Applied Regression Analysis and Other Multivariable Methods

Duxbury titles of related interest

Devore, *Probability and Statistics for Engineering and the Sciences, Fourth Edition*

DiIorio, *SAS Applications Programming: A Gentle Introduction*

Durrett, *Probability, Theory and Examples, Second Edition*

Elliott, *Learning SAS in the Computer Lab*

Hadi, *Matrix Algebra as a Tool*

Johnson, *Applied Multivariate Methods for Data Analysis*

Kuehl, *Statistical Principles of Research Design and Analysis*

Lapin, *Modern Engineering Statistics*

Ott, *Introduction to Statistical Methods and Data Analysis, Fourth Edition*

Pagano & Gauvreau, *Principles of Biostatistics*

Ramsey & Schafer, *Statistical Sleuth: A Course in Methods of Data Analysis*

Rao, *Statistical Research Methods in the Life Sciences*

Rosner, *Fundamentals of Biostatistics, Fourth Edition*

Scheaffer, Mendenhall, & Ott, *Elementary Survey Sampling, Fifth Edition*

Vining, *Statistical Methods for Engineers*

To order copies contact your local bookstore or call 1-800-354-9706.
For more information contact **Duxbury Press,** 511 Forest Lodge Road, Pacific Grove, CA,
or go to: **www.duxbury.com**

THIRD EDITION

Applied Regression Analysis and Other Multivariable Methods

David G. Kleinbaum
Emory University

Lawrence L. Kupper
University of North Carolina, Chapel Hill

Keith E. Muller
University of North Carolina, Chapel Hill

Azhar Nizam
Emory University

An Alexander Kugushev Book

 Duxbury Press
An Imprint of Brooks/Cole Publishing Company
I(T)P ®An International Thomson Publishing Company

Pacific Grove • Albany • Belmont • Bonn • Boston • Cincinnati • Detroit • Johannesburg • London
Madrid • Melbourne • Mexico City • New York • Paris • Singapore • Tokyo • Toronto • Washington

Sponsoring Editor: *Alexander Kugushev*
Assistant Editor: *Cynthia Mazow*
Marketing Team: *Carolyn Crockett, Deborah Petit,*
  and *Michele Mootz*
Editorial Assistant: *Rita Jaramillo*
Production Service: *The Book Company*
Manuscript Editor: *Steven Gray*

Illustrations: *Emspace Artwork*
Cover Design: *Stuart Paterson, Image House, Inc.*
Cover Photo: *Superstock*
Typesetting: *Omegatype Typography, Inc.*
Cover Printing: *Phoenix Color Corp.*
Printing and Binding: *R. R. Donnelley & Sons,*
  Crawfordsville

For more information, contact Duxbury Press at Brooks/Cole Publishing Company:

BROOKS/COLE PUBLISHING COMPANY
511 Forest Lodge Road
Pacific Grove, CA 93950
USA

International Thomson Publishing Europe
Berkshire House 168–173
High Holborn
London WC1V 7AA
England

Thomas Nelson Australia
102 Dodds Street
South Melbourne, 3205
Victoria, Australia

Nelson Canada
1120 Birchmount Road
Scarborough, Ontario
Canada M1K 5G4

International Thomson Editores
Seneca 53
Col. Polanco
11560 México, D. F., México

International Thomson Publishing GmbH
Königswinterer Strasse 418
53227 Bonn
Germany

International Thomson Publishing Asia
221 Henderson Road
#05–10 Henderson Building
Singapore 0315

International Thomson Publishing Japan
Hirakawacho Kyowa Building, 3F
2-2-1 Hirakawacho
Chiyoda-ku, Tokyo 102
Japan

Printed in the United States of America

10   9   8   7   6

Library of Congress Cataloging-in-Publication Data

Applied regression analysis and other multivariable methods. – 3rd
  ed. / David G. Kleinbaum . . . [et al.]
    p.   cm.
  Rev. ed. of: Applied regression analysis and other multivariable
methods / David G. Kleinbaum, Lawrence L. Kupper, Keith E. Muller.
2nd ed. c1988.
    "An Alexander Kugushev book."
    Includes bibliographical references and index.
    ISBN 0–534–20910–6
    1. Multivariate analysis.   2. Regression analysis.   I. Kleinbaum,
David G.   II. Kleinbaum, David G. Applied regression analysis and
other multivariable methods.
QA278.A665   1998
519.5'36–dc21   97-17655
                                              CIP

Dedicated to my wife, Edna Kleinbaum.

Dedicated to my wife Sandy, and to my son Mark.

To the memory of the outstanding teachers I have been privileged to know.

To My Loving Family: Janet, Zainab, Sohail, Saeeda, Nizam, Seema, Naveed, Nadir and Dahnish.

Preface

This is the second revision of our second-level statistics text, originally published in 1978 and first revised in 1987. As before, this text is intended primarily for advanced undergraduates, graduate students, and working professionals in the health, social, biological, and behavioral sciences who engage in applied research in their fields. The book may also provide professional statisticians with some new insights into the application of advanced statistical techniques to real-life problems.

We have attempted in this revision to retain the basic structure and flavor of the earlier two editions, while at the same time making changes to keep pace with current analytic practices and computer usage in applied research. Notable changes in the third edition, discussed in more detail later, include a fourth author (Azhar Nizam), some reorganization of topics (in Chapters 22–24), expanded coverage of some content areas (such as logistic regression, in Chapter 23), a new chapter (Chapter 21 on repeated measures ANOVA), some new exercises for the reader, and the integration of computer output (using the SAS package, primarily) into our discussion of examples in the main body of the text and as a component of exercises given at the end of each chapter. We have deleted from the previous editions chapters on discriminant analysis, factor analysis, and categorical data analysis. This decision was based on our finding from a survey of previous users of our text that these chapters were rarely used for classroom instruction and were largely out of date. At the same time, the chapters we have added to replace this material seem more relevant to current applied research practice.

In this revision, as in our previous versions, we emphasize the intuitive logic and assumptions that underlie the techniques covered, the purposes for which these techniques are designed, the advantages and disadvantages of the techniques, and valid interpretations based on the techniques. Although we describe the statistical calculations required for the techniques we cover, we rely on computer output (even more so in this revision than previously) to provide the results of such calculations, so the reader can concentrate on how to apply a given technique rather than on how to carry out the calculations. The mathematical formulas that we do present require no more than simple algebraic manipulations. Proofs are of secondary importance and are generally omitted. Neither calculus nor matrix algebra is used anywhere in the main text, although we have included an appendix on matrices for the interested reader.

The text is not intended to be a general reference work dealing with all the statistical techniques available for analyzing data involving several variables. Instead, we focus on the techniques we consider most essential for use in applied research. After becoming proficient with the material in this text, the reader should be able to benefit from more specialized discussions of applied topics not covered here.

The most notable features of this second revised edition are the following:

1. Regression analysis and analysis of variance are discussed in considerable detail and with pedagogical care that reflects the authors' extensive experience and insight as teachers of such material.

2. The relationship between regression analysis and analysis of variance is highlighted.

3. The connection between multiple regression analysis and multiple and partial correlation analysis is discussed in detail.

4. Several advanced topics are presented in a unique, nonmathematical manner, including the analysis of repeated measures data (a new topic in this edition), maximum likelihood methods, logistic regression (expanded into a new chapter), and Poisson regression (also expanded into a new chapter).

5. An up-to-date discussion of the issues and procedures involved in fine-tuning a regression analysis is presented in chapters on confounding and interaction in regression, regression diagnostics, and selecting the best model.

6. Numerous examples and exercises illustrate applications to real studies in a wide variety of disciplines. New exercises have been added to all chapters.

7. Representative computer results from packaged programs (primarily using the SAS package) are used to illustrate concepts in the body of the text, as well as to provide a basis for exercises for the reader. We have greatly expanded the quantity of computer results provided throughout the text. Whenever appropriate, we have used computer output to replace material in the previous edition that unnecessarily emphasized numerical calculations.

8. The complete set of data for most exercises is provided, along with related computer results. This allows the instructor to assign computer work based on available packaged programs. However, if the instructional objectives involve a minimum of computer work, the instructor can use the computer results to give the student practical experience in interpreting computer output based on the techniques described in the text.

9. The reorganization and expansion of the material on maximum likelihood methods into three chapters (22–24) provide a strong foundation for understanding the most widely used method for fitting mathematical models involving several variables.

10. A new chapter on methods for the analysis of repeated measures data (Chapter 21) extends the discussion of ANOVA methods to a rapidly developing area of statistical methodology for the analysis of correlated data.

For formal classroom instruction, the chapters fall naturally into three clusters: Chapters 4 through 16, on regression analysis; Chapters 17 through 20, on analysis of variance, with optional use of Chapter 21 to introduce the analysis of repeated measures data; and Chapters 22 through 24, on maximum likelihood methods and important applications involving logistic and Poisson regression modeling. For a first course in regression analysis, some of Chapters 11 through 16 may be considered too specialized. For example, Chapter 12 on regression diagnostics and Chapter 16 on selecting the best model might be used in a continuation course on regression modeling, which might also include some of the advanced topics covered in Chapters 21 through 24.

The Teaching Package

A data disk is bound into each copy of the book. This disk contains data for the problems; the data sets are formatted for SAS, StataQuest, Minitab, and in ASCII. A *Student Solutions Manual* contains complete solutions for all of the problems for which answers are given in Appendix D, and a *Solutions Manual,* available to adopting instructors, contains complete solutions to all problems in the book.

Acknowledgments

We wish to acknowledge several people who contributed to the preparation of this text. Drs. Kleinbaum and Kupper continue to be indebted to John Cassel and Bernard Greenberg, two mentors who have provided us with inspiration and the professional and administrative guidance that enabled us to gain the broad experience necessary to write this book. Dr. Muller adds his thanks to Bernard Greenberg. Dr. Kleinbaum also wishes to thank John Boring, Chair of the Epidemiology Department at Emory University for his strong support and encouragement and for his deep commitment to teaching excellence. Dr. Kupper wishes to thank Barry Margolin, Chair of the Biostatistics Department at the University of North Carolina for his leadership and support. Azhar Nizam wishes to thank the chair of his department, Dr. Vicki Hertzberg, Department of Biostatistics at Emory University.

We also wish to thank Edna Kleinbaum, Sandy Martin, Sally Muller, and Janet Nizam for their encouragement and support during the writing of this revision. We thank our many students and colleagues at Emory University and at the University of North Carolina for their helpful comments and suggestions. We also want to thank the reviewers: Robert J. Anderson, University of Illinois at Chicago; Alfred A. Bartolucci, The University of Alabama at Birmingham; Robert Cochran, University of Wyoming; Joseph L. Fleiss, Columbia University Medical Center; James E. Holstein, University of Missouri at Columbia; Robin H. Lock, St. Lawrence University; Frank P. Mathur, Cal Poly at Pomona; and Satya N. Mishra, University of South Alabama. Finally, we thank those persons responsible for publishing this book: Alex Kugushev, Jamie Sue Brooks, and Dusty Davidson.

Contents

11 CONFOUNDING AND INTERACTION IN REGRESSION   186

12 REGRESSION DIAGNOSTICS   212

13 POLYNOMIAL REGRESSION   281

14 DUMMY VARIABLES IN REGRESSION   317

15 ANALYSIS OF COVARIANCE AND OTHER METHODS FOR ADJUSTING CONTINUOUS DATA   361

16 SELECTING THE BEST REGRESSION EQUATION   386

17   ONE-WAY ANALYSIS OF VARIANCE   423

18   RANDOMIZED BLOCKS: SPECIAL CASE OF TWO-WAY ANOVA   484

19   TWO-WAY ANOVA WITH EQUAL CELL NUMBERS   516

24   POISSON REGRESSION ANALYSIS   687

A   APPENDIX A—TABLES   711

B   APPENDIX B—MATRICES AND THEIR RELATIONSHIP TO REGRESSION ANALYSIS   732

C   APPENDIX C—ANOVA INFORMATION FOR FOUR COMMON BALANCED REPEATED MEASURES DESIGNS   744

D   SOLUTIONS TO EXERCISES   758

INDEX   787

1

Concepts and Examples
of Research

1-1 Concepts

The purpose of most research is to assess relationships among a set of variables. *Multivariable*[1] *techniques* are concerned with the statistical analysis of such relationships, particularly when at least three variables are involved. Regression analysis, our primary focus, is one type of multivariable technique. Other techniques will also be described in this text. Choosing an appropriate technique depends on the purpose of the research and on the types of variables under investigation (a subject discussed in Chapter 2).

Research may be classified broadly into three types: *experimental, quasi-experimental,* or *observational.* Multivariable techniques are applicable to all such types, yet the confidence one may reasonably have in the results of a study can vary with the research type. In most types, one variable is usually taken to be a *response* or *dependent variable*—that is, a variable to be predicted from other variables. The other variables are called *predictor* or *independent variables.*

If observational units (subjects) are randomly assigned to levels of important predictors, the study is usually classified as an *experiment.* Experiments are the most controlled type of study; they maximize the investigator's ability to isolate the observed effect of the predictors from the distorting effects of other (independent) variables that might also be related to the response.

If subjects are assigned to treatment conditions without randomization, the study is called *quasi-experimental* (Campbell and Stanley 1963). Such studies are often more feasible

---

[1]The term *multivariable* is preferable to *multivariate.* Statisticians generally use the term *multivariate analysis* to describe a method in which several dependent variables can be considered simultaneously. Researchers in the biomedical and health sciences who are not statisticians, however, use this term to describe any statistical technique involving several variables, even if only one dependent variable is considered at a time. In this text we prefer to avoid the confusion by using the term *multivariable analysis* to denote the latter, more general description.

and less expensive than experimental studies, but they provide less control over the study situation.

Finally, if all observations are obtained without either randomization or artificial manipulation (i.e., allocation) of the predictor variables, the study is said to be *observational*. Experiments offer the greatest potential for drawing definitive conclusions, and observational studies the least; however, experiments are the most difficult studies to implement, and observational studies the easiest. A researcher must consider this trade-off between interpretive potential and complexity of design when choosing among types of studies (Kleinbaum, Kupper, and Morgenstern 1982, chap. 3).

To assess a relationship between two variables, one must measure both of them in some manner. Measurement inherently and unavoidably involves error. The need for statistical design and analysis emanates from the presence of such error. Traditionally, statistical inference has been divided into two kinds: estimation and hypothesis testing. *Estimation* refers to describing (i.e., quantifying) characteristics and strengths of relationships. *Testing* refers to specifying hypotheses about relationships, making statements of probability about the reasonableness of such hypotheses, and then providing practical conclusions based on such statements.

This text focuses on regression and correlation methods involving one response variable and one or more predictor variables. In these methods, a mathematical model is specified that describes how the variables of interest are related to one another. The model must somehow be developed from study data, after which inference-making procedures (e.g., testing hypotheses and constructing confidence intervals) are conducted about important parameters of interest. Although other multivariable methods will be discussed, regression techniques are emphasized for three reasons: they have wide applicability; they can be the most straightforward to implement; and many other, more complex statistical procedures can be better appreciated once regression methods are understood.

1-2 Examples

The examples that follow concern *real* problems from a variety of disciplines and involve variables to which the methods described in this book can be applied. We shall return to these examples later when illustrating various methods of multivariable analysis.

Example 1-1   *Study of the associations among the physician–patient relationship, perception of pregnancy, and the outcome of pregnancy, illustrating the use of regression analysis, discriminant analysis, and factor analysis.*

Thompson (1972) and Hulka and others (1971) looked at both the process and the outcomes of medical care in a cohort of 107 pregnant married women in North Carolina. The data were obtained through patient interviews, questionnaires completed by physicians, and a review of medical records. Several variables were recorded for each patient.

One research goal of primary interest was to determine what association, if any, existed between SATISfaction[2] with medical care and a number of variables meant to describe patient

---

[2]Capital letters denote the abbreviated variable name.

perception of pregnancy and the physician–patient relationship. Three perception-of-pregnancy variables measured the patient's WORRY during pregnancy, her desire (WANT) for the baby, and her concern about childBIRTH. Two other variables measured the physician–patient relationship in terms of informational communication (INFCOM) concerning prescriptions and affective communication (AFFCOM) concerning perceptions. Other variables considered were AGE, social class (SOCLS), EDUCation, and PARity.

Regression analysis was used to describe the relationship between scores measuring patient satisfaction with medical care and the preceding variables. From this analysis, variables found not to be related to SATIS could be eliminated, while those found to be associated with SATIS could be ranked in order of importance. Also, the effects of confounding variables such as AGE and SOCLS could be considered, to three ends: any associations found could not be attributed solely to such variables; measures of the strength of the relationship between SATIS and other variables could be obtained; and a functional equation predicting level of patient satisfaction in terms of the other variables found to be important in describing satisfaction could be developed.

Another question of interest in this study was whether patient perception of pregnancy and/or the physician–patient relationship was associated with COMPlications of pregnancy. A variable describing complications was defined so that the value 1 could be assigned if the patient experienced one or more complications of pregnancy and 0 if she experienced no complications. *Logistic regression analysis* was used to evaluate the relationship between complications of pregnancy and other variables. This method, like regression analysis, allows the researcher to determine and rank important variables that can distinguish between patients who have complications and patients who do not.

Example 1-2   *Study of the relationship between water hardness and sudden death, illustrating the use of regression analysis.*

Hamilton's (1971) study of the effects of environmental factors on mortality used 88 North Carolina counties as the observational units—in contrast to an earlier study, which used individual patients as the observational units. Hamilton's primary goal was to determine the relationship, if any, between the sudden death rate of residents of a county and the measure of water hardness for that county. However, the researcher also wanted to compare the mortality–water hardness relationship among four regions of the state and to determine whether this relationship was affected by other variables, such as the habits of the county coroner in recording deaths, the distance from the county seat to the main hospital, the per capita income, and the population per physician.

Regression analysis was used in this study. The four regions could have been compared by doing a separate analysis for each region and then comparing the results. Alternatively, though, the comparative analysis could have been performed in a single step by defining a number of artificial, or dummy, variables to represent the regions.

Example 1-3   *Comparative study of the effects of two instructional designs for teaching statistics, illustrating the use of analysis of covariance.*

Kleinbaum and Kleinbaum (1976) used a classroom experiment to compare two approaches for teaching probability to graduate students taking an introductory course in biostatistics. A class of 52 students was randomly split into two groups stratified on the basis of the students' fields of study. Both groups were taught in a lecture format by the same instructor.

One group, the control, was taught with the standard lecture method, in which an instructor (using chalk and a blackboard) lectured after a handout had been distributed. For the experimental group, 16 transparencies, systematically designed to fit a carefully defined set of objectives, were used along with a set of practice problems that were done and reviewed in class. Both groups were given the same pretests and posttests, as well as questionnaires to measure attitudes. Both received a copy of the objectives, the handout, and a homework assignment. The primary question of interest in this study was whether the instructional design for the experimental group was more effective than that for the control group, as measured by cognitive learning and attitudes.

In comparing the experimental and control groups, the researchers had to take into account pretest scores so that any differences found in posttest scores could not be attributed solely to differing levels of ability or knowledge between the two groups at the beginning of the study. One appropriate method of analysis for this situation was *analysis of covariance,* which showed that posttest scores, when adjusted for pretest scores, were significantly higher for the experimental group. ▣

▣ **Example 1-4** *Study of race and social influence in cooperative problem-solving dyads, illustrating the use of analysis of variance and analysis of covariance.*

James (1973) conducted an experiment on 140 seventh- and eighth-grade males to investigate the effects of two factors—race of the experimenter (E) and race of the comparison norm (N)—on social influence behaviors in three types of dyads: white–white; black–black; and white–black. Subjects played a game of strategy called Kill the Bull, in which 14 separate decisions must be made for proceeding toward a defined goal on a game board. In the game, each pair of players (dyad) must reach a consensus on a direction at each decision step, after which they signal the E, who then rolls a die to determine how far they can advance along their chosen path of six squares. Photographs of the current champion players (N) (either two black youths [black norm] or two white youths [white norm]) were placed above the game board.

Four measures of social influence activity were used as the outcome variables of interest. One of these, called performance output, was a measure of the number of times a given subject attempted to influence his dyad to move in a particular direction.

The major research question focused on the outcomes for biracial dyads. Previous research of this type had used only white investigators and implicit white comparison norms, and the results indicated that the white partner tended to dominate the decision making. James's study sought to determine whether such an "interaction disability," previously attributed to blacks, would be maintained, removed, or reversed when the comparison norm, the experimenter, or both were black. One approach to analyzing this problem was to perform a *two-way analysis of variance* on social-influence-activity difference scores between black and white partners, to assess whether such differences were affected by either the race of E or the race of N. No such significant effects were found, however, implying that neither E nor N influenced biracial dyad interaction. Nevertheless, through use of *analysis of covariance,* it was shown that, controlling for factors such as age, height, grade, and verbal and mathematical test scores, there was no statistical evidence of white dominance in any of the experimental conditions.

Furthermore, when combined output scores for both subjects in same-race dyads (white–white or black–black) were analyzed using a *three-way analysis of variance* (the three factors being race of dyad, race of E, and race of N), subjects in all-black dyads were found to be more verbally active (i.e., exhibited a greater tendency to influence decisions) under a black

E than under a white E; the same result was found for white dyads under a white E. This property is generally referred to in statistical jargon as a "race of dyad" by "race of E" interaction. The property continued to hold up after *analysis of covariance* was used to control for the effects of age, height, and verbal and mathematical test scores.

Example 1-5   *Study of the relationship of culture change to health, illustrating the use of factor analysis and analysis of variance.*

Patrick and others (1974) studied the effects of cultural change on health in the U.S. Trust Territory island of Ponape. Medical and sociological data were obtained on a sample of about 2,000 people by means of physical exams and a sociological questionnaire. This Micronesian island has experienced rapid Westernization and modernization since American occupation in 1945. The question of primary interest was whether rapid social and cultural change caused a rise in blood pressure and in the incidence of coronary heart disease. A specific hypothesis guiding the research was that persons with high levels of cultural ambiguity and incongruity and low levels of supportive affiliations with others have high levels of blood pressure and are at high risk for coronary heart disease.

A preliminary step in the evaluation of this hypothesis involved measuring three variables: attitude toward modern life; preparation for modern life; and involvement in modern life. Each of these variables was created by isolating specific questions from a sociological questionnaire. Then a *factor analysis*[3] determined how best to combine the scores on specific questions into a single overall score that defined the variable under consideration. Two cultural incongruity variables were then defined. One involved the discrepancy between attitude toward modern life and involvement in modern life; the other was defined as the discrepancy between preparation for modern life and involvement in modern life.

These variables were then analyzed to determine their relationship, if any, to blood pressure and coronary heart disease. Individuals with large positive or negative scores on either of the two incongruity variables were hypothesized to have high blood pressure and to be at high risk for coronary heart disease.

One approach to analysis involved categorizing both discrepancy scores into high and low groups. Then a *two-way analysis of variance* could be performed using blood pressure as the outcome variable. We will see later that this problem can also be described as a regression problem.

1-3 Concluding Remarks

The five examples described in section 1-2 indicate the variety of research questions to which multivariable statistical methods are applicable. In Chapter 2, we will provide a broad overview of such techniques; in the remaining chapters, we will discuss each technique in detail.

---

[3]Factor analysis was described in Chapter 24 of the second edition of this text, but this topic is not included as a topic in this (third) edition.

References

Campbell, D. T., and Stanley, J. C. 1963. *Experimental and Quasi-experimental Designs for Research.* Chicago: Rand McNally.

Hamilton, M. 1971. "Sudden Death and Water Hardness in North Carolina Counties in 1956–1964." Master's thesis, Department of Epidemiology, University of North Carolina, Chapel Hill.

Hulka, B. S.; Kupper, L. L.; Cassel, J. C.; and Thompson, S. J. 1971. "A Method for Measuring Physicians' Awareness of Patients' Concerns." *HSMHA Health Reports* 86: 741–51.

James, S. A. 1973. "The Effects of the Race of Experimenter and Race of Comparison Norm on Social Influence in Same Race and Biracial Problem-Solving Dyads." Ph.D. dissertation, Department of Clinical Psychology, Washington University, St. Louis, Mo.

Kleinbaum, D. G., and Kleinbaum, A. 1976. "A Team Approach for Systematic Design and Evaluation of Visually Oriented Modules." In J. R. O'Fallon and J. Service, eds., *Modular Instruction in Statistics—Report of ASA Study,* pp. 115–21. Washington, D.C.: American Statistical Association.

Kleinbaum, D. G.; Kupper, L. L.; and Morgenstern, H. 1982. *Epidemiologic Research.* Belmont, Calif.: Lifetime Learning Publications.

Patrick, R.; Cassel, J. C.; Tyroler, H. A.; Stanley, L.; and Wild, J. 1974. "The Ponape Study of Health Effects of Cultural Change." Paper presented at the annual meeting of the Society for Epidemiologic Research, Berkeley, Calif.

Thompson, S. J. 1972. "The Doctor–Patient Relationship and Outcomes of Pregnancy." Ph.D. dissertation, Department of Epidemiology, University of North Carolina, Chapel Hill.

2

Classification of Variables and the Choice of Analysis

2-1 Classification of Variables

Variables can be classified in a number of ways. Such classifications are useful for determining which method of data analysis to use. In this section we describe three methods of classification: by gappiness, by descriptive orientation, and by level of measurement.

2-1-1 Gappiness

In the classification scheme we call *gappiness,* we determine whether gaps exist between successively observed values of a variable (Figure 2-1). If gaps exist between observations, the variable is said to be *discrete;* if no gaps exist, the variable is said to be *continuous.* To speak more precisely, a variable is discrete if, between any two potentially observable values, a value exists that is not possibly observable. A variable is continuous if, between any two potentially observable values, another potentially observable value exists.

Examples of continuous variables are age, blood pressure, cholesterol level, height, and weight. Examples of discrete variables are sex (e.g., 0 if male and 1 if female), number of deaths, group identification (e.g., 1 if group A and 2 if group B), and state of disease (e.g., 1 if a coronary heart disease case and 0 if not a coronary heart disease case).

In analyses of actual data, the sampling frequency distributions for continuous variables are represented differently from those for discrete variables. Data on a continuous variable are usually *grouped* into class intervals, and a relative frequency distribution is determined by counting the proportion of observations in each interval. Such a distribution is usually represented by a histogram, as shown in Figure 2-2(a). Data on a discrete variable, on the other hand, are usually not grouped but are represented instead by a line chart, as shown in Figure 2-2(b).

Discrete variables can sometimes be treated for analysis purposes as continuous variables. This is possible when the values of such a variable, even though discrete, are not far apart and cover a wide range of numbers. In such a case, the possible values, although technically gappy, show such small gaps between values that a visual representation would approximate an interval (Figure 2-3).

7

FIGURE 2-1   Discrete versus continuous variables.

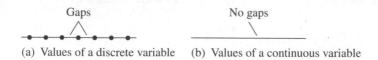

(a) Values of a discrete variable   (b) Values of a continuous variable

FIGURE 2-2   Sample frequency distributions of a continuous and a discrete variable.

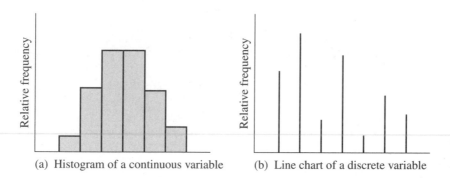

(a) Histogram of a continuous variable   (b) Line chart of a discrete variable

FIGURE 2-3   Discrete variable that may be treated as continuous.

Furthermore, a line chart, like the one in Figure 2-2(b), representing the frequency distribution of data on such a variable would probably show few frequencies greater than 1 and thus would be uninformative. As an example, the variable "social class" is usually measured as discrete; one popular measure of social class[1] takes on integer values between 11 and 77. When data on this variable are grouped into classes (e.g., 11–15, 16–20, etc.), the resulting frequency histogram gives a clearer picture of the characteristics of the variable than a line chart does. Thus, in this case, treating social class as a continuous variable is sometimes more useful than treating it as discrete.

Just as it is often useful to treat a discrete variable as continuous, some fundamentally continuous variables may be grouped into categories and treated as discrete variables in a given analysis. For example, the variable "age" can be made discrete by grouping its values into two categories, "young" and "old." Similarly, "blood pressure" becomes a discrete variable if it is categorized into "low," "medium," and "high" groups, or into deciles.

2-1-2 Descriptive Orientation

A second scheme for classifying variables is based on whether a variable is intended to _describe_ or _be described_ by other variables. Such a classification depends on the study objectives rather than on the inherent mathematical structure of the variable itself. If the variable under

---

[1]Hollingshead's "Two-Factor Index of Social Position," a description of which can be found in Green (1970).

investigation is to be described in terms of other variables, we call it a *response* or *dependent variable.* If we might be using the variable in conjunction with other variables to describe a given response variable, we call it a *predictor* or *independent variable.* Other variables may affect relationships but be of no intrinsic interest in a particular study. Such variables may be referred to as *control* or *nuisance variables* or, in some contexts, as *covariates* or *confounders.*

For example, in Thompson's (1972) study of the relationship between patient perception of pregnancy and patient satisfaction with medical care, the perception variables are independent and the satisfaction variable is dependent (Figure 2-4). Similarly, in studying the relationship of water hardness to sudden death rate in North Carolina counties, the water hardness index measured in each county is an independent variable, and the sudden death rate for that county is the dependent variable (Figure 2-5).

Usually, the distinction between independent and dependent variables is clear, as it is in the examples we have given. Nevertheless, a variable considered as dependent for purposes of evaluating one study objective may be considered as independent for purposes of evaluating a different objective. For example, in Thompson's study, in addition to determining the relation-

FIGURE 2-4   Descriptive orientation for Thompson's (1972) study of satisfaction with medical care.

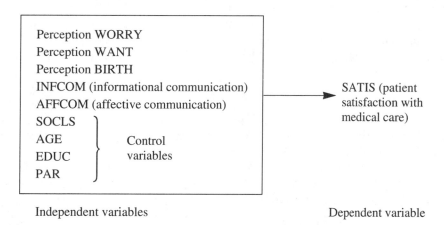

FIGURE 2-5   Descriptive orientation for Hamilton's (1971) study of water hardness and sudden death.

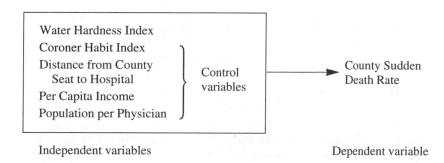

ship of perceptions as independent variables to patient satisfaction, the researcher sought to determine the relationships of social class, age, and education to perceptions treated as dependent variables.

2-1-3 Level of Measurement

A third classification scheme deals with the preciseness of measurement of the variable. There are three such levels: nominal, ordinal, and interval.

The numerically weakest level of measurement is the *nominal.* At this level, the values assumed by a variable simply indicate different categories. The variable "sex," for example, is nominal: by assigning the numbers 1 and 0 to denote male and female, respectively, we distinguish the two sex categories. A variable that describes treatment group is also nominal, provided that the treatments involved cannot be ranked according to some criterion (e.g., dosage level).

A somewhat higher level of measurement allows not only *grouping* into separate categories but also *ordering* of categories. This level is called *ordinal.* The treatment group may be considered ordinal if, for example, different treatments differ by dosage. In this case we could tell not only which treatment group an individual falls into but also who received a heavier dose of the treatment. Social class is another ordinal variable, since an ordering can be made among its different categories. For example, all members of the upper middle class are higher in some sense than all members of the lower middle class.

A limitation—perhaps debatable—in the preciseness of a measurement such as social class is the amount of information supplied by the magnitude of the differences between different categories. Thus, although upper middle class is higher than lower middle class, it is debatable *how much* higher.

A variable that can give not only an ordering but also a meaningful measure of the distance between categories is called an *interval* variable. To be interval, a variable must be expressed in terms of some standard or well-accepted physical unit of measurement. Height, weight, blood pressure, and number of deaths all satisfy this requirement, whereas subjective measures such as perception of pregnancy, personality type, prestige, and social stress do not.

An interval variable that has a scale with a true zero is occasionally designated as a *ratio* or *ratio-scale variable.* An example of a ratio-scale variable is the height of a person. Temperature is commonly measured in degrees Celsius, an interval scale. Measurement of temperature in degrees Kelvin is based on a scale that begins at absolute zero, and so is a ratio variable. An example of a ratio variable common in health studies is the concentration of a substance (e.g., cholesterol) in the blood.

Ratio-scale variables often involve measurement errors that follow a nonnormal distribution and are proportional to the size of the measurement. We will see in Chapter 5 that such proportional errors violate an important assumption of linear regression—namely, equality of error variance for all observations. Hence, the presence of a ratio variable is a signal to be on guard for a possible violation of this assumption. In Chapter 12 (on regression diagnostics), we will describe methods for detecting and dealing with this problem.

As with variables in other classification schemes, the same variable may be considered at one level of measurement in one analysis and at a different level in another analysis. Thus, "age" may be considered as interval in a regression analysis or, by being grouped into categories, as nominal in an analysis of variance.

The various levels of mathematical preciseness are cumulative. An ordinal scale possesses all the properties of a nominal scale plus ordinality. An interval scale is also nominal and ordi-

FIGURE 2-6   Overlap of variable classification.

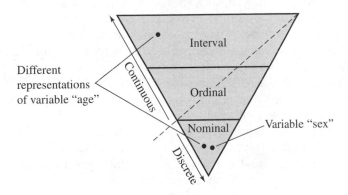

nal. The cumulativeness of these levels allows the researcher to drop back one or more levels of measurement in analyzing the data. Thus, an interval variable may be treated as nominal or ordinal for a particular analysis, and an ordinal variable may be analyzed as nominal.

2-2 Overlapping of Classification Schemes

The three classification schemes described in section 2-1 overlap in the sense that any variable can be labeled according to each scheme. "Social class," for example, may be considered as ordinal, discrete, and independent in a given study; "blood pressure" may be considered interval, continuous, and dependent in the same or another study.

The overlap between the level-of-measurement classification and the gappiness classification is shown in Figure 2-6. The diagram does not include classification into dependent or independent variables, because that dimension is entirely a function of the study objectives and not of the variable itself. In reading the diagram, one should consider any variable as being representable by some point within the triangle. If the point falls below the dashed line within the triangle, it is classified as discrete; if it falls above that line, it is continuous. Also, a point that falls into the area marked "interval" is classified as an interval variable; and similarly for the other two levels of measurement.

As Figure 2-6 indicates, any nominal variable must be discrete, but a discrete variable may be nominal, ordinal, or interval. Also, a continuous variable must be either ordinal or interval, although ordinal or interval variables may exist that are not continuous. For example, "sex" is nominal and discrete; "age" may be considered interval and continuous or, if grouped into categories, nominal and discrete; and "social class," depending on how it is measured and on the viewpoint of the researcher, may be considered ordinal and continuous, ordinal and discrete, or nominal and discrete.

2-3 Choice of Analysis

Any researcher faced with the need to analyze data requires a rationale for choosing a particular method of analysis. Four considerations should enter into such a choice: the purpose of the investigation; the mathematical characteristics of the variables involved; the statistical assumptions made about these variables; and how the data are collected (e.g., the sampling procedure). The first two considerations are generally sufficient to determine an appropriate analysis. However, the researcher must consider the latter two items before finalizing initial recommendations.

Here we focus on the use of variable classification, as it relates to the first two considerations noted at the beginning of this section, in choosing an appropriate method of analysis. Table 2-1 provides a rough guide to help the researcher in this choice when several variables are involved. The guide distinguishes among various multivariable methods. It considers the types of variable sets usually associated with each method and gives a general description of the purposes of each method. In addition to using the table, however, one must carefully check the statistical assumptions being made. These assumptions will be described fully later in the text. Table 2-2 shows how these guidelines can be applied to the examples given in Chapter 1.

Several methods for dealing with multivariable problems are *not* included in Table 2-1 or in this text—among them, nonparametric methods of analysis of variance, multivariate multiple regression, and multivariate analysis of variance (which are extensions of the corresponding methods given here that allow for *several* dependent variables), as well as methods of cluster analysis. In this book, we will cover only the multivariable techniques used most often by health and social researchers.

TABLE 2-1   Rough guide to multivariable methods.

Method	Classification of Variables		General Purpose
	Dependent	Independent	
Multiple regression analysis	Continuous	Classically all continuous, but in practice any type(s) can be used	To describe the extent, direction, and strength of the relationship between several independent variables and a continuous dependent variable
Analysis of variance	Continuous	All nominal	To describe the relationship between a continuous dependent variable and one or more nominal independent variables
Analysis of covariance	Continuous	Mixture of nominal variables and continuous variables (the latter used as control variables)*	To describe the relationship between a continuous dependent variable and one or more nominal independent variables, controlling for the effect of one or more continuous independent variables
Logistic regression analysis	Dichotomous	A mixture of various types can be used	To determine how one or more independent variables are related to the probability of the occurrence of one of two possible outcomes
Poisson regression analysis	Discrete	A mixture of various types can be used	To determine how one or more independent variables are related to the rate of occurrence of some outcome

* Generally, a *control variable* is a variable that must be considered before any relationships of interest can be quantified; this is because a control variable may be related to the variables of primary interest and must be taken into account in studying the relationships among the primary variables. For example, in describing the relationship between blood pressure and physical activity, we would probably consider "age" and "sex" as control variables because they are related to blood pressure and physical activity, and unless taken into account, could confound any conclusions regarding the primary relationship of interest.

TABLE 2-2   Application of Table 2-1 to examples in Chapter 1.

Study	Multivariable Method	Dependent Variable	Independent Variables	Purpose
Example 1-1	Multiple regression analysis	Patient satisfaction with medical care (SATIS), a continuous variable.	WANT, WORRY, BIRTH, INFCOM, AFFCOM, AGE, EDUC, SOCLS, PAR	To describe the relationship between SATIS and the variables WANT, WORRY, etc.
Example 1-1	Logistic regression analysis	Complications of pregnancy (0 = no, 1 = yes), a nominal variable.	WANT, WORRY, BIRTH, INFCOM, AFFCOM, AGE, EDUC, SOCLS, PAR	To determine whether and to what extent the independent variables are related to the probability of having pregnancy complications.
Example 1-2	Multiple regression analysis	Sudden death rate for a county in North Carolina during a 5-year period, a continuous variable.	Water hardness index for county, distance between hospital and county seat, per capita income, population per physician.	To describe the relationship between sudden death rate and the independent variables.
Example 1-3	Analysis of covariance	Posttest score, a continuous variable.	Pretest score, group designation (e.g., 1 = experimental, 0 = control)	To compare posttest scores for experimental and control groups, after adjusting for the possible effect of pretest scores.
Example 1-4	Analysis of covariance	Social influence activity score, a continuous variable.	Race of subject (e.g., 1 = white, 2 = black), age, height, etc.	To determine whether one racial group dominates the other in biracial dyads, after controlling for age, height, etc.
Example 1-4	Two-way analysis of variance	Social influence activity difference score between black and white partners in biracial dyads, a continuous variable.	Race of experimenter (e.g., 1 = white, 2 = black), race of comparison norm (e.g., 1 = white, 2 = black)	To determine whether the experimenter's race and the comparison norm's race have any effect on the difference score.
Example 1-5	Two-way analysis of variance	Systolic blood pressure (SBP), a continuous variable	Discrepancy between attitude toward and involvement in modern life, categorized as "high" or "low"; discrepancy between preparation for and involvement in modern life, categorized as "high" or "low."	To describe the relationship between nominal discrepancy scores and SBP.

References

Green, L. W. 1970. "Manual for Scoring Socioeconomic Status for Research on Health Behaviors." *Public Health Reports* 85: 815–27.

Hamilton, M. 1971. "Sudden Death and Water Hardness in North Carolina Counties in 1956–1964." Master's thesis, Department of Epidemiology, University of North Carolina, Chapel Hill.

Thompson, S. J. 1972. "The Doctor–Patient Relationship and Outcomes of Pregnancy." Ph.D. dissertation, Department of Epidemiology, University of North Carolina, Chapel Hill.

3

Basic Statistics: A Review

3-1 Preview

This chapter reviews the fundamental statistical concepts and methods that are needed to understand the more sophisticated multivariable techniques discussed in this text. Through this review, we shall introduce the statistical notation (using conventional symbols whenever possible) employed throughout the text.

The broad area associated with the word *statistics* involves the methods and procedures for collecting, classifying, summarizing, and analyzing data. We shall focus on the latter two activities here. The primary goal of most statistical analysis is to make *statistical inferences*—that is, to draw valid conclusions about a *population* of items or measurements based on information contained in a *sample* from that population.

A *population* is any set of items or measurements of interest, and a *sample* is any subset of items selected from that population. Any characteristic of that population is called a *parameter,* and any characteristic of the sample is termed a *statistic.* A statistic may be considered an estimate of some population parameter, and its accuracy of estimation may be good or bad.

Once sample data have been collected, it is useful, prior to analysis, to examine the data using tables, graphs, and *descriptive statistics,* such as the sample mean and the sample variance. Such descriptive efforts are important for representing the essential features of the data in easily interpretable terms.

Following such examination, statistical inferences are made through two related activities: *estimation* and *hypothesis testing.* The techniques involved here are based on certain assumptions about the probability pattern (or *distribution*) of the (*random*) variables being studied.

Each of the preceding key terms—*descriptive statistics, random variables, probability distribution, estimation,* and *hypothesis testing*—will be reviewed in the sections that follow.

3-2 Descriptive Statistics

A *descriptive statistic* may be defined as any single numerical measure computed from a set of data that is designed to describe a particular aspect or characteristic of the data set. The most common types of descriptive statistics are measures of *central tendency* and of *variability* (or *dispersion*).

The central tendency in a sample of data is the "average value" of the variable being observed. Of the several measures of central tendency, the most commonly used is the sample mean, which we denote by $\overline{X}$ whenever our underlying variable is called X. The formula for the sample mean is given by

$$\overline{X} = \frac{\sum_{i=1}^{n} X_i}{n}$$

where n denotes the sample size; $X_1, X_2,\ldots, X_n$ denote the n independent measurements on X; and $\sum$ denotes summation. The sample mean $\overline{X}$—in contrast to other measures of central tendency, such as the median or mode—uses in its computation all the observations in the sample. This property means that $\overline{X}$ is necessarily affected by the presence of extreme X-values, so it may be preferable to use the median instead of the mean. A remarkable property of the sample mean, which makes it particularly useful in making statistical inferences, follows from the *Central Limit Theorem,* which states that *whenever n is moderately large, $\overline{X}$ has approximately a normal distribution, regardless of the distribution of the underlying variable X.*

Measures of central tendency (such as $\overline{X}$) do not, however, completely summarize all features of the data. Obviously, two sets of data with the same mean can differ widely in appearance (e.g., an $\overline{X}$ of 4 results both from the values 4, 4, and 4 and from the values 0, 4, and 8). Thus, we customarily consider, in addition to $\overline{X}$, measures of variability, which tell us the extent to which the values of the measurements in the sample differ from one another.

The two measures of variability most often considered are the *sample variance* and the *sample standard deviation.* These are given by the following formulas when considering observations $X_1, X_2,\ldots, X_n$ on a single variable X:

$$\text{Sample variance} = S^2 = \frac{1}{n-1}\sum_{i=1}^{n} (X_i - \overline{X})^2 \tag{3.1}$$

$$\text{Sample standard deviation} = S = \sqrt{\frac{1}{n-1}\sum_{i=1}^{n} (X_i - \overline{X})^2} \tag{3.2}$$

The formula for S^2 describes variability in terms of an average of squared deviations from the sample mean—although $n - 1$ is used as the divisor instead of n, due to considerations that make S^2 a good estimator of the variability in the entire population.

A drawback to the use of S^2 is that it is expressed in squared units of the underlying variable X. To obtain a measure of dispersion that is expressed in the same units as X, we simply take the square root of S^2 and call it the sample standard deviation S. Using S in combination with $\overline{X}$ thus gives a fairly succinct picture of both the amount of spread and the center of the data, respectively.

When more than one variable is being considered in the same analysis (as will be the case throughout this text), we will use different letters and/or different subscripts to differentiate among the variables, and we will modify the notations for mean and variance accordingly. For example, if we are using X to stand for age and Y to stand for systolic blood pressure, we will denote the sample mean and the sample standard deviation for each variable as $(\overline{X}, S_X)$ and $(\overline{Y}, S_Y)$, respectively.

3-3 Random Variables and Distributions

The term *random variable* is used to denote a variable whose observed values may be considered outcomes of a stochastic or random experiment (e.g., the drawing of a random sample). The values of such a variable in a particular sample, then, cannot be anticipated with certainty before the sample is gathered. Thus, if we select a random sample of persons from some community and determine the systolic blood pressure (W), cholesterol level (X), race (Y), and sex (Z) of each person, then W, X, Y, and Z are four random variables whose particular realizations (or observed values) for a given person in the sample cannot be known for sure beforehand. In this text we shall denote random variables by capital italic letters.

The probability pattern that gives the relative frequencies associated with all the possible values of a random variable in a population is generally called the *probability distribution* of the random variable. We represent such a distribution by a table, graph, or mathematical expression that provides the probabilities corresponding to the different values or ranges of values taken on by a random variable.

Discrete random variables (such as the number of deaths in a sample of patients, or the number of arrivals at a clinic), whose possible values are countable, have (gappy) distributions that are graphed as a series of lines; the heights of these lines represent the probabilities associated with the various possible discrete outcomes (Figure 3-1(a)). *Continuous* random variables (such as blood pressure and weight), whose possible values are uncountable, have (nongappy) distributions that are graphed as smooth curves; an *area* under such a curve represents the probability associated with a *range of values* of the continuous variable (Figure 3-1(b)). We note in passing that the probability of a continuous random variable's taking on one particular value is 0, because there can be no area above a single point.

In the next two subsections, we will discuss two particular distributions of enormous practical importance: the binomial (which is discrete) and the normal (which is continuous).

FIGURE 3-1   Discrete and continuous distributions. $P(X = a)$ is read: "The probability that X takes the value a."

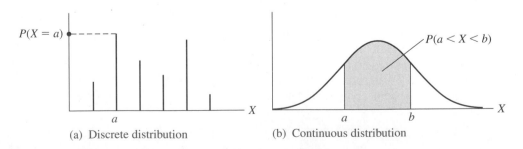

(a) Discrete distribution   (b) Continuous distribution

3-3-1 The Binomial Distribution

A *binomial* random variable describes the number of occurrences of a particular event in a series of n trials, under the following four conditions:

1. The n trials are identical.

2. The outcome of any one trial is independent of (i.e., is not affected by) the outcome of any other trial.

3. There are two possible outcomes of each trial: "success" (i.e., the event of interest occurs) or "failure" (i.e., the event of interest does not occur), with probabilities π and $1 - \pi$, respectively.

4. The probability of success, π, remains the same for all trials.

For example, the distribution of the number of lung cancer deaths in a random sample of $n = 400$ persons would be considered binomial only if the four conditions were all satisfied, as would the distribution of the number of persons in a sample of $n = 70$ who favor a certain form of legislation.

The two elements of the binomial distribution that one must specify to determine the precise shape of the probability distribution and to compute binomial probabilities are the sample size n and the parameter π. The usual notation for this distribution is, therefore, $B(n, \pi)$. If X has a binomial distribution, it is customary to write

$$X \frown B(n, \pi)$$

where $\frown$ stands for "is distributed as." The probability formula for this discrete random variable X is given by the expression

$$P(X = j) = {}_nC_j\, \pi^j (1 - \pi)^{n-j}, \quad j = 0, 1, \dots, n$$

where ${}_nC_j = n! / [j!\, (n - j)!]$ denotes the number of combinations of n distinct objects selected j at a time.

3-3-2 The Normal Distribution

The *normal distribution*, denoted as $N(\mu, \sigma)$, where μ and σ are the two parameters, is described by the well-known bell-shaped curve (Figure 3-2). The parameters μ (the mean) and σ (the standard deviation) characterize the center and the spread, respectively, of the distribution. We generally attach a subscript to the parameters μ and σ to distinguish among variables; that is, we often write

$$X \frown N(\mu_X, \sigma_X)$$

to denote a normally distributed X.

FIGURE 3-2   The normal distribution.

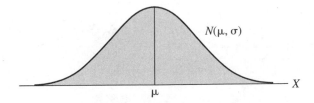

An important property of any normal curve is its *symmetry,* which distinguishes it from some other continuous distributions that we will discuss later. This symmetry property is quite helpful when using tables to determine probabilities or percentiles of the normal distribution.

Probability statements about a normally distributed random variable X that are of the form $P(a \leq X \leq b)$ require for computation the use of a single table (Table A-1 in Appendix A). This table gives the probabilities (or areas) associated with the *standard normal distribution,* which is a normal distribution with $\mu = 0$ and $\sigma = 1$. It is customary to denote a standard normal random variable by the letter Z, so we write

$$Z \sim N(0, 1)$$

To compute the probability $P(a \leq X \leq b)$ for an X that is $N(\mu, \sigma)$, we must transform (i.e., *standardize*) X to Z by applying the conversion formula

$$Z = \frac{X - \mu}{\sigma} \qquad (3.3)$$

to each of the elements in the probability statement about X, as follows:

$$P(a \leq X \leq b) = P\left(\frac{a - \mu}{\sigma} \leq Z \leq \frac{b - \mu}{\sigma} \right)$$

We then look up the equivalent probability statement about Z in the $N(0, 1)$ tables.

This rule also applies to the sample mean $\overline{X}$ whenever the underlying variable X is normally distributed or whenever the sample size is moderately large (by the Central Limit Theorem). But because the standard deviation of $\overline{X}$ is $\sigma/\sqrt{n}$, the conversion formula has the form

$$Z = \frac{\overline{X} - \mu}{\sigma/\sqrt{n}}$$

An inverse procedure for computing a probability for a range of values of X is to find a percentile of the distribution of X. A *percentile* is a value of the variable X *below which* the area under the probability distribution has a certain specified value. We denote the $(100p)$th percentile of X by X_p and picture it as in Figure 3-3, where p is the amount of area under the curve to the left of X_p. In determining X_p for a given p, we must again use the conversion formula (3.3). Since the procedure requires that we first determine Z_p and then convert back to X_p, however, we generally rewrite the conversion formula as

$$X_p = \mu + \sigma Z_p \qquad (3.4)$$

FIGURE 3-3   The $(100p)$th percentiles of X and Z.

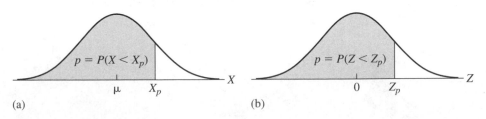

(a)

(b)

For example, if $\mu = 140$ and $\sigma = 40$, and we wish to find $X_{.95}$, the $N(0, 1)$ table first gives us $Z_{.95} = 1.645$, which we convert back to $X_{.95}$ as follows:

$$X_{.95} = 140 + (40)Z_{.95} = 140 + 40(1.645) = 205.8$$

Formulas (3.3) and/or (3.4) can also be used to approximate probabilities and percentiles for the binomial distribution $B(n, \pi)$ whenever n is moderately large (e.g., $n > 20$). Two conditions are usually required for this approximation to be accurate: $n\pi > 5$ and $n(1 - \pi) > 5$. Under such conditions, the mean and the standard deviation of the approximating normal distribution are

$$\mu = n\pi \quad \text{and} \quad \sigma = \sqrt{n\pi(1 - \pi)}$$

3-4 Sampling Distributions of t, χ^2, and F

The Student's t, chi-square (χ^2), and Fisher's F distributions are particularly important in statistical inference making.

The (*Student's*) *t distribution* (Figure 3-4(a)), which like the standard normal distribution is symmetric about 0, was originally developed to describe the behavior of the random variable

$$T = \frac{\overline{X} - \mu}{S/\sqrt{n}} \tag{3.5}$$

FIGURE 3-4   The t, χ^2, and F distributions.

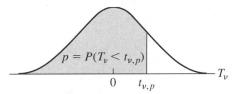

(a) Student's t distribution

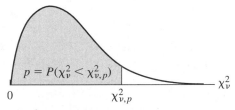

(b) χ^2 distribution

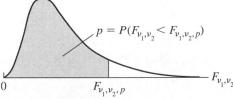

(c) F distribution

which represents an alternative to

$$Z = \frac{\overline{X} - \mu}{\sigma/\sqrt{n}}$$

whenever the population variance σ^2 is unknown and is estimated by S^2. The denominator of (3.5), $S/\sqrt{n}$, is the *estimated standard error of* $\overline{X}$. When the underlying distribution of X is normal, and when $\overline{X}$ and S^2 are calculated from a random sample from that distribution, then (3.5) has the *t distribution with* $n - 1$ *degrees of freedom,* where $n - 1$ is the quantity that must be specified in order to look up tabulated percentiles of this distribution. We denote all this by writing

$$T = \frac{\overline{X} - \mu}{S/\sqrt{n}} \frown t_{n-1}$$

It has generally been shown by statisticians that the t distribution is sometimes appropriate for describing the behavior of a random variable of the general form

$$T = \frac{\hat{\theta} - \mu_{\hat{\theta}}}{S_{\hat{\theta}}} \tag{3.6}$$

where $\hat{\theta}$ is any random variable that is normally distributed with mean $\mu_{\hat{\theta}}$ and standard deviation $\sigma_{\hat{\theta}}$, where $S_{\hat{\theta}}$ is the estimated standard error of $\hat{\theta}$, and where $\hat{\theta}$ and $S_{\hat{\theta}}$ are statistically independent. For example, when random samples are taken from two normally distributed populations with the same standard deviation (e.g., from $N(\mu_1, \sigma)$ and $N(\mu_2, \sigma)$), and we consider $\hat{\theta} = \overline{X}_1 - \overline{X}_2$ in (3.6), we can write

$$T = \frac{(\overline{X}_1 - \overline{X}_2) - (\mu_1 - \mu_2)}{S_p \sqrt{\frac{1}{n_1} + \frac{1}{n_2}}} \frown t_{n_1 + n_2 - 2}$$

where

$$S_p^2 = \frac{(n_1 - 1)S_1^2 + (n_2 - 1)S_2^2}{n_1 + n_2 - 2} \tag{3.7}$$

estimates the common variance σ^2 in the two populations. The quantity S_p^2 is called a *pooled sample variance,* since it is calculated by pooling the data from both samples in order to estimate the common variance σ^2.

The *chi-square* (or χ^2) *distribution* (Figure 3-4(b)) is a nonsymmetric distribution and describes, for example, the behavior of the nonnegative random variable

$$\frac{(n - 1)S^2}{\sigma^2} \tag{3.8}$$

where S^2 is the sample variance based on a random sample of size n from a normal distribution. The variable given by (3.8) has the chi-square distribution with $n - 1$ degrees of freedom:

$$\frac{(n - 1)S^2}{\sigma^2} \frown X_{n-1}^2$$

Because of the nonsymmetry of the chi-square distribution, both upper and lower percentage points of the distribution need to be tabulated, and such tabulations are solely a function of the degrees of freedom associated with the particular χ^2 distribution of interest. The chi-square distribution has widespread application in analyses of categorical data.

The *F distribution* (Figure 3-4(c)), which like the chi-square distribution is skewed to the right, is often appropriate for modeling the probability distribution of the ratio of independent estimators of two population variances. For example, given random samples of sizes n_1 and n_2 from $N(\mu_1, \sigma_1)$ and $N(\mu_2, \sigma_2)$, respectively, so that estimates S_1^2 and S_2^2 of σ_1^2 and σ_2^2 can be calculated, it can be shown that

$$\frac{S_1^2/\sigma_1^2}{S_2^2/\sigma_2^2} \tag{3.9}$$

has the *F* distribution with $n_1 - 1$ and $n_2 - 1$ degrees of freedom, which are called the *numerator* and *denominator* degrees of freedom, respectively. We write this as

$$\frac{S_1^2 \sigma_2^2}{S_2^2 \sigma_1^2} \frown F_{n_1-1, n_2-1}$$

The *F* distribution can also be related to the *t* distribution, when the numerator degrees of freedom equal 1; that is, the square of a variable distributed as Student's *t* with v degrees of freedom has the *F* distribution with 1 and v degrees of freedom. In other words,

$$T^2 \frown F_{1,v} \quad \text{if and only if} \quad T \frown t_v$$

Percentiles of the t, χ^2, and F distributions may be obtained from Tables A-2, A-3, and A-4 in Appendix A. The shapes of the curves that describe these probability distributions, together with the notation we will use to denote their percentile points, are given in Figure 3-4.

3-5 Statistical Inference: Estimation

Two general categories of statistical inference—estimation and hypothesis testing—can be distinguished by their differing purposes: estimation is concerned with quantifying the specific value of an unknown population parameter; hypothesis testing is concerned with making a decision about a hypothesized value of an unknown population parameter.

In estimation, which we focus on in this section, we wish to estimate an unknown parameter θ by using a random variable $\hat{\theta}$ ("theta hat," called a *point estimator* of θ). This point estimator takes the form of a formula or rule. For example,

$$\bar{X} = \frac{1}{n}\sum_{i=1}^{n} X_i \quad \text{or} \quad S^2 = \frac{1}{n-1}\sum_{i=1}^{n}(X_i - \bar{X})^2$$

tells us how to calculate a specific point estimate, given a particular set of data.

To estimate a parameter of interest (e.g., a population mean μ, a binomial proportion π, a difference between two population means $\mu_1 - \mu_2$, or a ratio of two population standard deviations σ_1/σ_2), the usual procedure is to select a random sample from the population or populations of interest, calculate the point estimate of the parameter, and then associate with this estimate a measure of its variability, which usually takes the form of a confidence interval for the parameter of interest.

As its name implies, a *confidence interval* (often abbreviated CI) consists of two random boundary points between which we have a certain specified *level of confidence* that the population parameter lies. More specifically, a 95% confidence interval for a parameter θ consists of lower and upper limits determined so that, in many repeated sets of samples of the same size, about 95% of all such intervals would be expected to contain the parameter θ. Care must be taken when interpreting such a confidence interval not to consider θ a random variable that either falls or does not fall in the calculated interval; rather, θ is a fixed (unknown) constant, and the random quantities are the lower and upper limits of the confidence interval, which vary from sample to sample.

We illustrate the procedure for computing a confidence interval with two examples—one involving estimation of a single population mean μ, and one involving estimation of the difference between two population means $\mu_1 - \mu_2$. In each case, the appropriate confidence interval has the following general form:

$$\left(\begin{array}{c}\text{Point estimate of}\\\text{the parameter}\end{array}\right) \pm \left[\left(\begin{array}{c}\text{Percentile of}\\\text{the } t \text{ distribution}\end{array}\right) \cdot \left(\begin{array}{c}\text{Estimated standard}\\\text{error of the estimate}\end{array}\right)\right] \qquad (3.10)$$

This general form also applies to confidence intervals for other parameters considered in the remainder of the text (e.g., those considered in multiple regression analysis).

Example 3-1   Suppose that we have determined the Quantitative Graduate Record Examination (QGRE) scores for a random sample of nine student applicants to a certain graduate department in a university, and we have found that $\overline{X} = 520$ and $S = 50$. If we wish to estimate with 95% confidence the population mean QGRE score (μ) for all such applicants to the department, and we are willing to assume that the population of such scores from which our random sample was selected is approximately normally distributed, the confidence interval for μ is given by the general formula

$$\overline{X} \pm t_{n-1,1-\alpha/2}\left(\frac{S}{\sqrt{n}}\right) \qquad (3.11)$$

which gives the $100(1 - \alpha)\%$ (small-sample) confidence interval for μ when σ is unknown. In our problem, $\alpha = 1 - .95 = .05$ and $n = 9$; therefore, by substituting the given information into (3.10), we obtain

$$520 \pm t_{8,0.975}\left(\frac{50}{\sqrt{9}}\right)$$

Since $t_{8,0.975} = 2.3060$, this formula becomes

$$520 \pm 2.3060\left(\frac{50}{\sqrt{9}}\right)$$

or

$$520 \pm 38.43$$

Our 95% confidence interval for μ is thus given by

$$(481.57, 558.43)$$

If we wanted to use this confidence interval to help determine whether 600 is a likely value for μ (i.e., if we were interested in making a decision about a specific value for μ), we would conclude that 600 is not a likely value, since it is not contained in the 95% confidence interval for μ just developed. This helps clarify the connection between estimation and hypothesis testing. ■

Example 3-2 Suppose that we wish to compare the change in health status of two groups of mental patients who are undergoing different forms of treatment for the same disorder. Suppose that we have a measure of change in health status based on a questionnaire given to each patient at two different times; and we are willing to assume that this measure of change in health status is approximately normally distributed and has the same variance in the populations of patients from which we selected our independent random samples. The data obtained are summarized as follows:

Group 1: $n_1 = 15, \overline{X}_1 = 15.1, S_1 = 2.5$
Group 2: $n_2 = 15, \overline{X}_2 = 12.3, S_2 = 3.0$

where the underlying variable X denotes the change in health status between time 1 and time 2.

A 99% confidence interval for the true mean difference $(\mu_1 - \mu_2)$ in health status change between these two groups is given by the following formula, which assumes equal population variances (i.e., $\sigma_1^2 = \sigma_2^2$):

$$(\overline{X}_1 - \overline{X}_2) \pm t_{n_1+n_2-2,1-\alpha/2} \, S_p \sqrt{\frac{1}{n_1} + \frac{1}{n_2}} \qquad (3.12)$$

where S_p is the pooled standard deviation derived from S_p^2, the pooled sample variance given by (3.7). Here we have

$$S_p^2 = \frac{(15 - 1)(2.5)^2 + (15 - 1)(3.0)^2}{15 + 15 - 2} = 7.625$$

so

$$S_p = \sqrt{7.625} = 2.76$$

Since $\alpha = .01$, our percentile in (3.12) is given by $t_{28,0.995} = 2.7633$. So the 99% confidence interval for $\mu_1 - \mu_2$ is given by

$$(15.1 - 12.3) \pm 2.7633(2.76)\sqrt{\frac{1}{15} + \frac{1}{15}}$$

which reduces to

$$2.80 \pm 2.78$$

yielding the following 99% confidence interval for $\mu_1 - \mu_2$:

(0.02, 5.58)

Since the value 0 is not contained in this interval, we conclude that there is evidence of a difference in health status change between the two groups.   ◾

3-6 Statistical Inference: Hypothesis Testing

Although closely related to confidence interval estimation, hypothesis testing has a slightly different orientation. When developing a confidence interval, we use our sample data to estimate what we think is a *likely* set of values for the parameter of interest. When performing a statistical test of a null hypothesis concerning a certain parameter, we use our sample data to *test* whether our estimated value for the parameter is *different enough* from the hypothesized value to support the conclusion that the null hypothesis is *unlikely* to be true.

The general procedure used in testing a statistical null hypothesis remains basically the same, regardless of the parameter being considered. This procedure (which we will illustrate by example) consists of the following seven steps:

1. Check the assumptions regarding the properties of the underlying variable(s) being measured that are needed to justify use of the testing procedure under consideration.

2. State the null hypothesis H_0 and the alternative hypothesis H_A.

3. Specify the significance level α.

4. Specify the test statistic to be used and its distribution under H_0.

5. Form the decision rule for rejecting or not rejecting H_0 (i.e., specify the rejection and nonrejection regions for the test).

6. Compute the value of the test statistic from the observed data.

7. Draw conclusions regarding rejection or nonrejection of H_0.

◾ **Example 3-3**   Let us again consider the random sample of nine student applicants with mean QGRE score $\bar{X} = 520$ and standard deviation $S = 50$. The department chairperson suspects that, because of the declining reputation of the department, this year's applicants are not quite as good quantitatively as those from the previous five years, for whom the average QGRE score was 600. If we assume that the population of QGRE scores from which our random sample has been selected is normally distributed, we can test the null hypothesis that the population mean score associated with this year's applicants is 600 versus the alternative hypothesis that it is less than 600. The *null hypothesis,* in mathematical terms, is $H_0: \mu = 600$, which asserts that the population mean μ for this year's applicants does not differ from what it has generally been in the past. The *alternative hypothesis* is stated as $H_A: \mu < 600$, which asserts that the QGRE scores, on average, have gotten worse.

We have thus far considered the first two steps of our testing procedure:

1. Assumptions: The variable QGRE score has a normal distribution, from which a random sample has been selected.

2. Hypotheses: H_0: $\mu = 600$; H_A: $\mu < 600$.

Our next step is to decide what error or probability we are willing to tolerate for incorrectly rejecting H_0 (i.e., making a Type I error, as discussed later in this chapter). We call this probability of making a Type I error the *significance level* α.[1]

We usually assign a value such as .1, .05, .025, or .01 to α. Suppose, for now, that we choose $\alpha = .025$. Then step 3 is

3. Use $\alpha = .025$.   *2.5% of the time reject null hypothesis when true*

Step 4 requires us to specify the test statistic that will be used to test H_0. In this case, with H_0: $\mu = 600$, we have

4. $T = \dfrac{\overline{X} - 600}{S/\sqrt{9}} \frown t_8$ under H_0.

Step 5 requires us to specify the decision rule that we will use to reject or not reject H_0. In determining this rule, we divide the possible values of T into two sets: the *rejection region* (or *critical region*), which consists of values of T for which we reject H_0; and the *nonrejection region,* which consists of those T-values for which we do not reject H_0. If our computed value of T falls in the rejection region, we conclude that the observed results deviate far enough from H_0 to cast considerable doubt on the validity of the null hypothesis.

In our example, we determine the critical region by choosing from t tables a point called the *critical point,* which defines the boundary between the nonrejection and rejection regions. The value we choose is

$$-t_{8,0.975} = -2.306$$

in which case the probability that the test statistic takes a value of less than -2.306 under H_0 is exactly $\alpha = .025$, the significance level (Figure 3-5). We thus have the following decision rule:

5. Reject H_0 if $T = \dfrac{\overline{X} - 600}{S/\sqrt{9}} < -2.306$; do not reject H_0 otherwise.

FIGURE 3-5   The critical point for Example 3-3.

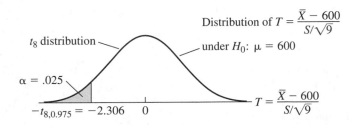

---

[1]Two types of errors can be made when performing a statistical test. A Type II error occurs if we fail to reject H_0 when H_0 is actually false. We denote the probability of a Type II error as β and call $(1 - \beta)$ the *power* of the test. For a fixed sample size, α and β for a given test are inversely related; that is, lowering one has the effect of increasing the other. In general, the power of any statistical test can be raised by increasing the sample size.

Now we simply apply the decision rule to our data by computing the observed value of T. In our example, since $\overline{X} = 520$ and $S = 50$, our computed T is

6. $T = \dfrac{\overline{X} - 600}{S/\sqrt{9}} = \dfrac{520 - 600}{50/3} = -4.8.$

The last step is to make the decision about H_0 based on the rule given in step 5:

7. Since $T = -4.8$, which lies below -2.306, we reject H_0 at significance level .025 and conclude that there is evidence that students currently applying to the department have QGRE scores *significantly lower* than 600.

In addition to performing the procedure just described, we often wish to compute a *P-value*, which quantifies *exactly how unusual the observed results would be if H_0 were true*. An equivalent way of describing the P-value is as follows: *The P-value gives the probability of obtaining a value of the test statistic that is at least as unfavorable to H_0 as the observed value* (Figure 3-6).

To get an idea of the approximate size of the P-value in this example, our approach is to determine from the table of the distribution of T under H_0 the two percentiles that bracket the observed value of T. In this case, the two percentiles are

$$-t_{8, 0.995} = -3.355 \quad \text{and} \quad -t_{8, 0.9995} = -5.041$$

Since the observed value of T lies between these two values, we conclude that the area we seek lies between the two areas corresponding to these two percentiles:

$$.0005 < P < .005$$

In interpreting this inequality, we observe that the P-value is *quite small,* indicating that we have observed a highly unusual result if H_0 is true. In fact, this P-value is so small as to lead us to reject H_0. Furthermore, the size of this P-value means that we would reject H_0 even for an α as small as .005.

For the general computation of a P-value, the appropriate P-value for a two-tailed test is twice that for the corresponding one-tailed test. If an investigator wishes to draw conclusions about a test on the basis of the P-value (e.g., in lieu of specifying α a priori), the following guidelines are recommended:

1. If P is small (less than .01), reject H_0.

2. If P is large (greater than .1), do not reject H_0.

3. If $.01 < P < .1$, the significance is borderline, since we reject H_0 for $\alpha = .1$ but not for $\alpha = .01$.

FIGURE 3-6   The P-value.

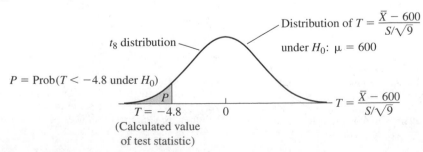

Notice that, if we actually do specify α a priori, we reject H_0 when $P < \alpha$.   ■

■ **Example 3-4**   We now look at one more worked example about hypothesis testing—this time involving a comparison of two means, μ_1 and μ_2. Consider the following data on health status change, which were discussed earlier:

$$\text{Group 1:}\quad n_1 = 15,\ \bar{X}_1 = 15.1,\ S_1 = 2.5$$
$$\text{Group 2:}\quad n_2 = 15,\ \bar{X}_2 = 12.3,\ S_2 = 3.0$$
$$(S_p = 2.76)$$

Suppose that we wish to test at significance level .01 whether the true average change in health status differs between the two groups. The steps required to perform this test are as follows:

1. Assumptions: We have independent random samples from two normally distributed populations. The population variances are assumed to be equal.

2. Hypotheses: H_0: $\mu_1 = \mu_2$; H_A: $\mu_1 \neq \mu_2$.

3. Use $\alpha = .01$.

4. $T = \dfrac{(\bar{X}_1 - \bar{X}_2) - 0}{S_p\sqrt{\dfrac{1}{n_1} + \dfrac{1}{n_2}}} \sim t_{28}$ under H_0.

5. Reject H_0 if $|T| \geq t_{28,0.995} = 2.763$; do not reject H_0 otherwise (Figure 3-7).

6. $T = \dfrac{(\bar{X}_1 - \bar{X}_2) - 0}{S_p\sqrt{\dfrac{1}{n_1} + \dfrac{1}{n_2}}} = \dfrac{15.1 - 12.3}{2.76\sqrt{\dfrac{1}{15} + \dfrac{1}{15}}} = 2.78.$

7. Since $T = 2.78$ exceeds $t_{28,0.995} = 2.763$, we reject H_0 at $\alpha = .01$ and conclude that there is evidence that the true average change in health status differs between the two groups.

The P-value for this test is given by the shaded area in Figure 3-8. For the t distribution with 28 degrees of freedom, we find that $t_{28,0.995} = 2.763$ and $t_{28,0.9995} = 3.674$. Thus, $P/2$ is given by the inequality

$$1 - .9995 < \frac{P}{2} < 1 - .995$$

so

$$.001 < P < .01 \quad ■$$

FIGURE 3-7   Critical region for the health status change example.

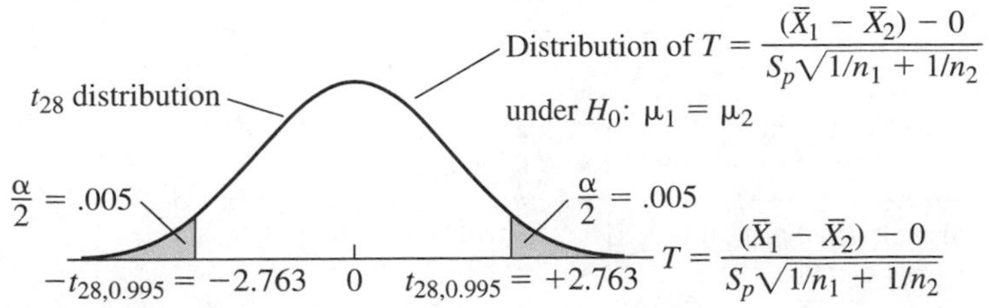

t_{28} distribution

Distribution of $T = \dfrac{(\bar{X}_1 - \bar{X}_2) - 0}{S_p\sqrt{1/n_1 + 1/n_2}}$ under H_0: $\mu_1 = \mu_2$

$\dfrac{\alpha}{2} = .005$

$\dfrac{\alpha}{2} = .005$

$-t_{28,0.995} = -2.763 \quad 0 \quad t_{28,0.995} = +2.763$

$T = \dfrac{(\bar{X}_1 - \bar{X}_2) - 0}{S_p\sqrt{1/n_1 + 1/n_2}}$

FIGURE 3-8   P-value for the health status change example.

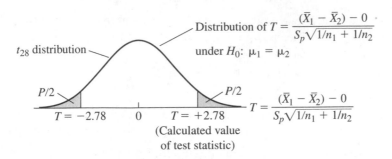

3-7 Error Rates, Power, and Sample Size

Table 3-1 summarizes the decisions that result from hypothesis testing. If the true state of nature is that the null hypothesis is true and if the decision is made that the null hypothesis is true, then a correct decision has been made. Similarly, if the true state of nature is that the alternative hypothesis is true and if the decision is made that the alternative is true, then a correct decision has been made. On the other hand, if the true state of nature is that the null hypothesis is true but the decision is made to choose the alternative, then a false positive error (commonly referred to as a _Type I error_) has been made. And if the true state of nature supports the alternative hypothesis but the decision is made that the null hypothesis is true, then a false negative error (commonly referred to as a _Type II error_) has been made.

Table 3-2 summarizes the probabilities associated with the outcomes of hypothesis testing just described. If the true state of nature corresponds to the null hypothesis but the alternative hypothesis is chosen, then a Type I error has been made, with probability denoted by the symbol α. Hence, the probability of making a correct choice of H_0 when H_0 is true must be $1 - \alpha$. In turn, if the actual state of nature is that the alternative hypothesis is true but the null hypothesis is chosen, then a Type II error has occurred, with probability denoted by β. In turn, $1 - \beta$ is the probability of choosing the alternative hypothesis when it is true, and this probability is often called the _power of the test._

When we design a research study, we would like to use statistical tests for which both α and β are small (i.e., for which there is a small chance of making either a Type I or a Type II error). For a given α, we can sometimes determine the sample size required in the study to ensure that β is no larger than some desired value for a particular alternative hypothesis of inter-

TABLE 3-1   Outcomes of hypothesis testing.

Hypothesis Chosen	True State of Nature	
	H_0	H_A
H_0	Correct decision	False negative decision (Type II error)
H_A	False positive decision (Type I error)	Correct decision

TABLE 3-2   Probabilities of outcomes
of hypothesis testing.

Hypothesis Chosen	True State of Nature	
	H_0	H_A
H_0	$1 - \alpha$	β
H_A	α	$1 - \beta$

est. Such a design consideration generally involves the use of a *sample size formula* pertinent to the research question(s). This formula usually requires the researcher to guess values for some of the unknown parameters to be estimated in the study (see Cohen 1977; Muller and Peterson 1984; Kupper and Hafner 1989).

For example, the classical sample size formula used for a one-sided test of H_0: $\mu_1 = \mu_2$ versus H_0: $\mu_2 > \mu_1$, when a random sample of size n is selected from each of two normal populations with common variance σ^2, is as follows:

$$n \geq \frac{2(Z_{1-\alpha} + Z_{1-\beta})^2 \sigma^2}{\Delta^2}$$

For chosen values of α, β, and σ^2, this formula provides the minimum sample size n required to detect a specified difference $\Delta = \mu_2 - \mu_1$ between μ_1 and μ_2 (i.e., to reject H_0: $\mu_2 - \mu_1 = 0$ in favor of H_A: $\mu_2 - \mu_1 = \Delta > 0$ with power $1 - \beta$). Thus, in addition to picking α and β, the researcher must guess the size of the population variance σ^2 and specify the difference Δ to be detected. An educated guess about the value of the unknown parameter σ^2 can sometimes be made by using information obtained from related research studies. To specify Δ intelligently, the researcher has to decide on the smallest population mean difference ($\mu_2 - \mu_1$) that is practically (as opposed to statistically) meaningful for the study.

For a fixed sample size, α and β are inversely related in the following sense. If one tries to guard against making a Type I error by choosing a small rejection region, the nonrejection region (and hence β) will be large. Conversely, protecting against a Type II error necessitates using a large rejection region, leading to a large value for α. Increasing the sample size generally decreases β; of course, α remains unaffected.

It is common practice to conduct several statistical tests using the same data set. If such a data-set-specific series of tests is performed and each test is based on a size α rejection region, the probability of making at least one Type I error will be much larger than α. This multiple-testing problem is pervasive and bothersome. One simple—but not optimal—method for addressing this problem is to employ the so-called *Bonferroni correction*. For example, if k tests are to be conducted and the overall Type I error rate (i.e., the probability of making at least one Type I error in k tests) is to be no more than α, then a rule of thumb is to conduct *each individual* test at a Type I error rate of α/k.

This simple adjustment ensures that the overall Type I error rate will (at least approximately) be no larger than α. In many situations, however, this correction leads to such a small rejection region for each individual test that the power of each test may be too low to detect important deviations from the null hypotheses being tested. Resolving this antagonism between Type I and Type II error rates requires a conscientious study design and carefully considered error rates for planned analyses.

Problems

1. **a.** Give two examples of discrete random variables.
   b. Give two examples of continuous random variables.

2. Name the four levels of measurement, and give an example of a variable at each level.

3. Assume that Z is a normal random variable with mean 0 and variance 1.
   a. $P(Z \geq -1) = ?$
   b. $P(Z \leq ?) = .20$

4. **a.** $P(\chi^2_7 \geq ?) = .01$
   b. $P(\chi^2_{12} \leq 14) = ?$

5. **a.** $P(T_{13} \leq ?) = .10$
   b. $P(|T_{28}| \geq 2.05) = ?$

6. **a.** $P(F_{6,24} \geq ?) = .05$
   b. $P(F_{5,40} \geq 2.9) = ?$

7. What are the **(a)** mean, **(b)** median, and **(c)** mode of the standard normal distribution?

8. An $F_{1,\nu}$ random variable can be thought of as the square of what kind of random variable?

9. Find the **(a)** mean, **(b)** median, and **(c)** variance for the following set of scores:
   $$\{0, 2, 5, 6, 3, 3, 3, 1, 4, 3\}$$
   d. Find the set of Z scores for the data.

10. Which of the following statements about descriptive statistics is correct?
    a. *All* of the data are used to compute the median.
    b. The mean should be preferred to the median as a measure of central tendency if the data are noticeably skewed.
    c. The variance has the same units of measurement as the original observations.
    d. The variance can never be 0.
    e. The variance is like an average of squared deviations from the mean.

11. Suppose that the weight W of male patients registered at a diet clinic has the normal distribution with mean 190 and variance 100.
    a. For a random sample of patients of size $n = 25$, the expression $P(\overline{W} < 180)$, in which $\overline{W}$ denotes the mean weight, is equivalent to saying $P(Z > ?)$. [*Note:* Z is a standard normal random variable.]
    b. Find the interval (a, b) such that $P(a < \overline{W} < b) = .80$ for the same random sample in part (a).

12. The limits of a 95% confidence interval for the mean μ of a normal population with unknown variance are found by adding to and subtracting from the sample mean a certain multiple of the estimated standard error of the sample mean. If the sample size on which this confidence interval is based is 28, the *multiple* referred to in the previous sentence is the number _____ .

13. A random sample of 32 persons attending a certain diet clinic was found to have lost (over a three-week period) an average of 30 pounds, with a sample standard deviation of 11. For these data, a 99% confidence interval for the true mean weight loss by all patients attending the clinic would have the limits (?, ?).

14. From two normal populations assumed to have the same variance, independent random samples of sizes 15 and 19 were drawn. The first sample (with $n_1 = 15$) yielded mean and standard deviation 111.6 and 9.5, respectively, while the second sample ($n_2 = 19$) gave mean and standard deviation 100.9 and 11.5, respectively. The estimated standard error of the difference in sample means is _____ .

15. For the data of Problem 14, suppose that a test of H_0: $\mu_1 = \mu_2$ versus H_A: $\mu_1 > \mu_2$ yielded a computed value of the appropriate test statistic equal to 2.55.
    a. What conclusions should be drawn for $\alpha = .05$?
    b. What conclusions should be drawn for $\alpha = .01$?

16. Test the hypothesis that average body weight is the same for two independent diagnosis groups from one hospital:

    Diagnosis group 1 data: $\{132, 145, 124, 122, 165, 144, 151\}$

    Diagnosis group 2 data: $\{141, 139, 172, 131, 150, 125\}$

    You may assume that the data are normally distributed, with equal variance in the two groups. What conclusion should be drawn, with $\alpha = .05$?

17. Independent random samples are drawn from two normal populations, which are assumed to have the same variance. One sample (of size 5) yields mean 86.4 and standard deviation 8.0, and the other sample (of size 7) has mean 78.6 and standard deviation 10. The limits of a 99% confidence interval for the difference in population means are found by adding to and subtracting from the difference in sample means a certain multiple of the estimated standard error of this difference. This *multiple* is the number _____ .

18. If a 99% confidence interval for $\mu_1 - \mu_2$ is 4.8 to 9.2, which of the following conclusions can be drawn *based on this interval*?
    a. Do not reject H_0: $\mu_1 = \mu_2$ at $\alpha = .05$ if the alternative is H_A: $\mu_1 \neq \mu_2$.
    b. Reject H_0: $\mu_1 = \mu_2$ at $\alpha = .01$ if the alternative is H_A: $\mu_1 \neq \mu_2$.
    c. Reject H_0: $\mu_1 = \mu_2$ at $\alpha = .01$ if the alternative is H_A: $\mu_1 < \mu_2$.
    d. Do not reject H_0: $\mu_1 = \mu_2$ at $\alpha = .01$ if the alternative is H_A: $\mu_1 \neq \mu_2$.
    e. Do not reject H_0: $\mu_1 = \mu_2 + 3$ at $\alpha = .01$ if the alternative is H_A: $\mu_1 \neq \mu_2 + 3$.

19. Assume that we gather data, compute a T, and reject the null hypothesis. If, in fact, the null hypothesis is true, we have made (a) _____ . If the null hypothesis is false, we have made (b) _____ . Assume instead that our data lead us to not reject the null hypothesis. If, in fact, the null hypothesis is true, we have made (c) _____ . If the null hypothesis is false, we have made (d) _____ .

20. Suppose that the critical region for a certain test of hypothesis is of the form $|T| \geq 2.5$ and the computed value of T from the data is -2.75. Which, if any, of the following statements is correct?
    a. H_0 should be rejected.
    b. The significance level α is the probability that, under H_0, T is either greater than 2.75 or less than -2.75.
    c. The non-rejection region is given by $-3.5 < T < 3.5$.
    d. The non-rejection region consists of values of T above 3.5 or below -3.5.
    e. The P-value of this test is given by the area to the right of $T = 3.5$ for the distribution of T under H_0.

21. Suppose that $\overline{X}_1 = 125.2$ and $\overline{X}_2 = 125.4$ are the mean systolic blood pressures for two samples of workers from different plants in the same industry. Suppose, further, that a test of H_0: $\mu_1 = \mu_2$ using these samples is rejected for $\alpha = .001$. Which of the following conclusions is most reasonable?
    a. There is a meaningful difference (clinically speaking) in population means but not a statistically significant difference.
    b. The difference in population means is both statistically and meaningfully significant.
    c. There is a statistically significant difference but not a meaningfully significant difference in population means.
    d. There is neither a statistically significant nor a meaningfully significant difference in population means.
    e. The sample sizes used must have been quite small.

22. The choice of an alternative hypothesis (H_A) should depend primarily on (choose all that apply):
    a. the data obtained from the study.
    b. what the investigator is interested in determining.
    c. the critical region.
    d. the significance level.
    e. the power of the test.

23. For each of the areas in the accompanying figure, labeled a, b, c, and d, select an answer from the following: α, $1 - \alpha$, β, $1 - \beta$.

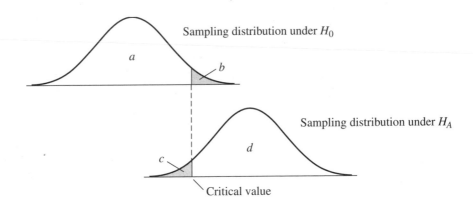

24. Suppose that H_0: $\mu_1 = \mu_2$ is the null hypothesis and that $.10 < P < .25$. What is the most appropriate conclusion?

25. Suppose that H_0: $\mu_1 = \mu_2$ is the null hypothesis and that $.005 < P < .01$. Which of the following conclusions is most appropriate?
    a. Do not reject H_0 because P is small.
    b. Reject H_0 because P is small.
    c. Do not reject H_0 because P is large.
    d. Reject H_0 because P is large.
    e. Do not reject H_0 at $\alpha = .01$.

References

Cohen, J. 1977. *Statistical Power Analysis for the Behavioral Sciences,* 2d ed. New York: Academic Press.

Kupper, L. L., and Hafner, K. B. 1989. "How Appropriate Are Popular Sample Size Formulas?" *The American Statistician* 43(2): 101–5.

Muller, K. E., and Peterson, B. L. 1984. "Practical Methods for Computing Power in Testing the Multivariate General Linear Hypothesis." *Computational Statistics and Data Analysis* 2: 143–58.

4

Introduction to Regression Analysis

4-1 Preview

Regression analysis is a statistical tool for evaluating the relationship of one or more independent variables $X_1, X_2, \ldots, X_k$ to a single, continuous dependent variable Y. It is most often used when the independent variables cannot be controlled, as when they are collected in a sample survey or other observational study. Nevertheless, it is equally applicable to more controlled experimental situations.

In practice, a regression analysis is appropriate for several possibly overlapping situations, including the following.

Application 1 You wish to *characterize the relationship* between the dependent and independent variables by determining the extent, direction, and strength of the association. For example ($k = 2$), in Thompson's (1972) study described in Chapter 1, one of the primary questions was to describe the extent, direction, and strength of the association between "patient satisfaction with medical care" (Y) and the variables "affective communication between patient and physician" (X_1) and "informational communication between patient and physician" (X_2).

Application 2 You *seek a quantitative formula* or equation to describe (e.g., predict) the dependent variable Y as a function of the independent variables $X_1, X_2, \ldots, X_k$. For example ($k = 1$), a quantitative formula may be desired for a study of the effect of dosage of a blood-pressure-reducing treatment (X_1) on blood pressure change (Y).

Application 3 You want to describe quantitatively or qualitatively the relationship between $X_1, X_2, \ldots, X_k$ and Y but *control for the effects of still other variables* $C_1, C_2, \ldots, C_p$, which you believe have an important relationship with the dependent variable. For example ($k = 2, p = 2$), a study of the epidemiology of chronic diseases might describe the relationship of blood pressure (Y) to smoking habits (X_1) and social class (X_2), controlling for age (C_1) and weight (C_2).

Application 4   You want to *determine which of several independent variables are important and which are not* for describing or predicting a dependent variable. You may want to control for other variables. You may also want to *rank* independent variables in their order of importance. In Thompson's (1972) study, for example ($k = 4$, $p = 2$), the researcher sought to determine for the dependent variable "satisfaction with medical care" (Y) which of the following independent variables were important descriptors: WORRY (X_1), WANT (X_2), INFCOM (X_3), and AFFCOM (X_4). It was also considered necessary to control for AGE (C_1) and EDUC (C_2).

Application 5   You want to *determine the best mathematical model* for describing the relationship between a dependent variable and one or more independent variables. Any of the previous examples can be used to illustrate this.

Application 6   You wish to *compare several derived regression relationships.* An example would be a study to determine whether smoking (X_1) is related to blood pressure (Y) in the same way for males as for females, controlling for age (C_1).

Application 7   You wish to *assess the interactive effects of two or more independent variables* with regard to a dependent variable. For example, you may wish to determine whether the relationship of alcohol consumption (X_1) to blood pressure level (Y) is different depending on smoking habits (X_2). In particular, the relationship between alcohol and blood pressure might be quite strong for heavy smokers but very weak for nonsmokers. If so, we would say that there is *interaction* between alcohol and smoking. Then, any conclusions about the relationship between alcohol and blood pressure must take into account whether—and possibly how much—a person smokes. More generally, if X_1 and X_2 interact in their joint effect on Y, then the relationship of either X variable to Y depends on the value of the other X variable.

Application 8   You want to *obtain a valid and precise estimate of one or more regression coefficients* from a larger set of regression coefficients in a given model. For example, you may wish to obtain an accurate estimate of the coefficient of a variable measuring alcohol consumption (X_1) in a regression model that relates hypertension status (Y), a dichotomous response variable, to X_1 and several other control variables (e.g., age and smoking status). Such an estimate may be used to quantify the effect of alcohol consumption on hypertension status after adjustment for the effects of certain control variables also in the model.

4-2 Association versus Causality

A researcher must be cautious about interpreting the results obtained from a regression analysis or, more generally, from any form of analysis seeking to quantify an association (e.g., via a correlation coefficient) among two or more variables. Although the statistical computations used to produce an estimated measure of association may be correct, the estimate itself may be biased. Such bias may result from the method used to select subjects for the study, from errors in the information used in the statistical analyses, or even from other variables that can account for the observed association but that have not been measured or appropriately considered in the analysis. (See Kleinbaum, Kupper, and Morgenstern 1982, for a discussion of validity in epidemiologic research.)

For example, if diastolic blood pressure and physical activity level were measured on a sample of individuals at a particular time, a regression analysis might suggest that, on the average, blood pressure decreases with increased physical activity; further, such an analysis may provide evidence (e.g., based on a confidence interval) that this association is of moderate strength and is statistically significant. If the study involved only healthy adults, however, or if physical activity level was measured inappropriately, or if such other factors as age, race, and sex were not correctly taken into account, the above conclusions might be rendered invalid or at least questionable.

Continuing with the preceding example, if the investigators were satisfied that the findings were basically valid (i.e., the observed association was not spurious), could they then conclude that a low level of physical activity is a cause of high blood pressure? The answer is an unequivocal *no*!

The finding of a "statistically significant" association in a particular study (no matter how well done) does not establish a causal relationship. To evaluate claims of causality, the investigator must consider criteria that are external to the specific characteristics and results of any single study.

It is beyond the scope of this text to discuss causal inference making. Nevertheless, we will briefly review some key ideas on this subject. Most strict definitions of causality (e.g., Blalock 1971; Susser 1973) require that a *change* in one variable (X) always *produce* a change in another variable (Y).[1] This suggests that, to demonstrate a cause–effect relationship between X and Y, *experimental proof* is required that a change in Y results from a change in X. Though it is needed, such experimental evidence is often impractical, infeasible, or even unethical to obtain, especially when considering risk factors (e.g., cigarette smoking or exposure to chemicals) that are potentially harmful to human subjects. Consequently, alternative criteria based on information *not* involving direct experimental evidence are typically employed when attempting to make causal inferences regarding variable relationships in human populations.

One school of thought regarding causal inference has produced a collection of procedures commonly referred to as *path analysis* (Blalock 1971) or *structural equations analysis* (Bollen 1989). To date, such procedures have been applied primarily to sociological and political science studies. Essentially, these methods attempt to assess causality indirectly, by eliminating competing causal explanations via data analysis and finally arriving at an acceptable causal model that is not obviously contradicted by the data at hand. Thus, these methods, rather than attempting to establish a particular causal theory directly, arrive at a final causal model through a process of elimination. In this procedure, literature relevant to the research question must be considered in order to postulate causal models; in addition, various estimated correlation ("path") coefficients must be compared by means of data analysis.

---

[1]An imperfect approximation to this ideal for real-world phenomena might be that, on the average, a change in Y is produced by a change in X.

A second, more widely used approach for making causal conjectures, particularly in the health and medical sciences, employs a judgmental (and more qualitative than quantitative) evaluation of the combined results from several studies, using a set of operational criteria generally agreed on as necessary (but not sufficient) for supporting a given causal theory. Efforts to define such a set of criteria were made in the late 1950s and early 1960s by investigators reviewing research on the health hazards of smoking. A list of general criteria for assessing the extent to which available evidence supports a causal relationship was formalized by Bradford Hill (1971), and this list has subsequently been adopted by many epidemiologic researchers. The list contains seven criteria:

1. *Strength of association.* The stronger an observed association appears over a series of different studies, the less likely it is that this association is spurious because of bias.

2. *Dose–response effect.* The value of the dependent variable (e.g., the rate of disease development) changes in a meaningful pattern (e.g., increases) with the dose (or level) of the suspected causal agent under study.

3. *Lack of temporal ambiguity.* The hypothesized cause precedes the occurrence of the effect. (The ability to establish this time pattern depends on the study design used.)

4. *Consistency of the findings.* Most or all studies concerned with a given causal hypothesis produce similar results. Of course, studies dealing with a given question may all have serious bias problems that can diminish the importance of observed associations.

5. *Biological and theoretical plausibility of the hypothesis.* The hypothesized causal relationship is consistent with current biological and theoretical knowledge. The current state of knowledge may nonetheless be insufficient to explain certain findings.

6. *Coherence of the evidence.* The findings do not seriously conflict with accepted facts about the outcome variable being studied (e.g., knowledge about the natural history of some disease).

7. *Specificity of the association.* The study factor (i.e., the suspected cause) is associated with only one effect (e.g., a specific disease). Many study factors have multiple effects, however, and most diseases have multiple causes.

Clearly, applying the above criteria to a given causal hypothesis is hardly a straightforward matter. Even if these criteria are all satisfied, a causal relationship cannot be claimed with complete certainty. Nevertheless, in the absence of solid experimental evidence, the use of such criteria may be a logical and practical way to address the issue of causality, especially with regard to studies on human populations.

4-3 Statistical versus Deterministic Models

Although *causality* cannot be established by statistical analyses, associations among variables can be well quantified in a *statistical* sense. With proper statistical design and analysis, an investigator can model the extent to which changes in independent variables are related to changes in dependent variables. However, *statistical models* developed by using regression or other multivariable methods must be distinguished from *deterministic models.*

The law of falling bodies in physics, for example, is a deterministic model that assumes an ideal setting: the dependent variable varies in a completely prescribed way according to a perfect (error-free) mathematical function of the independent variables.

Statistical models, on the other hand, allow for the possibility of error in describing a relationship. For example, in a study relating blood pressure to age, persons of the same age are unlikely to have exactly the same observed blood pressure. Nevertheless, with proper statistical methods, we might be able to conclude that, on the average, blood pressure increases with age. Further, appropriate statistical modeling can permit us to predict the expected blood pressure for a given age and to associate a measure of variability with that prediction. Through the use of probability and statistical theory, such statements take into account the uncertainty of the real world by means of measurement error and individual variability. Of course, because such statements are necessarily nondeterministic, they require careful interpretation. Unfortunately, such interpretation is often quite difficult to make.

4-4 Concluding Remarks

In this short chapter, we have informally introduced the general regression problem and indicated a variety of situations to which regression modeling can be applied. We have also cautioned the reader about the types of conclusions that can be drawn from such modeling efforts.

We now turn to the actual quantitative details involved in fitting a regression model to a set of data and then in estimating and testing hypotheses about important parameters in the model. In the next chapter, we will discuss the simplest form of regression model—a straight line. In subsequent chapters, we will consider more complex forms.

References

Blalock, H. M., Jr., ed. 1971. *Causal Models in the Social Sciences.* Chicago: Aldine Publishing.

Bollen, K. A. 1989. *Structural Equations with Latent Variables.* New York: John Wiley & Sons.

Hill, A. B. 1971. *Principles of Medical Statistics,* 9th ed. New York: Oxford University Press.

Kleinbaum, D. G.; Kupper, L. L.; and Morgenstern, H. 1982. *Epidemiologic Research: Principles and Quantitative Methods.* Belmont, Calif.: Lifetime Learning Publications.

Susser, M. 1973. *Causal Thinking in the Health Sciences.* New York: Oxford University Press.

Thompson, S. J. 1972. "The Doctor–Patient Relationship and Outcomes of Pregnancy." Ph.D. dissertation, Department of Epidemiology, University of North Carolina, Chapel Hill.

5

Straight-line
Regression Analysis

5-1 Preview

The simplest (but by no means trivial) form of the general regression problem deals with one dependent variable Y and one independent variable X. We have previously described the general problem in terms of k independent variables $X_1, X_2, \ldots, X_k$. Let us now restrict our attention to the special case $k = 1$ but denote X_1 as X to keep our notation as simple as possible. To clarify the basic concepts and assumptions of regression analysis, we find it useful to begin with a single independent variable. Furthermore, researchers often begin by looking at one independent variable at a time even when several independent variables are eventually jointly considered.

5-2 Regression with a Single Independent Variable

We begin this section by describing the statistical problem of finding the *curve* (straight line, parabola, etc.) that *best fits* the data, closely approximating the true (but unknown) relationship between X and Y.

5-2-1 The Problem

Given a sample of n individuals (or other study units, such as geographical locations, time points, or pieces of physical material), we observe for each a value of X and a value of Y. We thus have n pairs of observations that can be denoted by $(X_1, Y_1), (X_2, Y_2), \ldots, (X_n, Y_n)$, where the subscripts now refer to different individuals rather than different variables. Because these pairs may be considered as points in two-dimensional space, we can plot them on a graph. Such a graph is called a *scatter diagram*. For example, measurements of age and systolic blood pressure for 30 individuals might yield the scatter diagram given in Figure 5-1.

FIGURE 5-1   Scatter diagram of age and systolic blood pressure.

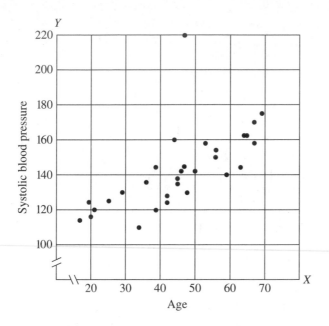

5-2-2 Basic Questions to Be Answered

Two basic questions must be dealt with in any regression analysis:

1. What is the most appropriate mathematical model to use—a straight line, a parabola, a log function, or what?

2. Given a specific model, what do we mean by and how do we determine the best-fitting model for the data? In other words, if our model is a straight line, how do we find the best-fitting line?

5-2-3 General Strategy

Several general strategies can be used to study the relationship between two variables by means of regression analysis. The most common of these is called the *forward method*. This strategy begins with a simply structured model—usually a straight line—and adds more complexity to the model in successive steps, if necessary. Another strategy, called the *backward method,* begins with a complicated model—such as a high-degree polynomial—and successively simplifies it, if possible, by eliminating unnecessary terms. A third approach uses *a model suggested from experience or theory,* which is revised either toward or away from complexity, as dictated by the data.

The strategy chosen depends on the type of problem and on the data; there are no hard-and-fast rules. The quality of the results often depends more on the skill with which a strategy is applied than on the particular strategy chosen. It is often tempting to try many strategies and then to use the results that provide the most "reasonable" interpretation of the relationship between the response and predictor variables. This exploratory approach demands particular care to ensure the reliability of any conclusions.

In Chapter 16, we will discuss in detail the issue of choosing a model-building strategy. For reasons discussed there, we often prefer the backward strategy. The forward method, however, corresponds more naturally to the usual development of theory from simple to complex. In some simple situations, forward and backward strategies lead to the same final model. In general, however, this is not the case!

Since it is the simplest method to understand and can therefore be used as a basis for understanding other methods, we begin by offering a step-by-step description of the forward strategy:

1. Assume that a straight line is the appropriate model. Later the validity of this assumption can be investigated.

2. Find the best-fitting straight line, which is the line among all possible straight lines that best agrees (as will be defined later) with the data.

3. Determine whether the straight line found in step 2 significantly helps to describe the dependent variable Y. Here it is necessary to check that certain basic statistical assumptions (e.g., normality) are met. These assumptions will be discussed in detail subsequently.

4. Examine whether the assumption of a straight-line model is correct. One approach for doing this is called *testing for lack of fit,* although other approaches can be used instead.

5. If the assumption of a straight line is found to be invalid in step 4, fit a new model (e.g., a parabola) to the data, determine how well it describes Y (i.e., repeat step 3), and then decide whether the new model is appropriate (i.e., repeat step 4).

6. Continue to try new models until an appropriate one is found.

A flow diagram for this strategy is given in Figure 5-2.

Since the usual (forward) approach to regression analysis with a single independent variable begins with the assumption of a straight-line model, we will consider this model first. Before describing the *statistical* methodology for this special case, let us review some basic straight-line mathematics. You may wish to skip the next section if you are already familiar with its contents.

FIGURE 5-2   Flow diagram of the forward method.

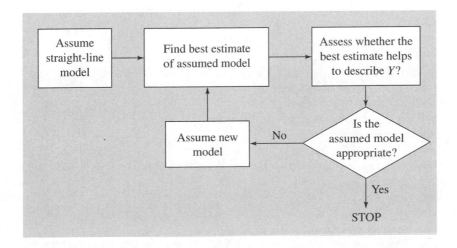

5-3 Mathematical Properties of a Straight Line

Mathematically, a straight line can be described by an equation of the form

$$y = \beta_0 + \beta_1 x \qquad (5.1)$$

We have used lowercase letters y and x, instead of capital letters, in this equation to emphasize that we are treating these variables in a purely mathematical, rather than statistical, context. The symbols β_0 and β_1 have constant values for a given line and are therefore not considered variables; β_0 is called the *y-intercept* of the line, and β_1 is called the *slope*. Thus, $y = 5 - 2x$ describes a straight line with intercept 5 and slope -2, whereas $y = -4 + 1x$ describes a different line with intercept -4 and slope 1. These two lines are shown in Figure 5-3.

The intercept β_0 is the value of y when $x = 0$. For the line $y = 5 - 2x$, $y = 5$ when $x = 0$. For the line $y = -4 + 1x$, $y = -4$ when $x = 0$. The slope β_1 is the amount of change in y for each 1-unit change in x. For any given straight line, this rate of change is always constant. Thus, for the line $y = 5 - 2x$, when x changes 1 unit from 3 to 4, y changes -2 units (the value of the slope) from $5 - 2(3) = -1$ to $5 - 2(4) = -3$; and when x changes from 1 to 2, also 1 unit, y changes from $5 - 2(1) = 3$ to $5 - 2(2) = 1$, also -2 units.

The properties of any straight line can be viewed graphically as in Figure 5-3. To graph a given line, plot any two points on the line and then connect them with a ruler. One of the two points often used is the y-intercept. This point is given by ($x = 0$, $y = 5$) for the line $y = 5 - 2x$ and by ($x = 0$, $y = -4$) for $y = -4 + 1x$. The other point for each line may be determined by arbitrarily selecting an x and finding the corresponding y. An x of 3 was used in our two examples. Thus, for $y = 5 - 2x$, an x of 3 yields a y of $5 - 2(3) = -1$; and for $y = -4 + 1x$, an x of 3 yields a y of $-4 + 1(3) = -1$. The line $y = 5 - 2x$ can then be drawn by connecting the points ($x = 0$, $y = 5$) and ($x = 3$, $y = -1$), and the line $y = -4 + 1x$ can be drawn from the points ($x = 0$, $y = -4$) and ($x = 3$, $y = -1$).

As Figure 5-3 illustrates, in the equation $y = 5 - 2x$, y decreases as x increases. Such a line is said to have *negative* slope. Indeed, this definition agrees with the sign of the slope -2 in

FIGURE 5-3   Straight-line plots.

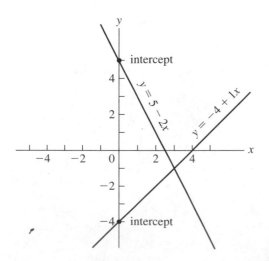

the equation. Conversely, the line $y - -4 + 1x$ is said to have *positive* slope, since y increases as x increases.

5-4 Statistical Assumptions for a Straight-line Model

Suppose that we have tentatively assumed a straight-line model as the first step in the forward method for determining the best model to describe the relationship between X and Y. We now wish to determine the best-fitting line. Certainly, we will have no trouble deciding what is meant by "best fitting" if the data allow us to draw a single straight line through every point in the scatter diagram. Unfortunately, this will never happen with real-life data. For example, persons of the same age are unlikely to have the same blood pressure, height, or weight.

Thus, the straight line we seek can only approximate the true state of affairs and cannot be expected to predict precisely each individual's Y from that individual's X. In fact, this need to approximate would exist even if we measured X and Y on the whole population of interest instead of on just a sample from that population. In addition, the fact that the line is to be determined from the sample data and not from the population requires us to consider the problem of how to estimate unknown population parameters.

What are these parameters? The ones of primary concern at this point are the intercept β_0 and the slope β_1 of the straight line of the general mathematical form of (5.1) that best fits the X-Y data for the entire population. To make inferences from the sample about this population line, we need to make five statistical assumptions covering existence, independence, linearity, homoscedasticity, and normality.

5-4-1 Statement of Assumptions

Assumption 1: Existence: *For any fixed value of the variable X, Y is a random variable with a certain probability distribution having finite mean and variance.* The (population) mean of this distribution will be denoted as $\mu_{Y|X}$ and the (population) variance as $\sigma^2_{Y|X}$. The notation "$Y \mid X$" indicates that the mean and the variance of the random variable Y depend on the value of X.

This assumption applies to any regression model, whether a straight line or not. Figure 5-4 illustrates the assumption. The different distributions are drawn vertically to correspond to different values of X. The dots denoting the mean values $\mu_{Y|X}$ at different X's have been connected to form the *regression equation,* which is the population model to be estimated from the data.

Assumption 2: Independence: *The Y-values are statistically independent of one another.* This assumption is appropriate in many, but not all, situations. In particular, Assumption 2 is usually violated when different observations are made on the same individual at different times. For example, if weight is measured on an individual at different times (i.e., longitudinally over time), we can expect that the weight at one time is related to the weight at a later time. As another example, if blood pressure is measured on a given individual longitudinally over time, we can expect the blood pressure value at one time to be in the same range as the blood pressure value at the previous or following time. When Assumption 2 is not satisfied, ignoring the dependency among the Y-values can often lead to invalid statistical conclusions.

FIGURE 5-4   General regression equation.

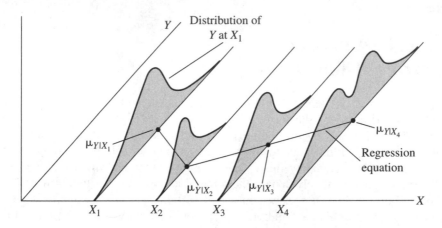

When the Y-values are not independent, special methods can be used to find the best-fitting model and to make valid statistical inferences. The method chosen depends on the characteristics of the response variable, the type of dependence, and the complexity of the problem. In Chapter 21, we describe a "mixed model" analysis of variance approach for designs involving repeated measurements on study subjects. In some cases, multivariate linear models are appropriate. See Morrison (1976) or Timm (1975) for a general introduction to multivariate linear models. More recently, Zeger and Liang (1986) introduced the "generalized estimating equations" (GEE) approach for analyzing correlated response data, and an excellent book on this very general and useful methodology is now available (Diggle, Liang, and Zeger 1994).

Assumption 3: Linearity: *The mean value of Y, $\mu_{Y|X}$, is a straight-line function of X.* In other words, if the dots denoting the different mean values $\mu_{Y|X}$ are connected, a straight line is obtained. This assumption is illustrated in Figure 5-5.

Using mathematical symbols, we can describe Assumption 3 by the equation

$$\mu_{Y|X} = \beta_0 + \beta_1 X \qquad (5.2)$$

where β_0 and β_1 are the intercept and the slope of this (population) straight line, respectively. Equivalently, we can express (5.2) in the form

$$Y = \beta_0 + \beta_1 X + E \qquad (5.3)$$

where E denotes a random variable that has mean 0 at fixed X (i.e, $\mu_{E|X} = 0$ for any X). More specifically, since X is fixed and not random, (5.3) represents the dependent variable Y as the sum of a constant term $(\beta_0 + \beta_1 X)$ and a random variable (E).[1] Thus, the probability distributions of Y and E differ only in the value of this constant term; that is, since E has mean 0, Y must have mean $\beta_0 + \beta_1 X$.

---

[1]The statement "X is fixed and not random" is often associated with the statement "X is measured without error." For our purposes, the practical implication of either statement for making statistical inferences from sample to population is that the *only* random component on the right-hand side of (5.3) is E.

FIGURE 5-5   Straight-line assumption.

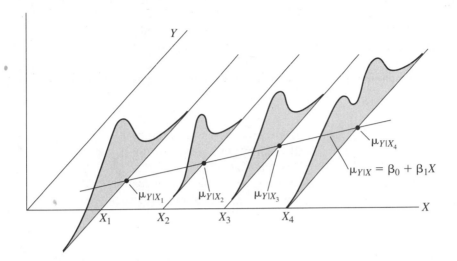

Equations (5.2) and (5.3) describe a *statistical* model. These equations should be distinguished from the *mathematical* model for a straight line described by (5.1), which does not consider Y as a random variable.

The variable E describes how distant an individual's response can be from the population regression line (Figure 5-6). In other words, what we observe at a given X (namely, Y) is in *error* from that expected on the average (namely, $\mu_{Y|X}$) by an amount E, which is random and varies from individual to individual. For this reason, E is commonly referred to as the *error component* in the model (5.3). Mathematically, E is given by the formula

$$E = Y - (\beta_0 + \beta_1 X)$$

or by

$$E = Y - \mu_{Y|X}$$

This concept of an error component is particularly important for defining a good-fitting line, since, as we will see in the next section, a line that fits data well ought to have small deviations (or errors) between what is observed and what is predicted by the fitted model.

FIGURE 5-6   Error component E.

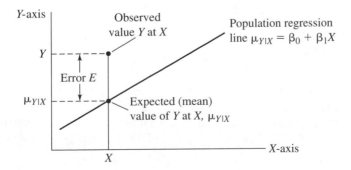

Assumption 4: Homoscedasticity: *The variance of Y is the same for any X.* (*Homo-* means "same," and *-scedastic* means "scattered.") An example of the violation of this assumption (called *heteroscedasticity*) is shown in Figure 5-5, where the distribution of Y at X_1 has considerably more spread than the distribution of Y at X_2. This means that $\sigma_{Y|X_1}^2$, the variance of Y at X_1, is greater than $\sigma_{Y|X_2}^2$, the variance of Y at X_2.

In mathematical terms, the homoscedastic assumption can be written as

$$\sigma_{Y|X}^2 \equiv \sigma^2$$

for all X. This formula is a short-hand way of saying that, since $\sigma_{Y|X_i}^2 = \sigma_{Y|X_j}^2$ for *any* two different values of X, we might as well simplify our notation by giving the common variance a single name—say, σ^2—that does not involve X at all.

A number of techniques of varying statistical sophistication can be used to determine whether the homoscedastic assumption is satisfied. Some of these procedures will be discussed in Chapter 12.

Assumption 5: Normal Distribution: *For any fixed value of X, Y has a normal distribution.* This assumption makes it possible to evaluate the statistical significance (e.g., by means of confidence intervals and tests of hypotheses) of the relationship between X and Y, as reflected by the fitted line.

Figure 5-5 provides an example in which this assumption is violated. In addition to the variances not being all equal in this figure, the distributions of Y at X_3 and at X_4 are not normal. The distribution at X_3 is skewed, whereas the normal distribution is symmetric. The distribution at X_4 is bimodal (two humps), whereas the normal distribution is unimodal (one hump). Methods for determining whether the normality assumption is tenable are described in Chapter 12.

If the normality assumption is not *badly* violated, the conclusions reached by a regression analysis in which normality is assumed will generally be reliable and accurate. This stability property with respect to deviations from normality is a type of *robustness*. Consequently, we recommend giving considerable leeway before deciding that the normality assumption is so badly violated as to require alternative inference-making procedures.

If the normality assumption is deemed unsatisfactory, the Y-values may be transformed by using a log, square root, or other function to see whether the new set of observations is approximately normal. Care must be taken when using such transformations to ensure that other assumptions, such as variance homogeneity, are not violated for the transformed variable. Fortunately, in practice such transformations usually help satisfy both the normality and variance homogeneity assumptions.

5-4-2 Summary and Comments

The assumptions of homoscedasticity and normality apply to the distribution of Y when X is fixed (i.e., Y given X), and not to the distributions of Y associated with different X-values. Many people find it more convenient to describe these two assumptions in terms of the error E. It is sufficient to say that the random variable E has a normal distribution with mean 0 and variance σ^2 for all observations. Of course, the linearity, existence, and independence assumptions must also be specified.

It is helpful to maintain distinctions among such concepts as *random variables, parameters,* and *point estimates.* The variable Y is a random variable, and an observation of it yields a particular value or "realization"; the variable X is a fixed (nonrandom), known variable. The constants β_0 and β_1 are parameters with unknown but specific values for a particular population. The variable E is a random, unobservable variable. Using some estimation procedure (e.g., least squares), one constructs point estimates $\hat{\beta}_0$ and $\hat{\beta}_1$ of β_0 and β_1, respectively. Once $\hat{\beta}_0$ and $\hat{\beta}_1$ are obtained, a point estimate of E at the value X is calculated as

$$\hat{E} = Y - \hat{Y} = Y - (\hat{\beta}_0 + \hat{\beta}_1 X)$$

The estimated error $\hat{E}$ is typically called a *residual.* If there are n (X, Y) pairs $(X_1, Y_1), (X_2, Y_2), \ldots,$ (X_n, Y_n), then there are n residuals $\hat{E}_i = Y_i - \hat{Y}_i = Y_i - (\hat{\beta}_0 + \hat{\beta}_1 X_i), i = 1, 2, \ldots, n.$

Some statisticians refer to a normally distributed random variable as having a Gaussian distribution. This terminology avoids confusing *normal* with its other meaning of "customary" or "usual"; it emphasizes the fact that the term *Gaussian* refers to a *particular* bell-shaped function; and it appropriately honors the mathematician Carl Gauss (1777–1855).

5-5 Determining the Best-fitting Straight Line

By far the simplest and quickest method for determining a straight line is to choose the line that can best be drawn by eye. Although this method often paints a reasonably good picture, it is extremely subjective and imprecise and is worthless for statistical inference. We now consider two analytical approaches for finding the best-fitting straight line.

5-5-1 The Least-squares Method

The *least-squares method* determines the best-fitting straight line as the line that *minimizes* the sum of squares of the lengths of the vertical-line segments (Figure 5-7) drawn from the observed data points on the scatter diagram to the fitted line. The idea here is that the smaller the deviations of observed values from this line (and consequently the smaller the sum of squares of these deviations), the closer or "snugger" the best-fitting line will be to the data.

In mathematical notation, the least-squares method is described as follows. Let $\hat{Y}_i$ denote the estimated response at X_i based on the fitted regression line; in other words, $\hat{Y}_i = \hat{\beta}_0 + \hat{\beta}_1 X_i$, where $\hat{\beta}_0$ and $\hat{\beta}_1$ are the intercept and the slope of the fitted line, respectively. The vertical distance between the observed point (X_i, Y_i) and the corresponding point $(X_i, \hat{Y}_i)$ on the fitted line is

FIGURE 5-7 Deviations of observed points from the fitted regression line.

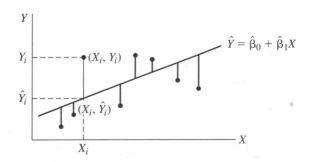

given by the absolute value $\left| Y_i - \hat{Y}_i \right|$ or $\left| Y_i - (\hat{\beta}_0 + \hat{\beta}_1 X_i) \right|$. The sum of the squares of all such distances is given by

$$\sum_{i=1}^{n} (Y_i - \hat{Y}_i)^2 = \sum_{i=1}^{n} (Y_i - \hat{\beta}_0 - \hat{\beta}_1 X_i)^2$$

The least-squares solution is defined to be the choice of $\hat{\beta}_0$ and $\hat{\beta}_1$ for which the sum of squares just described is a minimum. In standard jargon, $\hat{\beta}_0$ and $\hat{\beta}_1$ are termed the *least-squares estimates* of the parameters β_0 and β_1, respectively, in the statistical model given by (5.3).

The *minimum sum of squares* corresponding to the least-squares estimates $\hat{\beta}_0$ and $\hat{\beta}_1$ is usually called the *sum of squares about the regression line,* the *residual sum of squares,* or the *sum of squares due to error* (SSE). The measure SSE is of great importance in assessing the quality of the straight-line fit, and its interpretation will be discussed in section 5-6.

Mathematically, the essential property of the measure SSE can be stated in the following way. If β_0^* and β_1^* denote any other possible estimators of β_0 and β_1, we must have

$$\text{SSE} = \sum_{i=1}^{n} (Y_i - \hat{\beta}_0 - \hat{\beta}_1 X_i)^2 \leq \sum_{i=1}^{n} (Y_i - \beta_0^* - \beta_1^* X_i)^2$$

5-5-2 The Minimum-variance Method

The *minimum-variance method* is more classically statistical than the method of least squares, which can be viewed as a purely mathematical algorithm. In this second approach, determining the best fit becomes a statistical estimation problem. The goal is to find point estimators of β_0 and β_1 with good statistical properties. In this regard, under the previous assumptions, the best line is determined by the estimators $\hat{\beta}_0$ and $\hat{\beta}_1$ that are *unbiased* for their unknown population counterparts β_0 and β_1, respectively, and have minimum variance among all unbiased (linear) estimators of β_0 and β_1.

5-5-3 Solution to the Best-fit Problem

Fortunately, both the least-squares method and the minimum-variance method yield exactly the same solution, which we will state without proof.[2]

Let $\overline{Y}$ denote the sample mean of the observations on Y, and let $\overline{X}$ denote the sample mean of the values of X. Then the best-fitting straight line is determined by the formulas

$$\hat{\beta}_1 = \frac{\displaystyle\sum_{i=1}^{n} (X_i - \overline{X})(Y_i - \overline{Y})}{\displaystyle\sum_{i=1}^{n} (X_i - \overline{X})^2} \tag{5.4}$$

---

[2]Another general method of parameter estimation is called *maximum likelihood.* Under the assumption of a Gaussian distribution, the maximum-likelihood estimates of β_0 and β_1 are exactly the same as the least-squares and minimum-variance estimates. A general discussion of maximum-likelihood methods is given in Chapter 22.

$$\hat{\beta}_0 = \overline{Y} - \hat{\beta}_1 \overline{X} \tag{5.5}$$

In calculating $\hat{\beta}_0$ and $\hat{\beta}_1$, we recommend using a computer program for regression modeling from a convenient computer package. Many computer packages with regression programs are now available, the most popular of which are SAS, SPSS, BMDP, SYSTAT, and MINITAB. Other, more recently developed packages include EGRET, GLIM, MULTR, S+, SPIDA, STATA, and JMP, the latter two being available for MacIntosh computers. In this text, we will use SAS exclusively to present computer output, although we recognize that other packages may be preferred by particular users.

The least-squares line may generally be represented by

$$\hat{Y} = \hat{\beta}_0 + \hat{\beta}_1 X \tag{5.6}$$

or equivalently by

$$\hat{Y} = \overline{Y} + \hat{\beta}_1 (X - \overline{X}) \tag{5.7}$$

Either (5.6) or (5.7) may be used to determine predicted Y's that correspond to X's actually observed or to other X-values in the region of experimentation. Simple algebra can be used to demonstrate the equivalence of (5.6) and (5.7). The right-hand side of (5.7), $\overline{Y} + \hat{\beta}_1(X - \overline{X})$, can be written as $\overline{Y} + \hat{\beta}_1 X - \hat{\beta}_1 \overline{X}$, which in turn equals $\overline{Y} - \hat{\beta}_1 \overline{X} + \hat{\beta}_1 X$, which is equivalent to (5.6) since $\hat{\beta}_0 = \overline{Y} - \hat{\beta}_1 \overline{X}$.

Table 5-1 lists observations on systolic blood pressure and age for a sample of 30 individuals. The scatter diagram for this sample was presented in Figure 5-1. For this data set, the output from the SAS package's PROC REG routine is also shown.

TABLE 5-1   Observations on systolic blood pressure (SBP) and age for a sample of 30 individuals.

Individual (i)	SBP (Y)	Age (X)	Individual (i)	SBP (Y)	Age (X)
1	144	39	16	130	48
2	220	47	17	135	45
3	138	45	18	114	17
4	145	47	19	116	20
5	162	65	20	124	19
6	142	46	21	136	36
7	170	67	22	142	50
8	124	42	23	120	39
9	158	67	24	120	21
10	154	56	25	160	44
11	162	64	26	158	53
12	150	56	27	144	63
13	140	59	28	130	29
14	110	34	29	125	25
15	128	42	30	175	69

SAS Output for Data of Table 5-1

```
              REGRESSION OF SBP(Y) ON AGE (X)
                   Descriptive Statistics
Variables        Sum          X̄        Mean        Uncorrected SS
INTERCEP          30                     1                     30
X               1354          45.13333333              67894
Y               4276         142.53333333             624260


Variables        S²_X      Variance              Std Deviation
INTERCEP                       0                       0
X                          233.91264368          15.294202944
Y                          509.91264368          22.581245397
                 S²_Y
```

Analysis of Variance SSR

Sum of X^2 *dist* Mean SSE

Source	DF	Sum of Squares	Mean Square	F Value	Prob > F
Model	1	6394.02269 ←SSR	6394.02269	21.330	0.0001
Error	28	8393.44398 ←SSE	299.76586	$S²_{Y\mid X}$	r^2 (see Chapter 6)
C Total	29	14787.46667 ←SST			

					0.4324
Root MSE	$S_{Y\mid X}$ → 17.31375		R-square		0.4121
Dep Mean	142.53333		Adj R-sq		
C.V.	12.14716				

t-test statistics for tests of
$H_0: \beta_0 = 0$ and $H_0: \beta_1 = 0$
(vs. nondirectional alternatives)

Parameter Estimates

P-values for *t*-tests

Variable	DF	$\hat{\beta}_0$ Parameter Estimate	Standard $S_{\hat{\beta}_0}$ Error	T for H0: Parameter=0	Prob > |T|
INTERCEP	1	98.714718	10.00046756	9.871	0.0001
X	1	0.970870 $\hat{\beta}_1$	0.21021574 ← $S_{\hat{\beta}_1}$	4.618	0.0001

If $P < .05$

Reject H_0

The estimated line (5.6) is computed to be $\hat{Y} = 98.71 + 0.97X$ and is graphed as the solid line in Figure 5-8. This line reflects the clear trend that systolic blood pressure increases as age increases. Notice that one point, (47, 220), seems quite out of place with the other data; such an observation is often called an *outlier*. Because an outlier can affect the least-squares estimates, the determination of whether an outlier should be removed from the data is important. Usually this decision can be made only after thorough evaluation of the experimental conditions, the data collection process, and the data themselves. (See Chapter 12 for further discussion of the treatment of outliers.) If the decision is difficult, one can always determine the effect of removing the outlier by refitting the model to the remaining data. In this case, the resulting least-squares line is

$$\hat{Y} = 97.08 + 0.95X$$

and is shown on the graph in Figure 5-8 as the dashed line. As might be expected, this line is slightly below the one obtained by using all the data.

FIGURE 5-8 Best-fitting straight line to age systolic blood pressure data of Table 5-1.

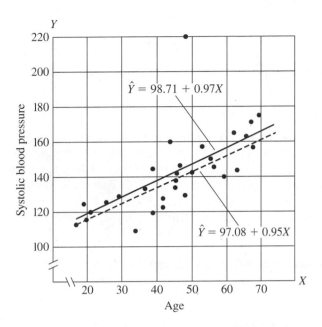

5-6 Measure of the Quality of the Straight-line Fit and Estimate of σ^2

Once the least-squares line is determined, we would like to evaluate whether the fitted line actually aids in predicting Y—and if so, to what extent. A measure that helps to answer these questions is provided by

$$\text{SSE} = \sum_{i=1}^{n} (Y_i - \hat{Y}_i)^2 \quad , \quad \frac{SSE}{\sigma^2} = \chi^2 (n-2)$$

where $\hat{Y}_i = \hat{\beta}_0 + \hat{\beta}_1 X_i$. Clearly, if SSE = 0, the straight line fits perfectly; that is, $Y_i = \hat{Y}_i$ for each i, and every observed point lies on the fitted line. As the fit gets worse, SSE gets larger, since the deviations of points from the regression line become larger.

Two possible factors contribute to the inflation of SSE. First, there may be a lot of variation in the data; that is, σ^2 may be large. Second, the assumption of a straight-line model may not be appropriate. It is important, therefore, to determine the separate effects of each of these components, since they address decidedly different issues with regard to the fit of the model. For the time being, we will assume that the second factor is not at issue. Thus, assuming that the straight-line model is appropriate, we can obtain an estimate of σ^2 by using SSE. Such an estimate is needed for making statistical inferences about the true (i.e., population) straight-line relationship between X and Y. This estimate of σ^2 is given by the formula

$$S_{Y|X}^2 = \frac{1}{n-2} \sum_{i=1}^{n} (Y_i - \hat{Y}_i)^2 = \frac{1}{n-2} \text{SSE} \qquad (5.8)$$

■ **Example 5-1**   From the computer output provided on p. 50, we find

$$S_{Y|X}^2 = 299.76586$$

Readers may wonder why $S_{Y|X}^2$ estimates σ^2—especially since, at first glance, (5.8) looks different from the formula usually used for the sample variance, $\sum_{i=1}^{n}(Y_i - \overline{Y})^2/(n-1)$. The latter formula is appropriate when the Y's are independent, with the same mean μ and variance σ^2. Since μ is unknown in this case (its estimate being, of course, $\overline{Y}$), we must divide by $n-1$ instead of n to make the sample variance an unbiased estimator of σ^2. To put it another way, we subtract 1 from n because to estimate σ^2 we had to estimate *one* other parameter first, μ.

If a straight-line model is appropriate, the population mean response $\mu_{Y|X}$ changes with X. For example, using the least-squares line (5.6) as an approximation to the population line for the age–systolic blood pressure data, the estimated mean of the Y's at $X = 40$ is approximately 138, whereas the estimated mean of the Y's at $X = 70$ is close to 167. Therefore, instead of subtracting $\overline{Y}$ from each Y_i when estimating σ^2, we should subtract $\hat{Y}_i$ from Y_i, because $\hat{Y}_i = \hat{\beta}_0 + \hat{\beta}_1 X_i$ is the estimate of $\mu_{Y|X_i}$. Furthermore, we subtract 2 from n in the denominator of our estimate, since the determination of $\hat{Y}_i$ requires the estimation of two parameters, β_0 and β_1.

When we discuss testing for lack of fit of the assumed model, we will show how it is possible to obtain an estimate of σ^2 that does not assume the correctness of the straight-line model. ■

5-7 Inferences About the Slope and Intercept

To assess whether the fitted line helps to predict Y from X, and to take into account the uncertainties of using a sample, it is standard practice to compute confidence intervals and/or test statistical hypotheses about the unknown parameters in the assumed straight-line model. Such confidence intervals and tests require, as described in section 5-4, the assumption that the random variable Y has a normal distribution at each fixed value of X. Working from this assumption, one can deduce that the estimators $\hat{\beta}_0$ and $\hat{\beta}_1$ are each normally distributed, with respective means β_0 and β_1 when (5.2) holds, and with easily derivable variances.[3] These estimators, together with estimates of their variances, can then be used to form confidence intervals and test statistics based on the t *distribution*.

---

[3]An important property that allows the normality assumption on Y to carry over to $\hat{\beta}_0$ and $\hat{\beta}_1$ is that these estimators are *linear functions* of the Y's. Such a function is defined by a formula of the structure

$$L = \sum_{i=1}^{n} c_i Y_i$$

or equivalently, $L = c_1 Y_1 + c_2 Y_2 + \cdots + c_n Y_n$, where the c_i's are constants not involving the Y's. A simple example of a linear function is $\overline{Y}$, which can be written as

$$\sum_{i=1}^{n} \frac{1}{n} Y_i$$

Here the c_i's equal $1/n$ for each i. The normality of $\hat{\beta}_0$ and $\hat{\beta}_1$ derives from a statistical theorem stating that linear functions of independent normally distributed observations are themselves normally distributed.

More specifically, to test the hypothesis $H_0: \beta_1 = \beta_1^{(0)}$, where $\beta_1^{(0)}$ is some hypothesized value for β_1, the test statistic used is

$$T = \frac{\hat{\beta}_1 - \beta_1^{(0)}}{S_{\hat{\beta}_1}} \qquad df = n - 2 \tag{5.9}$$

where

$$S_{\hat{\beta}_1} = \frac{S_{Y|X}}{S_x\sqrt{n-1}}$$

This test statistic has the t distribution with $n - 2$ degrees of freedom when H_0 is true. Here, $S_{Y|X}^2$ denotes the sample estimate of σ^2 defined by (5.8), and S_X is the sample standard deviation of the x's defined by (3.2) on p. 15. The denominator $S_{\hat{\beta}_1}$ in the test statistic is an estimate of the unknown standard error of the estimator $\hat{\beta}_1$, given by

$$\sigma_{\hat{\beta}_1} = \frac{\sigma}{S_x\sqrt{n-1}}$$

Thus, the test statistic (5.9) is the ratio of a normally distributed random variable minus its mean divided by an estimate of its standard error. Such a statistic has a t distribution for the kinds of situations encountered in this text.

Similarly, to test the hypothesis $H_0: \beta_0 = \beta_0^{(0)}$, we use the test statistic

$$T = \frac{\hat{\beta}_0 - \beta_0^{(0)}}{S_{Y|X}\sqrt{\dfrac{1}{n} + \dfrac{\overline{X}^2}{(n-1)S_X^2}}} \tag{5.10}$$

which also has the t distribution with $n - 2$ degrees of freedom when $H_0: \beta_0 = \beta_0^{(0)}$ is true. The denominator here estimates the standard error of $\hat{\beta}_0$, given by

$$\sigma_{\hat{\beta}_0} = \sigma\sqrt{\frac{1}{n} + \frac{\overline{X}^2}{(n-1)S_X^2}}$$

The reason why both test statistics (5.9) and (5.10) have $n - 2$ degrees of freedom is that both involve $S_{Y|X}^2$, which itself has $n - 2$ degrees of freedom and is the only random component in the denominator of both statistics.

In testing either of the preceding hypotheses at a significance level α, we should reject H_0 whenever any of the following occur:

$$\begin{cases} T \geq t_{n-2,1-\alpha} \text{ for an upper one-tailed test} & (\text{i.e., } H_A: \beta_1 > \beta_1^{(0)} \text{ or } H_A: \beta_0 > \beta_0^{(0)}) \\ T \leq -t_{n-2,1-\alpha} \text{ for a lower one-tailed test} & (\text{i.e., } H_A: \beta_1 < \beta_1^{(0)} \text{ or } H_A: \beta_0 < \beta_0^{(0)}) \\ |T| \geq t_{n-2,1-\alpha/2} \text{ for a two-tailed test} & (\text{i.e., } H_A: \beta_1 \neq \beta_1^{(0)} \text{ or } H_A: \beta_0 \neq \beta_0^{(0)}) \end{cases}$$

where $t_{n-2,1-\alpha}$ denotes the $100(1 - \alpha)\%$ point of the t distribution with $n - 2$ degrees of freedom. As an alternative to using a specified significance level, we may compute P-values based on the calculated value of the test statistic T.

Table 5-2 summarizes the formulas needed for performing statistical tests and computing confidence intervals for β_0 and β_1. Also given in this table are formulas for inference-making procedures concerned with prediction using the fitted line; these formulas are described in sections 5-9 and 5-10. Table 5-3 gives examples illustrating the use of each formula in Table 5-2, using the age–systolic blood pressure data previously considered.

TABLE 5-2 Confidence intervals, tests of hypotheses, and prediction intervals for straight-line regression analysis.

Parameter	$100(1-\alpha)\%$ Confidence Interval	H_0	Test Statistic (T)	Distribution Under H_0
β_1	$\hat{\beta}_1 \pm t_{n-2,1-\alpha/2}S_{\hat{\beta}_1}$	$\beta_1 = \beta_1^{(0)}$	$T = \dfrac{(\hat{\beta}_1 - \beta_1^{(0)})}{S_{\hat{\beta}_1}}$	t_{n-2}
β_0	$\hat{\beta}_0 \pm t_{n-2,1-\alpha/2}S_{\hat{\beta}_0}$	$\beta_0 = \beta_0^{(0)}$	$T = \dfrac{(\hat{\beta}_0 - \beta_0^{(0)})}{S_{\hat{\beta}_0}}$	t_{n-2}
$\mu_{Y\mid X_0}$	$\overline{Y} + \hat{\beta}_1(X_0 - \overline{X}) \pm t_{n-2,1-\alpha/2}S_{\hat{Y}_{X_0}}$	$\mu_{Y\mid X_0} = \mu_{Y\mid X_0}^{(0)}$	$T = \dfrac{\overline{Y} + \hat{\beta}_1(X_0 - \overline{X}) - \mu_{Y\mid X_0}^{(0)}}{S_{\hat{Y}_{X_0}}}$	t_{n-2}
Y_{X_0}[a]	$\overline{Y} + \hat{\beta}_1(X_0 - \overline{X})$ $\pm t_{n-2,1-\alpha/2}S_{Y\mid X}\sqrt{1 + \dfrac{1}{n} + \dfrac{(X_0 - \overline{X})^2}{(n-1)S_X^2}}$[b]			

Note: $\mu_{Y\mid X} = \beta_0 + \beta_1 X$ is the assumed true regression model.

$$\hat{\beta}_1 = \frac{\sum_1^n (X_i - \overline{X})(Y_i - \overline{Y})}{\sum_1^n (X_i - \overline{X})^2}$$

$$\hat{\beta}_0 = \overline{Y} - \hat{\beta}_1\overline{X}$$

$$\hat{Y} = \hat{\beta}_0 + \hat{\beta}_1 X = \overline{Y} + \hat{\beta}_1(X - \overline{X})$$

$t_{n-2,1-\alpha/2}$ is the $100(1-\alpha/2)\%$ point of the t distribution with $n-2$ degrees of freedom.

$$S_{Y\mid X}^2 = \frac{1}{n-2}\sum_1^n (Y_i - \hat{Y}_i)^2 \qquad S_{\hat{\beta}_1} = \frac{S_{Y\mid X}}{S_X\sqrt{n-1}}$$

$$S_Y^2 = \frac{1}{n-1}\sum_1^n (Y_i - \overline{Y})^2 \qquad S_{\hat{\beta}_0} = S_{Y\mid X}\sqrt{\frac{1}{n} + \frac{\overline{X}^2}{(n-1)S_X^2}}$$

$$S_X^2 = \frac{1}{n-1}\sum_1^n (X_i - \overline{X})^2 \qquad S_{\hat{Y}_{X_0}} = S_{Y\mid X}\sqrt{\frac{1}{n} + \frac{(X_0 - \overline{X})^2}{(n-1)S_X^2}}$$

[a]Single observation, not a parameter.
[b]Prediction interval for a new individual's Y.

5-8 Interpretations of Tests for Slope and Intercept

Researchers often make errors when interpreting the results of tests regarding the slope and the intercept. In this section we discuss conclusions that can be drawn based on nonrejection or rejection of the most common null hypotheses involving the slope and the intercept.[4] In our discussion we assume that the usual assumptions about normality, independence, and variance homogeneity are not violated. If these assumptions do not hold, any conclusions based on testing procedures developed under the assumptions are suspect.

---

[4]Statistically speaking, "nonrejection of H_0" really means "determination that there is insufficient evidence to reject H_0."

TABLE 5-3   Sample calculations of confidence intervals, tests of hypotheses, and prediction intervals for the age–systolic blood pressure data of Table 5-1.

Parameter	$100(1 - \alpha)\%$ Confidence Interval	H_0	Test Statistic (T)
β_1	For $\alpha = .05$: $0.97 \pm 2.0484(0.21)$ or $(0.54, 1.40)$	$\beta_1 = 0$	$T = \dfrac{(0.97 - 0)}{0.21} = 4.62$ Reject H_0 at $\alpha = .05$ (two-tailed test), since $t_{28, 0.975} = 2.0484$ ($P < .001$).
β_0	For $\alpha = .05$: $98.71 \pm (2.0484)(10.00)$ or $(78.23, 119.20)$	$\beta_0 = 75$	$T = \dfrac{(98.71 - 75)}{10.00} = 2.37$ Reject H_0 at $\alpha = .05$ (two-tailed test), since $t_{28, 0.975} = 2.0484$ ($.02 < P < .05$).
$\mu_{Y\mid 65}$[a]	For $\alpha = .10$: $142.53 + (0.97)(65 - 45.13) \pm (1.7011)(5.24)$ or $(152.89, 170.72)$	$\mu_{Y\mid 65} = 147$	$T = \dfrac{142.53 + (0.97)(65 - 45.13) - 147}{5.24}$ $= 2.82$ Reject H_0 at $\alpha = .10$ (two-tailed test), since $t_{28, 0.95} = 1.7011$ $(.001 < P < .01)$.
Y_{65}[b]	For $\alpha = .10$: $142.53 + (0.97)(65 - 45.13)$ $\pm (1.7011)(17.31)\sqrt{1 + \dfrac{1}{30} + \dfrac{(65 - 45.13)^2}{(30 - 1)(15.29)^2}}$ or $(131.04, 192.5)$		

Note: $n = 30$, $\hat{\beta}_0 = 98.71$, $\beta_1 = 0.97$, $\overline{Y} = 142.53$, $\overline{X} = 45.13$, $S_{Y\mid X} = 17.31$, $S_X = 15.29$,

$$S_{\hat{\beta}_1} = \frac{17.31}{15.29\sqrt{30 - 1}} = 0.21, \quad S_{\hat{\beta}_0} = 17.31\sqrt{\frac{1}{30} + \frac{(45.13)^2}{(30 - 1)(15.29)^2}} = 10.00, \quad S_{\hat{Y}_{X_0}} = 17.31\sqrt{\frac{1}{30} + \frac{(65 - 45.13)^2}{(30 - 1)(15.29)^2}} = 5.24$$

[a]Thus, $X_0 = 65$.

[b]Prediction interval at $X_0 = 65$.

5-8-1 Test for Zero Slope

The most important test of hypothesis dealing with the parameters of the straight-line model relates to whether the slope of the regression line differs significantly from zero or, equivalently, whether X helps to predict Y using a straight-line model. The appropriate null hypothesis for this test is H_0: $\beta_1 = 0$. Care must be taken in interpreting the result of the test of this hypothesis.

If we ignore for now the ever-present possibilities of making a Type I error (i.e., rejecting a true H_0) or a Type II error (i.e., not rejecting a false H_0), we can make the following interpretations:

1. If H_0: $\beta_1 = 0$ is not rejected, one of the following is true.
   a. For a true underlying straight-line model, X provides little or no help in predicting Y; that is, $\overline{Y}$ is essentially as good as $\overline{Y} + \hat{\beta}_1(X - \overline{X})$ for predicting Y (Figure 5-9(a)).
   b. The true underlying relationship between X and Y is *not* linear; that is, the true model may involve quadratic, cubic, or other more complex functions of X (Figure 5-9(b)).

<u>FIGURE 5-9</u>   Interpreting the test for zero slope.

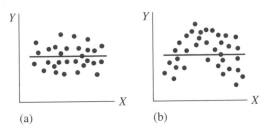

Examples when H_0: $\beta_1 = 0$ is rejected

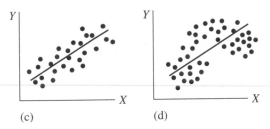

Combining (a) and (b), we can say that *not rejecting H_0: $\beta_1 = 0$ implies that a straight-line model in X is not the best model to use and does not provide much help for predicting Y.*

2. If H_0: $\beta_1 = 0$ is rejected, one of the following is true.
    a. X provides significant information for predicting Y; that is, the model $\overline{Y} + \hat{\beta}_1(X - \overline{X})$ is far better than the naive model $\overline{Y}$ for predicting Y (Figure 5-9(c)).
    b. A better model might have, for example, a curvilinear term (e.g., Figure 5-9(d)), although there is a definite linear component.

Combining (a) and (b), we can say that *rejecting H_0: $\beta_1 = 0$ implies that a straight-line model in X is better than a model that does not include X at all, although it may well represent only a linear approximation to a truly nonlinear relationship.*

An important point implied by these interpretations is that *whether or not the hypothesis H_0: $\beta_1 = 0$ is rejected, a straight-line model may not be appropriate; instead, some other curve may describe the relationship between X and Y better.*

5-8-2 Test for Zero Intercept

Another hypothesis sometimes tested involves whether the population straight line goes through the origin—that is, whether its Y-intercept β_0 is zero. The null hypothesis here is H_0: $\beta_0 = 0$. If this null hypothesis is not rejected, it may be appropriate to remove the constant from the model, provided that previous experience or a relevant theory suggests that the line may go through the origin and provided that observations are taken around the origin to improve the estimate of β_0. Forcing the fitted line through the origin merely because H_0: $\beta_0 = 0$ cannot be rejected may give a spurious appearance to the regression line. In any case this hypothesis is rarely of interest in most studies, because data are not usually gathered near the origin. For example, when dealing with age (X) and blood pressure (Y), we are not interested in knowing what happens at X = 0, and we rarely choose values of X near 0.

5-9 Inferences About the Regression Line $\mu_{Y|X} = \beta_0 + \beta_1 X$

In addition to making inferences about the slope and the intercept, we may also want to perform tests and/or compute confidence intervals concerning the regression line itself. More specifically, for a given $X = X_0$, we may want to find a confidence interval for $\mu_{Y|X_0}$, the mean value of Y at X_0.[5] We may also be interested in testing the hypothesis H_0: $\mu_{Y|X_0} = \mu_{Y|X_0}^{(0)}$, where $\mu_{Y|X_0}^{(0)}$ is some hypothesized value of interest.

The test statistic to use for the hypothesis H_0: $\mu_{Y|X_0} = \mu_{Y|X_0}^{(0)}$ is given by the formula

$$T = \frac{\hat{Y}_{X_0} - \mu_{Y|X_0}^{(0)}}{S_{\hat{Y}_{X_0}}} \qquad (5.11)$$

where $\hat{Y}_{X_0} = \hat{\beta}_0 + \hat{\beta}_1 X_0 = \bar{Y} + \hat{\beta}_1(X_0 - \bar{X})$ is the predicted value of Y at X_0 and

$$S_{\hat{Y}_{X_0}} = S_{Y|X}\sqrt{\frac{1}{n} + \frac{(X_0 - \bar{X})^2}{(n-1)S_X^2}} \qquad (5.12)$$

This test statistic, like those for slope and intercept, has the t distribution with $n - 2$ degrees of freedom when H_0 is true. The denominator $S_{\hat{Y}_{X_0}}$ is an estimate of the standard error of $\hat{Y}_{X_0}$, which is given by

$$\sigma_{\hat{Y}_{X_0}} = \sigma\sqrt{\frac{1}{n} + \frac{(X_0 - \bar{X})^2}{(n-1)S_X^2}}$$

The corresponding confidence interval for $\mu_{Y|X_0}$ at a given $X = X_0$ is given by the formula

$$\hat{Y}_{X_0} \pm t_{n-2, 1-\alpha/2} S_{\hat{Y}_{X_0}} \qquad (5.13)$$

In addition to drawing inferences about specific points on the regression line, researchers often find it useful to construct a confidence interval for the regression line over the entire range of X-values. The most convenient way to do this is to plot the upper and lower confidence limits obtained for several specified values of X and then to sketch the two curves that connect these points. Such curves are called *confidence bands* for the regression line. The confidence bands for the data of Table 5-1 are indicated in Figure 5-10.

Sketching confidence bands by hand calculator can be a painful job. Instead, we generally recommend using a computer program for regression analysis to compute confidence intervals for a range of X_0 values and then to plot these intervals on the same graph that contains the fitted regression line. A convenient way to choose X_0 values is to use $X_0 = \bar{X}$, $X_0 = \bar{X} \pm k$, $X_0 = \bar{X} \pm 2k$, $X_0 = \bar{X} \pm 3k$, etc., where k is chosen so that the range of X-values in the data is uniformly covered.

Example 5-2   For 90% confidence bands for our age–systolic blood pressure data, confidence interval formula (5.13) simplifies to:

$$142.53 + (0.97)(X_0 - 45.13) \pm 29.45\sqrt{0.033 + \frac{(X_0 - 45.13)^2}{6,783.48}} \qquad (5.14)$$

---

[5]The point $(X_0, \mu_{Y|X_0})$, of course, lies on the population regression line.

FIGURE 5-10   90% confidence and prediction bands for age–systolic blood pressure data of Table 5-1.

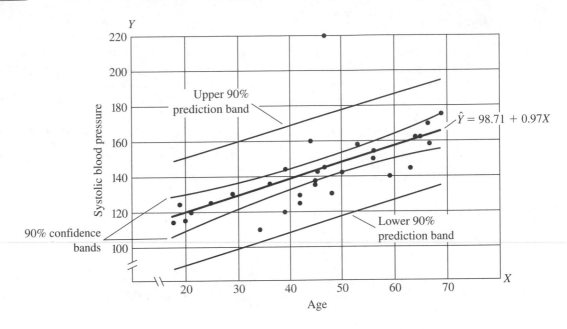

At $X_0 = \overline{X} = 45.13$, the formula simplifies to $142.53 \pm 29.45\sqrt{0.033}$, which yields a lower limit of 137.16 and an upper limit of 147.91. Notice that the minimum-width confidence interval is always obtained at $X_0 = \overline{X}$, since the second term under the square root sign in (5.14) is zero.

At $X_0 = \overline{X} \pm k$, the confidence interval formula becomes $142.53 \pm (0.97)k \pm 29.45\sqrt{0.033 + \frac{k^2}{6,783.48}}$. Thus, when $k = 10$, the confidence limits are 145.78 and 158.70 for $X_0 = \overline{X} + 10$, and 126.37 and 139.28 for $X_0 = \overline{X} - 10$. Figure 5-10 shows the 90% confidence bands for these data, together with the fitted model.

In SAS, PROC REG will compute 95% confidence intervals for $\mu_{Y|X}$, using the sample independent variable values of X_0. In the following output, the lower and upper 95% confidence limits for the systolic blood pressure–age example appear in the "Lower 95% Mean" and "Upper 95% Mean" columns, respectively.   ▪

SAS Output Showing Confidence and Prediction Intervals

REGRESSION OF SBP(Y) ON AGE(X)

		$\hat{Y}_{X_0}$	$S_{\hat{Y}_{X_0}}$	95% confidence intervals for $\mu_{Y\|X}$		95% prediction intervals	
Obs	Dep Var Y	Predict Value	Std Err Predict	Lower95% Mean	Upper95% Mean	Lower95% Predict	Upper95% Predict
1	144.0	136.6	3.414	129.6	143.6	100.4	172.7
2	220.0	144.3	3.185	137.8	150.9	108.3	180.4
3	138.0	142.4	3.161	135.9	148.9	106.4	178.5
4	145.0	144.3	3.185	137.8	150.9	108.3	180.4

```
          REGRESSION OF SBP(Y) ON AGE(X) (Continued)
      Dep Var   Predict   Std Err   Lower95%  Upper95%  Lower95%  Upper95%
Obs     Y       Value     Predict     Mean      Mean     Predict   Predict
 5    162.0     161.8     5.238      151.1     172.6     124.8     198.9
 6    142.0     143.4     3.166      136.9     149.9     107.3     179.4
 7    170.0     163.8     5.579      152.3     175.2     126.5     201.0
 8    124.0     139.5     3.229      132.9     146.1     103.4     175.6
 9    158.0     163.8     5.579      152.3     175.2     126.5     201.0
10    154.0     153.1     3.900      145.1     161.1     116.7     189.4
11    162.0     160.9     5.072      150.5     171.2     123.9     197.8
12    150.0     153.1     3.900      145.1     161.1     116.7     189.4
13    140.0     156.0     4.300      147.2     164.8     119.5     192.5
14    110.0     131.7     3.933      123.7     139.8     95.3551    168.1
15    128.0     139.5     3.229      132.9     146.1     103.4     175.6
16    130.0     145.3     3.218      138.7     151.9     109.2     181.4
17    135.0     142.4     3.161      135.9     148.9     106.4     178.5
18    114.0     115.2     6.706      101.5     129.0     77.1867    153.3
19    116.0     118.1     6.157      105.5     130.7     80.4909    155.8
20    124.0     117.2     6.338      104.2     130.1     79.3939    154.9
21    136.0     133.7     3.698      126.1     141.2     97.4003    169.9
22    142.0     147.3     3.322      140.5     154.1     111.1     183.4
23    120.0     136.6     3.414      129.6     143.6     100.4     172.7
24    120.0     119.1     5.977      106.9     131.3     81.5833    156.6
25    160.0     141.4     3.170      134.9     147.9     105.4     177.5
26    158.0     150.2     3.567      142.9     157.5     114.0     186.4
27    144.0     159.9     4.909      149.8     169.9     123.0     196.7
28    130.0     126.9     4.636      117.4     136.4     90.1549    163.6
29    125.0     123.0     5.283      112.2     133.8     85.9069    160.1
30    175.0     165.7     5.930      153.6     177.9     128.2     203.2
```

5-10 Prediction of a New Value of Y at X_0

We have just dealt with estimating the mean $\mu_{Y|X_0}$ at $X = X_0$. In practice, we may wish instead to estimate the response Y of a single individual, based on the fitted regression line; that is, we may want to predict an individual's Y given his $X = X_0$. The obvious point estimate to use in this case is $\hat{Y}_{X_0} = \hat{\beta}_0 + \hat{\beta}_1 X_0$. Thus, $\hat{Y}_{X_0}$ is used to estimate both the mean $\mu_{Y|X_0}$ and an individual's response Y at X_0.

Of course, some bounds (i.e., limits) must be placed on this estimate to take its variability into account. Here, however, we cannot say that we are constructing a confidence interval for Y, since Y is not a parameter; neither can we perform a test of hypothesis, for the same reason. The term used to describe the "hybrid limits" we require is the *prediction interval* (PI), which is given by

$$\overline{Y} + \hat{\beta}_1(X_0 - \overline{X}) \pm t_{n-2, 1-\alpha/2} S_{Y|X}\sqrt{1 + \frac{1}{n} + \frac{(X_0 - \overline{X})^2}{(n-1)S_X^2}} \tag{5.15}$$

We first note that an estimate of an individual's response should naturally have more variability than an estimate of a group's mean response. This is reflected by the extra term 1 under the square root sign in (5.15), which is not found in the square root part of the confidence interval formula for $\mu_{Y|X}$ (see (5.12) and (5.13)). To be more specific, in predicting an actual observed Y for a given individual, there are two sources of error operating: individual error as measured by σ^2 and the error in estimating $\mu_{Y|X_0}$ using $\hat{Y}_{X_0}$. More precisely, this can be expressed by the equation

$$\underbrace{Y - \hat{Y}_{X_0}}_{\substack{\text{Error in} \\ \text{predicting an} \\ \text{individual's} \\ Y \text{ at } X_0}} = \underbrace{(Y - \mu_{Y|X_0})}_{\substack{\text{Deviation of} \\ \text{individual's} \\ Y \text{ from true} \\ \text{mean at } X_0}} + \underbrace{(\mu_{Y|X_0} - \hat{Y}_{X_0})}_{\substack{\text{Deviation of} \\ \hat{Y}_{X_0} \text{ from true} \\ \text{mean at } X_0}}$$

This representation allows us to write the variance of an individual's predicted response at X_0 as

$$\text{Var } Y + \text{Var } \hat{Y}_{X_0} = \sigma^2 \left[1 + \frac{1}{n} + \frac{(X_0 - \overline{X})^2}{(n-1)S_X^2} \right]$$

This variance expression is estimated by replacing σ^2 by its estimate $S_{Y|X}^2$. This accounts for the expression on the right-hand side of the prediction interval in (5.15).

   Prediction bands, used to describe individual predictions over the entire range of X-values, may be determined in a manner analogous to that by which confidence bands are computed. Figure 5-10 gives 90% prediction bands for the age–systolic blood pressure data. As expected, the 90% prediction bands in this figure are wider than the corresponding 90% confidence bands.

   Once again, SAS can be used to compute 95% prediction intervals. In the previous computer output, the lower and upper limits for these intervals are given in the "Lower 95% Predict" and "Upper 95% Predict" columns. SAS uses the sample values of the independent variable X_0.

5-11 Assessing the Appropriateness of the Straight-line Model

   In section 5-1, we noted that the usual strategy for regression with a single independent variable is to assume that the straight-line model is appropriate. This assumption is then rejected if the data indicate that a more complex model is warranted.

   Many methods may be used to assess whether the straight-line assumption is reasonable. These will be discussed separately later. The basic techniques include tests for lack of fit and are understood most easily in terms of polynomial regression models (Chapter 13). Many regression diagnostics (Chapter 12) also help in evaluating the straight-line assumption, either explicitly or implicitly. With the linear model, the assumptions of linearity, homoscedasticity, and normality are so intertwined that they often are met or violated as a set.

Problems

SAS computer output is provided for most of the problems in this chapter. The same is true for many of the subsequent chapters. In some problems, using the output can significantly reduce the computational and programming effort required to answer the questions. Since the actual data are also often provided, however, instructors and students may choose to do their own computations and/or programming to perform the necessary analyses.

1. The accompanying table gives the dry weights (Y) of 11 chick embryos ranging in age from 6 to 16 days (X). Also given in the table are the values of the common logarithms of the weights (Z).

Age (X) (days)	6	7	8	9	10	11
Dry Weight (Y)	0.029	0.052	0.079	0.125	0.181	0.261
Log_{10} Dry Weight (Z)	-1.538	-1.284	-1.102	-0.903	-0.742	-0.583

Age (X) (days)	12	13	14	15	16
Dry Weight (Y)	0.425	0.738	1.130	1.882	2.812
Log_{10} Dry Weight (Z)	-0.372	-0.132	0.053	0.275	0.449

a. Observe the following two scatter diagrams. Describe the relationships between age (X) and dry weight (Y), and between age and $\log_{10}$ dry weight (Z).

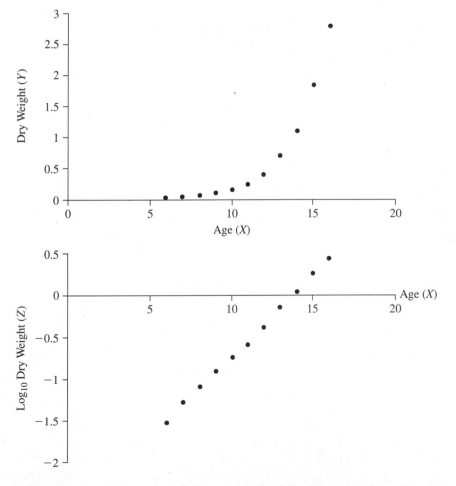

b. State the simple linear regression models for these two regressions: Y regressed on X, and Z regressed on X.

   c. Determine the least-squares estimates of each of the regression lines in part (b).

   d. Sketch each estimated line on the appropriate scatter diagram. Which of the two regression lines has the better fit? Based on your answers to parts (a)–(c), is it more appropriate to run a linear regression of Y on X, or of Z on X? Explain.

   e. For the regression that you chose as being more appropriate in part (d), find 95% confidence intervals for the true slope and intercept. Interpret each interval with regard to the null hypothesis that the true parameter is 0.

   f. For the regression that you chose as being more appropriate in part (d), find and sketch 95% confidence and prediction bands. Using your sketch, find and interpret an approximate 95% confidence interval on the mean response for an 8-day-old chick.

Edited SAS Output (PROC REG) for Problem 1

```
                    DRY WEIGHT (Y) Regressed on AGE (X)
                                    .
                          . [Portion of output omitted]
                                    .
          Root MSE            0.48185           R-square           0.7442
                          Parameter Estimates
                     Parameter         Standard       T for H0:
Variable    DF        Estimate           Error       Parameter=0    Prob > |T|
INTERCEP     1       -1.884527         0.52583537       -3.584        0.0059
X            1        0.235073         0.04594245        5.117        0.0006
        Dep Var    Predict    Std Err    Lower95%   Upper95%   Lower95%   Upper95%
Obs       Y         Value     Predict     Mean       Mean      Predict    Predict
 1      0.0290    -0.4741     0.272     -1.0889     0.1408    -1.7256     0.7774
 2      0.0520    -0.2390     0.234     -0.7690     0.2909    -1.4510     0.9730
 3      0.0790    -0.00395    0.200     -0.4570     0.4491    -1.1844     1.1765
 4      0.1250     0.2311     0.172     -0.1577     0.6200    -0.9262     1.3884
 5      0.1810     0.4662     0.152      0.1215     0.8109    -0.6770     1.6094
 6      0.2610     0.7013     0.145      0.3726     1.0299    -0.4372     1.8398
 7      0.4250     0.9363     0.152      0.5917     1.2810    -0.2069     2.0796
 8      0.7380     1.1714     0.172      0.7826     1.5603     0.0141     2.3287
 9      1.1300     1.4065     0.200      0.9535     1.8595     0.2261     2.5869
10      1.8820     1.6416     0.234      1.1116     2.1715     0.4296     2.8536
11      2.8120     1.8766     0.272      1.2618     2.4915     0.6252     3.1281

                 LOG10 DRY WEIGHT (Z) Regressed on AGE (X)
                                    .
                          . [Portion of output omitted]
                                    .
          Root MSE            0.02800           R-square           0.9983
                          Parameter Estimates
                     Parameter         Standard       T for H0:
Variable    DF        Estimate           Error       Parameter=0    Prob > |T|
INTERCEP     1       -2.689200         0.03055158      -88.022        0.0001
X            1        0.195881         0.00266930       73.383        0.0001
        Dep Var    Predict    Std Err    Lower95%   Upper95%   Lower95%   Upper95%
Obs       Z         Value     Predict     Mean       Mean      Predict    Predict
 1     -1.5376    -1.5139     0.016     -1.5496    -1.4782    -1.5866    -1.4412
 2     -1.2840    -1.3180     0.014     -1.3488    -1.2872    -1.3885    -1.2476
 3     -1.1024    -1.1222     0.012     -1.1485    -1.0958    -1.1907    -1.0536
 4     -0.9031    -0.9263     0.010     -0.9489    -0.9037    -0.9935    -0.8590
```

```
          LOG10 DRY WEIGHT (Z) Regressed on AGE (X) (continued)
                    Parameter Estimates  (continued)
        Dep Var   Predict   Std Err  Lower95%  Upper95%  Lower95%  Upper95%
 Obs      Z        Value    Predict    Mean      Mean     Predict   Predict
  5     -0.7423   -0.7304   0.009    -0.7504   -0.7104   -0.7968   -0.6640
  6     -0.5834   -0.5345   0.008    -0.5536   -0.5154   -0.6007   -0.4684
  7     -0.3716   -0.3386   0.009    -0.3587   -0.3186   -0.4050   -0.2722
  8     -0.1319   -0.1427   0.010    -0.1653   -0.1202   -0.2100   -0.0755
  9      0.0531    0.0531   0.012     0.0268    0.0795   -0.0154    0.1217
 10      0.2746    0.2490   0.014     0.2182    0.2798    0.1786    0.3194
 11      0.4490    0.4449   0.016     0.4092    0.4806    0.3722    0.5176
```

2. The following table gives the systolic blood pressure (SBP), body size (QUET),[6] age (AGE), and smoking history (SMK = 0 if nonsmoker, SMK = 1 if a current or previous smoker) for a hypothetical sample of 32 white males over 40 years old from the town of Angina.

Person	SBP	QUET	AGE	SMK	Person	SBP	QUET	AGE	SMK	Person	SBP	QUET	AGE	SMK
1	135	2.876	45	0	12	138	4.032	51	1	23	137	3.296	53	0
2	122	3.251	41	0	13	152	4.116	64	0	24	132	3.210	50	0
3	130	3.100	49	0	14	138	3.673	56	0	25	149	3.301	54	1
4	148	3.768	52	0	15	140	3.562	54	1	26	132	3.017	48	1
5	146	2.979	54	1	16	134	2.998	50	1	27	120	2.789	43	0
6	129	2.790	47	1	17	145	3.360	49	1	28	126	2.956	43	1
7	162	3.668	60	1	18	142	3.024	46	1	29	161	3.800	63	0
8	160	3.612	48	1	19	135	3.171	57	0	30	170	4.132	63	1
9	144	2.368	44	1	20	142	3.401	56	0	31	152	3.962	62	0
10	180	4.637	64	1	21	150	3.628	56	1	32	164	4.010	65	0
11	166	3.877	59	1	22	144	3.751	58	0					

a. On each of the accompanying scatter diagrams, sketch by eye a line that fits the data reasonably well. Comment on the relationships described.

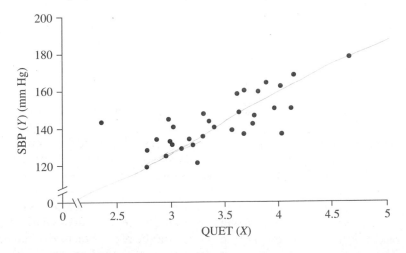

[6]QUET stands for "quetelet index," a measure of size defined by QUET = 100 (weight/height^2).

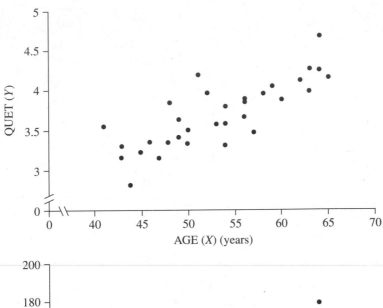

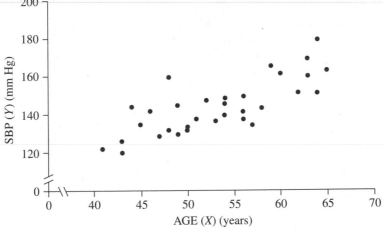

b. (1) Determine the least-squares estimates of the slope (β_1) and intercept (β_0) for the straight-line regression of SBP (Y) on QUET (X).

(2) Sketch the estimated regression line on the scatter diagram involving SBP and QUET. Compare this new line with the line you drew in part (a).

(3) Test the null hypothesis of zero slope; be sure to interpret the result.

(4) Based on your test in (3), would you conclude that blood pressure increases as body size increases?

(5) Find and sketch 95% prediction bands on the appropriate scatter diagram.

(6) Using your answer to part (b)(5), find an approximate 95% prediction interval for an individual with QUET = 3.4 (the sample mean value of QUET). Interpret your answer.

(7) Are any assumptions for straight-line regression clearly not satisfied in this example?

c. Repeat questions (1) through (4) in part (b) for the regression of QUET on AGE.

d. Repeat questions (1) through (4) in part (b) for the regression of SBP on AGE.

e. (1) Determine the least-squares estimates of the slope and intercept for the straight-line regression of SBP (Y) on SMK (X).

(2) Compare the value of $\hat{\beta}_0$ with the mean SBP for nonsmokers. Compare the values of $\hat{\beta}_0 + \hat{\beta}_1$ with the mean SBP for smokers. Explain the results of these comparisons.

(3) Test the null hypothesis that the true slope is 0; be sure to interpret the result.

(4) Is the test in part (e)(3) equivalent to the usual two-sample t test for the equality of two population means, assuming equal but unknown variances? Demonstrate your answer numerically.

Edited SAS Output (PROC REG) for Problem 2

```
                         SBP Regressed on QUET
                                    .
                                    .  [Portion of output omitted]
                                    .
          Root MSE           9.81160              R-square          0.5506
                         Parameter Estimates
                      Parameter        Standard      T for H0:
   Variable    DF     Estimate            Error    Parameter=0    Prob > |T|
   INTERCEP     1    70.576403      12.32186793          5.728        0.0001
   QUET         1    21.491669       3.54514702          6.062        0.0001

          Dep Var    Predict    Std Err    Lower95%    Upper95%
   Obs       SBP      Value     Predict     Predict     Predict      Residual
    1       135.0     132.4       2.650       111.6       153.1        2.6136
    2       122.0     140.4       1.861       120.1       160.8      -18.4458
    3       130.0     137.2       2.114       116.7       157.7       -7.2006
    4       148.0     151.6       2.086       131.1       172.0       -3.5570
    5       146.0     134.6       2.386       114.0       155.2       11.3999
    6       129.0     130.5       2.887       109.7       151.4       -1.5382
    7       162.0     149.4       1.912       129.0       169.8       12.5922
    8       160.0     148.2       1.837       127.8       168.6       11.7957
    9       144.0     121.5       4.181       99.6873      143.3       22.5313
   10       180.0     170.2       4.581       148.1       192.3        9.7667
   11       166.0     153.9       2.323       133.3       174.5       12.1004
   12       138.0     157.2       2.720       136.4       178.0      -19.2308
   13       152.0     159.0       2.955       138.1       180.0       -7.0361
   14       138.0     149.5       1.919       129.1       169.9      -11.5153
   15       140.0     147.1       1.787       126.8       167.5       -7.1297
   16       134.0     135.0       2.340       114.4       155.6       -1.0084
   17       145.0     142.8       1.758       122.4       163.1        2.2116
   18       142.0     135.6       2.279       115.0       156.1        6.4328
   19       135.0     138.7       1.981       118.3       159.2       -3.7265
   20       142.0     143.7       1.740       123.3       164.0       -1.6696
   21       150.0     148.5       1.857       128.2       168.9        1.4518
   22       144.0     151.2       2.053       130.7       171.7       -7.1917
   23       137.0     141.4       1.809       121.0       161.8       -4.4129
   24       132.0     139.6       1.918       119.1       160.0       -7.5647
   25       149.0     141.5       1.804       121.1       161.9        7.4796
   26       132.0     135.4       2.295       114.8       156.0       -3.4168
   27       120.0     130.5       2.890       109.6       151.4      -10.5167
   28       126.0     134.1       2.443       113.5       154.8       -8.1058
   29       161.0     152.2       2.151       131.7       172.8        8.7553
   30       170.0     159.4       3.001       138.4       180.3       10.6200
   31       152.0     155.7       2.533       135.0       176.4       -3.7264
   32       164.0     156.8       2.660       136.0       177.5        7.2420
```

(continued)

```
                    QUET Regressed on AGE
                              .
                              .  [Portion of output omitted]
                              .
     Root MSE              0.30131              R-square              0.6444
                       Parameter Estimates
                    Parameter        Standard      T for H0:
Variable    DF       Estimate           Error   Parameter=0     Prob > |T|
INTERCEP     1       0.386452      0.41769030         0.925         0.3622
AGE          1       0.057364      0.00777991         7.373         0.0001


                    SBP Regressed on AGE
                              .
                              .  [Portion of output omitted]
                              .
     Root MSE              9.24543              R-square              0.6009
                       Parameter Estimates
                    Parameter        Standard      T for H0:
Variable    DF       Estimate           Error   Parameter=0     Prob > |T|
INTERCEP     1      59.091625     12.81626145         4.611         0.0001
AGE          1       1.604500      0.23871593         6.721         0.0001


                    SBP Regressed on SMK
                              .
                              .  [Portion of output omitted]
                              .
     Root MSE             14.18082              R-square              0.0612
                       Parameter Estimates
                    Parameter        Standard      T for H0:
Variable    DF       Estimate           Error   Parameter=0     Prob > |T|
INTERCEP     1     140.800000      3.66147226        38.454         0.0001
SMK          1       7.023529      5.02349849         1.398         0.1723
```

3. For married couples with one or more offspring, a demographic study was conducted to determine the effect of the husband's annual income (at marriage) on the time (in months) between marriage and the birth of the first child. The following table gives the husband's annual income (INC) and the time between marriage and the birth of the first child (TIME) for a hypothetical sample of 20 couples.

INC	TIME	INC	TIME
5,775	16.20	4,608	9.70
9,800	35.00	24,210	20.00
13,795	37.20	19,625	38.20
4,120	9.00	18,000	41.25
25,015	24.40	13,000	44.00
12,200	36.75	5,400	9.20
7,400	31.75	6,440	20.00
9,340	30.00	9,000	40.20
20,170	36.00	18,180	32.00
22,400	30.80	15,385	39.20

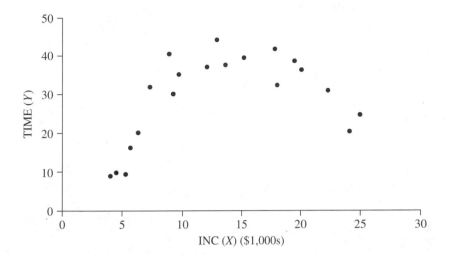

a. On the scatter diagram above, sketch by eye a line that fits the data reasonably well. Comment on the relationship between TIME (Y) and INC (X).

b. Determine the least-squares estimates of the slope (β_1) and intercept (β_0) for the straight-line regression of TIME (Y) on INC (X).

c. Draw the estimated regression line on the accompanying scatter diagram. Comment on how well this line fits the data.

d. Are any of the assumptions for straight-line regression clearly not satisfied in this example?

e. Test the null hypothesis that the true slope is 0. Interpret the results of this test.

f. Can you suggest a model that would describe the TIME–INC relationship better than a straight line does?

Edited SAS Output (PROC REG) for Problem 3

```
                     TIME (Y) Regressed on INC (X)
                                  .
                                  .   [Portion of output omitted]
                                  .

     Root MSE            10.49580              R-square            0.1853
                            Parameter Estimates
                       Parameter        Standard      T for H0:
     Variable   DF      Estimate           Error    Parameter=0    Prob > |T|
     INTERCEP    1     19.625748      5.21290778          3.765        0.0014
     X           1      0.000714      0.00035281          2.023        0.0582
```

4. A sociologist assigned to a correctional institution was interested in studying the relationship between intelligence and delinquency. A delinquency index (ranging from 0 to 50) was formulated to account for both the severity and the frequency of crimes committed, while intelligence was measured by IQ. The following table displays the delinquency index (DI) and IQ of a sample of 18 convicted minors.

DI (Y)	IQ (X)	DI (Y)	IQ (X)
26.20	110	22.10	92
33.00	89	18.60	116
17.50	102	35.50	85
25.25	98	38.00	73
20.30	110	30.00	90
31.90	98	19.70	104
21.10	122	41.10	82
22.70	119	39.60	134
10.70	120	25.15	114

a. Given that $\hat{\beta}_1 = -0.249$ and $\hat{\beta}_0 = 52.273$, draw the estimated regression line on the accompanying scatter diagram.

b. How do you account for the fact that $\hat{Y} = 52.273$ when IQ $= 0$, even though the delinquency index goes no higher than 50?

c. Find a 95% confidence interval for the true slope β_1, using the fact that $S_{Y|X} = 7.704$ and $S_X = 16.192$.

d. Interpret this confidence interval with regard to testing the null hypothesis of zero slope at the $\alpha = .05$ level.

e. Notice that the convicted minor with IQ $= 134$ and DI $= 39.6$ appears to be quite out of place in the data. Decide whether this outlier has any effect on your estimate of the IQ–DI relationship, by looking at the graph for the fitted line obtained when the outlier is omitted. (Note that $\hat{\beta}_0 = 70.846$ and $\hat{\beta}_1 = -0.444$ when the outlier is removed.)

f. Test the null hypothesis of zero slope when the outlier is removed, given that $S_{Y|X} = 4.933$, $S_X = 14.693$, and $n = 17$. (Use $\alpha = .05$.)

g. For these data, would you conclude that the delinquency index decreases as IQ increases?

5. Following the last congressional election, a political scientist attempted to investigate the relationship between campaign expenditures on television advertisements and subsequent voter turnout. The following table presents the percentage of total campaign expenditures delegated to television advertisements (TVEXP) and the percentage of registered voter turnout (VOTE) for a hypothetical sample of 20 congressional districts.

a. Determine the least-squares line of VOTE on TVEXP, and draw the estimated line on the accompanying scatter diagram.

b. Are any of the assumptions for straight-line regression clearly *not* satisfied in this example?

c. Test the null hypothesis that the true slope is 0; be sure to interpret your result.

d. Test the hypothesis $\mu_{Y|X_0} = 45$ for $X_0 = \bar{X} = 36.99$. Interpret your result.

e. Calculate the corresponding 95% confidence interval for $\mu_{Y|36.99}$, and interpret your result.

VOTE (Y)	TVEXP (X)	VOTE (Y)	TVEXP (X)
35.4	28.5	40.8	31.3
58.2	48.3	61.9	50.1
46.1	40.2	36.5	31.3
45.5	34.8	32.7	24.8
64.8	50.1	53.8	42.2
52.0	44.0	24.6	23.0
37.9	27.2	31.2	30.1
48.2	37.8	42.6	36.5
41.8	27.2	49.6	40.2
54.0	46.1	56.6	46.1

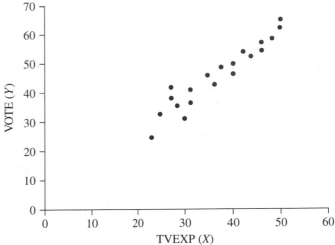

Edited SAS Output (PROC REG) for Problem 5

```
              VOTE (Y) Regressed on TVEXP (X)
                            .
                            .   [Portion of output omitted]
                            .

      Root MSE          3.33177          R-square       0.9101
                      Parameter Estimates
                     Parameter      Standard      T for H0:
      Variable   DF   Estimate        Error     Parameter=0    Prob > |T|
      INTERCEP    1   2.174067     3.30974362        0.657        0.5196
      X           1   1.176965     0.08718045       13.500        0.0001
```

6. A group of 13 children and adolescents (considered healthy) participated in a psychological study designed to analyze the relationship between age and average total sleep time (ATST). To obtain a measure for ATST (in minutes), recordings were taken on each subject on three consecutive nights and then averaged. The results obtained are displayed in the following table.

ATST (min/24 h)	AGE	ATST (min/24 h)	AGE
586.00	4.40	515.20	8.90
461.75	14.00	493.00	11.10
491.10	10.10	528.30	7.75
565.00	6.70	575.90	5.50
462.00	11.50	532.50	8.60
532.10	9.60	530.50	7.20
477.60	12.40		

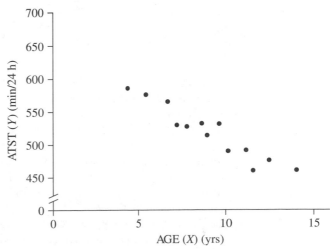

a. Determine the least-squares estimates of the slope and intercept for the straight-line regression of ATST (Y) on AGE (X). Draw the estimated line on the accompanying scatter diagram, and comment on the fit.

b. Are any of the assumptions for straight-line regression clearly *not* satisfied in this example?

c. Test the null hypothesis that the true slope is 0; be sure to interpret your result.

d. Obtain a 95% confidence interval for β_1. Interpret your result.

e. Would you reject the null hypothesis $H_0: \beta_1 = 0$ based on the confidence interval you calculated in part (d)? Explain.

f. Determine and sketch 95% confidence bands on the accompanying scatter diagram. Use your diagram to estimate the mean ATST when AGE = 10. Interpret your result.

Edited SAS Output (PROC REG) for Problem 6

```
                    ATST (Y) Regressed on AGE (X)
                                  .
                                  . [Portion of output omitted]
                                  .
        Root MSE          13.15238            R-square          0.9054
                         Parameter Estimates
                    Parameter         Standard        T for HO:
    Variable    DF   Estimate           Error      Parameter=0     Prob > |T|
    INTERCEP     1  646.483341      12.91772603         50.046        0.0001
    X            1  -14.041048       1.36811641        -10.263        0.0001

        Dep Var    Predict   Std Err    Lower95%   Upper95%
 Obs      Y         Value    Predict      Mean       Mean      Residual
  1      586.0      584.7     7.342       568.5      600.9       1.2973
  2      461.8      449.9     7.683       433.0      466.8      11.8413
  3      491.1      504.7     3.917       496.0      513.3     -13.5688
  4      565.0      552.4     4.869       541.7      563.1      12.5917
  5      462.0      485.0     4.947       474.1      495.9     -23.0113
  6      532.1      511.7     3.723       503.5      519.9      20.4107
  7      477.6      472.4     5.849       459.5      485.2       5.2257
  8      515.2      521.5     3.654       513.5      529.6      -6.3180
  9      493.0      490.6     4.595       480.5      500.7       2.3723
 10      528.3      537.7     4.063       528.7      546.6      -9.3652
 11      575.9      569.3     6.083       555.9      582.6       6.6424
 12      532.5      525.7     3.701       517.6      533.9       6.7697
 13      530.5      545.4     4.446       535.6      555.2     -14.8878
```

7. Several research workers associated with the Office of Highway Safety were evaluating the relationship between driving speed (MPH) and the distance a vehicle travels once brakes are applied (DIST). The results of 19 experimental tests are displayed in the following table.

a. Determine the least-squares estimates of the slope and intercept for each of the following straight-line regressions: Y_1(DIST) on X, and $Y_2(\sqrt{\text{DIST}})$ on X. Draw the estimated lines on the appropriate scatter diagrams.

b. Which of the two variable pairs mentioned in (a) seems to be better suited for straight-line regression?

c. For the variable pair Y_2 and X, test the hypothesis that the true slope is equal to 1 (use $\alpha = 0.01$). Be sure to interpret your result.

MPH (X)	DIST (Y_1)	$\sqrt{\text{DIST}}$ (Y_2)	MPH (X)	DIST (Y_1)	$\sqrt{\text{DIST}}$ (Y_2)
25.0	37.4	6.12	50.0	170.0	13.04
35.0	57.7	7.60	20.0	20.0	4.47
60.0	337.6	18.37	15.0	13.5	3.67
45.0	142.5	11.94	27.5	40.8	6.39
50.0	182.4	13.51	55.0	207.8	14.42
37.5	67.5	8.22	40.0	105.0	10.25
30.0	37.5	6.12	45.0	132.6	11.52
55.0	225.0	15.00	17.5	19.1	4.37
60.0	258.1	16.07	22.5	25.0	5.00
65.0	297.4	17.25			

d. Construct a 99% confidence interval for the true slope in part (c). Interpret your result.
e. For the same variable pair considered in parts (c) and (d), calculate and sketch 95% confidence bands on the appropriate scatter diagram. Using the confidence bands, estimate the mean value of Y_2 when $X = 45$. Interpret your result.

Edited SAS Output (PROC REG) for Problem 7

```
                        DIST (Y1) Regressed on MPH (X)
                                     .
                                     .  [Portion of output omitted]
                                     .

        Root MSE            31.64950          R-square          0.9106
                          Parameter Estimates
                      Parameter      Standard       T for H0:
     Variable   DF     Estimate        Error      Parameter=0     Prob > |T|
     INTERCEP    1   -122.344588   20.15623758       -6.070         0.0001
     X           1      6.227082    0.47318845       13.160         0.0001

                    SQRT(DIST) (Y2) Regressed on MPH (X)
                                     .
                                     .  [Portion of output omitted]
                                     .

        Root MSE             0.81097          R-square          0.9728
                          Parameter Estimates
                      Parameter      Standard       T for H0:
     Variable   DF     Estimate        Error      Parameter=0     Prob > |T|
     INTERCEP    1    -1.697124    0.51647104       -3.286         0.0044
     X           1     0.298775    0.01212469       24.642         0.0001

              Dep Var       Predict       Std Err      Lower95%      Upper95%
     Obs         Y2          Value        Predict        Mean          Mean
      1        6.1200        5.7723        0.258        5.2280        6.3165
      2        7.6000        8.7600        0.195        8.3492        9.1708
      3       18.3700       16.2294        0.308       15.5792       16.8796
      4       11.9400       11.7478        0.197       11.3328       12.1627
      5       13.5100       13.2416        0.224       12.7694       13.7139
      6        8.2200        9.5070        0.188        9.1103        9.9036
      7        6.1200        7.2661        0.220        6.8013        7.7310
      8       15.0000       14.7355        0.262       14.1819       15.2892
      9       16.0700       16.2294        0.308       15.5792       16.8796
     10       17.2500       17.7233        0.358       16.9671       18.4794
     11       13.0400       13.2416        0.224       12.7694       13.7139
     12        4.4700        4.2784        0.303        3.6389        4.9179
     13        3.6700        2.7845        0.353        2.0399        3.5292
     14        6.3900        6.5192        0.238        6.0171        7.0213
     15       14.4200       14.7355        0.262       14.1819       15.2892
     16       10.2500       10.2539        0.186        9.8613       10.6465
     17       11.5200       11.7478        0.197       11.3328       12.1627
     18        4.3700        3.5314        0.328        2.8403        4.2226
     19        5.0000        5.0253        0.280        4.4350        5.6157
```

8. The following table presents the starting annual salaries (SAL) of a group of 30 college
   graduates who have recently entered the job market, along with their cumulative grade-
   point averages (CGPA).
   a. Determine the least-squares estimates of the slope and intercept for the straight-line
      regression of SAL (Y) on CGPA (X). Draw the estimated line on the accompanying scat-
      ter diagram, and comment on the fit.
   b. Are any of the assumptions for straight-line regression clearly *not* satisfied in this example?

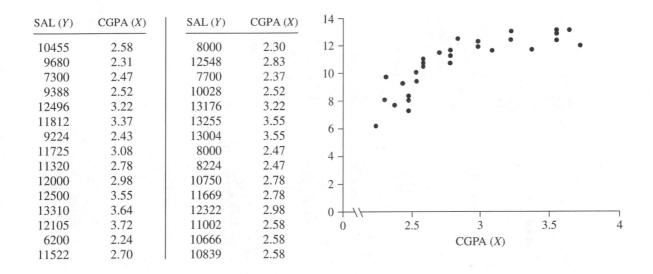

SAL (Y)	CGPA (X)	SAL (Y)	CGPA (X)
10455	2.58	8000	2.30
9680	2.31	12548	2.83
7300	2.47	7700	2.37
9388	2.52	10028	2.52
12496	3.22	13176	3.22
11812	3.37	13255	3.55
9224	2.43	13004	3.55
11725	3.08	8000	2.47
11320	2.78	8224	2.47
12000	2.98	10750	2.78
12500	3.55	11669	2.78
13310	3.64	12322	2.98
12105	3.72	11002	2.58
6200	2.24	10666	2.58
11522	2.70	10839	2.58

c. Obtain a 95% confidence interval for β_1.
d. Would you reject the null hypothesis $H_0: \beta_1 = 4000$ at the $\alpha = 0.05$ level?
e. Find and graph 95% confidence and prediction bands.
f. Would you reject the hypothesis $H_0: \mu_{Y|X} = 11{,}500$ at $X_0 = 2.75$?

Edited SAS Output (PROC REG) for Problem 8

SAL (Y) Regressed on CGPA (X)

[Portion of output omitted]

Root MSE   1124.71499   R-square   0.6845

Parameter Estimates

Variable	DF	Parameter Estimate	Standard Error	T for H0: Parameter=0	Prob > |T|
INTERCEP	1	435.923567	1337.8596742	0.326	0.7470
X	1	3630.561280	465.76874410	7.795	0.0001

Obs	Dep Var Y	Predict Value	Std Err Predict	Lower95% Mean	Upper95% Mean	Lower95% Predict	Upper95% Predict
1	10455.0	9802.8	238.000	9315.3	10290.3	7447.9	12157.7
2	9680.0	8822.5	320.503	8166.0	9479.0	6426.9	11218.1
3	7300.0	9403.4	267.579	8855.3	9951.5	7035.2	11771.6
4	9388.0	9584.9	253.279	9066.1	10103.8	7223.4	11946.5
5	12496.0	12126.3	271.602	11570.0	12682.7	9756.2	14496.4
6	11812.0	12670.9	321.696	12011.9	13329.9	10274.7	15067.2
7	9224.0	9258.2	279.889	8684.9	9831.5	6884.0	11632.3
8	11725.0	11618.1	234.171	11138.4	12097.7	9264.8	13971.3
9	11320.0	10528.9	207.134	10104.6	10953.2	8186.3	12871.5

(continued)

```
                    SAL (Y) Regressed on CGPA (X) (Continued)
          Dep Var   Predict    Std Err   Lower95%   Upper95%   Lower95%   Upper95%
   Obs       Y       Value     Predict     Mean       Mean     Predict    Predict
   10     12000.0   11255.0    215.685    10813.2    11696.8    8909.1    13600.9
   11     12500.0   13324.4    389.923    12525.7    14123.1    10886.0   15762.8
   12     13310.0   13651.2    426.130    12778.3    14524.1    11187.5   16114.9
   13     12105.0   13941.6    459.132    13001.1    14882.1    11453.2   16430.1
   14      6200.0    8568.4    346.167     7859.3     9277.5    6157.9    10978.9
   15     11522.0   10238.4    215.215     9797.6    10679.3    7892.8    12584.1
   16      8000.0    8786.2    324.093     8122.3     9450.1    6388.6    11183.8
   17     12548.0   10710.4    205.381    10289.7    11131.1    8368.4    13052.4
   18      7700.0    9040.4    299.581     8426.7     9654.0    6656.2    11424.6
   19     10028.0    9584.9    253.279     9066.1    10103.8    7223.4    11946.5
   20     13176.0   12126.3    271.602    11570.0    12682.7    9756.2    14496.4
   21     13255.0   13324.4    389.923    12525.7    14123.1    10886.0   15762.8
   22     13004.0   13324.4    389.923    12525.7    14123.1    10886.0   15762.8
   23      8000.0    9403.4    267.579     8855.3     9951.5    7035.2    11771.6
   24      8224.0    9403.4    267.579     8855.3     9951.5    7035.2    11771.6
   25     10750.0   10528.9    207.134    10104.6    10953.2    8186.3    12871.5
   26     11669.0   10528.9    207.134    10104.6    10953.2    8186.3    12871.5
   27     12322.0   11255.0    215.685    10813.2    11696.8    8909.1    13600.9
   28     11002.0    9802.8    238.000     9315.3    10290.3    7447.9    12157.7
   29     10666.0    9802.8    238.000     9315.3    10290.3    7447.9    12157.7
   30     10839.0    9802.8    238.000     9315.3    10290.3    7447.9    12157.7
```

9. In an experiment designed to describe the dose–response curve for vitamin K, individual rats were depleted of their vitamin K reserves and then fed dried liver for 4 days at different dosage levels.[7] The response of each rat was measured as the concentration of a clotting agent needed to clot a sample of its blood in 3 minutes. The results of the experiment on 12 rats are given in the following table; values are expressed in common logarithms for both dose and response.

   a. Determine the least-squares estimates of the slope (β_1) and the intercept (β_0) for the straight-line regression of Y on X.

Rat	$\log_{10}$ Concentration (Y)	$\log_{10}$ Dose (X)
1	2.65	0.18
2	2.25	0.33
3	2.26	0.42
4	1.95	0.54
5	1.72	0.65
6	1.60	0.75
7	1.55	0.83
8	1.32	0.92
9	1.13	1.01
10	1.07	1.04
11	0.95	1.09
12	0.88	1.15

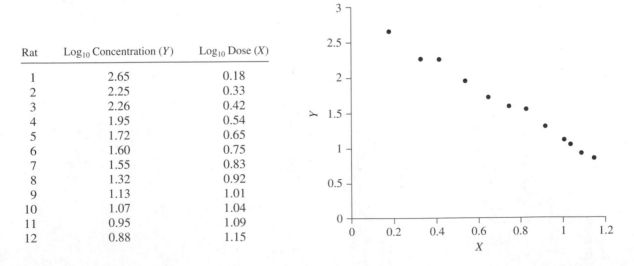

---

[7]Adapted from a study by Schønheyder (1936).

b. Draw the estimated regression line on the accompanying scatter diagram. How well does this line fit the data?

c. Determine and sketch 95% confidence bands based on the estimated regression line.

d. Convert the fitted straight line into an equation in the original units $Y' = 10^Y$ and $X' = 10^X$.

e. For the converted equation obtained in part (e), determine 99% confidence intervals for the true mean responses at the maximum and minimum doses used in the experiment.

f. If the values for X and Y on each rat are converted to their original units X' and Y', the following fitted straight line is obtained: $Y' = 237.16095 - 21.32117X'$. How would you evaluate whether using the variables X' and Y' is better or worse than using X and Y for the regression analysis?

Edited SAS Output (PROC REG) for Problem 9

```
          LOG10 (CONC) (Y) Regressed on LOG10 (DOSE) (X)

                           .
                           .  [Portion of output omitted]
                           .

        Root MSE             0.05589          R-square          0.9914
                          Parameter Estimates
                       Parameter     Standard      T for H0:
   Variable    DF      Estimate        Error     Parameter=0    Prob > |T|
   INTERCEP     1      2.936205      0.04230190      69.411        0.0001
   X            1     -1.785012      0.05266556     -33.893        0.0001

          Dep Var    Predict    Std Err    Lower95%    Upper95%
   Obs       Y        Value     Predict      Mean        Mean      Residual
    1      2.6500     2.6149      0.034      2.5397      2.6901      0.0351
    2      2.2500     2.3472      0.027      2.2869      2.4074     -0.0972
    3      2.2600     2.1865      0.023      2.1343      2.2387      0.0735
    4      1.9500     1.9723      0.019      1.9292      2.0154     -0.0223
    5      1.7200     1.7759      0.017      1.7384      1.8135     -0.0559
    6      1.6000     1.5974      0.016      1.5615      1.6334      0.00255
    7      1.5500     1.4546      0.017      1.4173      1.4920      0.0954
    8      1.3200     1.2940      0.019      1.2524      1.3355      0.0260
    9      1.1300     1.1333      0.021      1.0856      1.1811     -0.00334
   10      1.0700     1.0798      0.022      1.0297      1.1299     -0.00979
   11      0.9500     0.9905      0.024      0.9362      1.0449     -0.0405
   12      0.8800     0.8834      0.027      0.8236      0.9433     -0.00344
```

10. The susceptibility of catfish to a certain chemical pollutant was determined by immersing individual fish in 2 liters of an emulsion containing the pollutant and measuring the survival time in minutes.[8] The data in the following table give the common log of survival time (Y) and the common log of concentration (X) of the pollutant in parts per million for 18 fish.

a. Determine and draw the estimated straight line of Y regressed on X on the accompanying scatter diagram. Comment on the fit.

b. Test for the significance of the straight-line regression. Interpret your result.

c. Determine 95% confidence intervals for the true mean survival time $\mu_{Y|X}$ (where $Y = 10^Y$) at values of $X = 5$, 4.5, and 4. (Note $\overline{X} = 4.5$.) Interpret these intervals.

---

[8] Adapted from a study by Nagasawa, Osano, and Kondo (1964).

Fish	Log$_{10}$ Survival Time (Y)	Log$_{10}$ Concentration (X)
1	2.516	5.0
2	2.572	5.0
3	2.438	5.0
4	2.621	4.8
5	2.742	4.8
6	2.689	4.8
7	2.830	4.6
8	2.910	4.6
9	2.983	4.6
10	3.175	4.4
11	3.056	4.4
12	3.095	4.4
13	3.332	4.2
14	3.221	4.2
15	3.293	4.2
16	3.447	4.0
17	3.523	4.0
18	3.551	4.0

Edited SAS Output (PROC REG) for Problem 10

```
        LOG10 (SURV TIME) (Y) Regressed on LOG10 (CONC) (X)
                                    .
                                 .  [Portion of output omitted]
                                    .
Root MSE              0.05597              R-square            0.9766
                          Parameter Estimates

                       Parameter      Standard      T for H0:
Variable    DF         Estimate          Error   Parameter=0    Prob > |T|
INTERCEP     1         7.491095     0.17431396        42.975        0.0001
X            1        -0.998095     0.03862533       -25.840        0.0001

        Dep Var      Predict     Std Err   Lower95%    Upper95%
Obs        Y          Value      Predict      Mean        Mean     Residual
 1       2.5160       2.5006       0.023     2.4510      2.5502      0.0154
 .
 .  [Portion of output omitted]
 .
 9       2.9830       2.8999       0.014     2.8707      2.9290      0.0831
10       3.1750       3.0995       0.014     3.0703      3.1286      0.0755
 .
 .  [Portion of output omitted]
 .
18       3.5510       3.4987       0.023     3.4491      3.5483      0.0523
```

11. An experiment was conducted to determine the extent to which the growth rate of a certain fungus could be affected by filling test tubes containing the same medium at the same temperature with different inert gases.[9] Three such experiments were performed for each of six gases, and the average growth rate over these three tests was used as the response. The following table gives the molecular weight (X) of each gas used and the average growth rate (Y) in milliliters per hour for the three tests.

Gas	Average Growth Rate (Y)	Molecular Weight (X)
A	3.85	4.0
B	3.48	20.2
C	3.27	28.2
D	3.08	39.9
E	2.56	83.8
F	2.21	131.3

a. Find the least-squares estimates of slope and intercept for the straight-line regression of Y on X, and draw the estimated straight line on a scatter diagram for this data set.

b. Test for significant slope of the fitted straight line.

c. What information has not been used that might improve the sensitivity of the analysis?

d. What is the 90% confidence interval for the true average growth rate if the gas used has a molecular weight of 100?

e. Why would it be inappropriate to use the fitted line to estimate the growth rate of a gas whose molecular weight is 200?

f. Based on the choice of X-values used in this study, how would you criticize the accuracy of prediction obtained in this experiment by using the fitted straight line?

Edited SAS Output (PROC REG) for Problem 11

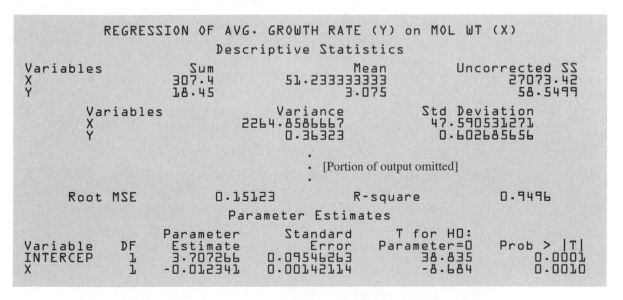

```
          REGRESSION OF AVG. GROWTH RATE (Y) on MOL WT (X)
                      Descriptive Statistics
Variables          Sum               Mean          Uncorrected SS
X                 307.4          51.233333333          27073.42
Y                 18.45              3.075             58.5499

         Variables              Variance         Std Deviation
            X                  2264.8586667        47.590531271
            Y                    0.36323            0.602685656

                                  .
                                  .   [Portion of output omitted]
                                  .

     Root MSE           0.15123          R-square          0.9496
                      Parameter Estimates

                  Parameter       Standard      T for H0:
Variable    DF    Estimate          Error     Parameter=0    Prob > |T|
INTERCEP    1     3.707266        0.09546263      38.835        0.0001
X           1    -0.012341        0.00142114      -8.684        0.0010
```

---

[9]Adapted from a study by Schreiner, Gregoine, and Lawrie (1962).

12. Consider the data in the following table.[10]

Age			Age		
Years	Months	Vocabulary Size	Years	Months	Vocabulary Size
0	8	0	3	0	896
0	10	1	3	6	1,222
1	0	3	4	0	1,540
1	3	19	4	6	1,870
1	6	22	5	0	2,072
1	9	118	5	6	2,289
2	0	272	6	0	2,562
2	6	446			

Note: Data are from M. E. Smith, "An Investigation of the Development of the Sentence and the Extent of Vocabulary in Young Children," *Studies in Child Welfare* (University of Iowa) 3(5) (1926).

a. First convert ages to decimal years (e.g., 1 year 6 months gives 1.5 years). Draw a scatter diagram for the variable pair vocabulary size (Y) and age (X).

b. Then calculate the least-squares estimates of the parameters of the regression line. Sketch this regression line on the scatter diagram.

c. Add a new observation to the vocabulary data, with values 0.00 years and 0 words. Plot this new point. Logically this value should be on the line of vocabulary growth. Is it near the fitted line from part (b)?

d. Recompute the least-squares estimates of the regression line to include the (0, 0) observation ($n = 16$). Sketch the new line distinctly on the scatter diagram.

e. Is either line fitted acceptably? If not, sketch your idea of the true relationship. If the data included observations through age 30 years, what would the extrapolated curve look like?

f. The data appear to be from one child. If this is true, what assumption of the least-squares approach is most likely violated, and why?

Edited SAS Output (PROC REG) for Problem 12

```
                    REGRESSION OF VOCAB (Y) ON AGE (X)
                              .
                              .  [Portion of output omitted]
                              .
        Root MSE          148.07090           R-square          0.9776
                         Parameter Estimates
                     Parameter         Standard       T for H0:
Variable   DF         Estimate            Error    Parameter=0    Prob > |T|
INTERCEP    1      -621.125954      74.08216150         -8.384        0.0001
X           1       526.718356      22.13536475         23.795        0.0001
```

---

[10]Taken from Bourne, Ekstrand, and Dominowski (1971, table 14.3).

```
              Y ON X, INCLUDING 16TH OBS.
                         .
                         .  [Portion of output omitted]
                         .
     Root MSE         205.90999         R-square         0.9558
                      Parameter Estimates
                    Parameter      Standard      T for H0:
  Variable    DF    Estimate       Error       Parameter=0    Prob > |T|
  INTERCEP     1   -496.775354    92.13221921     -5.392         0.0001
  X            1    494.893155    28.43144491     17.407         0.0001
```

13. The following table gives rat body weights (in grams) and latency to seizure (in minutes), following injection of 40 mg/kg of body weight of metrazol.

Latency	Weight	Latency	Weight
2.30	348	2.00	409
1.95	372	1.70	413
2.90	378	2.00	415
2.30	390	2.95	423
1.10	392	1.25	428
2.50	395	2.05	464
1.30	400	3.70	468

a. Draw a scatter diagram on graph paper, with latency plotted as a function of weight.
b. Determine the least-squares estimates of slope and intercept for the straight-line regression of latency on weight.
c. Test whether the slope is equal to 0. Use $\alpha = .10$.
d. Test whether the intercept is equal to 0. Use $\alpha = .10$.
e. Sketch the estimated regression line.
f. Distinctly sketch your choice for a regression line if it differs from that of part (e). Explain why you agree or disagree with part (e), noting whether any assumptions appear to be violated.

Edited SAS Output (PROC REG) for Problem 13

```
              REGRESSION OF LATENCY ON WEIGHT
                         .
                         .  [Portion of output omitted]
                         .
     Root MSE         0.72779          R-square         0.0519
                      Parameter Estimates
                    Parameter      Standard      T for H0:
  Variable    DF    Estimate       Error       Parameter=0    Prob > |T|
  INTERCEP     1    0.115983     2.50749283       0.046         0.9639
  WEIGHT       1    0.004983     0.00614559       0.811         0.4333
```

14. Stevens (1966), citing Dimmick and Hubbard (1939), reported data from 20 studies of the color perception of unitary hues. The wavelength (in millimeters) of light called green by

subjects in each experiment is given in the following table, along with the year the study was conducted.

Study	Wavelength	Date	Study	Wavelength	Date	Study	Wavelength	Date
1	532	1874	8	506	1909	15	500	1928
2	535	1884	9	509	1911	16	506	1931
3	495	1888	10	520	1912	17	528	1934
4	527	1890	11	514	1920	18	530	1935
5	505	1898	12	504	1922	19	512	1935
6	505	1898	13	515	1926	20	515	1939
7	503	1907	14	498	1927			

a. These data stimulate the question "Is there any linear trend over time in the wavelength of light called green?" Evaluate this question by finding the least-squares estimates of the straight-line regression functions for predicting wavelength from year. [*Hint:* Subtract 500 from the wavelengths and 1850 from the year.]

b. Find a 95% confidence interval for the true slope.

c. Draw a scatter diagram on graph paper.

Edited SAS Output (PROC REG) for Problem 14

```
              REGRESSION OF WAVELENGTH ON DATE
                          ·
                          · [Portion of output omitted]
                          ·
        Root MSE        12.17807          R-square        0.0310
                        Parameter Estimates
                      Parameter      Standard      T for H0:
    Variable   DF     Estimate         Error    Parameter=0   Prob > |T|
    INTERCEP    1     19.881714     9.52905925       2.086       0.0514
    DATE        1     -0.10933      0.14403292      -0.759       0.4576
```

15. The data listed in the following table are from a study by Benignus and others (1981). Blood and brain levels of toluene (a commonly used solvent) were measured following a 3-hour inhalation exposure to 50, 100, 500, or 1,000 parts per million (ppm) toluene (PPM_TOLU). Blood toluene (BLOODTOL) and brain toluene (BRAINTOL) are expressed in parts per million, weight in grams, and age in days.

Rat	BLOODTOL	BRAINTOL	PPM_TOLU	WEIGHT	AGE	LN_BLDTL	LN_BRNTL	LN_PPMTL
1	0.553	0.481	50	393	95	−0.593	−0.732	3.912
2	0.494	0.584	50	378	95	−0.706	−0.538	3.912
3	0.609	0.585	50	450	95	−0.495	−0.536	3.912
4	0.763	0.628	50	439	95	−0.270	−0.465	3.912
5	0.420	0.533	50	397	95	−0.868	−0.629	3.912
6	0.397	0.490	50	301	84	−0.923	−0.713	3.912
7	0.503	0.719	50	406	84	−0.687	−0.330	3.912
8	0.534	0.585	50	302	84	−0.628	−0.536	3.912
9	0.531	0.675	50	382	84	−0.633	−0.393	3.912

Rat	BLOODTOL	BRAINTOL	PPM_TOLU	WEIGHT	AGE	LN_BLDTL	LN_BRNTL	LN_PPMTL
10	0.384	0.442	50	355	84	−0.957	−0.816	3.912
11	0.215	0.492	50	405	85	−1.536	−0.709	3.912
12	0.552	0.859	50	405	85	−0.595	−0.152	3.912
13	0.420	0.650	50	387	85	−0.868	−0.431	3.912
14	0.324	0.528	50	358	85	−1.127	−0.639	3.912
15	0.387	0.546	50	311	85	−0.949	−0.605	3.912
16	1.036	1.262	100	355	86	0.035	0.233	4.605
17	1.065	1.584	100	440	86	0.063	0.460	4.605
18	1.084	1.773	100	421	86	0.081	0.573	4.605
19	0.944	1.307	100	370	86	−0.058	0.268	4.605
20	0.994	1.338	100	375	86	−0.006	0.291	4.605
21	1.146	1.180	100	368	83	0.136	0.166	4.605
22	1.167	1.108	100	321	83	0.154	0.103	4.605
23	0.833	0.939	100	359	83	−0.183	−0.063	4.605
24	0.630	0.909	100	367	83	−0.462	−0.095	4.605
25	0.955	1.078	100	363	83	−0.046	0.075	4.605
26	0.687	1.152	100	388	86	−0.376	0.141	4.605
27	0.723	1.796	100	404	86	−0.324	0.586	4.605
28	0.705	1.262	100	454	86	−0.349	0.233	4.605
29	0.696	1.865	100	389	86	−0.363	0.623	4.605
30	0.868	1.892	100	352	86	−0.142	0.638	4.605
31	8.223	19.843	500	367	83	2.107	2.988	6.215
32	10.604	24.450	500	406	83	2.361	3.197	6.215
33	12.085	29.297	500	371	83	2.492	3.377	6.215
34	7.936	18.098	500	408	83	2.071	2.896	6.215
35	11.164	25.196	500	305	83	2.413	3.227	6.215
36	10.289	18.266	500	391	84	2.331	2.905	6.215
37	11.140	19.486	500	396	84	2.411	2.970	6.215
38	9.647	18.479	500	347	84	2.267	2.917	6.215
39	13.343	21.920	500	372	84	2.591	3.087	6.215
40	11.292	20.861	500	331	84	2.424	3.038	6.215
41	7.524	22.130	500	365	85	2.018	3.097	6.215
42	10.783	18.301	500	348	85	2.378	2.907	6.215
43	8.595	17.038	500	416	85	2.151	2.835	6.215
44	9.616	22.423	500	344	85	2.263	3.110	6.215
45	11.956	15.452	500	398	85	2.481	2.738	6.215
46	30.274	44.900	1000	417	93	3.410	3.804	6.908
47	32.923	35.500	1000	351	93	3.494	3.570	6.908
48	28.619	30.800	1000	378	93	3.354	3.428	6.908
49	28.761	38.500	1000	338	93	3.359	3.651	6.908
50	25.402	31.500	1000	433	93	3.235	3.450	6.908
51	35.464	42.330	1000	342	85	3.569	3.745	6.908
52	32.706	34.030	1000	319	85	3.488	3.527	6.908
53	29.347	30.760	1000	440	85	3.379	3.426	6.908
54	26.481	32.360	1000	363	85	3.276	3.477	6.908
55	33.401	41.830	1000	336	85	3.509	3.734	6.908
56	39.541	54.930	1000	378	86	3.677	4.006	6.908
57	28.155	39.780	1000	420	86	3.338	3.683	6.908
58	25.629	49.290	1000	346	86	3.244	3.898	6.908
59	33.188	47.490	1000	413	86	3.502	3.861	6.908
60	33.505	42.660	1000	432	86	3.512	3.753	6.908

    a. Provide a scatter diagram, with BLOODTOL as the response and PPM_TOLU as the predictor.

    b. Compute least-squares estimates of the straight-line regression coefficients. Plot the line on the scatter diagram.

    c. Repeat part (a) for the natural logarithms LN_BLDTL and LN_PPMTL.

    d. Repeat part (b) for the natural logarithms.

    e. Which transformation leads to the best representation of the data? Note in your comments the validity of the regression assumptions.

Edited SAS Output (PROC REG) for Problem 15

```
              BLOODTOL REGRESSED ON PPM_TOLU
                              ⋮
                              ⋮  [Portion of output omitted]
                              ⋮
      Root MSE         2.85301         R-square        0.9497
                      Parameter Estimates
                    Parameter      Standard     T for H0:
   Variable    DF    Estimate         Error   Parameter=0    Prob > |T|
   INTERCEP     1   -2.546103    0.54253665        -4.693        0.0001
   PPM_TOLU     1    0.031959    0.00096570        33.094        0.0001

      SAS PROC REG OUTPUT FOR LN_BLDTL REGRESSED ON LN_PPMTL
                              ⋮
                              ⋮  [Portion of output omitted]
                              ⋮
      Root MSE         0.24144         R-square        0.9813
                      Parameter Estimates
                    Parameter      Standard     T for H0:
   Variable    DF    Estimate         Error   Parameter=0    Prob > |T|
   INTERCEP     1   -6.531583    0.14365100       -45.468        0.0001
   LN_PPMTL     1    1.430453    0.02592028        55.187        0.0001
```

16. Real estate prices depend, in part, on property size. The house size X (in hundreds of square feet) and house price Y (in thousands of dollars) of a random sample of houses in a certain county were recorded as in the following table.

House	1	2	3
X	18	20	25
Y	80	95	104

House	4	5	6	7
X	22	33	19	17
Y	110	175	85	89

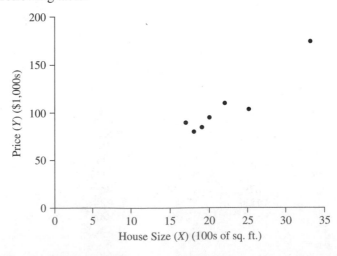

a. On the accompanying scatter diagram, sketch by eye a line that fits the data reasonably well. Comment on the relationship between house size and house price.
b. Determine the least-squares estimates of the slope (β_1) and the intercept (β_0) for the straight-line regression of Y on X.
c. Draw the estimated regression line on the scatter diagram. Comment on how well the line fits the data.
d. Test the null hypothesis that the true slope is 0. Interpret the results of this test.

Edited SAS Output (PROC REG) for Problem 16

```
               PRICE (Y) Regressed on HOUSE SIZE (X)
                                    .
                                    .   [Portion of output omitted]
                                    .
        Root MSE      10.70973          R-square         0.9091
                          Parameter Estimates

                      Parameter      Standard      T for H0:
    Variable    DF    Estimate         Error     Parameter=0    Prob > |T|
    INTERCEP     1   -17.364907     17.83512957     -0.974        0.3750
    X            1     5.581522      0.78953169      7.069        0.0009

         Dep Var     Predict    Std Err    Lower95%   Upper95%
    Obs     Y         Value     Predict      Mean       Mean     Residual
    1    80.0000     83.1025     5.134      69.9048    96.3002    -3.1025
    2    95.0000     94.2655     4.345      83.0964    105.4       0.7345
    3    104.0       122.2       4.690      110.1      134.2     -18.1731
    4    110.0       105.4       4.048      95.0231    115.8       4.5714
    5    175.0       166.8       9.582      142.2      191.5       8.1747
    6    85.0000     88.6840     4.690      76.6281    100.7      -3.6840
    7    89.0000     77.5210     5.654      62.9865    92.0554    11.4790
```

17. Sales revenue (Y) and advertising expenditure (X) data for a large retailer for the period 1988–1993 are given in the following table.

Year	1988	1989	1990
Sales ($millions)	4	8	2
Advertising ($millions)	2	5	0

Year	1991	1992	1993
Sales ($millions)	8	5	4
Advertising ($millions)	6	4	3

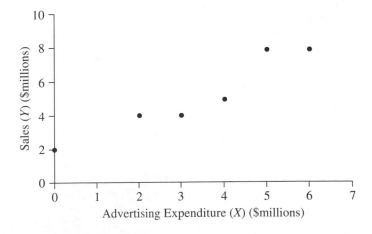

a. Does the plot of Y versus X suggest that a linear relationship exists between X and Y?

b. Calculate the least-squares estimates of the parameters of the regression line, and draw the estimated line on the accompanying scatter diagram. Does the line appear to fit the data well?

c. Find a 95% confidence interval for the slope parameter. Based on your interval, is sales revenue linearly related to advertising expenditure? Explain.

d. Would it be appropriate to use the estimated regression line in part (b) to estimate the sales for a new year in which an advertising expenditure of $10 million is planned? Why or why not?

Edited SAS Output (PROC REG) for Problem 17

```
            REGRESSION OF SALES (Y) ON ADV (X)

                      .
                      .   [Portion of output omitted]
                      .

        Root MSE      0.83023          R-square          0.9044
                       Parameter Estimates
                   Parameter        Standard      T for HO:
    Variable   DF   Estimate           Error    Parameter=0    Prob > |T|
    INTERCEP    1   1.642857      0.66566677          2.468        0.0691
    ADV         1   1.057143      0.17187442          6.151        0.0035
```

18. The production manager of a plant that manufactures syringes records the marginal cost at various levels of output for 14 randomly selected months. The data are shown below:

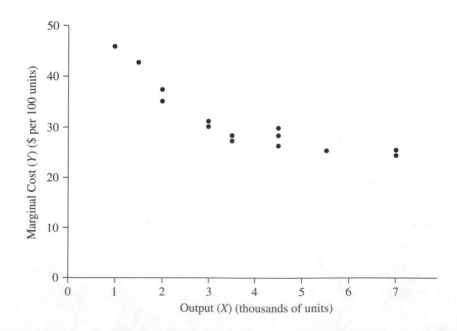

Marginal Cost Y ($ per 100 units)	Output X (thousands of units)
31.00	3.0
30.00	3.0
28.00	3.5
46.00	1.0
43.00	1.5
35.00	2.0
37.50	2.0
27.00	3.5
25.00	5.5
24.00	7.0
25.00	7.0
29.50	4.5
26.00	4.5
28.00	4.5

a. On the accompanying scatter diagram of marginal cost (Y) versus output (X), sketch by eye a line that fits the data reasonably well.

b. Find the estimated least-squares equation for the regression of marginal cost on output.

c. Sketch the estimated line on the scatter diagram. Does it seem to fit the data well?

d. Test the null hypothesis that the true slope is zero, at the $\alpha = .05$ significance level. Interpret your result.

e. Can you suggest a model that would describe the marginal cost–output relationship for this manufacturer better than a straight line does?

Edited SAS Output (PROC REG) for Problem 18

```
         MARGINAL COST (Y) Regressed on OUTPUT (X)
                          ⦙ [Portion of output omitted]
                          ⦙
      Root MSE        3.63338           R-square          0.7405
                        Parameter Estimates
                   Parameter       Standard      T for H0:
Variable    DF      Estimate          Error    Parameter=0     Prob > |T|
INTERCEP     1     42.842546     2.23377154         19.179         0.0001
X            1     -3.138965     0.53644247         -5.851         0.0001
```

19. The data shown in the following table were obtained from the 1990 Census.[11] Included is information on 26 randomly selected Metropolitan Statistical Areas (MSAs). Of interest are factors that potentially are associated with the rate of owner occupancy of housing units. Two variables are included in the data set: OWNEROCC = percentage of housing units that are owner-occupied (as opposed to renter-occupied); OWNCOST = median selected monthly ownership costs (in $).

a. Based on the accompanying scatter diagram of OWNEROCC versus OWNCOST, does there appear to be a linear relationship between these two variables?

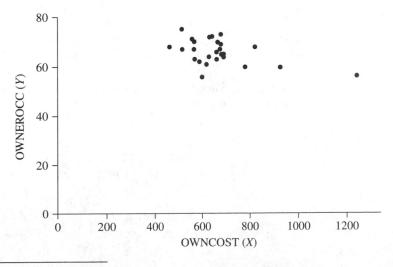

MSA	OWNEROCC (Y)	OWNCOST (X)
Abilene, TX	62	583
Burlington, NC	72	627
Daytona Beach, FL	72	636
Grand Rapids, MI	73	677
Laredo, TX	61	614
Louisville, KY–IN	67	561
Oklahoma City, OK	64	627
Pine Bluff, AR	67	513
San Francisco–Oakland–San Jose, CA	57	1234
Wichita Falls, TX	63	565
Albany, GA	56	597
Canton, OH	71	555
Des Moines, IA	67	673
Jacksonville, FL	65	687
Johnstown, PA	75	510
Medford, OR	66	660
Omaha, NE–IA	64	687
Provo–Orem, UT	63	659
Williamsburg, PA	70	564
Appleton–Oshkosh–Neenah, WI	70	663
Melbourne–Titusville–Palm Bay, FL	69	675
Redding, CA	65	682
Worcester, MA	60	918
Milwaukee–Racine, WI	60	777
Rochester, NY	68	818
St. Joseph, MO	68	457

b. State the model for the straight-line regression of OWNEROCC (Y) on OWNCOST (X). Determine the least-squares estimates for this regression line. Interpret the estimated values of the slope and the intercept in the context of the problem.

c. Sketch the estimated line on the scatter diagram and assess the fit.

d. Test for the significance of the slope parameter of the model in part (b). Interpret your result.

e. Determine a 95% confidence interval for the true slope in part (d). Interpret your result with regard to the test mentioned in part (d).

Edited SAS Output (PROC REG) for Problem 19

```
                    OWNEROCC Regressed on OWNCOST
                              .
                              .  [Portion of output omitted]
                              .
        Root MSE        4.41749           R-square          0.2207
                          Parameter Estimates
                      Parameter        Standard      T for H0:
     Variable   DF    Estimate         Error      Parameter=0    Prob > |T|
     INTERCEP    1    76.007640        3.94979480      19.243        0.0001
     OWNCOST     1    -0.015169        0.00581880      -2.607        0.0155
```

References

Benignus, V. A.; Muller, K. E.; Barton, C. N.; and Bittekofer, J. A. 1981. "Toluene Levels in Blood and Brain of Rats during and after Respiratory Exposure." *Toxicology and Applied Pharmacology* 61: 326–34.

Bourne, L. E.; Ekstrand, B. E.; and Dominowski, R. L. 1971. *The Psychology of Thinking.* Englewood Cliffs, N.J.: Prentice-Hall.

Diggle, P. J.; Liang, K. Y.; and Zeger, S. L. 1994. *Analysis of Longitudinal Data.* New York: Oxford University Press.

Dimmick, F. L., and Hubbard, M. R. 1939. "The Spectral Location of Psychologically Unique Yellow, Green, and Blue." *American Journal of Psychology* 52: 242.

Morrison, D. F. 1976. *Multivariate Statistical Methods.* New York: McGraw-Hill.

Nagasawa, S.; Osana, S.; and Kondo, K. 1964. "An Analytical Method for Evaluating the Susceptibility of Fish Species to an Agricultural Chemical." *Japanese Journal of Applied Enterological Zoology* 8: 118–22.

Schønheyder, F. 1936. "The Quantitative Determination of Vitamin K." *Biochemistry Journal* 30: 890–96.

Schreiner, H. R.; Gregoine, R. C.; and Lawrie, J. A. 1962. "New Biological Effects of the Gases on the Helium Group." *Science* 136: 653–54.

Siegel, S. 1956. *Nonparametric Statistics for the Behavioral Sciences.* New York: McGraw-Hill.

Smith, M. E. 1926. "An Investigation of the Development of the Sentence and the Extent of Vocabulary in Young Children." *Studies in Child Welfare* 3:5.

Stevens, S. S. 1966. *Handbook of Experimental Psychology.* New York: John Wiley & Sons.

Timm, N. H. 1975. *Multivariate Analysis with Applications in Education and Psychology.* Monterey, Calif.: Brooks/Cole.

Zeger, S. L., and Liang, K. Y. 1986. "Longitudinal Data Analysis for Discrete and Continuous Outcomes," *Biometrics* 42: 121–30.

6

The Correlation Coefficient and Straight-line Regression Analysis

6-1 Definition of r

The correlation coefficient is an often-used statistic that provides a measure of how two random variables are linearly associated in a sample and has properties closely related to those of straight-line regression. We define the *sample correlation coefficient r* for two variables X and Y by the formula

$$r = \frac{\sum_{i=1}^{n}(X_i - \overline{X})(Y_i - \overline{Y})}{\left[\sum_{i=1}^{n}(X_i - \overline{X})^2 \sum_{i=1}^{n}(Y_i - \overline{Y})^2\right]^{1/2}} = \frac{\text{SSXY}}{\sqrt{\text{SSX} \cdot \text{SSY}}} \tag{6.1}$$

where $\text{SSXY} = \sum_{i=1}^{n}(X_i - \overline{X})(Y_i - \overline{Y})$, $\text{SSX} = \sum_{i=1}^{n}(X_i - \overline{X})^2$, and $\text{SSY} = \sum_{i=1}^{n}(Y_i - \overline{Y})^2$.

An equivalent formula for r that illustrates its mathematical relationship to the least-squares estimate of the slope of a fitted regression line is[1]

$$r = \frac{S_X}{S_Y}\hat{\beta}_1 \tag{6.2}$$

■ **Example 6-1**   For the age–systolic blood pressure data in Table 5-1, r is 0.66. This value can be obtained from the SAS output on page 50 by taking the square root of "R–square = 0.4324." Alternatively, using (6.2), we have

$$r = \frac{15.29}{22.58}(0.97) = 0.66$$

---

[1] $S_X^2 = \dfrac{1}{n-1}\text{SSX}$ and $S_Y^2 = \dfrac{1}{n-1}\text{SSY}$ are the estimated sample variances of the X and Y variables, respectively.

Three important mathematical properties are associated with r:

1. The possible values of r range from -1 to 1.

2. r is a dimensionless quantity; that is, r is independent of the units of measurement of X and Y.

3. r is positive, negative, or zero as $\hat{\beta}_1$ is positive, negative, or zero; and vice versa. This property follows directly, of course, from (6.2). ■

6-2 *r* as a **Measure of Association**

In the statistical assumptions for straight-line regression analysis discussed earlier, we did not consider the variable X to be random. Nevertheless, it often makes sense to view the regression problem as one where both X and Y are random variables. The measure r can then be interpreted as an *index of linear association* between X and Y, in the following sense:

1. The more positive r is, the more positive the association is. This means that, when r is close to 1, an individual with a high value for one variable will likely have a high value for the other, and an individual with a low value for one variable will likely have a low value for the other (Figure 6-1(a)).

2. The more negative r is, the more negative the association is; that is, an individual with a high value for one variable will likely have a low value for the other when r is close to -1, and conversely (Figure 6-1(b)).

3. If r is close to 0, there is little, if any, *linear* association between X and Y (Figure 6-1(c) or 6-1(d)).[2]

By *association,* we mean the lack of statistical independence between X and Y. More loosely, the lack of an association means that the value of one variable cannot be reasonably anticipated from knowing the value of the other variable.

Since r is an index obtained from a *sample* of n observations, it can be considered as an estimate of an unknown population parameter. This unknown parameter, called the *population correlation coefficient,* is generally denoted by the symbol ρ_{XY} or more simply ρ (if it is clearly understood which two variables are being considered). We shall agree to use ρ unless confusion

FIGURE 6-1   Correlation coefficient as a measure of association.

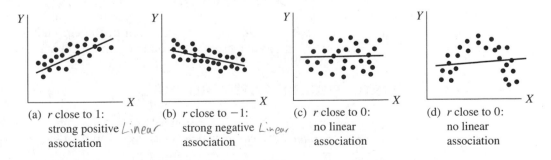

(a) r close to 1:
strong positive *Linear* association

(b) r close to -1:
strong negative *Linear* association

(c) r close to 0:
no linear association

(d) r close to 0:
no linear association

---

[2]Later we will see that a value of r close to 0 does not rule out a possible *nonlinear* association.

is possible. The parameter ρ_{XY} is defined as $\rho_{XY} = \sigma_{XY}/\sigma_X\sigma_Y$, where σ_X and σ_Y denote the population standard deviations of the random variables X and Y and where σ_{XY} is called the *covariance* between X and Y. The covariance σ_{XY} is a population parameter describing the average amount by which two variables covary. In actuality, it is the population mean of the random variable $\text{SSXY}/(n-1)$.

Figure 6-2 on the next page provides informative examples of scatter diagrams. Data were generated (via computer simulation) to have means and variances similar to those of the age–systolic blood pressure data of Chapter 5. The six samples observed here were produced by selecting 30 paired observations at random from each of six populations for which the population correlation coefficient ρ varied in value.

In Figure 6-2, the sample correlations range in value from .037 to .894. It should be clear that an eyeball analysis of the relative strengths of association is difficult, even though $n = 30$. For example, the difference between $r = .037$ in Figure 6-2(a) and $r = .220$ in Figure 6-2(b) is apparently due to the influence of just a few points. The study of so-called influential data points will be described in Chapter 12 on regression diagnostics.

In evaluating a scatter diagram, we find it helpful to include reference lines at the X and Y means, as in Figure 6-2(f). Roughly speaking, the proportions of observations in each quadrant reflect the strength of association. Notice that most of the observations in this figure are located in quadrants B and C, which are often referred to as the *positive quadrants*. Quadrants A and D are called the *negative quadrants*. When more observations are in positive quadrants than in negative quadrants, the sample correlation coefficient r is usually positive. On the other hand, if more observations are in negative quadrants, r is usually negative.

To understand why this is so, we need to examine the numerator part of equation (6.1)—namely,

$$\sum_{i=1}^{n} (X_i - \overline{X})(Y_i - \overline{Y})$$

(Notice that the denominator in (6.1) is simply a positive scaling factor ensuring that r is both dimensionless and satisfies the inequality $-1 \leq r \leq 1$.) The numerator describes how X and Y covary in terms of the n cross-products $(X_i - \overline{X})(Y_i - \overline{Y})$, where $i = 1, 2, \ldots, n$. For a given i, such a cross-product term is either positive or negative (or zero), depending on how X_i compares with $\overline{X}$ and how Y_i compares with $\overline{Y}$. In particular, if the ith observation (X_i, Y_i) is in quadrant B, then $X_i > \overline{X}$ and $Y_i > \overline{Y}$; hence, the product of $(X_i - \overline{X})$ and $(Y_i - \overline{Y})$ must be positive. Similarly, if (X_i, Y_i) is in quadrant C, $X_i < \overline{X}$ and $Y_i < \overline{Y}$, so $(X_i - \overline{X})(Y_i - \overline{Y})$ is again positive. Thus, observations in the positive quadrants B and C contribute positive values to the numerator of (6.1). Conversely, observations in the negative quadrants A and D contribute negative values to this numerator. So, roughly speaking, the sign of the correlation coefficient reflects the distribution of observations in these positive and negative quadrants.

6-3 The Bivariate Normal Distribution[3]

Another way of looking at straight-line regression is to consider X and Y as random variables having the *bivariate normal distribution*, which is a generalization of the *univariate nor-*

---

[3]This section is not essential for understanding the correlation coefficient as it relates to regression analysis.

FIGURE 6-2   Examples of a range of observed correlations between age and systolic blood pressure (SBP) for simulated data.

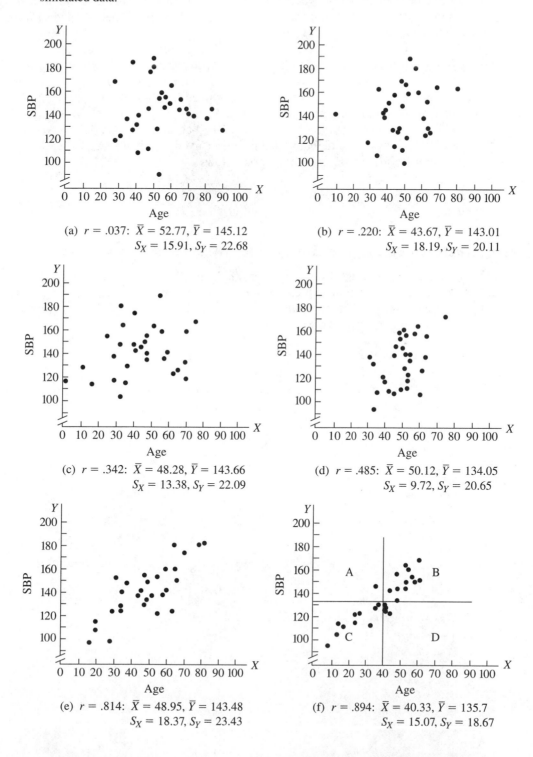

(a) $r = .037$: $\bar{X} = 52.77$, $\bar{Y} = 145.12$
$S_X = 15.91$, $S_Y = 22.68$

(b) $r = .220$: $\bar{X} = 43.67$, $\bar{Y} = 143.01$
$S_X = 18.19$, $S_Y = 20.11$

(c) $r = .342$: $\bar{X} = 48.28$, $\bar{Y} = 143.66$
$S_X = 13.38$, $S_Y = 22.09$

(d) $r = .485$: $\bar{X} = 50.12$, $\bar{Y} = 134.05$
$S_X = 9.72$, $S_Y = 20.65$

(e) $r = .814$: $\bar{X} = 48.95$, $\bar{Y} = 143.48$
$S_X = 18.37$, $S_Y = 23.43$

(f) $r = .894$: $\bar{X} = 40.33$, $\bar{Y} = 135.7$
$S_X = 15.07$, $S_Y = 18.67$

mal distribution. Just as the univariate normal distribution is described by a density function that appears as a bell-shaped curve when plotted in two dimensions, the bivariate normal distribution is described by a *joint density function* whose plot looks like a bell-shaped surface in three dimensions (Figure 6-3).

One property of the bivariate normal distribution that has implications for straight-line regression analysis is the following: If the bell-shaped surface is cut by a plane *parallel* to the YZ-plane and passing through a specific X-value, the curve, or *trace,* that results is a normal distribution. In other words, the distribution of Y for fixed X is univariate-normal. We call such a distribution the *conditional distribution* of Y at X, and we denote the corresponding random variable as Y_X. Let us denote the mean of this distribution as $\mu_{Y|X}$ and the variance as $\sigma^2_{Y|X}$. Then it follows from statistical theory that the mean and the variance, respectively, of Y_X can be written in terms of μ_X, μ_Y, σ^2_X, σ^2_Y, and ρ_{XY} as follows:

$$\mu_{Y|X} = \mu_Y + \rho_{XY}\frac{\sigma_Y}{\sigma_X}(X - \mu_X) \tag{6.3}$$

and

$$\sigma^2_{Y|X} = \sigma^2_Y(1 - \rho^2_{XY}) \tag{6.4}$$

Now suppose that we let $\beta_1 = \rho_{XY}(\sigma_Y/\sigma_X)$ and $\beta_0 = \mu_Y - \beta_1\mu_X$. Then (6.3) has been transformed into the familiar expression for a straight-line model; that is, $\mu_{Y|X} = \beta_0 + \beta_1 X$. Furthermore, if we substitute the estimators $\overline{X}$, $\overline{Y}$, S_X, S_Y, and r for their respective parameters μ_X, μ_Y, σ_X, σ_Y, and ρ_{XY} in (6.3), we obtain the formula

$$\hat{\mu}_{Y|X} = \overline{Y} + r\frac{S_Y}{S_X}(X - \overline{X})$$

The right-hand side of this equation is exactly equivalent to the expression for the least-squares straight line given in (5.7), since

$$\hat{\beta}_1 = r\frac{S_Y}{S_X}$$

FIGURE 6-3   The bivariate normal distribution.

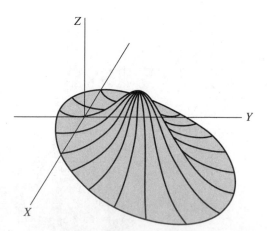

Thus, *the least-squares formulas for $\hat{\beta}_0$ and $\hat{\beta}_1$ can be developed by assuming that X and Y are random variables having the bivariate normal distribution and by substituting the usual estimates of μ_X, μ_Y, σ_X, σ_Y, and ρ_{XY} into the expression for $\mu_{Y|X}$, the conditional mean of Y given X.*

Our estimate $S_{Y|X}^2$ of $\sigma_{Y|X}^2$ can also be obtained by substituting the estimates S_Y^2 and r for σ_Y^2 and ρ_{XY} in (6.4). Thus, we obtain

$$S_{Y|X}^2 = S_Y^2(1 - r^2)$$

Finally, (6.4) can be algebraically manipulated into the form

$$\rho_{XY}^2 = \frac{\sigma_Y^2 - \sigma_{Y|X}^2}{\sigma_Y^2} \qquad (6.5)$$

This equation describes the square of the population correlation coefficient as the proportionate reduction in the variance of Y due to conditioning on X. The importance of (6.5) in describing the strength of the straight-line relationship will be discussed in the next section.

6-4 *r* and the Strength of the Straight-line Relationship

To quantify what we mean by the *strength* of the linear relationship between X and Y, we should first consider what our predictor of Y would be if we did not use X at all. The best predictor in this case would simply be $\overline{Y}$, the sample mean of the Y's. The sum of the squares of deviations associated with the naive predictor $\overline{Y}$ would then be given by the formula

$$SSY = \sum_{i=1}^{n} (Y_i - \overline{Y})^2$$

Now, if the variable X is of any value in predicting the variable Y, the residual sum of squares given by

$$SSE = \sum_{i=1}^{n} (Y_i - \hat{Y}_i)^2$$

should be considerably less than SSY. If so, the least-squares model $\hat{Y} = \hat{\beta}_0 + \hat{\beta}_1 X$ fits the data better than does the horizontal line $\hat{Y} = \overline{Y}$ (Figure 6-4). A quantitative measure of the improve-

FIGURE 6-4   Predictions of Y using and not using X.

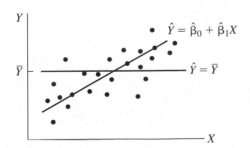

ment in the fit obtained by using X is given by the *square of the sample correlation coefficient r,* which can be written in the suggestive form

$$r^2 = \frac{SSY - SSE}{SSY} = \frac{SSR \ (regression)}{SST \ (total)} = 1 - \frac{SSE}{SST} \tag{6.6}$$

This quantity naturally varies between 0 and 1, since r itself varies between -1 and 1.

What interpretation can be given to the quantity r^2? To answer this question, we first note that the difference, or *reduction,* in SSY due to using X to predict Y may be measured by $(SSY - SSE)$, which is always nonnegative. Furthermore, the *proportionate reduction* in SSY due to using X to predict Y is this difference divided by SSY. Thus, r^2 measures the strength of the linear relationship between X and Y in the sense that it gives the proportionate reduction in the sum of the squares of vertical deviations obtained by using the least-squares line $\hat{Y} = \hat{\beta}_0 + \hat{\beta}_1 X$ instead of the naive model $\hat{Y} = \overline{Y}$ (the predictor of Y if X is ignored). The larger the value of r^2, the greater the reduction in SSE relative to $\sum_{i=1}^{n}(Y_i - \overline{Y})^2$, and the stronger the linear relationship between X and Y.

The largest value that r^2 can attain is 1, which occurs when $\hat{\beta}_1$ is nonzero and when SSE $= 0$ (i.e., when a perfect positive or negative straight-line relationship exists between X and Y). By "perfect" we mean that *all* the data points lie on the fitted straight line. In other words, when $Y_i = \hat{Y}_i$ for all i, we must have

$$SSE = \sum_{i=1}^{n}(Y_i - \hat{Y}_i)^2 = 0$$

so

$$r^2 = \frac{SSY - SSE}{SSY} = \frac{SSY}{SSY} = 1$$

Figure 6-5 illustrates examples of perfect positive and perfect negative linear association.

The smallest value that r^2 may take, of course, is 0. This value means that using X offers no improvement in predictive power; that is, SSE $=$ SSY. Furthermore, appealing to (6.2), we see that a correlation coefficient of 0 implies an estimated slope of 0 and consequently the absence of any linear relationship (although a nonlinear relationship is still possible).

Finally, one should *not* be led to a false sense of security by considering the magnitude of r, rather than of r^2, when assessing the strength of the linear association between X and Y. For example, when r is 0.5, r^2 is only 0.25, and it takes $r > 0.7$ to make $r^2 > 0.5$. Also, when r is 0.3, r^2 is 0.09, which indicates that only 9% of the variation in Y is explained with the help of X.

FIGURE 6-5   Examples of perfect linear association.

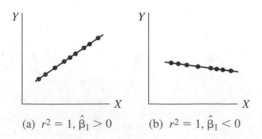

(a) $r^2 = 1, \hat{\beta}_1 > 0$ (b) $r^2 = 1, \hat{\beta}_1 < 0$

For the age–systolic blood pressure data, r^2 is 0.43, compared with an r of 0.66. The r^2 value also appears on the SAS output on p. 50.

6-5 What r Does Not Measure

Two common misconceptions about r (or, equivalently, about r^2) occasionally lead a researcher to make spurious interpretations of the relationship between X and Y. The correct notions are as follows:

1. r^2 *is not a measure of the magnitude of the slope of the regression line.* Even when the value of r^2 is high (i.e., close to 1), the magnitude of the slope $\hat{\beta}_1$ is not necessarily large. This phenomenon is illustrated in Figure 6-5. Notice that r^2 equals 1 in both parts, despite the fact that the slopes are different. Another way to understand this, using (6.2), is

$$\hat{\beta}_1^2 = \frac{S_Y^2}{S_X^2} \quad \text{when } r^2 = 1$$

    Thus, if two different sets of data have the same amount of X variation, but the first set has less Y variation than the second set, the magnitude of the slope for the first set is smaller than that for the second.

2. r^2 *is not a measure of the appropriateness of the straight-line model.* Thus, $r^2 = 0$ in parts (a) and (b) of Figure 6-6, even though no evidence of association between X

FIGURE 6-6   Examples showing that r^2 is not a measure of the appropriateness of the straight-line model.

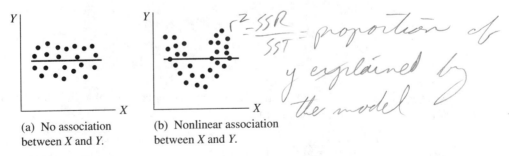

(a) No association between X and Y.

(b) Nonlinear association between X and Y.

Examples when r^2 is high

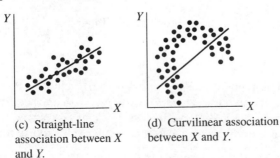

(c) Straight-line association between X and Y.

(d) Curvilinear association between X and Y.

and Y exists in (a) and strong evidence of a nonlinear association exists in (b). Conversely, r^2 is high in parts (c) and (d), even though a straight-line model is quite appropriate in (c) but not entirely appropriate in (d).

6-6 Tests of Hypotheses and Confidence Intervals for the Correlation Coefficient

Researchers interested in the association between two interval variables X and Y often wish to test the null hypothesis H_0: $\rho = 0$.

6-6-1 Test of H_0: $\rho = 0$

A test of H_0: $\rho = 0$ turns out to be mathematically equivalent to the test of the hypothesis H_0: $\beta_1 = 0$ described in section 5-8. This equivalence is suggested by the formulas $\beta_1 = \rho \sigma_Y / \sigma_X$ and $\hat{\beta}_1 = r S_Y / S_X$, which tell us, for example, that β_1 is positive, negative, or zero as ρ is positive, negative, or zero, and that an analogous relationship exists between $\hat{\beta}_1$ and r. The test statistic for the hypothesis H_0: $\rho = 0$ can be written entirely in terms of r and n, so we can perform the test without having to fit the straight line. This test statistic is given by the formula

$$T = \frac{r\sqrt{n - 2}}{\sqrt{1 - r^2}} \tag{6.7}$$

which has the t distribution with $n - 2$ degrees of freedom when the null hypothesis H_0: $\rho = 0$ (or equivalently, H_0: $\beta_1 = 0$) is true. Formula (6.7) yields exactly the same numerical answer as does (5.9), given by

$$T = \frac{\hat{\beta}_1 - \beta_1^{(0)}}{S_{\hat{\beta}_1}} \quad \text{when} \quad \beta_1^{(0)} = 0.$$

■ **Example 6-2**   For the age–systolic blood pressure data of Table 5-1, for which $r = 0.66$, the statistic in (6.7) is calculated as follows:

$$T = \frac{.66\sqrt{30 - 2}}{\sqrt{1 - (.66)^2}} = 4.62$$

which is the same value as obtained for the test for slope in Table 5-3.   ■

6-6-2 Test of H_0: $\rho = \rho_0$, $\rho_0 \neq 0$

A test concerning the null hypothesis H_0: $\rho = \rho_0$ ($\rho_0 \neq 0$) cannot be directly related to a test concerning β_1; moreover, the hypothesis H_0: $\rho = \rho_0$ ($\rho_0 \neq 0$) is not equivalent to the hypothesis H_0: $\beta_1 = \beta_1^{(0)}$ for some value $\beta_1^{(0)}$. Nevertheless, a test of H_0: $\rho = \rho_0$ ($\rho_0 \neq 0$) is meaningful when previous experience or theory suggests a particular value to use for ρ_0.

The test statistic in this case can be obtained by considering the distribution of the sample correlation coefficient r. This distribution happens to be symmetric, like the normal distribution, *only* when ρ is 0. When ρ is nonzero, the distribution of r is skewed. This lack of normality prevents us from using a test statistic of the usual form, which has a normally distributed estimator in the numerator and an estimate of its standard deviation in the denominator. But through an appropriate transformation, r can be changed into a statistic that is approximately normal. This transformation is called *Fisher's Z transformation.*[4] The formula for this transformation is

$$\frac{1}{2} \ln \frac{1 + r}{1 - r} \tag{6.8}$$

This quantity has approximately the normal distribution, with mean $\frac{1}{2} \ln \left[(1 + \rho)/(1 - \rho) \right]$ and variance $1/(n - 3)$ when n is not too small (e.g., $n \geq 20$). In testing the hypothesis $H_0: \rho = \rho_0$ ($\rho_0 \neq 0$), we can then use the test statistic

$$Z = \frac{\frac{1}{2} \ln \left[(1 + r)/(1 - r) \right] - \frac{1}{2} \ln \left[(1 + \rho_0)/(1 - \rho_0) \right]}{1/\sqrt{n - 3}} \tag{6.9}$$

This test statistic has approximately the standard normal distribution (i.e., $Z \frown N(0, 1)$) under H_0. To test $H_0: \rho = \rho_0$ ($\rho_0 \neq 0$), therefore, we use one of the following critical regions for significance level α:

$$Z \geq z_{1-\alpha} \qquad \text{(upper one-tailed alternative } H_A: \rho > \rho_0\text{)}$$
$$Z \leq -z_{1-\alpha} \qquad \text{(lower one-tailed alternative } H_A: \rho < \rho_0\text{)}$$
$$|Z| \geq z_{1-\alpha/2} \qquad \text{(two-tailed alternative } H_A: \rho \neq \rho_0\text{)}$$

where $z_{1-\alpha}$ denotes the $100(1 - \alpha)\%$ point of the standard normal distribution. Computation of Z can be aided by using Appendix Table A-5, which gives values of $\frac{1}{2} \ln \left[(1 + r)/(1 - r) \right]$ for given values of r.

■ **Example 6-3**   Suppose that from previous experience we can hypothesize that the true correlation between age and systolic blood pressure is $\rho_0 = 0.85$. To test the hypothesis $H_0: \rho = 0.85$ against the two-sided alternative $H_A: \rho \neq 0.85$, we perform the following calculations using $r = 0.66$, $\rho_0 = 0.85$, and $n = 30$:

$$\frac{1}{2} \ln \frac{1 + \rho_0}{1 - \rho_0} = \frac{1}{2} \ln \frac{1 + 0.85}{1 - 0.85} = 1.2561 \qquad \text{(from Table A-5)}$$

$$\frac{1}{2} \ln \frac{1 + r}{1 - r} = \frac{1}{2} \ln \frac{1 + 0.66}{1 - 0.66} = .7928 \qquad \text{(from Table A-5)}$$

$$Z = \frac{0.7928 - 1.2561}{1/\sqrt{30 - 3}} = -2.41$$

For $\alpha = .05$, the critical region is

$$|Z| \geq z_{.975} = 1.96$$

---

[4]Named after R. A. Fisher, who introduced it in 1925.

Since $|Z| = 2.41$ exceeds 1.96, the hypothesis $H_0: \rho_0 = 0.85$ is rejected at the .05 significance level. Further calculations show that the P-value for this test is $P = 0.0151$, which tells us that the result is not significant at $\alpha = .01$. ■

6-6-3 Confidence Interval for ρ

A $100(1 - \alpha)\%$ confidence interval for ρ can be obtained by using Fisher's Z transformation (6.8) as follows. First, compute a $100(1 - \alpha)\%$ confidence interval for the parameter $\frac{1}{2}\ln[(1 + \rho)/(1 - \rho)]$, using the formula

$$\frac{1}{2}\ln\frac{1 + r}{1 - r} \pm \frac{z_{1-\alpha/2}}{\sqrt{n - 3}} \qquad (6.10)$$

where $z_{1-\alpha/2}$ is as defined previously.

Denote the lower limit of the confidence interval (6.10) by L_Z, and the upper limit by U_Z; then use Appendix Table A-5 (in reverse) to determine the lower and upper confidence limits L_ρ and U_ρ for the confidence interval for ρ. In other words, determine L_ρ and U_ρ from the following formulas[5]:

$$L_Z = \frac{1}{2}\ln\frac{1 + L_\rho}{1 - L_\rho} \quad \text{and} \quad U_Z = \frac{1}{2}\ln\frac{1 + U_\rho}{1 - U_\rho}$$

■ **Example 6-4** Suppose that we seek a 95% confidence interval for ρ based on the age–systolic blood pressure data for which $r = 0.66$ and $n = 30$. A 95% confidence interval for $\frac{1}{2}\ln[(1 + \rho)/(1 - \rho)]$ is given by

$$\frac{1}{2}\ln\frac{1 + .66}{1 - .66} \pm \frac{1.96}{\sqrt{30 - 3}}$$

which is equal to

$$0.793 \pm 0.377$$

providing a lower limit of $L_Z = 0.416$ and an upper limit of $U_Z = 1.170$.

To transform these L_z and U_z values into lower and upper confidence limits for ρ, we determine the values of L_ρ and U_ρ that satisfy

$$0.416 = \frac{1}{2}\ln\frac{1 + L_\rho}{1 - L_\rho} \quad \text{and} \quad 1.170 = \frac{1}{2}\ln\frac{1 + U_\rho}{1 - U_\rho}$$

Using Table A-5, we see that a value of 0.416 corresponds to an r of about 0.394, so $L_\rho = 0.394$. Similarly, a value of 1.170 corresponds to an r of about 0.824, so $U_\rho = 0.824$. The 95% confidence interval for ρ thus has a lower limit of 0.394 and an upper limit of 0.824.

Notice that the interval (0.394, 0.824) does not contain the value 0.85, which agrees with the conclusion of the previous section that $H_0: \rho = 0.85$ is to be rejected at the 5% level (two-tailed test). ■

---

[5]L_ρ and U_ρ can also be calculated directly using the conversion formulas:

$$L_\rho = \frac{e^{2L_Z} - 1}{e^{2L_Z} + 1} \quad \text{and} \quad U_\rho = \frac{e^{2U_Z} - 1}{e^{2U_Z} + 1}$$

6-7 Testing for the Equality of Two Correlations

Suppose that independent random samples of sizes n_1 and n_2 are selected from two populations. Further, suppose that we wish to test H_0: $\rho_1 = \rho_2$ versus, say, H_A: $\rho_1 \neq \rho_2$. An appropriate test statistic can be developed based on the results given in section 6-6. In this section, we will also consider the situation where the sample correlations to be compared are calculated by using the same data set; in this case, the sample correlations are themselves "correlated."

6-7-1 Test of H_0: $\rho_1 = \rho_2$ Using Independent Random Samples

Let us assume that independent random samples of sizes n_1 and n_2 have been selected from two populations. For each population, the straight-line regression analysis assumptions given in Chapter 5, including that of normality, will hold.

An approximate test of H_0: $\rho_1 = \rho_2$ can be based on the use of Fisher's Z transformation. Let r_1 be the sample correlation calculated by using the n_1 observations from the first population, and let r_2 be defined similarly. Using (6.8), let

$$Z_1 = \frac{1}{2}\ln\frac{1 + r_1}{1 - r_1} \qquad (6.11)$$

and

$$Z_2 = \frac{1}{2}\ln\frac{1 + r_2}{1 - r_2} \qquad (6.12)$$

Appendix Table A-5 can be used to determine Z_1 and Z_2.
To test H_0: $\rho_1 = \rho_2$, we can compute the test statistic

$$Z = \frac{Z_1 - Z_2}{\sqrt{1/(n_1 - 3) + 1/(n_2 - 3)}} \qquad (6.13)$$

For large n_1 and n_2, this test statistic has (approximately) the standard normal distribution when H_0 is true. Hence, the following critical regions for significance level α should be used:

$$Z \geq z_{1-\alpha} \qquad \text{(upper one-tailed alternative } H_A: \rho_1 > \rho_2)$$
$$Z \leq -z_{1-\alpha} \qquad \text{(lower one-tailed alternative } H_A: \rho_1 < \rho_2)$$
$$|Z| \geq z_{1-\alpha/2} \qquad \text{(two-tailed alternative } H_A: \rho_1 \neq \rho_2)$$

To illustrate this procedure, let us test whether the data sets plotted in Figures 6-2(b) and 6-2(c) reflect populations with different correlations. In other words, we wish to test H_0: $\rho_1 = \rho_2$ versus the two-sided alternative H_A: $\rho_1 \neq \rho_2$.

For the data in Figure 6-2(b), $r_1 = 0.220$; for the Figure 6-2(c) data, $r_2 = 0.342$. Using Fisher's Z transformation and Table A-5, we can calculate Z_1 and Z_2 as

$$Z_1 = \frac{1}{2}\ln\frac{1 + r_1}{1 - r_1} = \frac{1}{2}\ln\frac{1 + .220}{1 - .220} = 0.2237$$

and

$$Z_2 = \frac{1}{2}\ln\frac{1 + r_2}{1 - r_2} = \frac{1}{2}\ln\frac{1 + .342}{1 - .342} = 0.3564$$

Then the test statistic (6.13) takes the value

$$Z = \frac{0.2237 - 0.3564}{\sqrt{1/(30 - 3) + 1/(30 - 3)}} = \frac{-0.1327}{0.2722} = -0.488$$

For $\alpha = .01$, the critical region is

$$|Z| \geq z_{.005} = 2.576$$

Since $|Z| = 0.488$ is less than 2.576, we cannot reject $H_0: \rho_1 = \rho_2$ at $\alpha = .01$.

6-7-2 Single Sample Test of $H_0: \rho_{12} = \rho_{13}$

Consider testing the null hypothesis that the correlation ρ_{12} of variable 1 with variable 2 is the same as the correlation ρ_{13} of variable 1 with variable 3. Let us assume that a single random sample of n subjects is selected and that the three sample correlations—r_{12}, r_{13}, and r_{23}—are calculated. Clearly these sample correlations are not independent, since they are computed using the same data set. Under the usual straight-line regression analysis assumptions, it can be shown (we omit the details) that an appropriate large-sample test statistic for testing $H_0: \rho_{12} = \rho_{13}$ is

$$Z = \frac{(r_{12} - r_{13})\sqrt{n}}{\sqrt{(1 - r_{12}^2)^2 + (1 - r_{13}^2)^2 - 2r_{23}^3 - (2r_{23} - r_{12}r_{13})(1 - r_{12}^2 - r_{13}^2 - r_{23}^2)}} \tag{6.14}$$

For large n, this test statistic has approximately the standard normal distribution under $H_0: \rho_{12} = \rho_{13}$ (Olkin and Siotani 1964; Olkin 1967).

■ **Example 6-5**   Assume that the weight, height, and age have been measured for each member of a sample of 12 nutritionally deficient children. Such a small sample brings into question the normal approximation involved in the use of (6.14). The data to be analyzed appear in Table 8-1 in Chapter 8. For these data, the three sample correlations are:

$$r_{12} = r_{(weight, height)} = 0.814$$
$$r_{13} = r_{(weight, age)} = 0.770$$
$$r_{23} = r_{(height, age)} = 0.614$$

We wish to test whether height and age are equally correlated with weight (i.e., $H_0: \rho_{12} = \rho_{13}$) versus the two-tailed alternative that they are not (i.e., $H_A: \rho_{12} \neq \rho_{13}$). Using (6.14), the test statistic takes the value

$$Z = \frac{(.814 - .770)\sqrt{12}}{\sqrt{\begin{array}{c}[1 - (.814)^2]^2 + [1 - (.770)^2]^2 - 2(.614)^3 \\ - [2(.614) - (.814)(.770)][1 - (.814)^2 - (.770)^2 - (.614)^2]\end{array}}}$$

$$= \frac{0.1524}{\sqrt{0.1968}} \neq 0.3435$$

It is clear that, for these data, we cannot reject the null hypothesis of equal correlation of weight with height and age.   ■

Problems

1. Using the data set of Problem 1 in Chapter 5, perform the following operations.
   a. Determine the sample correlation coefficients of (1) age with dry weight, and (2) age with $\log_{10}$ dry weight. Interpret your results.
   b. Using Fisher's Z transformation, obtain a 95% confidence interval for ρ based on each of the correlations obtained in part (a).
   c. For each straight-line regression, determine r^2 directly by squaring the r obtained in part (a); also determine r^2 from the computer output or from the formula $r^2 = (\text{SSY} - \text{SSE})/\text{SSY}$. Interpret your results.
   d. Based on the preceding results, which of the two regression lines provides the better fit? Explain. Does this agree with your earlier conclusion in Problem 1(d) of Chapter 5?

2. Examine the five pairs of data points given in the following table.

i	1	2	3	4	5
X_i	-2	-1	0	1	2
Y_i	4	1	0	1	4

   a. What is the mathematical relationship between X and Y?
   b. Show by computation that, for the straight-line regression of Y on X, $\hat{\beta}_1 = 0$.
   c. Show by computation that $r = 0$.
   d. Why is there apparently no relationship between X and Y, as indicated by the estimates of β_1 and ρ?

3. Consider the data in the following table.

i	1	2	3	4	5	6	7	8	9	10
X_i	1	1	1	2	2	2	3	3	3	20
Y_i	1	2	3	1	2	3	1	2	3	20

   a. Find the sample correlation coefficient r. Interpret your result.
   b. Show that the test statistic $T = \hat{\beta}_1/(S_{Y|X}/S_X\sqrt{n-1})$ for testing $H_0: \beta_1 = 0$ (based on a straight-line regression relationship between Y and X) is exactly equivalent to the test statistic $T' = r\sqrt{n-2}/\sqrt{1-r^2}$ for testing $H_0: \rho = 0$. [*Hint:* Use $\hat{\beta}_1 = rS_Y/S_X$ and $S_{Y|X}^2 = [(n-1)/(n-2)](S_Y^2 - \hat{\beta}_1^2 S_X^2).$]
   c. Using T', test $H_0: \rho = 0$ versus $H_A: \rho \neq 0$.
   d. Despite the conclusion you obtained in part (c), why should you be reluctant to conclude that the two variables are linearly related? ("A graph is worth a thousand words.")

4.–6. Answer the following questions concerning the straight-line regressions of Y on X referred to in parts (c), (d), and (e) of Problem 2 in Chapter 5.
   a. Determine r and r^2, and interpret your results.
   b. Find a 99% confidence interval for ρ, and interpret your result with regard to the test of $H_0: \rho = 0$ versus $H_A: \rho \neq 0$ at $\alpha = .01$.

7.–12. Answer the following questions for each of the data sets of Problems 3–8 in Chapter 5.
   a. Determine r and r^2 for each variable pair, and interpret your results.
   b. Test H_0: $\rho = 0$ versus H_A: $\rho \neq 0$, and interpret your findings.
   c. Find a 95% confidence interval for ρ. Interpret your result with regard to the test in part (b).

13. Suppose that, in a study on geographic variation in a certain species of beetle,[6] the mean tibia length (U) and the mean tarsus length (V) were obtained for samples of size 50 from each of 10 different regions spanning five southern states. Suppose further that the results were as given in the following table.

Region	1	2	3	4	5	6	7	8	9	10
U	7.500	7.164	7.512	8.544	7.380	7.860	7.836	8.100	7.584	7.344
V	1.680	1.596	1.680	1.908	1.632	1.752	1.776	1.860	1.692	1.680

   a. Determine the sample correlation coefficient between tarsus length and tibia length.
   b. Find a 99% confidence interval for ρ. Interpret your results with respect to the hypotheses: H_0: $\rho = 0$ versus H_A: $\rho \neq 0$.

14. In a sample of 23 young adult men, the correlation between total hemoglobin (THb) measured from venipuncture and measured from a finger needle puncture was 0.82. For a sample of 32 women of similar age, the correlation was 0.74. The two samples from each person were collected within 1 hour of each other. Assume that the straight-line regression assumptions hold.
   a. Test the hypothesis that the two population correlations are equal. Use a two-tailed test. What do you conclude?
   b. If the experimenter had planned to do so before collecting the data, a valid one-tailed test could have been conducted. With this assumption, repeat part (a) but use a one-tailed test to assess whether the correlation for women is lower than that for men. What do you conclude?
   c. Assume that the researcher had planned to conduct a one-tailed test of the hypothesis that the correlation for women is higher than that for men. What test should be conducted? What do you conclude?

15. A university admissions officer regularly administers a test to all entering freshmen. A new version of the test is marketed by the testing company. To evaluate the new form, the admissions officer has 121 freshmen take both the old and the new versions. After the end of the school year, the admissions officer correlates the two scores with each other and with the students' freshman grade point average (GPA). With 1 indicating the old version, 2 the new version, and G the GPA,

   $$r_{12} = .6969, \qquad r_{1G} = .5514, \qquad r_{2G} = .4188$$

Test the hypothesis that the two forms of the test are equally correlated with GPA. Use a two-tailed test with $\alpha = .05$. Assume that the straight-line regression assumptions hold.

---

[6] Adapted from a study by Sokal and Thomas (1965).

16.–25. Answer the following questions for each of the data sets of Problems 12(a), 12(c), 13, 14, 15(a), 15(c), 16, 17, 18, and 19 in Chapter 5.
   a. Determine r and r^2 for each variable pair, and interpret your results.
   b. Test H_0: $\rho = 0$ versus H_A: $\rho \neq 0$, and interpret your findings.
   c. Find a 95% confidence interval for ρ. Interpret your result with respect to the test in part (b).

References

Olkin, I. 1967. "Correlation Revisited." In Julian C. Stanley, ed., *Improving Experimental Design and Statistical Analysis.* Chicago: Rand McNally.

Olkin, I., and Siotani, M. 1964. "Asymptotic Distribution Functions of a Correlation Matrix." Stanford University Laboratory for Quantitative Research in Education, Report No. 6, Stanford, Calif.

Sokal, R. R., and Thomas, P. A. 1965. "Geographic Variation of *Pemphigus populitransversus* in Eastern North America: Stem Mothers and New Data on Alates." *University of Kansas Scientific Bulletin* 46: 201–52.

The Analysis-of-Variance Table

7-1 Preview

An overall summary of the results of any regression analysis, whether straight-line or not, can be provided by a table called an *analysis-of-variance (ANOVA) table*. This name derives primarily from the fact that the basic information in an ANOVA table consists of several estimates of variance. These estimates, in turn, can be used to answer the principal inferential questions of regression analysis. In the straight-line case, there are three such questions: (1) Is the true slope β_1 zero? (2) What is the strength of the straight-line relationship? (3) Is the straight-line model appropriate?

Historically, the name "analysis-of-variance table" was coined to describe the overall summary table for the statistical procedure known as *analysis of variance*. As we observed in Chapter 2 and will see later when discussing the ANOVA method, regression analysis and analysis of variance are closely related. More precisely, analysis-of-variance problems can be expressed in a regression framework. Thus, such a table can be used to summarize the results obtained from either method.

7-2 The ANOVA Table for Straight-line Regression

Various textbooks, researchers, and computer program printouts have slightly different ways of presenting the ANOVA table associated with straight-line regression analysis. This section describes the most common form.

The simplest version of the ANOVA table for straight-line regression is given in the accompanying SAS computer output, as applied to the age–systolic blood pressure data of Table 5-1. The mean-square term is obtained by dividing the sum of squares by its degrees of freedom.

The F statistic (i.e., F value in the output) is obtained by dividing the regression (i.e., model) mean square by the residual (i.e., error) mean square.

SAS Output for an ANOVA Table Based on Table 5-1 Data

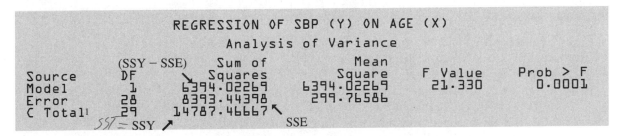

In Chapter 6, when describing the correlation coefficient, we observed in (6.6) that

$$r^2 = \frac{\text{SSY} - \text{SSE}}{\text{SSY}}$$

where $\text{SSY} = \sum_{i=1}^{n}(Y_i - \overline{Y})^2$ is the sum of the squares of deviations of the observed Y's from the mean $\overline{Y}$, and $\text{SSE} = \sum_{i=1}^{n}(Y_i - \hat{Y}_i)^2$ is the sum of squares of deviations of observed Y's from the fitted regression line. Since SSY represents the total variation of Y before accounting for the linear effect of the variable X, we usually call SSY the *total unexplained variation* or the *total sum of squares about* (or *corrected for*) *the mean*. Because SSE measures the amount of variation in the observed Y's that remains after accounting for the linear effect of the variable X, we usually call (SSY − SSE) the *sum of squares due to* (or *explained by*) *regression*. It turns out that (SSY − SSE) is mathematically equivalent to the expression

$$\sum_{i=1}^{n}(\hat{Y}_i - \overline{Y})^2$$

which represents the sum of squares of deviations of the predicted values from the mean $\overline{Y}$. We thus have the following mathematical result:

> Total unexplained variation = Variation due to regression
> + Unexplained residual variation

or

$$\sum_{i=1}^{n}(Y_i - \overline{Y})^2 = \sum_{i=1}^{n}(\hat{Y}_i - \overline{Y})^2 + \sum_{i=1}^{n}(Y_i - \hat{Y}_i)^2 \qquad (7.1)$$

---

[1]Corrected for the mean.

Equation (7.1), which is often called the *fundamental equation of regression analysis,* holds for any general regression situation. Figure 7-1 illustrates this equation.

The mean-square residual term is simply the estimate $S_{Y|X}^2$ presented earlier. If the true regression model is a straight line, then, as mentioned in section 5.6, $S_{Y|X}^2$ is an estimate of σ^2. On the other hand, the mean-square regression term (SSY $-$ SSE) provides an estimate of σ^2 only if the variable X does not help to predict the dependent variable Y—that is, only if the hypothesis H_0: $\beta_1 = 0$ is true. If in fact $\beta_1 \neq 0$, the mean-square regression term will be inflated in proportion to the magnitude of β_1 and will correspondingly overestimate σ^2.

It can be shown that the mean-square residual and mean-square regression terms are *statistically* independent of one another. Thus, if H_0: $\beta_1 = 0$ is true, the ratio of these terms represents the ratio of two independent estimates of the same variance σ^2. Under the normality assumption on the Y's, such a ratio has the F distribution, and this F statistic (with the value 21.330 in the accompanying SAS computer output) can be used to test the hypothesis H_0: "No significant straight-line relationship of Y on X" (i.e., H_0: $\beta_1 = 0$ or H_0: $\rho = 0$).

Fortunately, this way of testing H_0 is *equivalent* to using the two-sided t test previously discussed. This is so because, for v degrees of freedom,

$$F_{1,v} = T_v^2 \tag{7.2}$$

so

$$F_{1,v,\,1-\alpha} = t_{v,\,1-\alpha/2}^2 \tag{7.3}$$

The expression in (7.3) states that the $100(1-\alpha)\%$ point of the F distribution with 1 and v degrees of freedom is exactly the same as the square of the $100(1-\alpha/2)\%$ point of the t distribution with v degrees of freedom.

To illustrate the equivalence of the F and t tests, we can see from our age–systolic blood pressure example that $F = 21.33$ and $T^2 = 4.62^2 = 21.33$, where 4.62 is the figure obtained for T at the end of section 6-6-1. Also, it can be seen that $F_{1,28,\,0.95} = 4.20$ and that $t_{28,0.975}^2 = (2.05)^2 = 4.20$.

FIGURE 7-1   Variation explained and unexplained by straight-line regression.

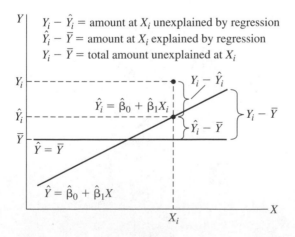

As thcsc cqualities establish, the critical region

$$|T| > t_{28,0.975} = 2.05$$

for testing H_0: $\beta_1 = 0$ against the two-sided alternative H_A: $\beta_1 \neq 0$ is exactly the same as the critical region

$$F > F_{1,28,0.95} = 4.20$$

Hence, if $|T|$ exceeds 2.05, then F will exceed 4.20. Similarly, if F exceeds 4.20, then $|T|$ will exceed 2.05. Thus, the null hypothesis H_0: $\beta_1 = 0$ (or equivalently, H_0: "No significant straight-line relationship of Y on X") is rejected at the $\alpha = .05$ level of significance.

An alternative but less common representation of the ANOVA table is given in Table 7-1. This table differs from the SAS output table only in that it splits up the total sum of squares corrected for the mean, SSY, into its two components: the *total uncorrected sum of squares,* $\sum_{i=1}^{n} Y_i^2$; and the *correction factor,* $(\sum_{i=1}^{n} Y_i)^2/n$. The relationship between these components is given by the equation

$$\sum_{i=1}^{n}(Y_i - \overline{Y})^2 = \sum_{i=1}^{n} Y_i^2 - \frac{\left(\sum_{i=1}^{n} Y_i\right)^2}{n}$$

conected $SS = SST = \sum_i y_i^2 - \frac{(\sum_i y)^2}{n}$

In the total (uncorrected) sum of squares $\sum_{i=1}^{n} Y_i^2$, the n observations on Y are considered before any estimation of the population mean of Y. The "Regression $\overline{Y}$" listed in Table 7-1 refers to the variability explained by using a model involving only β_0 (which is estimated by $\overline{Y}$). This is necessarily the same amount of variability as is explained by using only $\overline{Y}$ to predict Y, without attempting to account for the linear contribution of X to the prediction of Y. The "Regression $X \mid \overline{Y}$" describes the contribution of the variable X to predicting Y *over and above* that contributed by $\overline{Y}$ alone. Usually "Regression $X \mid \overline{Y}$" is written simply as "Regression X," the "given $\overline{Y}$" part being suppressed for notational simplicity. We will see more of this notation when we discuss multiple regression in subsequent chapters.

TABLE 7-1   Alternative ANOVA table for age–systolic blood pressure data of Table 5-1.

Source	Degrees of Freedom (df)	Sum of Squares (SS)	Mean Square (MS)	Variance Ratio (F)
Regression $\begin{cases} \overline{Y} \\ \\ X \mid \overline{Y} \end{cases}$	1	$\dfrac{\left(\sum_{i=1}^{n} Y_i\right)^2}{n} = 609{,}472.53$		
	1	6,394.02	6,394.02	21.33 $(P < .001)$
Residual	28	8,393.44	299.77	
Total	30	$\sum_{i=1}^{n} Y_i^2 = 624{,}260.00$		

Problems

1. Use the data set of Problem 1 in Chapter 5 to answer the following questions.
    a. Determine the ANOVA tables for the following regressions: (1) dry weight (Y) on age (X), and (2) $\log_{10}$ dry weight (Z) on age (X). The following results will be helpful in reducing computation time:

$$S^2_{Y|X} = 0.23218, \quad \text{SSY} = 8.16811 \qquad S^2_{Z|X} = 0.0007838, \quad \text{SSZ} = 4.2276839$$

    b. Use the tables in part (a) to perform the F test for the significance of each straight-line regression. Interpret your results.

2.–4. Answer the same questions as in parts (a) and (b) of Problem 1 for each regression of Y on X, using the data in parts (b), (c), and (d) of Problem 2 in Chapter 5. The following results will be helpful in reducing computation time:

SBP (Y) regressed on QUET (X): $S^2_{Y|X} = 96.26743, \quad \text{SSY} = 6{,}425.96875$

QUET (Y) regressed on AGE (X): $S^2_{Y|X} = 0.09079, \quad \text{SSY} = 7.65968$

SBP (Y) regressed on AGE (X): $S^2_{Y|X} = 85.47795$

5. Use the data of Problem 3 in Chapter 5 to answer the following questions.
    a.–b. Answer the same questions as in parts (a) and (b) of Problem 1 for the regression of TIME (Y) on INC (X). The following results will be helpful:

$$S^2_{Y|X} = 110.16190, \quad \text{SSY} = 2433.78137$$

    c. Compare the value of the test statistic F obtained in part (b) with the value of T^2, the square of the test statistic for testing $H_0: \beta_1 = 0$ that was required in part (g) of Problem 3 in Chapter 5.
    d. The p-values for the F test in part (b) and for the t test in part (g) of Problem 3 in Chapter 5 are the same. Intuitively, why does this make sense? [*Hint:* Compare the hypotheses for each of the tests.]

6.–10. Answer the same questions as in parts (a) and (b) of Problem 1 for each of the the regressions Y on X in Problems 5 through 8 and Problem 10 of Chapter 5. The following results will be helpful:

	$S^2_{Y|X}$	SSY
Chapter 5, Problem 5:	11.101	2,223.018
Chapter 5, Problem 6:	172.985	20,123.382
Chapter 5, Problem 7:	1,001.691	190,502.800
	0.658	410.531
Chapter 5, Problem 8:	1,264,983.805	112,278,032.670
Chapter 5, Problem 10:	0.003	2.142

11. A biologist wished to study the effects of the temperature of a certain medium on the growth of human amniotic cells in a tissue culture. Using the same parent batch, she conducted an experiment in which five cell lines were cultured at each of four temperatures. The procedure involved initially inoculating a fixed number (0.25 million) of cells into a fresh culture

flask and then, after 7 days, removing a small sample from the growing surface to use in estimating the total number of cells in the flask. The results are given in the following table, together with a computer printout for straight-line regression.

Number of Cells ($\times 10^{-6}$) after 7 Days (Y)	Temperature (X)	Number of Cells ($\times 10^{-6}$) after 7 Days (Y)	Temperature (X)
1.13	40	2.30	80
1.20	40	2.15	80
1.00	40	2.25	80
0.91	40	2.40	80
1.05	40	2.49	80
1.75	60	3.18	100
1.45	60	3.10	100
1.55	60	3.28	100
1.64	60	3.35	100
1.60	60	3.12	100

a. Complete the ANOVA table shown in the accompanying computer output.

b. Perform the F test for the significance of the straight-line regression of Y on X, and interpret your results.

Edited SAS Output (PROC REG) for Problem 11

```
         # OF CELLS (Y) Regressed on TEMPERATURE (X)
                   Analysis of Variance
                    Sum of          Mean
Source     DF      Squares         Square      F Value      Prob > F
Model      1      _____       _____     _____       0.0001
Error      18     0.36618
C Total    19    13.19690
            Root MSE        0.14263     R-square      0.9723
            Dep Mean        2.04500     Adj R-sq      0.9707
            C.V.            6.97454
                     Parameter Estimates
                  Parameter      Standard      T for H0:
Variable   DF     Estimate         Error     Parameter=0    Prob > |T|
INTERCEP   1     -0.462400     0.10481069       -4.412        0.0003
X          1      0.035820     0.00142629       25.114        0.0001
```

12.–21. Answer the same questions as in parts (a) and (b) of Problem 1 for each of the data sets in Problems 12(a), 12(c), 13, 14, 15(a), 15(c), 16, 17, 18, and 19 in Chapter 5. The following results will be useful:

	$S^2_{Y|X}$	SSY
Chapter 5, Problem 12(a):	21,924.992	12,699,346.400
Chapter 5, Problem 12(c):	42,398.926	13,439,939.000
Chapter 5, Problem 13:	0.530	6.704

	$S^2_{Y|X}$	SSY
Chapter 5, Problem 14:	384.701	3,770.900
Chapter 5, Problem 15(a):	8.140	9,386.654
Chapter 5, Problem 15(c):	0.058	180.911
Chapter 5, Problem 16:	114.698	6,305.714
Chapter 5, Problem 17:	0.689	28.833
Chapter 5, Problem 18:	13.201	610.429
Chapter 5, Problem 19:	19.514	600.962

Multiple Regression Analysis: General Considerations

8-1 Preview

Multiple regression analysis can be looked upon as an extension of straight-line regression analysis (which involves only one independent variable) to the situation where more than one independent variable must be considered. Several general applications of multiple regression analysis[1] were described in Chapter 4, and specific examples were given in Chapter 1. In this chapter we will describe the multiple regression method in detail, stating the required assumptions, describing the procedures for estimating important parameters, explaining how to make and interpret inferences about these parameters, and providing examples that illustrate how to use the techniques of multiple regression analysis. Dealing with several independent variables simultaneously in a regression analysis is considerably more difficult than dealing with a single independent variable, for the following reasons:

1. It is more difficult to choose the best model, since several reasonable candidates may exist.

2. It is more difficult to visualize what the fitted model looks like (especially if there are more than two independent variables), since it is not possible to plot either the data or the fitted model directly in more than three dimensions.

3. It is sometimes more difficult to interpret what the best-fitting model means in real-life terms.

4. Computations are virtually impossible without access to a high-speed computer and a reliable packaged computer program.

---

[1]We shall generally refer to multiple regression analysis simply as "regression analysis" throughout the remainder of the text.

8-2 Multiple Regression Models

One example of a multiple regression model is given by any second- or higher-order poly-nomial. Adding higher-order terms (e.g., an X^2 or X^3 term) to a model can be considered as equivalent to adding new independent variables. Thus, if we rename X as X_1 and X^2 as X_2, the second-order model

$$Y = \beta_0 + \beta_1 X + \beta_2 X^2 + E$$

can be rewritten as

$$Y = \beta_0 + \beta_1 X_1 + \beta_2 X_2 + E$$

Of course, in polynomial regression we have only one basic independent variable, the others being simple mathematical functions of this basic variable. In more general multiple regression problems, however, the number of basic independent variables may be greater than one. The general form of a regression model for k independent variables is given by

$$Y = \beta_0 + \beta_1 X_1 + \beta_2 X_2 + \cdots + \beta_k X_k + E$$

where $\beta_0, \beta_1, \beta_2, \ldots, \beta_k$ are the *regression coefficients* that need to be estimated. The *independent variables* $X_1, X_2, \ldots, X_k$ may all be separate basic variables, or some may be functions of a few basic variables.

Example 8-1   Suppose that we want to investigate how weight (WGT) varies with height (HGT) and age (AGE) for children with a particular kind of nutritional deficiency. The dependent variable here is $Y = $ WGT, and our two basic independent variables are $X_1 = $ HGT and $X_2 = $ AGE.[2]

Suppose that, as outlined in Example 6-5, a random sample consists of 12 children who attend a certain clinic. The WGT, HGT, and AGE data obtained for each child are given in Table 8-1.

In describing the relationship of WGT to HGT and AGE, we may want to consider the model

$$Y = \beta_0 + \beta_1 X_1 + \beta_2 X_2 + E$$

TABLE 8-1   WGT, HGT, and AGE of a random sample of 12 nutritionally deficient children.

Child	1	2	3	4	5	6	7	8	9	10	11	12
WGT (Y)	64	71	53	67	55	58	77	57	56	51	76	68
HGT (X_1)	57	59	49	62	51	50	55	48	42	42	61	57
AGE (X_2)	8	10	6	11	8	7	10	9	10	6	12	9

---

[2]Perhaps the main question associated with this type of study is whether the relationship for nutritionally deficient children is the same as that for "normal" children. To answer this question would require additional data on normal children and some kind of comparison of the models obtained for each group. Although we will learn how to deal with this kind of question in Chapter 14, we focus here on the methods needed to describe the relationship of weight to height and age for this single group of nutritionally deficient children.

if we are interested only in first-order terms. If we want to consider, in addition, the higher-order term X_1^2, our model is given by

$$Y = \beta_0 + \beta_1 X_1 + \beta_2 X_2 + \beta_3 X_3 + E$$

where $X_3 = X_1^2$. To consider all possible first- and second-order terms, we look at the model

$$Y = \beta_0 + \beta_1 X_1 + \beta_2 X_2 + \beta_3 X_3 + \beta_4 X_4 + \beta_5 X_5 + E$$

where $X_3 = X_1^2$, $X_4 = X_2^2$, and $X_5 = X_1 X_2$, or equivalently

$$Y = \beta_0 + \beta_1 X_1 + \beta_2 X_2 + \beta_3 X_1^2 + \beta_4 X_2^2 + \beta_5 X_1 X_2 + E$$

If we want to find the best predictive model, we might consider all of the preceding models (as well as some others) and then choose the best model according to some reasonable criterion.

We discuss the question of model selection in Chapter 16; the interpretation of product terms such as $X_1 X_2$ as interaction effects is explained in Chapter 11. For now our focus is on the methods used and the interpretations that can be made when the choice of independent variables to use in the model is not at issue. ■

8-3 Graphical Look at the Problem

When we are dealing with only one independent variable, our problem can easily be described graphically as that of finding the curve that best fits the scatter of points (X_1, Y_1), $(X_2, Y_2), \ldots, (X_n, Y_n)$ obtained on n individuals. Thus, we have a *two-dimensional* representation involving a plot of the form shown in Figure 8-1. Furthermore, the *regression equation* for this problem is defined as the path described by the mean values of the distribution of Y when X is allowed to vary.

When the number k of (basic) independent variables is two or more, the (graphical) dimension of the problem increases. The regression equation ceases to be a curve in two-dimensional space and becomes instead a *hypersurface in $(k + 1)$-dimensional space*. Obviously, we will not be able to represent in a single plot either the scatter of data points or the regression equation if more than two basic independent variables are involved. In the special case $k = 2$, as in the example just given where $X_1 = $ HGT, $X_2 = $ AGE, and $Y = $ WGT, the problem is to find the *surface* in three-dimensional space that best fits the scatter of points (X_{11}, X_{21}, Y_1),

FIGURE 8-1   Scatter plot for a single independent variable.

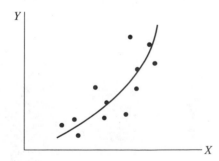

$(X_{12}, X_{22}, Y_2),\ldots, (X_{1n}, X_{2n}, Y_n)$, where (X_{1i}, X_{2i}, Y_i) denotes the X_1, X_2, and Y-values for the ith individual in the sample. The *regression equation* in this case, therefore, is the surface described by the mean values of Y at various combinations of values of X_1 and X_2; that is, corresponding to *each* distinct pair of values of X_1 and X_2 is a distribution of Y values with mean $\mu_{Y|X_1,X_2}$ and variance $\sigma^2_{Y|X_1,X_2}$.

Just as the simplest curve in two-dimensional space is a straight line, the simplest surface in three-dimensional space is a *plane,* which has the statistical model form $Y = \beta_0 + \beta_1 X_1 + \beta_2 X_2 + E$. Thus, finding the best-fitting plane is frequently the first step in determining the best-fitting surface in three-dimensional space when two independent variables are relevant, just as fitting the best straight line is the first step when one independent variable is involved. A graphical representation of a planar fit to data in the three-dimensional situation is given in Figure 8-2.

For the three-dimensional case, the least-squares solution that gives the best-fitting plane is determined by minimizing the sum of squares of the distances between the observed values Y_i and the corresponding predicted values $\hat{Y}_i = \hat{\beta}_0 + \hat{\beta}_1 X_{1i} + \hat{\beta}_2 X_{2i}$, based on the fitted plane. In other words, the quantity

$$\sum_{i=1}^{n} (Y_i - \hat{Y}_i)^2 = \sum_{i=1}^{n} (Y_i - \hat{\beta}_0 - \hat{\beta}_1 X_{1i} - \hat{\beta}_2 X_{2i})^2$$

is minimized to find the least-squares estimates $\hat{\beta}_0$ of β_0, $\hat{\beta}_1$ of β_1, and $\hat{\beta}_2$ of β_2.

How much can one learn by considering the independent variables in the multivariable problem separately? Probably the best answer is that we can learn something about what is going on, but there are too many separate (univariable) pieces of information to permit us to complete the (multivariable) puzzle. For example, consider the data previously given for $Y = \text{WGT}$, $X_1 = \text{HGT}$, and $X_2 = \text{AGE}$. If we plot separate scatter diagrams of WGT on HGT, WGT on AGE, and AGE on HGT, we get the results shown in Figure 8-3.

FIGURE 8-2   Best-fitting plane for three-dimensional data.

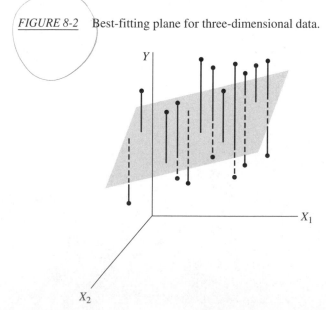

FIGURE 8-3   Separate scatter diagrams of WGT versus HGT, WGT versus AGE, and AGE versus HGT.

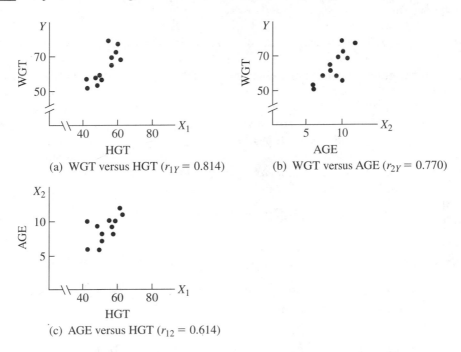

(a) WGT versus HGT ($r_{1Y} = 0.814$)

(b) WGT versus AGE ($r_{2Y} = 0.770$)

(c) AGE versus HGT ($r_{12} = 0.614$)

HGT is highly positively correlated with WGT ($r_{1Y} = 0.814$), as is AGE ($r_{2Y} = 0.770$). Thus, if we used each of these independent variables separately, we would likely find two separate, significant straight-line regressions. Does this mean that the best-fitting plane with both variables in the model together will also have significant predictive ability? The answer is probably yes. But what will the plane look like? This is difficult to say. We can get some idea of the difficulty if we consider the plot of HGT versus AGE in part (c), which reflects a positive correlation ($r_{12} = 0.614$). If, instead, these two variables were negatively correlated, we would expect a different orientation of the plane, although we could not clearly quantify either orientation. Thus, treating each independent variable separately does not help very much because the relationships between the independent variables themselves are not taken directly into account. The techniques of multiple regression, however, account for all these intercorrelations with regard to both estimation and inference making.

8-4   Assumptions of Multiple Regression

In the previous section we described the multiple regression problem in some generality and also hinted at some of the assumptions involved. We now state these assumptions somewhat more formally.

8-4-1   Statement of Assumptions

Assumption 1: Existence: *For each specific combination of values of the (basic) independent variables $X_1, X_2, \ldots, X_k$ (e.g., $X_1 = 57$, $X_2 = 8$ for the first child in Example 8-1),*

Y is a (univariate) random variable with a certain probability distribution having finite mean and variance.

Assumption 2: Independence: *The Y observations are statistically independent of one another.* As with straight-line regression, this assumption is usually violated when several Y observations are made on the same subject. Methods for dealing with regression modeling of correlated data include *repeated measures* ANOVA techniques (described in Chapter 21), *generalized estimating equations (GEE)* techniques (Zeger and Liang 1986; Diggle, Liang, and Zeger 1994), and *mixed model* techniques such as SAS's MIXED procedure, Release 6.07 (SAS Corporation 1992).

Assumption 3: Linearity: *The mean value of Y for each specific combination of X_1, $X_2,\ldots, X_k$ is a linear function[3] of $X_1, X_2,\ldots, X_k$.* That is,

$$\mu_{Y|X_1,X_2,\ldots,X_k} = \beta_0 + \beta_1 X_1 + \beta_2 X_2 + \cdots + \beta_k X_k \tag{8.1}$$

or

$$Y = \beta_0 + \beta_1 X_1 + \beta_2 X_2 + \cdots + \beta_k X_k + E \tag{8.2}$$

where E is the error component reflecting the difference between an individual's observed response Y and the true average response $\mu_{Y|X_1,X_2,\ldots,X_k}$. Some comments are in order regarding Assumption 3:

1. The surface described by (8.1) is called the *regression equation* (or *response surface* or *regression surface*).

2. If some of the independent variables are higher-order functions of a few basic independent variables (e.g., $X_3 = X_1^2$, $X_5 = X_1 X_2$), the expression $\beta_0 + \beta_1 X_1 + \beta_2 X_2 + \cdots + \beta_k X_k$ is nonlinear in the basic variables (hence the use of the word *surface* rather than *plane*).

3. Consonant with its meaning in straight-line regression, E is the amount by which any individual's observed response deviates from the response surface. Thus, E is the *error component* in the model.

Assumption 4: Homoscedasticity: *The variance of Y is the same for any fixed combination of $X_1, X_2,\ldots, X_k$.* That is,

$$\sigma_{Y|X_1,X_2,\ldots,X_k}^2 = \text{Var}\,(Y\mid X_1, X_2,\ldots, X_k) \equiv \sigma^2 \tag{8.3}$$

---

[3]The techniques of multiple regression that we will be describing are applicable as long as the model under consideration is *inherently linear* in the regression coefficients (regardless of how the independent variables are defined). For example, a model of the form $\mu_{Y|X} = \beta_0 e^{\beta_1 X}$ is inherently linear because it can be transformed into the equivalent form $\mu_{Y|X}^* = \beta_0^* + \beta_1 X$, where $\mu_{Y|X}^* = \ln \mu_{Y|X}$ and $\beta_0^* = \ln \beta_0$. However, the model $\mu_{Y|X_1,X_2} = e^{\beta_1 X_1} + e^{\beta_2 X_2}$ cannot be transformed directly into a form that is linear in β_1 and β_2; so estimating β_1 and β_2 requires the use of *nonlinear regression* procedures (see, e.g., Gallant 1975). A discussion of these procedures is beyond the scope of this text.

As before, this is called the assumption of homoscedasticity. An alternative (but equivalent) definition of homoscedasticity, based on (8.2), is that

$$\sigma^2_{E|X_1,X_2,\ldots,X_k} \equiv \sigma^2$$

This assumption may seem very restrictive. But variance heteroscedasticity needs to be considered only when the data show very obvious and significant departures from homogeneity. In general, mild departures do not have significant adverse effects on the results.

Assumption 5: Normality: *For any fixed combination of* $X_1, X_2,\ldots, X_k$, *the variable Y is normally distributed.* In other words,

$$Y \sim N(\mu_{Y|X_1,X_2,\ldots,X_k}, \sigma^2)$$

or equivalently,

$$E \sim N(0, \sigma^2) \tag{8.4}$$

This assumption is not necessary for the least-squares fitting of the regression model, but it is required in general for inference making. The usual parametric tests of hypotheses and confidence intervals used in a regression analysis are robust in the sense that only extreme departures of the distribution of Y from normality yield spurious results. (This statement is based on both theoretical and experimental evidence.) If the normality assumption does not hold, one typically seeks a transformation of Y—say, $\log Y$ or $\sqrt{Y}$—to produce a transformed set of Y observations that are approximately normal (see section 12-8-3). If the Y variable is either categorical or ordinal, however, alternative regression methods such as logistic regression (for binary Y's) or Poisson regression (for discrete Y's) are typically required (see Chapters 22 and 23).

8-4-2 Summary and Comments

Our assumptions for simple linear (i.e., straight-line) regression analysis can be generalized to multiple linear regression analysis. Here, homoscedasticity and normality apply to $Y \mid X_1, X_2,\ldots, X_k$, rather than to Y (i.e., to the conditional distribution of Y given $X_1, X_2,\ldots, X_k$, rather than to the so-called unconditional or marginal distribution of Y).

The assumptions for multiple linear regression analysis dictate that the random error component E have a normal distribution with mean 0 and variance σ^2. Of course, the linearity, existence, and independence assumptions must also hold.

Again, Y is an observable random variable, while $X_1, X_2,\ldots, X_k$ are fixed (nonrandom) known quantities. The constants $\beta_0, \beta_1,\ldots, \beta_k$ are unknown population parameters, and E is an unobservable random variable. If one estimates $\beta_0, \beta_1,\ldots, \beta_k$ with $\hat{\beta}_0, \hat{\beta}_1,\ldots, \hat{\beta}_k$, then an acceptable estimate of E_i for the i-th subject is

$$\hat{E}_i = Y_i - \hat{Y}_i = Y_i - (\hat{\beta}_0 + \hat{\beta}_1 X_{1i} + \cdots + \hat{\beta}_k X_{ki})$$

The estimated error $\hat{E}_i$ is usually called a *residual*.

The assumption of a Gaussian distribution is needed to justify the use of procedures of statistical inference involving the t and F distributions.

8-5 Determining the Best Estimate of the Multiple Regression Equation

As with straight-line regression, there are two basic approaches to estimating a multiple regression equation: the least-squares approach and the minimum-variance approach. In the straight-line case, both approaches yield the same solution. (We are assuming, as previously noted, that we already know the best form of regression model to use; that is, we have already settled on a fixed set of k independent variables $X_1, X_2, \ldots, X_k$. The problem of determining the best model form via algorithms for choosing the most important independent variables will be discussed in detail in Chapter 16.) The multiple regression model may also be fitted by using other statistical methodology, such as maximum likelihood (see Chapter 21). Under the assumption of a Gaussian distribution, the least-squares estimates of the regression coefficients are identical to the maximum-likelihood estimates.

8-5-1 Least-squares Approach

In general, the least-squares method chooses as the best-fitting model the one that minimizes the sum of squares of the distances between the observed responses and those predicted by the fitted model. Again, the better the fit, the smaller the deviations of observed from predicted values. Thus, if we let

$$\hat{Y} = \hat{\beta}_0 + \hat{\beta}_1 X_1 + \hat{\beta}_2 X_2 + \cdots + \hat{\beta}_k X_k$$

denote the fitted regression model, the sum of squares of deviations of observed Y-values from corresponding values predicted by using the fitted regression model is given by

$$\sum_{i=1}^{n} (Y_i - \hat{Y}_i)^2 = \sum_{i=1}^{n} (Y_i - \hat{\beta}_0 - \hat{\beta}_1 X_{1i} - \cdots - \hat{\beta}_k X_{ki})^2 \tag{8.5}$$

The least-squares solution then consists of the values $\hat{\beta}_0, \hat{\beta}_1, \ldots, \hat{\beta}_k$ (called the "least-squares estimates") for which the sum in (8.5) is a minimum. This minimum sum of squares is generally called the *residual sum of squares* (or equivalently, the *error sum of squares* or the *sum of squares about regression*); as in the case of straight line regression, it is referred to as the SSE.

8-5-2 Minimum-variance Approach

As in the straight-line case, the minimum-variance approach to estimating the multiple regression equation identifies as the best-fitting surface the one utilizing the minimum-variance (linear) unbiased estimates $\hat{\beta}_0, \hat{\beta}_1, \ldots, \hat{\beta}_k$ of $\beta_0, \beta_1, \ldots, \beta_k$, respectively.

8-5-3 Comments on the Least-squares Solutions

In this text we do not present matrix formulas for calculating the least-squares estimates $\hat{\beta}_0, \hat{\beta}_1, \ldots, \hat{\beta}_k$, since computer programs are readily available to perform the necessary calculations. Even so, we provide in Appendix B a discussion of matrices and their use in regression analysis; by using matrix mathematics, one can represent the general regression model and the associated least-squares methodology in compact form. Also, an understanding of the matrix formulation for regression analysis carries over to more complex modeling problems, such as those involving multivariate data (i.e., data relating to two or more dependent variables).

The least-squares solutions have several important properties:

1. Each of the estimates $\hat{\beta}_0, \hat{\beta}_1, \ldots, \hat{\beta}_k$ is a linear function of the Y-values. This linearity property makes determining the statistical properties of these estimates fairly straightforward. In particular, since the Y-values are assumed to be normally distributed and to be statistically independent of one another, each of the estimates $\hat{\beta}_0, \hat{\beta}_1, \ldots, \hat{\beta}_k$ will be normally distributed, with easily computable standard deviations.

2. The least-squares regression equation $\hat{Y} = \hat{\beta}_0 + \hat{\beta}_1 X_1 + \hat{\beta}_2 X_2 + \cdots + \hat{\beta}_k X_k$ is the unique linear combination of the independent variables $X_1, X_2, \ldots, X_k$ that has maximum possible correlation with the dependent variable. In other words, of all possible linear combinations of the form $b_0 + b_1 X_1 + b_2 X_2 + \cdots + b_k X_k$, the linear combination $\hat{Y}$ is such that the correlation

$$r_{Y,\hat{Y}} = \frac{\displaystyle\sum_{i=1}^{n}(Y_i - \overline{Y})(\hat{Y}_i - \overline{\hat{Y}})}{\sqrt{\displaystyle\sum_{i=1}^{n}(Y_i - \overline{Y})^2 \sum_{i=1}^{n}(\hat{Y}_i - \overline{\hat{Y}})^2}} \tag{8.6}$$

is a maximum, where $\hat{Y}_i$ is the predicted value of Y for the ith individual and $\overline{\hat{Y}}$ is the mean of the $\hat{Y}_i$'s. Incidentally, it is always true that $\overline{\hat{Y}} = \overline{Y}$; that is, the mean of the predicted values is equal to the mean of the observed values. The quantity $r_{Y,\hat{Y}}$ is called the *multiple correlation coefficient*.

3. Just as straight-line regression is related to the bivariate normal distribution, multiple regression can be related to the multivariate normal distribution. We will return to this point in section 10-4 of Chapter 10.

Example 8-2   For the data given in Table 8-1 on the variables $Y = \text{WGT}$, $X_1 = \text{HGT}$, and $X_2 = \text{AGE}$, the least-squares algorithm applied to the model

$$\text{WGT} = \beta_0 + \beta_1\text{HGT} + \beta_2\text{AGE} + \beta_3(\text{AGE})^2 + E$$

produces the estimated equation

$$\widehat{WGT} = 3.438 + 0.724\text{HGT} + 2.777\text{AGE} - 0.042(\text{AGE})^2$$

so

$$\hat{\beta}_0 = 3.438, \quad \hat{\beta}_1 = 0.724, \quad \hat{\beta}_2 = 2.777, \quad \hat{\beta}_3 = -0.042 \quad \blacksquare$$

8-6 The ANOVA Table for Multiple Regression

As with straight-line regression, an ANOVA table can be used to provide an overall summary of a multiple regression analysis. The particular form of an ANOVA table may vary, depending on how the contributions of the independent variables are to be considered (e.g., individually or collectively in some fashion). A simple form reflects the contribution that all independent variables considered collectively make to prediction. For example, consider Table 8-2, an ANOVA table based on the use of HGT, AGE, and $(AGE)^2$ as independent variables for the data of Table 8-1.

As before, the term $SSY = \sum_{i=1}^{n}(Y_i - \overline{Y})^2 = 888.25$ is called the *total sum of squares,* and this figure represents the total variability in the Y observations before accounting for the joint effect of using the independent variables HGT, AGE, and $(AGE)^2$. The term $SSE = \sum_{i=1}^{n}(Y_i - \hat{Y}_i)^2 = 195.19$ is the *residual sum of squares* (or the *sum of squares due to error*), which represents the amount of Y variation left unexplained after the independent variables have been used in the regression equation to predict Y. Finally, $SSY - SSE = \sum_{i=1}^{n}(\hat{Y}_i - \overline{Y})^2 = 693.06$ is called the *regression sum of squares* and measures the reduction in variation (or the variation explained) due to the independent variables in the regression equation. We thus have the familiar partition

Total sum of squares = Regression sum of squares + Residual sum of squares

or

$$\sum_{i=1}^{n}(Y_i - \overline{Y})^2 \quad = \quad \sum_{i=1}^{n}(\hat{Y}_i - \overline{Y})^2 \quad + \quad \sum_{i=1}^{n}(Y_i - \hat{Y}_i)^2$$

In Table 8-2, as in ANOVA tables for straight-line regression, the SS column identifies the various sums of squares. The df column gives the corresponding degrees of freedom: the regression degrees of freedom is k (the number of independent variables in the model); the residual degrees of freedom is $n - k - 1$; and the total degrees of freedom is $n - 1$. The MS column contains the mean-square terms, obtained by dividing the sum-of-squares terms by their corresponding degrees-of-freedom values. The F ratio is obtained by dividing the mean-square regression by the mean-square residual; the interpretation of this F ratio will be discussed in Chapter 9 on hypothesis testing.

The R^2 in Table 8-2 (with the value 0.7802) provides a quantitative measure of how well the fitted model containing the variables HGT, AGE, and $(AGE)^2$ predicts the dependent variable WGT. The computational formula for R^2 is

$$R^2 = \frac{SSY - SSE}{SSY} \tag{8.7}$$

The quantity R^2 lies between 0 and 1. If the value is 1, we say that the fit of the model is perfect. R^2 always increases as more variables are added to the model, but a very small increase in R^2 may be neither practically nor statistically important. Additional properties of R^2 are discussed in Chapter 10.

TABLE 8-2   ANOVA table for WGT regressed on HGT, AGE, and $(AGE)^2$.

Source	d.f.	SS	MS	F	R^2
Regression	$k = 3$	$SSY - SSE = 693.06$	231.02	9.47**	0.7802
Residual	$n - k - 1 = 8$	$SSE = 195.19$	24.40		
Total	$n - 1 = 11$	$SSY = 888.25$			

Note: The ** next to the computed F denotes significance at the 0.01 level; i.e., $P < 0.01$.

8-7 Numerical Examples

We conclude this chapter with some examples of the type of computer output to be expected from a typical regression program. This output generally consists of the values of the estimated regression coefficients, their estimated standard errors, the associated partial F (or T^2) statistics,[4] and an ANOVA table. For the data of Table 8-1, the six models that follow are by no means the only ones possible; for instance, no interaction terms were included.

Model 1 $\text{WGT} = \beta_0 + \beta_1\text{HGT} + E$

SAS Output for Regression of WGT on HGT

General Linear Models Procedure   *F statistic for overall test*

Source	DF	Sum of Squares	F Value	Pr > F
Model	1	588.92252318	19.67	0.0013
Error	10	299.32747682		
Corrected Total	11	888.25000000		

SSY → 888.25000000   SSE   SSY − SSE   *P-value for overall test ↑*

R-Square	C.V.	WGT Mean
0.663014	8.718857	62.7500000

R^2

[Portion of output omitted]

$\hat{\beta}_0$

Parameter	Estimate	T for H0: Parameter=0	P > |T|	Std Error of Estimate
INTERCEPT	6.189848707	0.48	0.6404	12.84874620
HGT	1.072230356 ← $\hat{\beta}_1$	4.44	0.0013	0.24173098

Model 2 $\text{WGT} = \beta_0 + \beta_2\text{AGE} + E$

SAS Output for Regression of WGT on AGE

General Linear Models Procedure

Source	DF	Sum of Squares	F Value	Pr > F
Model	1	526.39285714	14.55	0.0034
Error	10	361.85714286		
Corrected Total	11	888.25000000		

R-Square	C.V.	WGT Mean
0.592618	9.586385	62.7500000

[Portion of output omitted]

Parameter	Estimate	T for H0: Parameter=0	P > |T|	Std Error of Estimate
INTERCEPT	30.57142857	3.55	0.0053	8.61370526
AGE	3.64285714	3.81	0.0034	0.95511512

---

[4]Partial F statistics will be discussed in Chapter 9 on hypothesis testing.

Model 3 $WGT = \beta_0 + \beta_3(AGE)^2 + E$

SAS Output for Regression of WGT on $(AGE)^2$

```
                  General Linear Models Procedure
Source                  DF      Sum of Squares      F Value      Pr > F
Model                    1        521.93204725       14.25       0.0036
Error                   10        366.31795275
Corrected Total         11        888.25000000
          R-Square                  C.V.                  WGT Mean
          0.587596                9.645292              62.7500000
                              •  [Portion of output omitted]
                              •
                              T for H0:                      Std Error of
Parameter        Estimate     Parameter=0      P > |T|          Estimate
INTERCEPT      45.99764279         9.64         0.0001        4.76964028
AGESQ           0.20597161         3.77         0.0036        0.05456692
```

Model 4 $WGT = \beta_0 + \beta_1 HGT + \beta_2 AGE + E$

SAS Output for Regression of WGT on HGT and AGE

```
                  General Linear Models Procedure
Source                  DF      Sum of Squares      F Value      Pr > F
Model                    2        692.82260654       15.95       0.0011
Error                    9        195.42739346
Corrected Total         11        888.25000000
          R-Square                  C.V.                  WGT Mean
          0.779986                7.426048              62.7500000
                              •  [Portion of output omitted]
                              •
                              T for H0:                      Std Error of
Parameter        Estimate     Parameter=0      P > |T|          Estimate
INTERCEPT       6.553048251        0.60         0.5641       10.94482708
HGT             0.722037958        2.77         0.0218        0.26080506
AGE             2.050126352        2.19         0.0565        0.93722561
```

Test statistics and *P*-values for partial
tests on model parameters (see §9.3)

Model 5 $WGT = \beta_0 + \beta_1 HGT + \beta_3(AGE)^2 + E$

SAS Output for Regression of WGT on HGT and $(AGE)^2$

```
                  General Linear Models Procedure
Source                  DF      Sum of Squares      F Value      Pr > F
Model                    2        689.64995109       15.63       0.0012
Error                    9        198.60004891
Corrected Total         11        888.25000000
```

```
           General Linear Models Procedure (continued)
      R-Square                    C.V.                    WGT Mean
      0.776414                  7.486084                 62.7500000
                                     .
                                     .    [Portion of output omitted]
                                     .

                                    T for H0:                    Std Error of
      Parameter      Estimate     Parameter=0     P > |T|          Estimate
      INTERCEPT    15.11753900       1.28          0.2321       11.79690059
      HGT           0.72597651       2.76          0.0222        0.26333057
      AGESQ         0.11480164       2.14          0.0614        0.05373319
```

Model 6 $WGT = \beta_0 + \beta_1 HGT + \beta_2 AGE + \beta_3 (AGE)^2 + E$

SAS Output for Regression of WGT on HGT, AGE, and $(AGE)^2$

```
                    General Linear Models Procedure
      Source         DF      Sum of Squares    F Value      Pr > F
      Model           3        693.06046340      9.47       0.0052
      Error           8        195.18953660
      Corrected Total 11        888.25000000
            R-Square                    C.V.                 WGT Mean
            0.780254                  7.871718              62.7500000
                                         .
                                         .    [Portion of output omitted]
                                         .

                                    T for H0:                    Std Error of
      Parameter      Estimate     Parameter=0     P > |T|          Estimate
      INTERCEPT     3.438426001       0.10          0.9210       33.61081984
      HGT           0.723690241       2.61          0.0310        0.27696316
      AGE           2.776874563       0.37          0.7182        7.42727877
      AGESQ        -0.041706699      -0.10          0.9238        0.42240715
```

Although we will discuss model selection more fully in Chapter 16, it may already be clear from these results that model 4, involving HGT and AGE, is the best of the lot if we use R^2 and model simplicity as our criteria for selecting a model. The R^2-value of 0.7800 achieved by using this model is, for all practical purposes, the same as the maximum R^2-value of 0.7803 obtained by using all three variables.

Problems

1. The multiple regression relationships of SBP (Y) to AGE (X_1), SMK (X_2), and QUET (X_3) are studied using the data in Problem 2 of Chapter 5. Three regression models are considered, yielding least-squares estimates and ANOVA tables as shown in the accompanying SAS output.
   a. Use the model that includes all three independent variables ($X_1, X_2,$ and X_3) to answer the following questions. (1) What is the predicted SBP for a 50-year-old smoker with a

quetelet index of 3.5? (2) What is the predicted SBP for a 50-year-old nonsmoker with a quetelet index of 3.5? (3) For 50-year-old smokers, estimate the change in SBP corresponding to an increase in quetelet index from 3.0 to 3.5.

b. Using the computer output, determine and compare the R^2-values for the three models. If you use R^2 and model simplicity as the criteria for selecting a model, which of the three models appears to be the best?

Edited SAS Output (GLM PROC) for Problem 1

```
                    SBP Regressed on AGE
              General Linear Models Procedure
Source                  DF      Sum of Squares      F Value      Pr > F
Model                    1      3861.63037500        45.18       0.0001
Error                   30      2564.33837500
Corrected Total         31      6425.96875000
                              .
                              . [Portion of output omitted]
                              .
                                T for H0:
Parameter          Estimate    Parameter=0      P > |T|      Std Error of
                                                               Estimate
INTERCEPT        59.09162500       4.61          0.0001      12.81626145
AGE               1.60450000       6.72          0.0001       0.23871593

              SBP Regressed on AGE and SMK
              General Linear Models Procedure
Source                  DF      Sum of Squares      F Value      Pr > F
Model                    2      4689.68422867        39.16       0.0001
Error                   29      1736.28452133
Corrected Total         31      6425.96875000
                              .
                              . [Portion of output omitted]
                              .
                                T for H0:
Parameter          Estimate    Parameter=0      P > |T|      Std Error of
                                                               Estimate
INTERCEPT        48.04960299       4.32          0.0002      11.12955962
AGE               1.70915965       8.47          0.0001       0.20175872
SMK              10.29439180       3.72          0.0009       2.76810685

          SBP Regressed on AGE, SMK, and QUET
              General Linear Models Procedure
Source                  DF      Sum of Squares      F Value      Pr  F
Model                    3      4889.82569715        29.71       0.0001
Error                   28      1536.14305285
Corrected Total         31      6425.96875000
                              .
                              . [Portion of output omitted]
                              .
                                T for H0:
Parameter          Estimate    Parameter=0      P > |T|      Std Error of
                                                               Estimate
INTERCEPT        45.10319242       4.19          0.0003      10.76487511
AGE               1.21271462       3.75          0.0008       0.32381922
SMK               9.94556782       3.74          0.0008       2.65605655
QUET              8.59244866       1.91          0.0664       4.49868122
```

2. A psychiatrist wants to know whether the level of pathology (Y) in psychotic patients 6 months after treatment can be predicted with reasonable accuracy from knowledge of pretreatment symptom ratings of thinking disturbance (X_1) and hostile suspiciousness (X_2). The following table lists data collected on 53 patients.

Patient	Y	X_1	X_2	Patient	Y	X_1	X_2	Patient	Y	X_1	X_2
1	44	2.80	6.1	19	21	2.81	6.0	37	50	2.90	6.7
2	25	3.10	5.1	20	22	2.80	6.4	38	9	2.74	5.5
3	10	2.59	6.0	21	60	3.62	6.8	39	13	2.70	6.9
4	28	3.36	6.9	22	10	2.74	8.4	40	22	3.08	6.3
5	25	2.80	7.0	23	60	3.27	6.7	41	23	2.18	6.1
6	72	3.35	5.6	24	12	3.78	8.3	42	31	2.88	5.8
7	45	2.99	6.3	25	28	2.90	5.6	43	20	3.04	6.8
8	25	2.99	7.2	26	39	3.70	7.3	44	65	3.32	7.3
9	12	2.92	6.9	27	14	3.40	7.0	45	9	2.80	5.9
10	24	3.23	6.5	28	8	2.63	6.9	46	12	3.29	6.8
11	46	3.37	6.8	29	11	2.65	5.8	47	21	3.56	8.8
12	8	2.72	6.6	30	7	3.26	7.2	48	13	2.74	7.1
13	15	3.47	8.4	31	23	3.15	6.5	49	10	3.06	6.9
14	28	2.70	5.9	32	16	2.60	6.3	50	4	2.54	6.7
15	26	3.24	6.0	33	26	2.74	6.8	51	18	2.78	7.2
16	27	2.65	6.0	34	8	2.72	5.9	52	10	2.81	5.2
17	4	3.41	7.6	35	11	3.11	6.8	53	7	3.26	6.6
18	14	2.58	6.2	36	12	2.79	6.7				

a. The least-squares equation involving both independent variables is given by $\hat{Y} = -.0635 + 23.451X_1 - 7.073X_2$. Using this equation, determine the predicted level of pathology for a patient with pretreatment scores of 2.80 on thinking disturbance and 7.0 on hostile suspiciousness. How does this predicted value compare with the value actually obtained for patient 5?

b. Sums of squares are shown next for three regression models. Determine the R^2 values for each of these models. If you use R^2 and model simplicity as selection criteria, which model appears to be the best?

  Y regressed on X_1:   SSY = 13,791.1698,   SSE = 12,255.3128

  Y regressed on X_2:   SSE = 13,633.3225

  Y regressed on X_1 and X_2: SSE = 11,037.2985

3. The following table presents the weight (X_1), age (X_2), and plasma lipid levels of total cholesterol (Y) for a hypothetical sample of 25 patients suffering from hyperlipoproteinemia, before drug therapy.

a. Three estimated regression models, along with their sums-of-squares results, are as follows:

  $\hat{Y} = 77.983 + 0.417X_1 + 5.217X_2$ SSY = 145377.0400,   SSE = 42806.2254

  $\hat{Y} = 199.2975 + 1.622X_1$ SSE = 135145.3138

  $\hat{Y} = 102.5751 + 5.321X_2$ SSE = 43444.3743

  For each of these models, determine the predicted cholesterol level (Y) for patient 4, and compare these predicted cholesterol levels with the observed value. Comment on your findings.

Patient	Total Cholesterol (Y) (mg/100 ml)	Weight (X_1) (kg)	Age (X_2) (yr)	Patient	Total Cholesterol (Y) (mg/100 ml)	Weight (X_1) (kg)	Age (X_2) (yr)
1	354	84	46	14	254	57	23
2	190	73	20	15	395	59	60
3	405	65	52	16	434	69	48
4	263	70	30	17	220	60	34
5	451	76	57	18	374	79	51
6	302	69	25	19	308	75	50
7	288	63	28	20	220	82	34
8	385	72	36	21	311	59	46
9	402	79	57	22	181	67	23
10	365	75	44	23	274	85	37
11	209	27	24	24	303	55	40
12	290	89	31	25	244	63	30
13	346	65	52				

b. Determine R^2-values for each of the three models considered in part (a). If you use R^2 and model simplicity as selection criteria, which model appears to be the best predictive model?

4. A sociologist investigating the recent increase in the incidence of homicide throughout the United States studied the extent to which the homicide rate per 100,000 population (Y) is associated with the city's population size (X_1), the percentage of families with yearly income less than \$5,000 (X_2), and the rate of unemployment (X_3). Data are provided in the following table for a hypothetical sample of 20 cities.

City	Y	X_1 (thousands)	X_2	X_3	City	Y	X_1 (thousands)	X_2	X_3
1	11.2	587	16.5	6.2	11	14.5	7,895	18.1	6.0
2	13.4	643	20.5	6.4	12	26.9	762	23.1	7.4
3	40.7	635	26.3	9.3	13	15.7	2,793	19.1	5.8
4	5.3	692	16.5	5.3	14	36.2	741	24.7	8.6
5	24.8	1,248	19.2	7.3	15	18.1	625	18.6	6.5
6	12.7	643	16.5	5.9	16	28.9	854	24.9	8.3
7	20.9	1,964	20.2	6.4	17	14.9	716	17.9	6.7
8	35.7	1,531	21.3	7.6	18	25.8	921	22.4	8.6
9	8.7	713	17.2	4.9	19	21.7	595	20.2	8.4
10	9.6	749	14.3	6.4	20	25.7	3,353	16.9	6.7

a. Using the regression results presented next, determine the R^2-values for each two-variable model, and comment on which appears to be the best.

   Y regressed on X_1 and X_2: $SSY = 1855.2020$, $SSE = 537.4036$

   Y regressed on X_1 and X_3: $SSE = 431.9700$

   Y regressed on X_2 and X_3: $SSE = 367.3426$

b. When Y is regressed on X_1, X_2, and X_3, we get $SSE = 337.0571$. Determine and comment on the increase in R^2 in going from a model with just X_2 and X_3 to a model that includes all three independent variables.

c. Consider the model with independent variables X_2, X_3, and X_4, where $X_4 = X_2 X_3$.[5] The ANOVA table for this model, from SAS's GLM procedure, is given next. Does the addition of X_4 lead to a large improvement in fit over the model with X_2 and X_3 as the only independent variables? Explain.

Edited SAS Output (PROC GLM) for Problem 4

```
              Y Regressed on X2, X3, and X4
            General Linear Models Procedure
Source              DF      Sum of Squares    F Value     Pr > F
Model                3      1490.57127950      21.80      0.0001
Error               16       364.63072050
Corrected Total     19      1855.20200000
```

5. A panel of educators in a large urban community wanted to evaluate the effects of educational resources on student performance. They examined the relationship between twelfth-grade mean verbal SAT scores (Y) and the following independent variables for a random sample of 25 high schools: $X_1 = $ Per pupil expenditure (in dollars); $X_2 = $ Percentage of teachers with a master's degree or higher; and $X_3 = $ Pupil–teacher ratio. The sums of squares shown next can be used to summarize the key results from the regression of Y on X_1, X_2, and X_3:

$$\text{SSY} = 28{,}222.23 \qquad \text{SSE} = 2{,}248.23$$

a. Determine the ANOVA table for the regression of Y on X_1, X_2, and X_3.

b. Determine the R^2-value for the model in part (a). Based on this value, comment on whether the educational resources appear to be linearly associated with student performance.

6. A team of environmental epidemiologists used data from 23 counties to investigate the relationship between respiratory cancer mortality rates (Y) for a given year and the following three independent variables: $X_1 = $ Air pollution index for the county; $X_2 = $ Mean age (over 21) for the county; and $X_3 = $ Percentage of workforce in the county employed in a certain industry. The sums of squares shown next can be used to summarize the key results from the regression of Y on X_1, X_2, and X_3:

$$\text{SSY} = 2{,}387.653 \qquad \text{SSE} = 551.723$$

a. Determine the ANOVA table for the regression of Y on X_1, X_2, and X_3.

b. Determine the R^2-value for the model in part (a). Based on this value, comment on whether the pollution level, mean age, and percentage of people working in the particular industry appear to be linearly related (as a group), to respiratory cancer mortality rates.

---

[5]The coefficient of the product term X_4 measures an *interaction effect* associated with the variables X_2 and X_3, which concerns whether the relationship between Y and one of these two variables depends on the levels of the other variable. A more detailed discussion of the concept of interaction is given in Chapter 11.

7. In an experiment to describe the toxic action of a certain chemical on silkworm larvae,[6] the relationship of $\log_{10}$ dose and of $\log_{10}$ dose and $\log_{10}$ larva weight to $\log_{10}$ survival time was sought. The data, obtained by feeding each larva a precisely measured dose of the chemical in an aqueous solution and then recording the survival time (i.e., time until death), are given in the following table. Relevant computer results are also provided.

Larva	1	2	3	4	5	6	7	8
$\log_{10}$ survival time (Y)	2.836	2.966	2.687	2.679	2.827	2.442	2.421	2.602
$\log_{10}$ dose (X_1)	0.150	0.214	0.487	0.509	0.570	0.593	0.640	0.781
$\log_{10}$ weight (X_2)	0.425	0.439	0.301	0.325	0.371	0.093	0.140	0.406

Larva	9	10	11	12	13	14	15
$\log_{10}$ survival time (Y)	2.556	2.441	2.420	2.439	2.385	2.452	2.351
$\log_{10}$ dose (X_1)	0.739	0.832	0.865	0.904	0.942	1.090	1.194
$\log_{10}$ weight (X_2)	0.364	0.156	0.247	0.278	0.141	0.289	0.193

a. Compute R^2 for each of the three models. Which independent variable appears to be the single best predictor of survival time?

b. Which model involving one or both of the independent variables do you prefer? Why?

Edited SAS Output (PROC GLM) for Problem 7

```
                    Y Regressed on X1
              General Linear Models Procedure
Source                 DF      Sum of Squares      F Value       Pr > F
Model                   1         0.36327405        31.91        0.0001
Error                  13         0.14799288
Corrected Total        14         0.51126693
                              .
                              .  [Portion of output omitted]
                              .
                             T for H0:                      Std Error of
Parameter      Estimate     Parameter=0     Pr > |T|          Estimate
INTERCEPT     2.952199058        40.14        0.0001         0.07355501
X1           -0.549855934        -5.65        0.0001         0.09733756

                    Y Regressed on X2
              General Linear Models Procedure
Source                 DF      Sum of Squares      F Value       Pr > F
Model                   1         0.33667410        25.07        0.0002
Error                  13         0.17459283
Corrected Total        14         0.51126693
                              .
                              .  [Portion of output omitted]
                              .
                             T for H0:                      Std Error of
Parameter      Estimate     Parameter=0     Pr > |T|          Estimate
INTERCEPT     2.184705838        26.64        0.0001         0.08199583
X2            1.375578800         5.01        0.0002         0.27474016
```

---

[6]Adapted from a study by Bliss (1936).

```
                    Y Regressed on X1 and X2
                  General Linear Models Procedure
Source              DF       Sum of Squares    F Value      Pr > F
Model                2          0.46420205      59.18       0.0001
Error               12          0.04706489
Corrected Total     14          0.51126693
                                    .
                                    .   [Portion of output omitted]
                                    .

                               T for H0:                    Std Error of
Parameter         Estimate    Parameter=0    Pr > |T|        Estimate
INTERCEPT       2.588995674      30.97         0.0001       0.08360793
X1             -0.378480493      -5.70         0.0001       0.06637411
X2              0.874974778       5.07         0.0003       0.17248352
```

8. An experiment to evaluate the effects of certain variables on soil erosion was performed on 10-foot-square plots of sloped farmland subjected to 2 inches of artificial rain applied over a 20-minute period.[7] The data and related ANOVA table are as follows:

Plot	1	2	3	4	5	6	7	8	9	10	11
SL (Y)	27.1	35.6	31.4	37.8	40.2	39.8	55.5	43.6	52.1	43.8	35.7
SG (X_1)	0.43	0.47	0.44	0.48	0.48	0.49	0.53	0.50	0.55	0.51	0.48
LOBS (X_2)	1.95	5.13	3.98	6.25	7.12	6.50	10.67	7.08	9.88	8.72	4.96
PGC (X_3)	0.34	0.32	0.29	0.30	0.25	0.26	0.10	0.16	0.19	0.18	0.28

Note: SL denotes soil lost (in pounds/acre), SG denotes slope gradient of the plot, LOBS denotes length (in inches) of the largest opening of bare soil on any boundary, and PGC denotes percentage of ground cover.

Source	d.f.	SS
Regression	3	680.4912
Residual	7	16.0942
Total	10	696.5855

a. Compute R^2, and comment on the fit of the model.
b. The fitted model involving all three independent variables is given by
$\hat{Y} = -1.879 + 77.326X_1 + 1.559X_2 - 23.904X_3$. Compute and compare observed and predicted values of Y for plots 1, 5, and 7.

9. In a study by Yoshida (1961), the oxygen consumption of wireworm larva groups was measured at five temperatures. The rate of oxygen consumption per larva group (in milliliters per hour)—the dependent variable—was transformed to 0.5 less than the common logarithm.

---

[7]Adapted from a study by Packer (1951).

Another independent variable (other than temperature) of importance was larva group weight, which was also transformed to common logarithms. The data are given in the following table.

Oxygen Consumption (Y) (log ml/hr −0.5)	Larva Group Weight (X₁) (log cg)	Temperature (X₂) (°C)	Oxygen Consumption (Y) (log ml/hr −0.5)	Larva Group Weight (X₁) (log cg)	Temperature (X₂) (°C)
0.054	0.130	15.5	0.482	0.053	30.0
0.154	0.215	15.5	0.477	0.114	30.0
0.073	0.250	15.5	0.551	0.137	30.0
0.182	0.267	15.5	0.516	0.190	30.0
0.241	0.389	15.5	0.561	0.210	30.0
0.316	0.490	15.5	0.588	0.230	30.0
0.290	0.491	15.5	0.561	0.240	30.0
0.061	0.004	20.0	0.580	0.260	30.0
0.143	0.164	20.0	0.674	0.389	30.0
0.188	0.225	20.0	0.718	0.470	30.0
0.176	0.314	20.0	0.754	0.521	30.0
0.248	0.447	20.0	0.800	0.544	30.0
0.357	0.477	20.0	0.654	−0.004	35.0
0.403	0.505	20.0	0.744	0.033	35.0
0.342	0.537	20.0	0.711	0.049	35.0
0.335	−0.046	25.0	0.855	0.140	35.0
0.408	0.176	25.0	0.932	0.204	35.0
0.366	0.199	25.0	0.927	0.210	35.0
0.482	0.292	25.0	0.914	0.215	35.0
0.545	0.380	25.0	0.914	0.265	35.0
0.596	0.483	25.0	0.973	0.346	35.0
0.590	0.491	25.0	1.000	0.462	35.0
0.631	0.491	25.0	0.998	0.468	35.0
0.610	0.519	25.0			

a. The fitted multiple regression model containing both X_1 and X_2 is given by

$$\hat{Y} = -0.6838 + 0.5921X_1 + 0.0394X_2$$

On the basis of this fitted model, how much of a change in oxygen consumption would be predicted for a larva group with fixed weight X_1 if the temperature were increased from $X_2 = 20$ to $X_2 = 25$?

b. For a temperature of 20°C, compute and compare the predicted values of $\hat{Y}$ for weights of 0.250 and 0.500.

c. What is R^2 for each of the three models?

Edited SAS Output (PROC GLM) for Problem 9

```
                        Y Regressed on X1
                 General Linear Models Procedure
Source            DF      Sum of Squares      F Value      Pr > F
Model              1         0.06607190         0.89       0.3505
Error             45         3.33995010
Corrected Total   46         3.40602200
```

```
                    Y Regressed on X2
            General Linear Models Procedure
Source              DF      Sum of Squares      F Value     Pr > F
Model                1        2.77418353        197.58      0.0001
Error               45        0.63183847
Corrected Total     46        3.40602200
                Y Regressed on X1 and X2
            General Linear Models Procedure
Source              DF      Sum of Squares      F Value     Pr > F
Model                2        3.21119764        362.62      0.0001
Error               44        0.19482436
Corrected Total     46        3.40602200
```

10. Residential real estate prices depend, in part, on property size and number of bedrooms. The house size X_1 (in hundreds of square feet), number of bedrooms X_2, and house price Y (in thousands of dollars) of a random sample of houses in a certain county were observed. The data are listed in the following table.

House	1	2	3	4	5	6	7
House size (X_1)	18	20	25	22	33	19	17
Number of bedrooms (X_2)	3	3	4	4	5	4	3
House price (Y)	80	95	104	110	175	85	89

Use the accompanying output to answer the following questions.

a. Determine the least-squares estimates for the model in which house price is regressed on both house size and number of bedrooms. Find the predicted average price of a 2,500-square-foot home that has four bedrooms.

b. Compare the R^2-value for the regression in part (a) with the r^2 value in Problem 22 of Chapter 6. Does adding X_2 to a model that already contains X_1 appear to be useful in predicting the house price?

Edited SAS Output (PROC GLM) for Problem 10

```
            General Linear Models Procedure
Source              DF      Sum of Squares      F Value     Pr > F
Model                2       5733.32128961       20.03      0.0082
Error                4        572.39299611
Corrected Total      6       6305.71428571
                        .
                        .   [Portion of output omitted]
                        .
                             T for H0:
                          Parameter=0                   Std Error of
Parameter       Estimate  Parameter=0     Pr > |T|        Estimate
INTERCEPT     -16.09338521    -0.65         0.5494      24.64693805
X1              5.72178988     3.13         0.0352       1.82778969
X2             -1.17315175    -0.09         0.9344      13.38993701
```

11. Data on sales revenues Y, television advertising expenditures X_1, and print media advertising expenditures X_2 for a large retailer for the period 1988–1993 are given in the following table.

Year	1988	1989	1990	1991	1992	1993
Sales ($millions)	4.0	8.0	2.0	8.0	5.0	4.0
TV advertising ($millions)	1.5	4.5	0.0	5.0	3.0	1.5
Print advertising ($millions)	0.5	0.5	0.0	1.0	1.0	1.5

a. State the model for the regression of sales revenue on television advertising expenditure and print advertising expenditure.

Use the accompanying computer output to answer the following questions.

b. State the estimate of the model in part (a). What is the estimated change in sales revenue for every $1,000 increase in television advertising expenditure? What is the estimated change in sales revenue for every $1,000 increase in print advertising expenditure?

c. Find the R^2-value for the regression of Y on X_1 and X_2. Interpret your result.

d. Predict the sales for a year in which $5 million was spent on TV advertising, and $1 million was spent on print advertising.

Edited SAS Output (PROC GLM) for Problem 11

```
                    General Linear Models Procedure
Source                  DF        Sum of Squares       F Value      Pr > F
Model                    2          28.11853189         59.01       0.0039
Error                    3           0.71480144
Corrected Total          5          28.83333333
                                         .
                                         . [Portion of output omitted]
                                         .
                                  T for H0:                       Std Error of
Parameter       Estimate        Parameter=0       Pr > |T|           Estimate
INTERCEPT      2.104693141           4.99           0.0155         0.42196591
X1             1.241877256          10.42           0.0019         0.11913357
X2            -0.194945848          -0.44           0.6874         0.43944079
```

12. Radial keratotomy is a type of refractive surgery in which radial incisions are made in the cornea of myopic (nearsighted) patients in an effort to reduce their myopia. Theoretically, the incisions allow the curvature of the cornea to become less steep, thereby reducing the refractive error of the patient. This and other vision correction surgery have been growing in popularity in the 1980s and 1990s, both among the public and among ophthalmologists.

The Prospective Evaluation of Radial Keratotomy (PERK) study began in 1983 to investigate the effects of radial keratotomy. Lynn et al. (1987) examined the factors associated with the five-year postsurgical change in refractive error (Y, measured in diopters, D). Two independent variables under consideration were baseline refractive error (X_1, in diopters) and baseline curvature of the cornea (X_2, in diopters). [*Note:* Myopic patients have negative refractive errors. Patients who are far-sighted have positive refractive errors. Patients who are neither near nor far-sighted have zero refractive error.]

The accompanying computer output is based on data adapted from the PERK study. Use it to answer the following questions.

a. State the estimated least-squares equation for the regression of change in refractive error (Y) on baseline refractive error (X_1) and baseline curvature (X_2).

b. Using your answer to (a), give a point estimate for the change in refractive error for a patient who, at baseline, has a refractive error of $-8.00D$, and a corneal curvature of 44D. [*Note:* Myopic patients have negative refractive errors].

c. Find the R^2-value for the regression in (a), and comment on the fit of the model.

Edited SAS Output (PROC GLM) for Problem 12

```
                  General Linear Models Procedure
                        DF      Sum of Squares    F Value      Pr > F
Source
Model                    2        17.62277191       7.18       0.0018
Error                   51        62.63016929
Corrected Total         53        80.25294120
                                   .
                                   .  [Portion of output omitted]
                                   .

                                 T for H0:                  Std Error of
Parameter          Estimate    Parameter=0     Pr > |T|       Estimate
INTERCEPT        12.36001523        2.43         0.0187      5.08621177
X1               -0.29160125       -3.18         0.0025      0.09165295
X2               -0.22039615       -1.92         0.0605      0.11482391
```

13. In 1990, *Business Week* magazine compiled financial data on the 1,000 companies that had the biggest impact on the U.S. economy.[8] Data sampled from the top 500 companies in *Business Week*'s report are presented in the following table. In addition to the company name, four variables are shown:

1990 rank: Based on company's market value (share price on March 16, 1990, multiplied by available common shares outstanding).

1989 rank: Rank in 1989 compilation.

P-E ratio: Price-to-earnings ratio, based on 1989 earnings and March 16, 1990, share price.

Yield: Annual dividend rate as a percentage of March 16, 1990, share price.

Company	1990 Rank	1989 Rank	P-E Ratio	Yield	Company	1990 Rank	1989 Rank	P-E Ratio	Yield
AT&T	4	4	17	2.87	ITT	81	57	8	2.98
Merck	7	7	19	2.54	Humana	162	209	15	2.62
Boeing	27	41	24	1.72	Salomon	236	172	7	2.91
American Home Products	32	37	14	4.26	Walgreen	242	262	17	1.87
Walt Disney	33	42	23	0.41	Lincoln National	273	274	9	4.73
Pfizer	46	46	14	4.10	Citizens Utilities	348	302	21	0.00
MCI Communications	52	72	19	0.00	MNC Financial	345	398	6	5.46
Dunn & Bradstreet	55	48	15	4.27	Bausch & Lomb	354	391	15	1.99
United Telecommunications	63	93	22	2.61	Medtronic	356	471	16	1.10
Warner Lambert	77	91	17	2.94	Circuit City	497	514	14	0.33

[8]"The Business Week 1000," a special issue of *Business Week* magazine, April 13, 1990.

Answer the following questions about these data.

a. What is the estimated least-squares equation for the regression of yield (Y) on 1989 rank (X_2) and P-E ratio (X_3)?

b. Using your answer to part (a), give a point estimate for the yield for a company that had a 1989 ranking of 200 and a P-E ratio of 10.

c. Find the R^2-value for the regression in part (a), and comment on the fit of the model.

Edited SAS Output (PROC GLM) for Problem 13

```
        YIELD (Y) Regressed on 1989 RANK (X2) and P-E RATIO (X3)
                    General Linear Models Procedure
Dependent Variable: Y
                                Sum of              Mean
Source              DF          Squares            Square      F Value    Pr > F
Model                2       26.58272135       13.29136068      10.51     0.0011
Error               17       21.49757365        1.26456316
Corrected Total     19       48.08029500

        R-Square               C.V.            Root MSE            Y Mean
        0.552882             45.24353          1.124528           2.485500

                                         .    [Portion of output omitted]
                                         .

                                   T for H0:                        Std Error of
Parameter          Estimate      Parameter=0       Pr > |T|           Estimate
INTERCEPT        6.873525074          6.94           0.0001         0.99001534
X2              -0.004138085         -2.51           0.0224         0.00164749
X3              -0.234451675         -4.43           0.0004         0.05292277
```

14. This problem refers to the 1990 Census data presented in Problem 19 of Chapter 5. In addition to median selected monthly ownership costs (OWNCOST), another independent variable studied was the proportion of the total metropolitan statistical area (MSA) population living in urban areas (URBAN).

Use the accompanying computer output to answer the following questions about the regression of OWNEROCC on OWNCOST and URBAN.

a. What is the estimated least-squares equation for the regression of OWNEROCC on OWNCOST and URBAN?

b. Using your answer to part (a), give a point estimate for the rate of owner occupancy for an MSA for which the urban proportion is 0.73 and median selected monthly ownership cost is $513.

c. Find the R^2-value for the regression in part (a), and comment on the fit of the model.

Edited SAS Output (PROC REG) for Problem 14

```
            OWNEROCC Regressed on OWNCOST and URBAN
                    Analysis of Variance
                        Sum of              Mean
Source        DF        Squares            Square      F Value    Pr > F
Model          2       255.77850        127.88925       8.521     0.0017
Error         23       345.18304         15.00796
C Total       25       600.96154
```

```
        OWNEROCC Regressed on OWNCOST and URBAN (continued)
               Analysis of Variance (continued)
      Root MSE          3.87401      R-square        0.4256
      Dep Mean         65.96154      Adj R-sq        0.3757
      C.V.              5.87314
                      Parameter Estimates
                     Parameter      Standard      T for H0:
      Variable   DF   Estimate        Error     Parameter=0    Pr > |T|
      INTERCEP    1  86.758772    5.10721232        16.988      0.0001
      OWNCOST     1  -0.011130    0.00529413        -2.102      0.0467
      URBAN       1 -16.875462    5.89094370        -2.865      0.0088
```

References

Bliss, C. I. 1936. "The Size Factor in Action of Arsenic upon Silkworms' Larvae." *Journal of Experimental Biology* 13: 95–110.

Business Week. 1990. "The Business Week 1000, America's Most Valuable Companies." Special issue of *Business Week,* April 13, 1990.

Diggle, P. J.; Liang, K. Y.; and Zeger, S. L. 1994. *Analysis of Longitudinal Data.* New York: Oxford University Press.

Gallant, A. R. 1975. "Non-linear Regression." *American Statistician* 29: 73–81.

Lynn, M. J.; Waring, G. O.; Sperduto, R. D.; et al. 1987. "Factors Affecting Outcome and Predictability of Radial Keratotomy in the PERK Study." *Archives of Ophthalmology* 105: 42–51.

Packer, P. E. 1951. "An Approach to Watershed Protection Criteria." *Journal of Forestry* 49: 638–44.

SAS/STAT. 1992. *Software: Changes and Enhancements, Release 6.07.* SAS Technical Report P-229. SAS Institute, Inc. Cary, N.C.

Yoshida, M. 1961. "Ecological and Physiological Researches on the Wireworm, *Melanotus caudex* Lewis. Iwata." *Shizuoka Pref.,* Japan.

Zeger, S. L., and Liang, K. Y. 1986. "Longitudinal Data Analysis for Discrete and Continuous Outcomes." *Biometrics* 42: 121–30.

Testing Hypotheses
in Multiple Regression

9-1 Preview

Once we have fit a multiple regression model and obtained estimates for the various parameters of interest, we want to answer questions about the contributions of various independent variables to the prediction of Y. Such questions raise the need for three basic types of tests:

1. *Overall test.* Taken collectively, does the *entire set* of independent variables (or equivalently, the fitted model itself) contribute significantly to the prediction of Y?

2. *Test for addition of a single variable.* Does the addition of *one* particular independent variable of interest add significantly to the prediction of Y achieved by other independent variables already present in the model?

3. *Test for addition of a group of variables.* Does the addition of some *group* of independent variables of interest add significantly to the prediction of Y obtained through other independent variables already present in the model?

These questions are typically answered by performing statistical tests of hypotheses. The null hypotheses for the tests can be stated in terms of the unknown parameters (the regression coefficients) in the model. The form of these hypotheses differs depending on the question being asked. (In Chapter 10 we will look at alternative but equivalent ways to state such null hypotheses in terms of population correlation coefficients.)

In the sections that follow, we will describe the statistical test appropriate for each of the preceding questions. Each of these tests can be expressed as an F test; that is, the test statistic will have an F distribution when the stated null hypothesis is true. In some cases, the test may be equivalently expressed as a t test. (For a review of material concerning the F and t distributions, refer to Chapter 3.)

All F tests used in regression analyses involve a ratio of two independent estimates of variance—say, $F = \hat{\sigma}_0^2 / \hat{\sigma}^2$. Under the assumptions for the standard multiple linear regression analysis given earlier, the term $\hat{\sigma}_0^2$ estimates σ^2 if H_0 is true; the term $\hat{\sigma}^2$ estimates σ^2 whether H_0

is true or not. The specific forms that these variance estimates take will be described in subsequent sections. In general, each is a mean-square term that can be found in an appropriate ANOVA table. If H_0 is not true, then $\hat{\sigma}_0^2$ estimates some quantity larger than σ^2. Thus, we would expect a value of F that is close to $1 (= \sigma^2/\sigma^2)$ if H_0 is true, but larger than 1 if H_0 is not true. The larger the value of F, then, the likelier H_0 is to be untrue.

Another general characteristic of the tests to be discussed in this chapter is that *each test can be interpreted as a comparison of two models.* One of these models will be referred to as the *full* or *complete* model; the other will be called the *reduced* model (i.e., the model to which the complete model reduces under the null hypothesis).

As a simple example, consider the following two models:

$$Y = \beta_0 + \beta_1 X_1 + \beta_2 X_2 + E$$

and

$$Y = \beta_0 + \beta_1 X_1 + E$$

Under H_0: $\beta_2 = 0$, the larger (full) model reduces to the smaller (reduced) model. A test of H_0: $\beta_2 = 0$ is thus essentially equivalent to determining which of these two models is more appropriate.

As this example suggests, the set of independent variables in the reduced model (namely, X_1) is a subset of the independent variables in the full model (namely, X_1 and X_2). This is a characteristic common to all the basic types of tests to be described in this chapter. (More generally, this subset characteristic need not always be present. Suppose, for example, that we have H_0: $\beta_1 = \beta_2$. Then, the reduced model may be written as $Y = \beta_0 + \beta X + E$, with $\beta = \beta_1 = \beta_2$ and $X = X_1 + X_2$.)

9-2 Test for Significant Overall Regression

We now reconsider our first question, regarding an overall test for a model containing k independent variables—say,

$$Y = \beta_0 + \beta_1 X_1 + \beta_2 X_2 + \cdots + \beta_k X_k + E$$

The null hypothesis for this test may be generally stated as H_0: "All k independent variables considered together do not explain a significant amount of the variation in Y." Equivalently, we may state the null hypothesis as H_0: "There is no significant overall regression using all k independent variables in the model," or as H_0: $\beta_1 = \beta_2 = \cdots = \beta_k = 0$. Under this last version of H_0, the full model is reduced to a model that contains only the intercept term β_0.

To perform the test, we use the mean-square quantities provided in our ANOVA table (see Table 8-2 of Chapter 8). We calculate the F statistic

$$F = \frac{\text{MS regression}}{\text{MS residual}} = \frac{(\text{SSY} - \text{SSE})/k}{\text{SSE}/(n - k - 1)} \tag{9.1}$$

where $\text{SSY} = \sum_{i=1}^{n}(Y_i - \overline{Y})^2$ and $\text{SSE} = \sum_{i=1}^{n}(Y_i - \hat{Y}_i)^2$ are the total and error sums of squares, respectively. The computed value of F can then be compared with the critical point $F_{k,n-k-1, 1-\alpha}$, where α is the preselected significance level. We would reject H_0 if the computed F exceeded the critical point. Alternatively, we could compute the P-value for this test as the area under the curve of the $F_{k,n-k-1}$ distribution to the right of the computed F statistic. It can be shown that an equivalent expression for (9.1) in terms of R^2 is

$$F = \frac{R^2/k}{(1 - R^2)/(n - k - 1)} \qquad (9.2)$$

For the example summarized in Table 8-2, which concerns the regression of WGT on HGT, AGE, and $(AGE)^2$ for a sample of $n = 12$ children, we have $k = 3$, MS regression $= 231.02$, MS residual $= 24.40$, and $R^2 = 0.7802$, so that

$$F = \frac{231.02}{24.40} = \frac{0.7802/3}{(1 - 0.7802)/(12 - 3 - 1)} = 9.47$$

The critical point for $\alpha = .01$ is $F_{3,8,0.99} = 7.59$. Thus, we would reject H_0 at $\alpha = .01$, because the P-value is less than .01. (We usually denote $P < .01$ by putting a double ** next to the computed F, as in Table 8-2. When $.01 < P < .05$, we usually use only one *.)

In interpreting the results of this test, we can conclude that, based on the observed data, the set of variables HGT, AGE, and $(AGE)^2$ significantly help to predict WGT. This conclusion does not mean that *all three* variables are needed for significant prediction of Y; perhaps only one or two of them are sufficient. In other words, a more parsimonious model than the one involving all three variables may be adequate. To determine this requires further tests, to be described in the next section.

The mean-square residual term in the overall F test, which is the denominator of the F in (9.1), is given by the formula

$$\frac{1}{n - k - 1} \text{SSE} = \frac{1}{n - k - 1} \sum_{i=1}^{n} (Y_i - \hat{Y}_i)^2$$

This quantity provides an estimate of σ^2 under the assumed model. The mean-square regression term $\sum_{i=1}^{n}(\hat{Y}_i - \bar{Y})^2/k$, which is the numerator of the F in (9.1), provides an independent estimate of σ^2 only if the null hypothesis of no significant overall regression is true. Otherwise, the numerator overestimates σ^2 in direct proportion to the absolute values of the regression coefficients $\beta_1, \beta_2, \ldots, \beta_k$; this is why an F-value that is "too large" favors rejection of H_0. Thus, the F statistic (9.1) is the ratio of two independent estimates of the same variance only if the null hypothesis $H_0: \beta_1 = \beta_2 = \cdots = \beta_k = 0$ is true.

9-3 Partial F Test

Some important additional information regarding the fitted regression model can be obtained by presenting the ANOVA table as shown in Table 9-1. In this representation, we have partitioned the regression sum of squares into three components:

1. $\text{SS}(X_1)$: the sum of squares explained by using only $X_1 = $ HGT to predict Y.

2. $\text{SS}(X_2 \mid X_1)$: the extra sum of squares explained by using $X_2 = $ AGE in addition to X_1 to predict Y.

3. $\text{SS}(X_3 \mid X_1, X_2)$: the extra sum of squares explained by using $X_3 = (AGE)^2$ in addition to X_1 and X_2 to predict Y.

We can use the extra information in the table to answer the following questions:

1. Does $X_1 = $ HGT alone significantly aid in predicting Y?

2. Does the addition of $X_2 = $ AGE significantly contribute to the prediction of Y after we account (or control) for the contribution of X_1?

3. Does the addition of $X_3 = (\text{AGE})^2$ significantly contribute to the prediction of Y after we account for the contribution of X_1 and X_2?

To answer question 1, we simply fit the straight-line regression model, using $X_1 = $ HGT as the single independent variable. The value 588.92, therefore, is the regression sum of squares for this straight-line regression model. The SSE for this model can be obtained from Table 9-1 by adding 195.19, 103.90, and 0.24 together, which yields the sum of squares 299.33, having 10 degrees of freedom (i.e., $10 = 8 + 1 + 1$). The F statistic for testing whether there is significant straight-line regression when we use only $X_1 = $ HGT is then given by $F = (588.92/1)/(299.33/10) = 19.67$, which has a P-value of less than .01 (i.e., X_1 contributes significantly to the linear prediction of Y).

To answer questions 2 and 3, we must use what is called a *partial F test.* This test assesses whether the addition of any specific independent variable, <u>given others already in the model,</u> significantly contributes to the prediction of Y. The test, therefore, allows for the deletion of variables that do not help in predicting Y and thus enables one to reduce the set of possible independent variables to an economical set of "important" predictors.

9-3-1 The Null Hypothesis

Suppose that we wish to test whether adding a variable X^* significantly improves the prediction of Y, given that variables $X_1, X_2,..., X_p$ are already in the model. The null hypothesis may then be stated as H_0: "X^* does not significantly add to the prediction of Y, given that $X_1, X_2,..., X_p$ are already in the model," or equivalently, as H_0: $\beta^* = 0$ in the model $Y = \beta_0 + \beta_1 X_1 + \beta_2 X_2 + \cdots + \beta_p X_p + \beta^* X^* + E$.

As can be inferred from the second statement, the test procedure essentially compares two models: the *full* model contains $X_1, X_2,..., X_p$ and X^* as independent variables; the *reduced* model contains $X_1, X_2,..., X_p$, but not X^* (since $\beta^* = 0$ under the null hypothesis). The goal is to determine which model is more appropriate based on how much additional information X^* provides about Y over that already provided by $X_1, X_2,..., X_p$. In the next chapter, we shall see that an equivalent statement of H_0 can be given in terms of a partial correlation coefficient.

TABLE 9-1   ANOVA table for WGT regressed on HGT, AGE, and $(\text{AGE})^2$, containing components of the regression sum of squares.

Source		d.f.	SS	MS	F	R^2
Regression	X_1	1	588.92	588.92	19.67**	0.7802
	$X_2 \mid X_1$	1	103.90	103.90	4.78 $(.05 < P < .10)$	
	$X_3 \mid X_1, X_2$	1	0.24	0.24	0.01	
Residual		8	195.19	24.40		
Total		11	888.25			

9-3-2 The Procedure

To perform a partial F test involving a variable X^*, given that variables $X_1, X_2,\ldots, X_p$ are already in the model, we must first compute the extra sum of squares from adding X^*, given $X_1, X_2,\ldots, X_p$, which we place in our ANOVA table under the source heading "Regression $X^* \mid X_1, X_2,\ldots, X_p$." This sum of squares is computed by the formula

$$
\begin{bmatrix} \text{Extra sum of} \\ \text{squares from} \\ \text{adding } X^*, \text{ given} \\ X_1, X_2,\ldots, X_p \end{bmatrix} = \begin{bmatrix} \text{Regression sum} \\ \text{of squares when} \\ X_1, X_2,\ldots, X_p \\ \text{and } X^* \text{ are } all \\ \text{in the model} \end{bmatrix} - \begin{bmatrix} \text{Regression sum} \\ \text{of squares when} \\ X_1, X_2,\ldots, X_p \\ (\text{and } not \ X^*) \text{ are} \\ \text{in the model} \end{bmatrix} \tag{9.3}
$$

or, more compactly

$$
\boxed{\begin{aligned} \text{SS}(X^* \mid X_1, X_2,\ldots, X_p) &= \text{Regression SS}(X_1, X_2,\ldots, X_p, X^*) \\ &\quad - \text{Regression SS}(X_1, X_2,\ldots, X_p) \end{aligned}}
$$

[For any model, $\sum_{i=1}^{n}(Y_i - \overline{Y})^2$ can be split into two components—the regression sum of squares and the residual sum of squares. Therefore,

$$
\text{SS}(X^* \mid X_1, X_2,\ldots, X_p) = \text{Residual SS}(X_1, X_2,\ldots, X_p) - \text{Residual SS}(X_1, X_2,\ldots, X_p, X^*)
$$

is an equivalent expression.]

Thus, for our example,

$$
\begin{aligned} \text{SS}(X_2 \mid X_1) &= \text{Regression SS}(X_1, X_2) - \text{Regression SS}(X_1) \\ &= 692.82 - 588.92 \\ &= 103.90 \end{aligned}
$$

and

$$
\begin{aligned} \text{SS}(X_3 \mid X_1, X_2) &= \text{Regression SS}(X_1, X_2, X_3) - \text{Regression SS}(X_1, X_2) \\ &= 693.06 - 692.82 \\ &= 0.24 \end{aligned}
$$

To test the null hypothesis H_0: "The addition of X^* to a model already containing $X_1, X_2,\ldots, X_p$ does not significantly improve the prediction of Y," we compute

$$
F(X^* \mid X_1, X_2,\ldots, X_p) = \frac{\text{Extra sum of squares from adding } X^*, \text{ given } X_1, X_2,\ldots, X_p}{\substack{\text{Mean-square residual for the model} \\ \text{containing all the variables } X_1, X_2,\ldots, X_p, X^*}}
$$

or more compactly,

$$
\boxed{F(X^* \mid X_1, X_2,\ldots, X_p) = \frac{\text{SS}(X^* \mid X_1, X_2,\ldots, X_p)}{\text{MS residual } (X_1, X_2,\ldots, X_p, X^*)}} \tag{9.4}
$$

This F statistic has an F distribution with 1 and $n - p - 2$ degrees of freedom under H_0, so we should reject H_0 if the computed F exceeds $F_{1,n-p-2,1-\alpha}$. For our example, the partial F statistics are (from Table 9-1)

$$F(X_2 \mid X_1) = \frac{SS(X_2 \mid X_1)}{MS \text{ residual } (X_1, X_2)} = \frac{103.90}{(195.19 + 0.24)/9} = 4.78$$

and

$$F(X_3 \mid X_1, X_2) = \frac{SS(X_3 \mid X_1, X_2)}{MS \text{ residual } (X_1, X_2, X_3)} = \frac{0.24}{24.40} = 0.01$$

The quantity MS residual (X_1, X_2) can be obtained directly from the ANOVA table for only X_1 and X_2 or indirectly from the partitioned ANOVA table for $X_1, X_2,$ and X_3 by using the formula

$$MS \text{ residual } (X_1, X_2) = \frac{\text{Residual } SS(X_1, X_2, X_3) + SS(X_3 \mid X_1, X_2)}{8 + 1}$$

The statistic $F(X_2 \mid X_1) = 4.78$ has a P-value satisfying $.05 < P < .10$, since $F_{1,9,0.90} = 3.36$ and $F_{1,9,0.95} = 5.12$. Thus, we should reject H_0 at $\alpha = .10$ and conclude that the addition of X_2 after accounting for X_1 significantly adds to the prediction of Y at the $\alpha = .10$ level. At $\alpha = .05$, however, we would not reject H_0.

The statistic $F(X_3 \mid X_1, X_2)$ equals 0.01, so obviously H_0 should not be rejected regardless of the significance level; we therefore conclude that, once $X_1 = $ HGT and $X_2 = $ AGE are in the model, the addition of $X_3 = (\text{AGE})^2$ is superfluous.

9-3-3 The t Test Alternative

An equivalent way to perform the partial F test for the variable added last is to use a t test. (You may recall that an F statistic with 1 and $n - k - 1$ degrees of freedom is the square of a t statistic with $n - k - 1$ degrees of freedom.) The t test alternative focuses on a test of the null hypothesis $H_0: \beta^* = 0$, where β^* is the coefficient of X^* in the regression equation $Y = \beta_0 + \beta_1 X_1 + \beta_2 X_2 + \cdots + \beta_p X_p + \beta^* X^* + E$. The equivalent statistic for testing this null hypothesis is

$$T = \frac{\hat{\beta}^*}{S_{\hat{\beta}^*}} \tag{9.5}$$

where $\hat{\beta}^*$ is the corresponding estimated coefficient and $S_{\hat{\beta}^*}$ is the estimate of the standard error of $\hat{\beta}^*$, both of which are printed by standard regression programs.

In performing this test, we reject $H_0: \beta^* = 0$ if

$$\begin{cases} |T| > t_{n-p-2,1-\alpha/2} & \text{(two-sided test; } H_A: \beta^* \neq 0) \\ T > t_{n-p-2,1-\alpha} & \text{(upper one-sided test; } H_A: \beta^* > 0) \\ T < -t_{n-p-2,1-\alpha} & \text{(lower one-sided test; } H_A: \beta^* < 0) \end{cases}$$

It can be shown that a two-sided t test is equivalent to the partial F test described earlier. For example, in testing $H_0: \beta_3 = 0$ in the model $Y = \beta_0 + \beta_1 X_1 + \beta_2 X_2 + \beta_3 X_3 + E$ fit to the data in Table 8-1, we compute

$$T = \frac{\hat{\beta}_3}{S_{\hat{\beta}_3}} = \frac{-0.0417}{0.4224} = -0.10$$

Squaring, we get

$$T^2 = 0.01 = \text{partial } F(X_3 \mid X_1, X_2)$$

from Table 9-1.

9-3-4 Comments

An important general application of the partial F test concerns the control of extraneous variables (e.g., confounders, which will be discussed in Chapter 11). Consider, for example, a situation with one main study variable of interest, S, and p control variables $C_1, C_2,\ldots, C_p$. The effect of S on the outcome variable Y, controlling for $C_1, C_2,\ldots, C_p$, may be assessed by considering the model

$$Y = \beta_0 + \beta_1 C_1 + \beta_2 C_2 + \cdots + \beta_p C_p + \beta_{p+1} S + E$$

The appropriate null hypothesis is H_0: $\beta_{p+1} = 0$. The partial F statistic in this situation is given by $F(S \mid C_1, C_2,\ldots, C_p)$, using (9.4) with $X^* = S$ and $C_i = X_i$, for $i = 1, 2,\ldots, p$.

When several study variables (i.e., several S's) are involved, the task includes determining which of the S's are important and perhaps even rank-ordering them by their relative importance. Such a task amounts to finding a best model, a topic we will address in Chapter 16 (where the term *best* will be carefully defined). For now, we note that one strategy (detailed in Chapter 16) is to work backward by *deleting* S variables, one at a time, until a best model is obtained. This requires performing several partial F tests (as described in Chapter 16). If the starting model of interest is

$$Y = \beta_0 + \beta_1 C_1 + \cdots + \beta_p C_p + \beta_{p+1} S_1 + \cdots + \beta_{p+k} S_k + E$$

then the first backward step involves considering k partial F tests, $F(S_i \mid C_1, C_2,\ldots, C_p, S_1, S_2,\ldots, S_k$ except $S_i)$, where $i = 1, 2,\ldots, k$. The corresponding (separate) null hypotheses are H_0: $\beta_i = 0$, where $i = p + 1, p + 2,\ldots, p + k$. The usual backward procedure identifies the variable S_I associated with the smallest partial F value. This variable becomes the first to be deleted from the model, provided that its partial F is not significant. Then the elimination process starts all over again for the reduced model with S_I removed. Of course, if the smallest partial F value is significant, no S variables are deleted.

Each partial F test made at the first backward step weighs the contribution of a specific S variable, given that it is the last S variable to enter the model. It is therefore inappropriate to delete more than one S variable at this first step. For example, it is inappropriate to delete simultaneously *all* S variables from the model if all partial F's are nonsignificant at this first step. This is because, given that one particular S variable (say, S_I) is deleted, the remaining S variables may become important (based on consideration of their partial F's under the reduced model).

For example, suppose that we fit the model

$$Y = \beta_0 + \beta_1 C_1 + \beta_2 S_1 + \beta_3 S_2 + E$$

and obtain the following partial F results:

$$F(S_1 \mid C_1, S_2) = 0.01 \ (P = .90)$$
$$F(S_2 \mid C_1, S_1) = 0.85 \ (P = .25)$$

Then, S_1 is "less significant" than S_2, controlling for C_1 and the other S variable. Under the strategy of backward elimination, S_1 should be deleted before the elimination of S_2 is considered; however, to delete both S_1 and S_2 at this point would be incorrect. In fact, when considering the reduced model $Y = \beta_0 + \beta_1 C_1 + \beta_2 S_2 + E$, the partial F statistic $F(S_2 \mid C_1)$ may be highly significant. In other words, if S_1 is not significant given S_2 and C_1, and if S_2 is not significant given S_1 and C_1, it does not necessarily follow that S_2 is unimportant in a reduced model containing S_2 and C_1 but not S_1.

9-4 Multiple Partial *F* Test

This testing procedure addresses the more general problem of assessing the additional contribution of two or more independent variables over and above the contribution made by other variables already in the model. For the example involving $Y = \text{WGT}$, $X_1 = \text{HGT}$, $X_2 = \text{AGE}$, and $X_3 = (\text{AGE})^2$, we may be interested in testing whether the AGE variables, taken collectively, significantly improve the prediction of WGT given that HGT is already in the model. In contrast to the partial F test discussed in section 9-3, the multiple partial F test addresses the simultaneous addition of two or more variables to a model. Nevertheless, the test procedure is a straightforward extension of the partial F test.

9-4-1 The Null Hypothesis

We wish to test whether the addition of the k variables $X_1^*, X_2^*, \ldots, X_k^*$ significantly improves the prediction of Y, given that the p variables $X_1, X_2, \ldots, X_p$ are already in the model. The (full) model of interest is thus

$$Y = \beta_0 + \beta_1 X_1 + \beta_p X_p + \beta_1^* X_1^* + \cdots + \beta_k^* X_k^* + E$$

Then, the null hypothesis of interest may be stated as H_0: "$X_1^*, X_2^*, \ldots, X_k^*$ do not significantly add to the prediction of Y given that $X_1, X_2, \ldots, X_p$ are already in the model," or equivalently, H_0: $\beta_1^* = \beta_2^* = \cdots = \beta_k^* = 0$ in the (full) model.[1]

From the second version of H_0, it follows that the *reduced* model is of the form

$$Y = \beta_0 + \beta_1 X_1 + \beta_2 X_2 + \cdots + \beta_p X_p + E$$

(i.e., the X_i^* terms are dropped from the full model).

For the preceding example, the (full) model is

$$\text{WGT} = \beta_0 + \beta_1 \text{HGT} + \beta_1^* \text{AGE} + \beta_2^* (\text{AGE})^2 + E$$

The null hypothesis here is H_0: $\beta_1^* = \beta_2^* = 0$.

9-4-2 The Procedure

As in the case of the partial F test, we must compute the extra sum of squares due to the addition of the X_i^* terms to the model. In particular, we have

$$\text{SS}(X_1^*, X_2^*, \ldots, X_k^* \mid X_1, X_2, \ldots, X_p)$$
$$= \text{Regression SS}(X_1, X_2, \ldots, X_p, X_1^*, X_2^*, \ldots, X_k^*) - \text{Regression SS}(X_1, X_2, \ldots, X_p)$$
$$= \text{Residual SS}(X_1, X_2, \ldots, X_p) - \text{Residual SS}(X_1, X_2, \ldots, X_p, X_1^*, X_2^*, \ldots, X_k^*)$$

---

[1] In Chapter 10, an equivalent expression for this null hypothesis will be given in terms of a multiple partial correlation coefficient.

Using this extra sum of squares, we obtain the following F statistic:

$$F(X_1^*, X_2^*, \ldots, X_k^* \mid X_1, X_2, \ldots, X_p) = \frac{SS(X_1^*, X_2^*, \ldots, X_k^* \mid X_1, X_2, \ldots, X_p)/k}{MS \text{ residual } (X_1, X_2, \ldots, X_p, X_1^*, X_2^*, \ldots, X_k^*)} \quad (9.6)$$

This F statistic has an F distribution with k and $n - p - k - 1$ degrees of freedom under H_0: $\beta_1^* = \beta_2^* = \ldots = \beta_k^* = 0$.

In (9.6), we must divide the extra sum of squares by k, the number of regression coefficients specified to be zero under the null hypothesis of interest. This number k is also the numerator degrees of freedom for the F statistic. The denominator of the F statistic is the mean-square residual for the full model; its degrees of freedom is $n - (p + k + 1)$, which is $n - 1$ minus the number of variables in this model (namely, $p + k$).

An alternative way to write this F statistic is

$$F(X_1^*, X_2^*, \ldots, X_k^* \mid X_1, X_2, \ldots, X_p) = \frac{[\text{Regression SS(full)} - \text{Regression SS(reduced)}]/k}{MS \text{ residual (full)}}$$

$$= \frac{[\text{Residual SS(reduced)} - \text{Residual SS(full)}]/k}{MS \text{ residual (full)}}$$

Using the information in Table 9-1 involving WGT, HGT, AGE, and $(AGE)^2$, we can test H_0: $\beta_1^* = \beta_2^* = 0$ in the model $WGT = \beta_0 + \beta_1 HGT + \beta_1^* AGE + \beta_2^* (AGE)^2 + E$, as follows:

$$F(AGE, (AGE)^2 \mid HGT)$$

$$= \frac{\{\text{Regression SS}[HGT, AGE, (AGE)^2] - \text{Regression SS}(HGT)\}/2}{MS \text{ residual } [HGT, AGE, (AGE)^2]}$$

$$= \frac{[(588.92 + 103.90 + 0.24) - 588.92]/2}{24.40}$$

$$= 2.13$$

For $\alpha = .05$, the critical point is

$$F_{k, n-p-k-1, 0.95} = F_{2, 12-1-2-1, 0.95} = 4.46$$

so H_0 would not be rejected at $\alpha = .05$.

In the preceding calculation, we used the relationship

Regression SS$[HGT, AGE, (AGE)^2]$

= Regression SS(HGT) + Regression SS(AGE | HGT)

   + Regression SS$[(AGE)^2 \mid HGT, AGE)]$

= 588.92 + 103.90 + 0.24

Alternatively, we could form two ANOVA tables (Table 9-2)—one for the full and one for the reduced model—and then extract the appropriate regression and/or residual sum-of-square terms from these tables. More examples of partial F calculations are given at the end of this chapter.

9-4-3 Comments

Like the partial F test, the multiple partial F test is useful for assessing the importance of extraneous variables. In particular, it is often used to test whether a "chunk" (i.e., a group) of

TABLE 9-2   ANOVA tables for WGT regressed on HGT, AGE, and $(AGE)^2$.

Full Model					Reduced Model			
Source	d.f.	SS	MS		Source	d.f.	SS	MS
Regression (X_1, X_2, X_3)	3	693.06	231.02		Regression (X_1)	1	588.92	588.92
Residual	8	195.19	24.40		Residual	10	299.33	29.93
Total	11	888.25			Total	11	888.25	

variables having some trait in common is important when considered together. An example of a chunk is a collection of variables that are all of a certain order (e.g., $(AGE)^2$, HGT $\times$ AGE, and $(HGT)^2$ are all of order 2).

Another example is a collection of two-way product terms (e.g., X_1X_2, X_1X_3, X_2X_3); this latter group is sometimes referred to as a set of interaction variables (see Chapter 11). It is often of interest to assess the importance of interaction effects collectively before trying to consider individual interaction terms in a model. In fact, the initial use of such a chunk test can reduce the total number of tests to be performed, since variables may be dropped from the model as a group. This, in turn, helps provide better control of overall Type I error rates, which may be inflated due to multiple testing (Abt 1981; Kupper, Stewart, and Williams 1976).

9-5 Strategies for Using Partial F Tests

In applying the ideas presented in this chapter, readers will typically use a computer program to carry out the numerical calculations required. Therefore, we will briefly describe the computer output for typical regression programs. To help readers understand and use such output, we discuss two strategies for using partial F tests: *variables-added-in-order tests* and *variables-added-last tests*.

The accompanying computer output is from a typical regression computer program[2] for the model

$$\text{WGT} = \beta_0 + \beta_1\text{HGT} + \beta_2\text{AGE} + \beta_3(\text{AGE})^2 + E$$

The results here were computed with *centered* predictors[3] (see section 12-5-2), so $(\text{HGT} - 52.75)$, $(\text{AGE} - 8.833)$, and $(\text{AGE} - 8.833)^2$ were used, with mean HGT $= 52.75$ and mean AGE $= 8.833$. The computer output consists of five sections, labeled A through E. Section A provides the overall ANOVA table for the regression model. Computer output typically presents numbers with far more significant digits than can be justified. Section B provides a test for significant overall regression, the multiple R^2-value, the mean $(\overline{Y})$ of the dependent variable (WGT), the

---

[2]This particular output was produced by the SAS program GLM.

[3]*Centering* is the process of transforming a variable, say AGE, by subtracting from AGE its sample mean, e.g., 8.833 in the preceding data. The newly defined centered variable $(\text{AGE} - 8.833)$ will then have zero as its overall mean. Centering is frequently done to gain computational accuracy when estimating higher-order (e.g., polynomial) terms in one's model. See section 12-5-1 for further discussion.

WGT residual standard deviation or "root-mean-square error" (*s*), and the coefficient of variation $(100s/\overline{Y})$.

SAS Output for Data from Table 8-1 (HGT and AGE Centered)

```
                        General Linear Models Procedure
Dependent Variable: Weight        Body Weight in Pounds
Source                   DF          Sum of Squares        Mean Square⌐
Model                     3           693.06046340       231.02015447 │
Error                     8           195.18953660        24.39869208 ├A
Corrected Total          11           888.25000000                    ┘

Model F =               9.47                           PR > F = 0.0052⌐
R-Square              C.V.              Root MSE            Weight Mean ├B
0.780254             7.8717            4.93950322          62.75000000 ┘

Source         DF           Type I SS        F Value          Pr > F⌐
Height          1          588.92252318        24.14          0.0012 │
Age             1          103.90008336         4.26          0.0730 ├C
Age*Age         1            0.23785686         0.01          0.9238 ┘

Source         DF          Type III SS       F Value          Pr > F⌐
Height          1          166.58195495         6.83          0.0310 │
Age             1          101.80889273         4.17          0.0754 ├D
Age*Age         1            0.23785686         0.01          0.9238 ┘

                             T for H0:                     Std Error of⌐
Parameter       Estimate   Parameter=0      PR > T           Estimate  │
Intercept     62.88786380       31.51       0.0001         1.99570812  │
Height         0.72369024        2.61       0.0310         0.27696316  ├E
Age            2.04005621        2.04       0.0754         0.99869425  │
Age*Age       -0.04170670       -0.10       0.9238         0.42240715  ┘
```

Section C provides certain (Type I) tests for assessing the importance of each predictor in the model; section D provides a different set of (Type III) tests regarding these predictors; and section E provides yet a third set of (*t*) tests.

9-5-1 Basic Principles

Two methods (or strategies) are widely used for evaluating whether a variable should be included in a model: partial (Type I) *F* tests for variables added in order, and partial (Type III) *F* tests for variables added last.[4] For the first (variables-added-in-order) method, the following procedure is employed: (1) an order for adding variables one at a time is specified; (2) the significance of the (straight-line) model involving only the variable ordered first is assessed; (3) the significance of adding the second variable to the model involving only the first variable is assessed;

---

[4]Searle (1971), among others, refers to these methods as "ignoring" and "eliminating" tests, respectively.

Instead of considering variables added in order, it may be of interest to consider *variables deleted in order*. The latter strategy would apply, for example, in polynomial regression, where a backward selection algorithm is used to determine the proper degree of the polynomial. As can be seen from section C of the accompanying computer output, the same set of sums of squares is produced whether the variables are considered to be added in one order or deleted in the reverse order.

(4) the significance of adding the third variable to the model containing the first and second variables is assessed; and so on.

For the second (variables-added-last) method, the following procedure is used: (1) an initial model containing two or more variables is specified; (2) the significance of each variable in the initial model is assessed separately, as if it were the last variable to enter the model (i.e., if k variables are in the initial model, then k variables-added-last tests are conducted). In either method, each test is conducted using a partial F test for the addition of a single variable.

Variables-added-in-order tests can be illustrated with the weight example. One possible ordering is HGT first, followed by AGE, and then $(AGE)^2$. For this ordering, the smallest model considered is

$$WGT = \beta_0 + \beta_1 HGT + E$$

The overall regression F test of H_0: $\beta_1 = 0$ is used to assess the contribution of HGT. Next, the model

$$WGT = \beta_0 + \beta_1 HGT + \beta_2 AGE + E$$

is fit. The significance of adding AGE to a model already containing HGT is then assessed by using the partial $F(AGE \mid HGT)$. Finally, the full model is fit by using HGT, AGE, and $(AGE)^2$. The importance of the last variable is tested with the partial $F[(AGE)^2 \mid HGT, AGE]$. The tests used are those discussed in this chapter and summarized in Table 9-1. These are also the tests provided in section C of the earlier computer output (using Type I sums of squares). However, each test in Table 9-1 involves a different residual sum of squares, while those in the computer output use a common residual sum of squares. More will be said about this issue shortly.

To describe variables-added-last tests, consider again the full model

$$WGT = \beta_0 + \beta_1 HGT + \beta_2 AGE + \beta_3 (AGE)^2 + E$$

The contribution of HGT, when added last, is assessed by comparing the full model to the model with HGT deleted—namely,

$$WGT = \beta_0 + \beta_2 AGE + \beta_3 (AGE)^2 + E$$

The partial F statistic, based on (9.4), has the form $F[(HGT \mid AGE, (AGE)^2]$. The sum of squares for HGT added last is then the difference in the error sum of squares (or the regression sum of squares) for the two preceding models. Similarly, the reduced model with AGE deleted is

$$WGT = \beta_0 + \beta_1 HGT + \beta_3 (AGE)^2 + E$$

for which the corresponding partial F statistic is $F[AGE \mid HGT, (AGE)^2]$; and the reduced model with $(AGE)^2$ omitted is

$$WGT = \beta_0 + \beta_1 HGT + \beta_2 AGE + E$$

for which the partial F statistic is $F[(AGE)^2 \mid HGT, AGE]$. The three F statistics just described are provided in section D of the earlier computer output (using Type III sums of squares).[5]

---

[5]It is important to remember that the preceding computer output was based on the use of the *centered* predictor variables $(HGT - 52.75)$, $(AGE - 8.833)$, and $(AGE - 8.833)^2$, and not on the use of the original (uncentered) predictor variables HGT, AGE, and $(AGE)^2$. As a result, certain numerical results (specifically, some of those in sections D and E) for the centered predictors differ from those for the uncentered predictors (see, e.g., the computer model for Model 6 on page 123 in Chapter 8).

An important characteristic of variables-added-in-order sums of squares is that they decompose the regression sum of squares into a set of mutually exclusive and exhaustive pieces. For example, the sums of squares provided in section C of the computer output (588.922, 103.900, and 0.238) add to 693.060, which is the regression sum of squares given in section A. The variables-added-last sums of squares do not generally have this property (e.g., the sums of squares given in section D of the computer output do not add to 693.060).

Each of these two testing strategies has its own advantages, and the situation being considered determines which is preferable. For example, if all variables are considered to be of equal importance, the variables-added-last tests are usually preferred. Such tests treat all variables equally; and because the importance of each variable is assessed as if it were the last variable to enter the model, the order of entry is not a consideration.

In contrast, if the order in which the predictors enter the model is an important consideration, the variables-added-in-order testing approach may be better. An example where the entry order is important is one where main effects (e.g., X_1, X_2, and X_3) are forced into the model, followed by their cross-products (X_1X_2, X_1X_3, and X_2X_3) or so-called interaction terms (see Chapter 11). Such tests evaluate the contribution of a variable and adjust *only* for the variables just preceding it into the model.

9-5-2 Commentary

As discussed in the preceding subsection, section C of the earlier computer output provides variables-added-in-order tests, which are also given for the same data in Table 9-1. Section D of the computer output provides variables-added-last tests. Finally, section E provides *t* tests (which are equivalent to the variables-added-last *F* tests in section D), as well as regression coefficient estimates and their standard errors, for the *centered* predictor variables.

Table 9-3 gives an ANOVA table for the variables-added-last tests for the weight example. (We recommend that readers consider how this table was extracted from the earlier computer output.) The variables-added-last tests usually give a different ANOVA table from one based on the variables-added-in-order tests. A different residual sum of squares is used for each variables-added-in-order test in Table 9-1, while the same residual sum of squares (based on the three-variable model involving the *centered* versions of HGT, AGE, and $(AGE)^2$) is used for all the variables-added-last tests in Table 9-3.

An argument can be made that it is preferable to use the residual sum of squares for the three-variable model (i.e., the largest model containing all candidate predictors) for all tests. This is because the error variance σ^2 will *not* be correctly estimated by a model that ignores important predictors, but it will be correctly estimated (under the usual regression assumptions)

TABLE 9-3   ANOVA table for WGT regressed on HGT, AGE, and $(AGE)^2$, using variables-added-last tests.

Source	d.f.	SS	MS	F	R^2
$X_1 \mid X_2, X_3$	1	166.58	166.58	6.83*	.7802
$X_2 \mid X_1, X_3$	1	101.81	101.81	4.17	
$X_3 \mid X_1, X_2$	1	0.24	0.24	0.01	
Residual	8	195.19	24.40		
Total	11	888.25			

* Exceeds .05 critical value of 5.32 for *F* with 1 and 8 degrees of freedom.

by a model that contains all candidate predictors (even if some are not important). In other words, overfitting a model in estimating σ^2 is safer than underfitting it. Of course, extreme overfitting results in lost precision, but it still provides a valid estimate of residual variation. We generally prefer using the residual sum of squares based on fitting the "largest" model, although some statisticians disagree.

9-5-3 Models Underlying the Source Tables

Tables 9-4 and 9-5 present the models being compared based on the earlier computer output and the associated ANOVA tables. Table 9-4 summarizes the models and residual sums of squares needed to conduct variables-added-last tests for the full model containing the *centered* versions of HGT, AGE, and $(AGE)^2$. Table 9-5 lists the models that must be fitted to provide variables-added-in-order tests for the order of entry HGT, then AGE, and then $(AGE)^2$.

Table 9-6 details computations of regression sums of squares for both types of tests. For example, the first line, where $24,226.2637 = 24,421.4532 - 195.1895$, is the difference in the error sums of squares for models 1 and 5 given in Table 9-4. These results can then be used to produce any of the F tests given in Tables 9-1 and 9-2 and in the computer output.

TABLE 9-4 Variables-added-last regression models and residual sums of squares for data from Table 8-1.

Model No.	Model		SSE
1	WGT =	$\beta_1 HGT + \beta_2 AGE + \beta_3 (AGE)^2 + E$	24,421.4532
2	WGT = β_0	$+ \beta_2 AGE + \beta_3 (AGE)^2 + E$	361.7715
3	WGT = $\beta_0 + \beta_1 HGT$	$+ \beta_3 (AGE)^2 + E$	297.0071
4	WGT = $\beta_0 + \beta_1 HGT + \beta_2 AGE$	$+ E$	195.4274
5	WGT = $\beta_0 + \beta_1 HGT + \beta_2 AGE + \beta_3 (AGE)^2 + E$		195.1895

TABLE 9-5 Variables-added-in-order regression models and residual sums of squares for data from Table 8-1.

Model No.	Model		SSE
6	WGT =	$+ E$	48,139.0000
7	WGT = β_0	$+ E$	888.2500
8	WGT = $\beta_0 + \beta_1 HGT$	$+ E$	299.3275
4	WGT = $\beta_0 + \beta_1 HGT + \beta_2 AGE$	$+ E$	195.4274
5	WGT = $\beta_0 + \beta_1 HGT + \beta_2 AGE + \beta_3 (AGE)^2 + E$		195.1895

TABLE 9-6 Computations for regression sum of squares for data from Table 8-1.

		Regression SS	
Parameter	Variable	Added Last	In Order
β_0	Intercept	SSE(1) $-$ SSE(5) = 24,226.2637	SSE(6) $-$ SSE(7) = 47,250.7500
β_1	HGT	SSE(2) $-$ SSE(5) = 166.5820	SSE(7) $-$ SSE(8) = 588.9225
β_2	AGE	SSE(3) $-$ SSE(5) = 101.8176	SSE(8) $-$ SSE(4) = 103.9001
β_3	$(AGE)^2$	SSE(4) $-$ SSE(5) = 0.2379	SSE(4) $-$ SSE(5) = 0.2379

9-6 Tests Involving the Intercept

Inferences about the intercept β_0 are occasionally of interest in multiple regression analysis. A test of H_0: $\beta_0 = 0$ is usually carried out with an intercept-added-last test, although an intercept-added-in-order test is also feasible (where the intercept is the first term added to the model). Many computer programs provide only a t test involving the intercept. The t-test statistic for the intercept in the earlier computer output corresponds exactly to a partial F test for adding the intercept last. The two models being compared are

$$Y = \beta_0 + \beta_1 X_1 + \beta_2 X_2 + \cdots + \beta_k X_k + E$$

and

$$Y = \beta_1 X_1 + \beta_2 X_2 + \cdots + \beta_k X_k + E$$

The null hypothesis of interest is H_0: $\beta_0 = 0$ versus H_A: $\beta_0 \neq 0$. The test is computed as

$$F = \frac{(\text{SSE without } \beta_0 - \text{SSE with } \beta_0)/1}{\text{SSE with } \beta_0/(n - k - 1)}$$

This F statistic has 1 and $n - k - 1$ degrees of freedom and is equal to the square of the t statistic used for testing H_0: $\beta_0 = 0$. For the weight example involving the *centered* predictors, an intercept-added-last test is reported in the output on page 148 as a t test. The corresponding partial F equals $(31.51)^2 = 992.88$ and has 1 and 8 degrees of freedom.

An intercept-added-in-order test can also be conducted. In this case, the two models being compared are

$$Y = E$$

and

$$Y = \beta_0 + E$$

Again, the null hypothesis is H_0: $\beta_0 = 0$ versus the alternative H_A: $\beta_0 \neq 0$. The special nature of this test leads to the simple expression

$$F = \frac{n\overline{Y}^2}{\text{SSY}/(n - 1)}$$

which represents an F statistic with 1 and $n - 1$ degrees of freedom. This statistic involves SSY, the residual sum of squares from a model with just an intercept (such as model 7 in Table 9-5). Alternatively, the residual sum of squares from the "largest" model may be used. When we use the latter approach, the F statistic for the weight data becomes (see Table 9-5)

$$F = \frac{[\text{SSE}(6) - \text{SSE}(7)]/1}{\text{SSE}(5)/8}$$

$$= \frac{(48{,}139.00 - 888.25)/1}{195.1895/8}$$

$$= 1936.61$$

where SSE(5) denotes the residual (i.e., error) sum of squares for model 5, which is the largest of the models in Table 9-5, with 1 and 8 degrees of freedom. In general, using the residual from the largest model (with k predictors) gives $n - k - 1$ error degrees of freedom, so the F statistic is compared to a critical value with 1 and $n - k - 1$ degrees of freedom.

Problems

1. Use the data of Problem 1 in Chapter 8 to answer the following questions.
   a. Conduct the overall F tests for significant regression for the three models in Problem 1 of Chapter 8. Be sure to state the null and alternative hypotheses for each test, and interpret each result.
   b. Based on your results in part (a), which of the three models is best? Compare your answer here with your answer to Problem 1(b) in Chapter 8.

2. Use the information given in Problem 2 of Chapter 8 as well as the computer output given here to answer the following questions about the data from that problem.
   a. Conduct overall regression F tests for these three models: Y regressed on X_1 and X_2; Y regressed on X_1 alone; and Y regressed on X_2 alone. Based on your answers, how would you rate the importance of the two variables in predicting Y? Comment on how your answer compares with the answer to Problem 2(b) in Chapter 8.
   b. Provide variables-added-in-order tests for both variables, with thinking disturbances (X_1) added first. Use $\alpha = .05$.
   c. Provide variables-added-in-order tests for both variables, with hostile suspiciousness (X_2) added first. Use $\alpha = .05$.
   d. Provide a table of variables-added-last tests.
   e. What, if any, differences are present in the approaches in parts (a) through (d)?
   f. Which predictors appear to be necessary? Why?

Edited SAS Output (PROC GLM) for Problem 2

```
                    Y Regressed on X1 and X2
                 General Linear Models Procedure
Source                    DF       Sum of Squares    F Value      Pr > F
Model                      2       2753.87135941       6.24       0.0038
Error                     50      11037.29845191
Corrected Total           52      13791.16981132

   R-Square                       C.V.                        Y Mean
   0.199684                     65.45708                    22.6981132

Source          DF              Type I SS        F Value      Pr > F
X1               1            1535.85696503         6.96       0.0111
X2               1            1218.01439439         5.52       0.0228

Source          DF            Type III SS        F Value      Pr > F
X1               1            2596.02402494        11.76       0.0012
X2               1            1218.01439439         5.52       0.0228

                                   T for H0:                 Std Error
Parameter         Estimate        Parameter=0     Pr > |T|   of Estimate
INTERCEPT       -0.63535020          -0.03         0.9759     20.96833367
X1              23.45143521           3.43         0.0012      6.83850964
X2              -7.07260920          -2.35         0.0228      3.01092434
```

3. A psychologist examined the regression relationship between anxiety level (Y)—measured on a scale ranging from 1 to 50, as the average of an index determined at three points in a 2-week period—and the following three independent variables: $X_1 =$ Systolic blood pressure; $X_2 =$ IQ; and $X_3 =$ Job satisfaction (measured on a scale ranging from 1 to 25). The

following ANOVA table summarizes results obtained from a variables-added-in-order regression analysis on data involving 22 outpatients who were undergoing therapy at a certain clinic.

Source		d.f.	SS
Regression $\begin{cases} (X_1) \\ (X_2 \mid X_1) \\ (X_3 \mid X_1, X_2) \end{cases}$	(X_1)	1	981.326
	$(X_2 \mid X_1)$	1	190.232
	$(X_3 \mid X_1, X_2)$	1	129.431
Residual		18	442.292

a. Test for the significance of each independent variable as it enters the model. State the null hypothesis for each test in terms of regression coefficients.

b. Test for the significance of adding both X_2 and X_3 to a model already containing X_1. State the null hypothesis in terms of regression coefficients.

c. In terms of regression sums of squares, identify the test that corresponds to comparing the two models

$$Y = \beta_0 + \beta_1 X_1 + \beta_2 X_2 + \beta_3 X_3 + E$$

and

$$Y = \beta_0 + \beta_3 X_3 + E$$

Why can't this test be done by using the ANOVA table? Describe the appropriate test procedure.

d. Based on the tests made, what would you recommend as the most appropriate statistical model? Use $\alpha = .05$.

4. An educator examined the relationship between the number of hours devoted to reading each week (Y) and the independent variables social class (X_1), number of years of school completed (X_2), and reading speed (X_3), in pages read per hour. The following ANOVA table was obtained from a stepwise regression analysis on data for a sample of 19 women over 60.

Source		d.f.	SS
Regression $\begin{cases} (X_3) \\ (X_2 \mid X_3) \\ (X_1 \mid X_2, X_3) \end{cases}$	(X_3)	1	1,058.628
	$(X_2 \mid X_3)$	1	183.743
	$(X_1 \mid X_2, X_3)$	1	37.982
Residual		15	363.300

a. Test for the significance of each variable as it enters the model.

b. Test $H_0: \beta_1 = \beta_2 = 0$ in the model $Y = \beta_0 + \beta_1 X_1 + \beta_2 X_2 + \beta_3 X_3 + E$.

c. Why can't we test $H_0: \beta_1 = \beta_3 = 0$ by using the ANOVA table given? What formula would you use for this test?

d. Based on your results in parts (a) and (b), what is the most appropriate model to use?

5. An experiment was conducted regarding a quantitative analysis of factors found in high-density lipoprotein (HDL) in a sample of human blood serum. Three variables thought to be predictive of or associated with HDL measurement (Y) were the total cholesterol (X_1) and total triglyceride (X_2) concentrations in the sample, plus the presence or absence of a certain sticky component of the serum called sinking pre-beta, or SPB (X_3), coded as 0 if absent and 1 if present. The data obtained are shown in the following table.

a. Test whether X_1, X_2, or X_3 alone significantly helps to predict Y.

b. Test whether X_1, X_2, and X_3 taken together significantly helps to predict Y.

Y	X_1	X_2	X_3	Y	X_1	X_2	X_3	Y	X_1	X_2	X_3
47	287	111	0	63	339	168	1	36	318	180	0
38	236	135	0	40	161	68	1	42	270	134	0
47	255	98	0	59	324	92	1	41	262	154	0
39	135	63	0	56	171	56	1	42	264	86	0
44	121	46	0	76	265	240	1	39	325	148	0
64	171	103	0	67	280	306	1	27	388	191	0
58	260	227	0	57	248	93	1	31	260	123	0
49	237	157	0	57	192	115	1	39	284	135	0
55	261	266	0	42	349	408	1	56	326	236	1
52	397	167	0	54	263	103	1	40	248	92	1
49	295	164	0	60	223	102	1	58	285	153	1
47	261	119	1	33	316	274	0	43	361	126	1
40	258	145	1	55	288	130	0	40	248	226	1
42	280	247	1	36	256	149	0	46	280	176	1

c. Test whether the true coefficients of the product terms X_1X_3 and X_2X_3 are simultaneously zero in the model containing X_1, X_2, and X_3 plus these product terms. Specify the two models being compared, and state the null hypothesis in terms of regression coefficients. If this test is not rejected, what can you conclude about the relationship of Y to X_1 and X_2 when X_3 equals 1, compared to the same relationship when X_3 equals 0?

d. Test (at $\alpha = .05$) whether X_3 is associated with Y, after taking into account the combined contribution of X_1 and X_2. What does your result, together with your answer to part (c), tell you about the relationship of Y with X_1 and X_2 when SPB is present compared to the same relationship when it is absent?

e. What overall conclusion can you draw about the association of Y with the three independent variables for this data set? Specify the two models being compared, and state the appropriate null hypothesis in terms of regression coefficients.

Edited SAS Output (PROC GLM) for Problem 5

```
                          Y Regressed on X2
                    General Linear Models Procedure
Source            DF         Sum of Squares      F Value        Pr > F
Model              1           21.33970871         0.19         0.6687
Error             40         4592.27933891
Corrected Total   41         4613.61904762

   R-Square                      C.V.                          Y Mean
   0.004625                   22.43378                       47.7619048

Source            DF          Type I SS          F Value        Pr > F
X2                 1          21.33970871          0.19         0.6687

                          Y Regressed on X3
                            ⋮  [Portion of output omitted]
                            ⋮

Source            DF          Type I SS          F Value        Pr > F
X3                 1         735.20541126          7.58         0.0088
```

(continued)

```
           Y Regressed on X1, X2, X3, X1X3, and X2X3
               General Linear Models Procedure
                                 .
                                 . [Portion of output omitted]
                                 .
Source         DF          Type I SS        F Value        Pr > F
X1             1           46.23555746        0.45         0.5078
X2             1           89.14656809        0.86         0.3591
X3             1          684.36519731        6.62         0.0143
X1X3           1           48.69043235        0.47         0.4968
X2X3           1           26.12961110        0.25         0.6181
```

6. Use the results from Problem 3 in Chapter 8, as well as the computer output given here, to answer the following questions about the data from that problem.
   a. Conduct overall regression F tests for these three models: Y regressed on weight and age; Y regressed on weight alone; and Y regressed on age alone. Based on your answers, how would you rate the importance of the two variables in predicting Y? Comment on how your answer here compares with your answer to Problem 3(b) in Chapter 8.
   b. Provide variables-added-in-order tests for both variables, with weight (X_1) added first. Use $\alpha = .05$.
   c. Provide variables-added-in-order tests for both variables, with age (X_2) added first. Use $\alpha = .05$.
   d. Provide a table of variables-added-last tests for both weight and age.
   e. What, if any, differences are discernible in the results of the approaches in parts (a) through (d)?

Edited SAS Output (PROC GLM) for Problem 6

```
              TOT. CHOL (Y) Regressed on WEIGHT (X1)
                 General Linear Models Procedure
Source              DF      Sum of Squares    F Value      Pr > F
Model               1        10231.7261997     1.74        0.2000
Error               23      135145.3138003
Corrected Total     24      145377.0400000

              TOT. CHOL (Y) Regressed on AGE (X2)
                 General Linear Models Procedure
Source              DF      Sum of Squares    F Value      Pr > F
Model               1       101932.665711     53.96        0.0001
Error               23       43444.374289
Corrected Total     24      145377.040000

               Y Regressed on X1 and X2
                 General Linear Models Procedure
Source              DF      Sum of Squares    F Value      Pr > F
Model               2       102570.814650     26.36        0.0001
Error               22       42806.225350
Corrected Total     24      145377.040000
```

7. Use the results from Problem 4 in Chapter 8, as well as the computer output given here, to answer the following questions about the data from that problem.

   a. Conduct the overall regression F test for the model where Y is regressed on X_1, X_2, and X_3. Use $\alpha = .05$. Interpret your result.

   b. Provide variables-added-in-order tests for the order X_2, X_1, and X_3.

   c. Provide variables-added-in-order tests for the order X_3, X_1, and X_2.

   d. List all orders that can be tested, using the computer results below and in Problem 4 of Chapter 8. List all orders that *cannot* be so computed.

   e. Provide variables-added-last tests for X_1, X_2, and X_3.

   f. Provide the variables-added-last test for $X_4 \neq X_2X_3$, given X_2 and X_3 already in the model. Does X_4 significantly improve the prediction of Y, given that X_2 and X_3 are already in the model?

Edited SAS Output (PROC GLM) for Problem 7

```
                        Y Regressed on X2 and X1
                             ⋮ [Portion of output omitted]
                             ⋮
Source      DF          Type I SS        F Value         Pr > F
X2          1        1308.33947922        41.39          0.0001
X1          1           9.45893243         0.30          0.5915

Source      DF          Type III SS      F Value         Pr > F
X2          1        1309.44593073        41.42          0.0001
X1          1           9.45893243         0.30          0.5915
                             ⋮ [Portion of output omitted]
                             ⋮
                        Y Regressed on X3 and X1
                    General Linear Models Procedure
                             ⋮ [Portion of output omitted]
                             ⋮
Source      DF          Type I SS        F Value         Pr > F
X3          1        1387.59971830        54.61          0.0001
X1          1          35.63230678         1.40          0.2526

Source      DF          Type III SS      F Value         Pr > F
X3          1        1414.87954416        55.68          0.0001
X1          1          35.63230678         1.40          0.2526
                             ⋮ [Portion of output omitted]
                             ⋮
                        Y Regressed on X3 and X2
                    General Linear Models Procedure
                             ⋮ [Portion of output omitted]
                             ⋮
Source      DF          Type I SS        F Value         Pr > F
X3          1        1387.59971830        64.22          0.0001
X2          1         100.25968692         4.64          0.0459

Source      DF          Type III SS      F Value         Pr > F
X3          1         179.51992600         8.31          0.0103
X2          1         100.25968692         4.64          0.0459
                             ⋮ [Portion of output omitted]
                             ⋮
```

(continued)

```
               Y Regressed on X1, X2, and X3
             General Linear Models Procedure
                        :  [Portion of output omitted]
                        :

Source     DF          Type I SS        F Value       Pr > F
X1          1          8.35248092         0.40        0.5378
X2          1       1309.44593073        62.16        0.0001
X3          1        200.34652965         9.51        0.0071

Source     DF          Type III SS      F Value       Pr > F
X1          1         30.28553607         1.44        0.2480
X2          1         94.91291622         4.51        0.0497
X3          1        200.34652965         9.51        0.0071
```

8. The following ANOVA table is based on the data discussed in Problem 5 of Chapter 8. Use $\alpha = .05$.

Source		d.f.	SS
	(X_1)	1	18,953.04
Regression	$(X_3 \mid X_1)$	1	7,010.03
	$(X_2 \mid X_1, X_3)$	1	10.93
Residual		21	2,248.23
Total		24	28,222.23

a. Provide a test to compare the following two models:
$$Y = \beta_0 + \beta_1 X_1 + \beta_2 X_2 + \beta_3 X_3 + E$$
and
$$Y = \beta_0 + \beta_1 X_1 + E$$

b. Provide a test to compare the following two models:
$$Y = \beta_0 + \beta_1 X_1 + \beta_3 X_3 + E$$
and
$$Y = \beta_0 + E$$

c. State which two models are being compared in the computation
$$F = \frac{(18{,}953.04 + 7{,}010.03 + 10.93)/3}{2{,}248.23/21}$$

9. Residential real estate prices are thought to depend, in part, on property size and number of bedrooms. The house size X_1 (in hundreds of square feet), number of bedrooms X_2, and house price Y (in thousands of dollars) of a random sample of houses in a certain county were observed. The resulting data and some associated computer output were presented in Problem 10 of Chapter 8. Additional portions of the output are shown here. Use all of the output to answer the following questions.

a. Perform the overall F test for the regression of Y on both independent variables. Interpret your result.

b. Perform variables-added-in-order tests for both independent variables, with X_1 added first.

c. Perform variables-added-in-order tests for both independent variables, with X_2 added first.

d. Provide a table of variables-added-last tests.

e. Which predictors appear to be necessary? Why?

Edited SAS Output (PROC GLM) for Problem 9

```
              Y Regressed on X1 and X2
            General Linear Models Procedure
                           .
                           .  [Portion of output omitted]
                           .
Source      DF          Type I SS      F Value      Pr > F
X1           1       5732.22282609       40.06      0.0032
X2           1          1.09846352        0.01      0.9344

Source      DF          Type III SS     F Value      Pr > F
X1           1       1402.31533722        9.80      0.0352
X2           1          1.09846352        0.01      0.9344
```

10. Data on sales revenue Y, television advertising expenditures X_1, and print media advertising expenditures X_2 for a large retailer for the period 1988–1993 were given in Problem 11 of Chapter 8. Use the computer output for that problem, along with the additional portions of the output shown here, to answer the following questions.

a. Perform the overall F test for the regression of Y on both independent variables. Interpret your result.

b. Perform variables-added-in-order tests for both independent variables, with X_1 added first.

c. Perform variables-added-in-order tests for both independent variables, with X_2 added first.

d. Provide a table of variables-added-last tests.

e. Which predictors appear to be necessary? Why?

Edited SAS Output (PROC GLM) for Problem 10

```
              Y Regressed on X1 and X2
            General Linear Models Procedure
                           .
                           .  [Portion of output omitted]
                           .
Source      DF          Type I SS      F Value      Pr > F
X1           1         28.07164068      117.82      0.0017
X2           1          0.04689121        0.20      0.6874

Source      DF          Type III SS     F Value      Pr > F
X1           1         25.89125916      108.66      0.0019
X2           1          0.04689121        0.20      0.6874
```

11. Use the computer output for the radial keratotomy data of Problem 12 in Chapter 8, along with the additional output here, to answer the following questions.
    a. Perform the overall F test for the regression of Y on both independent variables. Interpret your result.
    b. Perform variables-added-in-order tests for both independent variables, with X_1 added first.
    c. Perform variables-added-in-order tests for both independent variables, with X_2 added first.
    d. Provide a table of variables-added-last tests.
    e. Which predictors appear to be necessary? Why?

Edited SAS Output (PROC GLM) for Problem 11

```
                    Y Regressed on X1 and X2
                General Linear Models Procedure
                              .
                              .   [Portion of output omitted]
                              .
Source       DF          Type I SS        F Value        Pr > F
X1            1         13.09841552         10.67         0.0020
X2            1          4.52435640          3.68         0.0605

Source       DF          Type III SS       F Value        Pr > F
X1            1         12.43080596         10.12         0.0025
X2            1          4.52435640          3.68         0.0605
```

12. Use the computer output for the *Business Week* magazine data of Problem 13 in Chapter 8, as well as the additional output here, to answer the following questions.
    a. Perform the overall F test for the regression of Y on both independent variables, X_2 and X_3. Interpret your result.
    b. Perform variables-added-in-order tests for both independent variables, with X_2 added first.
    c. Perform variables-added-in-order tests for both independent variables, with X_3 added first.
    d. Provide a table of variables-added-last tests.
    e. Which predictors appear to be necessary? Why?

Edited SAS Output (PROC GLM) for Problem 12

```
                    Y Regressed on X2 and X3
                General Linear Models Procedure
                              .
                              .   [Portion of output omitted]
                              .
Source   DF      Type I SS     Mean Square    F Value      Pr > F
X2        1     1.76498949     1.76498949       1.40       0.2537
X3        1    24.81773186    24.81773186      19.63       0.0004

Source   DF      Type III SS   Mean Square    F Value      Pr > F
X2        1     7.97795163     7.97795163       6.31       0.0224
X3        1    24.81773186    24.81773186      19.63       0.0004
```

13. This problem refers to the 1990 Census data presented in Problem 19 of Chapter 5 and in Problem 14 of Chapter 8.

Use the computer output from Problem 14 of Chapter 8, along with the additional output shown here, to answer the following questions about the regression of OWNEROCC on OWNCOST and URBAN.

a. Perform the overall F test for the regression of OWNEROCC on OWNCOST and URBAN. Interpret your result.

b. Perform variables-added-in-order tests for both independent variables, with OWNCOST added first.

c. Perform variables-added-in-order tests for both independent variables, with URBAN added first.

d. Provide a table of variables-added-last tests.

e. Which predictors appear to be necessary? Why?

Edited SAS Output (PROC GLM) for Problem 13

```
           OWNEROCC Regressed on OWNCOST and URBAN
                 General Linear Models Procedure

                         .
                         .  [Portion of output omitted]
                         .

Source    DF     Type I SS    Mean Square   F Value   Pr > F
OWNCOST    1   132.6203242    132.6203242      8.84   0.0068
URBAN      1   123.1581714    123.1581714      8.21   0.0088

Source    DF   Type III SS    Mean Square   F Value   Pr > F
OWNCOST    1    66.3336420     66.3336420      4.42   0.0467
URBAN      1   123.1581714    123.1581714      8.21   0.0088
```

References

Abt, K. 1981. "Problems of Repeated Significance Testing." *Controlled Clinical Trials* 1: 377–81.

Kupper, L. L.; Stewart, J. R.; and Williams, K. A. 1976. "A Note on Controlling Significance Levels in Stepwise Regression." *American Journal of Epidemiology* 103(1): 13–15.

Searle, S. R. 1971. *Linear Models.* New York: John Wiley & Sons.

Correlations: Multiple, Partial, and Multiple Partial

10-1 Preview

We saw in Chapter 5 that the following essential features of straight-line regression (excluding the quantitative prediction formula provided by the fitted regression equation) can also be described in terms of the correlation coefficient r. These features are summarized as follows:

1. The squared correlation coefficient r^2 measures the strength of the linear relationship between the dependent variable Y and the independent variable X. The closer r^2 is to 1, the stronger the linear relationship; the closer r^2 is to 0, the weaker the linear relationship.

2. $r^2 = (\text{SSY} - \text{SSE})/\text{SSY}$ is the proportionate reduction in the total sum of squares achieved by using a straight-line model in X to predict Y.

3. $r = \hat{\beta}_1(S_X/S_Y)$, where $\hat{\beta}_1$ is the estimated slope of the regression line.

4. r (or r_{XY}) is an estimate of the population parameter ρ (or ρ_{XY}), which describes the correlation between X and Y, both considered as random variables.

5. Assuming that X and Y have a bivariate normal distribution with parameters μ_X, μ_Y, σ_X^2, σ_Y^2, and ρ_{XY}, the conditional distribution of Y given X is $N(\mu_{Y|X}, \sigma_{Y|X}^2)$, where

$$\mu_{Y|X} = \mu_Y + \rho\frac{\sigma_Y}{\sigma_X}(X - \mu_X) \quad \text{and} \quad \sigma_{Y|X}^2 = \sigma_Y^2(1 - \rho^2)$$

Here, r^2 estimates ρ^2, which can be expressed as

$$\rho^2 = \frac{\sigma_Y^2 - \sigma_{Y|X}^2}{\sigma_Y^2}$$

6. *r* can be used as a general index of linear association between two random variables, in the following sense:

    a. The more highly positive *r* is, the more positive is the linear association; that is, an individual with a high value of one variable will likely have a high value of the other, and an individual with a low value of one variable will probably have a low value of the other.

    b. The more highly negative *r* is, the more negative is the linear association; that is, an individual with a high value of one variable will likely have a low value of the other, and conversely.

    c. If *r* is close to 0, there is little evidence of linear association, which indicates that there is a nonlinear association or no association at all.

This connection between regression and correlation can be extended to the multiple regression case, as we will discuss in this chapter. When several independent variables are involved, however, the essential features of regression are described not by a single correlation coefficient, as in the straight-line case, but by several correlations. These include a set of zero-order correlations such as *r*, plus a whole group of higher-order indices called *multiple correlations, partial correlations,* and *multiple partial correlations.*[1] These higher-order correlations allow us to answer many of the same questions that can be answered by fitting a multiple regression model. In addition, the correlation analog has been found to be particularly useful in uncovering spurious relationships among variables, identifying intervening variables, and making certain types of causal inferences.[2]

10-2 Correlation Matrix

When dealing with more than one independent variable, we can represent the collection of all zero-order correlation coefficients (i.e., the *r*'s between all possible pairs of variables) most compactly in *correlation matrix form.* For example, given $k = 3$ independent variables X_1, X_2, and X_3, and one dependent variable Y, there are $C_2^4 = 6$ zero-order correlations, and the correlation matrix has the general form

$$
\begin{array}{c}
 \\ Y \\ X_1 \\ X_2 \\ X_3
\end{array}
\begin{array}{cccc}
Y & X_1 & X_2 & X_3 \\
\begin{bmatrix}
1 & r_{Y1} & r_{Y2} & r_{Y3} \\
 & 1 & r_{12} & r_{13} \\
 & & 1 & r_{23} \\
 & & & 1
\end{bmatrix}
\end{array}
$$

Here, r_{Yj} ($j = 1, 2, 3$) is the correlation between Y and X_j, and r_{ij} ($i, j = 1, 2, 3$) is the correlation between X_i and X_j.

---

[1] The *order* of a correlation coefficient, as the term is used here, is the number of variables being controlled or adjusted for (see section 10-5).

[2] See Blalock (1971) and Bollen (1989) for discussions of techniques for causal inference using regression modeling.

For the data in Table 8-1, this matrix takes the form

$$
\begin{array}{c c}
 & \begin{array}{c c c c} \text{WGT} & \text{HGT} & \text{AGE} & (\text{AGE})^2 \end{array} \\
\begin{array}{c} \text{WGT} \\ \text{HGT} \\ \text{AGE} \\ (\text{AGE})^2 \end{array} &
\left[\begin{array}{c c c c}
1 & .814 & .770 & .767 \\
 & 1 & .614 & .615 \\
 & & 1 & .994 \\
 & & & 1
\end{array} \right]
\end{array}
$$

Taken separately, each of these correlations describes the strength of the linear relationship between the two variables involved. In particular, the correlations $r_{Y1} = .814$, $r_{Y2} = .770$, and $r_{Y3} = .767$ measure the strength of the linear association with the dependent variable WGT for each of the independent variables taken separately. As we can see, HGT ($r_{Y1} = .814$) is the independent variable with the strongest linear relationship to WGT, followed by AGE and then $(\text{AGE})^2$.

Nevertheless, these zero-order correlations do not describe (1) the overall relationship of the dependent variable WGT to the independent variables HGT, AGE, and $(\text{AGE})^2$ considered together; or (2) the relationship between WGT and AGE after controlling for[3] HGT; or (3) the relationship between WGT and the combined effects of AGE and $(\text{AGE})^2$ after controlling for HGT. The measure that describes relationship (1) is called the *multiple correlation coefficient* of WGT on HGT, AGE, and $(\text{AGE})^2$. The measure that describes relationship (2) is called the *partial correlation coefficient* between WGT and AGE controlling for HGT. Finally, the measure that describes relationship (3) is called the *multiple partial correlation coefficient* between WGT and the combined effects of AGE and (AGE)2 controlling for HGT.

Even though AGE and $(\text{AGE})^2$ are very highly correlated in our example, it is possible—if the general relationship of AGE to WGT is nonlinear—that $(\text{AGE})^2$ will be significantly correlated with WGT even after AGE has been controlled for. In fact, this is what happens in general when the addition of a second-order term in polynomial regression significantly improves the prediction of the dependent variable.

10-3 Multiple Correlation Coefficient

The *multiple correlation coefficient,* denoted as $R_{Y|X_1, X_2, \dots, X_k}$, is a measure of the overall *linear association* of one (dependent) variable Y with several other (independent) variables $X_1, X_2, \dots, X_k$. By "linear association" we mean that $R_{Y|X_1, X_2, \dots, X_k}$ measures the strength of the association between Y and the best-fitting linear combination of the X's, which is the least-squares solution $\hat{Y} = \hat{\beta}_0 + \hat{\beta}_1 X_1 + \hat{\beta}_2 X_2 + \dots + \hat{\beta}_k X_k$. In fact, no other linear combination of the X's will have as great a correlation with Y. Also, $R_{Y|X_1, X_2, \dots, X_k}$ is always nonnegative.

The multiple correlation coefficient is thus a direct generalization of the simple correlation coefficient r to the case of several independent variables. We have dealt with this measure up to now under the name R^2, which is the square of the multiple correlation coefficient.

---

[3]In this case the phrase "controlling for" pertains to determining the extent to which the variables WGT and AGE are related after removing the effect of HGT on WGT and AGE.

Two computational formulas provide useful interpretations of the multiple correlation coefficient $R_{Y|X_1,X_2,...,X_k}$ and its square:

$$R_{Y|X_1,X_2,...,X_k} = \frac{\sum_{i=1}^{n}(Y_i - \overline{Y})(\hat{Y}_i - \overline{\hat{Y}})}{\sqrt{\sum_{i=1}^{n}(Y_i - \overline{Y})^2 \sum_{i=1}^{n}(\hat{Y}_i - \overline{\hat{Y}})^2}}$$

and $\qquad\qquad\qquad\qquad\qquad\qquad\qquad\qquad\qquad\qquad\qquad$ (10.1)

$$R^2_{Y|X_1,X_2,...,X_k} = \frac{\sum_{i=1}^{n}(Y_i - \overline{Y})^2 - \sum_{i=1}^{n}(Y_i - \hat{Y}_i)^2}{\sum_{i=1}^{n}(Y_i - \overline{Y})^2} = \frac{\text{SSY} - \text{SSE}}{\text{SSY}}$$

where $\hat{Y}_i = \hat{\beta}_0 + \hat{\beta}_1 X_{1i} + \hat{\beta}_2 X_{2i} + \cdots + \hat{\beta}_k X_{ki}$ (the predicted value for the ith individual) and $\overline{\hat{Y}} = \sum_{i=1}^{n}\hat{Y}_i/n$. Formula (2), which we have seen several times before (as R^2), is most useful for assessing the fit of the regression model. Formula (1) indicates that $R_{Y|X_1,X_2,...,X_k} = r_{Y,\hat{Y}}$, the simple linear correlation between the observed values Y and the predicted values $\hat{Y}$.

As a numerical example, let us again consider the data of Table 8-1, where $X_1 = $ HGT, $X_2 = $ AGE, and $Y = $ WGT. Using only X_1 and X_2 in the model, the fitted regression equation is $\hat{Y} = 6.553 + 0.722X_1 + 2.050X_2$, and the observed and predicted values are as given in Table 10-1.

One can check that $\overline{Y} = \overline{\hat{Y}} = 62.75$. As we mentioned in Chapter 8, this is no coincidence: it is a mathematical fact that $\overline{Y} = \overline{\hat{Y}}$.

The following SAS output shows that the computed value of R^2 is 0.7800 for the model containing X_1 and X_2. This tells us that 78% of the variation in Y is explained by the regression model. The corresponding multiple correlation coefficient $R_{Y|X_1,X_2} = R$ is 0.8832, although this value is not shown in the output. The value 0.8832 can be obtained by taking the positive square root of 0.7800. In general, R is defined as the *positive* square root of R^2, so it always has a non-negative value.

TABLE 10-1   Observed and predicted values for the regression of WGT on HGT and AGE.

Child	1	2	3	4	5	6
observed	64	71	53	67	55	58
predicted	64.11	69.65	54.23	73.87	59.78	57.01

Child	7	8	9	10	11	12
observed	77	57	56	51	76	68
predicted	66.77	59.66	57.38	49.18	75.20	66.16

SAS Output for Regression of WGT on HGT and AGE

```
Model: MODEL1
Dependent Variable: WGT          Y
                        Analysis of Variance
                         Sum of            Mean
Source        DF         Squares          Square      F Value      Pr > F
Model          2        692.82261       346.41130      15.953      0.0011
Error          9        195.42739        21.71415
C Total       11        888.25000

        Root MSE          4.65984          R-square           0.7800
        Dep Mean         62.75000          Adj R-sq           0.7311
        C.V.              7.42605
```

10-4 Relationship of $R_{Y|X_1, X_2,...,X_k}$ to the Multivariate Normal Distribution[4]

An informative way of looking at the sample multiple correlation coefficient $R_{Y|X_1,X_2,...,X_k}$ is to consider it as an estimator of a population parameter characterizing the joint distribution of all the variables Y, X_1, X_2,..., X_k taken together. When we had two variables X and Y and assumed that their joint distribution was bivariate normal $N_2(\mu_Y, \mu_X, \sigma_Y^2, \sigma_X^2, \rho_{XY})$, we saw that r_{XY} estimated ρ_{XY}, which satisfied the formula $\rho_{XY}^2 = (\sigma_Y^2 - \sigma_{Y|X}^2)/\sigma_Y^2$, where $\sigma_{Y|X}^2$ was the variance of the conditional distribution of Y given X. Now, when we have k independent variables and one dependent variable, we get an analogous result if we assume that their joint distribution is *multivariate normal.* Let us now consider what happens with just two independent variables. In this case the *trivariate normal distribution* of Y, X_1, and X_2 can be described as

$$N_3(\mu_Y, \mu_{X_1}, \mu_{X_2}, \sigma_Y^2, \sigma_{X_1}^2, \sigma_{X_2}^2, \rho_{Y1}, \rho_{Y2}, \rho_{12})$$

where μ_Y, μ_{X_1}, and μ_{X_2} are the three (unconditional) means; σ_Y^2, $\sigma_{X_1}^2$, and $\sigma_{X_2}^2$ are the three (unconditional) variances; and ρ_{Y1}, ρ_{Y2}, and ρ_{12} are the three correlation coefficients. The *conditional distribution of Y given X_1 and X_2* is then a univariate normal distribution with a (conditional) mean denoted by $\mu_{Y|X_1,X_2}$ and a (conditional) variance denoted by $\sigma_{Y|X_1,X_2}^2$; we usually write this compactly as

$$Y \mid X_1, X_2 \frown N(\mu_{Y|X_1,X_2}, \sigma_{Y|X_1,X_2}^2)$$

In fact, it turns out that

$$\mu_{Y|X_1,X_2} = \mu_Y + \rho_{Y1|2}\frac{\sigma_{Y|X_2}}{\sigma_{X_1|X_2}}(X_1 - \mu_{X_1}) + \rho_{Y2|1}\frac{\sigma_{Y|X_1}}{\sigma_{X_2|X_1}}(X_2 - \mu_{X_2})$$

and

$$\sigma_{Y|X_1,X_2}^2 = (1 - \rho_{Y,\mu_{Y|X_1,X_2}}^2)\sigma_Y^2$$

where $\rho_{Y,\mu_{Y|X_1,X_2}}$ is the population correlation coefficient between the random variables Y and $\mu_{Y|X_1,X_2} = \beta_0 + \beta_1X_1 + \beta_2X_2$ (where we are considering X_1 and X_2 as random variables) and where $\rho_{Y1|2}$ and $\rho_{Y2|1}$ are partial correlations (to be discussed in section 10.5). Also, $\sigma_{Y|X_2}^2$, $\sigma_{Y|X_1}^2$,

---

[4]This section is not essential for the application-oriented reader.

$\sigma^2_{X_1 | X_2}$, and $\sigma^2_{X_2 | X_1}$ are the conditional variances, respectively, of Y given X_2, Y given X_1, X_1 given X_2, and X_2 given X_1.

The parameter $\rho_{Y, \mu_{Y|X_1,X_2}}$ is the *population analog of the sample multiple correlation coefficient* $R_{Y|X_1,X_2}$, and we write $\rho_{Y, \mu_{Y|X_1,X_2}}$ simply as $\rho_{Y|X_1,X_2}$. Furthermore, from the formula for $\sigma^2_{Y|X_1,X_2}$, we can confirm (with a little algebra) that

$$\rho^2_{Y|X_1,X_2} = \frac{\sigma^2_Y - \sigma^2_{Y|X_1,X_2}}{\sigma^2_Y}$$

which is the *proportionate reduction in the unconditional variance of Y due to conditioning on X_1 and X_2.*

Generalizing these findings to the case of k independent variables, we may summarize the characteristics of the multiple correlation coefficient $R_{Y|X_1,X_2,\ldots,X_k}$ as follows:

1. $R^2_{Y|X_1,X_2,\ldots,X_k}$ measures the proportionate reduction in the total sum of squares $\sum(Y_i - \overline{Y})^2$ to $\sum(Y_i - \hat{Y}_i)^2$ due to the multiple linear regression of Y on $X_1, X_2, \ldots, X_k$.

2. $R_{Y|X_1,X_2,\ldots,X_k}$ is the correlation $r_{Y,\hat{Y}}$ of the observed values (Y) with the predicted values ($\hat{Y}$), and this correlation is always nonnegative.

3. $R_{Y|X_1,X_2,\ldots,X_k}$ is an estimate of $\rho_{Y|X_1,X_2,\ldots,X_k}$, the correlation of Y with the true regression equations $\beta_0 + \beta_1 X_1 + \beta_2 X_2 + \cdots + \beta_k X_k$, where the X's are considered to be random.

4. $R^2_{Y|X_1,X_2,\ldots,X_k}$ is an estimate of the proportionate reduction in the unconditional variance of Y due to conditioning on $X_1, X_2, \ldots X_k$; that is, it estimates

$$\rho^2_{Y|X_1,X_2,\ldots,X_k} = \frac{\sigma^2_Y - \sigma^2_{Y|X_1,X_2,\ldots,X_k}}{\sigma^2_Y}$$

10-5 Partial Correlation Coefficient

The *partial correlation coefficient* is a measure of the strength of the linear relationship between two variables after we control for the effects of other variables. If the two variables of interest are Y and X, and the control variables are $Z_1, Z_2, \ldots, Z_p$, then we denote the corresponding partial correlation coefficient by $r_{YX|Z_1,Z_2,\ldots,Z_p}$. The order of the partial correlation depends on the number of variables that are being controlled for. Thus, *first-order* partials have the form $r_{YX|Z}$; *second-order* partials have the form $r_{YX|Z_1,Z_2}$; and in general, *p*th-order partials have the form $r_{YX|Z_1,Z_2,\ldots,Z_p}$.

For the three independent variables HGT, AGE, and $(AGE)^2$ in our example, the highest-order partial possible is second order. The values of most of the partial correlations that can be computed from this data set are given in Table 10-2.

The easiest way to obtain a partial correlation is to use a standard computer program. A computing formula that helps highlight the structure of the partial correlation coefficient will be given a little later. First, however, let us see how we can use the information in Table 10-2 to describe our data.

Looking back at our (zero-order) correlation matrix, we see that the variable most highly correlated with WGT is HGT ($r_{Y1} = .814$). Thus, of the three independent variables we are considering, HGT is the most important according to the strength of its linear relationship with WGT.

TABLE 10-2   Partial correlations for the WGT, HGT, and AGE data of Table 8-1.

Order	Controlling Variables	Form of Correlation	Computed Value
1	HGT	$r_{WGT, AGE \mid HGT}$	.589
1	HGT	$r_{WGT, (AGE)^2 \mid HGT}$	.580
1	HGT	$r_{AGE, (AGE)^2 \mid HGT}$	.991
1	AGE	$r_{WGT, HGT \mid AGE}$	.678
1	AGE	$r_{WGT, (AGE)^2 \mid AGE}$	.015
1	AGE	$r_{HGT, (AGE)^2 \mid AGE}$	.060
1	$(AGE)^2$	$r_{WGT, HGT \mid (AGE)^2}$	.677
1	$(AGE)^2$	$r_{WGT, AGE \mid (AGE)^2}$	.111
1	$(AGE)^2$	$r_{HGT, AGE \mid (AGE)^2}$	.022
2	HGT, AGE	$r_{WGT, (AGE)^2 \mid HGT, AGE}$	-0.035
2	HGT, $(AGE)^2$	$r_{WGT, AGE \mid HGT, (AGE)^2}$	.131
2	AGE, $(AGE)^2$	$r_{WGT, HGT \mid AGE, (AGE)^2}$	.679

After HGT, what is the next most important variable to the linear prediction of WGT? Since the first-order partial $r_{WGT, AGE \mid HGT} = .589$ is larger than $r_{WGT, (AGE)^2 \mid HGT} = .580$, it makes sense to conclude that AGE is next in importance, after we have accounted for HGT. (If we wanted to test the significance of this partial correlation coefficient, we would use a partial F test, as described in Chapter 9. We shall return to this point shortly.)

The only variable left to consider is $(AGE)^2$. But once we have accounted for HGT and AGE, does $(AGE)^2$ add anything to our knowledge of WGT? To answer this, we look at the second-order partial correlation $r_{WGT, (AGE)^2 \mid HGT, AGE} = -0.035$. Notice that the magnitude of this partial correlation is very small. Thus, we would be inclined to conclude that $(AGE)^2$ provides essentially no additional information about WGT once HGT and AGE have been used together as predictors. (This can be verified with a formal partial F-test.)

The procedure for selecting variables just described—starting with the most important variable and continuing step by step to add variables in descending order of importance while controlling for variables already selected, is called a _forward selection_ procedure. Alternatively, we could have handled the variable selection problem by working backward—starting with all the variables and deleting (step by step) variables that do not contribute much to the description of the dependent variable. We shall discuss procedures for selecting variables further in Chapter 16.

Output from SAS's REG procedure can be used to determine the partial correlations mentioned in the preceding discussion. The following output is for the model with WGT as the dependent variable, and HGT, AGE, and $(AGE)^2$ as the (uncentered) independent variables. The PCORR1 option was used in SAS to obtain the sequential or variables-added-in-order squared partial correlations (labeled "Squared Partial Corr Type I" on the output). The PCORR2 option was used to obtain the variables-added-last squared partial correlations (labeled "Squared Partial Corr Type II" on the output). The output reports that $r^2_{WGT, AGE \mid HGT} = 0.3471$. To determine $r_{WGT, AGE \mid HGT}$, the square root of 0.3471 is first taken. The sign of the partial correlation is then determined from the sign of the slope estimate for AGE in the regression of WGT on AGE and HGT (see output for model 4 on p. 122 in Chapter 8). Since the sign of the slope estimate for AGE is positive in this regression, the sign of $r_{WGT, AGE \mid HGT}$ is also positive. Hence, $r_{WGT, AGE \mid HGT} = +0.589$.

The other partial correlations can be determined similarly. For example, $r^2_{WGT,(AGE)^2|HGT,AGE}$ = 0.0012. Since the sign of the slope estimate for $(AGE)^2$, given that HGT and AGE are in the model, is negative, we have

$$r_{WGT,(AGE)^2|HGT,AGE} = -\sqrt{0.0012} = -0.035$$

SAS Output for Regression of WGT on HGT, AGE, and $(AGE)^2$

```
Model: MODEL1
Dependent Variable: WGT          Y
                    Analysis of Variance
                    Sum of          Mean
Source       DF     Squares         Square    F Value    Pr > F
Model         3     693.06046       231.02015  9.469     0.0052
Error         8     195.18954        24.39869
C Total      11     888.25000

        Root MSE        4.93950      R-square   0.7803
        Dep Mean       62.75000      Adj R-sq   0.6978
        C.V.            7.87172
                    Parameter Estimates
                    Parameter      Standard     T for H0:
Variable     DF     Estimate       Error        Parameter=0   Prob > |T|
INTERCEPT     1      3.438426      33.61081984    0.102        0.9210
HGT           1      0.723690       0.27696316    2.613        0.0310
AGE           1      2.776875       7.42727877    0.374        0.7182
AGESQ         1     -0.041707       0.42240715   -0.099        0.9238
                    Squared        Squared
              r²WGT,HGT  Partial    Partial
Variable     DF  Corr Type I   Corr Type II
INTERCEPT     1      .              .
HGT           1      0.66301438     0.46046181  ←  r²WGT,HGT|AGE,(AGE)²
AGE           1      0.34711175     0.01717277  ←  r²WGT,AGE|HGT,(AGE)²
AGESQ         1      0.00121711     0.00121711
        r²WGT,AGE|HGT       r²WGT,(AGE)²|HGT,AGE
```

10-5-1 Tests of Significance for Partial Correlations

Regardless of the procedure we use to select variables, we must decide at each step whether a particular partial correlation coefficient is significantly different from zero. We already described how to test for such significance in a slightly different context, in connection with the various uses of the ANOVA table in regression analysis. When we wanted to test whether adding a variable to the regression model was worthwhile, given that certain other variables were already in the model, we used a partial F test. It can be shown that this partial F test is exactly equivalent to a test of significance for the corresponding partial correlation coefficient. Thus, to test whether $r_{YX|Z_1,Z_2,...,Z_p}$ is significantly different from zero, we compute the corresponding partial $F(X \mid Z_1, Z_2,..., Z_p)$ and reject the null hypothesis if this F statistic exceeds an appropriate critical value of the $F_{1,n-p-2}$ distribution. For example, in testing whether $r_{WGT,(AGE)^2|HGT,AGE}$ is significant, we find that the partial $F[(AGE)^2 \mid HGT, AGE] = 0.010$ does not exceed $F_{1,12-2-2,0.90} = F_{1,8,0.90}$ = 3.46. Therefore, we conclude that this partial correlation is not significantly different from zero, so $(AGE)^2$ does not contribute to the prediction of WGT once we have accounted for HGT and AGE.

The null hypothesis for this test can be stated more formally by considering the population analog of the sample partial correlation coefficient $r_{YX|Z_1,Z_2,...,Z_p}$. This corresponding population parameter, usually denoted by $\rho_{YX|Z_1,Z_2,...,Z_p}$, is called the *population partial correlation coefficient*. The null hypothesis can then be stated as $H_0: \rho_{YX|Z_1,Z_2,...,Z_p} = 0$, and the associated alternative hypothesis as $H_A: \rho_{YX|Z_1,Z_2,...,Z_p} \neq 0$.

10-5-2 Relating the Test for Partial Correlation to the Partial F Test

The structure of the population partial correlation helps us relate this form of higher-order correlation to regression. For simplicity, let us consider this relationship for the special case of two independent variables. The formula for the square of $\rho_{YX_1|X_2}$ can be written as

$$\rho_{YX_1|X_2}^2 = \frac{\sigma_{Y|X_2}^2 - \sigma_{Y|X_1,X_2}^2}{\sigma_{Y|X_2}^2}$$

Thus, the square of the sample partial correlation $r_{YX_1|X_2}$ is an estimate of the proportionate reduction in the conditional variance of Y given X_2 due to conditioning on both X_1 and X_2.[5]

It then follows that an analogous formula for the squared sample partial correlation coefficient is

$$r_{YX_1|X_2}^2 = \frac{\left[\begin{array}{c}\text{Residual SS (using only } X_2 \text{ in the model)} \\ - \text{ Residual SS (using } X_1 \text{ and } X_2 \text{ in the model)}\end{array}\right]}{\text{Residual SS (using only } X_2 \text{ in the model)}}$$

$$= \frac{\left[\begin{array}{c}\text{Extra sum of squares due to adding } X_1 \text{ to the model,} \\ \text{given that } X_2 \text{ is already in the model}\end{array}\right]}{\text{Residual SS (using only } X_2 \text{ in the model)}}$$

(10.2)

It should be clear from the structure of (10.2) and from the discussion of partial F statistics in Chapters 8 and 9 why the test of $H_0: \rho_{YX_1|X_2} = 0$ is performed using $F(X_1 \mid X_2)$ as the test statistic.

10-5-3 Another Way of Describing Partial Correlations

Another way to compute a first-order partial correlation is to use the formula

$$r_{YX|Z} = \frac{r_{YX} - r_{YZ}r_{XZ}}{\sqrt{(1 - r_{YZ}^2)(1 - r_{XZ}^2)}}$$

(10.3)

---

[5]The partial correlation $\rho_{YX_1|X_2}$ can also be described as a zero-order correlation for a conditional bivariate distribution. If the joint distribution of Y, X_1, and X_2 is trivariate normal, the conditional joint distribution of Y and X_1, given X_2, is bivariate normal. The zero-order correlation between Y and X_1 for this conditional distribution is what we call $\rho_{YX_1|X_2}$; this is exactly the partial correlation between X_1 and Y, controlling for X_2.

For example, to compute $r_{\text{WGT,AGE}|\text{HGT}}$, we calculate

$$\frac{r_{\text{WGT,AGE}} - r_{\text{WGT,HGT}} r_{\text{AGE,HGT}}}{\sqrt{(1 - r^2_{\text{WGT,HGT}})(1 - r^2_{\text{AGE,HGT}})}} = \frac{.770 - (.814)(.614)}{\sqrt{[1 - (.814)^2][1 - (.614)^2]}}$$

$$= \frac{.770 - .500}{\sqrt{.337(.623)}} = .589$$

Notice that the first correlation in the numerator is the simple zero-order correlation between Y and X. The *control variable* Z appears in the second expression in the numerator (where it is correlated separately with each of the variables Y and X) and in both terms in the denominator. By using (10.3), we can interpret the partial correlation coefficient as an adjustment of the simple correlation coefficient to take into account the effect of the control variable. For example, if r_{YZ} and r_{XZ} have opposite signs, controlling for Z always increases r_{YX}. However, if r_{YZ} and r_{XZ} have the same sign, then $r_{YX|Z}$ could be either larger or smaller than r_{YX}.

To compute higher-order partial correlations, we simply reapply this formula using the appropriate next-lower-order partials. For example, the second-order partial correlation is an adjustment of the first-order partial, which, in turn, is an adjustment of the simple zero-order correlation. In particular, we have the following general formula for a second-order partial correlation:

$$r_{YX|Z,W} = \frac{r_{YX|Z} - r_{YW|Z} r_{XW|Z}}{\sqrt{(1 - r^2_{YW|Z})(1 - r^2_{XW|Z})}} = \frac{r_{YX|W} - r_{YZ|W} r_{XZ|W}}{\sqrt{(1 - r^2_{YZ|W})(1 - r^2_{XZ|W})}} \tag{10.4}$$

To compute $r_{\text{WGT,(AGE)}^2|\text{HGT,AGE}}$, for example, we have

$$\frac{r_{\text{WGT,(AGE)}^2|\text{HGT}} - r_{\text{WGT,AGE}|\text{HGT}} r_{\text{(AGE)}^2,\text{AGE}|\text{HGT}}}{\sqrt{(1 - r^2_{\text{WGT,AGE}|\text{HGT}})(1 - r^2_{\text{(AGE)}^2,\text{AGE}|\text{HGT}})}} = \frac{.580 - (.589)(.991)}{\sqrt{[1 - (.589)^2][1 - (.991)^2]}}$$

$$= -.035$$

10-5-4 Partial Correlation as a Correlation of Residuals of Regression

There is still another important interpretation concerning partial correlations. For the variables Y, X, and Z, suppose that we fit the two straight-line regression equations $Y = \beta_0 + \beta_1 Z + E$ and $X = \beta_0^* + \beta_1^* Z + E$. Let $\hat{Y} = \hat{\beta}_0 + \hat{\beta}_1 Z$ be the fitted line of Y on Z, and let $\hat{X} = \hat{\beta}_0^* + \hat{\beta}_1^* Z$ be the fitted line of X on Z. Then, the n pairs of deviations, or residuals, $(Y_i - \hat{Y}_i)$ and $(X_i - \hat{X}_i)$, $i = 1, 2, \ldots, n$, represent what remains unexplained after the variable Z has explained all the variation it can in the variables Y and X separately.

If we now correlate these n pairs of residuals (i.e., find $r_{Y-\hat{Y},X-\hat{X}}$), we obtain a measure that is independent of the effects of Z. It can be shown that *the partial correlation between Y and X, controlling for Z, can be defined as the correlation of the residuals of the straight-line regressions of Y on Z and of X on Z*; that is, $r_{YX|Z} = r_{Y-\hat{Y},X-\hat{X}}$.

10-5-5 Semipartial Correlations

An alternative form of partial correlation is sometimes considered. The partial correlation was just characterized as a correlation between Y adjusted for Z and X adjusted for Z. Some statisticians refer to this as a *full partial,* since both variables being correlated have been adjusted.

The *semipartial correlation* (or "part" correlation) may be characterized as the correlation between two variables when only one of the two has been adjusted for a third variable. For example, one may consider the semipartial correlation between Y and X with only X adjusted for Z or with only Y adjusted for Z. The first will be denoted by $r_{Y(X|Z)}$, and the second by $r_{X(Y|Z)}$. Thus, we have $r_{Y(X|Z)} = r_{Y,X-\hat{X}}$ and $r_{X(Y|Z)} = r_{Y-\hat{Y},X}$, where $\hat{X}$ and $\hat{Y}$ are obtained from straight-line regressions on Z.

Another way of describing these semipartials is in terms of zero-order correlations, as follows:

$$r_{Y(X|Z)} = \frac{r_{YX} - r_{YZ}r_{XZ}}{\sqrt{1 - r_{XZ}^2}} \tag{10.5}$$

and

$$r_{X(Y|Z)} = \frac{r_{YX} - r_{YZ}r_{XZ}}{\sqrt{1 - r_{YZ}^2}} \tag{10.6}$$

It is instructive to compare these formulas with formula (10.3) for the full partial. The numerator is the same in all three expressions: the partial covariance between Y and X adjusted for Z (with all three variables standardized to have variance 1). Clearly, then, these correlations all have the same sign; and if any one equals 0, they all do. For significance testing, it is appropriate to use the extra-sum-of-squares test described earlier.

These three correlations have different interpretations. They each describe the relationship between Y and X, but with adjustment for different quantities. The semipartial $r_{Y(X|Z)}$ is the correlation between Y and X with X adjusted for Z; the semipartial $r_{X(Y|Z)}$ is the correlation between Y and X with Y adjusted for Z. Finally, the full partial $r_{YX|Z}$ is the correlation between Y and X with both Y and X adjusted for Z.

Choosing the proper correlation coefficient depends on the relationship among the three variables $X, Y,$ and Z (the nuisance variable). Table 10-3 shows the four possible types of relationship. Case 1 involves assessing the relationship between X and Y without a nuisance variable present; here, the simple correlation r_{XY} should be used. Case 2 illustrates the situation where the nuisance variable Z is related to both X and Y, so that the use of $r_{YX|Z}$ is appropriate. In cases 3 and 4, the nuisance variable affects just one of the two variables X and Y. In these cases, semipartial correlations permit just one of two primary variables to be adjusted for the effects of a nuisance variable.

10-5-6 Summary of the Features of the Partial Correlation Coefficient

1. The partial correlation $r_{YX|Z_1, Z_2, ..., Z_p}$ measures the strength of the linear relationship between two variables X and Y while controlling for variables $Z_1, Z_2, ..., Z_p$.

TABLE 10-3   Possible relationships among variables X and Y and nuisance variable Z.

Case	Nuisance Relationship	Diagram	Preferred Correlation
1	Neither X nor Y affected by Z.	$X \leftrightarrow Y$	r_{XY}
2	Both X and Y affected by Z.	$X \leftrightarrow Y$ with Z below	$r_{YX \mid Z}$
3	Only X affected by Z.	$X \leftrightarrow Y$ with Z below	$r_{Y(X \mid Z)}$
4	Only Y affected by Z.	$X \leftrightarrow Y$ with Z below	$r_{X(Y \mid Z)}$

2. The square of the partial correlation $r_{YX \mid Z_1, Z_2, \ldots, Z_p}$ measures the proportion of the residual sum of squares that is accounted for by the addition of X to a regression model already involving $Z_1, Z_2, \ldots, Z_p$; that is,

$$r^2_{YX \mid Z_1, Z_2, \ldots, Z_p} = \frac{\left[\begin{array}{c}\text{Extra sum of squares due to adding} \\ X \text{ to a model already containing } Z_1, Z_2, \ldots, Z_p\end{array}\right]}{\text{Residual SS (using only } Z_1, Z_2, \ldots, Z_p \text{ in the model)}}$$

3. The partial correlation coefficient $r_{YX \mid Z_1, Z_2, \ldots, Z_p}$ is an estimate of the population parameter $\rho_{YX \mid Z_1, Z_2, \ldots, Z_p}$, which is the correlation between Y and X in the conditional joint distribution of Y and X given $Z_1, Z_2, \ldots, Z_p$. The square of this population partial correlation coefficient is given by the equivalent formula

$$\rho^2_{YX \mid Z_1, Z_2, \ldots, Z_p} = \frac{\sigma^2_{Y \mid Z_1, Z_2, \ldots, Z_p} - \sigma^2_{Y \mid X, Z_1, Z_2, \ldots, Z_p}}{\sigma^2_{Y \mid Z_1, Z_2, \ldots, Z_p}}$$

where $\sigma^2_{Y \mid Z_1, Z_2, \ldots, Z_p}$ is the variance of the conditional distribution of Y given Z_1, $Z_2, \ldots, Z_p$ (and where $\sigma^2_{Y \mid X, Z_1, Z_2, \ldots, Z_p}$ is similarly defined).

4. The partial F statistic $F(X \mid Z_1, Z_2, \ldots, Z_p)$ is used to test H_0: $\rho_{YX \mid Z_1, Z_2, \ldots, Z_p} = 0$.

5. The (first-order) partial correlation coefficient $r_{YX \mid Z}$ is an adjustment of the (zero-order) correlation r_{YX} that takes into account the effect of the control variable Z. This can be seen from the formula

$$r_{YX \mid Z} = \frac{r_{YX} - r_{YZ} r_{XZ}}{\sqrt{(1 - r^2_{YZ})(1 - r^2_{XZ})}}$$

Higher-order partial correlations are computed by reapplying this formula, using the next-lower-order partials.

6. The partial correlation $r_{YX \mid Z}$ can be defined as the correlation of the residuals of the straight-line regressions of Y on Z and of X on Z; that is, $r_{YX \mid Z} = r_{Y - \hat{Y}, X - \hat{X}}$.

10-6 Alternative Representation of the Regression Model

With the correlation analog to multiple regression, we can express the regression model $\mu_{Y|X_1, X_2, \ldots, X_k} = \beta_0 + \beta_1 X_1 + \beta_2 X_2 + \cdots + \beta_k X_k$ in terms of partial correlation coefficients and conditional variances. When $k = 3$, this representation takes the form

$$\mu_{Y|X_1, X_2, X_3} = \mu_Y + \rho_{Y1|23} \frac{\sigma_{Y|23}}{\sigma_{1|23}} (X_1 - \mu_{X_1}) \qquad (10.7)$$

$$+ \rho_{Y2|13} \frac{\sigma_{Y|13}}{\sigma_{2|13}} (X_2 - \mu_{X_2})$$

$$+ \rho_{Y3|12} \frac{\sigma_{Y|12}}{\sigma_{3|12}} (X_3 - \mu_{X_3})$$

where, for example, $\rho_{Y1|23} = \rho_{YX_1|X_2, X_3}$, and where we define

$$\beta_1 = \rho_{Y1|23} \frac{\sigma_{Y|23}}{\sigma_{1|23}}, \quad \beta_2 = \rho_{Y2|13} \frac{\sigma_{Y|13}}{\sigma_{2|13}}, \quad \beta_3 = \rho_{Y3|12} \frac{\sigma_{Y|12}}{\sigma_{3|12}}$$

Notice the similarity between this representation and the one for the straight-line case, where β_1 is equal to $\rho(\sigma_Y / \sigma_X)$. Also, here

$$\beta_0 = \mu_Y - \beta_1 \mu_{X_1} - \beta_2 \mu_{X_2} - \beta_3 \mu_{X_3}$$

An equivalent method to that of least squares for estimating the coefficients $\beta_0, \beta_1, \beta_2,$ and β_3 is to substitute for the population parameters in the preceding formulas the corresponding estimates

$$\hat{\mu}_Y = \overline{Y}, \quad \hat{\mu}_{X_1} = \overline{X}_1, \quad \hat{\mu}_{X_2} = \overline{X}_2, \quad \hat{\mu}_{X_3} = \overline{X}_3,$$

$$\hat{\beta}_1 = r_{Y1|23} \frac{S_{Y|23}}{S_{1|23}}, \quad \hat{\beta}_2 = r_{Y2|13} \frac{S_{Y|13}}{S_{2|13}}, \quad \hat{\beta}_3 = r_{Y3|12} \frac{S_{Y|12}}{S_{3|12}}$$

10-7 Multiple Partial Correlation

10-7-1 The Coefficient and Its Associated F Test

The *multiple partial correlation coefficient* is used to describe the overall relationship between a dependent variable and two or more independent variables while controlling for still other variables. For example, suppose that we consider the variable $X_1^2 = (HGT)^2$ and the product term $X_1 X_2 = HGT \times AGE$, in addition to the independent variables $X_1 = HGT$, $X_2 = AGE$, and $X_2^2 = (AGE)^2$. Our complete regression model is then of the form

$$Y = \beta_0 + \beta_1 X_1 + \beta_2 X_2 + \beta_{11} X_1^2 + \beta_{22} X_2^2 + \beta_{12} X_1 X_2 + E$$

We call such a model a *complete second-order model*, since it includes all possible variables up through second-order terms. For such a complete model, we frequently want to know whether any of the second-order terms are important—in other words, whether a first-order model involving only X_1 and X_2 (i.e., $Y = \beta_0 + \beta_1 X_1 + \beta_2 X_2 + E$) is adequate. There are two equivalent ways to represent this question as a hypothesis-testing problem: one is to test $H_0: \beta_{11} = \beta_{22} = \beta_{12} = 0$ (i.e., all second-order coefficients are zero); the other is to test the hypothesis $H_0: \rho_{Y(X_1^2, X_2^2, X_1 X_2)|X_1, X_2} = 0$, where $\rho_{Y(X_1^2, X_2^2, X_1 X_2)|X_1, X_2}$ is the population multiple partial correlation of Y with the second-order variables, controlling for the effects of the first-order variables. (In general, we write the multiple partial as $\rho_{Y(X_1, X_2, \ldots, X_k)|Z_1, Z_2, \ldots, Z_p}$.) This parameter is estimated by the sample multiple partial correlation $r_{Y(X_1^2, X_2^2, X_1 X_2)|X_1, X_2}$, which describes the overall multiple contribution of adding the second-order terms to the model after the effects of the first-order terms are partialed out or controlled for (hence the term *multiple partial*). Two equivalent formulas for $r_{Y(X_1^2, X_2^2, X_1 X_2)|X_1, X_2}^2$ are:

$$r_{Y(X_1^2, X_2^2, X_1 X_2)|X_1, X_2}^2 \qquad (10.8)$$

$$= \frac{\left[\begin{array}{c} \text{Residual SS (only } X_1 \text{ and } X_2 \text{ in the model)} \\ - \text{ Residual SS (all first- and second-order terms in the model)} \end{array} \right]}{\text{Residual SS (only } X_1 \text{ and } X_2 \text{ in the model)}}$$

$$= \frac{\left[\begin{array}{c} \text{Extra sum of squares due to the addition of the second-order terms } X_1^2, X_2^2, \text{ and} \\ X_1 X_2 \text{ to a model containing only the first-order terms } X_1 \text{ and } X_2 \end{array} \right]}{\text{Residual SS (only } X_1 \text{ and } X_2 \text{ in the model)}}$$

and

$$r_{Y(X_1^2, X_2^2, X_1 X_2)|X_1, X_2}^2 = \frac{R_{Y|X_1, X_2, X_1^2, X_2^2, X_1 X_2}^2 - R_{Y|X_1, X_2}^2}{1 - R_{Y|X_1, X_2}^2}$$

In most applications, estimating a multiple partial correlation is rarely of interest, so the preceding formulas are infrequently used. Nevertheless, there often is interest in testing hypotheses about a collection (or "chunk") of higher-order terms, and such tests involve the multiple partial correlation. For instance, to test either $H_0: \rho_{Y(X_1^2, X_2^2, X_1 X_2)|X_1, X_2} = 0$ (or equivalently, $H_0: \beta_{11} = \beta_{22} = \beta_{12} = 0$), we calculate the multiple partial F statistic given in general form by expression (9.6) in Chapter 9. For this example, the formula becomes

$$F(X_1^2, X_2^2, X_1 X_2 \mid X_1, X_2)$$

$$= \frac{[\text{Regression SS}(X_1, X_2, X_1^2, X_2^2, X_1 X_2) - \text{Regression SS}(X_1, X_2)]/3}{\text{MS residual}(X_1, X_2, X_1^2, X_2^2, X_1 X_2)}$$

$$= \frac{[\text{Residual SS}(X_1, X_2) - \text{Residual SS}(X_1, X_2, X_1^2, X_2^2, X_1 X_2)]/3}{\text{MS residual}(X_1, X_2, X_1^2, X_2^2, X_1 X_2)}$$

We would reject H_0 at the α level of significance if the calculated value of $F(X_1^2, X_2^2, X_1 X_2 \mid X_1, X_2)$ exceeded the critical value $F_{3, n-6, 1-\alpha}$.

In general, the null hypothesis $H_0: \rho_{Y(X_1, X_2, \ldots, X_k)|Z_1, Z_2, \ldots, Z_p} = 0$ is equivalent to the hypothesis $H_0: \beta_1^* = \beta_2^* = \ldots = \beta_k^* = 0$ in the model $Y = \beta_0 + \beta_1 Z_1 + \beta_2 Z_2 + \cdots + \beta_p Z_p + \beta_1^* X_1 + \beta_2^* X_2 + \cdots + \beta_k^* X_k + E$. The appropriate test statistic is the multiple partial F given by

$$F(X_1, X_2, \ldots, X_k \mid Z_1, Z_2, \ldots, Z_p)$$

which has the F distribution with k and $n - p - k - 1$ degrees of freedom under H_0.

The general F statistic given by (9.6) may be expressed in terms of squared multiple correlations (R^2 terms) involving the two models being compared. The general form for this alternative expression is

$$F = \frac{[R^2 \text{ (larger model)} - R^2 \text{ (smaller model)}]/[\text{regression df (larger model)} - \text{regression df (smaller model)}]}{[1 - R^2 \text{ (larger model)}]/\text{residual df (larger model)}} \qquad (10.9)$$

Concluding Remarks

As we have seen throughout this chapter, a regression F test is associated with many equivalent null hypotheses. As an example, the following null hypotheses all make the same statement but in different forms:

1. H_0: "adding variables to the smaller model to form the larger model does not significantly improve the prediction of Y"

2. H_0: "the population regression coefficients for the variables in the larger model but not in the smaller model are all equal to 0"

3. H_0: "the population multiple-partial correlation between Y and variables added to produce the larger model, controlling for the variables in the smaller model, is 0"

4. H_0: "the value of the population squared multiple correlation coefficient for the larger model is not greater than the value of that parameter for the smaller model"

The first two null hypotheses involve prediction, while the latter two involve association. The investigator, for interpretive purposes, can choose either group depending on whether his or her focus is on the predictive ability of the regression model or on a particular association of interest.

Problems

1. The correlation matrix obtained for the variables SBP (Y), AGE (X_1), SMK (X_2), and QUET (X_3), using the data from Problem 2 in Chapter 5, is given by

	SBP	AGE	SMK	QUET
SBP	1	.7752	.2473	.7420
AGE	.7752	1	−.1395	.8028
SMK	.2473	−.1395	1	−.0714
QUET	.7420	.8028	−.0714	1

a. Based on this matrix, which of the independent variables AGE, SMK, and QUET explains the largest proportion of the total variation in the dependent variable SBP?

b. Using the available computer outputs, determine the partial correlations $r_{SBP,SMK\,|\,AGE}$ and $r_{SBP,QUET\,|\,AGE}$.

c. Test for the significance of $r_{SBP,SMK\,|\,AGE}$, using the ANOVA results given in Problem 1 of Chapter 8. Express the appropriate null hypothesis in terms of a population partial correlation coefficient.

d. Determine the second-order partial $r_{SBP,QUET\,|\,AGE,SMK}$, and test for the significance of this partial correlation (again, using the computer output here and in Chapter 8, Problem 1).

e. Based on the results you obtained in parts (a) through (d), how would you rank the independent variables in terms of their importance in predicting Y? Which of these variables are relatively unimportant?

f. Compute the squared multiple partial correlation $r^2_{SBP(QUET,SMK)\,|\,AGE}$, using the output here and in Problem 1 of Chapter 8. Test for the significance of this correlation. Does this test result alter your decision in part (e) about which variables to include in the regression model?

Edited SAS Output (PROC REG) for Problem 1

```
                   SBP Regressed on AGE and SMK
                              .
                              . [Portion of output omitted]
                              .
                         Parameter Estimates
                    Parameter      Standard      T for HO:
Variable     DF     Estimate         Error    Parameter=0    Prob > |T|
INTERCEPT    1     48.049603     11.12955962        4.317        0.0002
AGE          1      1.709160      0.20175872        8.471        0.0001
SMK          1     10.294392      2.76810685        3.719        0.0009

                    Squared        Squared
                    Partial        Partial
Variable     DF  Corr Type I    Corr Type II
INTERCEPT    1        .
AGE          1     0.60094136     0.71219596
SMK          1     0.32291131     0.32291131

                   SBP Regressed on AGE and QUET
                              .
                              . [Portion of output omitted]
                              .
                         Parameter Estimates
                    Parameter      Standard      T for HO:
Variable     DF     Estimate         Error    Parameter=0    Prob > |T|
INTERCEPT    1     55.323436     12.53474644        4.414        0.0001
AGE          1      1.045157      0.38605667        2.707        0.0113
QUET         1      9.750732      5.40245598        1.805        0.0815

                    Squared        Squared
                    Partial        Partial
Variable     DF  Corr Type I    Corr Type II
INTERCEPT    1        .
AGE          1     0.60094136     0.20174580
QUET         1     0.10098584     0.10098584
```

(continued)

```
               SBP Regressed on AGE, SMK and QUET
                                 ·
                                 ·  [Portion of output omitted]
                                 ·

                      Parameter Estimates
                    Parameter        Standard        T for HO:
   Variable    DF   Estimate         Error          Parameter=0    Prob > |T|
   INTERCEPT    1   45.103192        10.76487511        4.190        0.0003
   AGE          1    1.212715         0.32381922        3.745        0.0008
   SMK          1    9.945568         2.65605655        3.744        0.0008
   QUET         1    8.592449         4.49868122        1.910        0.0664

                     Squared          Squared
                     Partial          Partial
   Variable    DF  Corr Type I     Corr Type II
   INTERCEPT    1        ·                ·
   AGE          1   0.60094136       0.33373458
   SMK          1   0.32291131       0.33366934
   QUET         1   0.11526997       0.11526997
```

2. An (equivalent) alternative to performing a partial F test for the significance of adding a new variable to a model while controlling for variables already in the model is to perform a t test using the appropriate partial correlation coefficient. If the dependent variable is Y, the independent variable of interest is X, and the controlling variables are $Z_1, Z_2, \ldots, Z_p$, then the t test for $H_0: \rho_{YX|Z_1,Z_2,\ldots,Z_p} = 0$ versus $H_A: \rho_{YX|Z_1,Z_2,\ldots,Z_p} \neq 0$ is given by the test statistic

$$T = r_{YX|Z_1,Z_2,\ldots,Z_p} \frac{\sqrt{n-p-2}}{\sqrt{1 - r^2_{YX|Z_1,Z_2,\ldots,Z_p}}}$$

which has a t distribution under H_0 with $n - p - 2$ degrees of freedom. The critical region for this test is therefore given by

$$|T| \geq t_{n-p-2,1-\alpha/2}$$

a. In a study of the relationship between water hardness and sudden death rates in $n = 88$ North Carolina counties, the following partial correlations were obtained:

$$r_{YX_3|X_1} = .124 \quad \text{and} \quad r_{YX_3|X_1,X_2} = .121$$

where

Y = Sudden death rate in county
X_1 = Distance from county seat to main hospital center
X_2 = Population per physician
X_3 = Water hardness index for county

Test separately the following hypotheses:

$$\rho_{YX_3|X_1} = 0 \quad \text{and} \quad \rho_{YX_3|X_1,X_2} = 0$$

b. An alternative way to form the ANOVA table associated with a regression analysis is to use partial correlation coefficients. For example, if three independent variables X_1, X_2, and X_3 are used, the ANOVA table is as shown next.

Source		d.f.	SS
Regression	X_1	1	$r^2_{YX_1}SSY$
	$X_2 \mid X_1$	1	$r^2_{YX_2 \mid X_1}(1 - r^2_{YX_1})SSY$
	$X_3 \mid X_1, X_2$	1	$r^2_{YX_3 \mid X_1, X_2}(1 - r^2_{YX_2 \mid X_1})(1 - r^2_{YX_1})SSY$
Residual		$n - 4$	$(1 - r^2_{YX_3 \mid X_1, X_2})(1 - r^2_{YX_2 \mid X_1})(1 - r^2_{YX_1})SSY$
Total		$n - 1$	$\sum_{i=1}^{n}(Y_i - \overline{Y})^2$

Determine the ANOVA table for the three independent variables in the water hardness study, using the following information:

$$r_{YX_1} = -.196, \quad r_{YX_2} = .033, \quad r_{X_1 X_2} = .038,$$

$$r_{YX_3 \mid X_1, X_2} = .121, \quad SSY = 21.05, \quad n = 88$$

c. In addition to the independent variables X_1, X_2, and X_3 already mentioned, the predictive abilities of the following independent variables were studied:

X_4 = Per capita income

X_5 = Coroner habit

Z_1 = 1 if county is in Piedmont area, and 0 otherwise

Z_2 = 1 if county is in Coastal Plains area, and 0 otherwise

Z_3 = 1 if county is in Tidewater area, and 0 otherwise

Furthermore, 25 first-order product (i.e., interaction) terms of the form $X_1 X_2$, $X_1 Z_1$, and so on, were also considered (excluding $Z_1 Z_2$, $Z_1 Z_3$, and $Z_2 Z_3$). The following three ANOVA tables were obtained.

Only X_1 Used

Source	d.f.	SS	MS
Regression	1	0.3846	0.3846
Residual	86	20.6703	0.2404

Only "Main Effects" Used

Source	d.f.	SS	MS
Regression	8	2.6853	0.3357
Residual	79	18.3696	0.2325

Main Effects Plus First-Order Interactions Used

Source	d.f.	SS	MS
Regression	33	7.4143	0.2247
Residual	54	13.6406	0.2258

Test whether adding all the interaction terms to the model significantly aids in predicting the dependent variable, after controlling for the main effects. State the null hypothesis for this test in terms of the appropriate multiple partial correlation coefficient.

d. Test whether significant overall prediction can be based on each of the three ANOVA tables. Determine the multiple R^2-values for each of the three tables.

e. What can you conclude from these results about the relationship of water hardness to sudden death?

3. Two variables X and Y are said to have a *spurious correlation* if their correlation solely reflects each variable's relationship to a third (antecedent) variable Z (and possibly to other variables). For example, the correlation between the total annual income (from all sources) of members of the U.S. Congress (Y) and the number of persons owning color television sets (X) is quite high. Simultaneously, however, a general upward trend has occurred in buying power (Z_1) and in wages of all types (Z_2), which would naturally be reflected in increased purchases of color TVs as well as in increased income of members of Congress. Thus, the high correlation between X and Y probably only reflects the influence of inflation on each of these two variables. Therefore, this correlation is spurious because it misleadingly suggests a relationship between color TV sales and the income of members of Congress.

a. How would you attempt to detect statistically whether a correlation between X and Y like the one described is spurious?

b. In a hypothetical study investigating socioecological determinants of respiratory morbidity for a sample of 25 communities, the following correlation matrix was obtained for four variables.

	Unemployment Level (X_1)	Average Temperature (X_2)	Air Pollution Level (X_3)	Respiratory Morbidity Rate (Y)
Unemployment level (X_1)	1	.51	.41	.35
Average temperature (X_2)	—	1	.29	.65
Air pollution level (X_3)	—	—	1	.50
Respiratory morbidity rate (Y)	—	—	—	1

(1) Determine the partial correlations $r_{YX_1|X_2}$, $r_{YX_1|X_3}$, and $r_{YX_1|X_2,X_3}$.

(2) Use the results in part (1) to determine whether the correlation of .35 between unemployment level (X_1) and respiratory morbidity rate (Y) is spurious. (Use the testing formula given in Problem 2 to make the appropriate tests.)

c. Describe a relevant example of spurious correlation in your field of interest. (Use only interval variables, and define them carefully.)

4. **a.** Using the information provided in Problem 2 of Chapter 9, determine the proportion of residual variation that is explained by the addition of X_2 to a model already containing X_1; that is, compute

$$Q = \frac{\text{Regression SS}(X_1, X_2) - \text{Regression SS}(X_1)}{\text{Residual SS}(X_1)}$$

b. How is the formula given in part (a) related to the partial correlation $r_{YX_2|X_1}$?

c. Test the hypothesis H_0: $\rho_{YX_2|X_1} = 0$, using both an F test and a two-sided t test. Check to confirm that these tests are equivalent.

5. Refer to Problem 7 of Chapter 9 to answer the following questions about the relationship of homicide rate (Y) to city population size (X_1), percentage of families with yearly incomes less than \$5,000 (X_2), and unemployment rate (X_3).

a. Determine the squared partial correlations $r^2_{YX_1|X_3}$ and $r^2_{YX_2|X_3}$ using the computer output here. Check the computation of $r^2_{YX_2|X_3}$ by means of an alternative formula, using the information that $r_{YX_2} = .8398$, $r_{YX_3} = .8648$, and $r_{X_2X_3} = .8154$.

b. Based on the results you obtained in part (a), which variable (if any) should next be considered for entry into the model, given that X_3 is already in the model?

c. Test H_0: $\rho_{YX_2|X_3} = 0$, using the t test described in Problem 2.

d. Determine the squared partial correlation $r^2_{YX_1|X_2,X_3}$ from the output here and/or in Problem 7 of Chapter 9, and then test H_0: $\rho_{YX_1|X_2,X_3} = 0$.

e. Determine the squared multiple partial correlation $r^2_{Y(X_1,X_2)|X_3}$, and test H_0: $\rho_{Y(X_1,X_2)|X_3} = 0$.

f. Based on the results you obtained in parts (a) through (e), what variables would you include in your final regression model? Use $\alpha = .05$.

Edited SAS Output (PROC REG) for Problem 5

```
                    Y Regressed on X3, X2 and X1
                       Analysis of Variance
                        Sum of          Mean
Source         DF      Squares         Square      F Value    Pr > F
Model           3    1518.14494      506.04831     24.022     0.0001
Error          16     337.05706       21.06607
C Total        19    1855.20200

          Root MSE          4.58978        R-square      0.8183
          Dep Mean         20.57000        Adj R-sq      0.7843
          C.V.             22.31297

                       Parameter Estimates
                    Parameter       Standard      T for H0:
Variable       DF   Estimate         Error      Parameter=0    Prob > |T|
INTERCEPT       1   -36.764925     7.01092577      -5.244        0.0001
X3              1     4.719821     1.53047547       3.084        0.0071
X2              1     1.192174     0.56165391       2.123        0.0497
X1              1     0.000763     0.00063630       1.199        0.2480

                    Squared         Squared
                    Partial         Partial
Variable       DF   Corr Type I   Corr Type II
INTERCEPT       1      .               .
X3              1   0.74795075      0.37280460
X2              1   0.21441231      0.21972110
X1              1   0.08244493      0.08244493
```

6. Using the ANOVA table given in Problem 8 of Chapter 9, which deals with the regression relationship of twelfth-grade mean verbal SAT scores (Y) to per pupil expenditures (X_1), percentage of teachers with advanced degrees (X_2), and pupil–teacher ratio (X_3), test the following null hypotheses.

a. H_0: $\rho_{YX_3|X_1} = 0$

b. H_0: $\rho_{YX_2|X_1,X_3} = 0$

c. H_0: $\rho_{Y(X_2,X_3)|X_1} = 0$

d. Based on these results, and assuming that X_1 is an important predictor of Y, what additional variables would you include in your regression model?

7. Using the following ANOVA table based on data in Problem 6 of Chapter 8 about the regression relationship of respiratory cancer mortality rates (Y) to air pollution index (X_1), mean age (X_2), and percentage of workforce employed in a certain industry (X_3), test the following hypotheses.

a. H_0: $\rho_{YX_2|X_1} = 0$

b. H_0: $\rho_{YX_3|X_1,X_2} - 0$

Source	d.f.	SS
X_1	1	1,523.658
$X_2 \mid X_1$	1	181.743
$X_3 \mid X_1, X_2$	1	130.529
Residual	19	551.723
Total	22	2,387.653

c. H_0: "The addition of X_2 and X_3 to a model already containing X_1 does not significantly improve the prediction of Y."

d. Based on these results, which variables are important predictors of Y? Use $\alpha = .05$.

e. State the results in part (a) in terms of equivalent tests of semipartial correlations.

8. Refer to the following ANOVA tables and to computer results given here (from data in Problem 8 of Chapter 8) to answer the following questions dealing with factors related to soil erosion.

a. Using the accompanying SAS output, compute $r_{YX_2 \mid X_1}$ and $r_{YX_3 \mid X_1}$.

b. Based on your results in part (a), which variable (if any) should next be entered into a regression model that already contains X_1?

c. Test H_0: $\rho_{YX_2 \mid X_1} = 0$, using the t test described in Problem 2.

d. Determine the squared multiple partial correlation $r^2_{Y(X_2, X_3) \mid X_1}$, and test H_0: $\rho_{Y(X_2, X_3) \mid X_1} = 0$.

Fitting X_1 First, Then Letting X_2 and X_3 Enter Stepwise

Source	d.f.	SS	Source	d.f.	SS
X_2	1	667.7279	X_1	1	640.4249
$X_3 \mid X_2$	1	5.8228	$X_2 \mid X_1$	1	32.7819
$X_1 \mid X_3, X_2$	1	6.9406	$X_3 \mid X_1, X_2$	1	7.2844
Residual	7	16.0942	Residual	7	16.0942

Edited SAS Output (PROC REG) for Problem 8

```
                    Y Regressed on X1 and X2
                    Analysis of Variance
                    Sum of           Mean
Source      DF      Squares          Square       F Value      Pr > F
Model        2      673.20680        336.60340    115.183      0.0001
Error        8       23.37865          2.92233
C Total     10      696.58545

          Root MSE         1.70948        R-square        0.9664
          Dep Mean        40.23636        Adj R-sq        0.9580
          C.V.             4.24860

                    Parameter Estimates
                    Parameter        Standard     T for H0:
Variable    DF      Estimate         Error        Parameter=0  Prob > |T|
INTERCEPT    1      -8.084810        20.06314836   -0.403      0.6975
X1           1      68.250681        49.84520939    1.369      0.2081
X2           1       2.293871         0.68488343    3.349      0.0101

                    Squared          Squared
                    Partial          Partial
Variable    DF      Corr Type I      Corr Type II
INTERCEPT    1         .                .
X1           1      0.91937735       0.18986133
X2           1      0.58371760       0.58371760
```

```
                  Y Regressed on X1 and X3
                    Analysis of Variance
                      Sum of         Mean
Source        DF     Squares        Square     F Value    Pr > F
Model          2    670.13094     335.06547    101.326    0.0001
Error          8     26.45452       3.30681
C Total       10    696.58545

        Root MSE         1.81846         R-square       0.9620
        Dep Mean        40.23636         Adj R-sq       0.9525
        C.V.             4.51946
                    Parameter Estimates
                    Parameter     Standard     T for H0:
Variable      DF     Estimate       Error     Parameter=0   Prob > |T|
INTERCEPT      1    -26.704505    16.62113471    -1.607       0.1468
X1             1    157.267135    28.44374038     5.529       0.0006
X3             1    -39.925950    13.32102838    -2.997       0.0171
                     Squared        Squared
                     Partial        Partial
Variable      DF   Corr Type I    Corr Type II
INTERCEPT      1        .              .
X1             1    0.91937735     0.79258760
X3             1    0.52894851     0.52894851
                Y Regressed on X1, X2 and X3
                        Correlation
Corr             X1             X2             X3             Y
X1           1.0000         0.9515        -0.8191         0.9588
X2           0.9515         1.0000        -0.8785         0.9791
X3          -0.8191        -0.8785         1.0000        -0.9038
Y            0.9588         0.9791        -0.9038         1.0000
                    Analysis of Variance
                      Sum of         Mean
Source        DF     Squares        Square     F Value    Pr > F
Model          3    680.49122     226.83041    98.657     0.0001
Error          7     16.09423       2.29918
C Total       10    696.58545

        Root MSE         1.51630         R-Square       0.9769
        Dep Mean        40.23636         Adj R-Sq       0.9670
        C.V.             3.76849
                    Parameter Estimates
                    Parameter     Standard     T for H0:
Variable      DF     Estimate       Error     Parameter=0   Prob > |T|
INTERCEPT      1     -1.879316    18.13419563    -0.104       0.9204
X1             1     77.325781    44.50547040     1.737       0.1259
X2             1      1.559100     0.73447037     2.123       0.0714
X3             1    -23.903783    13.42935668    -1.780       0.1183
                     Squared        Squared
                     Partial        Partial
Variable      DF   Corr Type I    Corr Type II
INTERCEPT      1        .              .
X1             1    0.91937735     0.30130747
X2             1    0.58371760     0.39162631
X3             1    0.31158432     0.31158432
```

9. Use the computer results from Problem 9 of Chapter 8 to answer the following questions.

    a. Test $H_0: \rho_{YX_1} = 0$ and $H_0: \rho_{YX_2} = 0$.

    b. Test H_0: $\rho_{YX_1|X_2} = 0$ and H_0: $\rho_{YX_2|X_1} = 0$.

    c. Based on your results in parts (a) and (b), which variables (if any) should be included in the regression model, and what is their order of importance?

10. Use the correlation matrix from Problem 1 to answer the following questions.

    a. Compute the semipartial $r_{SBP(SMK|AGE)}$.

    b. Compute the semipartial $r_{SMK(SBP|AGE)}$.

    c. Compare these correlations to the (full) partial $r_{SBP, SMK|AGE}$, computed in Problem 1.

11. For the data discussed in Problem 7, provide estimates of the following values.

    a. $r_{YX_1}^2$

    b. $R_{Y|X_1,X_2}^2$

    c. $R_{Y|X_1,X_2,X_3}^2$

    d. $r_{YX_3|X_1,X_2}^2$

    e. $r_{YX_2|X_1}^2$

12. Use the results in Problem 8 to answer the following questions.

    a. Provide an estimate of $r_{YX_1|X3,X2}^2$.

    b. What two models should you compare in testing whether the correlation in part (a) is zero in the population?

    c. Provide an estimate of $r_{YX_1}^2$.

    d. What does the difference between parts (a) and (c) say about the relationships among the three predictor variables?

    e. What two models should you compare in testing whether the correlations in parts (a) and (c) differ in the population?

13. The accompanying SAS computer output relates to the house price data of Problem 10 in Chapter 8. Use this output and, if necessary, the output associated with Problem 9 in Chapter 9 to answer the following questions.

    a. Determine $r_{Y|X_1,X_2}^2$, the squared multiple correlation between house price (Y) and the independent variables house size (X_1) and number of bedrooms (X_2).

    b. Determine $r_{YX_2|X_1}$, the partial correlation of Y with X_2 given that X_1 is in the model.

    c. Determine $r_{YX_1|X_2}$, the partial correlation of Y with X_1 given that X_2 is in the model.

    d. Using the t test technique of Problem 2, test H_0: $\rho_{YX_2|X_1} = 0$. Compare your test statistic value with the partial t test statistic for X_2 shown on the SAS output here.

    e. Using the t test technique of Problem 2, test H_0: $\rho_{YX_1|X_2} = 0$. Compare your test statistic value with the partial t test statistic for X_1 shown on the SAS output here.

    f. Based on your answers to parts (a) through (e), which variables (if any) should be included in the regression model, and what is their order of importance?

Edited SAS Output (PROC REG) for Problem 13

```
                   Y Regressed on X1 and X2
                     Analysis of Variance
                    Sum of            Mean
  Source     DF    Squares           Square      F Value    Pr > F
  Model       2   5733.32129      2866.66064      20.033     0.0082
  Error       4    572.39300       143.09825
  C Total     6   6305.71429
```

```
             Y Regressed on X1 and X2 (Continued)
               Analysis of Variance (Continued)
         Root MSE        11.96237        R-square        0.9092
         Dep Mean       105.42857        Adj R-sq        0.8638
         C.V.            11.34642
                     Parameter Estimates
                     Parameter        Standard      T for H0:
    Variable    DF    Estimate          Error      Parameter=0    Prob > |T|
    INTERCEPT    1   -16.093385      24.64693805      -0.653        0.5494
    X1           1     5.721790       1.82778969       3.130        0.0352
    X2           1    -1.173152      13.38993701      -0.088        0.9344

                      Squared          Squared
                      Partial          Partial
    Variable    DF  Corr Type I      Corr Type II
    INTERCEPT    1       .                .
    X1           1   0.90905210       0.71013795
    X2           1   0.00191540       0.00191540
```

14. The accompanying SAS computer output relates to the sales revenue data from Problem 11 in Chapter 8. Use this output and, if necessary, the output for Problem 10 in Chapter 9 to answer the following questions.

  a. Determine $r^2_{Y|X_1,X_2}$, the squared multiple correlation between sales revenue (Y) and the independent variables TV advertising expenditures (X_1) and print advertising expenditures (X_2).

  b.–f. For the sales revenue data, answer the questions posed in Problem 13, parts (b) through (f).

Edited SAS Output (PROC REG) for Problem 14

```
                   Y Regressed on X1 and X2
                     Analysis of Variance
                     Sum of           Mean
    Source     DF    Squares          Square      F Value     Pr > F
    Model       2   28.11853        14.05927      59.006      0.0039
    Error       3    0.71480         0.23827
    C Total     5   28.83333
         Root MSE         0.48813        R-square        0.9752
         Dep Mean         5.16667        Adj R-sq        0.9587
         C.V.             9.44760
                     Parameter Estimates
                     Parameter        Standard      T for H0:
    Variable    DF    Estimate          Error      Parameter=0    Prob > |T|
    INTERCEPT    1     2.104693       0.42196591       4.988        0.0155
    X1           1     1.241877       0.11913357      10.424        0.0019
    X2           1    -0.194946       0.43944079      -0.444        0.6874

                      Squared          Squared
                      Partial          Partial
    Variable    DF  Corr Type I      Corr Type II
    INTERCEPT    1       .                .
    X1           1   0.97358291       0.97313389
    X2           1   0.06156185       0.06156185
```

15. The accompanying SAS computer output relates to the radial keratotomy data from Problem 12 in Chapter 8. Use this output and, if necessary, the output from Problem 11 in Chapter 9 to answer the following questions.

    a. Determine $r^2_{Y|X_1,X_2}$, the squared multiple correlation between change in refractive error (Y) and the independent variables baseline refractive error (X_1) and baseline curvature (X_2).

    b–f. For the radial keratotomy data, answer the questions posed in Problem 13, parts (b) through (f).

Edited SAS Output (PROC REG) for Problem 15

```
                        Y Regressed on X1 and X2
                           Analysis of Variance
                             Sum of          Mean
Source          DF          Squares         Square     F Value      Pr > F
Model            2         17.62277        8.81139       7.175      0.0018
Error           51         62.63017        1.22804
C Total         53         80.25294

           Root MSE           1.10817          R-square          0.2196
           Dep Mean           3.83343          Adj R-sq          0.1890
           C.V.              28.90811

                           Parameter Estimates
                       Parameter        Standard      T for H0:
Variable        DF      Estimate          Error      Parameter=0    Prob > |T|
INTERCEPT        1     12.360015        5.08621177       2.430         0.0187
X1               1     -0.291601        0.09165295      -3.182         0.0025
X2               1     -0.220396        0.11482391      -1.919         0.0605

                        Squared          Squared
                        Partial          Partial
Variable        DF   Corr Type I      Corr Type II
INTERCEPT        1         .                .
X1               1     0.16321415       0.16560944
X2               1     0.06737232       0.06737232
```

16. The accompanying SAS computer output relates to the *Business Week* data from Problem 13 in Chapter 8. Use this output and, if necessary, the output from Problem 12 in Chapter 9 to answer the following questions.

    a. Determine $r^2_{Y|X_2,X_3}$, the squared multiple correlation between the yield (Y) and the independent variables 1989 rank (X_2) and P-E ratio (X_3).

    b.–f. For the *Business Week* data, answer the questions posed in Problem 13, parts (b) through (f).

Edited SAS Output (PROC REG) for Problem 16

```
       Yield (Y) Regressed on 1989 Rank (X2) and P-E Ratio (X3)
                                    :  [Portion of output omitted]
                                    :
                                 Squared          Squared
                                 Partial          Partial
       Variable         DF    Corr Type I      Corr Type II
       INTERCEPT         1         .                .
       X2                1     0.03670921       0.27066359
       X3                1     0.53584299       0.53584299
```

17. The accompanying SAS computer output relates to the 1990 Census data from Problem 14 in Chapter 8. Use this output and, if necessary, the output from Problem 13 in Chapter 9 to answer the following questions.

a. Determine $r^2_{Y|X_1,X_2}$, the squared multiple correlation between the rate of owner occupancy (Y = OWNEROCC) and the independent variables for monthly ownership costs (X_1 = OWNCOST) and proportion of population living in urban areas (X_2 = URBAN).

b.–f. For the Census data, answer the questions posed in Problem 13, parts (b) through (f).

Edited SAS Output (PROC REG) for Problem 17

```
              OWNEROCC Regressed on OWNCOST and URBAN
                        :  [Portion of output omitted]
                        :
                                  Squared              Squared
                                  Partial              Partial
          Variable       DF    Corr Type I         Corr Type II
          INTERCEPT       1
          OWNCOST         1    0.22068022          0.16119308
          URBAN           1    0.26296676          0.26296676
```

References

Blalock, H. M., Jr., ed. 1971. *Causal Models in the Social Sciences.* Chicago: Aldine Publishing.

Bollen, K. A. 1989. *Structural Equations with Latent Variables.* New York: John Wiley & Sons.

Confounding and Interaction in Regression

11-1 Preview

A regression analysis may have two different goals: to predict the dependent variable by using a set of independent variables; and to quantify the relationship of one or more independent variables to a dependent variable. The first of these goals focuses on finding a model that fits the observed data and predicts future data as well as possible, whereas the second pertains to producing accurate estimates of one or more regression coefficients in the model. The second goal is of particular interest when the research question concerns disease etiology, such as trying to identify one or more determinants of a disease or other health-related outcome.

Confounding and *interaction* are two methodological concepts relevant to attaining the second goal. In this chapter, we use regression terminology to describe these concepts. More general discussion of this subject can be found elsewhere (e.g., Kleinbaum, Kupper, and Morgenstern 1982) within the context of epidemiological research, which typically addresses etiologic questions involving the second goal. We begin here with a general overview of these concepts, after which we discuss the regression formulation of each concept separately. In Chapter 15, we describe a popular regression procedure, analysis of covariance (ANACOVA), which may be used to adjust or correct for problems of confounding. Subsequently, in Chapter 16, we briefly describe a strategy for obtaining a "best" regression model that incorporates the assessment of both confounding and interaction.

11-2 Overview

Both confounding and interaction involve the assessment of an association between two or more variables so that additional variables that may affect this association are accounted for. The measure of association chosen usually depends on the characteristics of the variables of interest. For example, if both variables are continuous, as in the classic regression context, the measure of association is typically a regression coefficient. The additional variables to be considered are synonymously referred to as *extraneous variables, control variables,* or *covariates.* The essential questions about these variables are whether and how they should be incorporated into a model that can be used to estimate the association of interest.

Suppose that we are conducting a study to assess whether physical activity level (PAL) is associated with systolic blood pressure (SBP), accounting (i.e., controlling) for AGE. The extraneous variable here is AGE. We need to determine whether we can ignore AGE in our analysis and still correctly assess the PAL–SBP association. In particular, we need to address the following two questions: (1) Is the estimate of the association between PAL and SBP meaningfully different depending on whether we ignore AGE? (2) Is the estimate of the association between PAL and SBP meaningfully different for different values of AGE? The first question is concerned with confounding; the second question, with interaction.

In general, *confounding exists if meaningfully different interpretations of the relationship of interest result when an extraneous variable is ignored or included in the data analysis.* In practice, the assessment of confounding requires a comparison between a *crude* estimate of an association (which ignores the extraneous variable(s) and an *adjusted* estimate of the association (which accounts in some way for the extraneous variables). If the crude and adjusted estimates are meaningfully different, confounding is present and one or more extraneous variables must be included in our data analysis. This definition does not require a statistical test but rather a comparison of estimates obtained from the data (see Kleinbaum, Kupper, and Morgenstern 1982, chap. 13, for further discussion of this point).

For example, a crude estimate of the relationship between PAL and SBP (ignoring AGE) is given by the regression coefficient—say, $\hat{\beta}_1$—of the variable PAL in the straight-line model that predicts SBP using just PAL. In contrast, an adjusted estimate is given by the regression coefficient $\hat{\beta}_1^*$ of the same variable PAL in the multiple regression model that uses both PAL and AGE to predict SBP. In particular, if PAL is defined dichotomously (e.g., PAL = 1 or 0 for high or low physical activity, respectively), then the crude estimate is simply the crude difference between the mean systolic blood pressures in each physical activity group, and the adjusted estimate represents an adjusted difference in these two mean systolic blood pressures that controls for AGE. In general, confounding is present if any meaningful difference exists between the crude and adjusted estimates.

Interaction is the condition where the relationship of interest is different at different levels (i.e., values) of the extraneous variable(s). In contrast to confounding, the assessment of interaction does not consider either a crude estimate or an (overall) adjusted estimate; instead, it focuses on describing the relationship of interest at different values of the extraneous variables. For example, in assessing interaction due to AGE in describing the PAL–SBP relationship, we must determine whether some description (i.e., estimate) of the relationship varies with different values of AGE (e.g., whether the relationship is strong at older ages and weak at younger ages). If the PAL–SBP relationship does vary with AGE, then we say that an AGE × (read "by") PAL interaction exists. To assess interaction, we may employ a statistical test in addition to subjective evaluation of the meaningfulness (e.g., clinical importance) of an estimated interaction effect. Again, for further discussion, see Kleinbaum, Kupper, and Morgenstern (1982).

When both confounding and interaction are considered for the same data set, using an overall (adjusted) estimate as a summary index of the relationship of interest tends to mask any (strong) interaction effects that may be present. For example, if the PAL–SBP association differs meaningfully at different values of AGE, using a single overall estimate, such as the regression coefficient of PAL in a multiple regression model containing both AGE and PAL, would hide this interaction finding. This illustrates the following important principle: *Interaction should be assessed before confounding is assessed; the use of a summary (adjusted) estimate that controls for confounding is recommended only when there is no meaningful interaction* (Kleinbaum, Kupper, and Morgenstern 1982, chap. 13).

A variable may manifest both confounding and interaction, neither, or only one of the two. But if strong interaction is found, an adjustment for confounding is inappropriate.

We are now ready to use regression terminology to address how these concepts can be employed, assuming a linear model and a continuous dependent variable. A regression analog for a dichotomous outcome variable might, for example, involve a logistic rather than a linear model. Logistic regression analysis is discussed in detail in Chapter 23. A more detailed discussion in which confounding and interaction are considered can be found in Kleinbaum, Kupper, and Morgenstern (1982, chaps. 20–24).

11-3 Interaction in Regression

In this section, we describe how two independent variables can interact to affect a dependent variable and how such an interaction can be represented by an appropriate regression model.

11-3-1 First Example

To illustrate the concept of interaction, let us consider the following simple example. Suppose that we wish to determine how two independent variables—temperature (T) and catalyst concentration (C)—jointly affect the growth rate (Y) of organisms in a certain biological system. Further, suppose that two particular temperature levels (T_0 and T_1) and two particular levels of catalyst concentration (C_0 and C_1) are to be examined, and that an experiment is performed in which an observation on Y is obtained for each of the four combinations of temperature–catalyst concentration level: (T_0, C_0), (T_0, C_1), (T_1, C_0), and (T_1, C_1). In statistical parlance, this experiment is called a *complete factorial experiment,* because observations on Y are obtained for all combinations of settings for the independent variables (or factors). The advantage of a factorial experiment is that any existing interaction effects can be detected and estimated efficiently.

Now, let us consider two graphs based on two hypothetical data sets for the experimental scheme just described. Figure 11-1(a) suggests that the rate of change[1] in the growth rate as a function of temperature remains the same regardless of the level of catalyst concentration; in other words, the relationship between Y and T does not in any way depend on C.

We are not saying that Y and C are unrelated, but that the relationship between Y and T does not vary as a function of C. When this is the case, we say that T and C do not interact or, equivalently, that there is no $T \times C$ interaction effect. Practically speaking, this means that we

---

[1] For readers familiar with calculus, the phrase "rate of change" is related to the notion of a derivative of a function. In particular, Figure 11-1(a) portrays a situation in which the partial derivative with respect to T of the response function relating the mean of Y to T and C is independent of C.

FIGURE 11-1   Graphs of noninteracting and interacting independent variables.

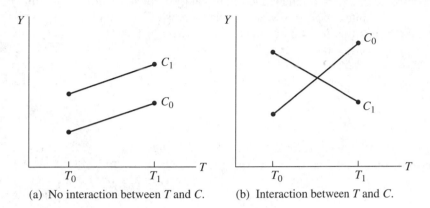

(a) No interaction between T and C.   (b) Interaction between T and C.

can investigate the effects of T and C on Y independently of one another and that we can legitimately talk about the separate effects (sometimes called the *main effects*) of T and C on Y.

One way to quantify the relationship depicted in Figure 11-1(a) is with a regression model of the form

$$\mu_{Y|T,C} = \beta_0 + \beta_1 T + \beta_2 C \qquad (11.1)$$

Here, the change in the mean of Y for a 1-unit change in T is equal to β_1, regardless of the level of C. In fact, changing the level of C in (11.1) has only the effect of shifting the straight line relating $\mu_{Y|T,C}$ and T either up or down, without affecting the value of the slope β_1, as seen in Figure 11-1(a). In particular, $\mu_{Y|T,C_0} = (\beta_0 + \beta_2 C_0) + \beta_1 T$ and $\mu_{Y|T,C_1} = (\beta_0 + \beta_2 C_1) + \beta_1 T$.

In general, then, we might say that no interaction is synonymous with parallelism, in the sense that the response curves of Y versus T for fixed values of C are parallel; in other words, these response curves (which may be linear or nonlinear) all have the same general shape, differing from one another only by additive constants independent of T (see, e.g., Figure 11-2).

In contrast, Figure 11-1(b) depicts a situation where the relationship between Y and T depends on C; in particular, Y appears to increase with increasing T when $C = C_0$ but to

FIGURE 11-2   Response curves illustrating no interaction between T and C.

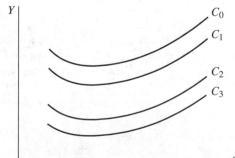

decrease with increasing T when $C = C_1$. In other words, the behavior of Y as a function of temperature cannot be considered independently of catalyst concentration. When this is the case, we say that T and C interact or, equivalently, that there is a $T \times C$ interaction effect. Practically speaking, this means that it does not make much sense to talk about the separate (or main) effects of T and C on Y, since T and C do not operate independently of one another in their effects on Y.

One way to represent such interaction effects mathematically is to use a regression model of the form

$$\mu_{Y|T,C} = \beta_0 + \beta_1 T + \beta_2 C + \beta_{12} TC \tag{11.2}$$

Here the change in the mean value of Y for a 1-unit change in T is equal to $\beta_1 + \beta_{12}C$, which clearly depends on the level of C. In other words, introducing a product term such as $\beta_{12}TC$ in a regression model of the type given in (11.2) is one way to account for the fact that two such factors as T and C do not operate independently of one another. For our particular example, when $C = C_0$, model (11.2) can be written as

$$\mu_{Y|T,C} = (\beta_0 + \beta_2 C_0) + (\beta_1 + \beta_{12}C_0)T$$

and when $C = C_1$, model (11.2) becomes

$$\mu_{Y|T,C} = (\beta_0 + \beta_2 C_1) + (\beta_1 + \beta_{12}C_1)T$$

In particular, Figure 11-1(b) suggests that the interaction effect β_{12} is negative, with the linear effect $(\beta_1 + \beta_{12}C_0)$ of T at C_0 being positive and the linear effect $(\beta_1 + \beta_{12}C_1)$ of T at C_1 being negative. A negative interaction effect is to be expected here, since Figure 11-1(b) suggests that the slope of the linear relationship between Y and T decreases (goes from positive to negative in sign) as C changes from C_0 to C_1. Of course, it is possible for β_{12} to be positive, in which case the interaction effect manifests itself as a larger positive value for the slope when $C = C_1$ than when $C = C_0$.

11-3-2 Interaction Modeling in General

As the preceding illustration suggests, interaction among independent variables can generally be described in terms of a regression model that involves product terms. Unfortunately, no precise rules exist for specifying such terms. For example, if interaction involving three variables X_1, X_2, and X_3 is of interest, one model to consider is

$$Y = \beta_0 + \beta_1 X_1 + \beta_2 X_2 + \beta_2 X_3 + \beta_4 X_1 X_2 + \beta_5 X_1 X_3 + \beta_6 X_2 X_3 \\ + \beta_7 X_1 X_2 X_3 + E \tag{11.3}$$

Here, the two-factor products of the form $X_i X_j$ are often referred to as *first-order interactions*; the three-factor products, like $X_1 X_2 X_3$, are called *second-order interactions*; and so on for higher-order products. The higher the order of interaction, the more difficult it becomes to interpret its meaning.

Model (11.3) is not the most general model possible for considering the three variables X_1, X_2, and X_3. Additional product terms such as $X_i X_j^2$, $X_i X_j^3$, $X_i^2 X_j^2$, and so on can be included. Nevertheless, there is a limit on the total number of such terms: a model with an intercept (β_0) term cannot contain more than $n - 1$ independent variables when n is the total number of observations in the data. Moreover, it may not even be possible to fit a model with fewer than $n - 1$

variables reliably if some of the variables (e.g., higher-order products) are highly correlated with other variables in the model, as would be the case when the model contains several interaction terms. This problem, called *collinearity,* is discussed in Chapter 12.

On the other hand, model (11.3) may be considered too general if one is focusing on particular interactions of interest. For example, if the purpose of one's study is to describe the relationship between X_1 and Y, controlling for the possible confounding and/or interaction effects of X_2 and X_3, the following simpler model may be of more interest than (11.3):

$$Y = \beta_0 + \beta_1 X_1 + \beta_2 X_2 + \beta_3 X_3 + \beta_4 X_1 X_2 + \beta_5 X_1 X_3 + E \qquad (11.4)$$

The terms $X_1 X_2$ and $X_1 X_3$ describe the interactions of X_2 and X_3, respectively, with X_1. In contrast, the term $X_2 X_3$, which is not contained in model (11.4), has no relevance to interaction involving X_1.

In using statistical testing to evaluate interaction for a given regression model, we have a number of available options. (A more detailed discussion of how to select variables is given in Chapter 16.) One approach is to test globally for the presence of any kind of interaction and then, if significant interaction is found, to identify particular interaction terms of importance by using other tests. For example, in considering model (11.3), we could first test $H_0: \beta_4 = \beta_5 = \beta_6 = \beta_7 = 0$, using the multiple partial F statistic

$$F(X_1 X_2, X_1 X_3, X_2 X_3, X_1 X_2 X_3 \mid X_1, X_2, X_3)$$

which has an $F_{4, n-8}$ distribution when H_0 is true. If this F statistic is found to be significant, individually important interaction terms may then be identified by using selected partial F tests.

A second way to assess interaction is to test for interaction in a hierarchical sequence, beginning with the highest-order terms and then proceeding sequentially through lower-order terms if higher-order terms are not significant. Using model (11.3), for example, we might first test $H_0: \beta_7 = 0$, which considers the second-order interaction, and then test $H_0: \beta_4 = \beta_5 = \beta_6 = 0$ in a reduced model (excluding the three-way product term $X_1 X_2 X_3$), if the result of the first test is nonsignificant.

11-3-3 Second Example

We present here an example of how to assess interaction by using SAS computer output. In Chapters 5 through 7, the age–systolic blood pressure example was used to illustrate the main principles and methods of straight-line regression analysis. These hypothetical data were shown, upon analysis, to support the commonly found epidemiological observation that blood pressure increases with age.

Another question that can be answered by such data is whether an interaction exists between age and gender: Does the slope of the straight-line relating systolic blood pressure to age significantly differ for males and for females? We will continue with our age–systolic blood pressure example to investigate this possible interaction.

In Chapter 5, we observed that the data point (47, 220) is an outlier quite distinct from the rest of the data. We will discard this data point in all further analyses; henceforth we will assume that all 29 remaining observations on age and systolic blood pressure considered previously were made on females and that a second sample of observations on age and systolic blood pressure was collected on 40 males. The data set for the 40 males is given in Table 11-1 and the data set for females appears in Table 5-1 in Chapter 5.

TABLE 11-1   Data on systolic blood pressure (SBP) and age for 40 males and 29 females, together with associated data for comparing two straight-line regression equations.

Male (i)	SBP (Y)	Age (X)	Male (i)	SBP (Y)	Age (X)	Male (i)	SBP (Y)	Age (X)
1	158	41	15	142	44	28	144	33
2	185	60	16	144	50	29	139	23
3	152	41	17	149	47	30	180	70
4	159	47	18	128	19	31	165	56
5	176	66	19	130	22	32	172	62
6	156	47	20	138	21	33	160	51
7	184	68	21	150	38	34	157	48
8	138	43	22	156	52	35	170	59
9	172	68	23	134	41	36	153	40
10	168	57	24	134	18	37	148	35
11	176	65	25	174	51	38	140	33
12	164	57	26	174	55	39	132	26
13	154	61	27	158	65	40	169	61
14	124	36						

The accompanying SAS computer output is given for the following regression model:

$$Y = \beta_0 + \beta_1 X + \beta_2 Z + \beta_3 XZ + E$$

where Z represents GENDER ($Z = 0$ if male, $Z = 1$ if female) and XZ is the interaction of AGE and GENDER.

SAS OUTPUT: SBP Regressed on Age (X), Gender (Z), and Interaction (XZ)

```
                  General Linear Models Procedure
                          Sum of          Mean
Source            DF      Squares         Square     F Value    Pr > F
Model              3     18010.329       6003.443     75.02     0.0001
Error             65      5201.439         80.022
Corrected Total   68     23211.768

      R-Square              C.V.          Root MSE        Y mean
      0.775914           6.014814          8.9455         148.72

                          Type I          Mean
Source            DF        SS            Square     F Value    Pr > F
X                  1     14951.255       14951.255    186.84     0.0001
Z                  1      3058.525        3058.525     38.22     0.0001
XZ                 1         0.549           0.549      0.01     0.9342

                          Type III        Mean
Source            DF        SS            Square     F Value    Pr > F
X                  1      7971.0071       7971.0071    99.61     0.0001
Z                  1       273.4433        273.4433     3.42     0.0691
XZ                 1         0.5494          0.5494     0.01     0.9342

                                     T for H0:                   Std Error
Parameter         Estimate        Parameter=0     Pr > |T|      of Estimate
INTERCEPT       110.0385285          23.23          0.0001      4.73610350
X                 0.9613526           9.98          0.0001      0.09632327
Z               -12.9614443          -1.85          0.0691      7.01172459
XZ               -0.0120301          -0.08          0.9342      0.14519328
```

From the output, we see that when $Z = 0$ (i.e., when GENDER = Male), the estimated model reduces to

$$Y = 110.04 + 0.96X + (-12.96)(0) + (-0.01)X(0)$$
$$= 110.04 + 0.96X$$

This is the estimated regression line for males. When $Z = 1$, the estimated model gives the estimated regression line for females:

$$Y = 110.04 + 0.96X + (-12.96)(1) + (-0.01)X(1)$$
$$= 97.08 + 0.95X$$

Plotting these two lines (Figure 11-3), we see that they appear to be almost parallel, indicating that there is probably no statistically significant interaction. We can confirm this lack of significance by inspecting the output: the partial F test for the significance of β_3, given that X and Z are in the model, has a p-value of 0.9342. Therefore, the slopes of the straight lines relating systolic blood pressure and age do not statistically significantly differ for males and for females: there is no interaction between age and gender in this situation.

11-3-4 Third Example

We now consider a study to assess physical activity level (PAL) as a predictor of systolic blood pressure (SBP), controlling for AGE and SEX. A model that allows for possible interactions of both AGE with PAL and SEX with PAL is given by

$$SBP = \beta_0 + \beta_1(PAL) + \beta_2(AGE) + \beta_3(SEX) + \beta_4(PAL \times AGE)$$
$$+ \beta_5(PAL \times SEX) + E$$

Notice the absence of any term involving AGE $\times$ SEX; such a term does not indicate interaction associated with the study variable of interest (PAL).

To assess interaction for this model, we might first perform a multiple partial F test of H_0: $\beta_4 = \beta_5 = 0$; if this test was found to be significant, we could conduct partial F tests to determine whether one or more of these product terms should be kept in the model. If the first test was found to be nonsignificant, we could then simplify the full model by removing these two product terms entirely, leaving the reduced model $SBP = \beta_0 + \beta_1(PAL) + \beta_2(AGE) + \beta_3(SEX) + E$. At this point the interaction phase of model building would be complete, and the next step would involve the assessment of confounding, as discussed in the next section.

FIGURE 11-3   Comparison by sex of straight-line regressions of systolic blood pressure on age.

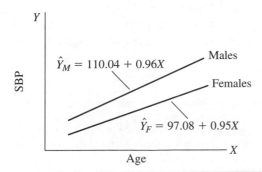

11-4 Confounding in Regression

We emphasized in section 11-1 that the assessment of confounding is questionable in the presence of interaction. Thus, our discussion of confounding here assumes throughout that no interaction is present.[2]

11-4-1 Controlling for One Extraneous Variable

Suppose that we are interested in describing the relationship between an independent variable T and a continuous dependent variable Y, taking into account the possible confounding effect of a third variable C. As described in section 11-2, the assessment of confounding requires us to compare a crude estimate of the T–Y relationship—which ignores the effect of the control variable (C)—with an estimate of the relationship that accounts (or controls) for this variable. This comparison can be expressed in terms of the following two regression models:

$$Y = \beta_0 + \beta_1 T + \beta_2 C + E \qquad (11.5)$$

and

$$Y = \beta_0 + \beta_1 T + E \qquad (11.6)$$

The assumption of no $T \times C$ interaction precludes the need to consider a product term of the form TC in these models.

From model (11.5), we can express the relationship between T and Y, adjusted for the variable C, in terms of the (partial) regression coefficient (β_1) of the T variable. The estimate of β_1 (which we will denote by $\hat{\beta}_{1|C}$), obtained from least-squares fitting of model (11.5), is an adjusted-effect measure in the sense that it gives the estimated change in Y per unit change in T after accounting for C (i.e., with C in the model).

A crude estimate of the T–Y relationship is the estimated coefficient of T (namely, $\hat{\beta}_1$) based on model (11-6)—a model that does not involve the variable C.

Thus, we have the following general rule for assessing the presence of confounding when only one independent variable is to be controlled: confounding is present if the estimate of the coefficient (β_1) of the study variable T meaningfully changes when the variable C is removed from model (11.5)—that is, if

$$\hat{\beta}_{1|C} \neq \hat{\beta}_1 \qquad (11.7)$$

where $\hat{\beta}_{1|C}$ denotes the (adjusted) estimate of β_1 using model (11.5), and $\hat{\beta}_1$ denotes the (crude) estimate of β_1 using model (11.6).

The $\neq$ sign in expression (11.7) indicates that a subjective decision is required as to whether the two estimates are meaningfully different; that is, we must determine subjectively whether the two estimates each describe a different interpretation of the T–Y association in question. A statistical test is neither required nor appropriate (Kleinbaum, Kupper, and Morgenstern 1982, chap. 13).

---

[2]It is possible, however, to assess confounding for variables that are not components of interaction terms. For example, if we consider the model $Y = \beta_0 + \beta_1 X_1 + \beta_2 X_2 + \beta_3 X_3 + \beta_4 X_1 X_3 + E$, where X_1 is the study variable of interest, we might wish to consider whether X_2 is a confounder, since it is not a component of $X_1 X_3$, the only interaction term in the model. For more realistic examples, see Kleinbaum, Kupper, and Morgenstern (1982, chap. 23).

As an example, suppose that Y denotes SBP, T denotes PAL, and C denotes AGE. For some set of data, suppose we found that

$$\hat{\beta}_{1|\text{AGE}} = 4.1 \quad \text{and} \quad \hat{\beta}_1 = 15.9$$

Then, we could conclude that a 1-unit change in PAL yields a 16-unit change in SBP when AGE is ignored, and that a 1-unit change in PAL yields only a 4.1-unit change in SBP when AGE is controlled; that is, the association between PAL and SBP is much weaker after controlling for AGE. (As a special case, if PAL is a 0–1 variable, then $\hat{\beta}_1$ gives the crude difference in mean systolic blood pressures between the two PAL groups, and $\hat{\beta}_{1|\text{AGE}}$ gives an adjusted [for AGE] difference in mean blood pressures.) Thus, we would treat AGE as a confounder and control for it in the analysis.

As another example, suppose that

$$\hat{\beta}_{1|\text{AGE}} = 6.2 \quad \text{and} \quad \hat{\beta}_1 = 6.1$$

Here, we would be inclined to say that AGE is not a confounder because there is no meaningful difference between the estimates 6.2 and 6.1. Unfortunately, an investigator may have to deal with much more problematic comparisons, such as $\hat{\beta}_{1|\text{AGE}} = 4.1$ versus $\hat{\beta}_1 = 5.5$. In comparing such estimates numerically, one must consider the clinical importance of the numerical difference between estimates, based on (a priori) knowledge of the variable(s) involved. For instance, since the coefficients 4.1 and 5.5 estimate, respectively, adjusted and crude differences in mean blood pressures between high and low PAL groups, it is important to decide whether a mean difference of 5.5 is clinically more important than a mean difference of 4.1. One approach to this problem is to control for any variable (as a confounder) that changes the crude effect estimate by some prespecified amount determined by clinical judgment.

One approach sometimes used (incorrectly) to assess confounding is, for example, to conduct a statistical test of H_0: $\beta_2 = 0$ in model (11.5). Such a test does not address confounding, but rather *precision;* that is, such a test evaluates whether significant additional variation in Y is explained by adding C to a model already containing T. An almost equivalent approach is to determine whether a confidence interval for β_1, the coefficient of T, is considerably narrower when C is in the model than when it is not. Precision is often an important issue to assess when considering extraneous factors, but it differs fundamentally from confounding. For etiologic questions, confounding, which concerns validity (i.e., do you have the right answer?), usually takes precedence over precision. Another reason for not focusing on β_2 is that, if $\hat{\beta}_2 \neq 0$, it does not follow that $\hat{\beta}_{1|C} \neq \hat{\beta}_1$. In other words, $\hat{\beta}_2 \neq 0$ is not a sufficient condition for confounding.[3]

What type of variables (i.e., covariates) should be considered for control as potential confounders? Although our answer is somewhat debatable, we consider that a list of eligible variables should be constructed based on prior knowledge and/or research about the relationship of the dependent variable to each covariate under consideration. In particular, we recommend that only variables known to be reasonably predictive of (i.e., associated with) the dependent variable should be considered as potential confounders and/or effect modifiers. In epidemiological terms, such variables are generally referred to as *risk factors* (Kleinbaum, Kupper, and Morgenstern 1982). The idea here is to restrict attention to controlling only the (previously

---

[3]Suppose that $n = 6$ and that we have the following data for (T, C, Y): $(1, 0, 4)$, $(1, 1, 5)$, $(1, 2, 6)$, $(0, 0, 1)$, $(0, 1, 2)$, and $(0, 2, 3)$. Then unweighted least-squares fitting gives $\hat{Y} = 1 + 3T + C$ when T and C are predictors, whereas $\hat{Y} = 2 + 3T$ when C is ignored. Thus, $\hat{\beta}_2 = 1 (\neq 0)$, yet there is no confounding since $\hat{\beta}_1 = 3 = \hat{\beta}_{1|C}$.

studied) extraneous variables that the investigator anticipates may account for the hypothesized relationship between T and Y currently being studied. To develop such a list, the investigators have to make a subjective decision.[4]

11-4-2 Controlling for Several Extraneous Variables

Suppose that we wish to describe the association between T and Y, taking into account several covariates $C_1, C_2, \ldots, C_p$. Similarly to our approach for handling one covariate, we can assess confounding by comparing a crude estimate of the T–Y relationship to some adjusted estimate. As before, the crude estimate can be defined in terms of a regression model like (11.6), which describes the relationship between T and Y while ignoring all covariates. To obtain the adjusted estimate, however, we must now consider an extended model defined as

$$Y = \beta_0 + \beta_1 T + \beta_2 C_1 + \beta_3 C_2 + \cdots + \beta_{p+1} C_p + E \qquad (11.8)$$

(Like model (11.5), model (11.8) assumes that there is no interaction involving T, since no product terms of the form TC_i are included.)

Using this model, we can define confounding that involves several variables as follows: Confounding is present if the estimate of the regression coefficient (β_1) of T in a regression model like (11.6), which ignores the variables $C_1, C_2, \ldots, C_p$, is meaningfully different from the corresponding estimate of β_1 based on a model like (11.8), which controls for $C_1, C_2, \ldots, C_p$—that is, if

$$\hat{\beta}_{1 \mid C_1, C_2, \ldots, C_p} \neq \hat{\beta}_1 \qquad (11.9)$$

where $\hat{\beta}_{1 \mid C_1, C_2, \ldots, C_p}$ denotes the (adjusted) estimate of β_1 using (11.8) and $\hat{\beta}_1$ is the (crude) estimate of β_1 using (11.6).

One problem with applying this definition is that it addresses the question of whether confounding is present without directly identifying specific variables to be controlled.[5] In other words, when confounding is deemed to be present based on (11.9), it may still be the case that only a subset of $C_1, C_2, \ldots, C_p$ is required for adequate control. How does one identify such a subset? More specifically, why bother to identify such a subset rather than simply to control for all variables $C_1, C_2, \ldots, C_p$?

One answer to the latter question is that, when addressing the control of covariates, we should consider the possible gains in *precision*[6] in addition to the control of confounding. In

---

[4]As a caveat to these recommendations, we note that certain variables usually referred to as *intervening variables* should not be considered as potential confounders (Kleinbaum, Kupper, and Morgenstern 1982). A variable C is said to intervene between T and Y if T causes C and then C causes Y. Controlling intervening variables may spuriously reduce or eliminate any manifestation in the data of a true association between T and Y.

[5]Another problem relates to assessing confounding when there are two or more study variables—say, T_1 and T_2—of interest. For this general situation, confounding may be defined to be present if (11.9) is satisfied for the coefficient of *any* study variable of interest, given a model containing all such study variables and all control variables. Unfortunately, this definition has the practical drawback of requiring the researcher to make a subjective decision for each study variable of interest.

[6]The term *precision* refers to the size of an estimator's variance or, equivalently, the narrowness of a confidence interval for the parameter being estimated. The smaller the variance of the estimator, the higher is the precision of the estimator. Equivalently, since the width of a confidence interval depends on the variance estimate, the narrower is the width of the confidence interval, the higher is the precision of the estimator.

particular, we may prefer a subset of C_i variables to the entire set because the subset may provide equivalent control of confounding (i.e., may give the same adjusted estimate) while providing greater precision in estimating the adjusted association of interest. However, there is no guarantee that precision will be increased by using a subset; in fact, precision may be reduced. In any case, confounding should take precedence over precision in the sense that no subset should be considered unless it gives the same adjusted-effect estimate as is obtained when the researcher controls for all C_i's.

To illustrate, suppose that $p = 5$; that is, we consider controlling for $C_1, C_2, \ldots, C_5$ using model (11.8). Suppose, too, that the estimate of β_1 takes on the following values, depending on which sets of $C_1, C_2, \ldots, C_5$ are controlled:

$$\hat{\beta}_{1 \mid C_1, C_2, \ldots, C_5} = 4.0, \quad \hat{\beta}_{1 \mid C_1, C_2} = 4.3, \quad \hat{\beta}_1 = 16.0$$

Then, because 16.0 is much different from 4.0, one can argue that confounding is present. Yet since 4.0 is not meaningfully different from 4.3, it can also be argued that C_3, C_4, and C_5 do not need to be controlled, since essentially the same (adjusted) estimate is obtained when we control only for C_1 and C_2 as when we adjust for all C_i's.

Thus, for this example, we have identified two sets of C_i variables that we can use for control. Which set do we choose? The answer depends on an evaluation of precision. One approach is to compare interval estimates for some parameter of interest—one interval being derived from a model that controls for C_1 and C_2 only, and the other interval coming from a model that controls for C_1 through C_5. The logical parameter for this example is the population regression coefficient β_1 of the variable T, when controlling for a particular set of C_i's. In other words, we may compare an interval estimate for β_1 when only C_1 and C_2 are controlled to a corresponding interval estimate for β_1 when C_1 through C_5 are controlled. The narrower interval of the two is the interval reflecting the greater precision. For example, if the two 95% interval estimates are (2.6, 7.4) for $\beta_{1 \mid C_1, C_2}$ and (1.7, 7.6) for $\beta_{1 \mid C_1, C_2, \ldots, C_5}$, then the former interval is narrower; in this case, some precision is gained by dropping C_3, C_4, and C_5 from the model.

An alternative, but not exactly equivalent, approach to evaluating precision is to perform a statistical test for the significance of the addition of C_3, C_4, and C_5 to a model containing T, C_1, and C_2. The null hypothesis for this test may be stated as $H_0: \beta_4 = \beta_5 = \beta_6 = 0$ in model (11.8), with $p = 5$. If this test is not significant, we could argue that retaining C_3, C_4, and C_5 does not provide additional precision (i.e., explanation of variance). This would indicate that only C_1 and C_2 should be controlled for greater precision.

Because this testing approach does not always lead to the same conclusion as the internal estimation approach, the investigator may need to choose between them. In most situations, however, both approaches lead to similar results.

How do we identify which set of C_i variables to control? We have seen, by example, that we must first identify a baseline-adjusted estimate (i.e., a "gold standard") against which to make comparisons. The ideal gold standard is the regression coefficient estimate that controls for all C_i's that are being considered. Then, any subset of C_i's that gives essentially the same adjusted estimate (i.e., an estimate that is not meaningfully different from the gold standard when only the C_i's in that subset are controlled) is a candidate set for control. Several such candidates may be possible (Kleinbaum, Kupper, and Morgenstern 1982, chap. 14).

Which set should we finally use? The answer, again, is based on precision: we should use the set that gives the greatest precision (e.g., the tightest confidence interval for the adjusted effect under study). (For "political" reasons—that is, to convince people that all variables have been controlled—it might be better to control for $C_1, C_2, \ldots, C_p$, unless some subset of C_i's leads to a large increase in precision.)

To illustrate, suppose that the candidate sets in Table 11-2 can be identified when $p = 5$ in model (11.8). All three proper subsets of C_1, C_2, C_3, C_4, and C_5 may be considered candidates for control, since they all give adjusted estimates that are roughly equal to the gold standard $\hat{\beta}_{1 \mid C_1, C_2, \ldots, C_5} = 4.8$. Of these candidates, the subset involving C_1, C_2, and C_4 gives the best precision (narrowest confidence interval); therefore, this subset can be used both to control confounding and to enhance precision.

11-4-3   An Example Revisited

In section 11-3-3 we considered a hypothetical study to assess the relationship between physical activity level (PAL) and systolic blood pressure (SBP) while controlling for both AGE and SEX. We considered a model that allows for possible interactions of AGE and SEX with PAL, and we described methods for testing for such interactions. Assuming that no significant interaction effects are found, the resulting reduced model is

$$SBP + \beta_0 + \beta_1(PAL) + \beta_2(AGE) + \beta_3(SEX) + E$$

Given this no-interaction model, we next assess confounding: Does the coefficient of PAL change when AGE and/or SEX is dropped from the model? To answer this, we can examine the estimate of the coefficient of PAL in four models—namely, one including both AGE and SEX, one involving either AGE or SEX but not both, and one involving neither. The gold standard model for comparison contains both control variables and PAL. Then, for example, if the estimate of β_1 changes considerably when at least one control variable is dropped from this gold standard model, we need to control for both AGE and SEX. However, if we obtain essentially the same estimate of β_1 (as obtained using the gold standard model) when only AGE is in the model, we do not need to retain SEX in the model to control for confounding. Even so, including the SEX variable in addition to AGE may increase or decrease precision. Thus, the decision as to whether to control for just AGE or for both AGE and SEX depends, for example, on a comparison of confidence intervals for β_1. If the confidence interval is considerably narrower when only AGE is controlled, we should not retain SEX in the model.

Finally, once we make a decision about which variables to control (i.e., which model is the best for providing a valid and precise estimate of the coefficient of PAL), we make statistical inferences about the true PAL–SBP relationship. Given the no-interaction model, this involves testing $H_0: \beta_1 = 0$ in the best model and then obtaining an interval estimate of β_1.

TABLE 11-2   An example of candidate sets for control.

Candidate Set	$\hat{\beta}_{1 \mid \text{Candidate Set}}$	95% Confidence Interval for $\beta_{1 \mid \text{Candidate Set}}$
C_1, C_2, C_3, C_4, C_5 (baseline)	4.8	(2.3, 7.2)
C_1, C_2	5.1	(2.6, 7.6)
C_1, C_4	4.6	(2.1, 7.0)
C_1, C_2, C_4	4.7	(2.6, 6.8)

11-5 Summary and Conclusions

Confounding and interaction are two methodological concepts that pertain to assessing a relationship between independent and dependent variables.

Interaction, which takes precedence over confounding, exists when the relationship of interest differs at different levels of extraneous (control) variables. In linear regression, interaction is evaluated by using statistical tests about product terms involving basic independent variables in the model.

Confounding, which is not evaluated with statistical testing, is present when the effect of interest differs depending on whether an extraneous variable is ignored or retained in the analysis. In regression terms, confounding is assessed by comparing crude versus adjusted regression coefficients from different models.

When several potential confounders are being considered, it may be worthwhile to identify nonconfounders that can be dropped from the model to gain precision; this may not be possible (i.e., precision may be lost by dropping variables) in some situations.

When there is strong interaction involving a certain extraneous variable, the assessment of confounding for that extraneous variable becomes irrelevant. Moreover, in such a situation, the assessment of confounding involving other extraneous variables, though possible, is quite complex and extremely subjective. Consequently, the assessment of confounding is not usually recommended when important interaction effects have been identified.

Problems

1. Consider the numerical examples given in section 8-7 of Chapter 8, involving assessment of the relationship of the independent variables HGT, AGE, and $(AGE)^2$ to the dependent variable WGT. Suppose that HGT is the independent variable of primary concern, so interest lies in evaluating the relationship of HGT to WGT, controlling for the possible confounding effects of AGE and $(AGE)^2$.

   a. Assuming that no interaction of any kind exists, state an appropriate regression model to use as the baseline (i.e., standard) for decisions about confounding.

   b. Using an appropriate (partial) regression coefficient as your measure of association, determine whether confounding exists due to AGE and/or $(AGE)^2$.

   c. Can $(AGE)^2$ be dropped from your initial model in part (a) because it is not needed to control adequately for confounding? Explain your answer (using a regression coefficient as your measure of association).

   d. Should $(AGE)^2$ be retained in the final model for the sake of precision? Explain.

   e. In light of both confounding and precision, what should be your final model? Why?

   f. How would you modify your initial model in part (a) to allow for assessing interactions?

   g. Regarding your answer to part (f), how would you test for interaction?

2. Consider the following computer results, which describe regression analyses involving two independent variables X_1 and X_2 and a dependent variable Y. Assume that your goal is to assess the relationship of X_1 with Y, controlling for the possible confounding effects of X_2.

Edited SAS Output (PROC REG) for Problem 2

```
                      Y Regressed on X1 and X2
                           Correlation
               CORR         X1            X2             Y
               X1        1.0000        0.5000        0.6527
               X2        0.5000        1.0000        0.9494
               Y         0.6527        0.9494        1.0000
                      Analysis of Variance
                         Sum of         Mean
Source         DF        Squares        Square     F Value    Prob > F
Model          2        268.00000     134.00000    108.875     0.0001
Error          13        16.00000       1.23077
C Total        15       284.00000

                              .
                              .  [Portion of output omitted]
                              .

                      Parameter Estimates
                    Parameter      Standard      T for H0:      Prob
Variable       DF   Estimate        Error      Parameter=0     > |T|
INTERCEP       1    5.000000      0.42365927      11.802       0.0001
      X1       1    2.000000      0.64051262       3.122       0.0081
      X2       1    7.000000      0.64051262      10.929       0.0001
                      Squared        Squared
                      Partial        Partial
Variable       DF   Corr Type I    Corr Type II
      X1       1    0.42605634     0.42857143
      X2       1    0.90184049     0.90184049
                       Y Regressed on X1
                      Analysis of Variance
                         Sum of         Mean
Source         DF        Squares        Square     F Value    Prob > F
Model          1        121.00000     121.00000     10.393     0.0061
Error          14       163.00000      11.64286
C Total        15       284.00000

                              .
                              .  [Portion of output omitted]
                              .

                      Parameter Estimates
                    Parameter      Standard      T for H0:      Prob
Variable       DF   Estimate        Error      Parameter=0     > |T|
INTERCEP       1    6.750000      1.20638184       5.595       0.0001
      X1       1    5.500000      1.70608156       3.224       0.0061
```

a. Using an appropriate regression coefficient as your measure of association, determine whether confounding exists. Explain.

b. Suppose that confounding was defined to require a comparison of crude versus adjusted (partial) correlation coefficients. What conclusion would you draw? Explain.

c. What is the moral of this example?

3. a–c. Consider the accompanying computer results, which describe regression analyses involving two independent variables X_1 and X_2 and a dependent variable Y (using a different data set from the one used in Problem 2). Answer the same questions as in Problem 2 for this new printout.

d. What does this example illustrate about using a test of the hypothesis $H_0: \beta_2 = 0$ to assess confounding?

Edited SAS Output (PROC REG) for Problem 3

```
                        Y Regressed on X1 and X2
                             Correlation
                CORR          X1            X2            Y
                X1         1.0000        0.0000        0.2649
                X2         0.0000        1.0000        0.9272
                Y          0.2649        0.9272        1.0000
                        Analysis of Variance
                        Sum of        Mean
Source       DF         Squares       Square      F Value      Prob > F
Model         2        106.00000     53.00000     33.125       0.0013
Error         5          8.00000      1.60000
C Total       7        114.00000
                              .
                              .    [Portion of output omitted]
                              .
                        Parameter Estimates
                     Parameter      Standard      T for H0:        Prob
Variable      DF     Estimate        Error       Parameter=0      > |T|
INTERCEP      1      5.000000       0.77459667      6.455         0.0013
      X1      1      2.000000       0.89442719      2.236         0.0756
      X2      1      7.000000       0.89442719      7.826         0.0005
                        Squared            Squared
                        Partial            Partial
Variable      DF     Corr Type I       Corr Type II
      X1      1      0.07017544        0.50000000
      X2      1      0.92452830        0.92452830
                           Y Regressed on X1
                        Analysis of Variance
                        Sum of        Mean
Source       DF         Squares       Square      F Value      Prob > F
Model         1          8.00000      8.00000      0.453        0.5260
Error         6        106.00000     17.66667
C Total       7        114.00000
                              .
                              .    [Portion of output omitted]
                              .
                        Parameter Estimates
                     Parameter      Standard      T for H0:        Prob
Variable      DF     Estimate        Error       Parameter=0      > |T|
INTERCEP      1      8.500000       2.10158670      4.045         0.0068
      X1      1      2.000000       2.97209242      0.673         0.5260
```

4. A regression analysis of data on $n = 53$ males considered the following variables:

$$Y = \text{SBPSL (estimated slope based on the straight-line regression of an individual's blood pressure over time)}$$

$$X_1 = \text{SBP1 (initial blood pressure)}$$

$$X_2 = \text{RW (relative weight)}$$

$$X_3 = X_1X_2 = \text{SR (product of SBP1 and RW)}$$

The accompanying computer printout was obtained by using a standard stepwise regression program (SPSS). Using this output, complete the following exercises.

a. Fill in the following ANOVA table for the fit of the model $Y = \beta_0 + \beta_1 X_1 + \beta_2 X_2 + \beta_3 X_3 + E$.

Source		d.f.	SS	MS
Regression $\begin{cases} X_1 \\ X_2 \mid X_1 \\ X_3 \mid X_1, X_2 \end{cases}$				
Residual				
Total		52		

b. Test H_0: $\rho_{YX_2 \mid X_1} = 0$.

c. Test H_0: "The addition of X_3 to the model, given that X_1 and X_2 are already in the model, is not significant."

d. Test H_0: $\rho_{Y(X_2, X_3) \mid X_1} = 0$.

e. Based on the tests in parts (b) through (d), what is the most appropriate regression model? Use $\alpha = .05$.

f. Based on the information provided, can you assess whether X_1 is a confounder of the X_2–Y relationship? Explain.

5. An experiment involved a quantitative analysis of factors found in high-density lipoprotein (HDL) in a sample of human blood serum. Three variables thought to be predictive of or associated with HDL measurement (Y) were the total cholesterol (X_1) and total triglyceride (X_2) concentrations in the sample, plus the presence or absence of a certain sticky component called sinking pre-beta, or SPB (X_3), which was coded as 0 if absent and 1 if present. The data obtained are shown in the following table and the accompanying computer results.

Edited SPSS Output for Problem 4

```
DEPENDENT VARIABLE.. SBPSL
VARIABLE(S) ENTERED ON STEP NUMBER 1.. SBP 1

MULTIPLE R        0.45834     ANALYSIS OF VARIANCE      DF      SUM OF SQUARES     MEAN SQUARE        F
R SQUARE          0.21007     REGRESSION                 1.          14.79083        14.79083     13.56308
STANDARD ERROR    1.04428     RESIDUAL                  51.          55.61661         1.09052
------VARIABLES IN THE EQUATION------           ------VARIABLES NOT IN THE EQUATION------

VARIABLE          B       BETA    STD ERROR B      F       VARIABLE  BETA IN  PARTIAL  TOLERANCE      F
SBP 1       -0.04660   -0.45834   0.01265       13.563     RW         0.23166  0.26007  0.99953     3.627
(CONSTANT)   5.10797                                       SR         0.23074  0.25953  0.99933     3.611

VARIABLE(S) ENTERED ON STEP NUMBER 2.. RW

MULTIPLE R        0.51332     ANALYSIS OF VARIANCE      DF      SUM OF SQUARES     MEAN SQUARE        F
R SQUARE          0.26350     REGRESSION                 2.          18.55240         9.27620      8.94435
STANDARD ERROR    1.01838     RESIDUAL                  50.          51.85504         1.03710
------VARIABLES IN THE EQUATION------           ------VARIABLES NOT IN THE EQUATION------

VARIABLE          B       BETA    STD ERROR B      F       VARIABLE  BETA IN  PARTIAL  TOLERANCE      F
SBP 1       -0.04817   -0.47382   0.01237       15.174     SR         0.04646  0.00450  0.00690     0.001
RW          -0.02252   -0.23166   0.01182        3.627
(CONSTANT)   5.38484

VARIABLE(S) ENTERED ON STEP NUMBER 3.. SR

MULTIPLE R        0.51334     ANALYSIS OF VARIANCE      DF      SUM OF SQUARES     MEAN SQUARE        F
R SQUARE          0.26352     REGRESSION                 3.          18.55345         6.18448      5.64409
STANDARD ERROR    1.02871     RESIDUAL                  49.          51.85399         1.05824
------VARIABLES IN THE EQUATION------           ------VARIABLES NOT IN THE EQUATION------

VARIABLE          B       BETA    STD ERROR B      F       VARIABLE  BETA IN  PARTIAL  TOLERANCE      F
SBP 1       -0.04798   -0.47193   0.01391       11.899
RW          -0.01801   -0.18527   0.14372       10.016
SR           0.00004    0.04646   0.00122        0.001
(CONSTANT)   5.36183
```

Y	X_1	X_2	X_3		Y	X_1	X_2	X_3
47	287	111	0		57	192	115	1
38	236	135	0		42	349	408	1
47	255	98	0		54	263	103	1
39	135	63	0		60	223	102	1
44	121	46	0		33	316	274	0
64	171	103	0		55	288	130	0
58	260	227	0		36	256	149	0
49	237	157	0		36	318	180	0
55	261	266	0		42	270	134	0
52	397	167	0		41	262	154	0
49	295	164	0		42	264	86	0
47	261	119	1		39	325	148	0
40	258	145	1		27	388	191	0
42	280	247	1		31	260	123	0
63	339	168	1		39	284	135	0
40	161	68	1		56	326	236	1
59	324	92	1		40	248	92	1
56	171	56	1		58	285	153	1
76	265	240	1		43	361	126	1
67	280	306	1		40	248	226	1
57	248	93	1		46	280	176	1

Edited SAS Output (PROC GLM) for Problem 5

```
                    General Linear Models Procedure
                         Y Regressed on X1
                                   Sum of
Source                   DF        Squares        F Value        Pr > F
Model                     1      46.23555746        0.40         0.5282
Error                    40    4567.38349016
Corrected Total          41    4613.61904762

              R-Square                  C.V.                     Y Mean
              0.010022               22.37289               47.7619048

                                   Type I
Source                   DF          SS           F Value        Pr > F
X1                        1      46.23555746        0.40         0.5282

                                  Type III
Source                   DF          SS           F Value        Pr > F
X1                        1      46.23555746        0.40         0.5282

                                 T for H0:                     Std Error
Parameter       Estimate      Parameter=0     Pr > |T|       of Estimate
INTERCEPT      52.47018272        6.92          0.0001        7.58057266
X1             -0.01758070       -0.64          0.5282        0.02762815

                         Y Regressed on X2
                                   Sum of
Source                   DF        Squares        F Value        Pr > F
Model                     1      21.33970871        0.19         0.6687
Error                    40    4592.27933891
Corrected Total          41    4613.61904762
```

```
                    Y Regressed on X2 (Continued)
                R-Square               C.V.                    Y Mean
                0.004625             22.43378               47.7619048
                                    Type I
Source              DF                 SS          F Value      Pr > F
X2                  1            21.33970871         0.19       0.6687
                                   Type III
Source              DF                 SS          F Value      Pr > F
X2                  1            21.33970871         0.19       0.6687
                                 T for H0:                    Std Error
Parameter        Estimate      Parameter=0    Pr > |T|      of Estimate
INTERCEPT       46.24519280       11.90         0.0001       3.88711502
X2               0.00978223        0.43         0.6687       0.02268966
                        Y Regressed on X3
                                   Sum of
Source              DF            Squares        F Value       Pr > F
Model               1          735.20541126       7.58         0.0088
Error               40        3878.41363636
Corrected Total     41        4613.61904762
                R-Square               C.V.                    Y Mean
                0.159355             20.61652               47.7619048
                                    Type I
Source              DF                 SS          F Value      Pr > F
X3                  1          735.20541126        7.58         0.0088
                                   Type III
Source              DF                 SS          F Value      Pr > F
X3                  1          735.20541126        7.58         0.0088
                                 T for H0:                    Std Error
Parameter        Estimate      Parameter=0    Pr > |T|      of Estimate
INTERCEPT       43.77272727       20.85         0.0001       2.09935424
X3               8.37727273        2.75         0.0088       3.04225332
                      Y Regressed on X1 and X2
                                   Sum of
Source              DF            Squares        F Value       Pr > F
Model               2          135.38212555       0.59         0.5595
Error               39        4478.23692207
Corrected Total     41        4613.61904762
                R-Square               C.V.                    Y Mean
                0.029344             22.43570               47.7619048
                                    Type I
Source              DF                 SS          F Value      Pr > F
X1                  1           46.23555746        0.40         0.5294
X2                  1           89.14656809        0.78         0.3837
                                   Type III
Source              DF                 SS          F Value      Pr > F
X1                  1          114.04241683        0.99         0.3251
X2                  1           89.14656809        0.78         0.3837
                                 T for H0:                    Std Error
Parameter        Estimate      Parameter=0    Pr > |T|      of Estimate
INTERCEPT       52.76400680        6.93         0.0001       7.60916422
X1              -0.03216055       -1.00         0.3251       0.03227093
X2               0.02328833        0.88         0.3837       0.02643061
```

(continued)

```
                 Y Regressed on X1 and X3
                                  Sum of
Source                 DF         Squares       F Value      Pr > F
Model                   2      783.16906883        3.99      0.0266
Error                  39     3830.44997878
Corrected Total        41     4613.61904762

               R-Square                C.V.                  Y Mean
               0.169752            20.74966              47.7619048

                                  Type I
Source                 DF             SS       F Value      Pr > F
X1                      1       46.23555746        0.47      0.4967
X3                      1      736.93351138        7.50      0.0092

                                 Type III
Source                 DF             SS       F Value      Pr > F
X1                      1       47.96365758        0.49      0.4888
X3                      1      736.93351138        7.50      0.0092

                               T for H0:               Std Error
Parameter        Estimate     Parameter=0   Pr > |T|   of Estimate
INTERCEPT      48.56350948         6.77      0.0001     7.17377750
X1             -0.01790642        -0.70      0.4888     0.02562390
X3              8.38720265         2.74      0.0092     3.06193225

                 Y Regressed on X2 and X3
                                  Sum of
Source                 DF         Squares       F Value      Pr > F
Model                   2      737.80689080        3.71      0.0334
Error                  39     3875.81215682
Corrected Total        41     4613.61904762

               R-Square                C.V.                  Y Mean
               0.159919            20.87216              47.7619048

                                  Type I
Source                 DF             SS       F Value      Pr > F
X2                      1       21.33970871        0.21      0.6457
X3                      1      716.46718209        7.21      0.0106

                                 Type III
Source                 DF             SS       F Value      Pr > F
X2                      1        2.60147955        0.03      0.8723
X3                      1      716.46718209        7.21      0.0106

                               T for H0:               Std Error
Parameter        Estimate     Parameter=0   Pr > |T|   of Estimate
INTERCEPT      43.26641927        11.44      0.0001     3.78286563
X2              0.00343683         0.16      0.8723     0.02124209
X3              8.32148668         2.69      0.0106     3.09921587

         Y Regressed on X1, X2, X3, X1X3, and X2X3
                                  Sum of
Source                 DF         Squares       F Value      Pr > F
Model                   5      894.56736631        1.73      0.1523
Error                  36     3719.05168130
Corrected Total        41     4613.61904762

               R-Square                C.V.                  Y Mean
               0.193897            21.28057              47.7619048
```

```
          Y Regressed on X1, X2, X3, X1X3, and X2X3 (Continued)
                                  Type I
Source               DF              SS          F Value      Pr > F
X1                    1        46.23555746         0.45       0.5078
X2                    1        89.14656809         0.86       0.3591
X3                    1       684.36519731         6.62       0.0143
X1X3                  1        48.69043235         0.47       0.4968
X2X3                  1        26.12961110         0.25       0.6181

                                 Type III
Source               DF              SS          F Value      Pr > F
X1                    1       154.23113774         1.49       0.2297
X2                    1        47.48756478         0.46       0.5021
X3                    1         1.99625065         0.02       0.8902
X1X3                  1        74.26755996         0.72       0.4021
X2X3                  1        26.12961110         0.25       0.6181

                                T for H0:                   Std Error
Parameter        Estimate    Parameter=0    Pr > |T|      of Estimate
INTERCEPT       52.33465853       5.65        0.0001        9.25548949
X1              -0.04955819      -1.22        0.2297        0.04055966
X2               0.03188430       0.68        0.5021        0.04702749
X3              -2 07933755      -0.14        0.8902       14.95830269
X1X3             0.05438698       0.85        0.4021        0.06414462
X2X3            -0.02821521      -0.50        0.6181        0.05610244
```

a. Test whether X_1, X_2, or X_3 alone significantly helps in predicting Y.

b. Test whether X_1, X_2, and X_3 taken together significantly help to predict Y.

c. Test whether the true coefficients of the product terms X_1X_3 and X_2X_3 are simultaneously zero in the model containing X_1, X_2, and X_3 plus these product terms. State the null hypothesis in terms of a multiple partial correlation coefficient. If this test is not rejected, what can you conclude about the relationship of Y to X_1 and X_2 when X_3 equals 1, as compared with when X_3 equals 0?

d. Using $\alpha = .05$, test whether X_3 is associated with Y, after the combined contribution of X_1 and X_2 is taken into account. State the appropriate null hypothesis in terms of a partial correlation coefficient. What does your result, together with your answer to part (c), tell you about the relationship of Y with X_1 and X_2 when SPB is present, as compared with when it is absent?

e. How would you determine whether X_1, X_2, or both X_1 and X_2 need to be retained in the model to control for confounding and possibly to enhance precision. Assume that no interaction occurs and that the study variable of interest is X_3.

f. Based on the information provided, can confounding of X_1 and/or X_2 be assessed in evaluating the relationship of X_3 to Y? Explain.

6. Use the computer output from Problem 10 in Chapter 8 and from Problem 16 in Chapter 5 to answer the following questions. (Assume that no interaction occurs between house size (X_1) and number of rooms (X_2).)

a. Does the number of rooms confound the relationship between house price (Y) and house size (X_1)?

b. Should the number of rooms be included in a model that already contains house size, based on considerations of precision?

    c. In light of your answers to parts (a) and (b), should the final model include both predictors? Explain.

7. Use the output from Problem 11 in Chapter 8 and the output given here to answer the following questions. (Assume that no interaction occurs between TV advertising expenditure (X_1) and print advertising expenditure (X_2).)

    a. Does the print advertising expenditure confound the relationship between sales (Y) and TV advertising expenditure (X_1)?

    b. Should print advertising expenditure be included in a model that already contains TV advertising expenditure, based on considerations of precision?

    c. In light of your answers to parts (a) and (b), should the final model include both types of advertising expenditures as predictors? Explain.

Edited SAS Output (PROC GLM) for Problem 7

```
                      Y Regressed on X1
                                .
                                .  [Portion of output omitted]
                                .
                              T for H0:
Parameter          Estimate   Parameter=0    Pr > |T|    Std Error
                                                         of Estimate
INTERCEPT       2.002227171        6.34       0.0032     0.31569714
X1              1.224944321       12.14       0.0003     0.10088866
```

8. Use the computer output from Problem 12 in Chapter 8 and the output given here to answer the following questions.

    a. State the model that relates change in refraction (Y) to baseline refraction (X_1), baseline curvature (X_2), and the interaction of X_1 and X_2. Is the partial test for the interaction significant?

    b. Is it appropriate to assess confounding, given your answer to part (a)? Explain.

    c. If your answer to part (b) is yes, does X_2 confound the relationship between Y and X_1?

    d. If your answer to part (b) is yes, does X_1 confound the relationship between Y and X_2?

    e. In light of your answers to parts (a) through (d), and on considerations of precision, which predictor(s) should be included in the model?

Edited SAS Output (PROC GLM) for Problem 8

```
                 Y Regressed on X1, X2, and X1*X2
                                .
                                .  [Portion of output omitted]
                                .
                              T for H0:
Parameter          Estimate   Parameter=0    Pr > |T|    Std Error
                                                         of Estimate
INTERCEPT      25.85215663        1.58       0.1199     16.33395716
X1              2.69854313        0.79       0.4339      3.41991591
X2             -0.52636869       -1.42       0.1613      0.37006537
X1X2           -0.06778472       -0.87       0.3861      0.07750534
```

```
                         Y Regressed on X1
                                  .
                                  . [Portion of output omitted]
                                  .

                              T for HO:                    Std Error
Parameter       Estimate      Parameter=0     Pr > |T|     of Estimate
INTERCEPT     2.619102900          6.35         0.0001     0.41260537
X1           -0.298729328         -3.16         0.0027     0.09463300

                         Y Regressed on X2
                                  .
                                  . [Portion of output omitted]
                                  .

                              T for HO:                    Std Error
Parameter       Estimate      Parameter=0     Pr > |T|     of Estimate
INTERCEPT    14.14611679          2.56         0.0135     5.52569396
X2           -0.23446227         -1.87         0.0674     0.12544713

                     Y Regressed on X1 and X2
Dependent Variable: Y

                                    Sum of
Source                   DF         Squares      F Value      Pr > F
Model                     2      17.53454762        7.02      0.0021
Error                    50      62.41955709
Corrected Total          52      79.95410472

                 R-Square                C.V.              Y mean
                 0.219308             29.22453          3.82320755

                                  .
                                  . [Portion of output omitted]
                                  .

                              T for HO:                    Std Error
Parameter       Estimate      Parameter=0     Pr > |T|     of Estimate
INTERCEPT    12.29334748          2.40         0.0204     5.13074580
X1           -0.29135396         -3.15         0.0027     0.09241113
X2           -0.21905406         -1.89         0.0644     0.11581741
```

9. Refer to the data in Problem 19 of Chapter 5. The relationship between the rate of owner occupancy of housing units (OWNEROCC) and the median monthly ownership costs (OWNCOST) was studied in that problem in connection with a random sample of data from 26 Metropolitan Statistical Areas (MSAs)[1]. The results of the regression of OWNEROCC on OWNCOST and a third variable, the median household income (INCOME) are presented next.

Use the output given here and in Problem 19 of Chapter 5 to answer the following questions.

a. Test whether OWNCOST and INCOME, taken together, significantly help to predict the rate of owner occupancy.

b. Test whether OWNCOST is associated with the rate of owner occupancy, after taking into account the contribution of INCOME.

c. Test whether INCOME is associated with the rate of owner occupancy, after taking into account the contribution of OWNCOST.

d. Should INCOME be included in the model to control for confounding? Explain your answer, including discussion of any assumptions you made in reaching your answer.

Edited SAS Output (PROC GLM) for Problem 9

```
              OWNEROCC Regressed on OWNCOST and INCOME
                 General Linear Models Procedure
Dependent Variable: OWNEROCC
                           Sum of              Mean
Source            DF       Squares            Square    F Value   Pr > F
Model              2    214.4833906       107.2416953      6.38   0.0062
Error             23    386.4781479        16.8033977
Corrected Total   25    600.9615385

              R-Square              C.V.        Root MSE      OWNEROCC Mean
              0.356900          6.214523        4.099195          69.96154
                           Type I              Mean
Source            DF          SS              Square    F Value   Pr > F
OWNCOST            1    132.6203242       132.6203242      7.89   0.0100
INCOME             1     81.8630664        81.8630664      4.87   0.0375
                           Type III            Mean
Source            DF          SS              Square    F Value   Pr > F
OWNCOST            1    192.6978217       192.6978217     11.47   0.0025
INCOME             1     81.8630664        81.8630664      4.87   0.0375
                                   T for H0:                      Std Error
Parameter         Estimate      Parameter=0     Pr > |T|       of Estimate
INTERCEPT       70.03201034           15.37       0.0001        4.55666725
OWNCOST         -0.03334031           -3.39       0.0025        0.00984532
INCOME           0.00064489            2.21       0.0375        0.00029217
```

10. Refer to the *Business Week* magazine data in Problem 13 of Chapter 8. The relationship between the yield (Y), 1989 rank (X_2), and P-E ratio (X_3) was studied in that problem, using a random sample of data from the magazine's compilation of information on the top 1,000 companies. The results of the regression of yield on 1989 rank, P-E ratio, and 1990 rank (X_1) are presented next. Use this output and the output from Problem 13 in Chapter 8 to answer the following questions.

Edited SAS Output (PROC GLM) for Problem 10

```
      YIELD (Y) Regressed on 1990 RANK (X1), 1989 RANK (X2),
                       and P-E RATIO (X3)
                 General Linear Models Procedure
Dependent Variable: Y
                           Sum of              Mean
Source            DF       Squares            Square    F Value   Pr > F
Model              3     29.97725832        9.99241944      8.83   0.0011
Error             16     18.10303668        1.13143979
Corrected Total   19     48.08029500

              R-Square              C.V.        Root MSE         Y Mean
              0.623483          42.79588        1.063692        .485500
```

```
         General Linear Models Procedure (Continued)
                     Type I              Mean
Source      DF         SS               Square      F Value    Pr > F
X1          1      1.82894871        1.82894871       1.62      0.2218
X2          1      0.00159198        0.00159198       0.00      0.9705
X3          1     28.14671763       28.14671763      24.88      0.0001
                    Type III             Mean
Source      DF         SS               Square      F Value    Pr > F
X1          1      3.39453697        3.39453697       3.00      0.1025
X2          1      1.35441258        1.35441258       1.20      0.2901
X3          1     28.14671763       28.14671763      24.88      0.0001
                                   T for HO:                  Std Error
Parameter          Estimate      Parameter=0    Pr > |T|    of Estimate
INTERCEPT         7.421187723        7.51         0.0001      0.98839344
X1               -0.013351684       -1.73         0.1025      0.00770835
X2                0.007620880        1.09         0.2901      0.00696539
X3               -0.261846279       -4.99         0.0001      0.05249866
```

a. Test whether 1989 rank, P-E ratio, and 1990 rank, taken together, significantly help to predict the yield.

b. Test whether 1990 rank is associated with the yield, after taking into account the contribution of 1989 rank and P-E ratio.

c. Should 1990 rank be included in the model to control for confounding? Explain your answer, including discussion of any assumptions you made in reaching your answer.

References

Kleinbaum, D. G.; Kupper, L. L.; and Morgenstern, H. 1982. *Epidemiologic Research.* Belmont, Calif.: Lifetime Learning Publications.

Nie, N., et al. 1975. *Statistical Package for the Social Sciences.* New York: McGraw-Hill.

Regression Diagnostics

12-1 Preview

This chapter provides a general overview of statistical techniques known as *regression diagnostics.*[1] Such techniques are employed to check the assumptions and to assess the accuracy of computations for a multiple regression analysis. We shall focus primarily on methods for analyzing residuals, assessing the influence of outliers, and assessing collinearity. All of these methods are essentially diagnostic tools. In addition, procedures for analyzing the data that solve, avoid, or help correct diagnosed problems are briefly discussed.

12-2 Simple Approaches to Diagnosing Problems in Data

In analyzing data, it is important to be familiar with their basic characteristics. Such familiarity helps avoid many errors. For example, it is essential to know the following:

1. The type of subject or experimental unit (e.g., small pine tree needles, elderly male humans)

2. The procedure for data collecting

3. The unit of measurement for each variable (e.g., kilograms, meters, square inches, cubic centimeters)

4. A plausible range of values and a typical value for each variable

---

[1] We use the term *regression diagnostics* much less restrictively than do many writers in the statistical literature. Our usage is intended to encourage the reader to examine the validity of all aspects of a regression analysis.

This knowledge, when combined with descriptive statistics computed for a set of data, can be used to detect errors in the data and pinpoint potential violations of the assumptions of a planned analysis. For regression analysis in particular, simple descriptive statistics computed on the response variable and the predictor variables can be extremely helpful for detecting potential violations of the assumptions. Using a comprehensive statistical computer package, a researcher can easily generate descriptive statistics for all variables in an analysis.

As a first step in detecting data base problems, we recommend listing the five largest and five smallest values for every variable. This is a simple but extremely powerful technique for examining data. It must be combined with knowledge of the type of subject, data collection procedure, measurement units, and plausible ranges. One can then immediately detect many data-recording errors, format errors in computer input, and some outliers. In addition, the extremes are a good indication of whether the data lie in the expected range. A listing of the data may be used to check for and correct errors, such as format errors in computer input; however, experience indicates that detecting individual data point problems is unlikely, particularly as the number of observations increases beyond 50.

A useful second step in detecting problems and assessing analysis assumptions is to calculate descriptive statistics. The choice of statistics depends on the scale of measurement (nominal, ordinal, or interval; see Chapter 3). Many good alternatives are available. One simple approach is to consider frequency tables for variables with few distinct values, and the mean, standard deviation, and minimum and maximum values for continuous variables. For continuous variables, more detailed information can usually be presented conveniently in some type of frequency histogram. It is often helpful to compute such descriptive statistics separately for important groups within the sample (such as males and females). Such descriptive statistics should be compared with what is expected from the study design and from scientific knowledge about the variables. Suspect and implausible data should be investigated more carefully.

More elaborate descriptive approaches can also be made part of a regression analysis, including correlations among pairs of variables and plots of the response as a function of each predictor. As will be discussed later, very high between-predictor correlations may signal collinearity problems. In other cases, the plots may indicate the presence of nonlinear relationships or troublesome unevenness in the distribution of the data.

We will illustrate many of the techniques described in this chapter with reference to two specific examples. The data for the first are taken from Lewis and Taylor (1967), where WEIGHT, HEIGHT, and AGE were all recorded for a sample of boys and girls; in this chapter, we will consider only the data from the 127 boys, and we will evaluate a model in which the response is the subject's WEIGHT and the predictors are HEIGHT, $(\text{HEIGHT})^2$, AGE, and $(\text{AGE})^2$. The data for the second example arise from a hypothetical calibration experiment in environmental engineering. The 17 values of the "concentration of a certain pollutant" (X) and the "instrument reading" (Y) are given in the following computer output, along with other "regression diagnostic" information to be discussed later.

In our first example, the first step in assessing the appropriateness of the regression assumptions is to consider descriptive statistics for the three variables WEIGHT, HEIGHT, and AGE. Frequency histograms and their close relatives, the schematic plot and the stem-and-leaf diagram (all to be described later in this chapter), are the most suitable tools for this task. Analyzing raw data can be helpful for detecting serious errors in individual observations or for suggesting an appropriate model. Analysis of residuals and other regression diagnostic procedures provide the most refined and accurate evaluation of model assumptions.

SAS Output Showing Residuals for the Calibration Experiment

						Studentized $\left(r_i = e_i/\sqrt{S^2(1-h_i)}\right)$	

Predictor* ↓ $\hat{Y}_i$ ↓ e_i ↓   residual ↓

X	Obs	Dep Var Y	Predict Value	Std Err Predict	Residual	Std Err Residual	Student Residual
0.0	1	10.7000	15.3941	1.909	-4.6941	3.641	-1.289
0.5	2	14.2000	17.2875	1.739	-3.0875	3.725	-0.829
1.0	3	16.7000	19.1809	1.577	-2.4809	3.797	-0.653
1.5	4	19.1000	21.0743	1.425	-1.9743	3.857	-0.512
2.0	5	24.9000	22.9676	1.287	1.9324	3.905	0.495
2.5	6	25.4000	24.8610	1.169	0.5390	3.942	0.137
3.0	7	32.3000	26.7544	1.077	5.5456	3.968	1.398
3.5	8	30.8000	28.6478	1.018	2.1522	3.983	0.540
4.0	9	39.6000	30.5412	0.997	9.0588	3.989	2.271
4.5	10	30.3000	32.4346	1.018	-2.1346	3.983	-0.536
5.0	11	37.2000	34.3279	1.077	2.8721	3.968	0.724
5.5	12	37.8000	36.2213	1.169	1.5787	3.942	0.401
6.0	13	37.5000	38.1147	1.287	-0.6147	3.905	-0.157
6.5	14	38.6000	40.0081	1.425	-1.4081	3.857	-0.365
7.0	15	42.6000	41.9015	1.577	0.6985	3.797	0.184
7.5	16	44.3000	43.7949	1.739	0.5051	3.725	0.136
8.0	17	37.2000	45.6882	1.909	-8.4882	3.641	-2.331

Dot-plot of jackknife residuals ↓   Jackknife $\left(r_{(-i)} = e_i/\sqrt{S^2_{(-i)}(1-h_i)}\right)$ residual ↓   Leverage (h_i) ↓

Obs	-2 -1 -0 1 2	Cook's D	Rstudent	Hat Diag H
1	| ** |	0.229	-1.3208	0.2157
2	| * |	0.075	-0.8197	0.1789
3	| * |	0.037	-0.6404	0.1471
4	| * |	0.018	-0.4989	0.1201
5	| |	0.013	0.4821	0.0980
6	| |	0.001	0.1322	0.0809
7	| ** |	0.072	1.4478	0.0686
8	| * |	0.010	0.5271	0.0613
9	| ****|	0.161	2.7089	0.0588
10	| * |	0.009	-0.5227	0.0613
11	| * |	0.019	0.7119	0.0686
12	| |	0.007	0.3890	0.0809
13	| |	0.001	-0.1522	0.0980
14	| |	0.009	-0.3543	0.1201
15	| |	0.003	0.1779	0.1471
16	| |	0.002	0.1311	0.1789
17	|**** |	0.747	-2.8204	0.2157

*SAS does not print the dependent variable values; here they have been added for completeness.

As mentioned earlier, plots are informative prior to regression analysis. The data from Lewis and Taylor (1967) are plotted in scatter diagrams that track the response variable WEIGHT on the vertical axis and one or the other of the two predictors HEIGHT and AGE on the horizontal axis (Figures 12-1 and 12-2, respectively). In both cases, a single observation (which is circled) appears isolated at the upper edge of the plot, making it highly suspect. A listing of the five largest and smallest values for each variable would also reveal the same point. Two possible interpretations of such an unusual observation are as follows: (1) the values for the observation were measured, recorded, or entered in the computer incorrectly; (2) the values are correct, and the observation must be evaluated for its effect on the analysis. The mere fact that an observation appears to be unusual when compared with the rest of the data does not automatically mean that it should be dropped. The observation appears below the main data cluster (but not nearly as suspiciously) in Figure 12-3, which plots HEIGHT against AGE. We will later

<u>*FIGURE 12-1*</u>   Children's body weight as a function of height; data from Lewis and Taylor (1967) ($n = 127$).

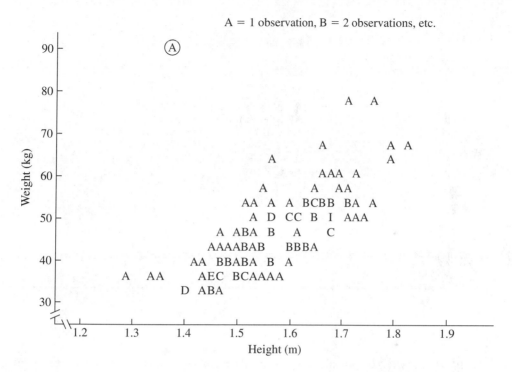

<u>*FIGURE 12-2*</u>   Children's body weight as a function of age; data from Lewis and Taylor (1967) ($n = 127$).

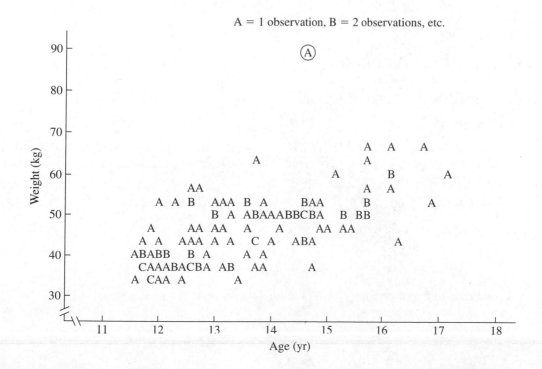

FIGURE 12-3   Children's height as a function of age; data from Lewis and Taylor (1967) ($n = 127$).

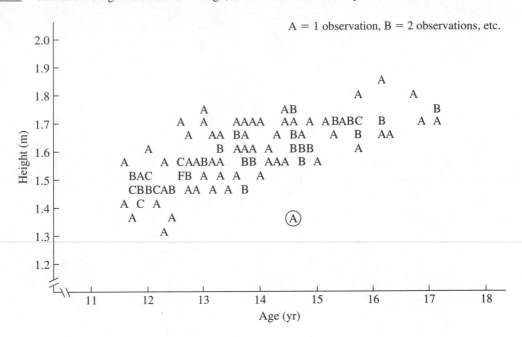

see how regression diagnostic procedures permit easier detection of such observations and indicate their influence on a regression analysis. Most of the regression diagnostics to be described are as easy to use with several predictors as with one or two.

12-3 Residual Analysis

12-3-1 Some Properties of Residuals

Analytic Properties

Given n observations $(Y_i, X_{i1}, X_{i2}, \ldots, X_{ik})$, where $i = 1, 2, \ldots, n$, recall that the methodology of regression analysis is concerned with the least-squares fitting of a model describing the observed response Y_i as

$$Y_i = \beta_0 + \beta_1 X_{i1} + \beta_2 X_{i2} + \cdots + \beta_k X_{ik} + E_i, \quad i = 1, 2, \ldots, n$$

in which E_i denotes the (unobserved) error term for the ith response. We write the fitted model as

$$\hat{Y} = \hat{\beta}_0 + \hat{\beta}_1 X_1 + \hat{\beta}_2 X_2 + \cdots + \hat{\beta}_k X_k$$

and then the predicted response at the ith data point is

$$\hat{Y}_i = \hat{\beta}_0 + \hat{\beta}_1 X_{i1} + \hat{\beta}_2 X_{i2} + \cdots + \hat{\beta}_k X_{ik}$$

With this framework in mind, we define the ith residual e_i as the difference between the observed value Y_i and the predicted value $\hat{Y}_i$—namely,

$$e_i = Y_i - \hat{Y}_i, \quad i = 1, 2,\ldots, n$$

In words, the $\{e_i\}$ reflect the amount of discrepancy between observed and predicted values that remains after the data have been fitted by the least-squares model. Each e_i represents an estimate of the corresponding unobserved error E_i. The usual assumptions (described in Chapter 8) made about the unobserved error terms $\{E_i\}$ for regression analysis are that they are independent, have a mean of zero, have a common variance σ^2, and follow a normal distribution (the normality assumption is required for performing parametric tests of significance). If the model is indeed appropriate for the data under analysis, it is reasonable to expect the *observed* residuals $\{e_i\}$ to exhibit properties in keeping with the stated assumptions. The basic strategy underlying the statistical procedure generally referred to as *residual analysis* is to assess the appropriateness of a model according to the behavior of the set of observed residuals. Our purpose here is to discuss methods for making such an assessment. The methodology discussed is applicable to any context in which a model is fitted and a set of residuals is produced (e.g., to analysis of variance and to nonlinear regression, as well as to multiple linear regression).

The n residuals $e_1, e_2,\ldots, e_n$ and the functions thereof possess the following characteristics:

1. The mean of the $\{e_i\}$ is 0:

$$\bar{e} = \frac{1}{n}\sum_{i=1}^{n} e_i = 0$$

2. The estimate of population variance computed from the sample of the n residuals is

$$S^2 = \frac{1}{n - k - 1}\sum_{i=1}^{n} e_i^2$$

which is exactly the residual mean square, $\mathrm{SSE}/(n - k - 1)$; S^2 is an unbiased estimator of σ^2 if the model involving $k + 1$ parameters is correct.

3. The $\{e_i\}$ are not independent random variables. This is obvious from the fact that the $\{e_i\}$ sum to zero. In general, if the number of residuals (n) is large relative to the number of independent variables (k), the dependency effect can, for all practical purposes, be ignored in any analysis of the residuals (see Anscombe and Tukey (1963) for a discussion of the effect of this dependency on graphical procedures involving residuals).

4. The quantity

$$z_i = \frac{e_i}{S}$$

is called a *standardized residual;* often it (rather than e_i) is examined in a residual analysis. Notice that, as with the $\{e_i\}$, the standardized residuals sum to 0 and hence are not independent. The standardized residuals have unit variance in the sense that

$$\frac{1}{n - k - 1}\sum_{i=1}^{n} z_i^2 = \frac{1}{n - k - 1}\sum_{i=1}^{n} \left(\frac{e_i}{S}\right)^2 = \frac{1}{S^2}\left(\frac{1}{n - k - 1}\sum_{i=1}^{n} e_i^2\right) = 1$$

5. The quantity

$$r_i = \frac{e_i}{S\sqrt{1 - h_i}} = \frac{z_i}{\sqrt{1 - h_i}}$$

is called a *studentized residual,* so named because it approximately follows a Student's t distribution with $n - k - 1$ degrees of freedom, if the data follow the usual assumptions for multiple regression (Chapter 8). The standard deviation of e_i is $S\sqrt{1 - h_i}$. The quantity h_i, the *leverage,*[2] is a measure of the importance of the ith observation in determining the model fit. Leverage values are such that $0 \le h_i \le 1$. Their role in helping diagnose regression problems is discussed in detail in Section 12-4. Studentized residuals have a mean near 0 (but not exactly 0), and a variance

$$\frac{1}{n - k - 1} \sum_{i=1}^{n} r_i^2$$

slightly larger than 1.

6. The quantity

$$r_{(-i)} = r_i \sqrt{\frac{S^2}{S_{(-i)}^2}} = \frac{e_i}{\sqrt{S_{(-i)}^2(1 - h_i)}} = r_i \sqrt{\frac{(n - k - 1) - 1}{(n - k - 1) - r_i^2}}$$

is called a *jackknife residual.* The quantity $S_{(-i)}^2$ is the residual variance computed with the ith observation deleted. Hence, the ith jackknife residual is standardized by a function of h_i and by a standard deviation based on $n - 1$ observations (deleting the ith observation). In contrast, a standardized residual involves S, the standard deviation based on all n observations. Jackknife residuals have a mean near 0 and a variance

$$\frac{1}{(n - k - 1) - 1} \sum_{i=1}^{n} r_{(-i)}^2$$

slightly greater than 1. If the usual assumptions are met, each jackknife residual exactly follows a t distribution with $(n - k - 1) - 1$ error degrees of freedom. If the standard regression assumptions are satisfied and approximately the same number of observations are made at all predictor values, then patterns in standardized, studentized, and jackknife residuals look very similar. As potential problems arise, however, studentized residuals and especially jackknife residuals make suspect values more obvious to the data analyst. For example, if the ith observation lies far from the rest of the data, $S_{(-i)}$ will tend to be much smaller than S, which in turn will make $r_{(-i)}$ large in comparison to r_i. Thus, $r_{(-i)}$ will tend to stand out more than r_i, further highlighting the outlier. Large h_i-values for high leverage observations lead to correspondingly larger $r_{(-i)}$-values than r_i-values.

7. Studentized residuals are distributed approximately, and jackknife residuals are distributed exactly, like t random variables under the usual assumptions. Furthermore, as error degrees of freedom ($n - k - 1$ for studentized and $n - k - 2$ for jackknife)

---

[2]The quantity h_i is the ith diagonal element of the ($n \times n$) matrix $\mathbf{X(X'X)^{-1}X'} = \mathbf{H}$, called the *hat matrix,* since $\hat{\mathbf{y}} = \mathbf{Hy}$, in which $\hat{\mathbf{y}} = (\hat{y}_1, \hat{y}_2,..., \hat{y}_n)'$ and $\mathbf{y} = (y_1, y_2,..., y_n)'$.

increase much above 30, the distributions of residuals can be approximated more and more closely by a standard normal (mean 0, variance 1) distribution. This information is helpful for evaluating the size of observed residuals by appealing to properties of a standard normal distribution. For example, if the residuals approximately represented a random sample from an $N(0, 1)$ distribution, no more than 5% of the residuals would be expected to exceed 1.96 in absolute value.

Residual Properties for the Weight Example

Using the WEIGHT, HEIGHT, and AGE data introduced previously, let us assume that a model has been fitted with WEIGHT as the response and HEIGHT, $(\text{HEIGHT})^2$, AGE, and $(\text{AGE})^2$ as predictors. We now analyze the residuals from that fitted model. Because the sample included 127 boys, there are 127 residuals. The 5 smallest and 5 largest residuals $\{e_i\}$ are $\{-11.5, -10.8, -8.3, -8.1, -7.8\}$ and $\{13.0, 14.8, 15.6, 18.4, 45.2\}$. The corresponding studentized residuals $\{r_i\}$ are $\{-1.74, -1.60, -1.22, -1.20, -1.19\}$ and $\{1.92, 2.30, 2.33, 2.70, 7.21\}$. The corresponding jackknife residuals $\{r_{(-i)}\}$ are $\{-1.76, -1.61, -1.22, -1.21, -1.19\}$ and $\{1.94, 2.34, 2.38, 2.78, 9.48\}$. Obviously, the largest value seems extraordinarily detached from the others.

With WEIGHT measured in kilograms, the residuals have variance 45.6, skewness 2.81, and kurtosis 15.33. *Skewness* indicates the degree of asymmetry of a distribution. Just as variance is the average squared deviation of observations about the mean, skewness is the average cubed deviation about the mean. To simplify comparisons between samples and to help account for estimation in small samples, skewness is usually computed as

$$\text{sk}(X) = \left(\frac{n}{n-2}\right)\left(\frac{1}{n-1}\right)\sum_{i=1}^{n}\left(\frac{X_i - \overline{X}}{S_X}\right)^3$$

The 2.81 value of skewness suggests that the normality assumption is questionable, since skewness is 0 for any symmetric distribution (such as a normal distribution). In addition, the positive value $(+2.81)$ indicates that relatively more values are above the mean than below it; the sample values are thus said to be "positively skewed." A negative value for skewness indicates that relatively more values are below the mean than above it.

Kurtosis indicates the heaviness of the tails relative to the middle of the distribution. Because kurtosis is the average of the fourth power of the deviations about the mean, it is always nonnegative. Standardized kurtosis may be computed as

$$\text{Kur}(X) = \left[\frac{n(n+1)}{(n-2)(n-3)}\right]\left(\frac{1}{n-1}\right)\sum_{i=1}^{n}\left(\frac{X_i - \overline{X}}{S_X}\right)^4$$

The term in brackets, which approaches 1.00 as n increases, helps to account for estimation based on a small sample. Since standardized kurtosis for a standard normal distribution is 3.0, this value is often subtracted from $\text{Kur}(X)$. The resulting statistic can be as small as -3 for flat distributions with short tails; it is 0 for a normal distribution; and it is positive for heavy-tailed distributions. Thus, the positive kurtosis value in our example suggests a distribution with tails heavier than in a normal distribution. Skewness and kurtosis statistics are highly variable in small samples and hence are often difficult to interpret. Since we have more than 100 observations in this example, however, these measures should be reasonably stable.

We have described four types of residuals for regression analysis. Which type is to be preferred? Since unstandardized and standardized residuals differ only by a multiplicative constant, they contain exactly the same information. For example, plots involving these two types of residuals have exactly the same shape. Standardized residuals (and studentized and jackknife residuals) have the advantage of being measured on a scale very similar to the one used for z scores. Consequently, the values of standardized, studentized, and jackknife residuals do not change if the Y variable is measured, for example, in inches rather than centimeters. Furthermore, one gains experience in interpretation that can easily be transferred across sets of data. If no problems with outliers are present, if $n - k - 1$ is not too small, and if all X-values are roughly equally represented (that is, approximately follow a uniform distribution), then standardized, studentized, and jackknife residuals are essentially the same. As a data set strays from these conditions, studentized residuals become preferable to standardized residuals, and eventually jackknife residuals become preferable to studentized residuals. Consequently, we emphasize the use of jackknife residuals in subsequent discussions.

12-3-2 Graphical Analysis of Residuals

Often the most direct and revealing way to examine a set of residuals is to make a series of plots of the residuals. Two basic kinds of plots are useful: one-dimensional and two-dimensional displays. The former employ only the properties and relationships of the observed residuals among themselves, while the latter consider the relationships of the residuals to other variables (such as the response and the predictors). In such a graphical analysis, a violation of a specific assumption (e.g., independence, model correctness, normality, or homogeneity of variance) sometimes is much more evident from one type of plot than from another. Here we discuss available types of plots and the interpretations that can be drawn from them.

One-dimensional Displays

The simplest plots are one-dimensional displays. Three kinds of one-dimensional displays of residuals are most useful: histograms (especially stem-and-leaf versions), schematic plots, and normal probability plots.

We shall first consider our second example—an instrument calibration experiment in environmental engineering. The 17 recorded values of the "concentration of a certain pollutant" (X) and the "instrument reading" (Y) are given in the computer output in section 12-2, along with the predicted values $\{\hat{Y}_i\}$, the residuals $\{e_i\}$, the studentized residuals $\{r_i\}$, and the jackknife residuals $\{r_{(-i)}\}$. All are based on the least-squares straight line $\hat{Y} = 15.39 + 3.79X_i$, for which $S = 4.11$.

First we examine a histogram of the jackknife residuals, shown in Figure 12-4(a). What should we look for in a figure of this type? Since we usually assume that $E_i \frown N(0, \sigma^2)$, we would expect (if the standard regression assumptions hold) the histogram of the $\{r_{(-i)}\}$ to reflect a random sample of 17 observations from a t distribution (with mean 0). Of course, as $n - k - 2$ increases much beyond 30, the distribution should reflect sampling from a standard normal distribution. For this example, with $n - k - 2 = 12$, we should see only slightly heavier tails than for an $N(0, 1)$ distribution. Otherwise, the picture should approximate the familiar bell-shaped curve.

For the calibration experiment, the residuals present a completely acceptable picture. As is customary for a histogram, the endpoints of the grouping intervals are indicated on the plot. For example, one observation (the largest) falls between 2.0 and 3.0, and six observations fall

FIGURE 12-4   Frequency histogram and stem-and-leaf diagram of jackknife residuals for the calibration experiment ($n = 17$).

```
 3 |  *               2 | 7
 2 |  *               1 | 4
 1 |  *******         0 | 1124557
 0 |  ******         -0 | 865542
-1 |  *             -1 | 3
-2 |  *             -2 | 8
-3 |
```

(a) Histogram.   (b) Stem-and-leaf diagram.

between -1.0 and 0.0. A frequency histogram can convey even more information if it is converted into a stem-and-leaf diagram, as in Figure 12-4(b). The lower (upper) endpoints of the grouping interval serve as *stems* for positive (negative) values. For example, the largest jackknife residual, rounded to two significant digits, is 2.7. Its stem is 2, the first significant digit. The *leaf* on the stem for the largest jackknife residual is 7, the second significant digit. The next largest value is read as 1.4. The third largest value is 0.7, then 0.5, 0.5, 0.4, and so on, down to the smallest value, -2.8. The figure may be compared with the original computer output, which lists all residuals.

The second kind of useful one-dimensional plot is a *schematic* plot. Figure 12-5 presents a schematic plot of the jackknife residuals for the calibration data ($n = 17$). A schematic plot is based entirely on the order of the values in the sample. *Quartiles* are the most important order-based statistics for the schematic plot. The *first quartile,* or 25th percentile, is the value at or

FIGURE 12-5   Schematic plot of jackknife residuals for the calibration experiment ($n = 17$).

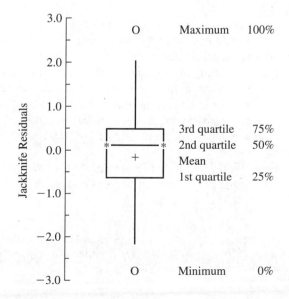

below which 25% of the data values lie; the *second quartile,* or 50th percentile (or median), is the value at or below which 50% of the data values lie; the *third quartile,* or 75th percentile, is the value at or below which 75% of the data values lie. The *interquartile range* (IQR), calculated as the value of the third quartile minus the value of the first quartile, is a measure of the spread of a distribution, like the variance. One important difference between the IQR and the variance, however, is illustrated by the fact that, whereas doubling the largest value in the sample would in general increase the variance dramatically, it would not change the IQR. For the jackknife residuals, the first, second, and third quartiles are approximately $-0.6, 0.1,$ and 0.5 respectively, with an interquartile range of 1.1. By way of comparison, it is easy to verify from Appendix Table A-1 that, for an $N(0, 1)$ distribution, the corresponding values are $-0.675, 0.0, +0.675,$ and 1.35, which compare quite closely (even though $n = 17$).

A schematic plot is sometimes called a *box-and-whisker plot,* or simply a *box plot,* due to its appearance. The box is outlined by three horizontal lines, which mark the values of the first quartile, the second quartile (the median, indicated by asterisks), and the third quartile (see Figure 12-5). The scale is determined by the units and range of the data. The mean is indicated by a + on the backbone of the plot. Since symmetric distributions have the same value for mean as for median, their schematic plots have the mean + in the middle of the median line (indicated by *———*) giving *—+—*. Symmetry also manifests itself in equal distances from the median to the other quartiles. The whiskers (vertical lines) extend from the box as far as the data extend up and down, to a limit of 1.5 IQRs (in the vertical direction). An O at the end of a whisker indicates an outside value, which occurs in about 1 observation out of 20 for a Gaussian sample (i.e., a sample from a normal distribution). An asterisk (*) at the end of a whisker indicates a detached value, which occurs in about 1 observation out of 200 in a Gaussian sample. A value is *outside* if it lies more than 1.5 IQRs beyond the box, and it is *detached* if it lies more than 3.0 IQRs beyond.

Referring to Figure 12-5, we see one positive outside value and one negative outside value. Recalling that we should expect to encounter slightly heavy tails and that jackknife residuals are sensitive to heavy tails, these two outsiders do not seem excessively bothersome. Even so, it is appropriate to examine such values more closely. With a little practice, one can extract a great deal of information about residuals from examining schematic plots. For more details about schematic plotting, see Tukey (1977).

Another graphical approach is to construct a normal or half-normal plot of the residuals (or their standardized, studentized, or jackknife counterparts) using normal probability paper. Figure 12-6 represents such a plot for the 17 jackknife residuals from the instrument calibration study. The plotting procedure first entails ordering the residuals from smallest to largest, followed by marking off the horizontal axis to include all the $r_{(-i)}$-values. The cumulative relative frequencies (i.e., i/n, where i indicates the ith ordered residual) up to each value are then plotted as ordinate values versus the $r_{(-i)}$-values as abscissas.[3] For example, the jackknife residual 1.45 is the second largest ($i = 16$) of the 17 residuals, so we plot $16/17$ (or 0.94) as the ordinate value corresponding to 1.45.

The cumulative relative frequencies for a normal distribution plot as a straight line. The line for the $N(0, 1)$ distribution is drawn on the graph and can be used as a yardstick in assess-

---

[3]To avoid treating a cumulative frequency of 1 (see Figure 12-6) and to provide better statistical estimation, one can plot $(i - 1/2)/n$, $i/(n + 1)$, or $(i - 3/8)/(n + 1/4)$, rather than i/n, versus residual values. Here, i denotes the ith ordered residual.

FIGURE 12-6   Normal probability plot of jackknife residuals for a linear
model fitted to data from the calibration experiment ($n = 17$).

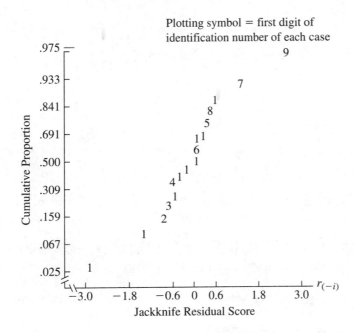

ing whether the scatter of points reflects any obvious deviation from normality. In our case, the stem-and-leaf diagram, the schematic plot, and the normal probability plot fail to suggest any blatant departures from the normality assumption—although such a statement is necessarily qualitative. More quantitative criteria are described in section 12-3-3.

Two-dimensional Displays

As mentioned earlier, plotting the observed response values against the predictor values is a good way to check the validity of regression assumptions. With a single predictor, one can plot $\hat{Y}$-values or residual values against X-values—for example, instrument readings versus pollutant levels for the calibration experiment. When many predictors are involved, the situation is more complex. For example, assume that air temperature is known to be important in determining the response of the instrument. Depending on the observed pattern of combinations of pollutant and temperature level, a plot of instrument reading against pollutant level might erroneously suggest, for example, heterogeneity of variance. In general, it is advisable to plot residuals not only versus each of the predictor variables but also versus the predicted responses, as well as plotting the observed responses versus the predicted responses.

The nature of predicted values helps explain the potential for obtaining misleading plots. For regression involving a single predictor,

$$\hat{Y} = \hat{\beta}_0 + \hat{\beta}_1 X$$

and for multiple regression,

$$\hat{Y} = \hat{\beta}_0 + \hat{\beta}_1 X_1 + \hat{\beta}_2 X_2 + \cdots + \hat{\beta}_k X_k$$

The predicted value $\hat{Y}$ represents the linear combination of X-variables that is most highly correlated with Y. For univariate regression,

$$r^2(Y, \hat{Y}) = r^2(Y, X)$$

This tells us that the strength of the (linear) relationship between Y and $\hat{Y}$ is the same as that between Y and X. For multiple regression, on the other hand,

$$r^2(Y, \hat{Y}) = R^2(Y \mid X_1, X_2, \ldots, X_k)$$

In general, however,

$$r^2(Y, \hat{Y}) \geq r^2(Y, X_j)$$

As a special case, for uncorrelated X-variables,

$$R^2(Y \mid X_1, X_2, \ldots, X_k) = r^2(Y, \hat{Y}) = r^2(Y, X_1) + r^2(Y, X_2) + \cdots + r^2(Y, X_k)$$

This helps demonstrate why the relationship between Y and any single X must be considered in light of all other X-variables. Unfortunately, even if all of the X-variables are mutually uncorrelated, consideration cannot be confined to single-predictor plots, since specific observations may constitute *multivariate* outliers (i.e., outliers if the X-variables are considered together).

One valid way to plot observed responses against predictors is to use a partial regression plot for each predictor. In such a case, we plot the response variable adjusted for $k - 1$ predictors against the predictor adjusted for the same $k - 1$ predictors. In particular, assume that the overall model of primary interest is

$$Y_i = \beta_0 + \beta_1 X_{i1} + \beta_2 X_{i2} + \cdots + \beta_k X_{ik} + E_i$$

To produce the partial regression plot for the kth predictor, we first fit two models:

$$Y_i = \beta_0 + \beta_1 X_{i1} + \beta_2 X_{i2} + \cdots + \beta_{k-1} X_{i(k-1)} + E_i$$
$$X_{ik} = \alpha_0 + \alpha_1 X_{i1} + \alpha_2 X_{i2} + \cdots + \alpha_{k-1} X_{i(k-1)} + E_i$$

Then we plot the residuals from the two models against each other; i.e., we plot $(Y_i - \hat{Y}_i)$ versus $(X_{ik} - \hat{X}_{ik})$ where $i = 1, 2, \ldots, n$. The best-fitting (i.e., least-squares) line for these paired residual data have an intercept of 0 and a slope of $\hat{\beta}_k$, the estimated regression coefficient based on fitting the original model. Notice the similarity of this process to the method of computing partial correlations presented in Chapter 10. The simple correlation between $(Y_i - \hat{Y}_i)$ and $(X_{ik} - \hat{X}_{ik})$ is the sample multiple partial correlation between Y and X_k, controlling for X_1 through X_{k-1}.

Of the various possible two-dimensional plots, some of the more useful graphs for checking assumptions in multiple regression involve plotting residuals (especially studentized or jackknife) versus predicted or predictor values. A few of the general patterns that may emerge from a plot of residuals versus predicted values are portrayed in Figure 12-7. Figure 12-7(a) illustrates the type of pattern to be expected when all basic assumptions hold: a horizontal band of points should be obtained with no hint of any systematic trends.

Figure 12-7(b) illustrates a systematic pattern to be expected when the data depart from linearity, indicating a need for curvilinear terms in the regression model. Naturally, different types of model inappropriateness result in different residual patterns.

Figure 12-7(c) represents the pattern to be expected when the error variance increases directly with $\hat{Y}$. It is possible, of course, to encounter situations where the error variance appears to be an even more complex function of the predicted value. In any case, transformations of the

FIGURE 12-7   Typical jackknife residual plots as a function of predicted value $\hat{Y}$ or time of data collection for hypothetical data.

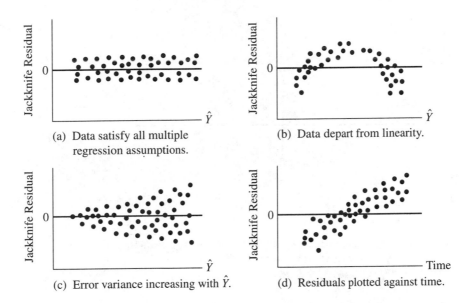

(a) Data satisfy all multiple regression assumptions.

(b) Data depart from linearity.

(c) Error variance increasing with $\hat{Y}$.

(d) Residuals plotted against time.

data often help eliminate such heteroscedasticity of the variance (see section 12-8-3). It is also helpful to take replicate observations at as many values of X as possible. This provides homogeneous (with respect to X) observation groups for which only variability due to "pure" error distinguishes the observations in a group. Observations associated with distinct predictor values vary—and their residuals vary—as a function of the *true* model. Without groups of replicates, any inappropriateness of the fitted model (sometimes called *misspecification*) is difficult to distinguish from simple heterogeneity of error variance. Such confusion could lead to the wrong choice of corrective actions.

Figure 12-7(d) is a plot of the jackknife residuals versus time; a linear time-related effect is clearly present. Whenever variables not included in the regression model (e.g., the variable "time" when the data are collected in a time sequence) may have a significant effect (such as that reflected in Figure 12-7(d) by a strong positive correlation with the residuals), it can be extremely informative to construct plots involving these variables.

For the calibration data, a plot of the jackknife residuals versus $\{\hat{Y}\}$ for the model $\hat{Y} = 15.39 + 3.79X$ is presented in Figure 12-8. The plot presents a systematic pattern similar to that of Figure 12-7(b), suggesting that introducing a quadratic term into the model will improve it. If we do this, the resulting model is $\hat{Y} = 10.00 + 8.10X - 0.54X^2$, with $S = 2.84$. A plot of jackknife residuals versus $\{\hat{Y}\}$ for this model is given in Figure 12-9. Although Figure 12-9 reflects considerable variation (due in part to the small sample size), the pattern of points very roughly reflects the horizontal band in Figure 12-7(a). Admittedly, some similarity to the pattern in Figure 12-7(c) is also evident.

Another way to examine the distribution of residuals is as follows. Since the studentized and jackknife residuals are assumed to represent a sample from a distribution that is approximately standard normal, we expect roughly 68% of these standardized residuals to lie in the interval $(-1.00, 1.00)$, about 95% to be contained in the interval $(-1.96, +1.96)$, and so on. If $n - k - 1$

FIGURE 12-8   Jackknife residuals as a function of the predicted values for the calibration experiment: linear model.

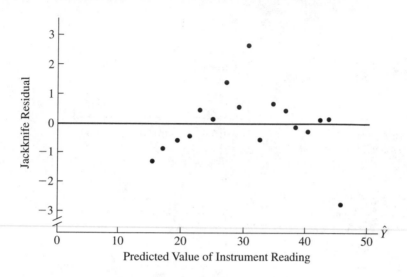

FIGURE 12-9   Jackknife residuals as a function of the predicted values for the calibration experiment: quadratic model.

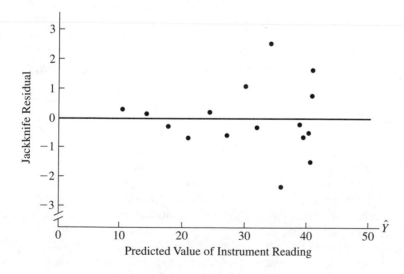

is small, with k = number of predictors, then 68% and 95% limits of the t distribution can be used, with $n - k - 1$ degrees of freedom for standardized and studentized, but $n - k - 2$ for jackknife. For these data, only two (i.e., 12%) of the seventeen jackknife residuals exceed 1.96 in absolute value. The same two jackknife residuals exceed the $n - k - 2 = 14$ degrees of freedom, two-tailed t critical value of 2.14 when $\alpha = .05$. Because the values of these two particular residuals are large (-2.82 and 2.71), the validity of the normality assumption might be questioned. Recall that the jackknife approach tends to highlight extreme residual values.

12-3-3 Significance Tests

More quantitative criteria for assessing the validity of the normality assumption can, of course, be based on the use of standard statistical testing procedures such as the chi-square and Kolmogorov–Smirnov tests (Stephens 1974). An additional significance test of interest is the Shapiro–Wilks (1965) test for normality, which is appropriate for small sample sizes—say, those less than 50. Since such tests are often discussed in a basic applied statistics course and can easily be conducted using standard computer programs, such discussion and computation will not be given here.

The analysis of residuals with respect to departures from normality is generally difficult because the distribution of a set of residuals is affected by several factors. For example, the residuals may exhibit a nonnormal pattern because of an inappropriate regression model, non-homogeneity of the variance, or too few residuals. In this last case, there is no pattern of sufficient stability to permit us to make a valid statistical inference about the nature of the underlying probability distribution. Hence, it is good practice to gather evidence concerning each of the possible types of departures from the assumptions and then to examine all evidence before making specific indictments. Following this strategy, we shall defer judgment about possible violation of the normality assumption in our calibration example until we have examined the possibility of other violations.

As mentioned earlier, a graphical analysis of a set of residuals is necessarily somewhat subjective. As demonstrated in the previous section, however, a careful graphical approach involving the simultaneous evaluation of several different types of residual plots often reveals whatever anomalies exist in the data. Nevertheless, on some occasions it is desirable to utilize statistical testing procedures to answer specific questions. We have already noted that standard goodness-of-fit tests are available for examining the normality assumption. In fact, statistical tests are available for examining each of the specific regression assumptions in question. For example, when a set of residuals is gathered in a time sequence, a nonparametric "runs test" (see, e.g., Siegel 1956, pp. 52–60) is frequently used to determine whether the time sequence of positive and negative residuals is sufficiently unusual to be viewed as more than just a random occurrence.

An additional method for assessing the validity of the independence assumption involves using Durbin–Watson (1951) statistics, which test the null hypothesis of independence (no autocorrelation) over time. An *autocorrelation* $r(Y_t, Y_{t-1})$ is the correlation between measurements taken at times t and $t - 1$. Assume, for example, that the data for our hypothetical instrument calibration experiment were collected in a particular order over time, such as one measurement per day. If the readings were slightly higher on each successive day (perhaps due to instrument drift), one would expect a positive autocorrelation among the residuals.

Tests for variance nonhomogeneity can be based on a comparison, by means of F statistics, of sample variances calculated from replicate observations obtained at each of a series of values of an independent variable (see, e.g., Bartlett 1947). However, Bartlett's test has been criticized for being overly sensitive to deviations from normality. Alternatively, to assess the amount of variance nonhomogeneity in terms of a suspected monotonic relationship between the error variance and an independent variable, one can test the significance of the Spearman rank correlation (again, see Siegel 1956, pp. 202–13) between the absolute value of the residual and the value of the independent variable.

12-4 Treating Outliers

An outlier among a set of residuals is much larger than the rest in absolute value, perhaps lying three or more standard deviations from the mean of the residuals. Clearly, the presence of such an extreme value can significantly affect the least-squares fitting of a model, so it is important to determine whether the analysis should be modified in some way (such as by deleting the observation in question). An outlier in the data may indicate special circumstances warranting further investigation (e.g., the presence of an unanticipated interaction effect). Therefore, we do not recommend immediately discarding the observation unless strong evidence indicates that it resulted from a mistake (e.g., an error in data recording or some other cause independent of the process under study, such as an obvious instrument malfunction). Statistical procedures for evaluating outliers are presented in the subsections that follow. Further discussion of some of these methods can be found in papers by Anscombe (1960), Anscombe and Tukey (1963), and Stevens (1984). See Tukey (1977) for a thorough discussion of outliers and their detection.

12-4-1 Definitions

Substantial differences exist among possible types of extreme values. An *outlier* is any rare or unusual observation that appears at one of the extremes of the data range. All regression observations—and hence, outliers in particular—may be evaluated as to three criteria: reasonableness, given knowledge of the variable; response extremeness; and predictor extremeness. The goal is to identify observations that significantly affect either the choice of variables in the model or the accuracy of estimations of the regression coefficients and associated standard errors. We shall discuss these concepts briefly and then describe procedures for quantifying them.

If an observation has been identified as an outlier, it should be checked for *plausibility*. As discussed earlier, the data analyst must be familiar with the basic characteristics of the data. Consider, as an example, body temperature as a response. The number 38.1 is plausible if the units are degrees Celsius and the subjects are humans. But if the units are degrees Fahrenheit, then 38.1 is an implausible value. More generally, one may classify any observation as being impossible, highly implausible, or plausible; it is then necessary to consider the importance of an observation in determining the choice of variables in the model, coefficient estimates, and associated statistics before deciding what, if any, action to take. The notion of importance includes *leverage* and *influence*—concepts we discuss in detail in section 12-4-2.

Traditionally, outliers among observations (the set of values of $Y, X_1, X_2, \ldots, X_k$) are detected by considering the residuals. But it may also be helpful to consider the location of a particular response value (Y) relative to the values of the other responses. An extreme response value may deserve attention to assess its plausibility and importance.

A more difficult task is to evaluate the extremeness of predictor values. The set $\{X_{i1}, X_{i2}, \ldots, X_{ik}\}$ represents a point in k-dimensional space. With two predictors, for example, the space is a plane and the points (X_{i1}, X_{i2}), where $i = 1, 2, \ldots, n$, are easily plotted. With more than two predictors, two-dimensional plots are not sufficient to locate outliers in the predictor set. Fortunately, quantitative methods can help solve this problem.

12-4-2 Detecting Outliers

Methods for detecting outliers have received a great deal of attention in the last two decades (see, e.g., Belsley, Kuh, and Welsch 1980; Cook and Weisberg 1982; and Stevens

1984). In this section, we describe three regression diagnostic statistics for evaluating outliers: jackknife residuals, leverages, and Cook's distance (a measure of influence). Other closely related statistics have been suggested. Our opinion is that *some* diagnostic statistics should be employed as part of any regression analysis. Most good regression computer programs provide a selection of diagnostics but require the user to request these diagnostics via options.

We have already discussed the utility of jackknife residuals in relation to standardized residuals. To understand their use in marking outliers, consider the components of the formula for the ith jackknife residual—namely,

$$r_{(-i)} = \frac{e_i}{S_{(-i)}\sqrt{1 - h_i}}$$

Three quantities merit discussion: e_i, $S_{(-i)}$, and h_i. Since

$$e_i = Y_i - \hat{Y}_i$$

the numerator of $r_{(-i)}$ reflects the extremeness of the ith observed response Y_i relative to the predicted value $\hat{Y}_i$.

Recall that the variance of the residuals is S^2; that is,

$$S^2 = \frac{1}{n - k - 1} \sum_{i=1}^{n} e_i^2$$

The S in the denominator of standardized residuals reflects the goodness-of-fit of the model and scales these residuals to have unit variance. In turn, $S_{(-i)}^2$ is a jackknifed estimate of the residual variance. Its purpose is to help prevent an outlier from masking its own effect (since its contribution to S^2 is ignored). To complete the dissection of $r_{(-i)}$, the best estimates of the variances of $\hat{Y}_i$ and $e_i = Y_i - \hat{Y}_i$ are, respectively,

$$S_{\hat{Y}_i}^2 = S^2 h_i \tag{12.1}$$

and

$$S_{(Y_i - \hat{Y}_i)}^2 = S^2(1 - h_i) \tag{12.2}$$

The leverage h_i is a measure of the geometric distance of the ith predictor point $(X_{i1}, X_{i2}, \ldots, X_{ik})$ from the center point $(\overline{X}_1, \overline{X}_2, \ldots, \overline{X}_k)$ of the predictor (X) space. Consequently, an observation may be associated with an outlier jackknife residual if the observation is an outlier in the response variable Y or in the predictor space of $X_1, X_2, \ldots, X_k$, or if it strongly affects the fit of the model (as reflected in the difference between S^2 and $S_{(-i)}^2$). Naturally, a combination of two or all three effects could yield a large jackknife residual.

It is easy to test whether a particular jackknife residual differs significantly from 0 (i.e., whether it has an extreme value not due to chance alone). Recall that a single jackknife residual exactly follows a t distribution with $n - k - 2$ degrees of freedom if the usual assumptions are met. A corrected significance level must be used to account for the fact that n tests (one for each observation) will be conducted. For example, if 50 subjects are observed and the testing is to be done at the .05 significance level, it is appropriate to require a P-value of $.05/50 = .001$. Since usually both extreme positive and extreme negative values are of concern, one may use $.025/50 = .0005$ for a two-tailed test. Using .025 rather than the corrected value would lead, on average, to falsely declaring two or three observations as outliers. Using this corrected significance level is an application of the *Bonferroni (α-splitting) correction*.

Consider, for example, the instrument calibration data in the computer output in section 12-3-2. Table A-8 in Appendix A provides two-tailed critical values for a useful range of n (number of observations) and k (number of predictors). For $\alpha = .05$, $n = 15$, and $k = 1$, Table A-8(a) gives a critical value of 3.65 for jackknife residuals, while Table A-8(b) gives a critical value of 2.61 for studentized residuals. Since $n = 17$ is not included in the table, interpolation should be used. For jackknife residuals, this interpolation gives

$$3.65 + \frac{(3.54 - 3.65)(20 - 17)}{(20 - 15)} = 3.584$$

The largest jackknife residual (in absolute value) is 2.82, so this value gives no cause for concern. Even the liberal $\alpha = .10$ $n = 15$ value of 3.27 is much larger than 2.82.

Another regression diagnostic is the set of *leverage values* $\{h_i\}$ introduced in our discussion of studentized residuals. Leverage measures the distance of an observation from the set of X-variable means—namely, from $\{\overline{X}_1, \overline{X}_2, \ldots, \overline{X}_k\}$. From equation (12.1), we know that

$$h_i = \frac{S_{\hat{Y}_i}^2}{S^2}$$

For the straight line model

$$Y_i = \beta_0 + \beta_1 X_i + E_i$$

the leverage value for the ith observation takes the special form

$$h_i = \frac{1}{n} + \frac{(X_i - \overline{X})^2}{(n-1)S_X^2}$$

in which

$$S_X^2 = \frac{1}{n-1} \sum_{i=1}^{n} (X_i - \overline{X})^2$$

The main component of the formula for leverage is the squared standardized distance of the X_i-value from the center (mean) of the set of X-values—namely, the squared z score

$$z_{X_i}^2 = \left(\frac{X_i - \overline{X}}{S_X} \right)^2$$

For simple linear regression involving a single predictor, then, leverage indicates the extremeness of an observation in the range of X-values.

More generally, for multiple regression, a leverage value measures the extremeness of an observation in the k-dimensional space of $X_1, X_2, \ldots, X_k$. For the special case in which all predictor variables $X_1, X_2, \ldots, X_k$ have mean 0 and are uncorrelated,

$$h_i = \frac{1}{n} + \sum_{j=1}^{k} \frac{X_{ij}^2}{(n-1)S_j^2}$$

in which

$$S_j^2 = \frac{1}{n-1} \sum_{i=1}^{n} X_{ij}^2$$

Leverages are related to an alternate regression diagnostic called *Mahalanobis distance* (Stevens 1984).

Interpretation of the size and extremeness of leverage values is simplified by the following properties. First, in general,

$$0 \le h_i \le 1$$

However, $h_i \ge 1/n$ if the regression model under consideration contains an intercept (i.e., a β_0 term). A leverage h_i of 1 indicates that $\hat{Y}_i = Y_i$, which ensures that the model is forced (levered) to fit the ith observed response exactly. Furthermore, with k being the number of predictors in the model $Y = \beta_0 + \beta_1 X_1 + \cdots + \beta_k X_k + E$,

$$\sum_{i=1}^{n} h_i = k + 1$$

Consequently, the average leverage value is

$$\bar{h} = \frac{k + 1}{n}$$

Hoaglin and Welsch (1978) recommended scrutinizing any observation for which $h_i > 2(k + 1)/n$. If the model is correct, the set of leverages for a sample has a frequency histogram that looks like a chi-square density. If the predictors follow a Gaussian distribution (i.e., each predictor has a normal distribution), then, for any single h_i,

$$F_i = \frac{[h_i - (1/n)]/k}{(1 - h_i)/(n - k - 1)}$$

follows an F distribution with k and $n - k - 1$ degrees of freedom under the null hypothesis that the ith observation is a random sample of size 1 from the Gaussian predictor population. Thus, a test of extreme leverage can be conducted by comparing F_i to an F critical value of $F_{k, (n-k-1), 1-\alpha/n}$. For example, with $\alpha = .05$ for a sample of 100 subjects, we may use a P-value of $.05/100 = .0005$ (using the Bonferroni correction) to avoid spuriously identifying too many leverages as outliers. Appendix Table A-9 provides leverage critical values corresponding to $F_{k, (n-k-1), 1-\alpha/n}$ for a useful range of k (the number of predictors) and n (the sample size). Even though the predictor values are assumed to be fixed values,[4] the F statistic just presented can offer a rough indication of troublesome observations.

The regression diagnostic usually referred to as *Cook's distance* is a measure of the *influence* of an observation. Cook's distance measures the extent to which the regression coefficients change when the particular observation in question is deleted. For the special case of mean 0, equal variance, and uncorrelated predictors, Cook's distance d_i for the ith observation is proportional to

$$\sum_{j=0}^{k} [\hat{\beta}_j - \hat{\beta}_{j(-i)}]^2 = [\hat{\beta}_0 - \hat{\beta}_{0(-i)}]^2 + [\hat{\beta}_1 - \hat{\beta}_{1(-i)}]^2 + \cdots + [\hat{\beta}_k - \hat{\beta}_{k(-i)}]^2$$

---

[4]We have assumed throughout our treatment of multiple regression that the predictor values are all fixed, known constants. Violation of this assumption can seriously bias estimates of variability and hence distort tests of hypotheses.

Here, $\hat{\beta}_j$ is the estimated regression coefficient for all the data, and $\hat{\beta}_{j(-i)}$ is the corresponding estimated regression coefficient with the ith observation deleted. If the predictors do not have mean 0 and equal variance and are not uncorrelated, then Cook's distance is proportional to a weighted sum of the terms $[\hat{\beta}_j - \hat{\beta}_{j(-i)}]^2$. For any set of data, Cook's distance d_i for the ith observation can be expressed in terms of leverages and studentized residuals as

$$d_i = \left(\frac{1}{k+1}\right)r_i^2\left(\frac{h_i}{1-h_i}\right) = \frac{e_i^2 h_i}{(k+1)S^2(1-h_i)^2}$$

These formulas show the close relationship of d_i to the leverage h_i and the studentized residual r_i. Clearly, d_i may be large either because the observation is extreme in the predictor space (i.e., h_i is near 1) or because the observation has a large studentized residual r_i.

Obviously, $d_i > 0$; and in general, d_i may be arbitrarily large. Cook and Weisberg (1982) suggested that any d_i value greater than 1 may deserve closer scrutiny. If the model is correct, d_i can be expected to be less than about 1.0. This simple approximation to a d_i-value keeps the regression coefficient estimates based on deleting the ith observation within a 50% confidence region of the original estimates. The value of 1 is based on the fact that d_i is roughly like an F random variable with k and $n - k - 1$ degrees of freedom. The question being addressed is whether the set of regression coefficients differs with the ith subject deleted from what it is with all subjects included. The suggestion corresponds to comparing d_i values to the median (50th percentile) values of an F random variable with k and $n - k - 1$ degrees of freedom.

Recently Muller and Chen Mok (1997) studied Cook's statistic, under the assumption that all predictors are Gaussian. They were motivated by their impression from various data analyses that comparing the statistic to the median of an F rarely highlights troublesome observations. They described computational forms for the exact distribution function, as well as a convenient F approximation, with accuracy increasing with sample size. They conducted computer simulations to verify the accuracy of their exact and approximate results, as well as the suggestion of Cook and Weisberg (1982). The simulations' results were consistent with the impression that the Cook and Weisberg rule for evaluating Cook's statistic proves very inaccurate. The approximation of Muller and Chen Mok worked well for evaluating a single value, but it performed inaccurately for evaluating the maximum value of a set, even with 200 observations. Hence, in Table A-10 we provide comparison values of $(N - k - 1)d_i$ for the maximum of N observations. The multiplication by the degrees of freedom simplifies the task of preserving numerical accuracy.

Table A-10 depends on assuming Gaussian predictors. Jensen and Ramirez (1996, 1997) discussed the contribution of Cook's statistic for fixed predictors at length. Muller and Chen Mok (1997) also briefly considered the issue. In considering fixed predictors, they recalled Obenchain's (1977) suggestion to ignore the statistic and concentrate on its two components, the residual and the leverage. Ultimately, the uncertainty surrounding the use of Cook's statistic reflects both the need for ongoing research in diagnostics and the impossibility of providing hard and fast rules for diagnostics.

Some observation must be the most extreme in every sample. It would be silly to delete automatically this most extreme observation or some cluster of extreme observations, based entirely on statistical testing procedures. The goal of regression diagnostics in evaluating outliers is to warn the data analyst to examine more closely such extreme observations. Scientific judgment is more important here than statistical tests, once influential observations have been flagged. Of course, deleting the most deviant observations always at least slightly improves—

and sometimes substantially improves—the fit of the model, but one must resist the temptation to polish the fit of the model by discarding troublesome data points.

12-4-3 Diagnostics for the Weight Example

One-dimensional plots are provided for jackknife residuals (Figure 12-10), leverages (Figure 12-11), and Cook's distance statistics (Figure 12-12) for the WEIGHT regression example. If the usual assumptions are correct, residuals should mimic a bell-shaped curve, while stem-and-leaf diagrams for both leverage and Cook's distance should look roughly like chi-square densities. Hence, the latter two diagrams typically have a long tail to the right and many observations bunched near 0. Figure 12-10 (the jackknife residual plot) looks acceptably bell-shaped, the important exception being the detached extreme value of 9.5. No significance test is needed to encourage us to examine this particular observation more closely. The critical value from Appendix Table A-8(a) is between 4.06 and 4.15 for $\alpha = .01$, $n = 127$, and $k = 4$. The leverage histogram also closely resembles the shape expected if all assumptions hold—an important exception being the detached value of .29. Since leverages are between 0 and 1, with a mean here of $(k + 1)/n = 5/127 = .039$, even the values in the neighborhood of .15 may deserve attention. (In fact, linear interpolation in Appendix Table A-9 gives a critical value of

FIGURE 12-10   Jackknife residuals for the four-predictor model of children's weights; data from Lewis and Taylor (1967) ($n = 127$).

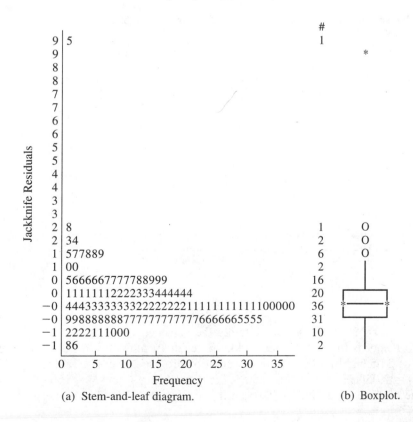

(a) Stem-and-leaf diagram.

(b) Boxplot.

<u>*FIGURE 12-11*</u> Leverages for the four-predictor model of children's weights; data
from Lewis and Taylor (1967) ($n = 127$).

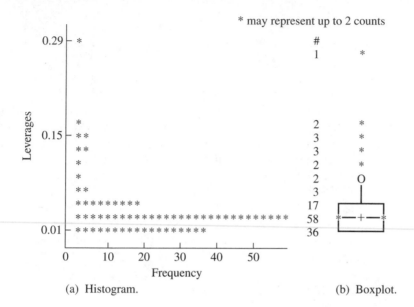

(a) Histogram.   (b) Boxplot.

<u>*FIGURE 12-12*</u> Cook's d_i statistics for the four-predictor model of children's weights; data from
Lewis and Taylor (1967) ($n = 127$).

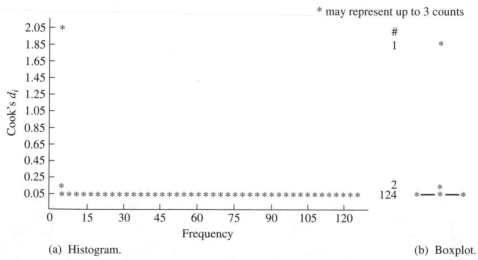

(a) Histogram.   (b) Boxplot.

.151 for $\alpha = .01$, $n = 127$, and $k = 4$.) Certainly the detached value of .29 is troublesome; as
expected, it corresponds to the detached jackknife residual value of 9.5. The maximum influence statistic of 2.05 highlights the same observation. To use Appendix Table A-10, compute
$d_i(n - k - 1) = 2.05(127 - 4 - 1) = 250.1$. For $\alpha = .01$, $k = 4$, and $N = 100$, the table yields
a critical value of 27.31; and for $N = 200$, a critical value of 31.26. Hence interpolation is not

needed to judge that the observation merits further consideration. The model includes not only HEIGHT and AGE as predictors, but also $(\text{HEIGHT})^2$ and $(\text{AGE})^2$. It is unlikely that all four predictors follow a Gaussian distribution, which is an assumption for using the table.

All regression diagnostics for the example indicate the need to consider one particular observation. In addition, a Kolmogorov–Smirnov test indicates that the residuals are not normally distributed. The particularly troublesome observation has variable values WEIGHT = 88.9 kg (195.6 lb), HEIGHT = 1.37 m (54 in.), and AGE = 14.5 yr. Although not impossible, this is an extremely unusual weight-and-height combination for a teenage boy. Consequently, for all subsequent analyses, this particular observation will be deleted.

It is informative to consider how the regression analysis changes as a result of deleting the observation in question. The R^2-value jumps from .52 (for $n = 127$ with the outlier) to .68 (for $n = 126$ without the outlier). The regression coefficients also change dramatically, as we would expect from the large Cook's distance statistic value of 2.05.

For an analysis excluding the outlier, one-dimensional plots are provided for jackknife residuals (Figure 12-13), leverage values (Figure 12-14), and influence values (Figure 12-15). The critical value ($n = 126$, $k = 4$) for jackknife residuals is approximately 3.64, with $\alpha = .05$. Hence, the maximum jackknife residual of 3.9 exceeds the .05 critical value but fails to exceed the .01 critical value of 4.06 for $n = 100$. The stem-and-leaf diagram of the jackknife residuals is approximately bell-shaped, and no obviously detached points are present.

The jackknife residual, leverage, and influence analyses each nominate different observations for attention. The jackknife residual of 3.9 corresponds to the observation WEIGHT =43.1 kg, HEIGHT = 1.54 m, and AGE = 13.67 yr; the leverage of .33 corresponds to WEIGHT = 35.8 kg, HEIGHT = 1.28 m, and AGE = 12.25 yr; and the influence statistic of 0.27 (Figure 12-15) corresponds to WEIGHT = 77.8 kg, HEIGHT = 1.71 m, and AGE = 17.1 yr. These WEIGHT–HEIGHT–AGE combinations are all plausible for teenage boys. The maximum leverage of .33 is

FIGURE 12-13   Jackknife residuals of children's weights without the outlier; data from Lewis and Taylor (1967) ($n = 126$).

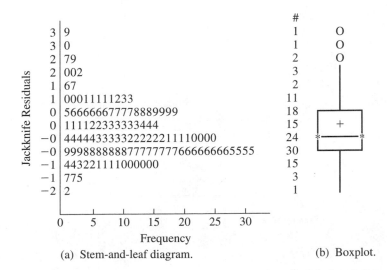

(a) Stem-and-leaf diagram.   (b) Boxplot.

FIGURE 12-14 Leverage values of children's weights without the outlier; data from Lewis and Taylor (1967) ($n = 126$).

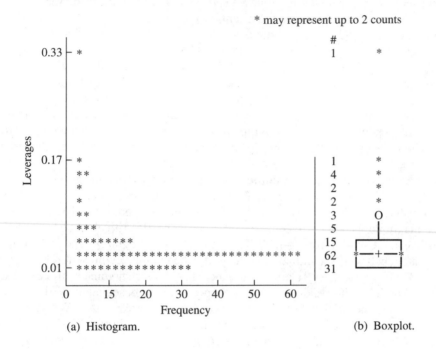

(a) Histogram.

(b) Boxplot.

FIGURE 12-15 Cook's d_i statistic of children's weights without the outlier; data from Lewis and Taylor (1967) ($n = 126$).

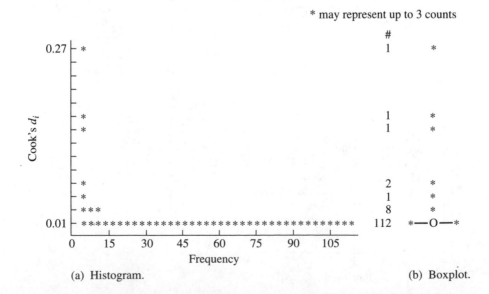

(a) Histogram.

(b) Boxplot.

much larger than a cutoff of about .17 (from Appendix Table A-9, for $\alpha = .05, n = 126$, and $k = 4$). The maximum influence statistic of 0.27 is below the suggested cutoff of 0.88 (from Appendix Table A-10) or the rule-of-thumb value 1. The small value of 0.27 for the maximum influence statistic reflects the fact that any single observation must be very unusual in order to change regression estimates noticeably if the assumptions are met and the error degrees-of-freedom value $n - k - 1$ is reasonably large. This is an excellent reason to avoid small samples! In addition, these numerical findings support *not* deleting plausible observations, such as the one with a 3.9 jackknife residual or the one with .33 leverage, since neither deletion would influence one's eventual conclusions.

In the next stage of analysis, the residuals of the model for predictor values surrounding these subjects would be considered. It is debatable whether the leverage and influence diagrams sufficiently resemble chi-square distributions and whether detached values are present. In any case, $R^2 = .68$, and the distribution of residuals (Figure 12-13) indicates that the fitted model is a reasonable one. For more details about interpreting regression diagnostic statistics, see Weisberg (1980), Cook and Weisberg (1982), Belsley, Kuh, and Welsch (1980), Stevens (1984), and Hoaglin and Welsch (1978).

12-5 Collinearity

In this section, we describe certain features of a regression analysis that can result in numerical problems that in turn yield inaccurate estimates of regression coefficients, variability, and P-values. These problems can be grouped loosely into one of two types: collinearity and scaling (including centering). Collinearity involves the relationship of the independent variables (predictors) to one another. Scaling pertains to the units in which the variables under study and their means are measured. Certain kinds of collinearity problems (e.g., those involving polynomial regression terms) can be expressed as scaling problems and, therefore, can easily be resolved. These concepts are expanded later, and methods for combating these problems are outlined.

12-5-1 Collinearity with Two Predictors

The problems emanating from collinearity can be illustrated with simple two-variable regression examples. Consider fitting the model

$$Y_i = \beta_0 + \beta_1 X_{i1} + \beta_2 X_{i2} + E_i$$

to produce $\hat{\beta}_0, \hat{\beta}_1$, and $\hat{\beta}_2$. In general, we can show that

$$\hat{\beta}_j = c_j \left[\frac{1}{1 - r^2(X_1, X_2)} \right]$$

for $j = 1$ or $j = 2$. Here c_j is a value that depends on the data, and $r^2(X_1, X_2)$ is the squared correlation between X_1 and X_2. In turn,

$$\hat{\beta}_0 = \overline{Y} - \hat{\beta}_1 \overline{X}_1 - \hat{\beta}_2 \overline{X}_2$$

so

$$\overline{Y} - \hat{\beta}_0 = \hat{\beta}_1\overline{X}_1 + \hat{\beta}_2\overline{X}_2 = \left[\frac{1}{1 - r^2(X_1, X_2)}\right](c_1\overline{X}_1 + c_2\overline{X}_2)$$

Here, $\overline{X}_1$ and $\overline{X}_2$ are the means of X_1 and X_2, respectively. These expressions tell us that $\hat{\beta}_1, \hat{\beta}_2$, and $\overline{Y} - \hat{\beta}_0$ are all proportional to $1/[1 - r^2(X_1, X_2)]$. More specifically, it is informative to consider fitting the model

$$Y_i = \theta_0 + \theta_1 X_{i1} + \theta_2 X_{i1} + E_i$$

In this case, a single variable, X_1, is included in the model twice. What are the estimates of the regression coefficients? Since $r^2(X_1, X_2) = r^2(X_1, X_1) = 1$, it follows that $1 - r^2(X_1, X_2) = 0$, and

$$\hat{\theta}_j = c_j'\left(\frac{1}{0}\right) = ?$$

From this we conclude that the estimates of the regression coefficients are indeterminate. Since the estimates of the variances of the regression coefficients are proportional to the "inflation factor" $1/[1 - r^2(X_1, X_2)]$, they are indeterminate, too. In turn, the P-values for tests about the coefficients are also indeterminate, since they involve the estimates just discussed. The preceding model can be rewritten in the form

$$Y_i = \theta_0 + (\theta_1 + \theta_2)X_{i1} + E_i$$

which establishes that an infinite number of pairs of θ_1 and θ_2 values add up to the same coefficient value; thus an estimate of the coefficient of X_1—say, $\widehat{\theta_1 + \theta_2}$—does not permit a unique determination of the individual estimates $\hat{\theta}_1$ and $\hat{\theta}_2$. Consequently, θ_1 and θ_2 cannot be estimated separately. For example, if the actual estimate for the slope of X_1 is 12.5, then $\hat{\theta}_1 = 12.5, \hat{\theta}_2 = 0$ are possible components, as are $\hat{\theta}_1 = 0, \hat{\theta}_2 = 12.5$ and $\hat{\theta}_1 = 6, \hat{\theta}_2 = 6.5$, and so on. In this extreme example, one of the predictor variables is a *perfect linear combination* of the other—namely,

$$X_1 = \alpha + \beta X_2 = 0 + (1)X_2 = X_2$$

Geometrically the points (X_{i1}, X_{i2}) all fall on the straight line $X_1 = X_2$; hence, the term *collinear* is applicable.

The following data provide a slightly more general example:

X_1	X_2
0	3
1	5
3	9
4	11
7	17

Plotting X_2 against X_1 yields a very simple picture, as shown in Figure 12-16. The plot illustrates that all of the data points (X_{i1}, X_{i2}) fall exactly on the straight line $X_2 = 3 + 2X_1$, which demonstrates that X_1 and X_2 are exactly collinear. Moreover, $r^2(X_1, X_2) = 1$ directly measures this perfect collinearity. The collinearity issue involves only the predictor variables and does not

FIGURE 12-16   Perfectly collinear set of pairs of variable values: $X_2 = 3 + 2X_1$.

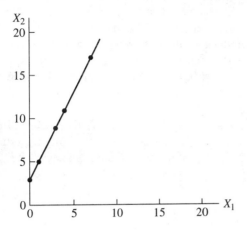

FIGURE 12-16   Perfectly collinear set of pairs of variable values: $X_2 = 3 + 2X_1$.

depend on the relationship between the response and any of the predictors. As $r^2(X_1, X_2)$ decreases, the collinearity problem between X_1 and X_2 becomes less severe, with the ideal situation occurring if X_1 and X_2 are uncorrelated.

As another example, consider fitting the quadratic model

$$Y_i = \beta_0 + \beta_1 X_i + \beta_2 X_i^2 + E_i$$

to the engineering calibration data given in the computer output in section 12-2. In this case, the size of $r^2(X_1, X_2) = r^2(X, X^2)$ is a measure of the degree of collinearity between X and X^2. The mean of the X-values (listed in the earlier computer output) is 4.0. Table 12-1 displays the effect of subtracting various constants from all of the X-values. The new values are computed as $X_i^* = (X_i - a)$ for various values of a between -4 and 4.1, including $a = 0$ (the original data). The mean of $X - a$ is shifted from the mean of X by a, while the variance of $X - a$ is the same

TABLE 12-1   Effects of subtracting a constant on $r[X^*, (X^*)^2]$ for the calibration example ($n = 17$), where $X^* = X - a$.

| | Mean | | Standard Deviation | | | | |
| | X^* | $(X^*)^2$ | X^* | $(X^*)^2$ | $\text{Cov}[X^*, (X^*)^2]$ | $r[X^*, (X^*)^2]$ | $\{1 - r^2[X^*, (X^*)^2]\}^{-1}$ |
a							
-4.0	8.0	70.00	2.52	40.77	102.00	.99	50.25
0.0	4.0	22.00	2.52	20.93	51.00	.96	12.76
1.0	3.0	15.00	2.52	16.12	38.25	.94	8.59
2.0	2.0	10.00	2.52	11.50	25.50	.88	4.43
3.0	1.0	7.00	2.52	7.47	12.75	.68	1.86
3.5	0.5	6.25	2.52	6.05	6.375	.42	1.21
3.7	0.3	6.09	2.52	5.71	3.825	.26	1.07
3.9	0.1	6.01	2.52	5.52	1.275	.09	1.01
centering → 4.0	0.0	6.00	2.52	5.50	0	.00	1.00
4.1	-0.1	6.01	2.52	5.52	-1.275	$-.09$	1.01

as that of X. Table 12-1 also lists the mean and the standard deviation of $(X - a)^2$ and the covariance (Cov) and correlation (r) between $X - a$ and $(X - a)^2$. The final column shows the associated value of the inflation factor $1/(1 - r^2)$. The squared correlation (r^2) ranges from .00 to .99, and the inflation factor ranges from 1 to approximately 50. The only change to the data is a constant shift in the X-values. Situations depicted across the rows range from near collinearity to no collinearity, depending on the value of a. The process of subtracting a constant from a variable is a form of scaling. In the example, if $a = \overline{X} = 4.0$, it is called *centering*. As can be seen, *centering in polynomial regression can substantially reduce collinearity*. More is said about this later in this chapter, and still more in Chapter 13.

In general, for a two-variable model, if $r^2(X_1, X_2)$ is nearly 1.0, then near collinearity is present. Although the coefficient estimates can be computed, they are highly unstable. In particular, this instability is reflected in large estimates of the coefficient variances, since such variance estimates are proportional to

$$\frac{1}{1 - r^2(X_1, X_2)}$$

As $r^2(X_1, X_2)$ gets closer to 1.0, this factor becomes large, thereby inflating the estimated variances of the regression coefficients. Later we will see that this factor is a special case of a variance inflation factor.

12-5-2   Collinearity Concepts

In this section, we generalize the discussion of the last subsection to treat any number of predictors. In addition, we describe methods for quantifying the degree of collinearity in fitting a regression model to a particular set of data.

As we observed in the examples just discussed, collinearity involves relationships among the predictor variables and does not directly involve the response variable. As such, one informative way to examine collinearity is to consider what happens if each predictor variable is treated as the response variable in a multiple regression model in which the independent variables are all of the remaining predictors. For k predictors, then, we have k such models. For example, with the four-predictor model

$$Y_i = \beta_0 + \beta_1 X_{i1} + \beta_2 X_{i2} + \beta_3 X_{i3} + \beta_4 X_{i4} + E_i$$

four models would be fitted as follows:

$$
\begin{aligned}
X_{i1} &= \alpha_{01} &&+ \alpha_{21}X_{i2} &&+ \alpha_{31}X_{i3} &&+ \alpha_{41}X_{i4} &&+ E_i \\
X_{i2} &= \alpha_{02} + \alpha_{12}X_{i1} && &&+ \alpha_{32}X_{i3} &&+ \alpha_{42}X_{i4} &&+ E_i \\
X_{i3} &= \alpha_{03} + \alpha_{13}X_{i1} &&+ \alpha_{23}X_{i2} && &&+ \alpha_{43}X_{i4} &&+ E_i \\
X_{i4} &= \alpha_{04} + \alpha_{14}X_{i1} &&+ \alpha_{24}X_{i2} &&+ \alpha_{34}X_{i3} && &&+ E_i
\end{aligned}
$$

To assess collinearity, we need to know the associated R^2-values based on fitting these four models—namely, $R^2(X_1 \mid X_2, X_3, X_4)$, $R^2(X_2 \mid X_1, X_3, X_4)$, $R^2(X_3 \mid X_1, X_2, X_4)$, and $R^2(X_4 \mid X_1, X_2, X_3)$. If any of these multiple R^2-values equals 1.0, a perfect collinearity is said to exist among the set of predictors. The term *collinearity* is used to indicate that one of the predictors is an *exact* linear combination of the others. As described in earlier examples, in the special case $k = 2$, perfect collinearity means that X_1 is a straight-line function of X_2—say, $X_1 = \alpha + \beta X_2$—so that the points (X_{i1}, X_{i2}) all fall on the line.

Consider, for example, predicting college grade point average (CGPA) on the basis of high school grade point average (HGPA) and College Board scores on the mathematics (MATH) and verbal (VERB) tests. The model is

$$CGPA_i = \beta_0 + \beta_1 HGPA_i + \beta_2 MATH_i + \beta_3 VERB_i + E_i$$

Now imagine trying to improve prediction by adding the total (combined) board scores (TOT = MATH + VERB) to create a new model:

$$CGPA_i = \alpha_0 + \alpha_1 HGPA_i + \alpha_2 MATH_i + \alpha_3 VERB_i + \alpha_4 TOT_i + E_i$$

This model has a perfect collinearity, which means that the parameters in the model cannot be estimated uniquely. To see this, begin by rewriting the model as

$$CGPA_i = \alpha_0 + \alpha_1 HGPA_i + \alpha_2 MATH_i + \alpha_3 VERB_i + \alpha_4 (MATH_i + VERB_i) + E_i$$

It follows that

$$CGPA_i = \alpha_0 + \alpha_1 HGPA_i + (\alpha_2 + \alpha_4)MATH_i + (\alpha_3 + \alpha_4)VERB_i + E_i$$

With this version of the model, consider choosing $\alpha_4 = 0$. Then $\alpha_2 = \beta_2$ and $\alpha_3 = \beta_3$ give the correct original model. Next, choose $\alpha_4 = 3$. Then $\alpha_2 = \beta_2 - 3$ and $\alpha_3 = \beta_3 - 3$ also give the correct original model. In fact, for any choice of α_4, we can choose values for α_2 and α_3 that provide the correct model. Since the α_4 parameter is irrelevant under the circumstances (it could best be set equal to 0), a model containing a perfect collinearity is sometimes said to be *overparameterized*.

Near collinearity arises if the multiple R^2 of one predictor with the remaining predictors is nearly 1. We saw an example of this in considering the squared correlation $r^2(X, X^2)$ between X and X^2 in section 12-5-1 (see Table 12-1, when $\alpha = 0$). For a more general model involving k predictors such as

$$Y_i = \beta_0 + \beta_1 X_{i1} + \beta_2 X_{i2} + \cdots + \beta_k X_{ik} + E_i$$

the multiple R^2-value of interest for the first predictor is $R^2(X_1 \mid X_2, X_3, \ldots, X_k)$, the multiple R^2-value of interest for the second predictor is $R^2(X_2 \mid X_1, X_3, \ldots, X_k)$, and so on. These quantities are generalizations of the statistic $r^2(X_1, X_2)$ for a $k = 2$ variable model. For convenience, we denote by R_j^2 the squared multiple correlation based on regressing X_j on the remaining $k - 1$ predictors.

The *variance inflation factor* (VIF) is often used to measure collinearity in a multiple regression analysis. It may be computed as

$$VIF_j = \frac{1}{1 - R_j^2}, \quad j = 1, 2, \ldots, k$$

The quantity VIF_j generalizes the variance inflation factor for a two-predictor model, $1/[1 - r^2(X_1, X_2)]$. Clearly, $VIF_j \geq 0$. As for the two-variable case, the estimates of the variances for the regression coefficients are proportional to the VIFs—namely,

$$S_{\hat{\beta}_j}^2 = c_j^*(VIF_j), \quad j = 1, 2, \ldots, k$$

This expression suggests that the larger value of VIF_j, the more troublesome is the variable X_j. A rule of thumb for evaluating VIFs is to be concerned with any value larger than 10.0. For VIF_j, this corresponds to $R_j^2 > .90$ or, equivalently, $R_j > .95$. Some people prefer to consider

$$Tolerance_j = \frac{1}{VIF_j} = 1 - R_j^2$$

The choice among R_j^2, $1 - R_j^2$, and VIF_j is a matter of personal preference since they all contain exactly the same information. As R_j^2 goes to 1.0, the tolerance $1 - R_j^2$ goes to 0 and VIF_j goes to infinity.

Because of its special nature, the intercept requires separate treatment in evaluating collinearity. For the general model involving k predictors,

$$Y_i = \beta_0 + \beta_1 X_{i1} + \beta_2 X_{i2} + \cdots + \beta_k X_{ik} + E_i$$

we find regression coefficient estimates $\hat{\beta}_0, \hat{\beta}_1, \ldots, \hat{\beta}_k$. The intercept estimate can be expressed simply as

$$\hat{\beta}_0 = \overline{Y} - (\hat{\beta}_1 \overline{X}_1 + \hat{\beta}_2 \overline{X}_2 + \cdots + \hat{\beta}_k \overline{X}_k) = \overline{Y} - \sum_{j=1}^{k} \hat{\beta}_j \overline{X}_j$$

Here, $\overline{Y} = \sum_{i=1}^{n} Y_i / n$ is the mean of the response values, and $\overline{X}_j = \sum_{i=1}^{n} X_{ij} / n$ is the mean of the values for predictor j. From this, we can deduce that the estimated intercept is affected by the VIF_j's, $j = 1, 2, \ldots, k$, since it is a function of the $\hat{\beta}_j$'s. The problem disappears if the means of all X_j's are 0 (e.g., if the predictor data are centered). In that case, $\overline{Y}$ is the estimated intercept.

In general, even if the predictor data are not centered, a variance inflation factor VIF_0 for $\hat{\beta}_0$ can be defined and is interpreted in the same way as VIF_j. First, define

$$VIF_0 = \frac{1}{1 - R_0^2}$$

Here R_0^2 is the generalized squared multiple correlation for the regression model

$$I_i = \alpha_1 X_{i1} + \alpha_2 X_{i2} + \cdots + \alpha_k X_{ik} + E_i$$

in which I_i is identically 1 (which may be thought of as the score for the intercept variable). No intercept is included in this model (as a predictor), which is why R_0^2 is called a *generalized* squared correlation. As with the VIF_j's, $VIF_0 \geq 0$, and

$$S_{\hat{\beta}_0}^2 = c_0^*(VIF_0)$$

Hence, the interpretation for VIF_0 is the same as for VIF_j, $j = 1, 2, \ldots, k$.

Controversy surrounds the treatment of the intercept in regression diagnostics. For some, it is simply another predictor; for others, it should be eliminated from discussion. We take a middle position, arguing that the model and the data at hand determine the role of the intercept. (See, for example, the discussions following Belsley 1984.) This leads us to discuss diagnostics both with and without the intercept included, corresponding to the cases with and without centering of the predictors and response variables.

The presence of collinearity or (more typically) near collinearity presents computational difficulties in calculating numerically reliable estimates of the R_j^2-values, tolerances, and the VIF_j's on the basis of standard regression procedures. This apparent impasse can be solved in at least three ways. The first way is to use computational algorithms that detect collinearity problems as they arise in the midst of the calculations. A discussion of such algorithms is beyond the scope of this text.

A second way to avoid the impasse is to scale the data appropriately. By *scaling,* we mean the choice of measurement unit (e.g., degrees Celsius versus degrees Fahrenheit) and the choice of measurement origin (e.g., degrees Celsius versus degrees Kelvin). We shall consider only linear changes in scale, such as

$$X_1 = \alpha + \beta X_2$$

$$C = \frac{5}{9}(F - 32) = -32\left(\frac{5}{9}\right) + \left(\frac{5}{9}\right)(F)$$

or

$$K = (-273) + (1)(C)$$

where C, F, and K are temperatures in degrees Celsius, degrees Fahrenheit, and degrees Kelvin, respectively.

Often scaling refers just to multiplying by a constant rather than also to adding a constant. An example of scaling is the conversion from feet to inches. One important case of adding a constant, a form of scaling, is *centering*. A set of scores $\{X_{ij}\}$ is centered by subtracting the mean $\overline{X}_j$ of the scores for each predictor from each individual score for that predictor, giving

$$X_{ij}^* = X_{ij} - \overline{X}_j$$

in which $\overline{X}_j = \sum_{i=1}^{n} X_{ij}/n$, for the jth predictor.

Computing standardized scores (z scores) is a closely related method of scaling. In particular, the standardized score corresponding to X_{ij} is

$$z_{ij} = \frac{X_{ij} - \overline{X}_j}{S_j}$$

in which $S_j^2 = \sum_{i=1}^{n}(X_{ij} - \overline{X}_j)^2/(n-1)$. Centered and standardized scores have mean 0, since $\sum_{i=1}^{n} X_{ij}^* = \sum_{i=1}^{n} z_{ij} = 0$. Also, $\sum_{i=1}^{n} z_{ij}^2/(n-1) = 1$, so the set $\{z_{ij}\}$ of standardized scores has a variance equal to 1. We shall delay the discussion of detecting and fixing scaling problems until later in this chapter. For now, we need the concept of scaling in order to understand the following discussion of other diagnostic statistics.

A third way to avoid the impasse created by collinearity and near collinearity is to use alternate computational methods to diagnose collinearity. One especially popular method for characterizing near and/or exact collinearities among the predictors involves computing the *eigenvalues* of the predictor variable *correlation matrix*. The eigenvalues are connected with the *principal component analysis* of the predictors. The principal components of the predictors are a set of new variables that are linear combinations of the original predictors. These components have two special properties: they are not correlated with each other; and each, in turn, has maximum variance, given that all are mutually uncorrelated. The principal components provide idealized predictor variables that still retain all of the same information as the original variables. The variances of these components (the new variables) are called *eigenvalues*. The larger the eigenvalue, the more important is the associated principal component in representing the information in the predictors. *As an eigenvalue approaches zero, the presence of a near collinearity among the original predictors is indicated.* The presence of an eigenvalue of exactly 0 means that a perfect linear dependency (i.e., an exact collinearity) exists among the predictors.

If a set of k predictor variables does *not* involve an *exact* collinearity, then k principal components are needed to reproduce exactly all of the information contained in the original variables.

If one of the predictors is a perfect linear combination of the others, then only $k - 1$ principal components are needed to provide all of the information in the original variables. *The number of zero (or near-zero) eigenvalues is the number of collinearities (or near collinearities) among the predictors.* Even a single eigenvalue near 0 presents a serious problem that must be resolved.

In using the eigenvalues to determine the presence of near collinearity, researchers usually employ three kinds of statistics: the *condition index* (CI), the *condition number* (CN), and the *variance proportions.*

Consider again the general linear model

$$Y_i = \beta_0 + \beta_1 X_{i1} + \beta_2 X_{i2} + \cdots + \beta_k X_{ik} + E_i$$

Often collinearity is assessed by considering only the k parameters β_1 through β_k and ignoring the intercept β_0. This is accomplished by centering the response and predictor variables, which corresponds to fitting the model

$$Y_i - \overline{Y} = \beta_1(X_{i1} - \overline{X}_1) + \beta_2(X_{i2} - \overline{X}_2) + \cdots + \beta_k(X_{ik} - \overline{X}_k) + E_i$$

Noting that $\hat{\beta}_0 = \overline{Y} - \sum_{j=1}^{k}\hat{\beta}_j\overline{X}_j$ in the original model, we observe that centering the predictors and the response forces the estimated intercept in that centered model to be 0. Hence, it can be dropped from that model.

It is also common to assess collinearity after centering and standardizing both the predictors and the response. This leads to the standardized model

$$\frac{Y_i - \overline{Y}}{S_Y} = \beta_1^*\frac{(X_{i1} - \overline{X}_1)}{S_1} + \beta_2^*\frac{(X_{i2} - \overline{X}_2)}{S_2} + \cdots + \beta_k^*\frac{(X_{ik} - \overline{X}_k)}{S_k} + E_i^*$$

The coefficients for this standardized model are often called the *standardized regression coefficients;* in particular,

$$\beta_j^* = \beta_j\left(\frac{S_j}{S_Y}\right), \quad j = 1, 2,\ldots, k$$

Table 12-2 summarizes an eigenanalysis of the predictor correlation matrix R_{xx} for a hypothetical four-predictor ($k = 4$) standardized regression model. With $k = 4$ predictors, four eigenvalues (denoted by λ's) exist. Later we will include the intercept in the analysis and have $k + 1$ eigenvalues. In the present case, $\lambda_1 = 2.0$, $\lambda_2 = 1.0$, $\lambda_3 = 0.6$, and $\lambda_4 = 0.4$. (The sum of the eigenvalues for a correlation matrix involving k predictors is *always* equal to k.) It is customary to list the eigenvalues from largest (λ_1) to smallest (λ_k). A condition index can be computed for each eigenvalue as

$$CI_j = \sqrt{\lambda_1/\lambda_j}$$

TABLE 12-2   Eigenanalysis of the predictor correlation matrix for the hypothetical four-variable model.

Variable	Eigenvalue	Condition Index	Variance Proportions			
			X_1	X_2	X_3	X_4
1	2.0	1.00	.09	.11	.08	.15
2	1.0	1.41	.32	.10	.25	.07
3	0.6	1.87	.40	.52	.12	.13
4	0.4	2.24	.19	.27	.55	.65

In particular, CI_3 for this example is $\sqrt{2.0/0.6} = 1.87$. The first (largest) eigenvalue always has an associated condition index (CI_1) of 1.0. The largest CI_j, called the *condition number*, always involves the largest (λ_1) and smallest (λ_k) eigenvalues. It is given by the formula

$$CN = \sqrt{\lambda_1 / \lambda_k}$$

For this example, $CN = \sqrt{2.0/0.4} = 2.24$. Since eigenvalues are variances, CI_j and CN are ratios of standard deviations (of principal components, the idealized predictors). Like VIFs, values of CI_j and CN are nonnegative, and larger values suggest potential near collinearity. Belsley, Kuh, and Welsch (1980) recommended interpreting a CN of 30 or more as reflecting moderate to severe collinearity, worthy of further investigation. Of course, such a CN may be associated with two or more CI_j's that are greater than or equal to 30.

A variance proportion indicates, for each predictor, the proportion of total variance of its estimated regression coefficient associated with a particular principal component. The variance proportions suggest collinearity problems if more than one predictor has a high variance proportion (loads highly) on a principal component having a high condition index. The presence of two or more proportions of at least .5 for such a component suggests a problem. One should definitely be concerned when two or more loadings greater than .9 appear on a component with a large condition index.

Table 12-2 includes variance proportions for the hypothetical four-predictor model under consideration. The entries represent a typical pattern of condition indices and proportions for a regression analysis with no major collinearity problems. Each column sums to a total proportion of 1.00 because each estimated regression coefficient has its own (total) variance partitioned among the four components. The last row involves two loadings greater than .5, corresponding to the smallest eigenvalue; but since the smallest eigenvalue has a CI of only 2.24 (far from the suggested warning level of 30.0 for moderate to severe collinearity), the proportion pattern does not indicate a major problem.

The intercept can play an important role in collinearity. In defining regression models, researchers follow the standard practice of including the intercept term β_0. The intercept is a regression coefficient for a variable that has a constant value of 1. Consequently, any variable with near-zero variance will be nearly a constant multiple of the intercept and hence will be nearly collinear with it. This problem can arise spuriously when variables are improperly scaled. To evaluate this possibility, the eigenanalysis discussed earlier can be modified to include the intercept by basing it on the scaled cross-products matrix.[5] The eigenanalysis including the intercept may suggest that this constant is nearly collinear with one or more predictors.

Centering may help decrease collinearity. From a purely theoretical perspective, statisticians disagree as to when centering helps regression calculations (for example, see Belsley 1991; Belsley 1984; Smith and Campbell 1980; and related comments in the same issues). For actual computations, centering can increase numerical accuracy in many situations. Polynomial regression (Chapter 13) is one situation where centering is recommended. In the example that follows, we numerically illustrate the effects of centering on procedures for diagnosing collinearity.

---

[5]The eigenanalysis would then be based on the $(k+1) \times (k+1)$ cross-products matrix $\mathbf{X'X}$ suitably scaled to have 1's on the diagonal, rather than on the $k \times k$ correlation matrix $\mathbf{R_{xx}}$. The sum of the eigenvalues for this scaled cross-products matrix is equal to $(k+1)$.

12-5-3 Example of Near Collinearity

The boys' WEIGHT data introduced on page 215 can be used to illustrate the use of collinearity diagnostics. First, consider the correlations among the predictor variables, as presented in Table 12-3. Since two correlations among predictors are greater than .99, one should expect trouble. These high correlations arise because $(AGE)^2$ is almost perfectly collinear with AGE, and $(HEIGHT)^2$ is essentially collinear with HEIGHT. In general, any between-predictor correlation above .9 (in absolute value) merits further attention.

Table 12-4 indicates that serious collinearity problems exist. This can be seen in several ways. First, two condition indices, CI_4 and CI_5, are much greater than 30. Second, the smallest eigenvalue (5×10^{-6}) is associated with maximum variance proportions (i.e., loadings) for HEIGHT and $(HEIGHT)^2$. The fact that the intercept loads highly (.73) on a problem component (one with an eigenvalue near 0) suggests to experienced analysts that centering may help to reduce the problem.

To evaluate the utility of centering for this example, we transformed the predictor variables as follows:

$$HEIGHT_* = HEIGHT - 1.577$$
$$(HEIGHT_*)^2 = (HEIGHT - 1.577)^2$$
$$AGE_* = AGE - 13.67$$
$$(AGE_*)^2 = (AGE - 13.67)^2$$

Because the mean HEIGHT (for $n = 126$) is 1.577 meters and the mean AGE is 13.67 years, the variables $HEIGHT_*$ and AGE_* are centered. This also means that the definitions of the parameters have changed, leading to different tests and interpretations. Next, we recomputed the same

TABLE 12-3   Correlations for the weight example, without centering ($n = 126$).

	HEIGHT	$(HEIGHT)^2$	AGE	$(AGE)^2$	WEIGHT
HEIGHT	1	.999	.774	.766	.790
$(HEIGHT)^2$		1	.777	.770	.794
AGE			1	.998	.699
$(AGE)^2$				1	.704

TABLE 12-4   Eigenanalysis of the scaled cross-products matrix for the model where WEIGHT is predicted from HEIGHT, $(HEIGHT)^2$, AGE, and $(AGE)^2$, without centering ($n = 126$).

Variable	Eigenvalue	Condition Index	Variance Proportions				
			Intercept	HEIGHT	$(HEIGHT)^2$	AGE	$(AGE)^2$
1	4.969	1.0	.00	.00	.00	.00	.00
2	0.025	14.1	.00	.00	.00	.00	.00
3	0.006	29.2	.00	.00	.00	.00	.00
4	2×10^{-5}	490.2	.27	.00	.00	.84	.83
5	5×10^{-6}	1,028.0	.73	1.00	1.00	.16	.17

regression using the transformed variables. As expected, the R^2-value remains .68. However, predictor correlations (Table 12-5) and collinearity diagnostics for the scaled cross-products matrix (Table 12-6) show radical improvement. The maximum between-predictor correlation in Table 12-5 has dropped from .999 to .774. The maximum condition index (the condition number) is now 3.7 in Table 12-6, compared with the value of 1,028 in Table 12-4. The only hint of a problem arises in the loadings for the principal component with the smallest eigenvalue (the last row); and even these variance proportions of .85 and .64 should not be bothersome, because the associated condition index is low (namely, 3.7).

The data and polynomial-type model we have just considered clearly illustrate a collinearity problem that could seriously affect the accuracy of regression calculations. In this example, simple centering of the original variables solved the problem, since the collinearity was due to high correlations among different powers of the same variable. A general treatment of numerical problems with polynomial regression models is described in Chapter 13.

12-5-4 Avoidable Collinearities

Problems with regression calculations due to collinearity problems may not be easy to detect. Therefore, researchers need to be aware of potential dangers when doing regression analyses. The following examples are chosen to highlight common occurrences in regression analyses that may lead to near collinearities (or even to exact linear dependencies) among the predictor variables.

In many situations, researchers wish to consider as predictors the powers of a continuous variable. An example would be the use of HEIGHT, $(HEIGHT)^2$, and $(HEIGHT)^3$ as predictors for the preceding weight example. These predictors, which are often called the *natural polynomials,* often lead to near collinearities when used carelessly. One solution to this problem, besides centering, is to use orthogonal polynomials. This topic is covered in detail in Chapter 13.

TABLE 12-5   Correlations for weight example, with centering of predictors ($n = 126$).

	HEIGHT	$(HEIGHT)^2$	AGE	$(AGE)^2$	WEIGHT
HEIGHT	1	−.067	.774	.221	.790
$(HEIGHT)^2$		1	.032	.372	.053
AGE			1	.430	.699
$(AGE)^2$				1	.396

TABLE 12-6   Eigenanalysis of the scaled cross-products matrix for the model where WEIGHT is predicted from HEIGHT, $(HEIGHT)^2$, AGE, and $(AGE)^2$, with centering ($n = 126$).

Variable	Eigenvalue	Condition Index	Variance Proportions				
			Intercept	HEIGHT	$(HEIGHT)^2$	AGE	$(AGE)^2$
1	2.353	1.0	.05	.01	.06	.01	.06
2	1.764	1.1	.02	.09	.02	.08	.00
3	0.376	2.5	.65	.05	.48	.03	.01
4	0.334	2.6	.03	.21	.44	.03	.53
5	0.172	3.7	.25	.64	.00	.85	.40

Collinearity problems can arise if particularly extreme data values are incorrectly included in the data set via errors in data collection. This data-handling problem can be detected by using the methods for analyzing outliers described earlier. Again, the reader is cautioned against blithely discarding troublesome observations.

The use of dummy variables may inadvertently introduce exact collinearities. Discussion of how to avoid this problem is deferred until Chapter 14.

Interaction terms generally create the atmosphere for collinearity problems, especially if such terms are overused. For example, if the predictors are AGE, HEIGHT, and AGE × HEIGHT, then a near collinearity may show up due to the close functional relationship between the product term and the two basic predictors AGE and HEIGHT. In general, fitting a model that contains several interaction terms almost guarantees that some collinearity problems will arise. An important special case in which interactions are not generally troublesome in this way is in ANOVA designs with equal or nearly equal numbers of subjects in each cell (Chapter 19).

A subtler form of near collinearity occurs with the following set of predictors: head-of-household income, education, number of years in work force, and age. Since these four variables tend to be highly positively correlated with one another, one of the four is likely to be nearly perfectly predicted from (some linear combination of) the remainder. This illustrates one of the most troublesome types of collinearity. Ideally one can avoid the problem by eliminating one or more of the variables. If this solution is unpalatable, certain esoteric methods of analysis (such as ridge regression) can be employed.

12-6 Scaling Problems

A general class of problems in regression analysis may arise due to improper scaling of the predictors and/or the response variable. Specifically, such problems involve the loss of computational accuracy. The resulting inaccuracy can sometimes be so great as to give coefficient estimates with the wrong sign. A scaling problem may occur if a predictor has too wide a range of values. For example, recording adult human body weight in grams might be problematic. Another scaling problem might occur with data for people who use vastly greater quantities of, say, vitamin C than the average person. Properly measured values of such vitamin C intake would still yield outliers, and their use could lead to modeling problems. Similarly, if the mean of a variable is large and has little variability, computational problems may arise. A simple example of this would involve recording human body temperature in degrees Kelvin (degrees Celsius minus 273). Most scaling problems can be avoided by using proper data validation and rescaling before performing regression analyses.

12-7 Treating Collinearity and Scaling Problems

Both near collinearity and bad scaling create numerical problems, including inaccurate computations of (1) estimates of the regression coefficients, (2) estimates of standard errors, and (3) hypothesis test statistics. In the worst cases, the analysis may change substantially if the data are input into a program in a different order or if a few observations are deleted. All reduce our confidence in the analysis.

The first step in treating numerical problems is to validate the data adequately before attempting any regression modeling. Appropriate validation procedures can detect most scaling problems and suggest solutions. For continuous variables, the numerical ranges of the predictor variables should be as similar as possible. For example, do not measure one set of weights in grams and another in tons. In general, numbers that are convenient to write and plot are preferred for data analysis. Typically these yield ranges such as 1 to 10 or 10 to 100.

The second step in treating numerical problems is to utilize regression diagnostics and collinearity diagnostics (e.g., eigenvalues and condition indices). These methods allow analysts to detect numerical problems, but they do not necessarily indicate a good solution.

The third step in treating numerical problems is to attempt to eliminate redundant variables. Ordering the variables by their importance is central to this task. More sophisticated techniques involve trying to formalize the selection of variables (which we discuss in Chapter 16; see also McCabe 1984). Analysis of the principal components may be applied to the process of reducing variables (see section 12.8).

12-8 Alternate Strategies of Analysis

When one or more basic assumptions of regression analysis are clearly not satisfied and/or when numerical problems are identified, the analyst may want to turn to other analytical strategies. In the subsections that follow, we list some alternate methods of analysis, briefly mention generalizations of linear regression that may be adequate in some applications, and briefly describe transformations to the data that may allow the use of multiple linear regression.

12-8-1 Alternate Approaches

If the analyst decides that the regression model used does not fit the data and cannot be made to fit via simple generalizations of linear regression (such as weighted least squares), other methods must be used. For situations where the response variable cannot be modeled as a linear function of the parameters, analysis methods do exist for nonlinear functions (Gallant 1975). If the problem concerns nonnormal distributions, methods of rank analysis (Conover and Iman 1981) or categorical data analysis (Agresti 1990) may be appropriate.

Some rank analysis methods may be thought of as being essentially two-step processes. The first step is to replace the original data with appropriate ranks. The second is to conduct linear regression analysis on the ranks. This approach cannot be presumed to work in any particular case; but a brief discussion of it is still helpful in understanding ranking methods.

12-8-2 Generalizations of Linear Regression

Survey of Methods

Generalizations of linear regression may be loosely grouped into exact and approximate methods. Exact methods have procedures of estimation and hypothesis testing with known properties for finite samples; approximate methods have only asymptotic results available and, therefore, must be used with caution in small samples. The generalized approximate methods include regression on principal components, ridge regression, and robust regression. Exact generalizations include multivariate techniques and exact weighted least squares (in which weights are known without error).

Regression on principal components and ridge regression are often recommended for treating collinearity problems. In the analysis of *principal components,* the original predictor variables are replaced by a set of mutually uncorrelated variables—the principal component scores. If necessary, components associated with eigenvalues near 0 are dropped from the analysis, thus eliminating the attendant collinearity problem. *Ridge regression* involves perturbing the eigenvalues of the original predictor variable cross-products matrix to "push" them away from 0, thus reducing the amount of collinearity. A detailed discussion of these methods is beyond the scope of this book (see, e.g., Gunst and Mason 1980). Both methods lead to biased regression estimates of the parameters under the assumed model. In addition, *P*-values for statistical tests may be optimistically small when based on such biased estimation methods.

Robust regression involves weighting or transforming the data so as to minimize the effects of extreme observations. The goal is to make the analysis more robust (i.e., less sensitive) to any particular observation and also less sensitive to the basic assumptions of regression analysis. (See Huber 1981, for a discussion of such procedures.)

Multivariate methods for the analysis of continuous responses include multivariate multiple regression, multivariate analysis of variance, canonical correlation analysis, and growth curve analysis, among others. In all these procedures, one can account for nonindependence among observations by explicitly modeling such nonindependence. The types of predictors and response variables that may be used in these methods differ, as do the hypotheses that are testable and methods by which the nonindependence is modeled. Many multivariate books are available; see, e.g., Timm (1975) and Morrison (1976).

Weighted Least-Squares Analysis

The *weighted least-squares* method of analysis is a modification of standard regression analysis procedures that is used when a regression model is to be fit to a set of data for which the assumptions of variance homogeneity and/or independence do not hold. We shall briefly describe here the weighted least-squares approach for dealing with variance heterogeneity. For discussions of the general method of weighted regression (which incorporate a discussion of treating nonindependence), see Draper and Smith (1981) and Neter, Wasserman, and Kutner (1983).

Weighted least-squares analysis can be used when the variance of Y varies for different values of the independent variable(s), provided that these variances (i.e., σ_i^2 for the ith observation on Y) are known or can be assumed to be of the form $\sigma_i^2 = \sigma^2 / W_i$, where the weights $\{W_i\}$ are known. The methodology then involves determining the regression coefficients $\hat{\beta}_0'$, $\hat{\beta}_1'$, ..., $\hat{\beta}_k'$ that minimize the expression

$$\sum_{i=1}^{n} W_i (Y_i - \hat{\beta}_0' - \hat{\beta}_1' X_{1i} - \hat{\beta}_2' X_{2i} - \cdots - \hat{\beta}_k' X_{ki})^2$$

where the weight W_i is given by $1/\sigma_i^2$ (when the $\{\sigma_i^2\}$ are known) or is exactly the W_i in the expression σ^2 / W_i (when this form applies).

The specific weighted least-squares solution for the straight-line regression case (i.e., $Y = \beta_0 + \beta_1 X + E$) is given by the formulas

$$\hat{\beta}_1' = \frac{\sum\limits_{i=1}^{n} W_i(X_i - \overline{X}')(Y_i - \overline{Y}')}{\sum\limits_{i=1}^{n} W_i(X_i - \overline{X}')^2}$$

and

$$\hat{\beta}_0' = \overline{Y}' - \hat{\beta}_1'\overline{X}'$$

in which

$$\overline{Y}' = \frac{\sum\limits_{i=1}^{n} W_i Y_i}{\sum\limits_{i=1}^{n} W_i} \quad \text{and} \quad \overline{X}' = \frac{\sum\limits_{i=1}^{n} W_i X_i}{\sum\limits_{i=1}^{n} W_i}$$

Under the usual normality assumption for the Y-variable, the same general procedures are applicable as are used in the unweighted case regarding t tests, F tests, and confidence intervals about the various regression parameters. For example, to test H_0: $\beta_1 = 0$, we may use the following test statistic:

$$T = \frac{\hat{\beta}_1' - 0}{S_{Y|X}'/S_X'\sqrt{n-1}} \sim t_{n-2} \quad \text{under } H_0$$

in which

$$S_{Y|X}'^2 = \frac{1}{n-2} \sum\limits_{i=1}^{n} W_i(Y_i - \hat{\beta}_0' - \hat{\beta}_1'X_i)^2$$

and

$$S_X'^2 = \frac{1}{n-1} \sum\limits_{i=1}^{n} W_i(X_i - \overline{X}')^2$$

12-8-3 Transformations

There are three primary reasons for using data transformations: (1) *to stabilize* the variance of the dependent variable, if the homoscedasticity assumption is violated; (2) *to normalize* (i.e., to transform to the normal distribution) the dependent variable, if the normality assumption is noticeably violated; (3) *to linearize* the regression model, if the original data suggest a model that is nonlinear in either the regression coefficients or the original variables (dependent or independent). Fortunately, the same transformation often helps to accomplish the first two goals and sometimes even the third, rather than achieving one goal at the expense of one or both of the other two.

A more complete discussion of the properties of various transformations can be found in Armitage (1971), Draper and Smith (1981), and Neter, Wasserman, and Kutner (1983). In addition, Box and Cox (1964) describe an approach to making an exploratory search for one of a family of transformations (see also Box and Cox 1984, and Carroll and Ruppert 1984). Nevertheless, we consider it useful to describe a few of the more commonly used transformations:

1. The *log transformation* ($Y' = \log Y$) is used (provided Y takes on only positive values) to stabilize the variance of Y, if the variance increases markedly with increasing Y; to normalize the dependent variable, if the distribution of the residuals for Y is positively skewed; and to linearize the regression model, if the relationship of Y to some independent variable suggests a model with consistently increasing slope.

2. The *square root transformation* ($Y' = \sqrt{Y}$) is used to stabilize the variance, if the variance is proportional to the mean of Y. This is particularly appropriate if the dependent variable has the Poisson distribution.

3. The *reciprocal transformation* ($Y' = 1/Y$) is used to stabilize the variance, if the variance is proportional to the fourth power of the mean of Y (which indicates that a huge increase in variance occurs above some threshold value of Y). This transformation minimizes the effect of large values of Y, since the transformed Y'-values for these values will be close to 0, and large increases in Y will cause only trivial decreases in Y'.

4. The *square transformation* ($Y' = Y^2$) is used to stabilize the variance, if the variance decreases with the mean of Y; to normalize the dependent variable, if the distribution of the residuals for Y is negatively skewed; and to linearize the model, if the original relationship with some independent variable is curvilinear downward (i.e., if the slope consistently decreases as the independent variable increases).

5. The *arcsin transformation* ($Y' = \arcsin \sqrt{Y} = \sin^{-1} \sqrt{Y}$) is used to stabilize the variance, if Y is a proportion or rate.

12-9 An Important Caution

All of the techniques described in this chapter involve checking the validity of the assumptions and estimates for a regression analysis. To that end, outlier analysis and collinearity diagnosis may suggest deleting observations or predictor variables in order to improve the quality of the model. Naturally, the observations and variables that are most at odds with the fitted model are the ones deleted. All of these techniques thus lead to underestimation of the true variability and give P-values that are optimistically small. The potential penalty associated with such improvements in fitting a model is biased results. The methods discussed in this chapter are safest to use with large samples and are least reliable with small samples. Unfortunately, of course, the potential influence of a single observation is largest in small samples. This is a strong argument against using small samples in regression analysis.

In some cases, a problem can be resolved only by evaluating a second sample of data. Picard and Cook (1984) discuss the issue of optimism in selecting a regression model and recommend considering split samples. In a split-sample design, some of the data are used for exploratory data analysis, and the rest are used for confirming the validity and reliability of the exploratory results. This technique is described in Chapter 16 in the context of selecting the best regression model.

handwritten annotation:
>library (MASS)
>help (studres)

Problems

You may use the computer output accompanying these problems to answer the questions posed; alternatively, you may use your own statistical software. Be aware that some of the data sets used in these problems have hidden problems: outliers, collinearities, and variables in need of transformation. In response to certain requests, some computer programs may either balk or blithely produce garbage. This state of affairs is realistic. Having "nice" data, of the type provided in most other problem sets, is not realistic, but it allows us to focus on the particular chapter topics.

1. Consider the data from Problem 1 of Chapter 5.
   a. Fit a univariate linear model, with dry weight as the response and age as the predictor.
   b. Provide a plot of studentized or jackknife residuals versus the predictor.
   c. Provide a frequency histogram and a schematic plot of the residuals.
   d. Report and interpret a test of whether any residuals have extreme values not due to chance alone.
   e. Do these analyses highlight any potentially troublesome observations? Why or why not?
   f. Why is it only approximately correct to test jackknife residuals for normality? What role does sample size play?

handwritten annotation: studentized ≟ jackknife

Edited SAS Output for Problem 1

Plot of JACKKNIFE by X. Legend: A = 1 obs, B = 2 obs, etc.

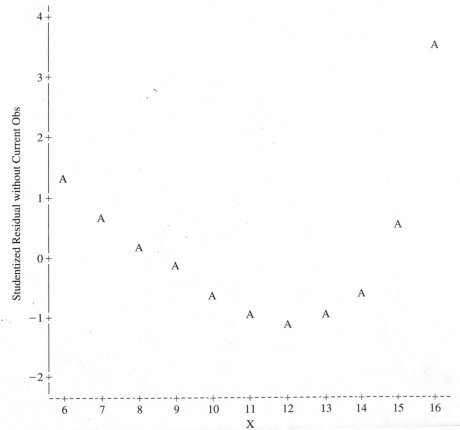

Univariate Procedure
•
• [Portion of output omitted]
•

Variable=JACKKNIFE		Studentized Residual without Current Obs	
Stem	Leaf	#	Boxplot
3	6	1	0
2			
1	3	1	|
0	257	3	+-----+-----+
−0	662	3	*----------*
−1	100	3	+----------+
	----+-----+-----+-----+		

2. a.–e. Repeat parts (a) through (e) of Problem 1, using $\log_{10}$ Dry weight as the response.
   f. Which approach—using the original data or using logarithms—leads to better-behaved residuals? Why?

Edited SAS Output for Problem 2

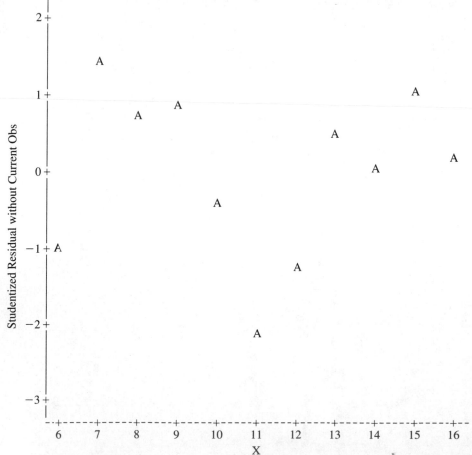

Plot of JACKKNIFE by X. Legend: A = 1 obs, B = 2 obs, etc.

Univariate Procedure

- [Portion of output omitted]

Variable= JACKKNIFE		Studentized Residual without Current Obs	
Stem	Leaf	#	Boxplot
1	5	1		
1	1	1	
0	89	2	+----------+
0	24	2	*----------*
-0	40	2	+
-0			
-1	30	2	+----------+
-1			
-2	2	1	

```
----+-----+-----+-----+
```

3.–14. a.–e. Repeat parts (a) through (e) of Problem 1, using the data from Problems 3 through 14 of Chapter 5.

15. Consider the data from Problem 15 of Chapter 5.

  a. Fit a univariate linear model, with BLOODTOL as the response and PPM_TOLU as the predictor.

  b. Provide a plot of studentized or jackknife residuals versus the predictor.

  c. Provide a frequency histogram and a schematic plot of the residuals.

  d. Report and interpret a test of whether any residuals have extreme values not due to chance alone.

  e. Do these analyses highlight any potentially troublesome observations? Why or why not?

16. a.–e. Repeat Problem 15 for LN_BLDTL as predicted by LN_PPMTL.

  f. Which approach—using the original data or using logarithms—leads to better-behaved residuals? Why?

17. Repeat Problem 15, replacing BLOODTOL with BRAINTOL.

18. Repeat Problem 16, replacing LN_BLDTL with LN_BRNTL.

19. The data in the following table come from an article by Bethel and others (1985). All subjects are asthmatics.

Subject	AGE (yr)	Sex	Height (cm)	Weight (kg)	FEV$_1$* (L)	Subject	AGE (yr)	Sex	Height (cm)	Weight (kg)	FEV$_1$* (L)
1	24	M	175	78.0	4.7	11	26	M	180	70.5	3.5
2	36	M	172	67.6	4.3	12	29	M	163	75.0	3.2
3	28	F	171	98.0	3.5	13	33	F	180	68.0	2.6
4	25	M	166	65.5	4.0	14	31	M	180	65.0	2.0
5	26	F	166	65.0	3.2	15	30	M	180	70.4	4.0
6	22	M	176	65.5	4.7	16	22	M	168	63.0	3.9
7	27	M	185	85.5	4.3	17	27	M	168	91.2	3.0
8	27	M	171	76.3	4.7	18	46	M	178	67.0	4.5
9	36	M	185	79.0	5.2	19	36	M	173	62.0	2.4
10	24	M	182	88.2	4.2						

* Forced expiratory volume in 1 second.

    a. Fit a model of FEV_1 as predicted by HEIGHT, WEIGHT, and AGE.

    b. Conduct variable-added-last tests for all predictors, and perform a test of the intercept.

    c. Determine variance inflation factors for each predictor.

    d. Determine a correlation matrix that includes all predictors and the response.

    e. Determine eigenvalues, condition indexes, and condition numbers for the correlation matrix (excluding the intercept).

    f. Determine eigenvalues, condition indexes, and condition numbers for the scaled cross-products matrix (including the intercept).

    g. Determine residuals (preferably studentized) and leverage values. Do any observations seem bothersome? Explain.

    h. Does there appear to be any problem with collinearity? Explain.

Edited SAS Output (PROC REG) for Problem 19

Correlation

CORR	HEIGHT	WEIGHT	AGE	FEV1
HEIGHT	1.0000	0.1291	0.2530	0.2257
WEIGHT	0.1291	1.0000	-0.1996	0.1807
AGE	0.2530	-0.1996	1.0000	-0.0552
FEV1	0.2257	0.1807	-0.0552	1.0000

Analysis of Variance

Source	DF	Sum of Squares	Mean Square	F Value	Prob > F
Model	3	1.11738	0.37246	0.440	0.7275
Error	15	12.68789	0.84586		
C Total	18	13.80526			

    . [Portion of output omitted]

Parameter Estimates

Variable	DF	Parameter Estimate	Standard Error	T for H0: Parameter=0	Prob > |T|
INTERCEP	1	-1.936831	5.72266521	-0.338	0.7397
HEIGHT	1	0.030148	0.03409922	0.884	0.3906
WEIGHT	1	0.011177	0.02151351	0.520	0.6110
AGE	1	-0.012643	0.03839506	-0.329	0.7465

Variable	DF	Variance Inflation
INTERCEP	1	0.00000000
HEIGHT	1	1.10810910
WEIGHT	1	1.08023527
AGE	1	1.13485151

Number	Eigenvalue	Condition Index	Var Prop INTERCEP	Var Prop HEIGHT	Var Prop WEIGHT	Var Prop AGE
1	3.95433	1.00000	0.0001	0.0001	0.0011	0.0021
2	0.03589	10.49649	0.0008	0.0005	0.1478	0.6272
3	0.00911	20.83984	0.0375	0.0293	0.8494	0.3407
4	0.0006744	76.57184	0.9617	0.9702	0.0017	0.0300

Collinearity Diagnostics(intercept adjusted)

Number	Eigenvalue	Condition Index	Var Prop HEIGHT	Var Prop WEIGHT	Var Prop AGE
1	1.26807	1.00000	0.2397	0.0605	0.4033
2	1.12400	1.06216	0.2641	0.5385	0.0135
3	0.60794	1.44425	0.4962	0.4011	0.5831

Obs	Dep Var FEV1	Predict Value	Std Err Predict	Residual	Std Err Residual	Student Residual
1	4.7000	3.9074	0.292	0.7926	0.872	0.909
2	4.3000	3.5490	0.362	0.7510	0.846	0.888
3	3.5000	3.9598	0.587	-0.4598	0.708	-0.650
4	4.0000	3.4837	0.395	0.5163	0.830	0.622
5	3.2000	3.4655	0.390	-0.2655	0.833	-0.319
6	4.7000	3.8232	0.434	0.8768	0.811	1.081
7	4.3000	4.2548	0.461	0.0452	0.796	0.057
8	4.7000	3.7299	0.255	0.9701	0.884	1.098
9	5.2000	4.0684	0.442	1.1316	0.806	1.403
10	4.2000	4.2325	0.461	-0.0325	0.796	-0.041
11	3.5000	3.9491	0.345	-0.4491	0.853	-0.527
12	3.2000	3.4489	0.454	-0.2489	0.800	-0.311
13	2.6000	3.8326	0.314	-1.2326	0.865	-1.426
14	2.0000	3.8244	0.342	-1.8244	0.854	-2.137
15	4.0000	3.8974	0.290	0.1026	0.873	0.118
16	3.9000	3.5540	0.448	0.3460	0.803	0.431
17	3.0000	3.8060	0.501	-0.8060	0.771	-1.045
18	4.5000	3.5968	0.643	0.9032	0.658	1.373
19	2.4000	3.5166	0.388	-1.1166	0.834	-1.339

Obs	-2-1-0 1 2	Cook's D	Rstudent	Hat Diag H	Cov Ratio
1	|*	0.023	0.9031	0.1006	1.1682
2	|*	0.036	0.8814	0.1545	1.2558
3	*|	0.073	-0.6367	0.4079	1.9855
4	|*	0.022	0.6084	0.1846	1.4560
5	|	0.006	-0.3089	0.1795	1.5630
6	|**	0.084	1.0879	0.2226	1.2253
7	|	0.000	0.0549	0.2514	1.7588
8	|**	0.025	1.1058	0.0766	1.0210
9	|**	0.148	1.4547	0.2314	0.9763
10	|	0.000	-0.0394	0.2516	1.7601
11	*|	0.011	-0.5137	0.1408	1.4234
12	|	0.008	-0.3017	0.2439	1.6982
13	**|	0.067	-1.4813	0.1163	0.8329
14	****|	0.183	-2.4749	0.1381	0.3581
15	|	0.000	0.1137	0.0994	1.4579
16	|	0.014	0.4189	0.2378	1.6449
17	**|	0.115	-1.0485	0.2967	1.3849
18	|**	0.450	1.4184	0.4883	1.5051
19	**|	0.097	-1.3788	0.1781	0.9635

20. Create a new variable, FEMALE, for the data in Problem 19, where FEMALE = 1 if sex is F and FEMALE = 0 if sex is M.

a.–h. Repeat parts (a) through (h) of Problem 19, adding FEMALE as a predictor.

Edited SAS Output (PROC REG) for Problem 20

```
                         Correlation
CORR          HEIGHT        WEIGHT          AGE        FEMALE          FEV1
HEIGHT        1.0000        0.1291       0.2530       -0.1563        0.2257
WEIGHT        0.1291        1.0000      -0.1996        0.1393        0.1807
AGE           0.2530       -0.1996       1.0000       -0.0156       -0.0552
FEMALE       -0.1563        0.1393      -0.0156        1.0000       -0.3476
FEV1          0.2257        0.1807      -0.0552       -0.3476        1.0000
                     Analysis of Variance
                          Sum of        Mean
Source        DF         Squares       Square      F Value      Prob > F
Model          4         2.72245      0.68061        0.860        0.5116
Error         14        11.08281      0.79163
C Total       18        13.80526
```

. [Portion of output omitted]

.

```
                      Parameter Estimates
                     Parameter       Standard      T for H0:         Prob
Variable      DF      Estimate          Error     Parameter=0        > |T|
INTERCEP       1     -0.718415     5.60191872         -0.128        0.8998
HEIGHT         1      0.021105     0.03359370          0.628        0.5400
WEIGHT         1      0.016417     0.02113526          0.777        0.4502
AGE            1     -0.009072     0.03722844         -0.244        0.8110
FEMALE         1     -0.819773     0.57571489         -1.424        0.1764
                     Variance
Variable      DF      Inflation
INTERCEP       1     0.00000000
HEIGHT         1     1.14917294
WEIGHT         1     1.11400386
AGE            1     1.14002376
FEMALE         1     1.05774385
```

```
                    Collinearity Diagnostics
                                       Var       Var       Var       Var       Var
                         Condition    Prop      Prop      Prop      Prop      Prop
Number   Eigenvalue         Index  INTERCEP    HEIGHT    WEIGHT       AGE    FEMALE
     1      4.15398       1.00000    0.0001    0.0001    0.0010    0.0019    0.0114
     2      0.80075       2.27763    0.0000    0.0000    0.0002    0.0007    0.9346
     3      0.03567      10.79146    0.0008    0.0005    0.1426    0.6288    0.0061
     4      0.00895      21.54913    0.0376    0.0278    0.8511    0.3352    0.0172
     5    0.0006537      79.71439    0.9615    0.9716    0.0052    0.0334    0.0308
                     Analysis of Variance
                          Sum of        Mean
Source        DF         Squares       Square      F Value      Prob > F
Model          6         3.18803      0.53134        0.601        0.7253
Error         12        10.61723      0.88477
C Total       18        13.80526
```

. [Portion of output omitted]

.

```
        Collinearity Diagnostics(intercept adjusted)
                                      Var      Var      Var      Var
                            Condition Prop     Prop     Prop     Prop
   Number   Eigenvalue       Index    HEIGHT   WEIGHT   AGE      FEMALE
       1     1.33567        1.00000    0.1929   0.0800   0.2648   0.1285
       2     1.12677        1.08876    0.2891   0.4885   0.0005   0.0100
       3     0.98490        1.16454    0.0012   0.0044   0.2703   0.6626
       4     0.55266        1.55461    0.5168   0.4271   0.4643   0.1989
```

Obs	Dep Var FEV1	Predict Value	Std Err Predict	Residual	Std Err Residual	Student Residual
1	4.7000	4.0378	0.297	0.6622	0.839	0.789
2	4.3000	3.6949	0.364	0.6051	0.812	0.745
3	3.5000	3.4257	0.681	0.0743	0.573	0.130
4	4.0000	3.6336	0.397	0.3664	0.796	0.460
5	3.2000	2.7965	0.602	0.4035	0.655	0.616
6	4.7000	3.8719	0.421	0.8281	0.784	1.057
7	4.3000	4.3448	0.451	-0.0448	0.767	-0.058
8	4.7000	3.8983	0.273	0.8017	0.847	0.947
9	5.2000	4.1564	0.432	1.0436	0.778	1.342
10	4.2000	4.3530	0.454	-0.1530	0.765	-0.200
11	3.5000	4.0021	0.336	-0.5021	0.824	-0.609
12	3.2000	3.6900	0.471	-0.4900	0.755	-0.649
13	2.6000	3.0778	0.611	-0.4778	0.647	-0.738
14	2.0000	3.8664	0.332	-1.8664	0.825	-2.261
15	4.0000	3.9642	0.284	0.0358	0.843	0.043
16	3.9000	3.6620	0.440	0.2380	0.773	0.308
17	3.0000	4.0796	0.521	-1.0796	0.721	-1.497
18	4.5000	3.7210	0.628	0.7790	0.630	1.236
19	2.4000	3.6241	0.383	-1.2241	0.803	-1.524

Obs	-2-1-0 1 2	Cook's D	Rstudent	Hat Diag H	Cov Ratio
1	| * |	0.016	0.7782	0.1112	1.2979
2	| * |	0.022	0.7331	0.1678	1.4214
3	| |	0.005	0.1251	0.5857	3.4750
4	| |	0.010	0.4467	0.1986	1.6750
5	| * |	0.064	0.6019	0.4583	2.3306
6	| **|	0.064	1.0614	0.2241	1.2321
7	| |	0.000	-0.0563	0.2564	1.9456
8	| * |	0.019	0.9431	0.0943	1.1488
9	| **|	0.111	1.3854	0.2362	0.9526
10	| |	0.003	-0.1930	0.2607	1.9314
11	| * |	0.012	-0.5952	0.1425	1.4768
12	| * |	0.033	-0.6350	0.2801	1.7270
13	| * |	0.097	-0.7259	0.4713	2.2460
14	**** |	0.165	-2.7344	0.1392	0.1735
15	| |	0.000	0.0410	0.1022	1.6124
16	| |	0.006	0.2977	0.2450	1.8545
17	|** |	0.234	-1.5745	0.3434	0.9217
18	| **|	0.303	1.2615	0.4979	1.6195
19	|*** |	0.106	-1.6083	0.1853	0.7176

21. Create the following three new interaction variables for the data in Problem 20:

FEMAGE = FEMALE * AGE

FEMHT   = FEMALE * HEIGHT

FEMWT = FEMALE * WEIGHT

a. Fit a model of FEV_1 as predicted by HEIGHT, WEIGHT, AGE, FEMALE, and the three newly specified interaction variables.

b. Conduct variable-added-last tests for all predictors, and perform a test of the intercept.

c. Determine variance inflation factors for each predictor.

d. Determine a correlation matrix that includes all predictors and the response.

e. Determine eigenvalues, condition indexes, and condition numbers for the correlation matrix (excluding the intercept).

f. Determine eigenvalues, condition indexes, and condition numbers for the scaled cross-products matrix (including the intercept).

g. Determine residuals (preferably studentized) and leverage values. Do any observations seem bothersome? Explain.

h. Does there appear to be any problem with collinearity? Explain.

i. Does the small number of females lead to difficulties in Problems 20 and 21? Suggest a simple solution for analyzing these data. Also, suggest a solution for future research.

Edited SAS Output (PROC REG) for Problem 21

```
                              Correlation
CORR            HEIGHT           WEIGHT             AGE           FEMALE
HEIGHT          1.0000           0.1291          0.2530          -0.1563
WEIGHT          0.1291           1.0000         -0.1996           0.1393
AGE             0.2530          -0.1996          1.0000          -0.0156
FEMALE         -0.1563           0.1393         -0.0156           1.0000
FEMAGE         -0.1166           0.1283          0.0065           0.9939
FEMHT          -0.1432           0.1375         -0.0083           0.9993
FEMWT          -0.1588           0.2561         -0.0218           0.9785
FEV1            0.2257           0.1807         -0.0552          -0.3476

CORR            FEMAGE            FEMHT           FEMWT             FEV1
HEIGHT         -0.1166          -0.1432         -0.1588           0.2257
WEIGHT          0.1283           0.1375          0.2561           0.1807
AGE             0.0065          -0.0083         -0.0218          -0.0552
FEMALE          0.9939           0.9993          0.9785          -0.3476
FEMAGE          1.0000           0.9973          0.9689          -0.3611
FEMHT           0.9973           1.0000          0.9772          -0.3522
FEMWT           0.9689           0.9772          1.0000          -0.3149
FEV1           -0.3611          -0.3522         -0.3149           1.0000
                          Analysis of Variance
                      Sum of            Mean
Source      DF       Squares           Square      F Value      Prob > F
Model        6       3.18803          0.53134        0.601        0.7253
Error       12      10.61723          0.88477
C Total     18      13.80526

                              .
                              .  [Portion of output omitted]
                              .

NOTE: Model is not full rank. Least-squares solutions for the
      parameters are not unique. Some statistics will be misleading.
      A reported DF of 0 or B means that the estimate is biased.

      The following parameters have been set to 0, since the
      variables are a linear combination of other variables as shown.
      FEMWT = -3610 * FEMALE -63.8571 * FEMAGE +32.1429 * FEMHT
```

Parameter Estimates

Variable	DF	Parameter Estimate	Standard Error	T for H0: Parameter=0	Prob > |T|
INTERCEP	1	-2.284006	6.31617975	-0.362	0.7239
HEIGHT	1	0.030437	0.03823938	0.796	0.4415
WEIGHT	1	0.014467	0.02827874	0.512	0.6182
AGE	1	-0.006540	0.04006671	-0.163	0.8730
FEMALE	B	6.200441	171.36441204	0.036	0.9717
FEMAGE	B	-0.098234	2.94520425	-0.033	0.9739
FEMHT	B	-0.024007	1.48871650	-0.016	0.9874
FEMWT	0	0	.	.	.

Variable	DF	Variance Inflation
INTERCEP	1	0.00000000
HEIGHT	1	1.33224307
WEIGHT	1	1.78436541
AGE	1	1.18147265
FEMALE	B	83849.187086
FEMAGE	B	21084.669881
FEMHT	B	188192.57055
FEMWT	0	.

Collinearity Diagnostics

Number	Eigenvalue	Condition Index	Var Prop INTERCEP	Var Prop HEIGHT	Var Prop WEIGHT	Var Prop AGE
1	5.54179	1.00000	0.0000	0.0000	0.0002	0.0007
2	2.37723	1.52683	0.0001	0.0001	0.0005	0.0017
3	0.04428	11.18748	0.0000	0.0000	0.0619	0.2262
4	0.02602	14.59356	0.0029	0.0024	0.0078	0.5150
5	0.00633	29.59364	0.0357	0.0131	0.5976	0.0927
6	0.00379	38.23872	0.0086	0.0238	0.3184	0.1446
7	0.0005639	99.13480	0.9527	0.9606	0.0136	0.0191
8	1E-12	2354100	0.0000	0.0000	0.0000	0.0000

Number	Var Prop FEMALE	Var Prop FEMAGE	Var Prop FEMHT	Var Prop FEMWT
1	0.0000	0.0000	0.0000	0.0000
2	0.0000	0.0000	0.0000	0.0000
3	0.0000	0.0000	0.0000	0.0000
4	0.0000	0.0000	0.0000	0.0000
5	0.0000	0.0000	0.0000	0.0000
6	0.0000	0.0000	0.0000	0.0000
7	0.0000	0.0000	0.0000	0.0000
8	1.0000	1.0000	1.0000	1.0000

NOTE: Singularities or near singularities caused grossly large variance calculations. To provide diagnostics, eigenvalues are inflated to a minimum of 1E-12.

Collinearity Diagnostics(intercept adjusted)

Number	Eigenvalue	Condition Index	Var Prop HEIGHT	Var Prop WEIGHT	Var Prop AGE
1	4.01994	1.00000	0.0016	0.0016	0.0001
2	1.24467	1.79714	0.2058	0.0209	0.4092
3	1.12543	1.88995	0.2020	0.3274	0.0247
4	0.58244	2.62715	0.4441	0.2596	0.5296
5	0.02303	13.21222	0.0822	0.3676	0.0157

(continued)

```
    Collinearity Diagnostics(intercept adjusted) (continued)
                                      Var       Var       Var
                           Condition  Prop      Prop      Prop
Number     Eigenvalue      Index      HEIGHT    WEIGHT    AGE
     6        0.00449    29.93449     0.0643    0.0231    0.0207
     7          1E-12     2004980     0.0000    0.0000    0.0000

                Var       Var       Var       Var
                Prop      Prop      Prop      Prop
Number     FEMALE    FEMAGE    FEMHT     FEMWT
     1     0.0000    0.0000    0.0000    0.0000
     2     0.0000    0.0000    0.0000    0.0000
     3     0.0000    0.0000    0.0000    0.0000
     4     0.0000    0.0000    0.0000    0.0000
     5     0.0000    0.0000    0.0000    0.0000
     6     0.0000    0.0000    0.0000    0.0000
     7     1.0000    1.0000    1.0000    1.0000
```

NOTE: Singularities or near singularities caused grossly large
      variance calculations. To provide diagnostics, eigenvalues
      are inflated to a minimum of 1E-12.

Obs	Dep Var FEV1	Predict Value	Std Err Predict	Residual	Std Err Residual	Student Residual
1	4.7000	4.0138	0.319	0.6862	0.885	0.776
2	4.3000	3.6936	0.387	0.6064	0.858	0.707
3	3.5000	3.5000	0.941	-335E-15	0.000	-0.000
4	4.0000	3.5525	0.446	0.4475	0.828	0.540
5	3.2000	3.2000	0.941	2.48E-13	0.000	0.000
6	4.7000	3.8765	0.482	0.8235	0.808	1.019
7	4.3000	4.4071	0.503	-0.1071	0.795	-0.135
8	4.7000	3.8479	0.304	0.8521	0.890	0.957
9	5.2000	4.2542	0.484	0.9458	0.806	1.173
10	4.2000	4.3744	0.515	-0.1744	0.787	-0.222
11	3.5000	4.0444	0.372	-0.5444	0.864	-0.630
12	3.2000	3.5725	0.533	-0.3725	0.775	-0.481
13	2.6000	2.6000	0.941	8.77E-14	.	.
14	2.0000	3.9322	0.395	-1.9322	0.854	-2.263
15	4.0000	4.0168	0.317	-0.0168	0.886	-0.019
16	3.9000	3.5969	0.507	0.3031	0.792	0.383
17	3.0000	3.9721	0.663	-0.9721	0.667	-1.457
18	4.5000	3.8021	0.673	0.6979	0.657	1.062
19	2.4000	3.6430	0.427	-1.2430	0.838	-1.483

Obs	-2-1-0	1 2	Cook's D	Rstudent	Hat Diag A	Cov Ratio	Dffits
1		*	0.011	0.7618	0.1152	1.4500	0.2749
2		*	0.015	0.6916	0.1688	1.6422	0.3117
3			0.000	-0.0000	1.0000	2.988E11	-0.0554
4		*	0.012	0.5238	0.2250	1.9967	0.2822
5			0.002	0.0000	1.0000	7.31E11	0.1005
6		**	0.053	1.0211	0.2622	1.3223	0.6088
7			0.001	-0.1291	0.2864	2.5496	-0.0818
8		*	0.015	0.9537	0.1044	1.1774	0.3257
9		**	0.071	1.1937	0.2653	1.0668	0.7172
10			0.003	-0.2126	0.2993	2.5501	-0.1389
11	*		0.011	-0.6137	0.1566	1.7225	-0.2645
12			0.016	-0.4648	0.3217	2.3657	-0.3201
13			.	.	1.0000	.	.
14	****		0.156	-2.8616	0.1760	0.0454	-1.3227
15			0.000	-0.0182	0.1138	2.0743	-0.0065
16			0.009	0.3685	0.2903	2.3775	0.2357
17			0.299	-1.5374	0.4968	0.9355	-1.5275
18		**	0.169	1.0685	0.5122	1.8881	1.0948
19	**		0.081	-1.5711	0.2060	0.5613	-0.8001

22. In an analysis of daily soil evaporation (EVAP), Freund (1979) identified the following predictor variables:

MAXAT = Maximum daily air temperature

MINAT = Minimum daily air temperature

AVAT = Integrated area under the daily air temperature curve (i.e., a measure of average air temperature)

MAXST = Maximum daily soil temperature

MINST = Minimum daily soil temperature

AVST = Integrated area under the soil temperature curve

MAXH = Maximum daily relative humidity

MINH = Minimum daily relative humidity

AVH = Integrated area under the daily humidity curve

WIND = Total wind, measured in miles per day

In addition, Freund provided the following overall ANOVA table and tests of coefficients.

Source	d.f.	SS	MS	F
Regression	10	8159.35	815.94	19.27
Residual	35	1482.76	42.36	
Total	45	9642.11		

Variable	$\hat{\beta}$	t	VIF
MAXAT	0.5011	0.88	8.828
MINAT	0.3041	0.39	8.887
AVAT	0.09219	0.42	22.21
MAXST	2.232	2.22	39.29
MINST	0.2049	0.19	14.08
AVST	0.7426	−2.12	52.36
MAXH	1.110	0.98	1.981
MINH	0.7514	1.54	25.38
AVH	−0.5563	−3.44	24.12
WIND	0.00892	0.97	1.985

a. Compute R^2, and provide a test of significance. Use $\alpha = .01$.

b. Compute R_j^2 for each predictor.

c. Which, if any, variables are implicated as possibly inducing collinearity?

d. Freund (1979) noted that "some of the coefficients have 'wrong' signs." Using your knowledge of evaporation, explain which coefficients have wrong signs, paying particular attention to those with extreme t-values.

23. The raw data for the study described in Problem 22, from Freund (1979), are given next.

a. Fit the model considered in Problem 22.

b. Perform a test of the intercept.

c. What numerical discrepancies do you observe between the output here and the data summary in Problem 22?

d. Determine a correlation matrix that includes all predictors and the response.

Observation	Month	Day	MAXST	MINST	AVST	MAXAT	MINAT	AVAT	MAXH	MINH	AVH	WIND	EVAP
1	6	6	84	65	147	85	59	151	95	40	398	273	30
2	6	7	84	65	149	86	61	159	94	28	345	140	34
3	6	8	79	66	142	83	64	152	94	41	388	318	33
4	6	9	81	67	147	83	65	158	94	50	406	282	26
5	6	10	84	68	167	88	69	180	93	46	379	311	41
6	6	11	74	66	131	77	67	147	96	73	478	446	4
7	6	12	73	66	131	78	69	159	96	72	462	294	5
8	6	13	75	67	134	84	68	159	95	70	464	313	20
9	6	14	84	68	161	89	71	195	95	63	430	455	31
10	6	15	86	72	169	91	76	206	93	56	406	604	38
11	6	16	88	73	178	91	76	208	94	55	393	610	43
12	6	17	90	74	187	94	76	211	94	51	385	520	47
13	6	18	88	72	171	94	75	211	96	54	405	663	45
14	6	19	88	72	171	92	70	201	95	51	392	467	45
15	6	20	81	69	154	87	68	167	95	61	448	184	11
16	6	21	79	68	149	83	68	162	95	59	436	177	10
17	6	22	84	69	160	87	66	173	95	42	392	173	30
18	6	23	84	70	160	87	68	177	94	44	392	76	29
19	6	24	84	70	168	88	70	169	95	48	398	72	23
20	6	25	77	67	147	83	66	170	97	60	431	183	16
21	6	26	87	67	166	92	67	196	96	44	379	76	37
22	6	27	89	69	171	92	72	199	94	48	393	230	50
23	6	28	89	72	180	94	72	204	95	48	394	193	36
24	6	29	93	72	186	92	73	201	94	47	386	400	54
25	6	30	93	74	188	93	72	206	95	47	385	339	44
26	7	1	94	75	199	94	72	208	96	45	370	172	41
27	7	2	93	74	193	95	73	214	95	50	396	238	45
28	7	3	93	74	196	95	70	210	96	45	380	118	42
29	7	4	96	75	198	95	71	207	93	40	365	93	50
30	7	5	95	76	202	95	69	202	93	39	357	269	48
31	7	6	84	73	173	96	69	173	94	58	418	128	17
32	7	7	91	71	170	91	69	168	94	44	420	423	20
33	7	8	88	72	179	89	70	189	93	50	399	415	15
34	7	9	89	72	179	95	71	210	98	46	389	300	42
35	7	10	91	72	182	96	73	208	95	43	384	193	44
36	7	11	92	74	196	97	75	215	96	46	389	195	41
37	7	12	94	75	192	96	69	198	95	36	380	215	49
38	7	13	96	75	195	95	67	196	97	24	354	185	53
39	7	14	93	76	198	94	75	211	93	43	364	466	53
40	7	15	88	74	188	92	73	198	95	52	405	399	21
41	7	16	88	74	178	90	74	197	95	61	447	232	1
42	7	17	91	72	175	94	70	205	94	42	380	275	44
43	7	18	92	72	190	95	71	209	96	44	379	166	44
44	7	19	92	73	189	96	72	208	93	42	372	189	46
45	7	20	94	75	194	95	71	208	93	43	373	164	47
46	7	21	96	76	202	96	71	208	94	40	368	139	50

e. Determine eigenvalues, condition indexes, and condition numbers for the correlation matrix (excluding the intercept).

f. Determine eigenvalues, condition indexes, and condition numbers for the scaled cross-products matrix (including the intercept).

g. Determine residuals (preferably studentized) and leverage values. Do any observations seem bothersome? Explain.

h. Does there appear to be any problem with collinearity? Explain.

24. Real estate prices depend, in part, on property size. The house size X (in hundreds of square feet) and house price Y (in $1,000s) of a random sample of houses in a certain county were observed. The data (which first appeared in Problem 16 of Chapter 5) are as follows:

X	18	20	25	22	33	19	17
Y	80	95	104	110	175	85	89

Edited SAS Output for Problem 24

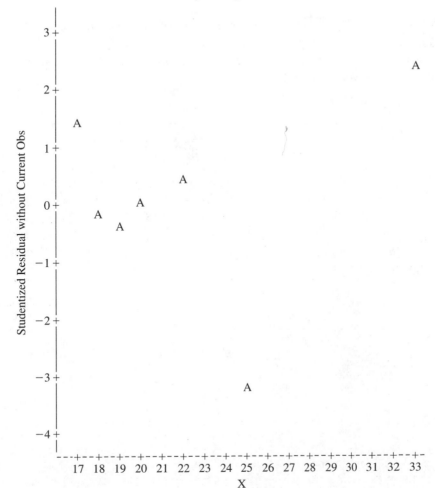

Plot of JACKKNIFE by X. Legend: A = 1 obs, B = 2 obs, etc.

Univariate Procedure

•

• [Portion of output omitted]

•

Variable-JACKKNIFE		Studentized Residual without Current Obs	
Stem	Leaf	#	Boxplot
2	4	1	I
1	4	1	+-----------+
0	14	2	*-----+-----*
-0	33	2	+-----------+
-1			
-2			
-3	1	1	0
	----+-----+-----+-----+		

a. Fit a univariate linear model, with house price Y as the response and house size X as the predictor.

b. Provide a plot of studentized or jackknife residuals versus the predictor.

c. Provide a frequency histogram and a schematic plot of the residuals.

d. Report and interpret a test of whether any residuals have extreme values.

e. Do these analyses highlight any potentially troublesome observations? Why or why not?

25. Data on sales revenues (Y) and advertising expenditures (X) for a large retailer for the period 1988-1993 are given in the following table.

Year	1988	1989	1990	1991	1992	1993
Sales Y ($millions)	4	8	2	8	5	4
Advertising X ($millions)	2	5	0	6	4	3

These data first appeared in Problem 17 of Chapter 5.

a.–e. Repeat parts (a) through (e) of Problem 24 for the current data set, using sales revenues Y as the response and advertising expenditures X as the predictor.

26. The production manager of a plant that manufactures syringes has recorded the marginal cost Y at various levels of output X for 14 randomly selected months. The data are shown in the following table.

Marginal Cost Y (per 100 units)	Output X (thousands of units)	Marginal Cost Y (per 100 units)	Output X (thousands of units)
31.00	3.0	27.00	3.5
30.00	3.0	25.00	5.5
28.00	3.5	24.00	7.0
46.00	1.0	25.00	7.0
43.00	1.5	29.50	4.5
35.00	2.0	26.00	4.5
37.50	2.0	28.00	4.5

These data first appeared in Problem 18 of Chapter 5.

Edited SAS Output for Problem 25

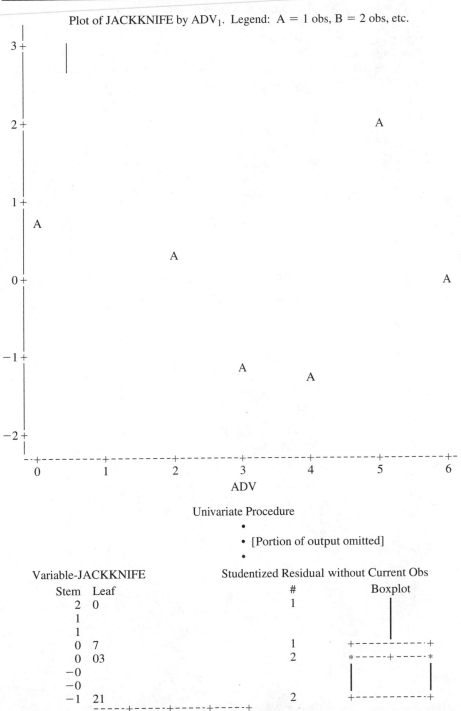

Plot of JACKKNIFE by ADV$_1$. Legend: A = 1 obs, B = 2 obs, etc.

Univariate Procedure

• [Portion of output omitted]

Variable-JACKKNIFE   Studentized Residual without Current Obs

Stem	Leaf	#	Boxplot
2	0	1	
1			
1			
0	7	1	
0	03	2	
−0			
−0			
−1	21	2	

a.–e. Repeat parts (a) through (e) of Problem 24 for the current data set, using marginal cost Y as the response and output X as the predictor.

Edited SAS Output for Problem 26

Plot of JACKKNIFE by OUTPUT. Legend: A = 1 obs, B = 2 obs, etc.

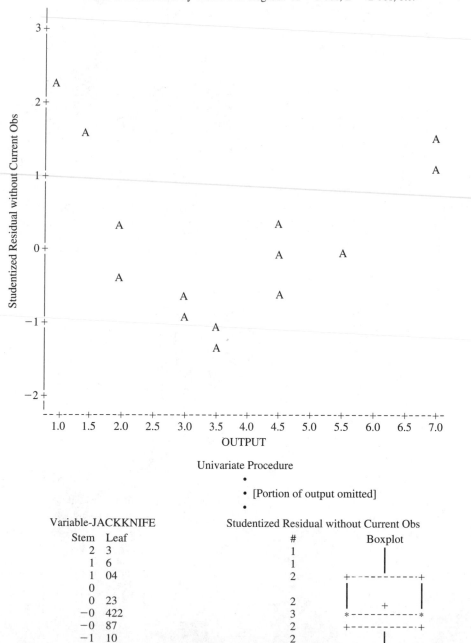

Univariate Procedure

• [Portion of output omitted]

Variable-JACKKNIFE			Studentized Residual without Current Obs	
Stem	Leaf		#	Boxplot
2	3		1	
1	6		1	
1	04		2	+----------+
0					
0	23		2		+	
−0	422		3	*----------*
−0	87		2	+----------+
−1	10		2	
−1	5		1	

27. Radial keratotomy is a type of refractive surgery in which radial incisions are made in the cornea of myopic (nearsighted) patients in an effort to reduce their myopia. Theoretically, the incisions allow the curvature of the cornea to become less steep, reducing the refractive error of the patient's vision. This and other types of vision correction surgery have been growing in popularity in the 1980s and 1990s among the public and among ophthalmologists.

The Prospective Evaluation of Radial Keratotomy (PERK) study was begun in 1983 to investigate the effects of radial keratotomy. Lynn et al. (1987) examined the factors associated with the five-year postsurgical change in refractive error (Y), measured in diopters. Two of the independent variables under consideration were baseline refractive error (X_1, in diopters) and baseline curvature of the cornea (X_2, in diopters). (Note: myopic patients have negative refractive errors. Patients who are far-sighted have positive refractive errors. Patients who are neither near nor far-sighted have zero refractive error.)

The computer output below is based on data adapted from the PERK study. These data first appeared in Chapter 8, Problem 12. Use the output to answer the following questions.

a. Fit a model of change in refractive error as predicted by baseline refractive error and baseline curvature.
b. Conduct variable-added-last tests for both predictors, and perform a test of the intercept.
c. Determine the variance inflation factors for each predictor.
d. Determine a correlation matrix that includes both predictors and the response.
e. Determine eigenvalues, condition indexes, and condition numbers for the correlation matrix (excluding the intercept).
f. Determine eigenvalues, condition indexes, and condition numbers for the scaled cross-products matrix (including the intercept).
g. Determine residuals and leverage values. Do any observations seem bothersome? Explain.
h. Does there appear to be any problem with collinearity? Explain.

Edited SAS Output (PROC REG) for Problem 27

```
                              Correlation
            CORR        X1             X2              Y
            X1      1.0000         0.0424        -0.4040
            X2      0.0424         1.0000        -0.2544
            Y      -0.4040        -0.2544         1.0000
                      Analysis of Variance
                       Sum of         Mean
      Source    DF    Squares        Square     F Value      Prob > F
      Model      2   17.62277       8.81139       7.175        0.0018
      Error     51   62.63017       1.22804
      C Total   53   80.25294

            Root MSE        1.10817     R-square        0.2196
            Dep Mean        3.83343     Adj R-sq        0.1890
            C.V.           28.90811
                       Parameter Estimates
                      Parameter       Standard      T for H0:        Prob
      Variable   DF   Estimate         Error       Parameter=0      > |T|
      INTERCEP    1  12.360015      5.08621177         2.430        0.0187
      X1          1  -0.291601      0.09165295        -3.182        0.0025
      X2          1  -0.220396      0.11482391        -1.919        0.0605
```

(continued)

```
                    Parameter Estimates (Continued)
                              Variance
Variable      DF           Inflation
INTERCEP       1          0.00000000
X1             1          1.00180088
X2             1          1.00180088
```

```
                     Collinearity Diagnostics

                                                Var         Var         Var
                              Condition        Prop        Prop        Prop
Number      Eigenvalue           Index      INTERCEP         X1          X2
    1          2.90058         1.00000        0.0001      0.0158      0.0001
    2          0.09898         5.41336        0.0014      0.9779      0.0015
    3        0.0004427        80.94123        0.9985      0.0063      0.9984
```

```
            Collinearity Diagnostics(intercept adjusted)

                                                     Var         Var
                                    Condition       Prop        Prop
        Number      Eigenvalue         Index          X1          X2
           1          1.04240        1.00000        0.4788      0.4788
           2          0.95760        1.04334        0.5212      0.5212
```

```
        Dep Var      Predict     Std Err                  Std Err     Student
Obs           Y        Value     Predict     Residual    Residual    Residual
  1      5.3750       4.1350       0.394       1.2400       1.036       1.197
  2      2.6250       3.7073       0.176      -1.0823       1.094      -0.989
  3      1.5000       3.2309       0.223      -1.7309       1.086      -1.594
  4      3.7500       3.9667       0.168      -0.2167       1.095      -0.198
  5      4.1250       3.7652       0.167       0.3598       1.095       0.328
  6      4.6250       4.4647       0.228       0.1603       1.084       0.148
  7      4.2500       5.3056       0.429      -1.0556       1.022      -1.033
  8      3.7500       1.9284       0.575       1.8216       0.947       1.923
  9      3.2500       3.5466       0.388      -0.2966       1.038      -0.286
 10      2.6250       3.3669       0.225      -0.7419       1.085      -0.684
 11      4.0000       3.5386       0.312       0.4614       1.063       0.434
 12      4.1250       4.2265       0.184      -0.1015       1.093      -0.093
 13      3.0000       3.0159       0.286      -0.0159       1.071      -0.015
 14      3.1250       3.5499       0.293      -0.4249       1.069      -0.398
 15      4.7500       4.6457       0.418       0.1043       1.026       0.102
 16      3.5000       3.8370       0.172      -0.3370       1.095      -0.308
 17      4.5000       4.0543       0.362       0.4457       1.047       0.426
 18      4.7500       3.9044       0.248       0.8456       1.080       0.783
 19      4.8750       4.6426       0.262       0.2324       1.077       0.216
 20      3.0000       3.6310       0.237      -0.6310       1.082      -0.583
 21      3.6250       4.0459       0.218      -0.4209       1.087      -0.387
 22      3.2500       3.4788       0.192      -0.2288       1.091      -0.210
 23      6.2500       3.9794       0.168       2.2706       1.095       2.073
 24      5.1250       4.0514       0.217       1.0736       1.087       0.988
 25      2.0000       3.1571       0.252      -1.1571       1.079      -1.072
 26      4.1300       3.0555       0.255       1.0745       1.078       0.996
 27      4.1250       3.3367       0.200       0.7883       1.090       0.723
 28      4.2500       3.9357       0.181       0.3143       1.093       0.287
 29      6.1250       4.9313       0.346       1.1937       1.053       1.134
 30      6.1250       3.9038       0.163       2.2212       1.096       2.026
 31      2.8750       3.4664       0.224      -0.5914       1.085      -0.545
 32      3.2500       4.8782       0.352      -1.6282       1.051      -1.549
 33      5.0000       4.0832       0.266       0.9168       1.076       0.852
 34      3.7500       3.2816       0.213       0.4684       1.087       0.431
 35      6.8750       4.4024       0.258       2.4726       1.078       2.294
```

```
Collinearity Diagnostics(intercept adjusted) (Continued)
```

Obs	Dep Var Y	Predict Value	Std Err Predict	Residual	Std Err Residual	Student Residual
36	2.3750	3.5047	0.181	-1.1297	1.093	-1.033
37	5.1250	4.3261	0.218	0.7989	1.086	0.735
38	4.3750	4.3028	0.210	0.0722	1.088	0.066
39	5.5000	3.8129	0.233	1.6871	1.083	1.557
40	2.2500	3.5445	0.273	-1.2945	1.074	-1.205
41	4.1250	3.4986	0.275	0.6264	1.074	0.583
42	3.8750	4.7244	0.283	-0.8494	1.071	-0.793
43	3.2500	4.5046	0.233	-1.2546	1.083	-1.158
44	2.7500	3.5428	0.187	-0.7928	1.092	-0.726
45	3.1250	3.3723	0.213	-0.2473	1.088	-0.227
46	2.6250	4.2043	0.180	-1.5793	1.093	-1.444
47	2.8750	3.7925	0.152	-0.9175	1.098	-0.836
48	5.2500	4.2388	0.298	1.0112	1.067	0.947
49	2.8750	3.7997	0.228	-0.9247	1.084	-0.853
50	2.1250	3.2733	0.241	-1.1483	1.082	-1.062
51	0.8750	3.4903	0.185	-2.6153	1.093	-2.394
52	3.5000	3.4745	0.184	0.0255	1.093	0.023
53	3.5000	3.2263	0.220	0.2737	1.086	0.252
54	4.3750	3.9205	0.154	0.4545	1.097	0.414

Obs	-2-1-0 1 2	Cook's D	Rstudent	Hat Diag H	Cov Ratio
1	| |**	0.069	1.2026	0.1267	1.1155
2	| * |	0.008	-0.9890	0.0253	1.0272
3	| *** |	0.036	-1.6196	0.0404	0.9486
4	| |	0.000	-0.1960	0.0230	1.0837
5	| |	0.001	0.3255	0.0228	1.0791
6	| |	0.000	0.1464	0.0425	1.1069
7	| ** |	0.063	-1.0339	0.1500	1.1717
8	| |***	0.454	1.9770	0.2693	1.1587
9	| |	0.004	-0.2831	0.1223	1.2033
10	| * |	0.007	-0.6802	0.0413	1.0768
11	| |	0.005	0.4305	0.0794	1.1400
12	| |	0.000	-0.0919	0.0276	1.0908
13	| |	0.000	-0.0147	0.0668	1.1372
14	| |	0.004	-0.3943	0.0700	1.1305
15	| |	0.001	0.1006	0.1423	1.2365
16	| |	0.001	-0.3051	0.0241	1.0814
17	| |	0.007	0.4221	0.1068	1.1755
18	| |*	0.011	0.7799	0.0501	1.0773
19	| |	0.001	0.2138	0.0558	1.1209
20	| * |	0.005	-0.5791	0.0459	1.0902
21	| |	0.002	-0.3841	0.0386	1.0941
22	| |	0.000	-0.2077	0.0299	1.0911
23	| |****	0.034	2.1450	0.0231	0.8342
24	| |*	0.013	0.9876	0.0383	1.0413
25	| ** |	0.021	-1.0739	0.0518	1.0451
26	| |*	0.019	0.9964	0.0531	1.0565
27	| |*	0.006	0.7198	0.0327	1.0636
28	| |	0.001	0.2849	0.0268	1.0851
29	| |**	0.046	1.1372	0.0977	1.0893
30	| |****	0.030	2.0925	0.0217	0.8433

(continued)

```
     Collinearity Diagnostics(intercept adjusted) (Continued)
                              Cook's              Hat Diag      Cov
     Obs  -2-1-0  1 2           D     Rstudent       H        Ratio
      31       *              0.004   -0.5412     0.0409     1.0873
      32    ***              0.090   -1.5715     0.1006     1.0211
      33         *           0.015    0.8498     0.0574     1.0785
      34                     0.002    0.4272     0.0370     1.0900
      35       ****          0.100    2.3986     0.0541     0.8092
      36    **               0.010   -1.0340     0.0268     1.0234
      37      *              0.007    0.7320     0.0388     1.0693
      38                     0.000    0.0657     0.0358     1.1004
      39       ***           0.037    1.5799     0.0442     0.9593
      40    **               0.031   -1.2109     0.0609     1.0361
      41      *              0.007    0.5796     0.0616     1.1083
      42     *               0.015   -0.7899     0.0653     1.0938
      43    **               0.021   -1.1619     0.0441     1.0249
      44     *               0.005   -0.7224     0.0283     1.0587
      45                     0.001   -0.2253     0.0369     1.0985
      46    **               0.019   -1.4603     0.0264     0.9616
      47     *               0.004   -0.8333     0.0187     1.0376
      48       *             0.023    0.9463     0.0721     1.0843
      49     *               0.011   -0.8504     0.0425     1.0616
      50    **               0.019   -1.0630     0.0472     1.0416
      51    ****             0.055   -2.5157     0.0279     0.7635
      52                     0.000    0.0231     0.0277     1.0914
      53                     0.001    0.2497     0.0395     1.1007
      54                     0.001    0.4107     0.0194     1.0713
```

28. In 1990, *Business Week* magazine compiled financial data on the 1,000 companies that had the biggest impact on the U.S. economy. Data from this compilation were presented in Problem 13 of Chapter 8. In addition to the company name, data on the following variables were shown in that problem:

1990 Rank: Based on company's market value (share price on March 16, 1990, multiplied by available common shares outstanding)

1989 Rank: Rank in 1989 compilation

P-E Ratio: Price-to-earnings ratio, based on 1989 earnings and March 16, 1990, share price

Yield: Annual dividend rate as a percentage of March 16, 1990, share price

The computer output given next is based on the *Business Week* magazine data. Use the output to answer the following questions:

a. Fit a model with yield (Y) as the response and 1990 rank (X_1), 1989 rank (X_2), and P-E ratio (X_3) as the predictors. State the estimated model.

b. Perform variables-added-last tests for each predictor.

c. Determine the variance inflation factors for each predictor.

d. Does there appear to be any problem with collinearity? Explain.

e. Examine the plot of studentized or jackknife residuals versus the predictor. Are any violations of linear regression assumptions obvious?

f. Determine residuals and leverage values. Do any observations seem bothersome? Explain.

Edited SAS Output (PROC REG) for Problem 28

```
Yield Regressed on 1990 Rank (X1), 1989 Rank (X2), and P-E Ratio (X3)
                          Analysis of Variance
                      Sum of        Mean
Source          DF     Squares      Square      F Value      Prob > F
Model           3      29.97726     9.99242     8.832        0.0011
Error           16     18.10304     1.13144
C Total         19     48.08030

          Root MSE         1.06369      R-square       0.6235
          Dep Mean         2.48550      Adj R-sq       0.5529
          C.V.            42.79588
                         Parameter Estimates
                   Parameter       Standard     T for H0:
Variable    DF     Estimate        Error        Parameter=0    Prob > |T|
INTERCEP    1       7.421188       0.98839344     7.508         0.0001
X1          1      -0.013352       0.00770835    -1.732         0.1025
X2          1       0.007621       0.00696539     1.094         0.2901
X3          1      -0.261846       0.05249866    -4.988         0.0001
                      Variance
Variable    DF      Inflation
INTERCEP    1       0.00000000
X1          1      23.11906214
X2          1      22.15953626
X3          1       1.21991445
          Dep Var    Predict    Std Err                Std Err    Student
Obs          Y        Value     Predict    Residual    Residual   Residual
 1        2.8700     2.9469     0.352      -0.0769      1.004      -0.077
 2        2.5400     2.4060     0.357       0.1340      1.002       0.134
 3        1.7200     1.0888     0.472       0.6312      0.953       0.662
 4        4.2600     3.6101     0.362       0.6499      1.000       0.650
 5        0.4100     1.2782     0.435      -0.8682      0.971      -0.894
 6        4.1000     3.4917     0.343       0.6083      1.007       0.604
 7        0          2.3005     0.325      -2.3005      1.013      -2.271
 8        4.2700     3.1250     0.319       1.1450      1.015       1.128
 9        2.6100     1.5282     0.404       1.0818      0.984       1.099
10        2.9400     2.6352     0.280       0.3048      1.026       0.297
11        2.9800     4.6793     0.532      -1.6993      0.921      -1.845
12        2.6200     2.9233     0.348      -0.3033      1.005      -0.302
13        2.9100     3.7481     0.665      -0.8381      0.830      -1.010
14        1.8700     1.7354     0.294       0.1346      1.022       0.132
15        4.7300     3.5077     0.403       1.2223      0.984       1.242
16        0         -0.4225     0.788       0.4225      0.714       0.591
17        5.4600     4.2769     0.610       1.1831      0.871       1.358
18        1.9900     1.7468     0.404       0.2432      0.984       0.247
19        1.1000     2.0679     0.735      -0.9679      0.769      -1.259
20        0.3300     1.0367     0.612      -0.7067      0.870      -0.812
                        Cook's              Hat Diag      Cov
Obs    -2-1-0 1 2        D       Rstudent      H         Ratio
 1       |     |       0.000     -0.0742     0.1095      1.4516
 2       |     |       0.001      0.1296     0.1124      1.4520
 3       |    *|       0.027      0.6499     0.1965      1.4417
 4       |    *|       0.014      0.6377     0.1158      1.3156
 5       |  * |        0.040     -0.8886     0.1673      1.2663
 6       |    *|       0.011      0.5916     0.1037      1.3170
```

(continued)

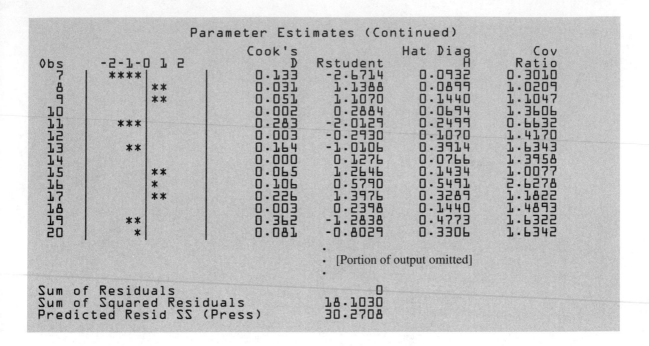

```
              Parameter Estimates (Continued)
                           Cook's              Hat Diag      Cov
Obs   -2-1-0 1 2              D    Rstudent        H        Ratio
  7   ****              0.133    -2.6714      0.0932      0.3010
  8            **       0.031     1.1388      0.0899      1.0209
  9            **       0.051     1.1070      0.1440      1.1047
 10                     0.002     0.2884      0.0694      1.3606
 11      ***            0.283    -2.0129      0.2499      0.6632
 12                     0.003    -0.2930      0.1070      1.4170
 13       **           0.164    -1.0106      0.3914      1.6343
 14                     0.000     0.1276      0.0766      1.3958
 15            **       0.065     1.2646      0.1434      1.0077
 16            *        0.106     0.5790      0.5491      2.6278
 17            **       0.226     1.3976      0.3289      1.1822
 18                     0.003     0.2398      0.1440      1.4893
 19       **           0.362    -1.2838      0.4773      1.6322
 20       *            0.081    -0.8029      0.3306      1.6342
```

. [Portion of output omitted]
.

```
Sum of Residuals                          0
Sum of Squared Residuals           18.1030
Predicted Resid SS (Press)         30.2708
```

Edited SAS Output for Problem 28

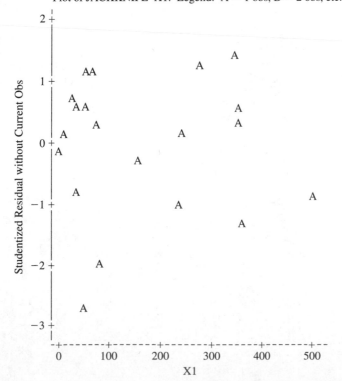

Plot of JACKKNIFE*X1. Legend: A = 1 obs, B = 2 obs, etc.

Plot of JACKKNIFE*X2. Legend: A = 1 obs, B = 2 obs, etc.

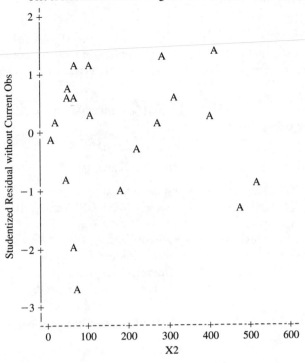

Plot of JACKKNIFE*X3. Legend: A = 1 obs, B = 2 obs, etc.

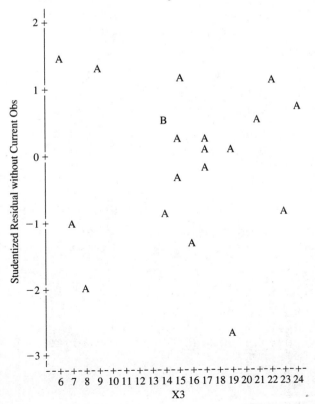

29. In Problem 19 of Chapter 5, data from the 1990 Census were shown for 26 randomly selected Metropolitan Statistical Areas (MSAs). Of interest are factors potentially associated with the rate of owner occupancy of housing units. Three variables included in the data set were as follows:

OWNEROCC: Proportion of housing units that are owner-occupied (as opposed to renter-occupied)

OWNCOST: Median selected monthly ownership costs, in $

URBAN: Proportion of population living in urban areas

Use the output given next to answer the following questions:

a. Fit a model with OWNEROCC as the response and OWNCOST and URBAN as the predictors. State the estimated model.

b. Perform variables-added-last tests for each predictor.

c. Determine the variance inflation factors for each predictor.

d. Does there appear to be any problem with collinearity? Explain.

e. Examine the plot of studentized or jackknife residuals versus the predictor. Are any violations of the assumptions of linear regression evident from the plot?

f. Determine residuals and leverage values. Do any observations seem bothersome? Explain.

Edited SAS Output (PROC REG) for Problem 29

```
                 OWNEROCC Regressed on OWNCOST and URBAN
                          Analysis of Variance
                        Sum of            Mean
Source       DF        Squares          Square      F Value      Prob > F
Model         2       255.77850       127.88925        8.521       0.0017
Error        23       345.18304        15.00796
C Total      25       600.96154

             Root MSE          3.87401      R-square         0.4256
             Dep Mean         65.96154      Adj R-sq         0.3757
             C.V.              5.87314

                          Parameter Estimates
                     Parameter         Standard      T for H0:
Variable    DF        Estimate            Error      Parameter=0     Prob > |T|
INTERCEP     1       86.758772       5.10721232         16.988         0.0001
OWNCOST      1       -0.011130       0.00529413         -2.102         0.0467
URBAN        1      -16.875462       5.89094370         -2.865         0.0088

                     Variance
Variable    DF      Inflation
INTERCEP     1     0.00000000
OWNCOST      1     1.07634548
URBAN        1     1.07634548

          Dep Var      Predict     Std Err                    Std Err      Student
Obs      OWNEROCC        Value     Predict      Residual      Residual     Residual
  1       62.0000      65.1719       1.107       -3.1719         3.713       -0.854
  2       72.0000      68.6629       1.087        3.3371         3.718        0.897
  3       72.0000      65.5199       0.825        6.4801         3.785        1.712
```

Parameter Estimates (Continued)

Obs	Dep Var OWNEROCC	Predict Value	Std Err Predict	Residual	Std Err Residual	Student Residual
4	73.0000	66.1494	0.777	6.8506	3.795	1.805
5	61.0000	64.2598	1.166	-3.2598	3.694	-0.882
6	67.0000	66.6296	0.968	0.3704	3.751	0.099
7	64.0000	64.6950	1.003	-0.6950	3.742	-0.186
8	67.0000	68.8092	1.095	-1.8092	3.716	-0.487
9	57.0000	56.8102	3.019	0.1898	2.427	0.078
10	63.0000	64.5294	1.362	-1.5294	3.627	-0.422
11	56.0000	66.5931	0.839	-10.5931	3.782	-2.801
12	71.0000	68.0833	0.952	2.9167	3.755	0.777
13	67.0000	64.5928	0.870	2.4072	3.775	0.638
14	65.0000	64.1798	0.913	0.8202	3.765	0.218
15	75.0000	74.1348	2.317	0.8652	3.104	0.279
16	66.0000	68.4097	1.135	-2.4097	3.704	-0.651
17	64.0000	63.9992	0.949	0.000769	3.756	0.000
18	63.0000	63.7493	1.096	-0.7493	3.716	-0.202
19	70.0000	71.3159	1.627	-1.3159	3.516	-0.374
20	70.0000	66.7189	0.806	3.2811	3.789	0.866
21	69.0000	63.7457	1.039	5.2543	3.732	1.408
22	65.0000	68.8408	1.349	-3.8408	3.632	-1.058
23	60.0000	63.7172	1.614	-3.7172	3.522	-1.056
24	60.0000	62.9960	1.050	-2.9960	3.729	-0.803
25	68.0000	65.7428	1.330	2.2572	3.639	0.620
26	68.0000	66.9432	1.493	1.0568	3.575	0.296

Obs	-2-1-0 1 2	Cook's D	Rstudent	Hat Diag H	Cov Ratio
1	| * |	0.022	-0.8492	0.0816	1.1294
2	| * |	0.023	0.8935	0.0787	1.1145
3	| ***|	0.046	1.7925	0.0454	0.7952
4	| ***|	0.046	1.9054	0.0402	0.7529
5	| * |	0.026	-0.8780	0.0906	1.1332
6	| |	0.000	0.0966	0.0624	1.2171
7	| |	0.001	-0.1818	0.0670	1.2192
8	| |	0.007	-0.4786	0.0799	1.2039
9	| |	0.003	0.0765	0.6075	2.9087
10	| |	0.008	-0.4140	0.1235	1.2737
11	|***** |	0.129	-3.3746	0.0469	0.3430
12	| * |	0.013	0.7698	0.0604	1.1229
13	| * |	0.007	0.6292	0.0505	1.1407
14	| |	0.001	0.2133	0.0556	1.2024
15	| |	0.014	0.2730	0.3578	1.7614
16	| * |	0.013	-0.6422	0.0858	1.1821
17	| |	0.000	0.0002	0.0600	1.2156
18	| |	0.001	-0.1974	0.0800	1.2355
19	| |	0.010	-0.3672	0.1764	1.3622
20	| * |	0.011	0.8610	0.0433	1.0813
21	| **|	0.051	1.4404	0.0720	0.9396
22	| ** |	0.051	-1.0604	0.1212	1.1196
23	| ** |	0.078	-1.0583	0.1736	1.1914
24	| * |	0.017	-0.7971	0.0735	1.1323
25	| * |	0.017	0.6119	0.1178	1.2314
26	| |	0.005	0.2897	0.1485	1.3267

.
: [Portion of output omitted]
.

Sum of Residuals	0
Sum of Squared Residuals	345.1830
Predicted Resid SS (Press)	396.8002

Plot of JACKKNIFE*OWNCOST. Legend: A = 1 obs, B = 2 obs, etc.

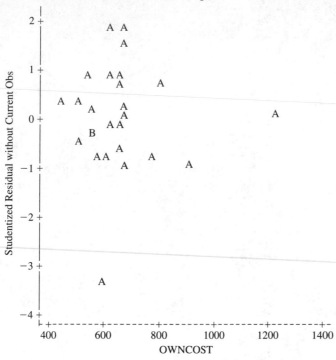

Plot of JACKKNIFE*URBAN. Legend: A = 1 obs, B = 2 obs, etc.

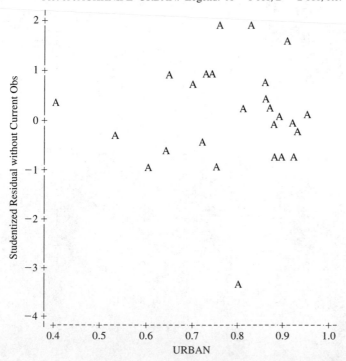

References

The following list includes sources not cited in Chapter 12 that nonetheless offer helpful discussions of topics covered here.

Agresti, A. 1990. *Categorical Data Analysis.* New York: John Wiley & Sons.

Anscombe, F. J. 1960. "Rejection of Outliers." *Technometrics* 2: 123–47.

Anscombe, F. J., and Tukey, J. W. 1963. "The Examination and Analysis of Residuals." *Technometrics* 5: 141–60.

Armitage, P. 1971. *Statistical Methods in Medical Research.* Oxford: Blackwell Scientific.

Barnett, V., and Lewis, T. 1978. *Outliers in Statistical Data.* New York: John Wiley & Sons.

Bartlett, M. S. 1947. "The Use of Transformations." *Biometrics* 3: 39–52.

Belsley, D. A. 1984. "Demeaning Conditioning Diagnostics Through Centering." *American Statistician* 38: 73–77.

———. 1991. *Conditioning Diagnostics, Collinearity and Weak Data in Regression.* New York: John Wiley & Sons.

Belsley, D. A.; Kuh, E.; and Welsch, R. E. 1980. *Regression Diagnostics: Identifying Influential Data and Sources of Collinearity.* New York: John Wiley & Sons.

Bethel, R. A.; Sheppard, D.; Geffroy, B.; Tam, E.; Nadel, J. A.; and Boushey, J. A. 1985. "Effect of 0.25 ppm Sulfur Dioxide on Airway Resistance in Freely Breathing, Heavily Exercising, Asthmatic Subjects." *American Review of Respiratory Diseases* 131: 659–61.

Box, G. E. P., and Cox, D. R. 1964. "An Analysis of Transformations." *Journal of the Royal Statistical Society* B26: 211–43 (with discussion, 244–52).

———. 1984. "An Analysis of Transformations Revisited, Rebuttal." *Journal of the American Statistical Association* 17: 209–10.

Carroll, R. J., and Ruppert, D. 1984. "Power Transformations When Fitting Theoretical Models to Data." *Journal of the American Statistical Association* 79: 321–28.

Conover, W. J., and Iman, R. L. 1981. "Rank Transformations as a Bridge Between Parametric and Nonparametric Statistics." *American Statistician* 35: 124–28.

Cook, R. D., and Weisberg, S. 1982. *Residuals and Influence in Regression.* New York: Chapman & Hall.

Draper, N. R., and Smith, H. 1981. *Applied Regression Analysis.* New York: John Wiley & Sons.

Durbin, J., and Watson, G. S. 1951. "Testing for Serial Correlation in Least Squares Regression." *Biometrika* 37: 409–28.

Freund, R. J. 1979. "Multicollinearity etc., Some 'New' Examples." *Proceedings of the Statistical Computing Section, American Statistical Association,* pp. 111–12.

Gallant, A. R. 1975. "Nonlinear Regression." *American Statistician* 29: 73–81.

Gunst, R. F., and Mason, R. L. 1980. *Regression Analysis and Its Application.* New York: Marcel Dekker.

Hackney, O. J., and Hocking, R. R. 1979. "Diagnostic Techniques for Identifying Data Problems in Multiple Linear Regression." *Proceedings of the Statistical Computing Section, American Statistical Association,* pp. 94–98.

Hinkley, D. V., and Runger, G. 1984. "The Analysis of Transformed Data." *Journal of the American Statistical Association,* 79: 302–20.

Hoaglin, D. C., and Welsch, R. E. 1978. "The Hat Matrix in Regression and ANOVA." *American Statistician* 32: 17–22.

Hocking, R. R. 1983. "Developments in Linear Regression Methodology: 1959–1982." *Technometrics* 25: 219–48.

Huber, P. J. 1981. *Robust Statistics.* New York: John Wiley & Sons.

Jensen, D. R., and Ramirez, D. E. 1996. "Computing the CDF of Cook's D_I Statistic." In A. Prat and E. Ripoll, eds., *Proceedings of the 12th Symposium in Computational Statistics,* pp. 65–66. Barcelona, Spain: Institut d'Estadística de Catalunya.

———. 1997. "Some Exact Properties of Cook's D_I." In C. R. Rao and N. Balakrishnan, eds., *Handbook of Statistics-16: Order Statistics and Their Applications.* Amsterdam: North Holland, in press.

Lewis, T., and Taylor, L. R. 1967. *Introduction to Experimental Ecology.* New York: Academic Press.

McCabe, G. P. 1984. "Principal Variables." *Technometrics* 26: 137–44.

Morrison, D. F. 1976. *Multivariate Statistical Methods.* New York: McGraw-Hill.

Mosteller, F., and Tukey, J. W. 1977. *Data Analysis and Regression.* Reading, Mass.: Addison-Wesley.

Muller, K. E., and Chen Mok, M. 1997. "The Distribution of Cook's Statistic." *Communications in Statistics: Theory and Methods* 26(3) pp. 525–546.

Neter, J.; Wasserman, W.; and Kutner, M. H. 1983. *Applied Linear Regression Models.* Homewood, Ill.: Richard D. Irwin.

Obenchain, R. L. 1977. Letter to the Editor. *Technometrics* 19: 348–49.

Picard, R. R., and Cook, R. D. 1984. "Cross-validation of Regression Models." *Journal of the American Statistical Association* 79: 575–83.

Shapiro, S. S., and Wilks, M. B. 1965. "An Analysis of Variance Test for Normality (Complete Samples)." *Biometrika* 52: 591–611.

Siegel, S. 1956. *Nonparametric Statistics for the Behavioral Sciences.* New York: McGraw-Hill.

Smith, G., and Campbell, F. 1980. "A Critique of Some Ridge Regression Methods." *Journal of the American Statistical Association* 75: 74–81.

Stephens, M. A. 1974. "EDF Statistics for Goodness of Fit and Some Comparisons." *Journal of the American Statistical Association* 69: 730–37.

Stevens, J. P. 1984. "Outliers and Influential Data Points in Regression Analysis." *Psychology Bulletin* 95: 334–44.

Timm, N. H. 1975. *Multivariate Analysis with Applications in Education and Psychology.* Belmont, Calif.: Wadsworth.

Tukey, J. W. 1977. *Exploratory Data Analysis.* Reading, Mass.: Addison-Wesley.

Weisberg, S. 1980. *Applied Linear Regressions.* New York: John Wiley & Sons.

13

Polynomial Regression

13-1 Preview

In this chapter, we focus on a special case of the multiple regression model, the *polynomial model,* which is often of interest when only *one basic* independent variable—say, X—is to be considered. We initially considered a straight-line model (Chapter 5) for this situation; however, we may wish to determine whether prediction can be improved significantly by increasing the complexity of the fitted straight-line model. The simplest extension of the straight-line model is the second-order polynomial, or *parabola,* which involves a second term, X^2, in addition to X. Adding high-order terms like X^2 and X^3, which are simple functions of a single basic variable, can be considered equivalent to adding new independent variables. Thus, if we rename X as X_1, and X^2 as X_2, the second-order model

$$Y = \beta_0 + \beta_1 X + \beta_2 X^2 + E$$

becomes

$$Y = \beta_0 + \beta_1 X_1 + \beta_2 X_2 + E$$

In general, polynomial models are special cases of the general multiple regression model. Since only one basic independent variable is being considered, however, any polynomial model can be represented by a curvilinear plot on a two-dimensional graph (rather than as a surface in higher-dimensional space). As mentioned in Chapter 5, when only one basic independent variable X is being considered, the fundamental goal is to find the curve that best fits the data so that the relationship between X and Y is appropriately described. Because a higher-order curve may be more appropriate than a straight line, considering fitting such (polynomial) curves is usually important.

We first consider methods for fitting and evaluating the second-order (parabolic) model, after which we consider higher-order polynomial models. Because these models are special cases of the general multiple regression model, the procedure for fitting these models and the methods for inference are essentially the same as those described more generally in Chapters 8 and 9. Since the independent variables in a polynomial model are functions of the same basic variable (X), they are inherently correlated. This, in turn, can lead to computational difficulties due to collinearity (Chapter 12). Fortunately, techniques such as centering and the use of orthogonal polynomials are available to help remedy such problems; these procedures are discussed later in this chapter. We shall also see that the use of orthogonal polynomials helps simplify hypothesis testing.

13-2 Polynomial Models

The most general kind of curve usually considered for describing the relationship between a single independent variable X and a response Y is called a *polynomial*. Mathematically, a polynomial of order k in x is an expression of the form

$$y = c_0 + c_1 x + c_2 x^2 + \cdots + c_k x^k$$

in which the c's and k (which must be a nonnegative whole number) are constants. We have already considered the simple polynomial corresponding to $k = 1$ (namely, the straight line having the form $y = c_0 + c_1 x$). The second-order polynomial corresponding to $k = 2$ (namely, the parabola) has the general form $y = c_0 + c_1 x + c_2 x^2$.

In going from a *mathematical* model to a *statistical* model, as we did in the straight-line case, we may write a parabolic model in either of the following forms:

$$\mu_{Y|X} = \beta_0 + \beta_1 X + \beta_2 X^2 \tag{13.1}$$

or

$$Y = \beta_0 + \beta_1 X + \beta_2 X^2 + E \tag{13.2}$$

In these equations, capital Y's and X's denote statistical variables; β_0, β_1, and β_2 denote the unknown parameters called *regression coefficients*; $\mu_{Y|X}$ denotes the mean of Y at a given X; and E denotes the error component, which represents the difference between the observed response Y at X and the true average response $\mu_{Y|X}$ at X.

If we tentatively assume that a parabolic model—as given by either (13.1) or (13.2)—is appropriate for describing the relationship between X and Y, we must then determine a specific estimated parabola that best fits the data. As in the straight-line case, this best-fitting parabola may be determined by employing the least-squares method.

13-3 Least-squares Procedure for Fitting a Parabola

The least-squares estimates of the parameters β_0, β_1, and β_2 in a parabolic model are chosen so as to minimize the sum of squares of deviations of observed points from corresponding points on the fitted parabola (Figure 13-1). Letting $\hat{\beta}_0$, $\hat{\beta}_1$, and $\hat{\beta}_2$ denote the least-squares esti-

FIGURE 13-1 Deviations of observed points from the least-squares parabola.

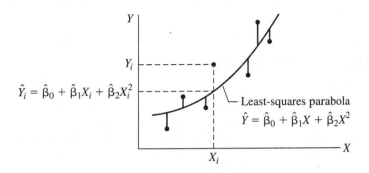

$$\hat{Y}_i = \hat{\beta}_0 + \hat{\beta}_1 X_i + \hat{\beta}_2 X_i^2$$

Least-squares parabola
$$\hat{Y} = \hat{\beta}_0 + \hat{\beta}_1 X + \hat{\beta}_2 X^2$$

mates of the unknown regression coefficients in the parabolic model (13.1), and letting $\hat{Y}$ denote the value of the predicted response at X, we can write the estimated parabola as

$$\hat{Y} = \hat{\beta}_0 + \hat{\beta}_1 X + \hat{\beta}_2 X^2 \tag{13.3}$$

The minimum sum of squares obtained by using this least-squares parabola is

$$\text{SSE} = \sum_{i=1}^{n} (Y_i - \hat{Y}_i)^2 = \sum_{i=1}^{n} (Y_i - \hat{\beta}_0 - \hat{\beta}_1 X_i - \hat{\beta}_2 X_i^2)^2 \tag{13.4}$$

As with the general regression model, we do not find it necessary to present the precise formulas for calculating the least-squares estimates $\hat{\beta}_0$, $\hat{\beta}_1$, and $\hat{\beta}_2$. These formulas are quite complex and become even more so for polynomials of orders higher than two. The researcher is not likely to employ such polynomial regression methods without using a packaged computer program, which can perform the necessary calculations and print the numerical results. (Appendix B contains a discussion of matrices and their relationship to regression analysis; by using matrix mathematics, we can compactly represent the general regression model and the associated least-squares methodology.)

▨ **Example 13-1** For the age–systolic blood pressure data of Table 5-1, with the outlier removed,[1] the least-squares estimates for the parabolic regression coefficients are computed to be

$$\hat{\beta}_0 = 113.41, \qquad \hat{\beta}_1 = 0.088, \qquad \hat{\beta}_2 = 0.010$$

The fitted model given by (13.3) then becomes

$$\hat{Y} = 113.41 + 0.088X + 0.010X^2 \tag{13.5}$$

This equation can be compared with the straight-line equation obtained in section 5-5 for these data with the outlier removed—namely,

$$\hat{Y} = 97.08 + 0.95X \tag{13.6}$$

---

[1]As described in section 5-5-3, the outlier corresponds to the data point ($X = 47$, $Y = 220$), for the second individual listed in Table 5-1.

A comparison of (13.5) and (13.6) reveals that the estimates of β_0 and β_1 differ in the two models, indicating that the estimation of β_2 affects the estimation of β_0 and β_1 in the quadratic model.

13-4 ANOVA Table for Second-order Polynomial Regression

As in the straight-line case, the essential results gathered from a second- or higher-order polynomial model can be summarized in an ANOVA table. The ANOVA table for a parabolic fit to the age–systolic blood pressure data of Table 5-1 (with the outlier removed) is given in Table 13-1.

The contents of Table 13-1 deserve comment. First, only variables-added-in-order tests are described. Natural variable orderings suggest themselves, either from the largest to the smallest power of the predictor or vice versa. Consequently, a variables-added-last test for each term should be avoided with polynomial models. Using variables-added-in-order tests aids in choosing the most parsimonious yet relevant model possible. Such tests should also utilize the residual from the largest model considered; this notion will be discussed more fully in section 13-10.

13-5 Inferences Associated with Second-order Polynomial Regression

Three basic inferential questions are associated with second-order polynomial regression:

1. Is the overall regression significant? That is, is more of the variation in Y explained by the second-order model than by ignoring X completely (and just using $\overline{Y}$)?

2. Does the second-order model provide significantly more predictive power than the straight-line model does?

3. Given that a second-order model is more appropriate than a straight-line model, should we add higher-order terms (X^3, X^4, etc.) to the second-order model?

TABLE 13-1   ANOVA table for a parabola fit to the age–systolic blood pressure data of Table 5-1, with the outlier removed.

Source		d.f.	SS	MS	F
Regression $\begin{cases} X \\ X^2 \mid X \end{cases}$	X	1	6,110.10	6,110.10	68.89
	$X^2 \mid X$	1	163.30	163.30	1.84
Residual		26	2,306.05	88.69	
Total (corrected)*		28	8,579.45		

Note: The residual from the largest model was used for all tests.

* $R^2 = .731$.

13-5-1 Test for Overall Regression and Strength of the Overall Parabolic Relationship

Determining whether the overall regression is significant involves testing the null hypothesis H_0: "There is no significant overall regression using X and X^2" (i.e., $\beta_1 = \beta_2 = 0$). The testing procedure used for this null hypothesis involves the overall F test described in Chapter 9—namely, computing

$$F = \frac{\text{MS regression}}{\text{MS residual}}$$

and then comparing the value of this F statistic with an appropriate critical point of the F distribution, which (in our example) has 2 and 26 degrees of freedom in the numerator and denominator, respectively. For $\alpha = .001$, we find that $F = 35.37 > F_{2,26,0.999} = 9.12$, so we reject the null hypothesis of nonsignificant overall regression ($P < .001$).

To obtain a quantitative measure of how well the second-order model predicts the dependent variable, we can use the squared multiple correlation coefficient (the multiple R^2). As with r^2 in straight-line regression, R^2 represents the proportionate reduction in the error sum of squares obtained by using X and X^2 instead of the naive predictor $\overline{Y}$. The formula for calculating R^2 is given by

$$R^2 \text{ (second-order model)} = \frac{\text{SSY} - \text{SSE (second-order model)}}{\text{SSY}} \tag{13.7}$$

For this example, $R^2 = .731$. The preceding F test (with $P < .001$) tells us that this R^2 is significantly different from 0. (It is possible, although not likely, that the overall F test for the second-order model will not lead to the rejection of H_0 even if the t test—or the equivalent F test—for significant regression of the straight-line model leads to rejection. This possibility arises because the loss of 1 degree of freedom in SSE in going from the straight-line model to the second-order model may result in a smaller computed F, coupled with an altered critical point of the F distribution. In our example, the computed F is reduced from 66.81 (68.89 for the variables-added-in-order test in Table 13-1) for the straight-line model to 35.37 for the second-order model, and the critical point of the F distribution for $\alpha = .001$ is reduced from $F_{1,27,0.999} = 13.6$ to $F_{2,26,0.999} = 9.12$.)

13-5-2 Test for the Addition of the X^2 Term to the Model

To answer the second question, about increased predictive power, we must perform a partial F test of the null hypothesis H_0: "The addition of the X^2 term to the straight-line model does not significantly improve the prediction of Y over and above that achieved by the straight-line model itself" (i.e., $\beta_2 = 0$). To test this null hypothesis, we compute the partial F statistic

$$F(X^2 \mid X) = \frac{(\text{Extra SS due to adding } X^2)/1}{\text{MS residual (second-order model)}} \tag{13.8}$$

and then compare this F value to an appropriate F percentage point (which is an $F_{1,26}$ value in our example). Since X^2 is the last variable added, this is a variables-added-last test. Alternatively, we could divide the estimated coefficient $\hat{\beta}_2$ by its estimated standard deviation to form a statistic that has a t distribution under H_0 with 26 degrees of freedom.

The ANOVA information needed to compute the F test for our example is given in Table 13-1. The extra sum of squares for $X^2 | X$ is 163.30 and is computed as the difference between the sum-of-squares regression values for the first- and second-order models. The partial F statistic (13.8) is then computed to be

$$F = \frac{163.30}{88.69} = 1.84$$

Since $F_{1,26,0.90} = 2.91$, we would not reject H_0 at the $\alpha = .10$ level. Furthermore, the P-value for this test satisfies the inequality $.10 < P < .25$. Thus, for this example, we conclude that adding a quadratic term to the straight-line model does not significantly improve prediction. Corroboration of the results of this partial F test is provided by a scatter diagram of the data (Figure 5-8 in Chapter 5), which offers no evidence of a parabolic relationship between X and Y. Finally, the conclusion is also supported by the small increase in R^2 when the X^2 term is added to the straight-line model. Since $r^2 = R^2 = .712$ for the straight-line model, and $R^2 = .731$ for the parabolic model, the increase in R^2 is $.731 - .712 = .019$.

13-5-3 Testing for Adequacy of the Second-order Model

The preceding analysis of the Table 5-1 data has shown that a straight-line model fits the data adequately, is significantly predictive of the response, and is preferable to a parabolic model. Consequently, it would be superfluous in this case to evaluate whether a model of an order higher than two would be significantly better than a straight-line model. Nevertheless, in general, any question of model adequacy can be addressed (with a *lack-of-fit* test) for any model (of any order) that is being considered at a given stage of an analysis. Any such lack-of-fit test can be characterized by a partial or multiple-partial F test for the addition of one or more terms to the model under study. More detailed discussion of lack-of-fit tests is provided in the subsequent sections.

13-6 Example Requiring a Second-order Model

We now turn to another hypothetical example to illustrate the methods of polynomial regression. This example will lead us to a different conclusion regarding the appropriateness of a second-order model.

Suppose that a laboratory study is undertaken to determine the relationship between the dosage (X) of a certain drug and weight gain (Y). Eight laboratory animals of the same sex, age, and size are selected and randomly assigned to one of eight dosage levels.[2]

The gain in weight (in dekagrams) is measured for each animal after a two-week time period during which all animals were subject to the same dietary regimen and general laboratory

---

[2]The study design here can certainly be criticized for not having more than one animal receive a particular dosage, as well as for involving such a small total sample size. Replication at each dosage would provide a reliable estimate of animal-to-animal variation in the data. However, for some laboratory studies, sufficient numbers of animals are not easily obtainable; cost and time are often limiting factors, as well. Finally, the data for this example have been contrived to simplify the analysis and to present a relationship that is clearly second-order in nature.

conditions. The data are given in Table 13-2, and a scatter diagram for these data in Figure 13-2. By simply eyeballing this diagram, one can see that a parabolic curve is a more appropriate model than a straight line. We shall now proceed to quantify this visual impression.

The complete ANOVA table, based on fitting a parabola to the data in Table 13-2, is given in Table 13-3. The equation for the least-squares parabola on which the ANOVA table is based has the form

$$Y = 1.35 - 0.41X + 0.17X^2$$

TABLE 13-2   Weight gain after two weeks as a function of dosage level.

Dosage level (X)	1	2	3	4	5	6	7	8
Weight gain (Y) (dag)	1	1.2	1.8	2.5	3.6	4.7	6.6	9.1

FIGURE 13-2   Scatter diagram of hypothetical data for the animal weight gain study.

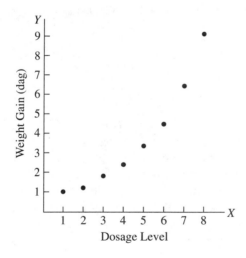

TABLE 13-3   Regression ANOVA table for the quadratic model fit to the weight gain data.

Source		d.f.	SS	MS	F
Regression	X	1	52.04	52.04	61.95
	$X^2 \mid X$	1	4.83	4.83	120.75
Residual		5	0.20	0.04	
Total (corrected)*		7	57.07		

* $R^2 = .997$.

Let us investigate the information contained in this ANOVA table. First, by combining the regression $X^2 \mid X$ sum of squares with the residual sum of squares, we can test whether there is a significant straight-line regression effect before we add the X^2 term to the model; in particular, the ANOVA table for straight-line regression derived from Table 13-3 is given in Table 13-4. The least-squares line is $\hat{Y} = -1.20 + 1.11X$. The null hypothesis of no significant linear regression is clearly rejected, since an F of 62.053 exceeds $F_{1,6,0.999} = 35.51$ ($P < .001$). Our next step is to examine the complete ANOVA table and decide whether adding the X^2 term significantly improves the prediction of Y over and above what is obtained from a simple straight-line model. In doing so, we are asking whether the increase in R^2 of .085 ($.997 - .912$), obtained by including the X^2 term in the model, significantly improves the fit. The appropriate test statistic to use in answering this question is the partial F statistic

$$F = \frac{(\text{Extra sum of squares due to adding } X^2)/1}{\text{MS residual (second-order model)}}$$

$$= \frac{4.83}{0.04} = 120.75$$

which exceeds $F_{1,5,0.999} = 47.18$ ($P < .001$). Therefore, adding the X^2 term to the model significantly improves prediction. As might be expected, a test for overall significant regression of the second-order model yields a highly significant F—namely,

$$F = \frac{\text{MS regression (second-order model)}}{\text{MS residual (second-order model)}}$$

$$= \frac{(52.04 + 4.83)/2}{0.04} = 710.88$$

Up to this point, we have concluded that a first-order (straight-line) model is not as good as a second-order model. We now need to determine whether adding higher-order terms to the second-order model is warranted. For example, we can add an X^3 term to the second-order model and then test whether prediction is significantly improved. Fitting this third-order model by least squares results in the ANOVA table given in Table 13-5. To test whether adding the third-order term significantly improves the fit, we calculate the following statistic:

$$F = \frac{(\text{Extra sum of squares due to adding } X^3)/1}{\text{MS residual (third-order model)}}$$

$$= \frac{0.14}{0.014} = 10.00$$

TABLE 13-4   Regression ANOVA table for the linear model fit to the weight gain data.

Source	d.f.	SS	MS	F
Regression (X)	1	52.04	52.04	62.05
Residual	6	5.03	0.84	
Total (corrected)*	7	57.07		

* $R^2 = .9118$.

TABLE 13-5   Regression ANOVA table for the third-order model fit to the weight gain data.

Source	d.f.	SS	MS	F
X	1	52.037	52.04	
Regression $X^2 \mid X$	1	4.835	4.84	
$X^3 \mid X, X^2$	1	0.141	0.14	10.00[a]
Residual	4	0.056	0.014	
Total (corrected)[b]	7	57.066		

[a] $.025 < P < .05$.

[b] $R^2 = .999$.

This F statistic has an F distribution with 1 and 4 degrees of freedom under H_0: "The addition of the X^3 term is not worthwhile" (i.e., $\beta_3 = 0$). Since $F_{1,4,0.95} = 7.71$ and $F_{1,4,0.975} = 12.22$, we have $.025 < P < .05$. This P-value would thus reject H_0 for $\alpha = .05$ but not for $\alpha = .025$. This makes the decision of whether to include the X^3 term in the model somewhat difficult. However, several other factors should be taken into consideration: (1) the R^2-value for the parabolic fit is very high (namely, .997); (2) the R^2-value only increases from .997 to .999 in going from a second-order model to a third-order model; (3) the scatter diagram clearly suggests a second-order curve; (4) when in doubt, the simpler model is preferable because it is easier to interpret. All things considered, then, it is most sensible to conclude that the second-order model is most appropriate.

In summary, for the data in Table 13-2, the best-fitting model is

$$\hat{Y} = 1.35 - 0.41X + 0.17X^2$$

with an R^2 of .997.

Finally, it is also valuable to have the standard deviations (or standard errors) of the estimated regression coefficients. These are difficult to compute by hand for models involving two or more predictors. However, all commonly used computer regression programs print the numerical values of the estimated coefficients and their estimated standard errors. For the second-order model fit to the data in Table 13-2, we obtain $S_{\hat{\beta}_1} = 0.141$ and $S_{\hat{\beta}_2} = 0.015$. For example, then, a $100(1 - \alpha)\%$ confidence interval for β_2 would be

$$\hat{\beta}_2 \pm t_{5,1-\alpha/2}S_{\hat{\beta}_2}$$

where the degrees of freedom for the appropriate critical t value are the degrees of freedom associated with the residual sum of squares in Table 13-3. In particular, a 95% confidence interval for β_2 in our example is

$$0.17 \pm (2.571)(0.015)$$

or (0.13, 0.21). This interval does not include 0, which agrees with our ANOVA table conclusion concerning the importance of the X^2 term in the quadratic model.

13-7 Fitting and Testing Higher-order Models

So far we have seen how the basic ideas of multiple regression may be applied to fitting and testing quadratic and cubic polynomial models. These same methods generalize to all higher-order polynomial models. Nevertheless, several related issues need to be discussed: the use of orthogonal polynomials, and strategies for choosing a polynomial model.

How large an order of polynomial model to consider depends on the problem being studied and the amount and type of data being collected. For studies in the biological and social sciences, one important consideration is whether the regression relationship can be described by a monotonic function (i.e., one that is always increasing or decreasing). If only monotonic functions are of interest, a second- or third-order model usually suffices (although monotonicity is not guaranteed, since, for example, some parabolas increase and then decrease). A large number of well-placed predictor values and a small error variance are needed to obtain reliable fits for models of higher order than cubic.

A more general consideration is the number of *bends* (more technically, *relative extrema*) in the polynomial curve that one wishes to fit. For example, a first-order model has no bends; a second-order model has no more than one bend, and each higher-order term adds another potential bend. In practice, fitting polynomial models of orders higher than three usually leads to models that are neither always decreasing nor always increasing. Substantial theoretical and/or empirical evidence should exist to support the employment of such complicated nonmonotonic models.

The quantity of data directly limits the maximum order of a polynomial that may be fit. Consider the weight gain data (Table 13-2). Given those eight distinct values, a polynomial of order seven would fit the eight points perfectly, giving an SSE value of 0 and an R^2-value of 1. (Nevertheless, because the fitted equation would have eight estimated parameters, no gain in parsimony is made over simply listing the eight data points.) *Generally, the maximum-order polynomial that may be fit is one less than the number of distinct X-values.* For example, consider the age–systolic blood pressure data in Table 5.1 (with the outlier removed). Of the 29 observations, 5 are replicates, which implies that 24 distinct X-values exist. Hence, a polynomial of order 23 could be fit to these data, although it would be absurd to consider fitting such a model.

13-8 Lack-of-fit Tests

Given that a polynomial model has been fitted and the estimated regression coefficients have been tested for their significance, how can one be confident that a model of order higher than the highest order tested is probably not needed? A lack-of-fit (LOF) test can be used to address this question. Conceptually, a lack-of-fit test evaluates a model more complex than the one under primary consideration.

The classical lack-of-fit test can be applied only if there are replicate observations. The term *replicate* means that an experimental unit (subject) has the same X-value as another experimental unit. With n total observations, if d X-values are distinct, the number of replicates is $r = n - d$. Recall that a polynomial curve of order $d - 1$ can pass through exactly d distinct points. A classical lack-of-fit test compares the fit of a polynomial of order $d - 1$ with the fit of the polynomial model currently under consideration.

For the age–systolic blood pressure example (with the outlier removed), 5 X-values out of the total of 29 involve replicates (i.e., $r = 5$). These are listed in Table 13-6. (Notice that these replicate data in Table 13-6 call into question the validity of the variance homogeneity assumption.) In a classical lack-of-fit test of these data, we consider a polynomial of order $d - 1 = n - r - 1 = 29 - 5 - 1 = 23$, which is the higher order of the two polynomial models being compared. The lower-order polynomial is the model of primary interest, such as the second-order model fit to the age–systolic blood pressure data. For our example, the two models would be

$$Y = \beta_0 + \beta_1 X + \beta_2 X^2 + E$$

and

$$Y = \beta_0 + \beta_1 X + \beta_2 X^2 + \beta_3 X^3 + \cdots + \beta_{23} X^{23} + E$$

Table 13-7 contains the ANOVA information needed to compare these two models. The F test for such a comparison is a multiple partial F test of the null hypothesis H_0: $\beta_j = 0$, $j = 3, 4, \ldots, 23$. (Trying to fit the larger 23rd-degree polynomial directly would lead to serious collinearity problems; a possible alternative method of fitting, described in section 13-9, involves the use of orthogonal polynomials.)

The ANOVA framework for the classical lack-of-fit test (see Table 13-7) partitions the residual sum of squares (SSE) for the model whose fit is being questioned into two components: a pure-error sum of squares SS_{PE} (with degrees of freedom df_{PE}), and a lack-of-fit sum of squares SS_{LOF}. The test statistic, which is equivalent to the multiple partial F test just described, can be written as $F = MS_{LOF}/MS_{PE}$. In actuality, SS_{PE} is the error sum of squares for the higher-degree polynomial model being compared, of order $d - 1 = n - r - 1$. The quantity SS_{LOF} is the extra sum of squares due to the addition to the lower-order model of all higher-order terms needed to construct the higher-order model. In the example under consideration,

$$SS_{LOF} = \text{Regression } SS(X, X^2, \ldots, X^{23}) - \text{Regression } SS(X, X^2)$$

and

$$SS_{PE} = \text{Residual } SS(X, X^2, \ldots, X^{23})$$

TABLE 13-6   Replicates and pure-error estimates from repeated observations for the age–systolic blood pressure data of Table 5-1.

X	Y	SS*	d.f.
39	144, 120	288.0	1
42	124, 128	8.0	1
45	138, 135	4.5	1
56	154, 150	8.0	1
67	170, 158	72.0	1
		380.5 (SS_{PE})	5

* The sum of squares for a given X is calculated by using the formula $\sum_m (Y_{mx} - \overline{Y}_x)^2$, where Y_{mx} is the mth observation of Y at $X = x$ and $\overline{Y}_x$ is the mean of all replicates at $X = x$.

TABLE 13-7   Regression ANOVA table for the classical lack-of-fit test of the second-order polynomial fit to the age–systolic blood pressure data of Table 5-1.

Source	d.f.
X	1
$X^2 \mid X$	1
Residual $\begin{cases} \text{Lack of fit } (X^3, X^4,\dots, X^{23} \mid X, X^2) \\ \text{Pure error} \end{cases}$	21
	5
Total	28

Note: The lack-of-fit test statistic is given by

$$F = F(X^3, X^4,\dots, X^{23} \mid X, X^2)$$

$$= \frac{[\text{Regression SS}(X, X^2,\dots, X^{23}) - \text{Regression SS}(X, X^2)]/21}{\text{Residual SS}(X, X^2,\dots, X^{23})/5}$$

$$= \frac{\text{MS}_{\text{LOF}}}{\text{MS}_{\text{PE}}}$$

Under H_0: "No lack of fit of second-order model," this F statistic should have an F distribution with 21 and 5 degrees of freedom.

With the availability of standard computer regression packages, using a multiple partial F test for lack of fit can be less computationally cumbersome than identifying replicate observations in order to compute SS_{PE} directly. Orthogonal polynomials *must* be used to avoid serious inaccuracy in the multiple regression computations leading to SS_{LOF}.

13-9 Orthogonal Polynomials

In Chapter 12, we illustrated collinearity problems that can arise in work with polynomial models, and we demonstrated that centering the predictor helped remedy such problems for a second-order polynomial model. A more sophisticated approach is needed for higher-order models. The polynomials we have discussed so far have all been *natural polynomials*. This terminology derives from the fact that each of the independent variables (X, X^2, X^3, etc.) in a polynomial of the form

$$Y = \beta_0 + \beta_1 X + \beta_2 X^2 + \cdots + \beta_k X^k + E$$

is a simple polynomial by itself. Another method of fitting polynomial models involves *orthogonal polynomials,* which constitute a new set of independent variables that are defined in terms of the simple polynomials but have more complicated structures. In this section we explain the method and describe its advantages and disadvantages. The basic motivation for using orthogonal polynomials is to avoid the serious collinearity inherent in using natural polynomials.

In section 13-2, we alluded to the natural polynomial model

$$Y = \beta_0 + \beta_1 X + \beta_2 X^2 + \cdots + \beta_k X^k + E$$

The simple polynomials $X, X^2, \ldots, X^k$ are the predictors. Orthogonal polynomial variables are new predictor variables that consist of linear combinations of these simple polynomials. Denoting orthogonal polynomials as $X_1^*, X_2^*, \ldots, X_k^*$, we can write them as linear combinations of the form

$$X_1^* = a_{01} + a_{11}X$$
$$X_2^* = a_{02} + a_{12}X + a_{22}X^2$$
$$\vdots$$
$$X_k^* = a_{0k} + a_{1k}X + a_{2k}X^2 + \cdots + a_{kk}X^k$$

where the a's are constants that relate the X^*'s to the original predictors. Each linear combination has the form of a polynomial. It can be shown that each simple polynomial can be written as a linear combination of the X^*'s as follows:

$$X = b_{01} + b_{11}X_1^*$$
$$X^2 = b_{02} + b_{12}X_1^* + b_{22}X_2^*$$
$$\vdots$$
$$X^k = b_{0k} + b_{1k}X_1^* + b_{2k}X_2^* + \cdots + b_{kk}X_k^*$$

where the b's are constants. With no loss of information, we can write either

$$Y = \beta_0 + \beta_1 X + \beta_2 X^2 + \cdots + \beta_k X^k + E$$

or

$$Y = \beta_0^* + \beta_1^* X_1^* + \beta_2^* X_2^* + \cdots + \beta_k^* X_k^* + E$$

The parameters $\{\beta_j^*\}$ in the latter model differ numerically from those in the former. Moreover, whereas the simple polynomials are highly correlated with one another, the orthogonal polynomials (via a judicious choice of the a's) are pairwise uncorrelated.[3] However, although the parameters and predictors in the two models have very different interpretations it can be shown that the multiple R^2-values and the overall regression F tests obtained by fitting these two models are exactly the same.

In summary, the orthogonal polynomial values represent a recoding of the original predictors with two desirable basic properties: *the orthogonal polynomial variables contain exactly the same information as the simple polynomial variables,* and *the orthogonal polynomial variables are uncorrelated with each other.* The first property means that all questions about the natural polynomial model can be answered by using the orthogonal polynomial model. For example, one can assess the overall strength of the regression relationship, conduct the overall regression F test, and even compute partial F tests while using the orthogonal polynomial model. The second property, zero pairwise correlation, completely eliminates any collinearity.

With regard to partial F tests, it can be shown that the partial F test of $H_0: \beta_i^* = 0$ for the orthogonal polynomial model of order k is equivalent to the partial F test of $H_0: \beta_i = 0$ in the reduced natural polynomial model

$$Y = \beta_0 + \beta_1 X + \beta_2 X^2 + \cdots + \beta_i X^i + E$$

---

[3] $r_{x_i^*, x_j^*} = 0$ for all $i \neq j$.

for any $i \leq k$. Thus, only one orthogonal polynomial model has to be fit (of the highest order—say, k—of interest) in order to test sequentially for the significance of each simple polynomial term in the original natural polynomial model. This may be summarized by saying that the k tests of $H_0: \beta_i^* = 0$, for $i = 1, 2, \ldots, k$, using the orthogonal polynomial model are equivalent to the k variables-added-in-order tests of $H_0: \beta_i = 0$, for $i = 1, 2, \ldots, k$, using an appropriate sequence of k natural polynomial models.

As an illustration, consider a third-order natural polynomial model

$$Y = \beta_0 + \beta_1 X + \beta_2 X^2 + \beta_3 X^3 + E$$

Suppose that it is of interest to determine a best model by proceeding backward, starting with a test of $H_0: \beta_3 = 0$ in the preceding cubic model. If this test is nonsignificant, one would proceed to a test of $H_0: \beta_2 = 0$ in a (reduced) parabolic model; in turn, if $H_0: \beta_2 = 0$ is not rejected, one would perform a test of $H_0: \beta_1 = 0$ in a (reduced) straight-line model. One way to conduct these three tests would be to fit three separate natural polynomial models (cubic, parabolic, and straight-line models) and then to perform partial F (or t) tests—starting with the cubic model—to assess the significance of the highest-order term in each model, stopping at any point if significance is obtained. Unfortunately, collinearity will generally compromise reliable fitting of the cubic and (without centering) quadratic models. To avoid this problem, one can fit a *single* third-order orthogonal polynomial model, and then test sequentially $H_0: \beta_3^* = 0$, $H_0: \beta_2^* = 0$, and finally $H_0: \beta_1^* = 0$, stopping at any point if significance is obtained. This approach generally ensures computational accuracy.

13-9-1 Transforming from Natural to Orthogonal Polynomials

In this section we describe a method for expressing orthogonal polynomials in terms of the original simple polynomials. This procedure involves specifying the a's in the earlier equations relating the X^*'s to powers of X. To illustrate the method, assume that the scientist who collected the data in Table 13-2 decided to try to confirm the conclusions made in the first study by collecting new data. The same eight dosage values, from 1 to 8, were used. Three new observations were taken at each dosage, giving a total of 24 observations. Since eight distinct X-values are available, the highest-order polynomial that can be fitted is of the seventh order. To demonstrate the use of orthogonal polynomials and an associated lack-of-fit test, let us consider fitting third- and seventh-order polynomials to these data.

Table 13-8 provides the data from the new study and also gives the values of the $X^2, X^3, \ldots, X^7$ terms (which are labeled DOSE2, DOSE3, $\ldots$, DOSE7 in the table). These are the natural polynomial values. Table 13-9 includes the same response (Y) values, but the predictor (X) values have been replaced by orthogonal polynomial values (labeled ODOSE1, ODOSE2, $\ldots$, ODOSE7). These values were obtained by using Table A-7 in Appendix A as follows.

Table A-7 can be used only if the predictor (X) values are equally spaced and if the same number of observations (i.e., replicates) occur at each value. If either of these conditions is not satisfied, Table A-7 cannot be used. In that case, a reasonable alternative is to use a computer program (such as the ORPOL function in SAS PROC IML) to calculate the appropriate orthogonal polynomial values.

In describing the transition from Table 13-8 to Table 13-9, we begin by noting that the variable "dosage" has eight distinct, equally spaced values and that three replicates are taken

TABLE 13-8   Observed weight gains and simple polynomial values.

Observation	WGTGAIN (Y)	DOSE1 (X)	DOSE2 (X^2)	DOSE3 (X^3)	DOSE4 (X^4)	DOSE5 (X^5)	DOSE6 (X^6)	DOSE7 (X^7)
1	0.9	1	1	1	1	1	1	1
2	1.1	2	4	8	16	32	64	128
3	1.6	3	9	27	81	243	729	2187
4	2.3	4	16	64	256	1024	4096	16384
5	3.5	5	25	125	625	3125	15625	78125
6	5.0	6	36	216	1296	7776	46656	279936
7	6.6	7	49	343	2401	16807	117649	823543
8	8.7	8	64	512	4096	32768	262144	2097152
9	0.9	1	1	1	1	1	1	1
10	1.1	2	4	8	16	32	64	128
11	1.6	3	9	27	81	243	729	2187
12	2.1	4	16	64	256	1024	4096	16384
13	3.4	5	25	125	625	3125	15625	78125
14	4.5	6	36	216	1296	7776	46656	279936
15	6.7	7	49	343	2401	16807	117649	823543
16	8.6	8	64	512	4096	32768	262144	2097152
17	0.8	1	1	1	1	1	1	1
18	1.2	2	4	8	16	32	64	128
19	1.4	3	9	27	81	243	729	2187
20	2.2	4	16	64	256	1024	4096	16384
21	3.2	5	25	125	625	3125	15625	78125
22	4.8	6	36	216	1296	7776	46656	279936
23	6.7	7	49	343	2401	16807	117649	823543
24	8.8	8	64	512	4096	32768	262144	2097152

Note: The linear predictor variable X is called DOSE1. The quadratic simple polynomial X^2 is called DOSE2, etc.; in general, the term X^i is called DOSEi.

at each of these values. Hence, Table A-7 can be used. In Table A-7, k indicates the number of distinct values of X—in our case, 8. The row in Table A-7 labeled "Linear" for $k = 8$ consists of the entries $(-7, -5, -3, -1, 1, 3, 5, 7)$. These eight entries will be used to replace the eight original values for the linear term X. Thus, a subject who received a dosage of 1 (as did the first subject) is given a linear orthogonal polynomial score of -7; a subject who received a dosage of 2 is given a linear orthogonal polynomial score of -5; and so on, with a dosage of 8 corresponding to a score of 7. This pattern is repeated in Table 13-10 twice more for the replicate observations. Analogous operations yield the remaining columns appearing in Table 13-10. For example, the eight quadratic entries for $k = 8$ $(7, 1, -3, -5, -5, -3, 1, 7)$ give the orthogonal polynomial values for the variable ODOSE2 in Table 13-10, which replace the original eight values for the quadratic term X^2 (given in the corresponding column of Table 13-9).

Finally, the rightmost column of Table A-7 gives the sum of squared values $\left(\sum p_i^2\right)$ for each row in the table. For $k = 8$, dividing each linear orthogonal polynomial score by $\sqrt{168}$, each quadratic score by $\sqrt{168}$, each cubic score by $\sqrt{264}$, and so on, makes the variance of

TABLE 13-9   Observed weight gains and orthogonal polynomial values.

Observation	WGTGAIN	ODOSE1	ODOSE2	ODOSE3	ODOSE4	ODOSE5	ODOSE6	ODOSE7
1	0.9	-7	7	-7	7	-7	1	-1
2	1.1	-5	1	5	-13	23	-5	7
3	1.6	-3	-3	7	-3	-17	9	-21
4	2.3	-1	-5	3	9	-15	-5	35
5	3.5	1	-5	-3	9	15	-5	-35
6	5.0	3	-3	-7	-3	17	9	21
7	6.6	5	1	-5	-13	-23	-5	-7
8	8.7	7	7	7	7	7	1	1
9	0.9	-7	7	-7	7	-7	1	-1
10	1.1	-5	1	5	-13	23	-5	7
11	1.6	-3	-3	7	-3	-17	9	-21
12	2.1	-1	-5	3	9	-15	-5	35
13	3.4	1	-5	-3	9	15	-5	-35
14	4.5	3	-3	-7	-3	17	9	21
15	6.7	5	1	-5	-13	-23	-5	-7
16	8.6	7	7	7	7	7	1	1
17	0.8	-7	7	-7	7	-7	1	-1
18	1.2	-5	1	5	-13	23	-5	7
19	1.4	-3	-3	7	-3	-17	9	-21
20	2.2	-1	-5	3	9	-15	-5	35
21	3.2	1	-5	-3	9	15	-5	-35
22	4.8	3	-3	-7	-3	17	9	21
23	6.7	5	1	-5	-13	-23	-5	-7
24	8.8	7	7	7	7	7	1	1

Note: The linear orthogonal polynomial predictor variable is called ODOSE1, the quadratic orthogonal polynomial variable is called ODOSE2, etc.; in general, the *i*th-order variable is called ODOSE*i*.

each set of orthogonal polynomial scores (i.e., each set of row entries) equal to 1. This helps in two ways. First, numerical accuracy is improved by the avoidance of scaling problems. Second, the estimated standard errors of the resulting estimated regression coefficients are all equal, which simplifies the task of comparing and interpreting such regression coefficients.

13-9-2 Regression Analysis with Orthogonal Polynomials

Let us now consider the regression ANOVA tables based on fitting third- and seventh-order polynomial models to the weight gain data in Tables 13-9 and 13-10. Table 13-11 summarizes the results for the third-order natural polynomial model. Clearly, the model fits extremely well ($R^2 = .998$). Furthermore, the variable-added-in-order tests give *P*-values of .0001, .0001, and .3995 for linear, quadratic, and cubic tests, respectively. These results argue persuasively for a second-order model. Since the simple polynomials X, X^2, and X^3 are correlated, the variable-added-last tests are more difficult to interpret. Nevertheless, the cubic test again gives $P = .3995$, since it is the last variable to enter; the linear effect is not significant

TABLE 13-10   Third-order natural polynomial model for the weight gain data $(N = 24)$.

Source	d.f.	SS	MS
Model[a]	3	162.7122	56.5707
Error	20	0.3274	0.0164
Total (corrected)[b]	23	170.0396	

Source	d.f.	Variables-Added-in-Order SS	F
DOSE1 (X)	1	155.6667	9491.87 $(P = .0001)$
DOSE2 $(X^2\,|\,X)$	1	14.0334	855.70 $(P = .0001)$
DOSE3 $(X^3\,|\,X, X^2)$	1	0.0121	0.74 $(P = .3995)$

Source	d.f.	Variables-Added-Last SS	F
DOSE1 $(X\,|\,X^2, X^3)$	1	0.0395	2.41 $(P = 0.1359)$
DOSE2 $(X^2\,|\,X, X^3)$	1	0.1662	10.15 $(P = 0.0046)$
DOSE3 $(X^3\,|\,X, X^2)$	1	0.0121	0.74 $(P = 0.3995)$

Parameter	Estimate	T for H_0: Parameter $= 0$	$S_{\hat{\beta}_j}$
Intercept	1.0261	5.64 $(P = .0001)$	0.182
DOSE1	−0.2559	−1.55 $(P = .1359)$	0.165
DOSE2	0.1316	3.19 $(P = .0046)$	0.041
DOSE3	0.0026	0.86 $(P = .3995)$	0.003

[a] $F = 3,455.65$ $(P = .0001)$.

[b] $R^2 = .998075$.

$(P = .1359)$, but the quadratic effect is $(P = .0046)$. When using variable-added-last tests, analysts should proceed backward, starting with the cubic model and including in the final model all lower-order terms below the highest term deemed significant. Then, the series of variable-added-last tests also supports the choice of the second-order model.

The preference for a second-order model is further supported by comparing R^2-values for the second- and third-order models. It can be shown, using the variable-added-in-order SS entries in Table 13-11, that the R^2-value for the second-order model is

$$R^2 = \frac{155.6667 + 14.0334}{170.040} = .998$$

which is the same (within round-off error) as the R^2-value for the third-order model.

The P-values for the t test are the same as the P-values for the variable-added-last F test (.1359, .0046, and .3995, respectively, for linear, quadratic, and cubic); this is because these t tests are equivalent to the variable-added-last F tests (indeed, $T_{20}^2 = F_{1,20}$). All such tests use the residual from the largest model, which has 20 degrees of freedom.

TABLE 13-11 Third-order orthogonal polynomial model for the weight gain data ($N = 24$).

Source	d.f.	SS	MS
Model[a]	3	169.7122	56.5707
Error	20	0.3274	0.0164
Total (corrected)[b]	23	170.0396	

Source	d.f.	Variables-Added-in-Order SS	F
ODOSE1	1	155.6667	9508.98 ($P = .0001$)
ODOSE2	1	14.0334	857.23 ($P = .0001$)
ODOSE3	1	0.0121	0.74 ($P = .3995$)

Source	d.f.	Variables-Added-Last SS	F
ODOSE1	1	155.6667	9508.98 ($P = .0001$)
ODOSE2	1	14.0334	857.23 ($P = .0001$)
ODOSE3	1	0.0121	0.74 ($P = .3995$)

Parameter	Estimate	T for H_0: Parameter $= 0$	$S_{\hat{\beta}_j^*}$
Intercept	3.6542	139.92 ($P = .0001$)	0.0261
ODOSE1	7.2034	97.51 ($P = .0001$)	0.0739
ODOSE2	2.1628	29.28 ($P = .0001$)	0.0739
ODOSE3	0.0636	0.86 ($P = .3995$)	0.0739

[a] $F = 3,455.65$ ($P = .0001$).

[b] $R^2 = .998075$.

For comparison, consider the results of fitting the third-order orthogonal polynomial model, as summarized in Table 13-11. The most important difference between Tables 13-10 and 13-11 is that the two types of sum of squares (namely, variable-added-in-order and variable-added-last) are equal in the latter. Consequently, the corresponding partial F test results are the same and coincide with the t test results. Furthermore, the standard errors of the estimated linear, quadratic, and cubic regression coefficients are identical. As a result, the estimated regression coefficients can be compared directly.

The preceding results, whether from analysis using natural or orthogonal polynomials, indicate that we need not consider any polynomial model higher than second order. Nevertheless, it is instructive to use these data to illustrate particular problems that arise if the model order does exceed two. Specifically, we address the (hidden) problem of collinearity. The existence of this collinearity problem can be demonstrated by considering the predictor correlations given in Table 13-12 for both the simple polynomials and their centered counterparts—that is, for $(X - \overline{X})$, $(X - \overline{X})^2$,..., as well as for X, X^2.... We first focus on the correlations among the linear, quadratic, and cubic terms in a third-order model; this information is contained in the upper left corner of each array (above and to the left of the dashed lines). The existence of a

TABLE 13-12   Predictor correlations for the weight gain data ($N = 24$).

Simple Polynomials

	X	X^2	X^3	X^4	X^5	X^6	X^7
X	1	.976	.932	.887	.846	.810	.779
X^2		1	.988	.963	.935	.908	.882
X^3			1	.993	.978	.960	.941
X^4				1	.996	.986	.973
X^5					1	.997	.990
X^6						1	.998
X^7							1

Centered Simple Polynomials

	$(X - \overline{X})$	$(X - \overline{X})^2$	$(X - \overline{X})^3$	$(X - \overline{X})^4$	$(X - \overline{X})^5$	$(X - \overline{X})^6$	$(X - \overline{X})^7$
$(X - \overline{X})$	1	0	.926	0	.855	0	.813
$(X - \overline{X})^2$		1	0	.969	0	.932	0
$(X - \overline{X})^3$			1	0	.985	0	.965
$(X - \overline{X})^4$				1	0	.992	0
$(X - \overline{X})^5$					1	0	.996
$(X - \overline{X})^6$						1	0
$(X - \overline{X})^7$							1

collinearity problem is suggested by the presence of three very high correlations (all above .93) for the noncentered predictor data. Centering X helps reduce collinearity: Two of the three correlations become 0 after centering. Turning to the complete array, we again see, for the noncentered polynomials, that the smallest off-diagonal correlation is .779 and that four are greater than .990. As before, centering X helps reduce collinearity substantially, since correlations between odd and even powers are then all 0. Nevertheless, the remaining correlations are high, with two correlations greater than .99. The analogous correlation array using orthogonal polynomials would have all zeros off the diagonal.

In Table 13-13, the collinearity problem suggested by Table 13-12 can be examined further. Predictor correlation matrix eigenvalues (see Chapter 12) are reported here for third- and seventh-order natural, centered, and orthogonal polynomial models. For the third-order natural polynomial model, the condition number (CN; see Chapter 12) is 70.8, suggesting a collinearity problem. If the X-variable is centered (i.e., if the dosage mean $\overline{X} = 4.5$ is subtracted from the dosage X before the second- and third-order terms are computed), the condition number is reduced to 5.1, indicating no severe collinearity problem. Unfortunately, as shown by the correlation arrays in Table 13-12, such centering does not solve the collinearity problem for higher-order models. For a seventh-order model based on centered dosage data, the condition number is a disturbing 430.0. And the condition number for an uncentered, seventh-order natural polynomial model is an extremely alarming 569,664. In contrast, the condition number for both the third-order and the seventh-order orthogonal polynomial models is 1.0. This will be true for any fitted orthogonal polynomial model.

The preceding condition numbers lead us to recommend using only orthogonal polynomials (i.e., we rule out using natural polynomials) to conduct lack-of-fit tests. Table 13-14 sum-

TABLE 13-13   Eigenvalues of predictor correlation matrices for
polynomial models for the weight gain data
($N = 24$).

Third-order Model

| Eigenvalue | Natural Polynomial | | Orthogonal Polynomial |
	Uncentered	Centered	
1	2.931	1.926	1.000
2	0.069	1.000	1.000
3	6×10^{-4}	0.074	1.000
CN	70.8	5.1	1.0

Seventh-order Model

| Eigenvalue | Natural Polynomial | | Orthogonal Polynomial |
	Uncentered	Centered	
1	6.638	3.773	1.000
2	0.348	2.929	1.000
3	0.013	0.221	1.000
4	3×10^{-4}	0.071	1.000
5	4×10^{-6}	0.006	1.000
6	2×10^{-8}	5×10^{-4}	1.000
7	2×10^{-11}	2×10^{-5}	1.000
CN	569,664.0	430.0	1.0

marizes the ANOVA results based on fitting a seventh-order orthogonal polynomial model to the $N = 24$ dosage–weight gain observations. The lower-order polynomial results (up to, say, order three) remain virtually unchanged. This follows from the fact that the quadratic model fits so well and from the properties of orthogonal polynomials. After the .0001 P-values for the linear and quadratic terms, the next smallest P-value is .0958 for the fourth-order term. As before, a second-order polynomial model seems most appropriate.

As discussed in section 13-8, the lack-of-fit test for the second-order model may be performed by using a multiple partial F statistic of the form

$$F = \frac{[\text{regression SS (seventh order)} - \text{regression SS (second order)}]/(7 - 2)}{\text{SSE (seventh order)}/(24 - 1 - 7)}$$

The actual value of the statistic in our example is

$$F = \frac{(169.7795 - 169.6999)/5}{.2600/16} = 0.98$$

With 5 and 16 degrees of freedom for $\alpha = .25$, the critical F value is 1.48. Hence, we fail to reject the null hypothesis of no lack of fit; i.e., a second-order model provides a good description of these data.

TABLE 13-14   Seventh-order orthogonal polynomial model for the weight gain
data ($N = 24$).

Source	d.f.	SS	MS
Model[a]	7	169.7795	24.2542
Error	16	0.2600	0.0162
Total (corrected)[b]	23	170.0396	

Source	d.f.	Variables-Added-in-Order SS	F
ODOSE1	1	155.6667	9579.98 ($P = .0001$)
ODOSE2	1	14.0334	863.59 ($P = .0001$)
ODOSE3	1	0.0121	0.75 ($P = .4003$)
ODOSE4	1	0.0509	3.13 ($P = .0958$)
ODOSE5	1	0.0000	0.00 ($P = .9924$)
ODOSE6	1	0.0036	0.22 ($P = .6420$)
ODOSE7	1	0.0128	0.79 ($P = .3871$)

Source	d.f.	Variables-Added-Last SS	F
ODOSE1	1	155.6667	9579.48 ($P = .0001$)
ODOSE2	1	14.0334	863.59 ($P = .0001$)
ODOSE3	1	0.0121	0.75 ($P = .4003$)
ODOSE4	1	0.0509	3.13 ($P = .0958$)
ODOSE5	1	0.0000	0.00 ($P = .9924$)
ODOSE6	1	0.0036	0.22 ($P = .6420$)
ODOSE7	1	0.0128	0.79 ($P = .3871$)

Parameter	Estimate	T for H_0: Parameter $= 0$	$S_{\hat{\beta}_j^*}$
Intercept	3.6541	140.43 ($P = .0001$)	0.0260
ODOSE1	7.2033	97.87 ($P = .0001$)	0.0735
ODOSE2	2.1628	29.39 ($P = .0001$)	0.0735
ODOSE3	0.0635	0.86 ($P = .4003$)	0.0735
ODOSE4	-0.1302	-1.77 ($P = .0958$)	0.0735
ODOSE5	0.0007	0.01 ($P = .9924$)	0.0735
ODOSE6	-0.0348	0.47 ($P = .6420$)	0.0735
ODOSE7	-0.0654	-0.89 ($P = .3871$)	0.0735

[a] $F = 1{,}492.57$ ($P = .0001$).

[b] $R^2 = .998075$.

13-10 Strategies for Choosing a Polynomial Model

In our discussion of polynomial models, we have sometimes started with the smallest model, involving only a linear term, and sequentially added increasing powers of X. This is a forward-selection model-building strategy. Although it is a natural approach, it can produce misleading results for inference-making procedures.

With a forward-selection strategy, one usually tests for the importance of a candidate predictor by comparing the extra regression sum of squares for the addition of that predictor to the residual mean square. This residual mean square is based on fitting a model containing the candidate (predictor) variable and the variables already in the model. The corresponding partial F statistic is of the form

$$F(X^i \mid X, X^2, \ldots, X^{i-1}) = \frac{SS(X^i \mid X, X^2, \ldots, X^{i-1})/1}{MS \text{ residual}(X^i \mid X, X^2, \ldots, X^i)}$$

when one is testing for the importance of X^i in a polynomial model already containing terms through X^{i-1}. The mean-square residual in the preceding expression is not based on terms of order higher than X^i, even though such terms may actually belong in the final model.

The forward-selection testing approach just described can lead to underfitting the data (i.e., the forward-selection algorithm is likely to quit too soon, thereby choosing a model of an order lower than is actually required). The reason for this problem is that, when the computations proceed forward, the residual mean-square error estimate of σ^2 at any step will be biased upward if the polynomial model at that step is of too low a degree. Since this causes the denominator in the earlier partial F expression to be too large, the F statistic itself may be too small and hence nonsignificant, thus stopping the forward selection algorithm prematurely.

The underfitting bias can be avoided by using a backward elimination strategy (see Chapter 16), where the partial F test at each backward step involves the residual mean error for the full (or largest) model fitted. When using this backward elimination approach, however, one may overfit the data (i.e., choose a final model of order slightly higher than required). Fortunately, the residual mean-square error estimate from the full model is still a valid (unbiased) estimate of σ^2. Consequently, using this estimate in the denominator of the partial F test at any backward step is a statistically valid procedure. By slightly overfitting the data, one loses some statistical power; but usually this loss is negligible.

Thus, to fit polynomial models, we generally recommend a backward elimination strategy for selecting variables, using in all partial F tests the mean-square error estimate based on the largest-order polynomial model fitted. To implement this strategy, we recommend first choosing the full model to be of third order or lower, to simplify interpretation and to enhance computational accuracy. Second, proceeding backward in a stepwise fashion starting with the largest power term, one should sequentially delete nonsignificant terms, stopping at the first power term that is significant; this term and all lower-order terms should be retained in the final model. Third, one should conduct a multiple partial F test for lack of fit. Fourth, the residual analysis methods of Chapter 12 should be applied, as in all regression modeling. Of particular use for polynomial regression is a plot of jackknife residuals against X. The need for a higher-order model often appears as a nonlinear trend in the residuals.

Problems

1. In an environmental engineering study of a certain chemical reaction, the concentrations of 18 separately prepared solutions were recorded at different times (three measurements at each of six times). The natural logarithms of the concentrations were also computed. The data recorded are reproduced in the following table.

Solution Number (i)	TIME (X_i) (hr)	Concentration (Y_i) (mg/ml)	Ln Concentration $(\ln Y_i)$
1	6	0.029	−3.540
2	6	0.032	−3.442
3	6	0.027	−3.612
4	8	0.079	−2.538
5	8	0.072	−2.631
6	8	0.088	−2.430
7	10	0.181	−1.709
8	10	0.165	−1.802
9	10	0.201	−1.604
10	12	0.425	−0.856
11	12	0.384	−0.957
12	12	0.472	−0.751
13	14	1.130	0.122
14	14	1.020	0.020
15	14	1.249	0.222
16	16	2.812	1.034
17	16	2.465	0.902
18	16	3.099	1.131

a. Plot on separate sheets of graph paper:
   (1) Concentration (Y) versus time (X).
   (2) Natural logarithm of concentration $(\ln Y)$ versus time (X).
b. Using the accompanying computer output, obtain the following:
   (1) The estimated equation of the straight-line (degree 1) regression of Y on X.
   (2) The estimated equation of the quadratic (degree 2) regression of Y on X.
   (3) The estimated equation of the straight-line (degree 1) regression of $\ln Y$ on X.
   (4) Plots of each of these fitted equations on their respective scatter diagrams.
c. Based on the accompanying computer output, complete the following table for the straight-line regression of Y on X.

Source		d.f.	SS	MS	F
Regression		1			
Residual	Lack of fit	4			
	Pure error	12			
Total		17			

d. Based on the accompanying computer output, complete the following ANOVA table.

Source		d.f.	SS	MS	F
Regression	Degree 1 (X)	1			
	Degree 2 $(X^2 \mid X)$	1			
Residual	Lack of fit	3			
	Pure error	12			
Total		17			

e. Determine and compare the proportions of the total variation in Y that are explained by the straight-line regression on X and by the quadratic regression on X.

f. Carry out F tests for the significance of the straight-line regression of Y on X and for the adequacy of fit of the estimated regression line.

g. Carry out an overall F test for the significance of the quadratic regression of Y on X, a test for the significance of the addition of X^2 to the model, and an F test for the adequacy of fit of the estimated quadratic model.

h. For the straight-line regression of $\ln Y$ on X, carry out F tests for the significance of the overall regression and for the adequacy of fit of the straight-line model.

i. What proportion of the variation in $\ln Y$ is explained by the straight-line regression of $\ln Y$ on X? Compare this result with the result you obtained in part (e) for the quadratic regression of Y on X.

j. A fundamental assumption in regression analysis is variance homoscedasticity.

    (1) Examine the scatter diagrams constructed in part (a), and state why taking natural logarithms of the concentrations helps with regard to the assumption of variance homogeneity.

    (2) Is the straight-line regression of $\ln Y$ on X better for describing this set of data than the quadratic regression of Y on X? Explain.

k. What key assumption about the data would be in question if, instead of 18 different solutions, there were only 3 different solutions, each of which was analyzed at the six different time points?

Edited SAS Output (PROC REG and PROC RSREG) for Problem 1

Straight-line Regression of Y on X

Variable	DF	Parameter Estimate	Standard Error	T for H0: Parameter=0	Prob > |T|
INTERCEP	1	-1.931797	0.42857949	-4.507	0.0004
X	1	0.245971	0.03720921	6.610	0.0001

Quadratic Regression of Y on X

Regression	Degrees of Freedom	Type I Sum of Squares	R-Square	F-Ratio	Prob > F
Linear	1	12.705408	0.7320	255.1	0.0000
Quadratic	1	3.905067	0.2250	78.421	0.0000
Crossproduct	0	0	0.0000	.	.
Total Regress	2	16.610475	0.9570	166.8	0.0000

Residual	Degrees of Freedom	Sum of Squares	Mean Square	F-Ratio	Prob > F
Lack of Fit	3	0.514457	0.171486	8.852	0.0023
Pure Error	12	0.232482	0.019374		
Total Error	15	0.746939	0.049796		

Parameter	Degrees of Freedom	Parameter Estimate	Standard Error	T for H0: Parameter=0	Prob > |T|
INTERCEPT	1	3.172052	0.603016	5.260	0.0001
X	1	-0.781023	0.116989	-6.676	0.0000
X*X	1	0.046682	0.005271	8.856	0.0000

```
              Straight-line Regression of ln Y on X
                                      .
                                      .  [Portion of output omitted]
                                      .

                       Parameter         Standard        T for H0:         Prob
Variable      DF       Estimate           Error         Parameter=0       >  |T|
INTERCEP       1      -6.209556         0.07703798        -80.604         0.0001
X              1       0.451167         0.00668843         67.455         0.0001

              Quadratic Regression of ln Y on X
                                      .
                                      .  [Portion of output omitted]
                                      .

              Degrees        Type I
                of           Sum of
Regression    Freedom        Squares       R-Square       F-Ratio       Prob > F
Linear           1          42.745786       0.9965         4277.0         0.0000
Quadratic        1           0.000396       0.0000         0.0396         0.8448
Crossproduct     0               0          0.0000            .              .
Total Regress    2          42.746182       0.9965         2138.5         0.0000

              Degrees
                of           Sum of          Mean
Residual      Freedom        Squares        Square        F-Ratio       Prob > F
Lack of Fit      3           0.027439       0.009146        0.896         0.4714
Pure Error      12           0.122475       0.010206
Total Error     15           0.149914       0.009994
```

2. With the addition of five pairs of observations—(18,000, 39.2), (22,400, 27.9), (24,210, 22.3), (5,400, 11.7), and (9,340, 32.5)—to the data in Problem 3 in Chapter 5, the accompanying computer output is obtained for the regression of TIME (Y) on INC (X).

a. Using the accompanying computer output, complete the following ANOVA table for the straight-line regression of TIME (Y) on INC (X).

Source		d.f.	SS	MS	F
Regression		1			
Residual	Lack of fit	18			
	Pure error	5			
Total		24			

b. Using the accompanying computer output, complete the following ANOVA table for the quadratic regression of TIME (Y) on INC (X).

Source		d.f.	SS	MS	F
Regression	Degree 1 (X)	1			
	Degree 2 ($X^2 \mid X$)	1			
Residual	Lack of fit	17			
	Pure error	5			
Total		24			

    c. On the scatter diagram of the data for this problem, plot the fitted straight-line (degree 1) equation and the fitted quadratic (degree 2) equation.

    d. Calculate and compare the R^2-values obtained for the straight-line, quadratic, and cubic fits.

    e. Carry out F tests for the significance of the straight-line regression and for the adequacy of fit of the straight-line model.

    f. Carry out F tests for the significance of the quadratic regression, of the addition of the quadratic term to the model, and of the adequacy of fit of the quadratic model.

    g. Which model is most appropriate: straight-line, quadratic, or cubic?

Edited SAS Output (PROC REG and PROC RSREG) for Problem 2

```
              Linear Regression of TIME (Y) on INC (X)
                                   .
                                   . [Portion of output omitted]
                                   .
                        Parameter Estimates

                     Parameter       Standard      T for H0:        Prob
Variable    DF        Estimate          Error    Parameter=0       > |T|
INTERCEP     1       20.176551     4.60655277          4.380      0.0002
X            1        0.000612     0.00030000          2.040      0.0530
              Quadratic Regression of TIME (Y) on INC (X)
                                   .
                                   . [Portion of output omitted]
                                   .
                        Parameter Estimates

                     Parameter       Standard      T for H0:        Prob
Variable    DF        Estimate          Error    Parameter=0       > |T|
INTERCEP     1      -19.866022     3.90293672         -5.090      0.0001
X            1        0.007871     0.00064060         12.287      0.0001
X2           1    -0.000000253     0.00000002        -11.521      0.0001
                Cubic Regression of TIME (Y) on INC (X)
                                   .
                                   . [Portion of output omitted]
                                   .
                        Parameter Estimates

                     Parameter       Standard      T for H0:        Prob
Variable    DF        Estimate          Error    Parameter=0       > |T|
INTERCEP     1      -35.292777     8.14733679         -4.332      0.0003
X            1        0.012227     0.00214424          5.702      0.0001
X2           1    -0.000000593     0.00000016        -.3.656      0.0015
X3           1   7.807155E-12     0.00000000          2.115      0.0466
                Regression of TIME (Y) on INC (X)
                                   .
                                   . [Portion of output omitted]
                                   .
                     Degrees            Type I
                         of             Sum of
Regression          Freedom            Squares
Linear                    1         442.914529
Quadratic                 1        2100.080668
Cubic                     1          61.100193
```

```
         Regression of TIME (Y) on INC (X) (Continued)
              Degrees
                of              Sum of              Mean
Residual      Freedom          Squares             Square
Lack of Fit      16          271.748723         16.984300
Pure Error        5           15.201250          3.040250
```

3. a.–g. For the data on DIST (Y) and MPH (X) in Problem 7 in Chapter 5, use the following information to answer the same questions as in parts (a) through (g) of Problem 2.

Degree 1 fit: $\hat{Y} = -122.345 + 6.227X$

Degree 2 fit: $\hat{Y} = 32.901 - 3.051X + 0.1176X^2$

Degree 3 fit: $\hat{Y} = 114.621 - 10.620X + 0.3247X^2 - 0.00173X^3$

Source		d.f.	SS	MS
	X	1	173,473.96	173,473.96
Regression	$X^2 \mid X$	1	10,515.44	10,515.44
	$X^3 \mid X, X^2$	1	415.19	415.19
Residual	Lack of fit	11	2,664.15	242.20
	Pure error	4	3,433.93	858.48
Total		18	190,502.67	

4. For the data on VOTE (Y) and TVEXP (X) in Problem 5 in Chapter 5, you found that the straight-line model was adequate. Using the accompanying computer output for quadratic regression, do the following.

a. Plot the fitted straight-line model and the fitted quadratic model on the scatter diagram for the data of this problem.

b. Determine the change in R^2 in going from a degree 1 to a degree 2 model.

c. Test for the significance of the addition of the X^2 term to the model.

d. Assess whether the results in parts (a) through (c) contradict your earlier conclusion about the adequacy of fit of the straight-line model.

Edited SAS Output (PROC REG and PROC RSREG) for Problem 4

```
            Straight-line Regression of Y on X
                          .
                          .  [Portion of output omitted]
                          .

      Root MSE       3.33177      R-square      0.9101
      Dep Mean      45.71000      Adj R-sq      0.9051
      C.V.           7.28894
```

(continued)

```
            Straight-line Regression of Y on X (Continued)
                         Parameter Estimates
                    Parameter         Standard        T for HO:        Prob
Variable    DF      Estimate          Error         Parameter=0       > |T|
INTERCEP    1       2.174067          3.30974362         0.657        0.5196
X           1       1.176965          0.08718045        13.500        0.0001
                 Quadratic Regression of Y on X
                                       .
                                       . [Portion of output omitted]
                                       .

              Degrees          Type I
              of               Sum of
Regression    Freedom          Squares       R-Square     F-Ratio     Prob > F
Linear        1              2023.205063      0.9101       174.3       0.0000
Quadratic     1                 2.450153      0.0011       0.211       0.6518
Crossproduct  0                 0             0.0000         .            .
Total Regress 2              2025.655215      0.9112       87.241      0.0000

            Degrees of         Sum of          Mean
Residual    Freedom            Squares         Square      F-Ratio     Prob > F
Lack of Fit    12            166.802785     13.900232      2.274       0.1874
Pure Error      5             30.560000      6.112000
Total Error    17            197.362785     11.609576

            Degrees of       Parameter       Standard      T for HO:      Prob
Parameter   Freedom          Estimate        Error        Parameter=0    > |T|
INTERCEPT      1             9.656782        16.636082        0.580       0.5692
X              1             0.750772         0.931996        0.806       0.4316
X*X            1             0.005746         0.012508        0.459       0.6518
```

5. For the regression of PCI (Y) on YNG (X) for the African countries considered in Problem 2 in Chapter 14, use the accompanying information to do the following.

   a. Plot the estimated straight-line and quadratic models on the scatter diagram for the data of the African countries.

   b. Test for the significance of the straight-line regression and for the adequacy of fit of the straight-line model.

   c. Test for the significance of the addition of the X^2 term to the model.

   d. Which model is more appropriate, the straight-line model or the quadratic model?

   Degree 1 fit: $\hat{Y} = 893.57 - 17.276X$

   Degree 2 fit: $\hat{Y} = 732.05 - 9.203X - 0.0996X^2$

Source		d.f.	SS	MS
Regression	X	1	153,784.8	153,784.8
	$X^2 \mid X$	1	88.3	88.3
Residual	Lack of fit	15	2,773.9	184.9
	Pure error	8	911.5	113.9
Total		25	157,558.5	

6. For the data given in Problem 11 in Chapter 7, which concerns the relationship between the temperature (X) of a certain medium and the growth (Y) of human amniotic cells in a

Edited SPSS Output for Problem 6

```
DEPENDENT VARIABLE.. Y
VARIABLE(S) ENTERED ON STEP NUMBER 1..  X

MULTIPLE R         0.98603      ANALYSIS OF VARIANCE   DF    SUM OF SQUARES    MEAN SQUARE         F
R SQUARE           0.97225      REGRESSION             1.        12.83072        12.83072     630.71608
ADJUSTED R SQUARE  0.97071      RESIDUAL              18.        0.36618          0.02034
STANDARD ERROR     0.14263

-------VARIABLES IN THE EQUATION-------      -----VARIABLES NOT IN THE EQUATION-------
VARIABLE     B       BETA    STD ERROR B     F            VARIABLE   BETA IN   PARTIAL   TOLERANCE      F
X         0.03582  0.98603   0.00143     630.716          XX        0.84502   0.64297   0.01606   11.981
(CONSTANT)-0.46240

VARIABLE(S) ENTERED ON STEP NUMBER 2..  XX

MULTIPLE R         0.99183      ANALYSIS OF VARIANCE   DF    SUM OF SQUARES    MEAN SQUARE         F
R SQUARE           0.98372      REGRESSION             2.        12.98210        6.49105      513.73362
ADJUSTED R SQUARE  0.98181      RESIDUAL              17.        0.21480          0.01264
STANDARD ERROR     0.11241

-------VARIABLES IN THE EQUATION-------      -----VARIABLES NOT IN THE EQUATION-------
VARIABLE     B       BETA    STD ERROR B     F            VARIABLE   BETA IN   PARTIAL   TOLERANCE      F
X         0.00537  0.14782   0.00887      0.367
XX        0.00022  0.84502   0.00006     11.981
(CONSTANT) 0.49460

MAXIMUM STEP REACHED
```

Note: B stands for the regression coefficient $\hat{\beta}$, XX stands for X^2, and you can ignore for now the terms "BETA," "PARTIAL," and "TOLERANCE." Also,

$$\text{adjusted } R^2 = R^2 - \left(\frac{k}{n-k-1}\right)(1-R^2).$$

From *Statistical Package for the Social Sciences* by Nie et al. Copyright © 1975 by McGraw-Hill, Inc. Used with permission of McGraw-Hill Book Company and Dr. Norman Nie, President, SPSS Inc.

tissue culture, researchers wish to evaluate whether a parabolic model is more appropriate than a straight-line model. Use the accompanying computer output to answer the following questions.

   a. Plot the fitted straight-line model and the fitted quadratic model on the same scatter diagram.
   b. Test for the significance of adding the X^2 term to the model.
   c. Determine the change in R^2 in going from a straight-line to a parabolic model.
   d. How do your results in parts (b) and (c) compare with the results in Problem 12 in Chapter 7 for the earlier test of adequacy of fit of the straight-line model?
   e. Which model is more appropriate—straight line or parabolic?

7. The skin response in rats to different concentrations of a newly developed vaccine was measured in an experiment, resulting in the data, models, and computer output that follow.

Concentration (X) (ml/l)	0.5	0.5	1.0	1.0	1.5	1.5	2.0	2.0	2.5	2.5	3.0	3.0
Skin response (Y) (mm)	13.90	13.81	14.08	13.99	13.75	13.60	13.32	13.39	13.45	13.53	13.59	13.64

   a. Plot the straight-line, quadratic, and cubic equations on the scatter diagram for this data set.
   b. Test sequentially for significant straight-line fit, for significant addition of X^2, and for significant addition of X^3 to the model.
   c. Which of the three models do you recommend, and why? (*Note*: You might also wish to consider R^2 for each model.)

Edited SAS Output for Problem 7

```
                    Linear Regression of Y on X
                       Analysis of Variance
                      Sum of          Mean
Source       DF      Squares         Square      F Value      Prob > F
Model         1      0.28440        0.28440        8.218        0.0168
Error        10      0.34609        0.03461
C Total      11      0.63049
                       Parameter Estimates
                    Parameter        Standard        T for H0:         Prob
Variable     DF     Estimate          Error        Parameter=0        > |T|
INTERCEP      1    13.986333        0.12246336        114.208         0.0001
X             1    -0.180286        0.06289138         -2.867         0.0168
                   Quadratic Regression of Y on X
                       Analysis of Variance
                      Sum of          Mean
Source       DF      Squares         Square      F Value      Prob > F
Model         2      0.35362        0.17681        5.747        0.0246
Error         9      0.27688        0.03076
C Total      11      0.63049
```

```
           Quadratic Regression of Y on X (Continued)
                        Parameter Estimates
                  Parameter        Standard      T for H0:        Prob
Variable    DF    Estimate         Error        Parameter=0      > |T|
INTERCEP    1     14.270500        0.22186125       64.322       0.0001
X           1     -0.606536        0.29029538       -2.089       0.0663
X2          1      0.121786        0.08119290        1.500       0.1679
                    Cubic Regression of Y on X
                        Analysis of Variance
                  Sum of           Mean
Source      DF    Squares          Square       F Value       Prob > F
Model       3     0.52218          0.17406       12.857        0.0020
Error       8     0.10831          0.01354
C Total     11    0.63049
                        Parameter Estimates
                  Parameter        Standard      T for H0:        Prob
Variable    DF    Estimate         Error        Parameter=0      > |T|
INTERCEP    1     13.361667        0.29664990       45.042       0.0001
X           1      1.679974        0.67600948        2.485       0.0378
X2          1     -1.392937        0.43263967       -3.220       0.0122
X3          1      0.288519        0.08176643        3.529       0.0077
```

8. This problem uses the data presented in the table in Problem 12 in Chapter 5. Use $\alpha = .05$.

a. Use a computer program to fit a natural polynomial cubic model for predicting vocabulary size as a function of age in years. Provide estimated regression coefficients.

b. Using variables-added-in-order tests, determine the best model.

c. Report variables-added-last tests. Explain any differences between your results here and those you obtained in part (b).

d. Report appropriate collinearity diagnostics for the model, and evaluate them. Include predictor correlations.

e. Plot jackknife (or studentized) residuals against predicted values for the best model based on your results in part (b). Provide a frequency histogram or schematic plot of the residuals, and comment on it.

f. Compare your results here with those you obtained in Problem 12 in Chapter 5.

9. a.–e. Repeat Problem 8, parts (a) through (e), after centering the predictor (age).

f. Compare the results you get here to those you obtained in Problem 8.

10. a.–e. Repeat Problem 8, (a) through (e), but use orthogonal polynomials.

f. Compare the results you get here to those you obtained in Problems 8 and 9. (*Hint:* Table A-7 cannot be used.)

11. This problem uses the data from Problem 13 in Chapter 5. Use $\alpha = .10$.

a. Use a computer program to fit a quadratic natural polynomial model for predicting latency as a function of weight minus average weight.

b.–e. Repeat parts (b) through (e) from Problem 8 for this analysis.

f. Compare your results here to those you obtained in Chapter 5.

12. This problem uses the data from Problem 15 in Chapter 5.

    a. Using a computer program, fit a cubic polynomial model with BLOODTOL as the response and PPM_TOLU as the predictor. Center PPM_TOLU. Provide the prediction equation.

    b.–e. Repeat parts (b) through (e) from Problem 8 for this analysis.

    f. For each predictor value, compute an estimate of variance of the response variable. Are these estimates approximately equal?

    g. Compare your results here to those you obtained in Chapter 5.

13. a. For the data from Problem 15 in Chapter 5, specify the orthogonal polynomial codings needed for coding PPM_TOLU linear, quadratic, and cubic terms. Repeat part (a) of Problem 12, but use the orthogonal coding. (*Hint:* Table A-7 cannot be used.)

    b.–e. Repeat parts (b) through (e) from Problem 8 for this analysis.

14. To promote safe driving habits and to better protect its customers, an insurance company offers a discount of between 5% and 20% on renewal insurance premiums to customers who have completed a defensive driving course. The following table of data shows the number of customers who have applied for the discount at various discount levels, over a period of 12 months.

Month	Discount (X, %)	Number of Renewing Customers Applying for Discount
1	5	485
2	5	1,025
3	5	1,056
4	10	1,020
5	10	1,149
6	10	1,100
7	15	1,800
8	15	1,805
9	15	1,725
10	20	2,225
11	20	2,325
12	20	2,650

Use the accompanying computer output to answer the following questions.

    a. Determine the estimated equation of the straight-line regression of the number of customers applying for the discount (Y) on the discount level (X).

    b. Determine the estimated equation of the quadratic regression of the number of customers applying for the discount (Y) on the discount level (X).

    c. Plot both estimated lines, along with a scatter plot of the data. Which regression appears to fit better?

    d. Conduct variables-added-in-order tests for the model in part (b).

    e. Carry out tests for the significance of the straight-line regression in part (a) and for the adequacy of fit of the estimated regression line.

    f. Carry out tests for the significance of the quadratic regression in part (b) and for the adequacy of fit of the estimated regression line.

    g. Based on the results from parts (a) through (f), which of the two regressions appears to be more appropriate for predicting the number of customers that apply for the discount?

Edited SAS Output (PROC REG and PROC RSREG) for Problem 14

```
                    Straight-line Regression of Y on X
                         Analysis of Variance
                       Sum of          Mean
Source         DF      Squares         Square       F Value     Prob > F
Model          1      4246956.15     4246956.15      90.183      0.0001
Error          10    470928.76667    47092.87667
C Total        11    4717884.9167

         Root MSE          217.00893       R-square      0.9002
         Dep Mean         1530.41667       Adj R-sq      0.8902
         C.V.               14.17973

                         Parameter Estimates

                     Parameter        Standard      T for H0:        Prob
Variable       DF     Estimate          Error      Parameter=0      > |T|
INTERCEP       1    200.166667     153.44848756       1.304        0.2213
X              1    106.420000      11.20629307       9.496        0.0001

                    Quadratic Regression of Y on X
                         Analysis of Variance
                       Sum of          Mean
Source         DF      Squares         Square       F Value     Prob > F
Model          2      4360446.9      2180223.45      54.896      0.0001
Error          9     357438.01667    39715.33519
C Total        11    4717884.9167

         Root MSE          199.28707       R-square      0.9242
         Dep Mean         1530.41667       Adj R-sq      0.9074
         C.V.               13.02175

                         Parameter Estimates

                     Parameter        Standard      T for H0:        Prob
Variable       DF     Estimate          Error      Parameter=0      > |T|
INTERCEP       1    686.416667     320.30914634       2.143        0.0607
X              1      9.170000      58.44244028       0.157        0.8788
X2             1      3.890000       2.30116884       1.690        0.1252

                           Variance
Variable       DF       Inflation
INTERCEP       1      0.00000000
X              1     32.25000000
X2             1     32.25000000

                    Quadratic Regression of Y on X
                                  .
                                  .    [Portion of output omitted]
                                  .

                Degrees      Type I
                  of         Sum of
Regression     Freedom      Squares      R-Square     F-Ratio     Prob > F
Linear            1        4246956       0.9002       106.9        0.0000
Quadratic         1         113491       0.0241        2.858       0.1252
Crossproduct      0              0       0.0000          .            .
Total Regress     2        4360447       0.9242        54.896      0.0000
```

(continued)

```
            Quadratic Regression of Y on X (Continued)
            Degrees
               of      Sum of       Mean
Residual     Freedom   Squares      Square    F-Ratio    Prob > F
Lack of Fit     1       39990       39990      1.008      0.3448
Pure Error      8      317448       39681
Total Error     9      357438       39715
```

15. Refer to Problem 14. Using the computer output for that problem, and the accompanying output here, answer the following questions.

   a. Determine the variance inflation factors for the estimated model in part (b) of Problem 14. Does collinearity appear to be a problem?

   b. Determine the estimated equation of the quadratic regression of the number of customers applying for the discount (Y) on the centered discount level (Z).

   c. Determine the variance inflation factors for the estimated model in part (b) of this problem. Does collinearity appear to be a problem?

   d. Conduct variables-added-in-order tests for the model in part (b).

   e. Carry out tests for the significance of the quadratic regression in part (b) and for the adequacy of fit of the estimated regression line.

   f. Based on the results from Problem 14 and parts (a) through (e) of this problem, which of the two regressions appears to be more appropriate for predicting the number of customers that apply for the discount?

Edited SAS Output (PROC REG and PROC RSREG) for Problem 15

```
               Quadratic Regression of Y on Z
                    Analysis of Variance
                      Sum of          Mean
Source     DF        Squares         Square      F Value    Prob > F
Model       2       4360446.9      2180223.45     54.896     0.0001
Error       9      357438.01667    39715.33519
C Total    11      4717884.9167

         Root MSE       199.28707      R-square     0.9242
         Dep Mean      1530.41667      Adj R-sq     0.9074
         C.V.            13.02175

                    Parameter Estimates
                     Parameter      Standard      T for H0:        Prob
Variable   DF        Estimate        Error      Parameter=0       > |T|
INTERCEP    1      1408.854167    92.09168729      15.298        0.0001
Z           1       106.420000    10.29113990      10.341        0.0001
Z2          1         3.890000     2.30116884       1.690        0.1252

                      Variance
Variable   DF        Inflation
INTERCEP    1       0.00000000
Z           1       1.00000000
Z2          1       1.00000000
```

```
              Quadratic Regression of Y on Z
             Degrees        Type I
                of          Sum of
Regression   Freedom        Squares      R-Square    F-Ratio     Prob > F
Linear          1          4246956        0.9002      106.9       0.0000
Quadratic       1           113491        0.0241      2.858       0.1252
Crossproduct    0                0        0.0000        .           .
Total Regress   2          4360447        0.9242      54.896      0.0000

             Degrees
                of          Sum of         Mean
Residual     Freedom        Squares       Square     F-Ratio     Prob > F
Lack of Fit     1            39990         39990      1.008       0.3448
Pure Error      8           317448         39681
Total Error     9           357438         39715
```

16. Columbus Airlines introduced a new, specially discounted line of air fares in 1995. Annual ticket revenues Y (in $1,000s) are shown in the following table, along with the time period X (in months, with $X = 1$ for January 1995).

Month	Ticket Revenues
1	34.9
2	38.8
3	41.5
4	45.1
5	48.3
6	51.2
7	56.6
8	59.9
9	65.4

Use the accompanying computer output to answer the following questions.

a. Determine the estimated equation for the straight-line regression of the ticket revenues (Y) on month (X).

b. Determine the estimated equation for the quadratic regression of ticket revenues (Y) on month (X).

c. Plot both estimated lines, along with a scatter plot of the data. Which regression appears to fit better?

d. Conduct variables-added-in-order tests for the model in part (b).

e. Carry out tests for the significance of the straight-line regression in part (a).

f. Carry out tests for the significance of the quadratic regression in part (b).

g. Examine the variance inflation factors for the estimated model in part (b). Does there appear to be a problem with collinearity?

h. Based on the results from parts (a) through (g), which of the two regression models appears to be more appropriate for predicting ticket revenues?

Edited SAS Output (PROC REG) for Problem 16

Straight-line Regression of Y on X
Analysis of Variance

Source	DF	Sum of Squares	Mean Square	F Value	Prob > F
Model	1	818.44267	818.44267	858.563	0.0001
Error	7	6.67289	0.95327		
C Total	8	825.11556			

Root MSE	0.97636	R-square	0.9919	
Dep Mean	49.07778	Adj R-sq	0.9908	
C.V.	1.98940			

Parameter Estimates

Variable	DF	Parameter Estimate	Standard Error	T for H0: Parameter=0	Prob > |T|
INTERCEP	1	30.611111	0.70930574	43.156	0.0001
X	1	3.693333	0.12604694	29.301	0.0001

Quadratic Regression of Y on X
Analysis of Variance

Source	DF	Sum of Squares	Mean Square	F Value	Prob > F
Model	2	822.91956	411.45978	1124.211	0.0001
Error	6	2.19599	0.36600		
C Total	8	825.11556			

Root MSE	0.60498	R-square	0.9973	
Dep Mean	49.07778	Adj R-sq	0.9965	
C.V.	1.23269			

Parameter Estimates

Variable	DF	Parameter Estimate	Standard Error	T for H0: Parameter=0	Prob > |T|
INTERCEP	1	32.821429	0.76978509	42.637	0.0001
X	1	2.487706	0.35345534	7.038	0.0004
X2	1	0.120563	0.03447183	3.497	0.0129

Variable	DF	Variance Inflation
INTERCEP	1	0.00000000
X	1	20.48051948
X2	1	20.48051948

Dummy Variables in Regression

14-1 Preview

To this point we have considered only continuous variables as predictors, but the methods of regression analysis can be generalized to treat categorical predictors as well. The generalization is based entirely on the use of dummy variables, the central idea of this chapter.

By using dummy variables, we can broaden the application of regression analysis. In particular, dummy variables allow us to employ regression analysis to produce the same information obtained by such seemingly distinct analytical procedures as analysis of covariance (Chapter 15) and analysis of variance (Chapters 17 through 20).

In this chapter we focus on one important application of dummy variables: for comparing several regression equations by use of a single multiple regression model. We also describe an alternative method that can be used if only two equations are being compared.

14-2 Definitions

A *dummy,* or *indicator, variable* is any variable in a regression equation that takes on a finite number of values so that different categories of a nominal variable can be identified. The term *dummy* reflects the fact that the values taken on by such variables (usually values like 0, 1, and -1) do not indicate meaningful measurements but rather the categories of interest.

Examples of dummy variables include the following:

$$X_1 = \begin{cases} 1 & \text{if treatment A is used} \\ 0 & \text{otherwise} \end{cases}$$

$$X_2 = \begin{cases} 1 & \text{if subject is female} \\ -1 & \text{if subject is male} \end{cases}$$

$$Z_1 = \begin{cases} 1 & \text{if residence is in western United States} \\ 0 & \text{if residence is in central United States} \\ -1 & \text{if residence is in eastern United States} \end{cases}$$

$$Z_2 = \begin{cases} 0 & \text{if residence is in western United States} \\ 1 & \text{if residence is in central United States} \\ -1 & \text{if residence is in eastern United States} \end{cases}$$

The variable X_1 indicates a nominal variable describing the category "treatment group" (either treatment A or not treatment A); the variable X_2 indexes the levels of the nominal variable "sex"; and variables Z_1 and Z_2 work in tandem to describe the nominal variable "geographical residence." In the last case, the three categories of geographical residence are described by the following combination of the two variables Z_1 and Z_2:

Residence in western United States: $Z_1 = 1, Z_2 = 0$
Residence in central United States: $Z_1 = 0, Z_2 = 1$
Residence in eastern United States: $Z_1 = -1, Z_2 = -1$

14-3 Rule for Defining Dummy Variables

The following simple rule should always be applied to avoid collinearity in defining a dummy variable for regression analysis: *If the nominal independent variable of interest has k categories, then exactly k − 1 dummy variables must be defined to index these categories, provided that the regression model contains a constant term (i.e., an intercept β_0). If the regression model does not contain an intercept, then k dummy variables are needed to index the k categories of interest.* For example, given $k = 3$ categories, the number of dummy variables should be $k - 1 = 2$ for a model containing an intercept. If an intercept is not included in an overall regression model designed to compare several regression equations, however, the dummy variables can be defined so that each of the regression equations derived from the overall model has its own intercept. Thus, using an intercept in the overall model generally depends on how the investigator prefers to code the dummy variables.

Applying this rule raises several significant points:

1. If an intercept is used in the regression equation, proper definition of the $k - 1$ dummy variables automatically indexes all k categories.

2. If k dummy variables are used to describe a nominal variable with k categories in a model containing an intercept, all the coefficients in the model cannot be uniquely estimated because collinearity is present.

3. The $k - 1$ dummy variables for indexing the k categories of a given nominal variable can be properly defined in many different ways. For example, two equivalent ways to describe the nominal variable "geographical residence" (represented earlier by Z_1 and Z_2) are

$$Z_1^* = \begin{cases} 1 & \text{if residence is in western United States} \\ 0 & \text{otherwise} \end{cases}$$

$$Z_2^* = \begin{cases} 1 & \text{if residence is in central United States} \\ 0 & \text{otherwise} \end{cases}$$

and

$$Z_1' = \begin{cases} 1 & \text{if residence is in western United States} \\ 0 & \text{otherwise} \end{cases}$$

$$Z_2' = \begin{cases} 1 & \text{if residence is in eastern United States} \\ 0 & \text{otherwise} \end{cases}$$

Among the coding schemes available for regression, we recommend the method often referred to as *reference cell coding*, which uses $k - 1$ dummy variables as suggested earlier. Each variable takes on only values of 1 and 0, and each variable indicates group membership (1 for a specific group, 0 otherwise). Some computer programs use 1 and -1 for coding. Since the choice of coding scheme affects analysis and interpretation, it is important to specify which coding method is being used.

We now illustrate how to use dummy variables to compare two or more regression models. We begin by considering two straight-line models, and then we extend the discussion to comparisons of more than two multiple regression models.

14-4 Comparing Two Straight-line Regression Equations: An Example

In Chapter 11, the age–systolic blood pressure example was used to assess whether there is a significant relationship between blood pressure and age. Here, we use the data to investigate the common observations that males tend to have higher blood pressure than females of similar age. To do this, we must compare the straight-line regression of systolic blood pressure versus age for females against the corresponding regression for males. The entire data set for this example was presented in Table 11-1. Table 14-1 also provides the information needed to compare the two fitted straight lines. For each data set, this information consists of the sample size (n), the intercept ($\hat{\beta}_0$), the slope ($\hat{\beta}_1$), the sample mean $\overline{X}$ of the X's, the sample mean $\overline{Y}$ of the Y's, the sample variance S_X^2 of the X's, and the residual mean-square error ($S_{Y|X}^2$). To distinguish between male and female data, we have used the subscripts M and F, respectively. Thus, n_M, $\hat{\beta}_{0M}$, $\hat{\beta}_{1M}$, and $S_{Y|X_M}^2$ denote the sample size, intercept, slope, and mean-square error for the male data, whereas n_F, $\hat{\beta}_{0F}$, $\hat{\beta}_{1F}$, and $S_{Y|X_F}^2$ denote the corresponding information for the female data.

The least-squares lines are then given as follows:[1]

Males: $\hat{Y}_M = 110.04 + 0.96X$

Females: $\hat{Y}_F = 97.08 + 0.95X$

TABLE 14-1

Group	n	$\hat{\beta}_0$	$\hat{\beta}_1$	$\overline{X}$	$\overline{Y}$	S_X^2	$S_{Y|X}^2$
Males	40	110.04	0.96	46.93	155.15	221.15	71.90
Females	29	97.08	0.95	45.07	139.86	242.14	91.46

---

[1]It can be shown that the straight-line model for males, like that for females, is appropriate based on a lack-of-fit test.

FIGURE 14-1   Comparison by sex of straight-line regressions of systolic blood pressure on age.

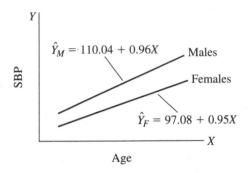

These lines are sketched in Figure 14-1. It can be seen from the figure that the male line lies completely above the female line. This fact alone supports the contention that males have higher blood pressure than females over the age range being considered. Nevertheless, it is necessary to explore statistically whether the observed differences between the regression lines could have occurred by chance. In other words, to be statistically precise when comparing two regression lines, we must consider the sampling variability of the data by using statistical test(s) and/or confidence interval(s). The sections that follow describe a number of statistical procedures for dealing with this comparison problem.

14-5 Questions for Comparing Two Straight Lines

There are three basic questions to consider when comparing two straight-line regression equations:

1. Are the two slopes the same or different (regardless of whether the intercepts are different)?[2]

2. Are the two intercepts the same or different (regardless of whether the slopes are different)?

3. Are the two lines coincident (that is, the same), or do they differ in slope and/or intercept?

Situations pertaining to these three questions are illustrated in Figure 14-2.

For our particular age–systolic blood pressure example, concluding that the lines are parallel (Figure 14-2(a)) is equivalent to finding that one sex has a consistently higher systolic blood pressure than the other at all ages, but that the rate of change with respect to age is the same for both sexes. If we conclude that the two lines have a common intercept but different slopes (Figure 14-2(b)), we find that the two sexes begin at an early age with the same average blood pressure but that the blood pressure changes with respect to age at different rates for each sex. If the

---

[2]If the two slopes are not different, we say that the two lines are *parallel*.

<u>FIGURE 14-2</u> Possible conclusions from comparing two
straight-line regressions.

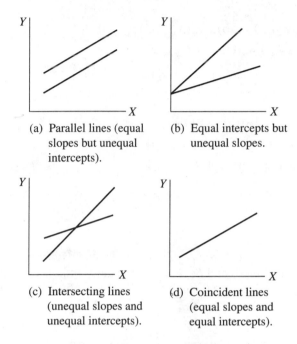

(a) Parallel lines (equal
slopes but unequal
intercepts).

(b) Equal intercepts but
unequal slopes.

(c) Intersecting lines
(unequal slopes and
unequal intercepts).

(d) Coincident lines
(equal slopes and
equal intercepts).

two lines have different slopes and different intercepts[3] (Figure 14-2(c)), it means that the relationship between age and systolic blood pressure differs for the two sexes with regard to both the origins and the rates of change. Furthermore, if the lines intersect in the range of X-values of interest, this indicates that at early ages one sex has a higher average systolic blood pressure than the other, but at later ages the other sex does.

14-6 Methods of Comparing Two Straight Lines

There are two general approaches to answering the earlier three questions related to comparing two straight lines:

Method I   Treat the male and female data separately by fitting the two separate regression equations

$$Y_M = \beta_{0M} + \beta_{1M} X + E \tag{14.1}$$

and

$$Y_F = \beta_{0F} + \beta_{1F} X + E \tag{14.2}$$

and then make appropriate two-sample t tests.

---

[3]It is possible for two lines to have unequal slopes and unequal intercepts and yet not intersect within the range of X-values of interest. This is illustrated by our example in Figure 14-1.

Method II   Define the dummy variable Z to be 0 if the subject is male and 1 if female. Thus, for the n_M observations on males, $Z = 0$; and for the n_F observations on females, $Z = 1$. Our data will then be of the form:

Males: $(X_{1M}, Y_{1M}, 0), (X_{2M}, Y_{2M}, 0), \ldots, (X_{n_M M}, Y_{n_M M}, 0)$

Females: $(X_{1F}, Y_{1F}, 1), (X_{2F}, Y_{2F}, 1), \ldots, (X_{n_F F}, Y_{n_F F}, 1)$

Then, for the combined data above, the single multiple regression model

$$Y = \beta_0 + \beta_1 X + \beta_2 Z + \beta_3 XZ + E \qquad (14.3)$$

yields the following two models for the two values of Z:

$$\begin{cases} Z = 0: & Y_M = \beta_0 + \beta_1 X + E \\ Z = 1: & Y_F = (\beta_0 + \beta_2) + (\beta_1 + \beta_3)X + E \end{cases}$$

This allows us to write the regression coefficients for the separate models for method I in terms of the coefficients of model (14.3) as follows:

$$\beta_{0M} = \beta_0, \quad \beta_{0F} = \beta_0 + \beta_2, \quad \beta_{1M} = \beta_1, \quad \beta_{1F} = \beta_1 + \beta_3$$

Thus, *model (14.3) incorporates the two separate regression equations within a single model* and allows for different slopes (β_1 for males and $\beta_1 + \beta_3$ for females) and different intercepts (β_0 for males and $\beta_0 + \beta_2$ for females). We now consider the details involved in making statistical inferences for these two methods.

14-7 Method I: Using Separate Regression Fits to Compare Two Straight Lines

14-7-1 Testing for Parallelism

From (14.1) and (14.2), we can conclude that the appropriate null hypothesis for comparing the slopes (i.e., for conducting a test of parallelism) is given by

$$\boxed{H_0: \beta_{1M} = \beta_{1F}}$$

When the null hypothesis $H_0: \beta_{1M} = \beta_{1F}$ is true, the two regression lines simplify to $Y_M = \beta_{0M} + \beta_1 X + E$ for males and $Y_F = \beta_{0F} + \beta_1 X + E$ for females, where $\beta_1 (= \beta_{1M} = \beta_{1F})$ is the common slope. An estimate of this common slope β_1 is given by the following formula, which is a weighted average of the two separate slope estimates:

$$\hat{\beta}_1 = \frac{(n_M - 1)S_{X_M}^2 \hat{\beta}_{1M} + (n_F - 1)S_{X_F}^2 \hat{\beta}_{1F}}{(n_M - 1)S_{X_M}^2 + (n_F - 1)S_{X_F}^2}$$

Notice that $\hat{\beta}_1$ equals the slope computed by fitting a straight line to the pooled data.

Any of the following three alternative hypotheses can be used:

$$H_A: \begin{cases} \beta_{1M} > \beta_{1F} & \text{(one sided)} \\ \beta_{1M} < \beta_{1F} & \text{(one sided)} \\ \beta_{1M} \neq \beta_{1F} & \text{(two sided)} \end{cases}$$

The test statistic for evaluating parallelism is then given by

$$T = \frac{\hat{\beta}_{1M} - \hat{\beta}_{1F}}{S_{\hat{\beta}_{1M} - \hat{\beta}_{1F}}} \qquad (14.4)$$

where

$\hat{\beta}_{1M} =$ Least-squares estimate of the slope β_{1M}, using the n_M observations (on males)

$\hat{\beta}_{1F} =$ Least-squares estimate of the slope β_{1F}, using the n_F observations (on females)

$S_{\hat{\beta}_{1M} - \hat{\beta}_{1F}} =$ Estimate of the standard deviation of the estimated difference between slopes $(\hat{\beta}_{1M} - \hat{\beta}_{1F})$

This standard deviation involves pooling and summing the estimated variances of the slopes of the fitted regression lines.[4] It is equal to the square root of the following variance:

$$S^2_{\hat{\beta}_{1M} - \hat{\beta}_{1F}} = S^2_{P,Y|X} \left[\frac{1}{(n_M - 1)S^2_{X_M}} + \frac{1}{(n_F - 1)S^2_{X_F}} \right] \qquad (14.5)$$

where

$$S^2_{P,Y|X} = \frac{(n_M - 2)S^2_{Y|X_M} + (n_F - 2)S^2_{Y|X_F}}{n_M + n_F - 4} \qquad (14.6)$$

is a pooled estimate of σ^2 based on combining residual mean-square errors for males and females, and where

$S^2_{Y|X_M} =$ Residual mean-square error for the male data

$S^2_{Y|X_F} =$ Residual mean-square error for the female data

$S^2_{X_M} =$ Variance of the X's for the male data

$S^2_{X_F} =$ Variance of the X's for the female data

The test statistic given by (14.4) will, under the usual regression assumptions, be distributed as a Student's t with $n_M + n_F - 4$ degrees of freedom when H_0 is true. We then have the following critical regions for different hypotheses and significance level α:

$$\begin{cases} T \geq t_{n_M + n_F - 4, 1 - \alpha} & \text{for } H_A: \beta_{1M} > \beta_{1F} \\ T \leq -t_{n_M + n_F - 4, 1 - \alpha} & \text{for } H_A: \beta_{1M} < \beta_{1F} \\ |T| > t_{n_M + n_F - 4, 1 - \alpha/2} & \text{for } H_A: \beta_{1M} \neq \beta_{1F} \end{cases}$$

---

[4]This pooling is valid only if the variance of Y_M—say, σ^2_M—is equal to the variance of Y_F—say, σ^2_F—that is, only if the assumption of homogeneity of error variance holds.

The associated $100(1 - \alpha)\%$ confidence interval for $\beta_{1M} - \beta_{1F}$ is of the form

$$(\hat{\beta}_{1M} - \hat{\beta}_{1F}) \pm t_{n_M+n_F-4,1-\alpha/2} S_{\hat{\beta}_{1M}-\hat{\beta}_{1F}}$$

■ **Example**   Using the data given in Table 14-1, we can compute the estimates $S^2_{P,Y|X}$, $S^2_{\hat{\beta}_{1M}-\hat{\beta}_{1F}}$, $S^2_{\hat{\beta}_{1M}}$, and $S^2_{\hat{\beta}_{1F}}$ as follows:

$$S^2_{P,Y|X} = \frac{(n_M - 2)S^2_{Y|X_M} + (n_F - 2)S^2_{Y|X_F}}{n_M + n_F - 4} = \frac{38(71.90) + 27(91.46)}{40 + 29 - 4}$$

$$= \frac{5{,}201.62}{65} = 80.02$$

$$S^2_{\hat{\beta}_{1M}-\hat{\beta}_{1F}} = S^2_{P,Y|X}\left[\frac{1}{(n_M - 1)S^2_{X_M}} + \frac{1}{(n_F - 1)S^2_{X_F}}\right]$$

$$= 80.02\left[\frac{1}{39(221.15)} + \frac{1}{28(242.14)}\right]$$

$$= 0.021$$

$$S^2_{\hat{\beta}_{1M}} = \frac{S^2_{Y|X_M}}{(n_M - 1)S^2_{X_M}} = \frac{71.90}{39(221.15)} = 0.0083$$

$$S^2_{\hat{\beta}_{1F}} = \frac{S^2_{Y|X_F}}{(n_F - 1)S^2_{X_F}} = \frac{91.46}{28(242.14)} = 0.0135$$

The test statistic (14.4) is then computed as

$$T = \frac{\hat{\beta}_{1M} - \hat{\beta}_{1F}}{S_{\hat{\beta}_{1M}-\hat{\beta}_{1F}}} = \frac{0.96 - 0.95}{\sqrt{0.021}} = \frac{0.01}{0.145} = 0.069$$

For this test statistic, the critical value for a two-sided test (i.e., $H_A: \beta_{1M} \neq \beta_{1F}$) with $\alpha = .05$ is given by

$$t_{65,0.975} = 1.9964$$

Since $|T| = 0.069$ does not exceed 1.9964, we do not reject H_0. Thus, we conclude that there is insufficient evidence to permit us to reject the hypothesis of parallelism (namely, that the lines for males and females have the same slope).   ■

14-7-2 Comparing Two Intercepts

We now describe how to use separate regression fits to determine whether both straight lines have the same intercept, regardless of the two slopes. The null hypothesis in this case is given by

$$H_0: \beta_{0M} = \beta_{0F}$$

If H_0: $\beta_{0M} = \beta_{0F}$ is true, the two regression lines simplify to $Y_M = \beta_0 + \beta_{1M}X + E$ and $Y_F = \beta_0 + \beta_{1F}X + E$, where $\beta_0 (= \beta_{0M} = \beta_{0F})$ is the common intercept. An estimate of this common intercept β_0 is given by the equation

$$\hat{\beta}_0 = \frac{n_M\hat{\beta}_{0M} + n_F\hat{\beta}_{0F}}{n_M + n_F}$$

which is the weighted average of the two separate intercept estimates.

The test statistic in this case is given by:

$$T = \frac{\hat{\beta}_{0M} - \hat{\beta}_{0F}}{S_{\hat{\beta}_{0M} - \hat{\beta}_{0F}}} \qquad (14.7)$$

where $\hat{\beta}_{0M}$ and $\hat{\beta}_{0F}$ are the intercept estimates for males and females, respectively, and where $S^2_{\hat{\beta}_{0M} - \hat{\beta}_{0F}}$ estimates the variance of the estimated difference between the intercepts by means of the formula

$$S^2_{\hat{\beta}_{0M} - \hat{\beta}_{0F}} = S^2_{P, Y|X} \left[\frac{1}{n_M} + \frac{1}{n_F} + \frac{\overline{X}^2_M}{(n_M - 1)S^2_{X_M}} + \frac{\overline{X}^2_F}{(n_F - 1)S^2_{X_F}} \right] \qquad (14.8)$$

The statistic T given in (14.7) will have the t distribution with $n_M + n_F - 4$ degrees of freedom when H_0: $\beta_{0M} = \beta_{0F}$ is true. We therefore have the following critical regions for different hypotheses and significance level α:

$$\begin{cases} T \geq t_{n_M+n_F-4, 1-\alpha} & \text{for } H_A: \beta_{0M} > \beta_{0F} \\ T \leq -t_{n_M+n_F-4, 1-\alpha} & \text{for } H_A: \beta_{0M} < \beta_{0F} \\ |T| \geq t_{n_M+n_F-4, 1-\alpha/2} & \text{for } H_A: \beta_{0M} \neq \beta_{0F} \end{cases}$$

The associated $100(1 - \alpha)\%$ confidence interval for $\beta_{0M} - \beta_{0F}$ is

$$(\hat{\beta}_{0M} - \hat{\beta}_{0F}) \pm t_{n_M+n_F-4, 1-\alpha/2}S_{\hat{\beta}_{0M} - \hat{\beta}_{0F}}$$

■ **Example**   For the data in Table 14-1 and the value of $S^2_{P, Y|X}$ obtained in section 14-7-1, the estimates $S^2_{\hat{\beta}_{0M} - \hat{\beta}_{0F}}$, $S^2_{\hat{\beta}_{0M}}$, and $S^2_{\hat{\beta}_{0F}}$ are computed as follows:

$$S^2_{\hat{\beta}_{0M} - \hat{\beta}_{0F}} = S^2_{P, Y|X} \left[\frac{1}{n_M} + \frac{1}{n_F} + \frac{\overline{X}^2_M}{(n_M - 1)S^2_{X_M}} + \frac{\overline{X}^2_F}{(n_F - 1)S^2_{X_F}} \right]$$

$$= 80.02 \left[\frac{1}{40} + \frac{1}{29} + \frac{(46.93)^2}{39(221.15)} + \frac{(45.07)^2}{28(242.14)} \right]$$

$$= 80.02(0.0250 + 0.0345 + 0.2554 + 0.2996)$$

$$= 49.17$$

$$S^2_{\hat{\beta}_{0M}} = S^2_{Y|X_M}\left[\frac{1}{n_M} + \frac{\bar{X}^2_M}{(n_M - 1)S^2_{X_M}}\right] = 71.90\left[\frac{1}{40} + \frac{(46.93)^2}{39(221.15)}\right]$$

$$= 71.90(0.0250 + 0.2554) = 20.16$$

$$S^2_{\hat{\beta}_{0F}} = S^2_{Y|X_F}\left[\frac{1}{n_F} + \frac{\bar{X}^2_F}{(n_F - 1)S^2_{X_F}}\right] = 91.46\left[\frac{1}{29} + \frac{(45.07)^2}{28(242.14)}\right]$$

$$= 91.46(0.0345 + 0.2996) = 30.56$$

From these results, we compute the T statistic of (14.7) as

$$T = \frac{\hat{\beta}_{0M} - \hat{\beta}_{0F}}{S_{\hat{\beta}_{0M}-\hat{\beta}_{0F}}} = \frac{110.04 - 97.08}{\sqrt{49.17}} = \frac{12.96}{7.01} = 1.85$$

For a two-sided test ($H_A:\beta_{0M} \neq \beta_{0F}$) with $\alpha = .05$, we find that $|T| = 1.85$ does not exceed $t_{65,0.975} = 1.9964$ (i.e., $.05 < P < .1$). Thus, the null hypothesis of common intercepts is not rejected at $\alpha = .05$, but it is rejected at $\alpha = .1$. ■

14-7-3 Testing for Coincidence from Separate Straight-line Fits

Two straight lines are coincident if their slopes and their intercepts are equal. In considering the male–female regression equations given by (14.1) and (14.2), we may conclude that the null hypothesis of coincidence is therefore equivalent to testing H_0: $\beta_{0M} = \beta_{0F}$ and $\beta_{1M} = \beta_{1F}$ *simultaneously*. If so, the two regression models both reduce to the general form

$$Y = \beta_0 + \beta_1 X + E$$

where $\beta_0 (= \beta_{0M} = \beta_{0F})$ and $\beta_1 (= \beta_{1M} = \beta_{1F})$ are the common intercept and slope, respectively. The estimates of the common slope β_1 and common intercept β_0 are obtained by pooling all observations on males and females together and determining the usual least-squares slope and intercept estimates using the pooled data set.

A preferred way to test the null hypothesis of coincident lines is to employ a multiple regression model involving dummy variables. Another, generally less efficient (e.g., not as powerful) procedure is often convenient when separate models are fit. In practice, this procedure frequently yields the same conclusion as is obtained from using dummy variables.

Using separate regression fits, we perform both the test of H_0: $\beta_{0M} = \beta_{0F}$ of equal intercepts and the test of H_0: $\beta_{1M} = \beta_{1F}$ of equal slopes. If either one or both of these null hypotheses are rejected, we can conclude that statistical evidence indicates that the two lines do not coincide. If neither is rejected, we must conclude that no evidence of noncoincidence exists in the data.

A valid criticism of this testing procedure, which calls into question its power, is that it involves two separate tests rather than a single test. This fact raises two difficulties:

1. The procedure does not precisely test for coincidence.

2. If α is the significance level of each separate test, the overall significance level for the two tests combined is greater than α; that is, there is more chance of rejecting a true H_0 (i.e., of making a Type I error).

One reasonable (but fairly conservative) way to get around the second difficulty is to use $\alpha/2$ for each separate test to guarantee an overall significance level of no more than α (the Bonferroni correction). Nevertheless, using $\alpha/2$ for each test is conservative (i.e., makes it harder to reject either H_0), thus making it difficult to detect a real difference between the two lines.

With regard to the first difficulty, even if both tests are not rejected, it is still possible (although unlikely) that the two lines do not coincide. This is so because each separate test (e.g., the test of $H_0: \beta_{1M} = \beta_{1F}$) allows the remaining parameters (β_{0M} and β_{0F}) to be unequal. In other words, the test for equal slopes does not assume equal intercepts; nor does the test for equal intercepts assume equal slopes. The multiple regression procedure, which involves using a single model containing a dummy variable for group status, avoids this drawback and permits testing for common slope and common intercept *simultaneously*.

Example   We saw in sections 14-7-1 and 14-7-2 that the null hypothesis of equal slopes (regardless of the intercepts) was not rejected ($P > .40$), and that the null hypothesis of equal intercepts (regardless of the slopes) was associated with a P-value of between .05 and .1. Putting these two facts together, we would be inclined to support the conclusion that there is no *strong* evidence for noncoincidence. As we shall see shortly, the more appropriate test procedure involving a single model yields a different conclusion.   ■

14-8 Method II: Using a Single Regression Equation to Compare Two Straight Lines

Another approach for comparing regression equations uses a single multiple regression model that contains one or more dummy variables to distinguish the groups being compared. The model for comparing two straight lines is given by (14.3), which we restate here:

$$Y = \beta_0 + \beta_1 X + \beta_2 Z + \beta_3 XZ + E$$

where $Y = \text{SBP}$, $X = \text{AGE}$, and Z is a dummy variable indicating gender (1 if female, 0 if male). For the data in Table 14-1 ($n_M = 40$, $n_F = 29$), the fitted model is

$$\hat{Y} = 110.04 + 0.96X - 12.96Z - 0.012XZ$$

which yields the following separate straight-line equations:

$$Z = 0: \quad \hat{Y}_M = 110.04 + 0.96X$$
$$Z = 1: \quad \hat{Y}_F = 97.08 + 0.95X$$

These two straight-line equations are identical to those obtained in section 14-4 by fitting separate regressions.

Table 14-2 provides ANOVA results needed to answer statistical inference questions about this model. This table provides variables-added-in-order tests for the fitted regression equation and allows us to perform appropriate tests for parallelism, for equal intercepts, and for coincidence.

14-8-1 Test of Parallelism: Single-model Approach

Referring again to the dummy variable model of (14.3), we know the null hypothesis that the two regression lines are parallel is equivalent to $H_0: \beta_3 = 0$. If $\beta_3 = 0$, then the slope for

TABLE 14-2   Three models for the age–systolic blood pressure example.

Source	d.f.	SS	MS	F
Regression (X)	1	14,951.25	14,951.25	121.27
Residual	67	8,260.51	123.29	
Regression (X, Z)	2	18,009.78	9,004.89	114.25
Residual	66	5,201.99	78.82	
Regression (X, Z, XZ)	3	18,010.33	6,003.44	75.02
Residual	65	5,201.44	80.02	

females, $\beta_{1F} = \beta_1 + \beta_3$, simplifies to β_1, which is the slope for males (i.e., the two lines are parallel). The test statistic for testing H_0: $\beta_3 = 0$ is the partial F statistic (or equivalent t test) for the significance of the addition of the variable XZ to a model already containing X and Z.[5]

In our example, this test statistic is computed as follows:

$$F(XZ \mid X, Z) = \frac{\text{Regression SS}(X, Z, XZ) - \text{Regression SS}(X, Z)}{\text{MS residual}(X, Z, XZ)}$$

$$= \frac{18,010.33 - 18,009.78}{80.02}$$

$$= 0.007 \qquad (P = .9342)$$

This F statistic, with 1 and 65 degrees of freedom, is extremely small (P is very large); so we do not reject H_0 and, therefore, have no statistical basis for believing that the two lines are not parallel. This was the same decision we reached on the basis of separate regression fits. In fact, the F computed here is (theoretically) the square of the corresponding T computed when using separate straight-line fits in section 14-7-1, although the numerical answers may not exactly agree due to round-off errors.

14-8-2 Test of Equal Intercepts: Single-model Approach

The hypothesis that the two intercepts are equal, allowing for unequal slopes, is equivalent to H_0: $\beta_2 = 0$ for the overall model (14.3). The test compares the overall model

$$Y = \beta_0 + \beta_1 X + \beta_2 Z + \beta_3 XZ + E$$

to the reduced model

$$Y = \beta_0 + \beta_1 X + \beta_3 XZ + E$$

This is a variables-added-last test considering Z, the sex group dummy variable.[6] Another approach involves a variables-added-in-order test comparing

$$Y = \beta_0 + \beta_1 X + \beta_2 Z + E$$

---

[5] If H_0 is not rejected by the test of H_0: $\beta_3 = 0$, model (14.3) can be revised to eliminate the β_3 term. This revised (or reduced) model becomes $Y = \beta_0 + \beta_1 X + \beta_2 Z + E$, which has the form of an analysis-of-covariance model (see Chapter 15).

[6] The partial F statistic for the variables-added-last test for equal intercepts, $F(Z \mid X, XZ)$, is the square of the statistic given by (14.7) for fitting separate straight-line models.

to the reduced model

$$Y = \beta_0 + \beta_1 X + E$$

The latter test presumes equal slopes, so it is essentially a test for coincidence, assuming parallelism. Not surprisingly, neither test is uniformly preferred (see section 14-10). As discussed in Chapters 9 and 13, we recommend using the residual from the full model, (14.3), for either test.

For the example under consideration, we opt for the second approach, since the slope test was not significant. The variables-added-in-order test for H_0: $\beta_2 = 0$ is then computed as follows:

$$F(Z \mid X) = \frac{\text{Regression SS}(Z, X) - \text{Regression SS}(X)}{\text{MS residual}(X, Z, XZ)}$$
$$= \frac{18{,}009.78 - 14{,}951.25}{80.02}$$
$$= 38.22$$

The preceding test statistic is modified from the usual partial F statistic in that we are now using for the denominator the mean-square residual for the full model, which contains X, Z, and XZ; the usual partial F statistic would use the mean-square residual for the model containing only X and Z.

Table 14-3 indicates that, with 1 and 65 degrees of freedom, $P < .0001$. Hence the intercepts are judged to be different for the male and female straight-line models.

14-8-3 Test of Coincidence: Single-model Approach

The hypothesis that the two regression lines coincide is H_0: $\beta_2 = \beta_3 = 0$. When both β_2 and β_3 are 0, the model for females, $Y_F = (\beta_0 + \beta_2) + (\beta_1 + \beta_3)X + E$, reduces to $Y_M = \beta_0 + \beta_1 X + E$, the model for males (i.e., the two lines coincide). The test of H_0: $\beta_2 = \beta_3 = 0$ is thus a multiple-partial F test, since it involves a subset of regression coefficients.[7] The two models being compared are therefore

$$Y = \beta_0 + \beta_1 X + \beta_2 Z + \beta_3 XZ + E$$

and

$$Y = \beta_0 + \beta_1 X + E$$

TABLE 14-3   ANOVA table for method II for the age–systolic blood pressure example.

Source	d.f.	SS	MS	F	P
X (AGE)	1	14,951.25	14,951.25	186.84	.0001
Z (SEX)$\mid X$	1	3,058.52	3,058.52	38.22	.0001
$X \times Z \mid X, Z$	1	0.55	0.55	0.01	.9342
Residual	65	5,201.44	80.02		
Total (corrected)	68	23,211.76			

[7]If the test for coincidence is not rejected, model (14.3) can be reduced to the form $Y = \beta_0 + \beta_1 X + E$.

For our example, the information in either Table 14-2 or Table 14-3 leads to the following computation:

$$F(XZ, Z \mid X) = \frac{[\text{Regression SS}(X, Z, XZ) - \text{Regression SS}(X)]/2}{\text{MS residual}(X, Z, XZ)}$$

$$= \frac{(18,010.33 - 14,951.25)/2}{80.02}$$

$$= 19.1$$

Comparing this F with $F_{2,65,0.999} = 7.72$, we reject H_0 with $P < .001$ and conclude that very strong evidence exists that the two lines are *not* coincident. This conclusion contradicts our earlier conclusion (section 14-7-3) based on the results from separate tests for equal slopes and equal intercepts.

14-9 Comparison of Methods I and II

Does the method that uses dummy variables differ from the method that fits two separate regression equations? And if so, is one of the methods preferable to the other?

In deciding whether one method is preferable, we first observe that the two methods yield exactly the same estimated regression coefficients for the two straight-line models. That is, if we fit the model $Y = \beta_0 + \beta_1 X + \beta_2 Z + \beta_3 XZ + E$ by the least-squares method to obtain estimated coefficients $\hat{\beta}_0, \hat{\beta}_1, \hat{\beta}_2,$ and $\hat{\beta}_3$, the straight-line equations obtained by setting Z equal to 0 and to 1 in this estimated model will be the same as those obtained by fitting the two straight lines separately. In particular, if $\hat{\beta}_{0M}, \hat{\beta}_{0F}, \hat{\beta}_{1M},$ and $\hat{\beta}_{1F}$ denote the estimated regression coefficients based on separate regression fits, then $\hat{\beta}_{0M} = \hat{\beta}_0, \hat{\beta}_{0F} = \hat{\beta}_0 + \hat{\beta}_2, \hat{\beta}_{1M} = \hat{\beta}_1,$ and $\hat{\beta}_{1F} = \hat{\beta}_1 + \hat{\beta}_3$. As for statistical tests involving regression coefficients estimated by the two methods, the following two points are valid:

1. *The tests for parallel lines are exactly equivalent*; that is, the T statistic with $n_1 + n_2 - 4$ degrees of freedom computed for testing $H_0: \beta_3 = 0$ in the dummy variable model is exactly the same as the T statistic given by (14.4) for testing $H_0: \beta_{1M} = \beta_{1F}$ based on fitting two separate models.

2. *The tests for coincident lines differ,* and the one using the dummy variable model is generally preferable. The approach using separate regressions tests $H_0: \beta_{1M} = \beta_{1F}$ and $H_0: \beta_{0M} = \beta_{0F}$ separately and then rejects the null hypothesis of coincident lines if either or both null hypotheses are rejected. This is exactly equivalent to performing two separate tests of $H_0: \beta_2 = 0$ and $H_0: \beta_3 = 0$ and using the same decision rule for the dummy variable approach; but it is not equivalent to testing the single null hypothesis $H_0: \beta_2 = \beta_3 = 0$ (i.e., testing whether β_2 and β_3 are both simultaneously 0).

14-10 Testing Strategies and Interpretation: Comparing Two Straight Lines

Several strategies can be used to identify a best model for comparing two straight lines. Strategies for more general situations are described in Chapter 16. We prefer a backward strategy for most situations—that is, starting with the largest model of interest and then trying to reduce the model through a sequence of hypothesis tests. A flow diagram of this strategy for comparing two straight lines is given in Figure 14-3. In this case, the largest model to be considered is model (14.3), which contains X, Z, and XZ as independent variables. To reduce the

FIGURE 14-3   Comparing two straight lines: Backward testing strategy.

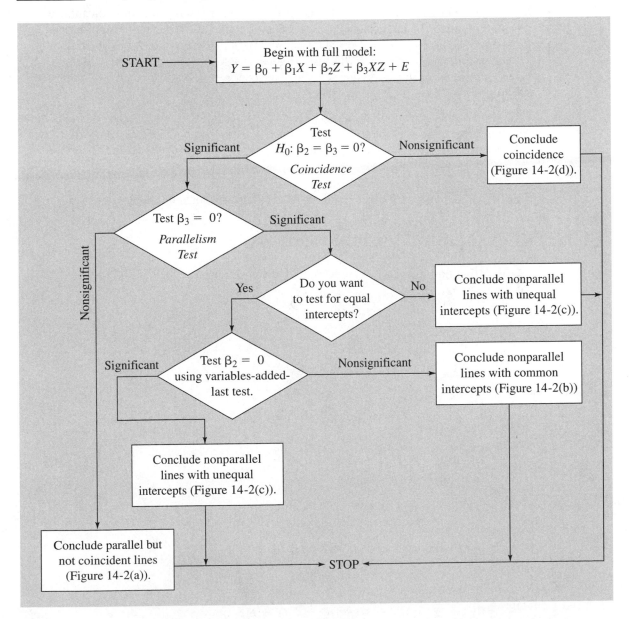

model, we perform tests for coincidence, then for parallelism, and then for equal intercepts, as follows:

1. If the test for coincidence is nonsignificant, we stop further testing and conclude that the best model is $Y = \beta_0 + \beta_1 X + E$ (i.e., coincident lines, Figure 14-2(d)).

2. If the test for coincidence is significant and the test for parallelism is nonsignificant, the data support a finding of parallel but noncoincident lines (Figure 14-2(a)).

3. If the test for coincidence is significant and the test for parallelism is significant, we might not even be interested in a test for equal intercepts; if we are, the appropriate test procedure involves the variables-added-last statistic $F(Z \mid X, XZ)$, which does not assume parallel lines. If this test produces a significant result, we would argue for Figure 14-2(c); if it produces a nonsignificant result, we would tend to support Figure 14-2(b).

Applying this strategy to the age–systolic blood pressure data, we would conclude, based on the preceding tests, that the test for coincidence is significant and the test for parallelism is nonsignificant. Our overall conclusion, therefore, is that the best model has the form

$$Y = \beta_0 + \beta_1 X + \beta_2 Z + E$$

In words, we assume parallel (and noncoincident) lines, as shown in Figure 14-2(a).

14-11 Other Dummy Variable Models

Two other dummy variable models could have been used instead of (14.3):

$$Y = \beta_0^* + \beta_1^* X + \beta_2^* Z^* + \beta_3^* X Z^* + E \qquad (14.9)$$

for which

$$Z^* = \begin{cases} 1 & \text{if subject is male} \\ -1 & \text{if subject is female} \end{cases}$$

and

$$Y = \beta_{0M} Z_1' + \beta_{0F} Z_2' + \beta_{1M} X Z_1' + \beta_{1F} X Z_2' + E \qquad (14.10)$$

for which

$$Z_1' = \begin{cases} 1 & \text{if subject is male} \\ 0 & \text{if subject is female} \end{cases}$$

$$Z_2' = \begin{cases} 0 & \text{if subject is male} \\ 1 & \text{if subject is female} \end{cases}$$

For model (14.9), the separate regression equations are

$$Z^* = 1: \quad Y_M = (\beta_0^* + \beta_2^*) + (\beta_1^* + \beta_3^*)X + E$$

and

$$Z^* = -1: \quad Y_F = (\beta_0^* - \beta_2^*) + (\beta_1^* - \beta_3^*)X + E$$

The test for parallel lines is equivalent to testing H_0: $\beta_3^* = 0$; the test for equal intercepts is equivalent to testing H_0: $\beta_2^* = 0$; and the test for coincident lines is equivalent to testing H_0: $\beta_2^* = \beta_3^* = 0$.[8]

For model (14.10), the separate regression equations are

$$Z_1' = 1, Z_2' = 0: \quad Y_M = \beta_{0M} + \beta_{1M}X + E$$

and

$$Z_1' = 0, Z_2' = 1: \quad Y_F = \beta_{0F} + \beta_{1F}X + E$$

The test for parallel lines here is equivalent to testing H_0: $\beta_{1M} = \beta_{1F}$ (not necessarily equal to 0); the test for equal intercepts is equivalent to testing H_0: $\beta_{0M} = \beta_{0F}$ (not necessarily equal to 0); and the test for coincident lines is equivalent to testing H_0: $\beta_{1M} = \beta_{1F}$ and $\beta_{0M} = \beta_{0F}$ simultaneously.[9] The test of H_0: $\beta_{1M} = \beta_{1F}$ and $\beta_{0M} = \beta_{0F}$ differs from previously discussed multiple-partial F tests because the coefficients under this H_0 are not equal to 0. The testing procedure in such a case is given as follows:

1. Reduce the model according to the specifications under the null hypothesis; for the test of coincidence, for example, the full model (14.10) becomes

$$Y = \beta_{0M}(Z_1' + Z_2') + \beta_{1M}(XZ_1' + XZ_2') + E$$
$$= \beta_{0M} + \beta_{1M}X + E \text{ since } Z_1' + Z_2' = 1$$

2. Find the residual sum of squares for this reduced model.

3. Compute the following F statistic:

$$F = \frac{[\text{Residual SS (reduced model)} - \text{Residual SS (full model)}]/v^*}{\text{MS residual (full model)}}$$

where v^* is the number of independent linear parametric functions specified to be 0 under H_0 (in our case $v^* = 2$, since the null hypothesis of coincidence specifies that $\beta_{1M} - \beta_{1F} = 0$ and that $\beta_{0M} - \beta_{0F} = 0$).

4. Test H_0 using F tables with v^* and $n - 4$ degrees of freedom, where 4 is the number of parameters in the full model.

Model (14.10) does not include an overall intercept. Care must be taken in using this particular model with most computer programs (which generally include an overall intercept by default). Collinearity (often labeled "LESS THAN FULL RANK MODEL") may occur.

---

[8]We can express the coefficients of model (14.9) in terms of the regression coefficients for the separate male and female models:

$$\beta_0^* = \frac{\beta_{0M} + \beta_{0F}}{2}, \quad \beta_1^* = \frac{\beta_{1M} + \beta_{1F}}{2}, \quad \beta_2^* = \frac{\beta_{0M} - \beta_{0F}}{2}, \quad \beta_3^* = \frac{\beta_{1M} - \beta_{1F}}{2}$$

[9]The coefficients of model (14.10) are identical to the correspondingly labeled coefficients for the separate male and female models.

14-12 Comparing Four Regression Equations

Suppose that we wish to compare the separate multiple regressions of systolic blood pressure on age and weight for four social class groups. For each individual in each social class group, we observe values of the variables $Y = \text{SBP}$, $X_1 = \text{AGE}$, and $X_2 = \text{WEIGHT}$. Further, let us suppose that there are n_i individuals in the ith social class (SC) group, $i = 1, 2, 3, 4$. We begin by defining three dummy variables Z_1, Z_2, and Z_3:

$$Z_1 = \begin{cases} 1 & \text{if SC2 member} \\ 0 & \text{otherwise} \end{cases} \qquad Z_2 = \begin{cases} 1 & \text{if SC3 member} \\ 0 & \text{otherwise} \end{cases}$$

$$Z_3 = \begin{cases} 1 & \text{if SC4 member} \\ 0 & \text{otherwise} \end{cases}$$

The complete model to be used (if no interaction between AGE and WEIGHT is considered) is given as follows:

$$\begin{aligned} Y = {} & \beta_0 + \beta_1 X_1 + \beta_2 X_2 + \beta_3 Z_1 + \beta_4 Z_2 + \beta_5 Z_3 + \beta_6 X_1 Z_1 + \beta_7 X_2 Z_1 \\ & + \beta_8 X_1 Z_2 + \beta_9 X_2 Z_2 + \beta_{10} X_1 Z_3 + \beta_{11} X_2 Z_3 + E \end{aligned} \qquad (14.11)$$

For each particular social class, model (14.11) specializes as follows:

$$\begin{aligned} \text{SC1}(Z_1 = Z_2 = Z_3 = 0): &\quad Y = \beta_0 + \beta_1 X_1 + \beta_2 X_2 + E \\ \text{SC2}(Z_1 = 1, Z_2 = Z_3 = 0): &\quad Y = (\beta_0 + \beta_3) + (\beta_1 + \beta_6) X_1 + (\beta_2 + \beta_7) X_2 + E \\ \text{SC3}(Z_1 = Z_3 = 0, Z_2 = 1): &\quad Y = (\beta_0 + \beta_4) + (\beta_1 + \beta_8) X_1 + (\beta_2 + \beta_9) X_2 + E \\ \text{SC4}(Z_1 = Z_2 = 0, Z_3 = 1): &\quad Y = (\beta_0 + \beta_5) + (\beta_1 + \beta_{10}) X_1 + (\beta_2 + \beta_{11}) X_2 + E \end{aligned}$$

14-12-1 Tests of Hypotheses

The following hypotheses involving the parameters in model (14.11) are of interest.

1. *All four regression equations are parallel* (i.e., test H_0: $\beta_6 = \beta_7 = \beta_8 = \beta_9 = \beta_{10} = \beta_{11} = 0$). When H_0 is true, the models for each social class reduce to

$$\begin{aligned} \text{SC1:} &\quad Y = \beta_0 + \beta_1 X_1 + \beta_2 X_2 + E \\ \text{SC2:} &\quad Y = (\beta_0 + \beta_3) + \beta_1 X_1 + \beta_2 X_2 + E \\ \text{SC3:} &\quad Y = (\beta_0 + \beta_4) + \beta_1 X_1 + \beta_2 X_2 + E \\ \text{SC4:} &\quad Y = (\beta_0 + \beta_5) + \beta_1 X_1 + \beta_2 X_2 + E \end{aligned}$$

Thus, the coefficients of X_1 and X_2 are the same for each social class when H_0 is true (i.e., the four regression equations are said to be parallel). The test statistic for testing H_0 is given by the multiple partial

$$F(X_1 Z_1, X_2 Z_1, X_1 Z_2, X_2 Z_2, X_1 Z_3, X_2 Z_3 \mid X_1, X_2, Z_1, Z_2, Z_3)$$

which has 6 and $n_1 + n_2 + n_3 + n_4 - 12$ degrees of freedom.

2. *All four regression equations are coincident* (i.e., test H_0: $\beta_3 = \beta_4 = \beta_5 = \beta_6 = \beta_7 = \beta_8 = \beta_9 = \beta_{10} = \beta_{11} = 0$)? When H_0 is true, all four social class models reduce to the form

$$Y = \beta_0 + \beta_1 X_1 + \beta_2 X_2 + E$$

The test statistic is the multiple partial

$$F(Z_1, Z_2, Z_3, X_1 Z_1, X_1 Z_2, X_2 Z_1, X_2 Z_2, X_1 Z_3, X_2 Z_3 \mid X_1, X_2)$$

which has 9 and $n_1 + n_2 + n_3 + n_4 - 12$ degrees of freedom.

14-12-2   An Alternate Dummy Variable Model

Another dummy variable coding scheme for comparing the four social class groups begins by defining

$$Z_1^* = \begin{cases} -1 & \text{if SC1 member} \\ 1 & \text{if SC2 member} \\ 0 & \text{if SC3 member} \\ 0 & \text{if SC4 member} \end{cases} \qquad Z_2^* = \begin{cases} -1 & \text{if SC1 member} \\ 0 & \text{if SC2 member} \\ 1 & \text{if SC3 member} \\ 0 & \text{if SC4 member} \end{cases}$$

$$Z_3^* = \begin{cases} -1 & \text{if SC1 member} \\ 0 & \text{if SC2 member} \\ 0 & \text{if SC3 member} \\ 1 & \text{if SC4 member} \end{cases}$$

Then we use the model

$$\begin{aligned} Y = {} & \beta_0^* + \beta_1^* X_1 + \beta_2^* X_2 + \beta_3^* Z_1^* + \beta_4^* Z_2^* + \beta_5^* Z_3^* + \beta_6^* X_1 Z_1^* + \beta_7^* X_2 Z_1^* \\ & + \beta_8^* X_1 Z_2^* + \beta_9^* X_2 Z_2^* + \beta_{10}^* X_1 Z_3^* + \beta_{11}^* X_2 Z_3^* + E \end{aligned} \qquad (14.12)$$

For the preceding dummy variable coding, the four regression equations for the four social classes based on model (14.12) are

SC1 $(Z_1^* = Z_2^* = Z_3^* = -1)$:
$$Y = (\beta_0^* - \beta_3^* - \beta_4^* - \beta_5^*) + (\beta_1^* - \beta_6^* - \beta_8^* - \beta_{10}^*)X_1 + (\beta_2^* - \beta_7^* - \beta_9^* - \beta_{11}^*)X_2 + E$$
SC2 $(Z_1^* = 1, Z_2^* = Z_3^* = 0)$:
$$Y = (\beta_0^* + \beta_3^*) + (\beta_1^* + \beta_6^*)X_1 + (\beta_2^* + \beta_7^*)X_2 + E$$
SC3 $(Z_1^* = 0, Z_2^* = 1, Z_3^* = 0)$:
$$Y = (\beta_0^* + \beta_4^*) + (\beta_1^* + \beta_8^*)X_1 + (\beta_2^* + \beta_9^*)X_2 + E$$
SC4 $(Z_1^* = Z_2^* = 0, Z_3^* = 1)$:
$$Y = (\beta_0^* + \beta_5^*) + (\beta_1^* + \beta_{10}^*)X_1 + (\beta_2^* + \beta_{11}^*)X_2 + E$$

So the null hypotheses to be tested in connection with parallelism and coincidence (using appropriate multiple partial F tests) are

Parallelism: H_0: $\beta_6^* = \beta_7^* = \beta_8^* = \beta_9^* = \beta_{10}^* = \beta_{11}^* = 0$

Coincidence: H_0: $\beta_3^* = \beta_4^* = \beta_5^* = \beta_6^* = \beta_7^* = \beta_8^* = \beta_9^* = \beta_{10}^* = \beta_{11}^* = 0$

14-13 Comparing Several Regression Equations Involving Two Nominal Variables

Suppose that we want to compare eight regression equations of SBP (Y) on AGE (X_1) and WEIGHT (X_2), corresponding to the eight combinations of SEX (Q) and social class (SC) groups. Then the following regression model can be used:

$$
\begin{aligned}
Y = {} & \beta_0 + \beta_1 X_1 + \beta_2 X_2 + \beta_3 Z_1 + \beta_4 Z_2 + \beta_5 Z_3 + \beta_6 Q + \beta_7 Z_1 Q + \beta_8 Z_2 Q \\
& + \beta_9 Z_3 Q + \beta_{10} X_1 Z_1 + \beta_{11} X_2 Z_1 + \beta_{12} X_1 Z_2 + \beta_{13} X_2 Z_2 + \beta_{14} X_1 Z_3 \\
& + \beta_{15} X_2 Z_3 + \beta_{16} X_1 Q + \beta_{17} X_2 Q + \beta_{18} X_1 Z_1 Q + \beta_{19} X_2 Z_1 Q \\
& + \beta_{20} X_1 Z_2 Q + \beta_{21} X_2 Z_2 Q + \beta_{22} X_1 Z_3 Q + \beta_{23} X_2 Z_3 Q + E
\end{aligned}
\tag{14.13}
$$

in which the dummy variables are defined as

$$
Z_1 = \begin{cases} 1 & \text{if SC2 member} \\ 0 & \text{otherwise} \end{cases}
\qquad
Z_2 = \begin{cases} 1 & \text{if SC3 member} \\ 0 & \text{otherwise} \end{cases}
$$

$$
Z_3 = \begin{cases} 1 & \text{if SC4 member} \\ 0 & \text{otherwise} \end{cases}
\qquad
Q = \begin{cases} 1 & \text{if subject is male} \\ 0 & \text{if subject is female} \end{cases}
$$

For each SEX–SC combination, we have

SC1–male: $\quad Y = (\beta_0 + \beta_6) + (\beta_1 + \beta_{16})X_1 + (\beta_2 + \beta_{17})X_2 + E$

SC2–male: $\quad Y = (\beta_0 + \beta_3 + \beta_6 + \beta_7) + (\beta_1 + \beta_{10} + \beta_{16} + \beta_{18})X_1$
$\qquad\qquad\quad + (\beta_2 + \beta_{11} + \beta_{17} + \beta_{19})X_2 + E$

SC3–male: $\quad Y = (\beta_0 + \beta_4 + \beta_6 + \beta_8) + (\beta_1 + \beta_{12} + \beta_{16} + \beta_{20})X_1$
$\qquad\qquad\quad + (\beta_2 + \beta_{13} + \beta_{17} + \beta_{21})X_2 + E$

SC4–male: $\quad Y = (\beta_0 + \beta_5 + \beta_6 + \beta_9) + (\beta_1 + \beta_{14} + \beta_{16} + \beta_{22})X_1$
$\qquad\qquad\quad + (\beta_2 + \beta_{15} + \beta_{17} + \beta_{23})X_2 + E$

SC1–female: $\quad Y = \beta_0 + \beta_1 X_1 + \beta_2 X_2 + E$

SC2–female: $\quad Y = (\beta_0 + \beta_3) + (\beta_1 + \beta_{10})X_1 + (\beta_2 + \beta_{11})X_2 + E$

SC3–female: $\quad Y = (\beta_0 + \beta_4) + (\beta_1 + \beta_{12})X_1 + (\beta_2 + \beta_{13})X_2 + E$

SC4–female: $\quad Y = (\beta_0 + \beta_5) + (\beta_1 + \beta_{14})X_1 + (\beta_2 + \beta_{15})X_2 + E$

The model therefore includes SEX × SC interaction, but not WEIGHT × AGE interaction. The following hypotheses concerning the parameters in model (14.13) are of interest.

1. *Male and female regression equations are parallel (controlling for SC).* This is a test of

$$
H_0: \beta_{16} = \beta_{17} = \beta_{18} = \beta_{19} = \beta_{20} = \beta_{21} = \beta_{22} = \beta_{23} = 0
$$

When this null hypothesis is true, the eight equations above reduce to

$$
\begin{cases}
\text{SC1–male:} & Y = (\beta_0 + \beta_6) + \beta_1 X_1 + \beta_2 X_2 + E \\
\text{SC1–female:} & Y = \beta_0 + \beta_1 X_1 + \beta_2 X_2 + E
\end{cases}
$$

$$
\begin{cases}
\text{SC2–male:} & Y = (\beta_0 + \beta_3 + \beta_6 + \beta_7) + (\beta_1 + \beta_{10})X_1 + (\beta_2 + \beta_{11})X_2 + E \\
\text{SC2–female:} & Y = (\beta_0 + \beta_3) + (\beta_1 + \beta_{10})X_1 + (\beta_2 + \beta_{11})X_2 + E
\end{cases}
$$

$$\begin{cases} \text{SC3--male:} & Y = (\beta_0 + \beta_4 + \beta_6 + \beta_8) + (\beta_1 + \beta_{12})X_1 + (\beta_2 + \beta_{13})X_2 + E \\ \text{SC3--female:} & Y = (\beta_0 + \beta_4) + (\beta_1 + \beta_{12})X_1 + (\beta_2 + \beta_{13})X_2 + E \end{cases}$$

$$\begin{cases} \text{SC4--male:} & Y = (\beta_0 + \beta_5 + \beta_6 + \beta_9) + (\beta_1 + \beta_{14})X_1 + (\beta_2 + \beta_{15})X_2 + E \\ \text{SC4--female:} & Y = (\beta_0 + \beta_5) + (\beta_1 + \beta_{14})X_1 + (\beta_2 + \beta_{15})X_2 + E \end{cases}$$

Thus, within any specific social class group, the male and female regression equations are parallel (since they have the same X_1 and X_2 coefficients).

2. *Male and female regression equations are coincident (controlling for SC).* This is a test of

$$H_0: \beta_6 = \beta_7 = \beta_8 = \beta_9 = \beta_{16} = \beta_{17} = \beta_{18} = \beta_{19} = \beta_{20} = \beta_{21} = \beta_{22} = \beta_{23} = 0$$

3. *All four social class equations are parallel (controlling for SEX).* This is a test of

$$H_0: \beta_{10} = \beta_{11} = \beta_{12} = \beta_{13} = \beta_{14} = \beta_{15} = \beta_{18} = \beta_{19} = \beta_{20} = \beta_{21} = \beta_{22} = \beta_{23} = 0$$

When this hypothesis is true, the eight equations reduce to

$$\begin{cases} \text{SC1--male:} & Y = (\beta_0 + \beta_6) + (\beta_1 + \beta_{16})X_1 + (\beta_2 + \beta_{17})X_2 + E \\ \text{SC2--male:} & Y = (\beta_0 + \beta_3 + \beta_6 + \beta_7) + (\beta_1 + \beta_{16})X_1 + (\beta_2 + \beta_{17})X_2 + E \\ \text{SC3--male:} & Y = (\beta_0 + \beta_4 + \beta_6 + \beta_8) + (\beta_1 + \beta_{16})X_1 + (\beta_2 + \beta_{17})X_2 + E \\ \text{SC4--male:} & Y = (\beta_0 + \beta_5 + \beta_6 + \beta_9) + (\beta_1 + \beta_{16})X_1 + (\beta_2 + \beta_{17})X_2 + E \end{cases}$$

$$\begin{cases} \text{SC1--female:} & Y = \beta_0 + \beta_1 X_1 + \beta_2 X_2 + E \\ \text{SC2--female:} & Y = (\beta_0 + \beta_3) + \beta_1 X_1 + \beta_2 X_2 + E \\ \text{SC3--female:} & Y = (\beta_0 + \beta_4) + \beta_1 X_1 + \beta_2 X_2 + E \\ \text{SC4--female:} & Y = (\beta_0 + \beta_5) + \beta_1 X_1 + \beta_2 X_2 + E \end{cases}$$

Thus, within any given sex group, all four regression equations have the same coefficients for X_1 and X_2.

4. *All four social class equations are coincident (controlling for SEX).* This is a test of

$$H_0: \beta_3 = \beta_4 = \beta_5 = \beta_7 = \beta_8 = \beta_9 = \beta_{10} = \beta_{11} = \beta_{12} = \beta_{13} = \beta_{14} = \beta_{15}$$
$$= \beta_{18} = \beta_{19} = \beta_{20} = \beta_{21} = \beta_{22} = \beta_{23} = 0$$

5. *All eight regression equations are parallel.* This is a test of

$$H_0: \text{``}\beta_{10} \text{ through } \beta_{23} \text{ are simultaneously 0.''}$$

When this hypothesis is true, the eight equations are all of the form

$$Y = \beta_{0(i)} + \beta_1 X_1 + \beta_2 X_2 + E \qquad \text{for } i = 1, 2, \dots, 8$$

(i.e., the eight models differ only in intercept).

6. *All eight regression equations are coincident.* This is a test of

$$H_0: \text{``}\beta_3 \text{ through } \beta_{23} \text{ are simultaneously 0.''}$$

7. *There is no interaction effect between SEX and SC.* This is a test of

$$H_0: \beta_7 = \beta_8 = \beta_9 = \beta_{18} = \beta_{19} = \beta_{20} = \beta_{21} = \beta_{22} = \beta_{23} = 0$$

From the form of the complete model (14.13), each of the coefficients in H_0 corresponds to a product term of the general form Z_iQ or X_iZ_iQ, which involves the product of a social class variable and the sex variable.

Problems

1. Using the data from Problem 2 in Chapter 5 and/or the SAS output given here, answer the following questions about the separate straight-line regressions of SBP on QUET for smokers (SMK = 1) and nonsmokers (SMK = 0).

   a. Determine the least-squares line of SBP (Y) on QUET (X) separately for smokers and nonsmokers.

   b. Test H_0: "The slopes are the same for the populations of smokers and nonsmokers being sampled," versus H_A: "Nonsmokers have a more positive slope."

   c. Test H_0: "The intercepts are the same for the populations of smokers and nonsmokers being sampled," versus H_A: "The intercepts are different."

   d. Test H_0: "The straight lines coincide for the populations of smokers and nonsmokers being sampled," versus H_A: "The straight lines do not coincide."

Edited SAS Output (PROC REG) for Problem 1

```
            SBP Regressed on QUET for Nonsmokers
                  Descriptive Statistics
Variables            Sum                Mean          Uncorrected SS
QUET              52.174          3.4782666667          183.93641
SBP                 2112               140.8               299700

       Variables                Variance          Std Deviation
       QUET              0.1758089238          0.4192957474
       SBP             166.45714286           12.901827113

                 Analysis of Variance
                       Sum of          Mean
Source       DF        Squares         Square     F Value    Prob > F
Model         1      1702.83966     1702.83966     35.275     0.0001
Error        13       627.56034       48.27387
C Total      14      2330.40000

         Root MSE         6.94794      R-square      0.7307
         Dep Mean       140.80000      Adj R-sq      0.7100
         C.V.             4.93462

                 Parameter Estimates
                    Parameter       Standard      T for H0:        Prob
Variable     DF      Estimate          Error     Parameter=0      > |T|
INTERCEP      1     49.311759     15.50814374        3.180        0.0072
QUET          1     26.302825      4.42865240        5.939        0.0001
```

```
              SBP Regressed on QUET for Smokers
                   Descriptive Statistics
Variables            Sum              Mean        Uncorrected SS
QUET              57.941       3.4082941176         202.639313
SBP                 2513        147.82352941             375183

        Variables          Variance        Std Deviation
          QUET          0.3224589706        0.567854709
          SBP           231.40441176        15.211982506

                   Analysis of Variance
                     Sum of           Mean
Source        DF     Squares         Square     F Value    Prob > F
Model          1   2088.16977     2088.16977     19.403      0.0005
Error         15   1614.30082      107.62005
C Total       16   3702.47059

        Root MSE        10.37401       R-square       0.5640
        Dep Mean       147.82353       Adj R-sq       0.5349
        C.V.             7.01783

                   Parameter Estimates
                   Parameter       Standard      T for H0:       Prob
Variable      DF    Estimate         Error     Parameter=0      > |T|
INTERCEP       1   79.255330     15.76836893        5.026       0.0002
QUET           1   20.118041      4.56719317        4.405       0.0005
```

2. A topic of major concern to demographers and economists is the effect of a high fertility rate on per capita income. The first two accompanying tables display values of per capita income (PCI) and population percentage under age 15 (YNG) for a hypothetical sample of developing countries in Latin America and Africa, respectively. The third table summarizes the results of straight-line regressions of PCI (Y) on YNG (X) for each group of countries.

a.–d. Repeat parts (a) through (d) of Problem 1 for the straight-line regressions of PCI (Y) on YNG (X) for Latin American and African countries.

Latin American Countries

YNG (X)	PCI (Y)	YNG (X)	PCI (Y)	YNG (X)	PCI (Y)
32.2	788	44.0	292	35.0	685
47.0	202	44.0	321	47.4	220
34.0	825	43.0	300	48.0	195
36.0	675	43.0	323	37.0	605
38.7	590	40.0	484	38.4	530
40.9	408	37.0	625	40.6	480
45.0	324	39.0	525	35.8	690
45.4	235	44.6	340	36.0	685
42.2	338	33.0	765		

African Countries

YNG (X)	PCI (Y)	YNG (X)	PCI (Y)	YNG (X)	PCI (Y)
34.0	317	41.0	188	39.0	225
36.0	270	42.0	166	39.0	232
38.2	208	45.0	132	37.0	260
43.0	150	36.0	290	37.0	250
44.0	105	42.6	160	46.0	92
44.0	128	33.0	300	45.6	110
46.0	85	33.0	320	42.0	180
48.0	75	47.0	85	38.8	235
40.0	210	47.0	75		

Summary of Separate Straight-Line Fits

Location	n	$\hat{\beta}_0$	$\hat{\beta}_1$	$\overline{X}$	$\overline{Y}$	S_X^2	$S_{Y|X}^2$	r
Latin America	26	2170.67	−42.0	40.277	478.846	21.633	1391.756	−.983
Africa	26	897.519	−17.39	40.892	186.462	20.244	188.919	−.985

e. Test H_0: "The population correlation coefficients are equal for the two groups of countries under study." Use $\alpha = .05$. Does your conclusion here clash with your findings regarding the equality of slopes? [*Hint:* See section 6-7.]

For each of the preceding tests, assume that the alternative hypothesis is two-sided.

3. A team of anthropologists and nutrition experts investigated the influence of protein content in diet on the relationship between AGE and height (HT) for New Guinean children. The two accompanying tables display values of HT (in centimeters) and AGE for a hypothetical sample of children with protein-rich and protein-poor diets, respectively.

Protein-Rich Diet

AGE (X)	0.2	0.5	0.8	1.0	1.0	1.4	1.8	2.0	2.0	2.5	2.5	3.0	2.7
HT (Y)	54	54.3	63	66	69	73	82	83	80.3	91	93.2	94	94

Protein-Poor Diet

AGE (X)	0.4	0.7	1.0	1.0	1.5	2.0	2.0	2.4	2.8	3.0	1.3	1.8	0.2	3.0
HT (Y)	52	55	61	63.4	66	68.5	67.9	72	76	74	65	69	51	77

Summary of Separate Straight-line Fits

Diet	n	$\hat{\beta}_0$	$\hat{\beta}_1$	$\overline{X}$	$\overline{Y}$	S_X^2	$S_{Y|X}^2$	r
Protein-rich	13	50.324	16.009	1.646	76.677	0.808	5.841	.937
Protein-poor	14	51.225	8.686	1.650	65.557	0.873	4.598	.969

a.–d. Repeat parts (a) through (d) of Problem 1 for the straight-line regressions of HT (Y) on AGE (X) for the two diets. (Consider a two-sided alternative in each case.)

e. Test whether the population correlation coefficient for children with a protein-rich diet differs significantly from that for children with a protein-poor diet. (Consider a two-sided alternative.)

4. For the data involving regression of DI (Y) on IQ (X) in Problem 4 in Chapter 5, assume that the sample of 17 observations (with the outlier removed) consists of males only. Now suppose that another sample of observations on DI (Y) and IQ (X) has been obtained for 14 females. The information needed to compare the straight-line regression equations for males and females is given in the following table.

Sex Group	n	$\hat{\beta}_0$	$\hat{\beta}_1$	$\overline{X}$	$\overline{Y}$	S_X^2	$S_{Y|X}^2$	r
Males	17	70.846	−0.444	101.411	25.812	215.882	24.335	−.807
Females	19	61.871	−0.438	101.053	17.579	175.497	16.692	−.825

a.–d. Repeat parts (a) through (d) of Problem 1 for the straight-line regressions of DI (Y) on IQ (X) for the two sexes. (Consider two-sided alternatives.)

e. Perform a two-sided test of whether the population correlation coefficients for males and females are equal.

5. Assume that the data involving the regression of VOTE (Y) on TVEXP (X) in Problem 5 in Chapter 5 came from congressional districts in New York. Now, suppose that researchers selected a second sample of 17 congressional districts in California and recorded the same information for it. The following table provides the information needed to compare the straight-line regression equations for New York and California.

Location	n	$\hat{\beta}_0$	$\hat{\beta}_1$	$\overline{X}$	$\overline{Y}$	S_X^2	$S_{Y|X}^2$	r
New York	20	2.174	1.177	36.99	45.71	76.870	11.101	.954
California	17	8.030	1.036	36.371	45.706	97.335	13.492	.945

a.–d. Repeat parts (a) through (d) of Problem 1 for the straight-line regression of VOTE (Y) on TVEXP (X) for the two states. Consider the one-sided alternative H_A: $\beta_{1(CAL)} < \beta_{1(NY)}$ for the test for slope, and consider the one-sided alternative H_A: $\beta_{0(CAL)} < \beta_{0(NY)}$ for the test for intercept.

e. Perform a two-sided test of whether the correlation coefficients for New York and California are equal.

6. The data in the following table represent four-week growth rates for depleted chicks at different dosage levels of vitamin B, by sex.[10]

---

[10]Adapted from a study by Clark, Lechyeka, and Cook (1940).

Males		Females	
Growth Rate (Y)	Log$_{10}$ Dose (X)	Growth Rate (Y)	Log$_{10}$ Dose (X)
17.1	0.301	18.5	0.301
14.3	0.301	22.1	0.301
21.6	0.301	15.3	0.301
24.5	0.602	23.6	0.602
20.6	0.602	26.9	0.602
23.8	0.602	20.2	0.602
27.7	0.903	24.3	0.903
31.0	0.903	27.1	0.903
29.4	0.903	30.1	0.903
30.1	1.204	28.1	0.903
28.6	1.204	30.3	1.204
34.2	1.204	33.0	1.204
37.3	1.204	35.8	1.204
33.3	1.505	32.6	1.505
31.8	1.505	36.1	1.505
40.2	1.505	30.5	1.505

Use the information provided in the next table to answer the following questions.

a. Determine the dose–response straight lines separately for each sex, and plot them on the same graph.

b. Test whether the slopes for males and females differ.

c. Find a 99% confidence interval for the true difference between the male and female slopes.

d. Test for coincidence of the two straight lines.

Sex Group	n	$\hat{\beta}_0$	$\hat{\beta}_1$	$\bar{X}$	$\bar{Y}$	S_X^2	$S_{Y|X}^2$	r
Males	16	14.178	14.825	0.922	27.84	0.187	10.5553	.807
Females	16	15.656	12.735	0.903	27.16	0.1812	8.4719	.788

7. The results in the first of the following tables were obtained in a study of the amount of energy metabolized by two similar species of birds under constant temperature.[11] Information based on separate straight-line fits to each data set is summarized in the second table.

a. Plot the least-squares straight lines for each species on the same graph.

b. Test whether the two lines are parallel.

c. Test whether the two lines have the same intercept.

d. Give a 95% confidence interval for the true difference between the mean amounts of energy metabolized by each species at 15°C. [*Hint:* Use a confidence interval of the form

$$\left(\hat{Y}_{15}^A - \hat{Y}_{15}^B\right) \pm t_{n_A+n_B-4,\,1-\alpha/2}\sqrt{S_{\hat{Y}_{15}^A}^2 + S_{\hat{Y}_{15}^B}^2}$$

---

[11]Adapted from a study by Davis (1955).

	Species A		Species B	
	Calories (Y)	Temperature (X) (°C)	Calories (Y)	Temperature (X) (°C)
	36.9	0	41.1	0
	35.8	2	40.6	2
	34.6	4	38.9	4
	34.3	6	37.9	6
	32.8	8	37.0	8
	31.7	10	36.1	10
	31.0	12	36.3	12
	29.8	14	34.2	14
	29.1	16	33.4	16
	28.2	18	32.8	18
	27.4	20	32.0	20
	27.8	22	31.9	22
	25.5	24	30.7	24
	24.9	26	29.5	26
	23.7	28	28.5	28
	23.1	30	27.7	30

Species	n	$\hat{\beta}_0$	$\hat{\beta}_1$	$\overline{X}$	$\overline{Y}$	S_X^2	$S_{Y|X}^2$	r
A	16	36.579	−0.4528	15.00	29.79	90.6667	0.1662	−.9959
B	16	40.839	−0.4368	15.00	34.29	90.6667	0.1757	−.9953

where $\hat{Y}_{15}^i$ is the predicted value at 15°C for species i (i = A, B), and $S_{\hat{Y}_{15}^i}^2$ is the estimated variance of $\hat{Y}_{15}^i$ given by the general formula (for $X = X_0$)

$$S_{\hat{Y}_{X_0}^i}^2 = S_{P,Y|X}^2 \left[\frac{1}{n_i} + \frac{(X_0 - \overline{X})^2}{(n_i - 1)S_{X_i}^2} \right]$$

in your calculations.]

8. In Problem 1, separate straight-line regressions of SBP on QUET were compared for smokers (SMK = 1) and nonsmokers (SMK = 0).

a. Define a single multiple regression model that uses the data for both smokers and non-smokers and that defines straight-line models for each group with possibly differing intercepts and slopes. Define the intercept and slope for each straight-line model in terms of the regression coefficients of the single regression model.

b. Using the accompanying computer results, determine and plot on graph paper the two fit-ted straight lines obtained from the fit of the regression model.

c. Test the following null hypotheses:

H_0: "The two lines are parallel."

H_0: "The two lines are coincident."

For each of these tests, state the appropriate null hypothesis in terms of the regression coefficients of the regression model.

d. Compare your answers in parts (b) and (c) with those you obtained by fitting separate re-gressions in Problem 1.

Edited SAS Output (PROC REG) for Problem 8

```
          Regression of SBP on QUET, SMK, and QUET × SMK
                      Analysis of Variance
                      Sum of           Mean
Source       DF       Squares         Square      F Value    Prob > F
Model         3     4184.10759     1394.70253     17.419      0.0001
Error        28     2241.86116       80.06647
C Total      31     6425.96875

          Root MSE        8.94799      R-square       0.6511
          Dep Mean      144.53125     Adj R-sq       0.6137
          C.V.            6.19104

                      Parameter Estimates
                      Parameter      Standard      T for H0:        Prob
Variable     DF       Estimate        Error      Parameter=0       > |T|
INTERCEP      1      49.311759     19.97234635      2.469          0.0199
QUET          1      26.302825      5.70349238      4.612          0.0001
SMK           1      29.943571     24.16355268      1.239          0.2256
QUETSMK       1      -6.184785      6.93170667     -0.892          0.3799

Variable     DF       Type I SS
INTERCEP      1         668457
QUET          1     3537.945739
SMK           1      582.420754
QUETSMK       1       63.741095
```

9. Use the computer results shown next to compare the separate regressions of SBP on AGE and QUET for smokers and nonsmokers (based on the data from Problem 2 in Chapter 5), as follows.
   a. State the appropriate regression model, incorporating both equations for smokers and nonsmokers.
   b. Determine the fitted regression equations for smokers and nonsmokers separately, using the fitted regression model.
   c. Test for parallelism of the two models, stating the null hypothesis in terms of appropriate regression coefficients.
   d. Test for coincidence of the two models, stating the null hypothesis in terms of regression coefficients.

Edited SAS Output (PROC REG) for Problem 9

```
          SBP Regressed on AGE, QUET, SMK, AGESMK, and QUETSMK
                      Analysis of Variance
                      Sum of           Mean
Source       DF       Squares         Square      F Value    Prob > F
Model         5     4915.63040      983.12608     16.924      0.0001
Error        26     1510.33835       58.08994
C Total      31     6425.96875
```

```
SBP Regressed on AGE, QUET, SMK, AGESMK, and QUETSMK (Continued)
                    Analysis of Variance (Continued)
         Root MSE        7.62168     R-square      0.7650
         Dep Mean      144.53125     Adj R-sq      0.7198
         C.V.            5.27338
                         Parameter Estimates
                     Parameter      Standard     T for H0:        Prob
    Variable   DF    Estimate         Error    Parameter=0       > |T|
    INTERCEP    1    48.612699    17.01536941        2.857       0.0083
    AGE         1     1.028915     0.50176589        2.051       0.0505
    QUET        1    10.451041     9.13014561        1.145       0.2628
    SMK         1    -0.537436    23.23003664       -0.023       0.9817
    AGESMK      1     0.437325     0.71279040        0.614       0.5449
    QUETSMK     1    -3.706822    10.76763124       -0.344       0.7334

    Variable   DF    Type I SS
    INTERCEP    1       668457
    AGE         1  3861.630375
    QUET        1   258.961870
    SMK         1   769.233452
    AGESMK      1    18.920333
    QUETSMK     1     6.884366
```

10. The results presented in the following tables are based on data from a study by Gruber (1970) to determine how and to what extent changes in blood pressure over time depend on initial blood pressure (at the beginning of the study), the sex of the individual, and the relative weight of the individual. Data were collected on $n = 104$ persons, for which the $k = 7$ independent variables were examined by multiple regression. The following variables were used in the study:

Y = SBPSL (estimated slope based on the straight-line regression of an individual's blood pressure over time)[12]

X_1 = SBP1 (initial systolic blood pressure)

X_2 = SEX (male = 1, female = -1)

X_3 = RW (relative weight)

$X_4 = X_1 X_2, \qquad X_5 = X_1 X_3$

$X_6 = X_2 X_3, \qquad X_7 = X_1 X_2 X_3$

Use only the ANOVA table to answer the following questions.

a. Determine the form of the separate fitted regression models of SBPSL (Y) on SBP1 (X_1), RW (X_3), and SBP1 $\times$ RW (X_5) for each sex in terms of the regression coefficients of the fitted regression model.

---

[12]This choice of dependent variable can be criticized because observations of an individual's blood pressure taken over time are *not* independent, thus violating Assumption 2 in Chapter 8. A statistic preferable to the one Gruber used could be obtained by weighted regression (see Chapter 12) or by growth curve analysis (see Allen and Grizzle 1969).

Variable	$\hat{\beta}$	$S_{\hat{\beta}}$	Partial F $(\hat{\beta}^2/S_{\hat{\beta}}^2)$
X_1(SBP1)	-0.045	0.00762	34.987
X_2(SEX)	0.695	0.86644	0.643
X_3(RW)	0.027	0.07049	0.149
$X_4(X_1X_2)$	-0.0029	0.00762	0.145
$X_5(X_1X_3)$	-0.00018	0.00062	0.084
$X_6(X_2X_3)$	-0.0092	0.07049	0.017
$X_7(X_1X_2X_3)$	0.00022	0.00062	0.125
(Intercept)	4.667		

Source		d.f.	SS	MS	F
Overall regression		7	37.148	5.307	$\dfrac{5.307}{76.246/96} = 6.68^{**}$
Regression	X_1	1	24.988	24.988	$\dfrac{24.988}{88.406/102} = 28.83^{**}$
	$X_2 \mid X_1$	1	7.886	7.886	$\dfrac{7.886}{80.520/101} = 9.89^{**}$
	$X_3 \mid X_1, X_2$	1	1.057	1.057	$\dfrac{1.057}{79.463/100} = 1.33$
	$X_4 \mid X_1, X_2, X_3$	1	0.020	0.020	$\dfrac{0.020}{79.443/99} = 0.025$
	$X_5 \mid X_1, X_2, X_3, X_4$	1	0.254	0.254	$\dfrac{0.254}{79.189/98} = 0.314$
	$X_6 \mid X_1, X_2, X_3, X_4, X_5$	1	2.844	2.844	$\dfrac{2.844}{76.345/97} = 3.613$
	$X_7 \mid X_1, X_2, X_3, X_4, X_5, X_6$	1	0.099	0.099	$\dfrac{0.099}{76.246/96} = 0.125$
Residual		96	76.246	0.794	
Total (corrected)		103	113.394		

Note: $F_{7,96,0.95} = 2.11$, $F_{1,96,0.95} = 3.95$, $F_{1,102,0.95} = 3.94$.

b. Why can't you use this ANOVA table to test whether the two regression equations are either parallel or coincident? Describe the appropriate testing procedure in each case.

11. For the data from Problem 2 in this chapter, address the following questions, using the information provided in the accompanying SAS output.

    a. State the regression model that incorporates the straight-line models for each group of countries.

    b. Determine and plot the separate fitted straight lines, based on the fitted regression model given in part (a).

    c. Test for parallelism of the two straight lines.

    d. Test for coincidence of the two straight lines.

    e. Compare your results in parts (b) through (d) to those obtained in Problem 2.

Edited SAS Output (PROC REG) for Problem 11

```
                  Regression of PCI on YNG, Z, and YNG x Z
                         Analysis of Variance
                          Sum of             Mean
Source        DF         Squares            Square      F Value    Prob > F
Model          3      2218613.5581      739537.85268    935.724     0.0001
Error         48        37936.21118        790.33773
C Total       51      2256549.7692

          Root MSE        28.11295      R-square       0.9832
          Dep Mean       332.65385      Adj R-sq       0.9821
          C.V.             8.45111

                         Parameter Estimates
                         Parameter      Standard      T for H0:        Prob
Variable      DF         Estimate         Error      Parameter=0      > |T|
INTERCEP       1       897.518610     51.39769142       17.462        0.0001
YNG            1       -17.388529      1.24965131      -13.915        0.0001
Z              1      1273.151485     71.01247197       17.929        0.0001
YNGZ           1       -24.616267      1.73867221      -14.158        0.0001

Variable      DF       Type I SS
INTERCEP       1        5754246
YNG            1        1089772
Z              1         970417
YNGZ           1         158424
```

12. Using the information given here, answer the same questions as in Problem 11 about the regression of height (Y) on age (X) for children in one of two diet categories. (This problem is based on the data in Problem 3.)

Edited SAS Output (PROC REG) for Problem 12

```
             Height Regressed on Age, Z, and Age x Z Where Z = 1
                       if Protein Rich, 0 Otherwise
                         Analysis of Variance
                          Sum of             Mean
Source        DF         Squares            Square      F Value    Prob > F
Model          3      4174.20665        1391.40222      267.981     0.0001
Error         23       119.42001           5.19217
C Total       26      4293.62667

          Root MSE         2.27863      R-square       0.9722
          Dep Mean        70.91111      Adj R-sq       0.9686
          C.V.             3.21337

                         Parameter Estimates
                         Parameter      Standard      T for H0:        Prob
Variable      DF         Estimate         Error      Parameter=0      > |T|
INTERCEP       1        51.225175      1.27112426       40.299        0.0001
AGE            1         8.686041      0.67620919       12.845        0.0001
Z              1        -0.901476      1.86193670       -0.484        0.6329
AGEZ           1         7.322927      0.99647349        7.349        0.0001
```

(continued)

```
        Height Regressed on Age, Z, and Age x Z Where Z = 1
             if Protein Rich, 0 Otherwise (Continued)
                 Parameter Estimates (Continued)
Variable     DF       Type I SS
INTERCEP     1            135766
AGE          1       3053.348351
Z            1        840.452385
AGEZ         1        280.405918
```

13. In Gruber's (1970) study of $n = 104$ individuals (discussed in Problem 10), the relationship between blood pressure change (SBPSL) and relative weight (RW), controlling for initial blood pressure (SBP1), was compared for three different geographical backgrounds and for three different psychosocial orientations, using the following 15 variables:

$$Y = \text{SBPSL}$$
$$X_1 = \text{SBP1 (initial blood pressure)}$$
$$X_2 = \text{R (1 if rural background, 0 if town, } -1 \text{ if urban)}$$
$$X_3 = \text{T (1 if town background, 0 if rural, } -1 \text{ if urban)}$$
$$X_4 = \text{TD (1 if traditional orientation, 0 if transitional, } -1 \text{ if modern)}$$
$$X_5 = \text{TN (1 if transitional orientation, 0 if traditional, } -1 \text{ if modern)}$$
$$X_6 = \text{RW (relative weight)}$$
$$X_7 = \text{T} \times \text{TD}$$
$$X_8 = \text{T} \times \text{TN}$$
$$X_9 = \text{R} \times \text{TD}$$
$$X_{10} = \text{R} \times \text{TN}$$
$$X_{11} = \text{R} \times \text{TD} \times \text{RW}$$
$$X_{12} = \text{R} \times \text{TN} \times \text{RW}$$
$$X_{13} = \text{T} \times \text{TD} \times \text{RW}$$
$$X_{14} = \text{T} \times \text{TN} \times \text{RW}$$

A standard stepwise-regression program was run using these data, yielding the following ANOVA table (variables were forced to enter in the order presented) based on the model

$$Y = \beta_0 + \beta_1 X_1 + \beta_2 X_2 + \cdots + \beta_{14} X_{14} + E$$

a. Using this regression model, determine the form of the nine fitted regression equations corresponding to the nine possible combinations of background with orientation (i.e., $R = 1$ and $TD = 1$, $R = 0$ and $TD = 1$, $R = -1$ and $TD = 1$, etc.). [*Note:* Each of the nine equations will be of the form $\hat{Y} = \hat{\beta}_0^* + \hat{\beta}_1^*(\text{SBP1}) + \hat{\beta}_2^*(\text{RW})$.]

b. Test the null hypothesis that the nine regression equations determined in part (a) are parallel. State the null hypothesis in terms of the regression coefficients of the original 14-variable regression model.

c. Test the hypothesis H_0: "The three regression equations corresponding to the three backgrounds (rural, town, and urban) are parallel (but not necessarily coincident)," against the alternative H_A: "They are not parallel."

d. Set up the formula for testing H_0: "The nine regression equations dealt with in part (a) are coincident," against H_A: "They are not coincident." State the null hypothesis in terms of the coefficients in the regression equation.

Source		d.f.	SS	F
	X_1	1	24.9878	28.830
	$X_2 \mid X_1$	1	0.5218	0.600
	$X_3 \mid X_1, X_2$	1	0.0057	0.006
	$X_4 \mid X_1, X_2, X_3$	1	1.0520	1.199
	$X_5 \mid X_1, X_2, X_3, X_4$	1	1.1116	1.271
	$X_6 \mid X_1{-}X_5$	1	0.8321	0.951
Regression	$X_7 \mid X_1{-}X_6$	1	0.2919	0.331
	$X_8 \mid X_1{-}X_7$	1	1.6601	1.902
	$X_9 \mid X_1{-}X_8$	1	0.5843	0.667
	$X_{10} \mid X_1{-}X_9$	1	0.2266	0.257
	$X_{11} \mid X_1{-}X_{10}$	1	1.1916	1.355
	$X_{12} \mid X_1{-}X_{11}$	1	2.0853	2.407
	$X_{13} \mid X_1{-}X_{12}$	1	1.5915	1.854
	$X_{14} \mid X_1{-}X_{13}$	1	0.0208	0.024
Residual		89	77.2303	
Total		103	113.3934	

14. The Environmental Protection Agency conducted an experiment to assess the characteristics of sampling procedures designed to measure the suspended particulate concentration (X) in a particular city. At each of two distinct locations (designated as location 1 and location 2), two identical sampling units were set up side by side, and readings were taken on each of 10 days. The data are given here in tabular form, where X_{ij1} and X_{ij2} are, respectively, the measured concentration for samplers 1 and 2 at location i on day j ($i = 1, 2$; $j = 1, 2, \ldots, 10$). Researchers hypothesized that the inherent variation in the observations depends on the level of concentration being measured. To quantify this hypothesis, they proposed to fit a model of the form

$$\left| d_{ij} \right| = \beta_0 + \beta_1 Z + \beta_2 (X_{ij1} + X_{ij2}) + \beta_3 (X_{ij1} + X_{ij2})Z + E$$

where Z is 1 if the observation pertains to location 1 and is 0 otherwise, and where $\left| d_{ij} \right| = \left| X_{ij1} - X_{ij2} \right|$.

a. Using the results provided in the following table, determine and interpret the fitted straight-line relationship between $\left| d_{ij} \right|$ and $(X_{ij1} + X_{ij2})$ at each location.

	Location 1			Location 2		
Day	X_{1j1}	X_{1j2}	$d_{1j} = X_{1j1} - X_{1j2}$	X_{2j1}	X_{2j2}	$d_{2j} = X_{2j1} - X_{2j2}$
1	4	3	1	6	5	1
2	8	6	2	3	1	2
3	12	16	−4	1	2	−1
4	1	1	0	10	12	−2
5	7	6	1	17	17	0
6	11	8	3	4	7	−3
7	14	10	4	8	6	2
8	10	12	−2	12	12	0
9	2	2	0	10	9	1
10	15	20	−5	20	19	1

    b. Test for parallelism of the two lines.

    c. Test for coincidence of the two lines.

    d. How would you test whether level of concentration is significantly related to inherent variation in the observations for at least one of the two locations?

Edited SAS Output (PROC REG) for Problem 14

```
       ABS(d) Regressed on X (=X1+X2), Z, and X x Z Where Z = 1 if
                       Location 1, 0 Otherwise
                       Analysis of Variance
                       Sum of          Mean
Source        DF       Squares        Square        F Value      Prob > F
Model          3      31.33952       10.44651       19.873       0.0001
Error         16       8.41048        0.52565
C Total       19      39.75000

            Root MSE      0.72502      R-square      0.7884
            Dep Mean      1.75000      Adj R-sq      0.7487
            C.V.         41.42974

                       Parameter Estimates
                       Parameter      Standard      T for H0:        Prob
Variable      DF       Estimate         Error      Parameter=0      > |T|
INTERCEP       1       2.008561      0.43196210       4.650        0.0003
X              1      -0.039147      0.02022626      -1.935        0.0708
Z              1      -2.427916      0.61773149      -3.930        0.0012
XZ             1       0.195061      0.03022850       6.453        0.0001

Variable      DF       Type I SS
INTERCEP       1      61.250000
X              1       4.834820
Z              1       4.616603
XZ             1      21.888102
```

    15. A biologist compared the effect of temperature of each of two media on the growth of human amniotic cells in a tissue culture. The data shown in the following table were obtained.

Medium A				Medium B			
No. of Cells $\times 10^{-6}$ (Y)	Temperature (°F) (X)	No. of Cells $\times 10^{-6}$ (Y)	Temperature (°F) (X)	No. of Cells $\times 10^{-6}$ (Y)	Temperature (°F) (X)	No. of Cells $\times 10^{-6}$ (Y)	Temperature (°F) (X)
1.13	40	2.30	80	0.98	40	2.20	80
1.20	40	2.15	80	1.05	40	2.10	80
1.00	40	2.25	80	0.92	40	2.20	80
0.91	40	2.40	80	0.90	40	2.30	80
1.05	40	2.49	80	0.89	40	2.26	80
1.75	60	3.18	100	1.60	60	3.10	100
1.45	60	3.10	100	1.45	60	3.00	100
1.55	60	3.28	100	1.40	60	3.13	100
1.64	60	3.35	100	1.50	60	3.20	100
1.60	60	3.12	100	1.56	60	3.07	100

a. Assuming that a parabolic model is appropriate for describing the relationship between Y and X for each medium, provide a single regression model that incorporates two separate parabolic models, one corresponding to each medium.

b. Use the SAS output provided here to determine and plot the separate fitted parabolas for each medium. [*Note:* $Z = 0$ for medium A and $Z = 1$ for medium B.]

c. Test for "parallelism" of the two parabolas.

d. Test for coincidence of the two parabolas.

e. Is it possible to test whether a quadratic term should be included in the model for each medium, using only the SAS output? Explain.

Edited SAS Output (PROC REG) for Problem 15

```
        Y Regressed on X, X2, Z, ZX, and ZX2 Where Z = 1
                  if Medium B, 0 Otherwise
                   Analysis of Variance

                    Sum of         Mean
Source      DF      Squares        Square     F Value     Prob > F
Model        5     26.06864       5.21373     583.965      0.0001
Error       34      0.30356       0.00893
C Total     39     26.37220

          Root MSE      0.09449      R-square      0.9885
          Dep Mean      1.99275      Adj R-sq      0.9868
          C.V.          4.74163

                   Parameter Estimates

                    Parameter      Standard     T for H0:        Prob
Variable    DF      Estimate       Error        Parameter=0      > |T|
INTERCEP     1      0.494600       0.24256234       2.039        0.0493
X            1      0.005370       0.00745505       0.720        0.4763
X2           1      0.000217       0.00005282       4.118        0.0002
Z            1     -0.143700       0.34303495      -0.419        0.6779
ZX           1      0.001235       0.01054303       0.117        0.9074
ZX2          1     -0.000008750    0.00007470      -0.117        0.9074

Variable    DF      Type I SS
INTERCEP     1     158.842103
X            1      25.668613
X2           1       0.290702
Z            1       0.109202
ZX           1       0.000000500
ZX2          1       0.000122
```

16. Answer the following questions, using the accompanying SPSS output, which is based on data on the growth rates (Y) of depleted chicks at different (log) dosage levels (X) of vitamin B for males and females.

a. Define a single multiple regression model that incorporates different straight-line models for males and females.

b. Plot the fitted straight lines for each sex on graph paper.

c. Test for parallelism.

d. Test for coincidence.

Edited SPSS Output for Problem 16

```
VARIABLE(S) ENTERED ON STEP NUMBER 1.. X

MULTIPLE R         0.88583     ANALYSIS OF VARIANCE    DF    SUM OF SQUARES    MEAN SQUARE        F
R SQUARE           0.78470     REGRESSION              1.      1041.33728      1041.33728    109.33735
ADJUSTED R SQUARE  0.77752     RESIDUAL               30.       285.72230         9.52408
STANDARD ERROR     3.08611

-------VARIABLES IN THE EQUATION-------          ------VARIABLES NOT IN THE EQUATION------

VARIABLE       B        BETA    STD ERROR B      F       VARIABLE   BETA IN    PARTIAL    TOLERANCE    F
X          13.85501   0.88583     1.32502    109.337     Z          0.01293    0.02784    0.99791    0.022
(CONSTANT) 14.72828                                      XZ         0.04475    0.08801    0.83267    0.226


VARIABLE(S) ENTERED ON STEP NUMBER 2.. Z (= 1 if MALE, = 0 if FEMALE)

MULTIPLE R         0.88592     ANALYSIS OF VARIANCE    DF    SUM OF SQUARES    MEAN SQUARE        F
R SQUARE           0.78486     REGRESSION              2.      1041.55874       520.77937     52.89862
ADJUSTED R SQUARE  0.77002     RESIDUAL               29.       285.50084         9.84486
STANDARD ERROR     3.13765

-------VARIABLES IN THE EQUATION-------          ------VARIABLES NOT IN THE EQUATION------

VARIABLE       B        BETA    STD ERROR B      F       VARIABLE   BETA IN    PARTIAL    TOLERANCE    F
X          13.84577   0.88524     1.34856    109.413     XZ         0.19139    0.15355    0.13847    0.676
Z          10.16655   0.01293     1.11049      0.022
(CONSTANT) 14.65352


VARIABLE(S) ENTERED ON STEP NUMBER 3.. XZ

MULTIPLE R         0.88678     ANALYSIS OF VARIANCE    DF    SUM OF SQUARES    MEAN SQUARE        F
R SQUARE           0.78639     REGRESSION              3.      1048.28988       349.42996     35.09721
ADJUSTED R SQUARE  0.76743     RESIDUAL               28.       278.76970         9.95606
STANDARD ERROR     3.15532

-------VARIABLES IN THE EQUATION-------          ------VARIABLES NOT IN THE EQUATION------

VARIABLE       B         BETA    STD ERROR B      F      VARIABLE   BETA IN    PARTIAL    TOLERANCE    F
X          12.73533    0.81424    1.91389     44.278
Z          -1.88944   -0.14670    2.73852      0.476
XZ          2.23020    0.19139    2.71233      0.676
(CONSTANT) 15.65624
```

From Nie et al., *Statistical Package for the Social Sciences.* Copyright © 1975 by McGraw-Hill Book Company and Dr. Norman Nie. Used with permission of McGraw-Hill Book Company and Dr. Norman Nie, President, SPSS Inc.

17. Market research was conducted for a national retail company to compare the relationship between sales and advertising during the warm Spring and Summer seasons as compared with the cool Fall and Winter seasons. The data shown in the following table were collected over a period of several years.

Season (Warm = 0, Cool = 1)	Advertising Expenditure ($millions)	Sales Revenue ($millions)	Season (Warm = 0, Cool = 1)	Advertising Expenditure ($millions)	Sales Revenue ($millions)
0	17.0	156.1	1	10.0	131.0
0	12.5	142.6	1	13.8	136.8
0	20.5	166.8	1	15.0	141.5
0	16.0	155.4	1	19.5	151.8
0	15.0	150.5	1	17.0	148.3
0	14.5	147.5	1	12.5	133.3
0	17.5	156.9	1	14.5	138.0
0	12.5	138.8	1	12.5	135.9
0	11.5	134.3	1	12.0	132.0

a. Identify a single regression model that uses the data for both warm and cool seasons and that defines straight-line models relating sales revenue (Y) to advertising expenditure (X) for each season.

b. Using the computer output given next, determine and plot the fitted straight lines for each season.

c. Test whether the straight lines for cool and warm seasons coincide.

d. Test the following hypotheses:

H_0: The lines are parallel.

H_A: The lines are not parallel.

e. In light of your answers to parts (c) and (d), comment on differences and similarities in the sales–advertising expenditure relationship between cooler and warmer seasons.

Edited SAS Output (PROC REG) for Problem 17

```
        Sales Revenue (Y) Regressed on Adv. Expenditure (X),
                      Season (Z), and XZ
                    Descriptive Statistics
Variables           Sum                   Mean          Uncorrected SS
INTERCEP            18                     1                 18
X                   263.8                 14.65555556        4002.94
Z                   9                      0.5               9
XZ                  126.8                  7.0444444444      1851.44
Y                   2597.5                144.3055556       376638.73
```

(continued)

```
          Sales Revenue (Y) Regressed on Adv. Expenditure (X),
                    Season (Z), and XZ (Continued)
                   Descriptive Statistics (Continued)
         Variables                Variance           Std Deviation
         INTERCEP                        0                       0
         X                   8.0473202614           2.8367799106
         Z                   0.2647058824           0.5144957554
         XZ                  56.36496732            7.5076605757
         Y                   106.17937908          10.304337877

                         Analysis of Variance
                          Sum of           Mean
     Source      DF       Squares          Square     F Value    Prob > F
     Model        3      1755.87993       585.29331   166.650     0.0001
     Error       14        49.16951         3.51211
     C Total     17      1805.04944

          Root MSE          1.87406       R-square      0.9728
          Dep Mean        144.30556       Adj R-sq      0.9669
          C.V.              1.29868

                         Parameter Estimates
                        Parameter       Standard       T for H0:        Prob
     Variable    DF     Estimate          Error      Parameter=0       > |T|
     INTERCEP     1     96.830446       3.56515723       27.160       0.0001
     X            1      3.484861       0.23058405       15.113       0.0001
     Z            1      7.172687       4.88169856        1.469       0.1639
     XZ           1     -1.019784       0.32745582       -3.114       0.0076

     Variable    DF     Type I SS
     INTERCEP     1       374834
     X            1     1461.750224
     Z            1      260.067028
     XZ           1       34.062678
```

18. A testing laboratory studies and compares the relationship between tire tread wear per 1,000 miles (Y) and average driving speed (X) for two competing tires. The data shown in the following table were collected for a random sample of 20 tires.

   a. Identify a single regression model that uses the data for both tire types and that defines straight-line models relating tread wear (Y) to average speed (X) for each tire.

Tire	Tread Wear Per 100 Miles of Travel (% of tread thickness)	Average Speed (mph)	Tire	Tread Wear Per 100 Miles of Travel (% of tread thickness)	Average Speed (mph)
A	0.5	65	B	0.7	65
A	0.4	55	B	0.7	64
A	0.6	70	B	0.8	69
A	0.6	68	B	0.7	66
A	0.6	70	B	0.9	70
A	0.5	66	B	0.9	69
A	0.4	55	B	0.5	58
A	0.3	50	B	0.6	62
A	0.5	64	B	0.6	64
A	0.6	68	B	0.8	68

b. Using the computer output given next, determine and plot the fitted straight lines for each tire type.

c. Test whether the straight lines for the two tire types coincide.

d. Test the following hypotheses:

H_0: The lines are parallel.

H_A: The lines are not parallel.

e. In light of your answers to parts (c) and (d), comment on differences and similarities in the tread wear–average speed relationship for the two tire types.

Edited SAS Output (PROC REG) for Problem 18

```
    Tread Wear (Y) Regressed on Avg. Speed (X), Tire (Z), and XZ
                (Z = 0 for Tire A, Z = 1 for Tire B)
                      Descriptive Statistics
Variables              Sum            Mean         Uncorrected SS
INTERCEP                20              1                 20
X                     1286           64.3              83302
Z                       10            0.5                 10
XZ                     655          32.75              43027
Y                     12.2           0.61               7.94

        Variables              Variance          Std Deviation
        INTERCEP                  0                    0
        X                   32.221052632         5.6763590999
        Z                    0.2631578947         0.512989176
        XZ                1135.5657895          33.698157063
        Y                    0.0262105263         0.1618966532

                       Analysis of Variance
                      Sum of          Mean
Source        DF      Squares        Square      F Value    Prob > F
Model          3      0.47861        0.15954     131.640     0.0001
Error         16      0.01939        0.00121
C Total       19      0.49800

          Root MSE        0.03481      R-square       0.9611
          Dep Mean        0.61000      Adj R-sq       0.9538
          C.V.            5.70697

                       Parameter Estimates
                      Parameter       Standard      T for H0:        Prob
Variable      DF      Estimate         Error      Parameter=0       > |T|
INTERCEP       1     -0.407518       0.10313225      -3.951         0.0011
X              1      0.014382       0.00162509       8.850         0.0001
Z              1     -1.082121       0.22917196      -4.722         0.0002
XZ             1      0.019353       0.00351783       5.501         0.0001

Variable      DF      Type I SS
INTERCEP       1      7.442000
X              1      0.295057
Z              1      0.146875
XZ             1      0.036678
```

19. A random sample of data were collected on residential sales in a large city. The following table shows the sales price Y (in \$1,000s), area X_1 (in hundreds of square feet), number of bedrooms X_2, total number of rooms X_3, age X_4 (in years), and location (dummy variables Z_1 and Z_2, defined as follows: $Z_1 = Z_2 = 0$ for intown; $Z_1 = 1$, $Z_2 = 0$ for inner suburbs; $Z_1 = 0$, $Z_2 = 1$ for outer suburbs) of each house.

House	Y	X_1	X_2	X_3	X_4	Z_1	Z_2
1	84.0	13.8	3	7	10	1	0
2	93.0	19.0	2	7	22	0	1
3	83.1	10.0	2	7	15	0	1
4	85.2	15.0	3	7	12	0	1
5	85.2	12.0	3	7	8	0	1
6	85.2	15.0	3	7	12	0	1
7	85.2	12.0	3	7	8	0	1
8	63.3	9.1	3	6	2	0	1
9	84.3	12.5	3	7	11	0	1
10	84.3	12.5	3	7	11	0	1
11	77.4	12.0	3	7	5	1	0
12	92.4	17.9	3	7	18	0	0
13	92.4	17.9	3	7	18	0	0
14	61.5	9.5	2	5	8	0	0
15	88.5	16.0	3	7	11	0	0
16	88.5	16.0	3	7	11	0	0
17	40.6	8.0	2	5	5	0	0
18	81.6	11.8	3	7	8	0	1
19	86.7	16.0	3	7	9	1	0
20	89.7	16.8	2	7	12	0	0
21	86.7	16.0	3	7	9	1	0
22	89.7	16.8	2	7	12	0	0
23	75.9	9.5	3	6	6	0	1
24	78.9	10.0	3	6	11	1	0
25	87.9	16.5	3	7	15	1	0
26	91.0	15.1	3	7	8	0	1
27	92.0	17.9	3	8	13	0	1
28	87.9	16.5	3	7	15	1	0
29	90.9	15.0	3	7	8	0	1
30	91.9	17.8	3	8	13	0	1

a. Identify a single regression model that uses the data for all three locations and that defines straight-line models relating sales price (Y) to area (X_1) for each location.

b. Using the computer output given next, determine and plot the fitted straight lines for each location.

c. Test whether the straight lines for the three locations coincide.

d. Test the following hypotheses:

H_0: The lines are parallel.

H_A: The lines are not parallel.

e. In light of your answers to parts (c) and (d), comment on differences and similarities in the sales price–area relationship for the three locations.

Edited SAS Output (PROC REG) for Problem 19

```
       Sales Price (Y) Regressed on Area (X1), Location (Z), and XZ
                         Descriptive Statistics
Variables             Sum              Mean           Uncorrected SS
INTERCEPT              30                1                     30
X1                   423.9            14.13                6276.55
Z1                     7            0.2333333333                7
Z2                    15               0.5                    15
X1Z1                 100.8             3.36                1490.94
Z1Z2                 204.2         6.806666667             2914.06
Y                   2504.9         83.49666667           212704.81

       Variables          Variance          Std Deviation
       INTERCEP              0                    0
       X1              9.891137931          3.145017954
       Z1              0.1850574713         0.4301830672
       Z2              0.2586206897         0.5085476277
       X1Z1            39.732827586         6.3033980983
       X1Z2            52.556505747         7.2495865915
       Y               122.55205747         11.070323278
                    Analysis of Variance
                      Sum of          Mean
Source        DF      Squares         Square      F Value    Prob > F
Model          5    3158.41441      631.68288      38.323     0.0001
Error         24     395.59526       16.48314
C Total       29    3554.00967

          Root MSE      4.05994       R-square      0.8887
          Dep Mean     83.49667       Adj R-sq      0.8655
          C.V.          4.86240
                    Parameter Estimates
                    Parameter       Standard     T for H0:          Prob
Variable      DF     Estimate         Error    Parameter=0         > |T|
INTERCEP       1     8.968611      6.07754231       1.476         0.1530
X1             1     4.806990      0.39734913      12.098         0.0001
Z1             1    52.122387     11.22484137       4.643         0.0001
Z2             1    48.558241      7.79709969       6.228         0.0001
X1Z1           1    -3.201206      0.75896494      -4.218         0.0003
X1Z2           1    -2.803086      0.52980716      -5.291         0.0001

Variable      DF      Type I SS
INTERCEP       1       209151
X1             1     2271.713503
Z1             1        0.016309
Z2             1      336.002950
X1Z1           1       89.282266
X1Z2           1      461.399382
```

20. In Problem 19 in Chapter 5 and Problem 14 in Chapter 8, data from the 1990 Census for 26 randomly selected Metropolitan Statistical Areas (MSAs) were discussed. Of interest were factors potentially associated with the rate of owner occupancy of housing units. The following three variables were included in the data set:

OWNEROCC: Proportion of housing units that are owner-occupied (as opposed to renter-occupied)

OWNCOST: Median selected monthly ownership costs, in $

URBAN: Proportion of population living in urban areas

It is also of interest to see whether the owner occupancy rate–ownership cost relationship is different in metropolitan areas where 75% or more of the population lives in urban areas than in metropolitan areas where less than 75% of the population lives in urban areas. For this purpose, the following additional variable is defined:

$$X_1 = \begin{cases} 1 \text{ if proportion of population living in urban areas} \geq 0.75 \\ 0 \text{ otherwise} \end{cases}$$

a. State a single regression model that defines straight-line models relating OWNEROCC to OWNCOST both for MSAs with high percentages ($\geq 75\%$) of urban populations and for MSAs with lower percentages of urban populations ($< 75\%$).

b. Using the computer output given next, test whether the straight lines for the two MSA types described in part (a) coincide. [*Note:* In the accompanying SAS output, the variable OWNCOST has been centered to avoid collinearity problems.]

c. Test the following hypotheses:

H_0: The lines are parallel.

H_A: The lines are not parallel.

d. In light of your answers to parts (b) and (c), comment on differences and similarities in the owner occupancy rate–ownership cost relationship for the two types of MSA.

Edited SAS Output (PROC REG) for Problem 20

```
      OWNEROCC Regressed on OWNCOST, X1, and X1*OWNCOST (X1OWNCOS)
                     Analysis of Variance
                      Sum of         Mean
Source      DF      Squares        Square     F Value     Prob > F
Model        3    214.19496      71.39832       4.061       0.0194
Error       22    386.76658      17.58030
C Total     25    600.96154
                                      .
                                      . [Portion of output omitted]
                                      .

           OWNEROCC Regressed on OWNCOST and X1
                     Analysis of Variance
                      Sum of         Mean
Source      DF      Squares        Square     F Value     Prob > F
Model        2    213.49860     106.74930       6.337       0.0064
Error       23    387.46294      16.84621
C Total     25    600.96154
                                      .
                                      . [Portion of output omitted]
                                      .

                     Parameter Estimates
                   Parameter      Standard      T for H0:        Prob
Variable    DF     Estimate          Error    Parameter=0       > |T|
INTERCEP     1    68.665453     1.47336083        46.605       0.0001
OWNCOST      1    -0.012668     0.00552565        -2.293       0.0314
X1           1    -3.905669     1.78250451        -2.191       0.0388
```

21. This problem involves the PERK study data discussed in Problem 12 in Chapter 8. Suppose that we wish to compare the relationship between change in refraction five years after surgery and baseline refractive error, for males and females. To this end, we define the following dummy variable:

$$Z = \begin{array}{l} 1 \text{ if patient is male} \\ 0 \text{ otherwise} \end{array}$$

a. State a single regression model that defines straight-line models relating change in refraction five years after surgery and baseline refractive error for both males and females.

b. Using the computer output given next, test whether the straight lines for males and females coincide.

c. Test the following hypotheses:

H_0: The lines are parallel.

H_A: The lines are not parallel.

d. In light of your answers to parts (b) and (c), comment on differences and similarities in the change in refraction–baseline refractive error relationship for males and females.

Edited SAS Output (PROC REG) for Problem 21

```
                    Y Regressed on X1, Z, and X1Z
                       Analysis of Variance
                     Sum of        Mean
Source       DF      Squares       Square      F Value    Prob > F
Model         3     19.65170      6.55057       5.405      0.0027
Error        50     60.60124      1.21202
C Total      53     80.25294

          Root MSE     1.10092      R-square      0.2449
          Dep Mean     3.83343      Adj R-sq      0.1996
          C.V.        28.71896

                       Parameter Estimates
                     Parameter     Standard     T for H0:       Prob
Variable     DF      Estimate       Error      Parameter=0     > |T|
INTERCEP      1      3.178210     0.46488413      6.837        0.0001
X1            1     -0.201008     0.10785972     -1.864        0.0683
Z             1     -1.995126     0.88972206     -2.242        0.0294
X1Z           1     -0.383826     0.20265951     -1.894        0.0640
                     Y Regressed on X1, Z
                       Analysis of Variance
                     Sum of        Mean
Source       DF      Squares       Square      F Value    Prob > F
Model         2     15.30414      7.65207       6.009      0.0045
Error        51     64.94880      1.27351
C Total      53     80.25294

          Root MSE     1.12850      R-square      0.1907
          Dep Mean     3.83343      Adj R-sq      0.1590
          C.V.        29.43835

                       Parameter Estimates
                     Parameter     Standard     T for H0:       Prob
Variable     DF      Estimate       Error      Parameter=0     > |T|
INTERCEP      1      2.752647     0.41716942      6.598        0.0001
X1            1     -0.309731     0.09360197     -3.309        0.0017
Z             1     -0.412878     0.31372310     -1.316        0.1940
```

References

Allen, D. M., and Grizzle, J. E. 1969. "Analysis of Growth and Dose Response Curves." *Biometrics* 25: 357–82.

Armitage, P. 1971. *Statistical Methods in Medical Research.* Oxford: Blackwell Scientific Publications.

Clark, M. F.; Lechyeka, M.; and Cook, C. A. 1940. "The Biological Assay of Riboflavia." *Journal of Nutrition* 20: 133–44.

Davis, E. A., Jr. 1955. "Seasonal Changes in the Energy Balance of the English Sparrow." *Auk.,* 72(4): 385–411.

Gruber, F. J. 1970. "Industrialization and Health." Ph.D. dissertation, Department of Epidemiology, University of North Carolina, Chapel Hill, N.C.

Nie, N., et al. 1975. *Statistical Package for the Social Sciences.* New York: McGraw-Hill.

15

Analysis of Covariance and Other Methods for Adjusting Continuous Data

15-1 Preview

In Chapter 11 we discussed the issue of controlling for extraneous variables when assessing an association of interest. Three reasons for considering control are to assess *interaction,* to correct for *confounding,* and to increase the *precision* in estimating the association of interest. In regression the usual approach for carrying out such control is to fit a regression model that contains as independent variables not only the study factors (exposure variables) of interest, but also extraneous variables considered to be important (and perhaps even product terms involving these variables). The focus then becomes one of determining the effects of the study factors on the response variable *adjusted* for the presence of the control variables in the model.

In this chapter we describe how to carry out this process of adjustment by using a popular procedure for regression modeling called analysis of covariance (ANACOVA). This technique involves a multiple regression model in which the study factors of interest are all treated as nominal variables, whereas the variables being controlled—that is, the *covariates*—may be measurements on any measurement scale. As discussed in Chapter 14, nominal variables are incorporated into regression models by means of dummy variables. Thus, the general ANACOVA model usually contains a mixture of dummy variables and other types of variables, and the dependent variable is considered continuous. In using the ANACOVA model, we also assume that there is no interaction of covariates with study variables, although (as discussed later) this assumption should be assessed in the analysis.

In addition to considering ANACOVA, we briefly review other regression-type methods for controlling for extraneous factors.

15-2 Adjustment Problem

Suppose, as in the example in section 14-4, that we are considering a sample of observations made on the dependent variable systolic blood pressure and the independent variable age for $n_F = 29$ women and $n_m = 40$ men.

Two questions are often of interest in analyses of such data:

1. Is the true straight-line relationship between blood pressure and age (given that a straight-line model is adequate) the same for male and female populations?

2. Do the mean blood pressure levels for the male and female groups differ significantly after one takes into account (i.e., after one adjusts for or controls for) the possible confounding effect of there being differing age distributions in the two groups?

Although the statistical techniques required to answer these questions are related, the questions nevertheless differ in emphasis: The first focuses on a comparison of straight-line regression equations, whereas the second focuses on a comparison of the mean blood pressure levels in the two groups.

We have already considered question 1 (in Chapter 14) through use of the model

$$Y = \beta_0 + \beta_1 X + \beta_2 Z + \beta_3 XZ + E \qquad (15.1)$$

where Z is a dummy variable identifying the sex group ($Z = 1$ if female and $Z = 0$ if male). By using appropriate tests of hypotheses about the parameters in this model, we may reach one of three important conclusions regarding question 1:

1. The lines are coincident (i.e., $\beta_2 = \beta_3 = 0$).

2. The lines are parallel but not coincident (i.e., $\beta_3 = 0$, but $\beta_2 \neq 0$).

3. The lines are not parallel ($\beta_3 \neq 0$).

These conclusions greatly influence the answer to question 2. If conclusion 1 is appropriate, we say that the two sex groups do not differ in mean blood pressure level after the effect of age is controlled. If conclusion 2 holds, we say that the sex group associated with the higher straight line has a higher mean blood pressure level at all ages. If conclusion 3 is valid, we must look closer at the orientation of the two straight lines: if they do not intersect in the age range of interest, we say that the sex group associated with the higher curve has a higher mean blood pressure level at each age; if they do intersect in the age range of interest, we say that one sex group has a higher mean blood pressure level than the other group at lower ages and a lower mean blood pressure level at higher ages (i.e., there is an age–sex group interaction effect).

Thus, by considering question 1 as described, we may draw reasonable inferences about the relationship between the true average blood pressure levels in the two groups as a function of age. Nevertheless, we must take additional statistical considerations into account to estimate the true adjusted mean difference and the adjusted means for each group. In this regard, question 2 raises the problem of determining an appropriate method for adjusting the sample mean blood pressure levels to take into account the effect of age, as well as the problem of providing a statistical test to compare these adjusted mean scores.

In the example we are considering, age is a factor known to be strongly associated with blood pressure, and the two groups, as sampled, may have widely different age distributions. Without adjusting the sample mean values to reflect any difference in the age distributions in the two groups, we could not determine (e.g., through the use of a two-sample t test based on the unadjusted sample means) whether a significant difference was attributable solely to age or to other factors. With adjustment, however, we could determine whether any findings were solely attributable, for example, to the fact that the females in the sample were older than the males (or vice versa).

15-3 Analysis of Covariance

The usual statistical technique for handling the adjustment problem described in section 15-2 is called the *analysis of covariance*. In this approach, we fit a regression model of the form

$$Y = \beta_0 + \beta_1 X + \beta_2 Z + E \qquad (15.2)$$

where X (age) is referred to as the *covariate* and Z is a dummy variable that indexes the two groups to be compared ($Z = 1$ if female, $Z = 0$ if male). This model assumes—in contrast to model (15.1), which contains a $\beta_3 XZ$ term—that the regression lines for males and females are parallel. Under this model, *the adjusted mean scores for males and females are defined to be the predicted values obtained by evaluating the model at $Z = 0$ and $Z = 1$, when X is set equal to the overall mean age for the two groups. A partial F test of the hypothesis $H_0: \beta_2 = 0$ is then used to determine whether these adjusted mean scores are significantly different.*

In computing these adjusted scores, we need to consider the two straight lines obtained by fitting model (15.2):

$$\begin{array}{lll}
\text{Males } (Z = 0): & \hat{Y}_M = \hat{\beta}_0 + \hat{\beta}_1 X \\
\text{Females } (Z = 1): & \hat{Y}_F = (\hat{\beta}_0 + \hat{\beta}_2) + \hat{\beta}_1 X
\end{array} \qquad (15.3)$$

Explicit formulas for the estimated coefficients in (15.3) are

$$\hat{\beta}_1 = \frac{(n_M - 1)S_{X_M}^2 \hat{\beta}_{1M} + (n_F - 1)S_{X_F}^2 \hat{\beta}_{1F}}{(n_M - 1)S_{X_M}^2 + (n_F - 1)S_{X_F}^2}$$

$$\hat{\beta}_0 = \overline{Y}_M - \hat{\beta}_1 \overline{X}_M \qquad (15.4)$$

$$\hat{\beta}_0 + \hat{\beta}_2 = \overline{Y}_F - \hat{\beta}_1 \overline{X}_F$$

where $\hat{\beta}_{1M}$ and $\hat{\beta}_{1F}$ are the estimated slopes based on separate straight-line fits for males and females, $S_{X_M}^2$ and $S_{X_F}^2$ are the sample variances (on X) for males and females, $\overline{Y}_M$ and $\overline{Y}_F$ are the mean blood pressures for the male and female samples, and $\overline{X}_M$ and $\overline{X}_F$ are the mean ages for the male and female samples, respectively. Notice that $\hat{\beta}_1$ is a weighted average of the slopes $\hat{\beta}_{1M}$ and $\hat{\beta}_{1F}$, which are estimated separately from the male and female data sets.

Based on (15.3) and (15.4), two alternative formulas for computing adjusted mean scores can be used:

Sex	Z	Adjusted Score	Formula 1	Formula 2	
Male	0	$\overline{Y}_M \text{ (adj)}$	$\hat{\beta}_0 + \hat{\beta}_1 \overline{X}$	$\overline{Y}_M - \hat{\beta}_1(\overline{X}_M - \overline{X})$	(15.5)
Female	1	$\overline{Y}_F \text{ (adj)}$	$(\hat{\beta}_0 + \hat{\beta}_2) + \hat{\beta}_1 \overline{X}$	$\overline{Y}_F - \hat{\beta}_1(\overline{X}_F - \overline{X})$	

In this table, $\overline{X}$ is the overall mean age for the combined data on males and females:

$$\overline{X} = \frac{n_M \overline{X}_M + n_F \overline{X}_F}{n_M + n_F}$$

Formula 1 is useful when model (15.2) has been estimated directly using a standard multiple regression program; formula 2, on the other hand, can be used without resorting to multiple regression procedures, although separate straight lines must be fitted to the male and female data sets.

Given that the parallel straight-line assumption of model (15.2) is appropriate (we discuss this assumption in section 15-4 and 15-6-2), the two formulas provide a comparison of the mean blood pressure levels for the two sex groups as if they both had the same age distribution. In this regard, the covariance approach just described attempts artificially to equate the age distributions by treating both sex groups as if they had the same mean age, the best estimate of which is $\overline{X}$. The adjusted scores, then, represent the predicted $\hat{Y}$-values for each fitted line at $\overline{X}$, the assumed common mean age. This is depicted graphically in Figure 15-1.

That the partial F test of H_0: $\beta_2 = 0$ addresses the question of whether a significant difference exists between the adjusted means follows because, from formula 1 in (15.5), the difference in the two adjusted mean scores is exactly equal to β_2; that is,

$$\hat{\beta}_2 = \overline{Y}_F(\text{adj}) - \overline{Y}_M(\text{adj}) = [(\hat{\beta}_0 + \hat{\beta}_2) + \hat{\beta}_1\overline{X}] - (\hat{\beta}_0 + \hat{\beta}_1\overline{X})$$

■ **Example**   The least-squares fitting of the model (15.2) for the male–female data we have been discussing yields the following estimated model:

$$\hat{Y} = 110.29 + 0.96X - 13.51Z$$

The separate fitted equations for males and females, respectively, are

Males ($Z = 0$): $\hat{Y}_M = 110.29 + 0.96X$

Females ($Z = 1$): $\hat{Y}_F = 96.78 + 0.96X$

The adjusted mean scores obtained from these equations, using formula 1 of (15.5) with $\overline{X} = 46.14$, are

$$\overline{Y}_M(\text{adj}) = 110.29 + 0.96(46.14) = 154.40$$

$$\overline{Y}_F(\text{adj}) = 96.78 + 0.96(46.14) = 140.89$$

FIGURE 15-1   Adjusted mean systolic blood pressure (SBP) scores for males and females, controlling for age by using analysis of covariance.

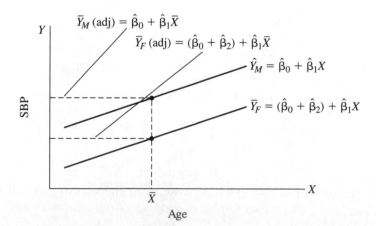

A comparison of these adjusted mean scores with the unadjusted means yields the following statistics:

Sex	Unadjusted $\overline{Y}$	Adjusted
Male	155.15	154.40
Female	139.86	140.89

Notice that the adjusted mean for males is slightly lower than the unadjusted mean for males, whereas the female adjusted mean is slightly higher than its unadjusted counterpart. The direction of these changes accurately reflects the fact that, in this sample, the males are somewhat older on the average ($\overline{X}_M = 46.93$) than the females ($\overline{X}_F = 45.07$). Using the adjusted mean scores, in effect, removes the influence of age on the comparison of mean blood pressures by considering what the mean blood pressures in the two groups would be if both groups had the same mean age ($\overline{X} = 46.14$).

But whether adjusted or not, the mean blood pressure for males in this example appears to be considerably higher than that for females. In fact, the covariance adjustment has done little to change this impression: the discrepancy between the male and female groups is 15.29 using unadjusted mean scores and 13.51 using adjusted mean scores. To test whether this difference in adjusted mean scores is significant, we use the partial F test of the hypothesis $H_0: \beta_2 = 0$ based on model (15.2), which can be computed from the following analysis-of-variance presentation:

Source		SS	d.f.	MS
Reduced model ($\beta_2 = 0$)	Regression (X)	14,951.25	1	14,951.25
	Residual	8,260.51	67	123.29
Complete model (15.2)	Regression (X, Z)	18,009.78	2	9,004.89
	Residual	5,201.99	66	78.82

From this presentation, we can obtain the appropriate partial F statistic as follows:

$$F(Z \mid X) = \frac{\text{Regression SS}(X, Z) - \text{Regression SS}(X)}{\text{MS residual}(X, Z)}$$

$$= \frac{18{,}009.78 - 14{,}951.25}{78.82}$$

$$= 38.80$$

which has 1 and 66 degrees of freedom. The P-value for this test satisfies $P < .001$, so we reject H_0 and conclude that the two adjusted scores differ significantly.

15-4 Assumption of Parallelism: A Potential Drawback

A potential problem raised by using the ANACOVA involves the assumption of parallelism of the regression lines. In certain applications, these regression lines may have different slopes. In such cases, the parallelism assumption is invalid and the covariance method of adjustment just described should be avoided. *To guard against applying the covariance method of adjustment incorrectly, we recommend conducting a test for parallelism before proceeding with the ANACOVA.* This amounts to testing $H_0: \beta_3 = 0$ for the complete model (15.1). We saw in Chapter 14 that the parallelism hypothesis ($H_0: \beta_3 = 0$) is not rejected for the age–systolic blood pressure data. This result supports the use of the ANACOVA model for these data (given, of course, that the assumption of variance homogeneity also holds).

If the test for parallelism supports the conclusion that the regression lines are not truly parallel, what, if anything, should be done about adjustment? Usually no adjustment at all should be made to the sample means, since any such adjustment would be misleading; that is, a direct comparison of means is not appropriate when the true difference between the mean blood pressure levels in the two groups varies with age (i.e., when there is an age–sex interaction). In this case, since the main feature of the data is that the two regression lines describe very different relationships between age and blood pressure, an analysis that allows two separate regression lines to be fitted (without assuming parallelism) and that quantifies how the lines differ is sufficient.

15-5 Analysis of Covariance: Several Groups and Several Covariates

In the example discussed in previous sections, we compared two groups and adjusted for the single covariate, age. In general, ANACOVA may be used to provide adjusted scores when there are several (say, k) groups and when it is necessary to adjust simultaneously for several (say, p) covariates. The regression model describing this general situation is written

$$Y = \beta_0 + \beta_1 X_1 + \beta_2 X_2 + \cdots + \beta_p X_p + \beta_{p+1} Z_1 + \beta_{p+2} Z_2$$
$$+ \cdots + \beta_{p+k-1} Z_{k-1} + E \tag{15.6}$$

This model includes p covariates $X_1, X_2,\ldots, X_p$ and k groups, which are represented by the $k - 1$ dummy variables $Z_1, Z_2,\ldots, Z_{k-1}$. As discussed in Chapter 14, we have some leeway in defining these dummy variables; but for our purposes we assume that the Z's are defined as follows:

$$Z_j = \begin{cases} 1 & \text{if group } j \\ 0 & \text{otherwise} \end{cases} \quad j = 1, 2,\ldots, k - 1$$

The fitted regression equations for the k different groups are then determined by specifying the appropriate combinations of values for the Z's. These are

Group 1 ($Z_1 = 1$, other $Z_j = 0$):
$$\hat{Y}_1 = (\hat{\beta}_0 + \hat{\beta}_{p+1}) + \hat{\beta}_1 X_1 + \hat{\beta}_2 X_2 + \cdots + \hat{\beta}_p X_p$$

Group 2 ($Z_2 = 1$, other $Z_j = 0$):
$$\hat{Y}_2 = (\hat{\beta}_0 + \hat{\beta}_{p+2}) + \hat{\beta}_1 X_1 + \hat{\beta}_2 X_2 + \cdots + \hat{\beta}_p X_p$$

$$\vdots \qquad\qquad\qquad\qquad \vdots \tag{15.7}$$

Group $k - 1$ ($Z_{k-1} = 1$, other $Z_j = 0$):
$$\hat{Y}_{k-1} = (\hat{\beta}_0 + \hat{\beta}_{p+k-1}) + \hat{\beta}_1 X_1 + \hat{\beta}_2 X_2 + \cdots + \hat{\beta}_p X_p$$

Group k (all $Z_j = 0$):
$$\hat{Y}_k = \hat{\beta}_0 + \hat{\beta}_1 X_1 + \hat{\beta}_2 X_2 + \cdots + \hat{\beta}_p X_p$$

From (15.7) we see that the corresponding coefficients of the covariates $X_1, X_2,\ldots, X_p$ in each of the k equations are identical. Thus, these regression equations represent "parallel" hypersurfaces in $(p + 1)$-space, which is a natural generalization of the situation for the single covariate case. The adjusted mean score for a particular group is then computed as the predicted Y-value obtained by evaluating the fitted equation for that group at the mean values $\overline{X}_1, \overline{X}_2,\ldots,$ $\overline{X}_p$ of the p covariates, based on the combined data for all k groups:

$$\overline{Y}_j(\text{adj}) = (\hat{\beta}_0 + \hat{\beta}_{p+j}) + \hat{\beta}_1\overline{X}_1 + \cdots + \hat{\beta}_p\overline{X}_p \qquad j = 1, 2,\ldots, k - 1$$
$$\overline{Y}_k(\text{adj}) = \hat{\beta}_0 + \hat{\beta}_1\overline{X}_1 + \cdots + \hat{\beta}_p\overline{X}_p \qquad\qquad\qquad (15.8)$$

To determine whether the k adjusted mean scores differ significantly from one another, we test the null hypothesis

$$H_0: \beta_{p+1} = \beta_{p+2} = \cdots = \beta_{p+k-1} = 0$$

using a multiple partial F test with $k - 1$ and $n - p - k$ degrees of freedom, based on model (15.6). If H_0 is rejected, we conclude that significant differences exist among the adjusted means (although we cannot, without further inspection, determine where the major differences are).

Example   In the Ponape study (Patrick et al. 1974) of the effect of rapid cultural change on health status, one research goal was to determine whether blood pressure was associated with a measure of the strength of a Ponapean male's prestige in the modern (i.e., western) part of his culture relative to the traditional culture. A measure of prestige discrepancy (PD) was developed and then measured by questionnaire for each of 550 Ponapean males. Of particular interest was whether a higher prestige discrepancy score corresponded to higher blood pressure. The effects of the covariates age and body size were adjusted or controlled for in considering these questions.

To perform this analysis, the researchers categorized prestige discrepancy into three groups:

Group 1: Modern prestige much higher than traditional prestige

Group 2: Modern prestige not much different from traditional prestige

Group 3: Traditional prestige much higher than modern prestige

Then an ANACOVA was carried out with diastolic blood pressure (DBP) as the dependent variable and with AGE and quetelet index (QUET, a measure of body size) as the two covariates. An analysis of covariance model in this case is given by

$$\text{DBP} = \beta_0 + \beta_1\text{AGE} + \beta_2\text{QUET} + \beta_3 Z_1 + \beta_4 Z_2 + E \qquad (15.9)$$

where

$$Z_1 = \begin{cases} 1 & \text{if group 1} \\ 0 & \text{otherwise} \end{cases} \quad \text{and} \quad Z_2 = \begin{cases} 1 & \text{if group 2} \\ 0 & \text{otherwise} \end{cases}$$

A test of the parallelism assumption implicit in model (15.9) considers the null hypothesis H_0: $\beta_5 = \beta_6 = \beta_7 = \beta_8 = 0$ in the full model

$$\text{DBP} = \beta_0 + \beta_1\text{AGE} + \beta_2\text{QUET} + \beta_3 Z_1 + \beta_4 Z_2 + \beta_5 Z_1\text{AGE} + \beta_6 Z_2\text{AGE}$$
$$+ \beta_7 Z_1\text{QUET} + \beta_8 Z_2\text{QUET} + E$$

This multiple-partial F test (with 4 and 541 degrees of freedom) was not rejected using the Ponape data, thus supporting the use of model (15.9).

TABLE 15-1 ANACOVA example using Ponape study data.

DBP	PD Group 1 ($n_1 = 87$)	PD Group 2 ($n_2 = 383$)	PD Group 3 ($n_3 = 80$)	P Value for F Test of H_0: $\beta_3 = \beta_4 = 0$*
Unadjusted	71.68	68.16	68.55	$.001 < P < .005$
Adjusted	72.07	68.02	68.80	

* Using model (15.9) with 2 and 545 d.f.

The researchers then determined adjusted DBP mean scores for the three groups by substituting the values of the overall means $\overline{AGE}$ and $\overline{QUET}$ into the fitted equations for the three groups, as follows:

$$\overline{DBP}_1(adj) = (\hat{\beta}_0 + \hat{\beta}_3) + \hat{\beta}_1\overline{AGE} + \hat{\beta}_2\overline{QUET}$$

$$\overline{DBP}_2(adj) = (\hat{\beta}_0 + \hat{\beta}_4) + \hat{\beta}_1\overline{AGE} + \hat{\beta}_2\overline{QUET}$$

$$\overline{DBP}_3(adj) = \hat{\beta}_0 + \hat{\beta}_1\overline{AGE} + \hat{\beta}_2\overline{QUET}$$

The test for equality of these adjusted means was based on use of a multiple partial F test of the null hypothesis H_0: $\beta_3 = \beta_4 = 0$ under model (15.9). Table 15-1 summarizes these calculations.

The results indicate the presence of highly significant differences among the adjusted mean blood pressure scores, with group 1 having a somewhat higher adjusted mean blood pressure than the other two groups. Despite the statistical significance found, however, the adjusted mean blood pressure of 72.07 for group 1 is close enough (clinically speaking) to the adjusted means for the other groups to cast doubt on the clinical significance of these results. ■

15-6 Comments and Cautions

15-6-1 Rationale for Adjustment

ANACOVA adjusts for disparities in covariate distributions over groups by artificially assuming that all groups have the same set of mean covariate values. For example, if age and weight are the covariates and two groups are being compared, the ANACOVA adjustment procedure treats both groups as if they had the same mean age and the same mean weight.

The ANACOVA adjustment procedure is equivalent to assuming a common covariate *distribution* based on the combined sample over all groups. That is, not only are the means assumed to be equal, but the entire distribution of the covariates in the combined sample is assumed to be the same as the distribution of the covariates in each group. The adjusted score for any group can be expressed as the average over the combined sample of the predicted scores for that group; that is,

$$\hat{Y}_j(adj) = \frac{1}{n}\sum_{k=1}^{n} \hat{Y}_{jk} \qquad j = 1, 2, \ldots, k$$

where $\hat{Y}_j(adj)$ is the adjusted score for group j defined by (15.8), n is the number of subjects in the combined sample, and $\hat{Y}_{jk}$ is the predicted response in the jth group, based on the set of equations (15.7) and the covariate values of the kth individual in the combined sample. For example,

$$\hat{Y}_{1k} = (\hat{\beta}_0 + \hat{\beta}_{p+1}) + \hat{\beta}_1X_{1k} + \hat{\beta}_2X_{2k} + \cdots + \hat{\beta}_pX_{pk}$$

is the predicted group 1 response for the kth individual in the combined sample, where $(X_{1k}, X_{2k}, \ldots, X_{pk})$ are the covariate values for that individual. Thus, the adjusted score for a given group can be obtained by artificially assuming that all n persons in the combined sample constitute the given group; then the covariate distribution of the group is assumed to be that of the combined sample.

Thus the method of adjustment using ANACOVA corrects the disparity in covariate distributions over groups by assuming a common distribution (not just a common set of means).

15-6-2 The Parallelism Assumption

As stated earlier, the ANACOVA method is inappropriate when the relationship between the covariates and the response is not the same in each group. Such nonparallelism or interaction might be reflected, for example, in a finding that males have higher blood pressures than females at older ages and that females have higher blood pressures than males at younger ages. Consequently, using a standard ANACOVA could lead to adjusted (for age) mean scores for each (gender) group that are roughly equal; this would create the misleading impression that there was little difference between male and female blood pressures, when, in fact, there were large differences in certain age categories. This example illustrates why we recommend that no method of adjustment be used in the presence of interaction; instead, the nature of the interaction itself should be quantified. The method for assessing interaction in this context, as previously indicated, requires first that the ANACOVA model (15.6) be expanded to include product terms between covariates and group variables and second that the coefficients of these product terms be tested for significance. The extended ANACOVA model that allows for such interaction terms has the form

$$Y = \beta_0 + \sum_{i=1}^{p} \beta_i X_i + \sum_{j=1}^{k-1} \beta_{p+j} Z_j + \sum_{i=1}^{p} \sum_{j=1}^{k-1} \gamma_{ij} X_i Z_j + E \qquad (15.10)$$

In this model, a chunk test for parallelism would test H_0: All $\gamma_{ij} = 0$, and would involve a multiple partial F statistic of the form

$$F(\text{All } X_i Z_j \text{ product terms} \mid \text{All } X_i \text{ and } Z_j \text{ terms})$$

which would have an F distribution under H_0 with $p(k-1)$ and $n - 1 - p - (k-1) - p(k-1)$ degrees of freedom. The use of ANACOVA-adjusted scores is appropriate only when the preceding test for interaction yields the conclusion that interaction is not significant.

15-6-3 Validity and Precision

As discussed in Chapter 11, validity and precision are two reasons to consider controlling covariates. In using ANACOVA, validity is achieved by adjusting for confounding, thereby obtaining an estimate of association that would have been distorted (biased) if the covariate(s) of interest had been ignored in the analysis.

Although validity should be the first consideration, it is possible to find no confounding in one's data and still control for one or more covariates to gain precision. To assess precision, we can consider either the variances of the estimators of the association(s) of interest (the smaller these variances, the greater the precision) or confidence intervals for the association(s) of interest (the narrower the confidence intervals, the greater the precision). For example, consider an ANACOVA model involving two groups—say, males and females—with systolic blood pressure as the dependent variable and age as the only covariate of interest. If the age distribution for males were identical to that for females in the data, then age would not be a confounder. Nevertheless, because age is strongly positively associated with systolic blood pressure, precision will probably be increased by adjusting for age, even though confounding (i.e., validity) is not at issue.

15-6-4 Alternatives to ANACOVA

When adjusting for covariates, we can fit a model to contain the covariates and the study variables of primary interest (and perhaps even product terms, so interaction can be assessed) without having to make the study variables categorical. A best model would then have to be derived by means of criteria for selecting variables (see Chapter 16). If, for such a model, significant interaction is found, it is inappropriate (as noted earlier with regard to categorical study variables) to derive adjusted scores. Moreover, even in the absence of interaction, it is impossible to obtain adjusted scores for groups, if the study variables are not defined categorically.

Nevertheless, predicted values based on the best regression model can be treated as adjusted values, since the covariates are being taken into account in the modeling process. Furthermore, adjusted scores for distinct values of the study variables can be obtained by computing predicted values using the overall mean covariate values in the best model, as was done for ANACOVA by using categorical study variables. For example, if we determine that the best model relating systolic blood pressure to age and weight is

$$\widehat{\text{SBP}} = \hat{\beta}_0 + \hat{\beta}_1 \text{AGE} + \hat{\beta}_2 \text{WEIGHT}$$

then we can compute adjusted blood pressure scores for persons weighing 150 and 175 pounds (controlling for age) as

$$\widehat{\text{SBP}}_{150}(\text{adj}) = \hat{\beta}_0 + \hat{\beta}_1 \overline{\text{AGE}} + 150\hat{\beta}_2$$

and

$$\widehat{\text{SBP}}_{175}(\text{adj}) = \hat{\beta}_0 + \hat{\beta}_1 \overline{\text{AGE}} + 175\hat{\beta}_2$$

These adjusted values are similar to group adjusted scores: the value of 150 can be thought of as representing a group of people whose weights are all close to 150; and a similar interpretation can be given to the 175-pound value.

In a more restrictive alternative to ANACOVA, *all* variables—even covariates—are treated as categorical. If we distinguish the set of dummy variables defining the covariates from the set of dummy variables defining the study variables, we can treat the regression model under consideration as a two-way analysis-of-variance model with unequal cell numbers (see Chapter 20). This alternative, however, might be inappropriate if the underlying scales of measurement of some of the covariates are actually noncategorical (e.g., are continuous). If, for example, the inherently continuous variable age is categorized into three age groups, a completely categorical model based on that arbitrary categorization might lead to different results from those obtained by using a model that treats age continuously.

15-7 Summary

In this chapter, we have examined the most popular approach to controlling for covariates, ANACOVA. To use ANACOVA, the study variables of interest must be treated as categorical variables, whereas the covariates are not so restricted. ANACOVA also requires the assumption that there is no interaction between covariates and study variables. This assumption can be checked by testing for the significance of appropriate product terms in an extended ANACOVA model. If the test for interaction yields a finding of nonsignificance, adjusted scores for different groups can be obtained by substituting the mean covariate values for the combined sample into the group-specific ANACOVA model. If the test for interaction is significant, adjusted scores should not be used; instead, the nature of the interaction should be described.

Problems

1. Problem 8 in Chapter 14 involved comparing straight-line regression fits of SBP on QUET for smokers and nonsmokers, and you demonstrated that these straight lines could be considered parallel. Use your results based from that problem and the computer output given next (and the fact that the overall sample mean value $\overline{\text{QUET}} = 3.441$) to address the following issues.

   a. State the appropriate ANACOVA regression model to use for comparing the mean blood pressures in the two smoking categories, controlling for QUET.

   b. Determine the adjusted SBP means for smokers and nonsmokers. Compare these values to the unadjusted mean values

   $$\overline{\text{SBP}}(\text{smokers}) = 147.823 \qquad \text{and} \qquad \overline{\text{SBP}}(\text{nonsmokers}) = 140.800$$

   c. Test whether the true adjusted mean blood pressures in the two groups are equal. State the null hypothesis in terms of the regression coefficients in the ANACOVA model given in part (a).

   d. Obtain a 95% confidence interval for the true difference in adjusted SBP means.

Edited SAS Output (PROC REG) for Problem 1

```
                    SBP Regressed on QUET and SMK
                        Analysis of Variance
                       Sum of           Mean
Source      DF        Squares          Square    F Value    Prob > F
Model        2     4120.36649      2060.18325     25.913      0.0001
Error       29     2305.60226        79.50353
C Total     31     6425.96875

        Root MSE        8.91647    R-square       0.6412
        Dep Mean      144.53125    Adj R-sq       0.6165
        C.V.            6.16924

                        Parameter Estimates
                     Parameter       Standard     T for H0:        Prob
Variable    DF        Estimate          Error    Parameter=0      > |T|
INTERCEP     1       63.876033     11.46811185        5.570       0.0001
QUET         1       22.115604      3.22995641        6.847       0.0001
SMK          1        8.571015      3.16670062        2.707       0.0113
```

2. a.–d. Answer the same questions as in parts (a) through (d) in Problem 1 regarding an analysis of covariance designed to control for *both* AGE and QUET. [*Note:* $\overline{AGE} = 53.250$.] Use the results from Problem 9 in Chapter 14.

   e. Is it necessary to control for *both* AGE and QUET as opposed to controlling just one of the two covariates?

Edited SAS Output (PROC REG) for Problem 2

```
              SBP Regressed on SMK, AGE, and QUET
                     Analysis of Variance
                        Sum of         Mean
Source       DF        Squares        Square     F Value    Prob > F
Model         3      4889.82570     1629.94190    29.710      0.0001
Error        28      1536.14305       54.86225
C Total      31      6425.96875

         Root MSE         7.40691     R-square      0.7609
         Dep Mean       144.53125     Adj R-sq      0.7353
         C.V.             5.12478

                     Parameter Estimates
                     Parameter      Standard      T for H0:       Prob
Variable     DF      Estimate        Error      Parameter=0      > |T|
INTERCEP      1     45.103192     10.76487511       4.190       0.0003
SMK           1      9.945568      2.65605655       3.744       0.0008
AGE           1      1.212715      0.32381922       3.745       0.0008
QUET          1      8.592449      4.49868122       1.910       0.0664

Variable     DF      Type I SS
INTERCEP      1        668457
SMK           1       393.098162
AGE           1      4296.586067
QUET          1       200.141468
```

3. In an experiment conducted at the National Institute of Environmental Health Sciences, the absorption (or uptake) of a chemical by a rat on one of two different diets, I or II, was known to be affected by the weight (or size) of the rat. A completely randomized design utilizing four rats on each diet was employed in the experiment, and the initial weight of each rat was recorded so that the diets could be compared after the researchers adjusted for the effect of initial weight. The data for the experiment are given in the following table.

Initial Weight (X)	3	1	4	4	5	2	3	2
Diet (Z)	I	I	I	I	II	II	II	II
Response (Y)	14	13	14	15	16	15	15	14

   a. Using the initial weight as a covariate, state the ANACOVA regression model for comparing the two diets (set $Z = -1$ if diet I is used, and $Z = 1$ if diet II is used).
   b. Use the computer results given next to determine the adjusted mean responses for each diet, controlling for initial weight.
   c. Use the ANACOVA regression model defined in part (a) to test whether the two diets differ significantly.

d. Test whether the two diets differ significantly, completely ignoring the covariate. How do the two testing procedures compare?

e. Determine a 95% confidence interval for the true difference in the adjusted mean responses.

Edited SAS Output (PROC REG) for Problem 3

```
   Regression of Response (Y) on Initial Weight (X) and Diet (Z)
                      Analysis of Variance
                      Sum of          Mean
Source         DF     Squares         Square      F Value      Prob > F
Model           2     5.00000         2.50000      12.500        0.0113
Error           5     1.00000         0.20000
C Total         7     6.00000

           Root MSE        0.44721       R-square       0.8333
           Dep Mean       14.50000       Adj R-sq       0.7667
           C.V.            3.08423

                      Parameter Estimates
                      Parameter        Standard       T for H0:       Prob
Variable       DF     Estimate         Error          Parameter=0     > |T|
INTERCEP        1     13.000000        0.41833001      31.076         0.0001
X               1      0.500000        0.12909944       3.873         0.0117
Z               1      0.500000        0.15811388       3.162         0.0250

Variable       DF     Type I SS
INTERCEP        1     1682.000000
X               1        3.000000
Z               1        2.000000
```

4. A political scientist developed a questionnaire to determine political tolerance scores (Y) for a random sample of faculty members at her university. She wanted to compare mean scores adjusted for age (X) in each of three categories: full professors, associate professors, and assistant professors. The data results are given in the accompanying tables (the higher the score, the more tolerant the individual).

Group 1: Full Professors $(Z1 = 1, Z2 = 0)$

Age (X)	65	61	47	52	49	45	41	41	40	39
Tolerance (Y)	3.03	2.70	4.31	2.70	5.09	4.02	3.71	5.52	5.29	4.62

Group 2: Associate Professors $(Z1 = 0, Z2 = 1)$

Age (X)	34	31	30	35	49	31	42	43	39	49
Tolerance (Y)	4.62	5.22	4.85	4.51	5.12	4.47	4.50	4.88	5.17	5.21

Group 3: Assistant Professors $(Z1 = Z2 = 1)$

Age (X)	26	33	48	32	25	33	42	30	31	27
Tolerance (Y)	5.20	5.86	4.61	4.55	4.47	5.71	4.77	5.82	3.67	5.29

a. State an ANACOVA regression model that can be used to compare the three groups, controlling for age.

b. What model should be used to check whether the ANACOVA model in part (a) is appropriate? Carry out the appropriate test. Use $\alpha = .01$.

c. Using ANACOVA, determine adjusted mean tolerance scores for each group, and test whether these differ significantly from one another. Also, compare the adjusted means with the unadjusted means. [*Note:* $\overline{X}$(overall) = 39.667, $\overline{Y}$(group 1) = 4.10, $\overline{Y}$(group 2) = 4.86, and $\overline{Y}$(group 3) = 5.00.]

Edited SAS Output (PROC REG) for Problem 4

```
             Tolerance (Y) Regressed on Age (X), Z1, and Z2

                                  .
                                  .  [Portion of output omitted]
                                  .

                          Parameter Estimates
                     Parameter        Standard        T for H0:        Prob
 Variable    DF       Estimate           Error      Parameter=0       > |T|
 INTERCEP     1       6.183716      0.61188805           10.106      0.0001
 X            1      -0.036352      0.01740518           -2.089      0.0467
 Z1           1      -0.339812      0.41456861           -0.820      0.4199
 Z2           1       0.063572      0.33234326            0.191      0.8498

             Regression of Y on X, Z1, Z2, XZ1, and XZ2

                          Analysis of Variance
                     Sum of            Mean
 Source      DF      Squares          Square     F Value      Prob > F
 Model        5     10.20934         2.04187       5.020        0.0027
 Error       24      9.76275         0.40678
 C Total     29     19.97210

         Root MSE        0.63779      R-square        0.5112
         Dep Mean        4.64967      Adj R-sq        0.4093
         C.V.           13.71699

                          Parameter Estimates
                     Parameter        Standard        T for H0:        Prob
 Variable    DF       Estimate           Error      Parameter=0       > |T|
 INTERCEP     1       5.427065      0.98483339            5.511      0.0001
 X            1      -0.013213      0.02947890           -0.448      0.6580
 Z1           1       2.784902      1.51590599            1.837      0.0786
 Z2           1      -1.223434      1.50992943           -0.810      0.4258
 XZ1          1      -0.072474      0.03778586           -1.918      0.0671
 XZ2          1       0.030220      0.04164509            0.726      0.4751

 Variable    DF       Type I SS
 INTERCEP     1     648.582003
 X            1       6.204949
 Z1           1       0.624940
 Z2           1       0.018469
 XZ1          1       3.146784
 XZ2          1       0.214202
```

5. A psychological experiment was performed to determine whether in problem-solving dyads containing one male and one female, "influencing" behavior depended on the sex of the experimenter. The problem for each dyad was a strategy game called "Rope a Steer," which required 20 separate decisions about which way to proceed toward a defined goal on a game board. For each subject in the dyad, a verbal-influence activity score was derived as a function of the number of statements made by the subject to influence the dyad to move in a particular direction. The difference in verbal-influence activity scores within a dyad was denoted as the variable VIAD, which was then used as the dependent variable in an ANACOVA designed to control for the effects of differing IQ scores of the male and female in each dyad. The relevant data are given in the following tables.

Group 1: Male Experimenter (Z = 0)

VIAD	−10	−4	9	−15	−15	5	−8	−4	−1	13
IQ_M	115	112	106	123	125	105	115	122	138	110
IQ_F	100	110	108	135	115	112	121	132	135	126

Group 2: Female Experimenter (Z = 1)

VIAD	8	−5	2	−7	15	−10	−3	10	2	4
IQ_M	120	130	110	113	102	141	120	113	114	102
IQ_F	141	128	104	98	106	130	128	105	107	111

a. State an ANACOVA model appropriate for these data.
b. Determine adjusted mean VIAD scores for each group and compare these to the unadjusted means for each group.
c. Test whether the adjusted mean scores differ significantly.
d. Find a 95% confidence interval for the true difference in adjusted mean scores.

Edited SAS Output (PROC REG) for Problem 5

```
            VIAD Regressed on IQM, IQF, and Z
                 Descriptive Statistics
Variables       Sum          Mean        Uncorrected SS
INTERCEP         20            1               20
IQM            2336         116.8           275060
IQF            2352         117.6           279924
Z                10           0.5               10
VIAD            -14          -0.7             1518
                            .  [Portion of output omitted]
                            .
```

(continued)

```
             VIAD Regressed on IQM, IQF, and Z (Continued)
                      Parameter Estimates
                      Parameter       Standard      T for H0:        Prob
Variable      DF      Estimate        Error         Parameter=0      > |T|
INTERCEP      1       44.590006       17.85878842       2.497        0.0238
IQM           1       -0.693012        0.19389273      -3.574        0.0025
IQF           1        0.281086        0.15966746       1.760        0.0974
Z             1        5.196104        3.13292211       1.659        0.1167
                           .
                           . [Portion of output omitted]
                           .

       Regression of VIAD on IQM, IQF, Z, IQM x Z, and IQF x Z
                      Analysis of Variance
                      Sum of          Mean
Source        DF      Squares         Square        F Value     Prob > F
Model         5       746.73852       149.34770      2.746       0.0623
Error         14      761.46148        54.39011
C Total       19     1508.20000

          Root MSE             7.37496      R-square       0.4951
          Dep Mean            -0.70000      Adj R-sq       0.3148
          C.V.             -1053.56640

                      Parameter Estimates
                      Parameter       Standard      T for H0:        Prob
Variable      DF      Estimate        Error         Parameter=0      > |T|
INTERCEP      1       43.990541       29.67269111       1.483        0.1604
IQM           1       -0.734946        0.31358043      -2.344        0.0344
IQF           1        0.327234        0.25687820       1.274        0.2234
Z             1        6.038753       38.43472793       0.157        0.8774
IQMZ          1        0.076127        0.41737246       0.182        0.8579
IQFZ          1       -0.082647        0.34324514      -0.241        0.8132

Variable      DF      Type I SS
INTERCEP      1         9.800000
IQM           1       502.258505
IQF           1       109.792500
Z             1       131.467153
IQMZ          1         0.067092
IQFZ          1         3.153274
```

6. An experiment was conducted to compare the effects of four different drugs (A, B, C, and D) in delaying atrophy of denervated muscles. A certain leg muscle in each of 48 rats was deprived of its nerve supply by surgical severing of the appropriate nerves. The rats were then put randomly into four groups, and each group was assigned treatment with one of the drugs. After 12 days, four rats from each group were killed and the weight W (in grams) of the denervated muscle was obtained, as listed in the following table.

Theoretically, atrophy should be measured as the loss in weight of the muscle, but the initial weight of the muscle could not have been obtained without killing the rat. Consequently, the initial total body weight X (in grams) of the rat was measured. It was assumed that this figure is closely related to the initial weight of the muscle.

Drugs A and C were large and small dosages, respectively, of atropine sulfate. Drug B was quinidine sulfate. Drug D acted as a control; it was simply a saline solution and could not have had any effect on atrophy. Use ANACOVA to compare the effects of the four drugs, controlling for initial total body weight (X). Use the results given in the SAS output shown next to perform your analysis. [*Note:* $\overline{X} = 226.125$.]

Drug A ($Z1 = 1$, ($Z2 = Z3 = 0$)		Drug B ($Z2 = 1$, ($Z1 = Z3 = 0$)		Drug C ($Z3 = 1$, ($Z1 = Z2 = 0$)		Drug D ($Z1 = Z2 = Z3 = 0$)	
X	W	X	W	X	W	X	W
198	0.34	233	0.41	204	0.57	186	0.81
175	0.43	250	0.87	234	0.80	286	1.01
199	0.41	289	0.91	211	0.69	245	0.97
224	0.48	255	0.87	214	0.84	215	0.87

Edited SAS Output (PROC REG) for Problem 6

```
            W Regressed on X, Z1, Z2, and Z3
                  Analysis of Variance
                    Sum of        Mean
Source       DF    Squares       Square      F Value     Prob > F
Model         4    0.61847       0.15462      10.715       0.0009
Error        11    0.15873       0.01443
C Total      15    0.77720

         Root MSE        0.12013     R-square      0.7958
         Dep Mean        0.70500     Adj R-sq      0.7215
         C.V.           17.03922

                  Parameter Estimates
                  Parameter      Standard      T for H0:        Prob
Variable     DF    Estimate         Error     Parameter=0      > |T|
INTERCEP      1    0.172827      0.30373303        0.569       0.5808
X             1    0.003185      0.00127783        2.493       0.0299
Z1            1   -0.391700      0.09540843       -4.106       0.0017
Z2            1   -0.225651      0.09020099       -2.502       0.0294
Z3            1   -0.135054      0.08775570       -1.539       0.1521

Variable     DF    Type I SS
INTERCEP      1    7.952400
X             1    0.320154
Z1            1    0.201770
Z2            1    0.062364
Z3            1    0.034177
```

7. Trough urine samples were analyzed for sodium content for each of two collection periods, one before and one after administration of Mercuhydrin, for each of 30 dogs. The experimenter used 7 dogs as a control group for the study; these dogs were not administered the drug, but their urine samples were collected for two similar time periods.

Experimental Group ($Z = 1$)			Experimental Group ($Z = 1$)		
Animal	First Collection Period (X) ([Na], mM/l)	Second Collection Period (Y) ([Na], mM/l)	Animal	First Collection Period (X) ([Na], mM/l)	Second Collection Period (Y) ([Na], mM/l)
1	17.5	22.1	18	6.3	12.7
2	9.4	12.0	19	9.7	17.1
3	10.0	15.2	20	7.1	9.5
4	7.4	23.1	21	7.2	11.0
5	8.8	9.8	22	5.3	8.2
6	18.9	26.9	23	14.3	15.8
7	10.8	11.1	24	7.9	9.7
8	8.8	13.6	25	14.1	14.7
9	8.8	12.8	26	12.8	17.0
10	9.2	7.5	27	12.8	20.2
11	8.1	8.1	28	10.7	13.9
12	10.3	27.5	29	5.9	11.8
13	10.1	11.2	30	3.8	9.0
14	7.3	11.0			
15	11.1	15.3	Mean	9.7	13.9
16	9.4	11.5	S.D.	3.3	5.3
17	8.2	8.4			

Control Group ($Z = 0$)		
Animal	First Collection Period (X) ([Na], mM/l)	Second Collection Period (Y) ([Na], mM/l)
1	11.1	9.4
2	5.1	5.9
3	6.5	14.8
4	17.2	15.5
5	11.8	23.4
6	6.6	7.3
7	4.1	8.2
Mean	8.9	12.1
S.D.	4.7	6.4

a. Use ANACOVA (computer results are given next), with the before measure as the covariate, to find the adjusted mean sodium contents for the experimental and control groups.

b. Test whether the two adjusted means differ significantly.

c. What alternative testing approach (involving a t test) could be used for these data? Carry out this test. [*Hint:* Variances of before–after differences for each group are 16.78517 (experimental) and 25.915 (control).]

d. Are the two testing methods (t test versus covariance analysis) equivalent in this problem? Explain by comparing regression models appropriate for each method.

e. What do the results of a lack-of-fit test, based on using the computer output given, indicate about the appropriateness of the covariance model used in parts (a) and (b)?

Dummy of Computer Results for Problem 7

					ANOVA		
Multiple $R^2 = .406$							
Variable	$\hat{\beta}$	S.D. of $\hat{\beta}$	$\hat{\beta}/$S.D.	Source	d.f.	MS	
X	0.96155	0.20418	4.709	X	1	434.4861	
Z	1.06435	1.83729	0.5793	Added by Z	1	6.3764	
(intercept)	3.49992			Lack of fit	30	21.0913	
				Pure error	4	3.3179	

8. Consider again the data in Problem 4.
   a. How would you compute appropriate cross-product variables to allow testing of whether an interaction exists between age and faculty rank?
   b. State the associated regression model.
   c. Using a computer, fit the model. Provide estimates of the coefficients.
   d. Provide a multiple partial F test for interaction, controlling for age and faculty rank. Use $\alpha = .05$.
   e. Does the F test in part (d) indicate that the ANACOVA is valid?

9. This problem involves data from Problem 15 in Chapter 5. Treat LN_BRNTL as the dependent variable, and dosage level of toluene (PPM_TOLU) as a categorical predictor (four levels). The experimenter wished to explore the possibility of using WEIGHT as a control variable.
   a. State the appropriate ANACOVA model. Use the 50-ppm exposure group as the reference group in coding dummy variables.
   b. Use a computer program to fit the model and provide estimated regression coefficients.
   c. Provide adjusted mean estimates. Compare these with the unadjusted means. (Do not perform any tests.)
   d. Test whether the adjusted means are all equal. Use $\alpha = .05$. State the null hypothesis in terms of the regression coefficients.

10. Repeat Problem 9, using LN_BLDTL as the dependent variable.

11. a. How would you compute appropriate cross-product terms for testing the interaction of WEIGHT and dosage for Problem 9?
    b. State the appropriate regression model.
    c. State the null hypothesis (of no interaction) in terms of regression coefficients.

12. a. For the model fitted in Problem 9, compute adjusted predicted scores and residuals.
    b. Plot residuals against predicted scores.
    c. Compute estimates of the residual variances separately for each dosage group. Are they approximately equal?
    d. Do these diagnostics suggest any problems?

13. Refer to the residential sales data in Problem 19 in Chapter 14. Use the computer output given below to answer the following questions.
    a. State an ANACOVA regression model that can be used to compare outer suburbs with other locations (intown and inner suburbs), controlling for age. [*Hint:* Let $Z = 0$ if intown or inner suburbs, 1 if outer suburbs.]
    b. Identify the model that should be used to check whether the ANACOVA model in part (a) is appropriate. Carry out the appropriate test.

c. Using ANACOVA, determine adjusted mean sales prices for the two locations, and test whether they differ significantly from one another. [*Note:* Mean house age $= 10.8667$; unadjusted mean sales price for intown and inner suburbs $= 82.1867$; unadjusted mean sales price for outer location $= 84.8067$.]

Edited SAS Output (PROC REG) for Problem 13

```
 Sales Price (Y) Regressed on Age (X4), Location (Z), and X4 x Z
                    Analysis of Variance
                   Sum of          Mean
Source       DF    Squares        Square     F Value    Prob > F
Model         3   1646.14368    548.71456     7.478      0.0009
Error        26   1907.86599     73.37946
C Total      29   3554.00967

         Root MSE        8.56618    R-square     0.4632
         Dep Mean       83.49667    Adj R-sq     0.4012
         C.V.           10.25931

                    Parameter Estimates
                   Parameter      Standard     T for H0:        Prob
Variable     DF    Estimate        Error      Parameter=0      > |T|
INTERCEP      1    55.457061    6.85946863       8.085        0.0001
X4            1     2.372450    0.57631064       4.117        0.0003
Z             1    17.905876    8.90572622       2.011        0.0548
X4Z           1    -1.279100    0.76285683      -1.677        0.1056

Variable     DF     Type I SS
INTERCEP      1       209151
X4            1   1324.847351
Z             1    114.997101
X4Z           1    206.299224

     Sales Price (Y) Regressed on Age (X4), and Location (Z)
                    Analysis of Variance
                   Sum of          Mean
Source       DF    Squares        Square     F Value    Prob > F
Model         2   1439.84445    719.92223     9.194      0.0009
Error        27   2114.16521     78.30242
C Total      29   3554.00967

                    Parameter Estimates
                   Parameter      Standard     T for H0:        Prob
Variable     DF    Estimate        Error      Parameter=0      > |T|
INTERCEP      1    63.681896    4.95305242      12.857        0.0001
X4            1     1.642435    0.39005384       4.211        0.0003
Z             1     3.933948    3.24618128       1.212        0.2361
```

14. A company wishes to compare three different point-of-sale promotions for its snack foods. The three promotions are:

Promotion 1: Buy two items, get a third free.

Promotion 2: Mail in a rebate for $1.00 with any $2.00 purchase.

Promotion 3: Buy reduced-price multi-packs of each snack food.

The company is interested in the average increase in sales volume due to the promotions. Fifteen grocery stores were selected in a targeted market, and each store was randomly assigned one of the promotion types. During the month-long run of the promotions, the company collected data on increase in sales volume (Y, in hundreds of units) at each store, to be gauged against average monthly sales volume (X, in hundreds of units) prior to the promotions. Let $Z_1 = 1$ if promotion 1, or 0 otherwise. Let $Z_2 = 1$ if promotion type 2, or 0 otherwise. The sample data are shown in the following table.

Store	Promotion	Y	X
1	1	12	39
2	1	23	42
3	2	11	23
4	3	17	39
5	3	15	37
6	3	18	31
7	1	12	36
8	2	19	38
9	3	21	33
10	1	13	44
11	1	7	26
12	2	5	20
13	2	8	32
14	3	17	36
15	2	19	29

a. State an ANACOVA regression model for comparing the three promotion types, controlling for average prepromotion monthly sales.

b. Identify the model that should be used to check whether the ANACOVA model in part (a) is appropriate. Carry out the appropriate test.

c. Using ANACOVA, determine adjusted mean increases in sales volume for the three promotions, and test whether they differ significantly from one another. [*Note:* Mean prepromotional average sales volume = 33.6667; unadjusted mean increases in sales volume were 13.4 for promotion 1, 12.4 for promotion 2, and 17.6 for promotion 3.]

Edited SAS Output (PROC REG) for Problem 14

```
  Increase in Sales Volume (Y) Regressed on Prepromotional Average
   Monthly Sales Volume (X) and Promotion Type (Z1, Z2, XZ1, XZ2)
                        Analysis of Variance
                       Sum of          Mean
 Source      DF        Squares        Square     F Value     Prob > F
 Model        5      219.94330      43.98866      2.252       0.1369
 Error        9      175.79003      19.53223
 C Total     14      395.73333
```

(continued)

```
Increase in Sales Volume (Y) Regressed on Prepromotional Average
  Monthly Sales Volume (X) and Promotion Type (Z1, Z2, XZ1, XZ2)
                          (Continued)
                      Parameter Estimates

                Parameter          Standard       T for H0:          Prob
Variable   DF    Estimate           Error      Parameter=0          > |T|
INTERCEP   1    31.921569       24.43508414           1.306        0.2238
X          1    -0.406863        0.69190391          -0.588        0.5710
Z1         1   -39.962733       27.16860093          -1.471        0.1754
Z2         1   -36.295838       26.03369419          -1.394        0.1967
XZ1        1     0.980156        0.75946321           1.291        0.2290
XZ2        1     0.997506        0.75757331           1.317        0.2205

Variable   DF     Type I SS
INTERCEP   1    3139.266667
X          1     115.601870
Z1         1      61.212205
Z2         1       6.851693
XZ1        1       2.413906
XZ2        1      33.863626

 Increase in Sales Volume (Y) Regressed on Prepromotional Average
      Monthly Sales Volume (X) and Promotion Type (Z1, Z2)
                    Analysis of Variance
                   Sum of           Mean
Source    DF       Squares         Square      F Value       Prob > F
Model      3      183.66577       61.22192      3.176         0.0674
Error     11      212.06757       19.27887
C Total   14      395.73333

                    Parameter Estimates
                Parameter          Standard       T for H0:          Prob
Variable   DF    Estimate           Error      Parameter=0          > |T|
INTERCEP   1     0.300449        7.58359834           0.040        0.9691
X          1     0.491465        0.20809575           2.362        0.0377
Z1         1    -5.281222        2.81445171          -1.876        0.0874
Z2         1    -1.858041        3.11671580          -0.596        0.5631

Variable   DF     Type I SS
INTERCEP   1    3139.266667
X          1     115.601870
Z1         1      61.212205
Z2         1       6.851693
```

15. In Problem 19 in Chapter 5 and Problem 14 in Chapter 8, data from the 1990 Census for 26 randomly selected Metropolitan Statistical Areas (MSAs) were discussed. Of interest were factors potentially associated with the rate of owner occupancy of housing units. The following three variables were included in the data set:

OWNEROCC: Proportion of housing units that are owner-occupied (as opposed to renter-occupied)

OWNCOST: Median selected monthly ownership costs, in $

URBAN: Proportion of population living in urban areas

It is also of interest to see whether the average owner occupancy rate differs in metropolitan areas where 75% or more of the population lives in urban areas from metropolitan areas where less than 75% of the population lives in urban areas, while controlling for ownership costs. For this purpose, the following additional variable is defined:

$Z = 1$ if proportion of population living in urban areas ≥ 0.75

0 otherwise

a. State an ANACOVA regression model that can be used to compare the two MSA types, controlling for average monthly ownership costs.

b. State the model that should be used to check whether the ANACOVA model in part (a) is appropriate. Using the computer output given next, carry out the appropriate test. [*Note:* In the SAS output, the variable OWNCOST has been centered to avoid collinearity problems.]

c. Using ANACOVA methods and the accompanying output, determine adjusted mean owner occupancy rates for the two types of MSAs, and test whether they significantly differ from one another. [*Note:* The unadjusted average owner occupancy rate for MSAs where at least 75% of the population lives in urban areas = 64.5%; the unadjusted average owner occupancy rate for MSAs where less than 75% of the population lives in urban areas = 69.25%.]

Edited SAS Output for Problem 15

```
        OWNEROCC Regressed on OWNCOST, Z, and OWNCOST*Z
                    Analysis of Variance
                      Sum of          Mean
Source      DF       Squares         Square      F Value     Prob > F
Model        3      214.19496       71.39832       4.061       0.0194
Error       22      386.76658       17.58030
C Total     25      600.96154

                    .  [Portion of output omitted]
                    .
                    .

            OWNEROCC Regressed on OWNCOST and Z
                    Analysis of Variance
                      Sum of          Mean
Source      DF       Squares         Square      F Value     Prob > F
Model        2      213.49860      106.74930       6.337       0.0064
Error       23      387.46294       16.84621
C Total     25      600.96154

                    .  [Portion of output omitted]
                    .
                    .

                    Parameter Estimates
                    Parameter       Standard      T for H0:        Prob
Variable    DF       Estimate         Error      Parameter=0       > |T|
INTERCEP     1      68.665453      1.47336083       46.605         0.0001
OWNCOST      1      -0.01266&      0.00552565       -2.293         0.0314
Z            1      -3.905669      1.78250451       -2.191         0.0388
```

16. This problem refers to the radial keratotomy study data from Problem 12 in Chapter 8. Suppose that we wish to compare the average change in refraction for males and females, controlling for baseline refractive error and baseline curvature. To this end, we define the following dummy variable:

$$Z = \begin{array}{ll} 1 & \text{if patient is male} \\ 0 & \text{otherwise} \end{array}$$

a. State an ANACOVA regression model that can be used to compare the mean change in refraction (Y) for males and females, controlling for baseline refractive error (X_1) and baseline curvature (X_2).

b. State the model that should be used to check whether the ANACOVA model in part (a) is appropriate. Using the computer output given next, carry out the appropriate test.

c. Using ANACOVA methods and the SAS output, determine adjusted mean changes in refractive error for males and females, and test whether they differ significantly from one another. [*Note:* Unadjusted average change in refraction for males = 3.64 diopters; unadjusted average change in refraction for females = 3.965 diopters; overall mean baseline refraction = -4.03 diopters; overall mean baseline curvature = 44.02 diopters.]

Edited SAS Output for Problem 16

```
              Y Regressed on Z, X1, X2, X1Z, and X2Z
                      Analysis of Variance

                        Sum of        Mean
Source       DF        Squares       Square     F Value    Prob > F
Model         5       24.71516      4.94303      4.272      0.0027
Error        48       55.53778      1.15704
C Total      53       80.25294

         Root MSE        1.07566      R-square     0.3080
         Dep Mean        3.83343      Adj R-sq     0.2359
         C.V.           28.05993

                      Parameter Estimates

                     Parameter       Standard     T for H0:       Prob
Variable     DF      Estimate          Error     Parameter=0      > |T|
INTERCEP      1     13.065628      6.97300361         1.874      0.0671
Z             1     -0.914275     10.01056704        -0.091      0.9276
X1            1     -0.202204      0.10538796        -1.919      0.0610
X2            1     -0.223889      0.15755988        -1.421      0.1618
X1Z           1     -0.358378      0.19863985        -1.804      0.0775
X2Z           1     -0.024258      0.22572002        -0.107      0.9149

                Y Regressed on Z, X1, and X2
                      Analysis of Variance

                        Sum of        Mean
Source       DF        Squares       Square     F Value    Prob > F
Model         3       20.90031      6.96677      5.869      0.0016
Error        50       59.35263      1.18705
C Total      53       80.25294
```

```
              Y Regressed on Z, X1, and X2 (Continued)
                 Analysis of Variance (Continued)
         Root MSE        1.08952    R-square       0.2604
         Dep Mean        3.83343    Adj R-sq       0.2161
         C.V.           28.42156
                      Parameter Estimates
                     Parameter      Standard     T for H0:        Prob
Variable      DF     Estimate         Error    Parameter=0       > |T|
INTERCEP       1    13.719864     5.06712978          2.708      0.0093
Z              1    -0.508595     0.30607853         -1.662      0.1028
X1             1    -0.303822     0.09040997         -3.360      0.0015
X2             1    -0.247698     0.11408078         -2.171      0.0347
```

Reference

Patrick, R.; Cassel, J. C.; Tyroler, H. A.; Stanley, L.; and Wild, J. 1974. "The Ponape Study of the Health Effects of Cultural Change." Paper presented to annual meeting of Society for Epidemiologic Research, Berkeley, California.

Selecting the Best
Regression Equation

16-1 Preview

The general problem to be discussed in this chapter is as follows: we have one response variable Y and a set of k predictor variables $X_1, X_2,\ldots, X_k$; and we want to determine the *best* (*most important* or *most valid*) subset of the k predictors and the corresponding *best-fitting* regression model for describing the relationship between Y and the X's.[1] What exactly we mean by "best" depends in part on our overall goal for modeling. Two such goals were described in Chapter 11 and are briefly reviewed now.

One goal is to find a model that provides the best prediction of Y, given $X_1, X_2,\ldots, X_k$, for some new observation or for a batch of new observations. In practice, we emphasize estimating the regression of Y on the X's (see Chapter 8)—$E(Y \mid X_1, X_2,\ldots, X_k)$—which expresses the mean of Y as a function of the predictors. Using this goal, we may say that our best model is *reliable* if it predicts well in a new sample. The details of the model may be of little or no consequence, such as whether we include any particular variable or what magnitude or sign we use for its regression coefficient. For example, in considering a sample of systolic blood pressures, we may simply wish to predict blood pressure (SBP) as a function of demographic variables like AGE, RACE, and GENDER. We may not care which variables are in the model or how they are defined, as long as the final model obtained gives the best prediction possible.

Alongside the question of prediction is the question of validity—that is, of obtaining accurate estimates for one or more regression coefficient parameters in a model and then making inferences about these parameters of interest. The goal here is to quantify the relationship between one or more independent variables of interest and the dependent variable, controlling for the other variables. For example, we might wish to describe the relationship of SBP to AGE, controlling for RACE and GENDER. In this case, we are focusing on regression coefficients

---

[1]Hocking (1976) provided an exceptionally thorough and well-written review of this topic. His presentation is at a technical level somewhat higher than that of this text.

involving AGE (including functions of AGE); although RACE and GENDER may remain in the model for control purposes, their regression coefficients are not of primary interest.

In this chapter we focus on strategies for selecting the best model when the primary goal of analysis is prediction. In the last section we briefly mention a strategy for modeling in situations where the validity of the estimates of one or more regression coefficients is of primary importance (see also Chapter 11).

16-2 Steps in Selecting the Best Regression Equation

To select the best regression equation, carry out the following steps:

1. Specify the maximum model (defined in section 16-3) to be considered.

2. Specify a criterion for selecting a model.

3. Specify a strategy for selecting variables.

4. Conduct the specified analysis.

5. Evaluate the reliability of the model chosen.

By following these steps, you can convert the fuzzy idea of finding the best predictors of Y into simple, concrete actions. Each step helps to ensure reliability and to reduce the work required. Specifying the maximum model forces the analyst to state the analysis goal clearly, recognize the limitations of the data at hand, and describe the range of plausible models explicitly. All computer programs for selecting models require that a maximum model be specified. In doing so, the analyst should consider all available scientific knowledge. In turn, specifying the criterion for selecting a model and the strategy for applying that criterion (i.e., for selecting variables) simplifies and speeds the analysis process. Finally, whether the primary goal is prediction or validity, the reliability of the chosen model must be demonstrated.

We illustrate the recommended process of analysis via two examples. The first of these was introduced in Chapter 8. The data (given in Table 8-1) are the hypothetical results of measuring a response variable weight (WGT) and predictor variables height (HGT) and age (AGE) on 12 children. The second example uses real data from a study involving more than 200 subjects (Lewis and Taylor 1967, discussed in Chapter 12).

16-3 Step 1: Specifying the Maximum Model

The *maximum model* is defined to be the largest model (the one having the most predictor variables) considered at any point in the process of model selection. All other possible models can be created by deleting predictor variables from the maximum model. A model created by deleting predictors from the maximum model is called a *restriction* of the maximum model.

Throughout this chapter we assume that the maximum model with k variables or some restriction of it with $p \le k$ variables is the *correct model* for the population. An important implication of this assumption is that the *population* squared multiple correlation for the maximum model—namely, $\rho^2(Y \mid X_1, X_2, \ldots, X_k)$—is no larger than that for the correct model (which may have fewer variables); as a result, adding more predictors to the correct model does not increase the *population* squared multiple correlation for the correct model. In turn, for the sample at hand, $R^2(Y \mid X_1, X_2, \ldots, X_k)$ for the maximum model is always at least as large as the corresponding R^2 for any subset model. However, other sample-based criteria may not necessarily suggest that the largest model is best.

To illustrate, consider the data from Table 8-1, reproduced in the following table.

Child	1	2	3	4	5	6	7	8	9	10	11	12
Y (WGT)	64	71	53	67	55	58	77	57	56	51	76	68
X_1 (HGT)	57	59	49	62	51	50	55	48	42	42	61	57
X_2 (AGE)	8	10	6	11	8	7	10	9	10	6	12	9

One possible model (not necessarily the maximum one) to consider is given by

$$Y = \beta_0 + \beta_1 X_1 + \beta_2 X_2 + E$$

This model allows for only a linear (planar) relationship between the response (WGT) and the two predictors (HGT and AGE). Nevertheless, the nature of growth suggests that the relationship, although monotonic in both HGT and AGE, may well be nonlinear. This implies that at least one quadratic (squared) term may have to be included in the model. Since HGT and AGE are highly correlated, and since the sample size is very small,[2] only $(\text{AGE})^2$ will be considered. Should the interaction between AGE and HGT be considered? Should transformations of the predictors and/or the response be considered (e.g., replacing AGE and $(\text{AGE})^2$ with log (AGE))? The limitations of the data lead us to define the maximum model as

$$Y = \beta_0 + \beta_1 X_1 + \beta_2 X_2 + \beta_3 X_3 + E$$

with $X_1 = \text{HGT}$, $X_2 = \text{AGE}$, and $X_3 = (\text{AGE})^2$.

The most important reason for choosing a large maximum model is to minimize the chance of making a Type II (false negative) error. In a regression analysis, a Type II error corresponds to omitting a predictor that has a truly nonzero regression coefficient in the population. Other reasons for considering a large maximum model are the wishes to include the following elements:

1.   All conceivable basic predictors

2.   High-order powers of basic predictors $((\text{AGE})^2, (\text{HGT})^2)$

3.   Other transformations of predictors (log (AGE), $1/\text{HGT}$)

4.   Interactions among predictors (e.g., AGE $\times$ HGT), including two-way and higher-order interactions

5.   All possible "control" variables, as well as their powers and interactions

Recall that overfitting a model (including variables in the model with truly zero regression coefficients in the population) will not introduce bias when population regression coefficients are estimated, if the usual regression assumptions are met. We must be careful, however, to

---

[2]Because the sample size here is so small, it almost precludes making reliable conclusions from any regression analysis. This small sample size is used to simplify our discussion.

ensure that overfitting does not introduce harmful collinearity (Chapter 12). Underfitting (i.e., leaving important predictors out of the final model), too, will introduce bias in the estimated regression coefficients.

There are also important reasons for working with a small maximum model. With a prediction goal, the need for reliability strongly argues for a small maximum model; and with a validity goal, we want to focus on a few important variables. In either case, we wish to avoid a Type I (false positive) error. In a regression analysis, a Type I error corresponds to including a predictor that has a zero population regression coefficient. The desire for parsimony is another important reason for choosing a small maximum model: practically unimportant but statistically significant predictors can greatly confuse the interpretation of regression results; and complex interaction terms are particularly troublesome in this regard.

The particular sample of data to be analyzed imposes certain constraints on the choice of the maximum model. In general, the smaller the sample size, the smaller the maximum model should be. The idea here is that a larger number of independent observations are needed to estimate reliably a larger number of regression coefficients. This notion has led to various guidelines about the size of a maximum model. The most basic constraint is that the error degrees of freedom must be positive. Symbolically, we require

$$\text{d.f. error} = n - k - 1 > 0$$

which is equivalent to the constraint

$$n > k + 1$$

As always, n is the number of observations and k is the number of predictors, giving $k + 1$ regression coefficients (including the intercept). With negative error degrees of freedom, the model has at least one perfect collinearity; consequently, unique estimates of coefficients and variances cannot be computed. With zero error degrees of freedom ($n = k + 1$), unique estimates of the coefficients can be computed, but unique estimates of the variances cannot (recall that $\hat{\sigma}^2 = \text{SSE/d.f. error}$). Furthermore, even if the population correlation is 0, $R^2 = 1.00$ when d.f. error $= 0$, which reflects the fact that the model exactly fits the observed data (i.e., SSE $= 0$). This is the most extreme example of the positive bias in R^2. In such a situation, we have exchanged n values of Y for n estimated regression coefficients. Hence we have gained nothing, since the dimensionality of the problem remains the same.

The question then arises as to how many error degrees of freedom are needed. Simple rules can be suggested to ensure that the estimates for a *single* model are reliable, but these rules are inadequate when applied to a series of models. Later in this chapter split-sample approaches will be offered as possible ways to assess reliability.

The weakest requirement is for a minimum of approximately 10 error degrees of freedom—namely,

$$n - k - 1 \geq 10$$

or

$$n \geq 10 + k + 1$$

Another suggested rule of thumb for regression is to have at least 5 (or 10) observations per predictor—namely, to require that

$$n \geq 5k \qquad (\text{or } n \geq 10k)$$

Assume, for example, that we wish to consider a maximum model involving 30 predictors. To have 10 error degrees of freedom, we need a sample of size 41, and $n \geq 5k$ demands a sample of size 150. Split-sample approaches, discussed later in this chapter, may reduce the required sample size substantially.

Another constraint on the maximum model involves the amount of variability present in the predictor values, considered either individually or jointly. If a predictor has the same value for all subjects, it obviously cannot be used in any model. For example, consider the variable GENDER of child (male $= 1$, female $= 0$) as a candidate predictor for the weight example. If all of the subjects are male, then GENDER $= 1$ for all subjects, the sample variance for the variable is 0.00, and it is perfectly collinear with the intercept variable. Clearly, if all subjects are of one gender, comparisons of gender cannot be made.

Similarly, consider the variables GENDER, RACE (white $= 0$, black $= 1$), and their interaction (GENRACE $=$ GENDER $\times$ RACE). If no black females are represented in the data, the 1 degree of freedom interaction effect cannot be estimated. If a race–sex cell is nearly empty, the estimated interaction coefficient may be very unstable (i.e., have a large variance).

Polynomial terms (e.g., X^2 and X^3) and other transformations merit particular consideration when the maximum model is being specified. For the weight example, we might wish to consider AGE, $(AGE)^2$, and exp(AGE) as possible predictors. Near collinearities (Chapter 12) can lead to very unstable and often uninterpretable results when we try to find the best model. (See Marquardt and Snee 1975, for further discussion of this topic.) Consequently, we should attempt to reduce such collinearity if possible. Centering, if applicable, almost always helps increase numerical accuracy, as does multiplying variables by various constants (a special form of scaling; see Chapter 12) to produce nearly equal variances for all predictors (say, variances around 1).

If the collinearity problem cannot be overcome, then we have three recourses: conduct separate analyses for each form of X (for example, one analysis using X and X^2 and then a separate analysis using exp(X)); eliminate some variables; or impose structure (e.g., a fixed order of tests) on the testing procedure. If we do nothing about severe collinearity, the estimated regression coefficients in the best model may be highly unstable (i.e., have high variances) and may be quite far from the true parameter values.

16-4 Step 2: Specifying a Criterion for Selecting a Model

The second step in selecting the best model is to specify the selection criterion. A *selection criterion* is an index that can be computed for each candidate model and used to compare models. Thus, given one particular selection criterion, candidate models can be ordered from best to worst. This helps automate the process of choosing the best model. As we shall see, this selection-criterion-specific process may not find the *best* model in a global sense. Nonetheless, using a specific selection criterion can substantially reduce the work involved in finding a *good* model.

Obviously, the selection criterion should be related to the goal of the analysis. For example, if the goal is reliable prediction of future observations, the selection criterion should be somewhat liberal, to avoid missing useful predictors. Distinctions can be drawn among numerical differences, statistically significant differences, and scientifically important differences. Numerical differences may or may not correspond to significant or important differences. With sensitive tests (typical of analyses of large samples), significant differences may or may not correspond to important differences. And, with insensitive tests (typical of analyses of very small samples), important differences may not be significant.

Many selection criteria for choosing the best model have been suggested. Hocking (1976), for example, reviewed eight candidates. We consider four criteria: R_p^2, F_p, MSE(p), and C_p. But before discussing these, we must define the notation needed to understand them. All four criteria attempt to compare two model equations: the maximum model with k predictors and a restricted model with p predictors ($p \leq k$). The maximum model is

$$Y = \beta_0 + \beta_1 X_1 + \beta_2 X_2 + \cdots + \beta_p X_p + \beta_{p+1} X_{p+1} + \cdots + \beta_k X_k + E \qquad (16.1)$$

and the reduced model (a restriction of the maximum model) is

$$Y = \beta_0 + \beta_1 X_1 + \beta_2 X_2 + \cdots + \beta_p X_p + E \qquad (16.2)$$

Let SSE(k) be the error sum of squares for the k-variable model, let SSE(p) be the error sum of squares for the p-variable model, and so on. Also, $SSY = \sum_{i=1}^{n}(Y_i - \overline{Y})^2$ is the total (corrected) sum of squares for the response Y. Often $p = k - 1$, which is the case when we evaluate the addition or deletion of a single variable. We assume that the $k - p$ variables under consideration for addition or deletion are denoted $X_{p+1}, X_{p+2}, \ldots, X_k$, for notational convenience.

The sample squared multiple correlation R^2 is a natural candidate for deciding which model is best, so we will discuss this criterion first. The multiple R^2 for the p-variable model is

$$R_p^2 = R^2(Y \mid X_1, X_2, \ldots, X_p) = 1 - \frac{\text{SSE}(p)}{\text{SSY}} \qquad (16.3)$$

Unfortunately, R_p^2 has three potentially misleading characteristics. First, it tends to overestimate ρ_p^2, the corresponding population value. Second, adding predictors, even useless ones, can never decrease R_p^2. In fact, adding variables invariably increases R_p^2, at least slightly. Finally, R_p^2 is always largest for the maximum model, even though a better model may be obtained by deleting some (or even many) variables. The reduced (or restricted) model may be better because it may sacrifice only a negligible amount of predictive strength while substantially simplifying the model.

Another reasonable criterion for selecting the best model is the F test statistic for comparing the full and restricted models. This statistic, F_p, can be expressed in terms of sums of squares for error (SSEs) as

$$F_p = \frac{[\text{SSE}(p) - \text{SSE}(k)]/(k - p)}{\text{SSE}(k)/(n - k - 1)} = \frac{[\text{SSE}(p) - \text{SSE}(k)]/(k - p)}{\text{MSE}(k)} \qquad (16.4)$$

This statistic may be compared to an F distribution with $k - p$ and $n - k - 1$ degrees of freedom. The criterion F_p tests whether SSE(p) − SSE(k), the difference between the residual sum of squares for the p-variable model and the residual sum of squares for the maximum model, differs significantly from 0. If F_p is *not* significant, we can use the smaller (p-variable) model and achieve roughly the same predictive ability as that yielded by the full model. Hence, a reasonable rule for selecting variables is to retain p variables if F_p is not significant and if p is as small as possible. An often-used special case of F_p occurs when $p = k - 1$, in which case F_p is a test of H_0: $\beta_k = 0$ in the full model.

A third criterion to consider in selecting the best model is the estimated error variance for the p-variable model—namely,

$$\text{MSE}(p) = \frac{\text{SSE}(p)}{n - p - 1} \qquad (16.5)$$

In considering one particular model, we earlier symbolized this estimate as S^2. The quantity $MSE(p)$ is an inviting choice for a selection criterion, since we wish to find a model with small residual variance.

A less obvious candidate for a selection criterion involving $SSE(p)$ is Mallow's C_p:

$$C_p = \frac{SSE(p)}{MSE(k)} - [n - 2(p + 1)] \qquad (16.6)$$

The C_p criterion helps us to decide how many variables to put in the best model, since it achieves a value of approximately $p + 1$ if $MSE(p)$ is roughly equal to $MSE(k)$ (i.e., if the correct model is of size p). Knowing the correct model size greatly aids in choosing the best model.

The criteria F_p, R_p^2, $MSE(p)$, and C_p are intimately related. For example, the F_p test can be expressed in terms of multiple squared correlations as follows:

$$F_p = \frac{(R_k^2 - R_p^2)/(k - p)}{(1 - R_k^2)/(n - k - 1)} \qquad (16.7)$$

and the C_p statistic is the following simple function of the F_p statistic:

$$C_p = (k - p)F_p + (2p - k + 1) \qquad (16.8)$$

The reason to consider more than one criterion is that no single criterion is always best. In practice, the alternatives can lead to different model choices. In the remainder of the chapter we shall consider all of the criteria mentioned, at least to some extent. An important aspect of our discussion will be a demonstration of the limitations of R_p^2 as the sole criterion for selecting a model. We favor C_p because it tends to simplify the decision about how many variables to retain in the final model.

16-5 Step 3: Specifying a Strategy for Selecting Variables

The third step in choosing the best model is to specify the strategy for selecting variables. Such a strategy is concerned with determining how many variables and also which particular variables should be in the final model. Traditionally, such strategies have focused on deciding whether a single variable should be added to a model (a forward selection method) or whether a single variable should be deleted from a model (a backward elimination method). As computers became more powerful, methods for considering more than one variable per step became practical (by generalizing single-variable methods to deal with sets, or *chunks*, of variables). Before discussing these strategies in detail, we consider an algorithm for evaluating models that, if feasible to conduct, should be the method of choice.

16-5-1 All Possible Regressions Procedure

Whenever practical, the *all possible regressions procedure* is to be preferred over any other variable selection strategy. It is the only method guaranteed to find the model having the largest R_p^2, the smallest $MSE(p)$, and so on. This strategy is not always used because the amount of calculation necessary becomes impractical when the number of variables k in the maximum model is large. The all possible regressions procedure requires that we fit each possible regression equation associated with each possible combination of the k independent variables. For our example, we must fit the seven models corresponding to the following seven sets of independent variables: HGT; AGE; $(AGE)^2$; HGT and AGE; HGT and $(AGE)^2$; AGE and $(AGE)^2$; and HGT, AGE, and $(AGE)^2$. For k independent variables, the number of models to be fitted would be $2^k - 1$; for example, if $k = 10$, then $2^{10} - 1 = 1,023$.

Once all $2^k - 1$ models have been fitted, we assemble the fitted models into sets involving from 1 to k variables and then order the models within each set according to some criterion (e.g., R_p^2, F_p, MSE(p), or C_p).

For our data, a summary of the results of the all possible regressions procedure appears in Table 16-1. From the table, the leaders (in terms of R_p^2-values) in each of the sets involving one, two, and three variables are

One-variable set: HGT with $R_1^2 = .6630$

Two-variable set: HGT, AGE with $R_2^2 = .7800$

Three-variable set: HGT, AGE, $(AGE)^2$ with $R_3^2 = .7802$

Of the three models (models 1, 4, and 7, respectively, in Table 16-1), model 4, involving HGT and AGE, should clearly be our choice, since its R^2-value is essentially the same as that for model 7 and is much higher than the value for model 1. Thus, our choice of the best regression equation based on the all possible regressions procedure with R_p^2 as the criterion is

$$WGT = 6.553 + 0.722HGT + 2.050AGE$$

Now consider the partial F statistics in Table 16-1. For a given variable in a given model, the associated partial F statistic assesses the contribution made by that variable to the prediction of Y (WGT) over and above the contributions made by other variables already in the given model. For example, for model 4, which involves only X_1 and X_2, the partial F for X_2 is $F(X_2 \mid X_1) = 4.785$; but for model 7, which includes X_1, X_2, and X_3, the partial F for X_2 is $F(X_2 \mid X_1, X_3) = 0.140$.

TABLE 16-1   Summary of results of all possible regressions procedure.

Model	No. of Variables	Variables Used	Estimated Coefficients				Partial F Statistics			Overall F Statistic	R_p^2	MSE(p)	C_p
			$\hat{\beta}_0$	$\hat{\beta}_1$	$\hat{\beta}_2$	$\hat{\beta}_3$	X_1	X_2	X_3				
1	1	HGT (X_1)	6.190	1.073			19.67**			19.67**	.6630	29.93	4.27
2	1	AGE (X_2)	30.571		3.643			14.55**		14.55**	.5926	36.18	6.83
3	1	$(AGE)^2$ (X_3)	45.998			0.206			14.25**	14.25**	.5876	36.63	7.01
4	2	HGT, AGE	6.553	0.722	2.050		7.665*	4.785		15.95**	.7800	21.71	2.01
5	2	HGT, $(AGE)^2$	15.118	0.726		0.115	7.601*		4.565	15.63**	.7764	22.07	2.14
6	2	AGE, $(AGE)^2$	32.404		3.205	0.025		0.113	0.002	6.55*	.5927	40.20	8.83
7	3	HGT, AGE, $(AGE)^2$	3.438	0.724	2.777	−0.042	6.827*	0.140	0.010	9.47**	.7802	24.40	4.00

These partial F's are variables-added-last tests, each based on the MSE for the corresponding model being fit for that row. Such tests must be treated with some caution. Any test based on a model with fewer terms than the correct model will be biased, perhaps substantially, because the test involves using biased error terms. Furthermore, so many tests are computed that the Type I error rate should be higher than the nominal rate α.

Overall F tests are also affected by the estimate of the error variance used. In variable selection, this estimate is important because biased tests conducted early in a stepwise algorithm may stop the procedure prematurely and miss important predictors. As calculated in Table 16-1, each overall F statistic is based on the corresponding MSE listed in the same row. If the MSE for the largest model, number 7, had been used in the denominator of each overall F test instead, each resulting test statistic would have been an F_p statistic. Such a statistic involves a comparison of the R_p^2-value for a reduced model (models 1 through 6) and the R_k^2-value for the largest model (model 7). For example, for model 4,

$$F_p = \frac{(R_k^2 - R_p^2)/(k - p)}{(1 - R_k^2)/(n - k - 1)} = \frac{(.7802 - .7800)/(3 - 2)}{(1 - .7802)/(12 - 3 - 1)} = 0.007$$

In general, such a test is a multiple partial F test. The small F-value here indicates that the predictive abilities of the maximum model and of model 4 do not differ significantly.

Table 16-1 also provides C_p-values. The value of C_p is expected to be close to $p + 1$ if the correct model—or a larger model that contains the correct model—is considered. If important predictors are omitted, C_p should be larger than $p + 1$. Also, if $F_p < 1$, then $C_p < (p + 1)$; this can occur when R_p^2 is close enough to R_k^2 in value. Users prefer models with C_p not too far from $p + 1$. Therefore, for a model with one variable, C_p is compared with the value 2.0; for two variables, with the value 3.0; and so on. For this example, no one-variable model has a C_p-value near 2.0. For the only three-variable model, C_p is exactly 4.00. The full model with k predictors is guaranteed to have C_p exactly equal to $k + 1$; to see this, examine equation (16.6), the formula for C_p. The two-variable model with AGE and $(\text{AGE})^2$ has a C_p-value much greater than 3, while models 4 and 5 have C_p-values near the minimum possible C_p-value of

$$\frac{(n - k - 1)\text{MSE}(k)}{\text{MSE}(k)} - [n - 2(p + 1)] = (2p - k + 1)$$

which equals 2 when $k = 3$ and $p = 2$. (We can see from (16.8) that this lower bound for C_p is attained when $F_p = 0$, and that it can be negative.) Such a value is better than the value of $p + 1 = 3$.

The all possible regressions procedure was presented first, since it is preferred whenever practical. It alone is guaranteed to find the best model, in the sense that any selection criterion will be numerically optimized for the particular sample under study. Naturally, using it does not guarantee finding the correct (or population) model. In fact, in many situations, several reasonable candidates for the best model can be found, with different selection criteria suggesting different best models. Furthermore, such findings may vary from sample to sample, even though all the samples are chosen from the same population. Consequently, the choice of the best model may vary from sample to sample. These considerations motivate the discussion in section 16-7 of evaluating the reliability of the chosen regression equation.

As mentioned earlier, the all possible regressions algorithm is often impractical, since $2^k - 1$ models must be fitted when k candidate predictors are being evaluated. Many methods have been suggested as computationally feasible alternatives for approximating the all possible regressions procedure. Although these methods are not guaranteed to find the best model, they can (with careful use) glean essentially all the information from the data needed to choose the best model.

16-5-2 Backward Elimination Procedure

In the *backward elimination procedure,* we proceed as follows:

1. *Determine the fitted regression equation containing all independent variables* (i.e., the estimated maximum model). From the accompanying SAS computer output, we obtain

$$\widehat{\text{WGT}} = 3.438 + 0.724\text{HGT} + 2.777\text{AGE} - 0.042(\text{AGE})^2$$

2. *Determine the partial F statistic for every variable in the model as though it were the last variable to enter, and determine the p-values associated with the test statistics.* The partial F statistics and *p*-values are indicated in the Step 0 portion of the SAS output. (Recall that the partial F statistics test whether adding the last variable to the model significantly helps predict the dependent variable, given that the other variables are already in the model.)

3. *Focus on the lowest observed partial F statistic (or, equivalently, on the highest p-value).* From the Step 0 portion of the SAS output, we see that we should focus on $(\text{AGE})^2$.

4. *Compare the p-value with a preselected significance level (say, 10%) and decide whether to remove the variable under consideration. If the variable is dropped, recompute the regression equation for the remaining variables, and repeat backward elimination procedure steps 2, 3, and 4. If the variable is not dropped, the backward elimination procedure ends, and the selected model consists of variables remaining in the model.* From the Step 1 portion of the SAS output, we see that $(\text{AGE})^2$ is dropped and the regression is recomputed. With $(\text{AGE})^2$ out of the picture, the smallest partial F statistic is 4.78, with an associated *p*-value of .0565. The *p*-value is less than .10, so we stop here with this model, which is the same model that we arrived at using the all possible regressions procedure:

$$\widehat{\text{WGT}} = 6.553 + 0.722\text{HGT} + 2.050\text{AGE}$$

Edited SAS Output (PROC REG) for Backward Elimination Procedure

```
        Backward Elimination Procedure for Dependent Variable WGT
Step 0   All Variables Entered   R-square = 0.78025383
         C(p) = 4.00000000
                             Sum of              Mean              Prob
                    DF       Squares            Square       F     > F
Regression           3     693.06046340      231.02015447    9.47   0.005
Error                8     195.18953660       24.39869208
Total               11     888.25000000
```

(continued)

```
              Backward Elimination Procedure for Dependent
                      Variable WGT (Continued)
                                                       Partial F      P-value
Step 0   (Continued)                                   statistics        ↓
                                              Type II     ↓
                Parameter         Standard     Sum of               Prob
Variable        Estimate          Error        Squares      F        > F
INTERCEP        3.43842600       33.61081984   0.25534520   0.01    0.9210
HGT             0.72369024        0.27696316 166.58195495   6.83    0.0310
AGE             2.77687456        7.42727877   3.41051231   0.14    0.7182
AGE2           -0.04170670        0.42240715   0.23785686   0.01    0.9238
Bounds on condition number:    89.96948,    543.7942
------------------------------------------------------------------------
Step 1   Variable AGE2 Removed   R-square = 0.77998605
         C(p) = 2.00974875
                                 Sum of         Mean                 Prob
              DF                 Squares        Square       F        > F
Regression     2             692.82260654   346.41130327   15.95   0.0011
Error          9             195.42739346    21.71415483
Total         11             888.25000000
                                              Type II
                Parameter         Standard     Sum of               Prob
Variable        Estimate          Error        Squares      F        > F
INTERCEP        6.55304825       10.94482708   7.78416178   0.36    0.5641
HGT             0.72203796        0.26080506 166.42974940   7.66    0.0218
AGE             2.05012635        0.93722561 103.90008336   4.78    0.0565
Bounds on condition number:     1.604616,    6.418463
------------------------------------------------------------------------
All variables left in the model are significant at the 0.1000 level.
Summary of Backward Elimination Procedure for Dependent Variable WGT
         Variable    Number   Partial    Model                        Prob
Step     Removed       In      R**2      R**2      C(p)        F        > F
  1        AGE2         2      0.0003    0.7800    2.0097    0.0097    0.923
```

16-5-3 Forward Selection Procedure

In the *forward selection procedure,* we proceed as follows:

1. *Select as the first variable to enter the model the variable most highly correlated with the dependent variable, and then fit the associated straight-line regression equation. If the overall F test for this regression is not significant, stop and conclude that no independent variables are important predictors. If the test is significant, include this variable in the model, and proceed to step 2 of the procedure.* From the accompanying SAS output, we see that the highest squared correlation is for X_1, HGT. The straight-line regression equation relating WGT and HGT is

$$\widehat{\text{WGT}} = 6.190 + 1.072\text{HGT}$$

2. *Determine the partial F statistic and p-value associated with each remaining variable, based on a regression equation containing that variable and the variable ini-*

tially selected. For our data, the statistics are shown in the Step 2 portion of the SAS output.

3. *Focus on the variable with the largest partial F statistic (i.e., the variable with the smallest p-value).* From the Step 2 portion of the SAS output, we see that AGE has the largest partial *F* statistic, with a *p*-value of .0565.

4. *Test for the significance of the partial F statistic associated with the variable from Step 3 of the forward selection procedure. If this test is significant, add the new variable to the regression equation. If it is not significant, use in the model only the variable added in 1. above.* For our data, since the *p*-value for AGE is less than .10, we add AGE to get the following two-variable model:

$$\widehat{\text{WGT}} = 6.553 + 0.722\text{HGT} + 2.050\text{AGE}$$

5. *At each subsequent step, determine the partial F statistics for the variables not yet in the model, and then add to the model the variable with the largest partial F value (if it is statistically significant). At any step, if the largest partial F is not significant, no more variables are included in the model and the process is terminated.* For our example we have already added HGT and AGE to the model. We now check to see whether we should add $(\text{AGE})^2$. The partial *F* for $(\text{AGE})^2$, controlling for HGT and AGE, is .01, with a *p*-value of .9238. This value is not statistically significant at $\alpha = .10$. Again, we have arrived at the same two-variable model chosen via the previously discussed methods.

Edited SAS Output (PROC REG) for Forward Selection Procedure

```
     Forward Selection Procedure for Dependent Variable WGT
                  Statistics for Entry: Step 0
                          DF = 1,10
                                 Model
         Variable    Tolerance    R**2             F        Prob > F
         HGT         1.000000     0.6630       19.6749        0.0013
         AGE         1.000000     0.5926       14.5470        0.0034
         AGE2        1.000000     0.5876       14.2481        0.0036
  Step 1  Variable HGT Entered   R-square = 0.66301438
          C(p) = 4.26817716
                              Sum of          Mean                 Prob
                 DF          Squares        Square         F        > F
  Regression      1       588.92252318   588.92252318   19.67     0.0013
  Error          10       299.32747682    29.93274768
  Total          11       888.25000000
                                              Type II
                 Parameter      Standard      Sum of                Prob
  Variable       Estimate       Error         Squares        F       > F
  INTERCEP       6.18984871    12.84874620    6.94680569    0.23    0.6404
  HGT            1.07223036     0.24173098  588.92252318   19.67    0.0013
  Bounds on  condition number:          1,            1
  ------------------------------------------------------------------------
```

(continued)

```
Forward Selection Procedure for Dependent Variable WGT (Continued)
                 Statistics for Entry: Step 2
                          DF = 1,9
                                Model
       Variable      Tolerance     R**2              F        Prob > F
       AGE           0.623202      0.7800         4.7849        0.0565
       AGE2          0.621230      0.7764         4.5647        0.0614
Step 2  Variable AGE Entered    R-square = 0.77998605
        C(p) = 2.00974875
                             Sum of           Mean                 Prob
                  DF         Squares          Square         F      > F
Regression        2       692.82260654    346.41130327     15.95  0.0011
Error             9       195.42739346     21.71415483
Total            11       888.25000000
                                            Type II
                 Parameter      Standard     Sum of              Prob
Variable         Estimate        Error       Squares       F      > F
INTERCEP        6.55304825    10.94482708   7.78416178    0.36   0.5641
HGT             0.72203796     0.26080506 166.42974940    7.66   0.0218
AGE             2.05012635     0.93722561 103.90008336    4.78   0.0565
Bounds on condition number:     1.604616,    6.418463
--------------------------------------------------------------------
                 Statistics for Entry: Step 3
                          DF = 1,8
                                Model
       Variable      Tolerance     R**2              F        Prob > F
       AGE2          0.011115      0.7803         0.0097        0.9238
  No other variable met the 0.1000 significance level for entry
                          into the model.
Summary of Forward Selection Procedure for Dependent Variable WGT
          Variable  Number Partial    Model                         Prob
Step      Entered     In    R**2      R**2       C(p)       F        > F
  1        HGT         1    0.6630    0.6630     4.2682   19.6749   0.0013
  2        AGE         2    0.1170    0.7800     2.0097    4.7849   0.0565
```

16-5-4 Stepwise Regression Procedure

Stepwise regression is a modified version of forward regression that permits reexamination, at every step, of the variables incorporated in the model in previous steps. A variable that entered at an early stage may become superfluous at a later stage because of its relationship with other variables subsequently added to the model. To check this possibility, at each step we make a partial F test for each variable currently in the model, as though it were the most recent variable entered, irrespective of its actual entry point into the model. The variable with the smallest nonsignificant partial F statistic (if there is such a variable) is removed; the model is refitted with the remaining variables; the partial F's are obtained and similarly examined; and so on. The whole process continues until no more variables can be entered or removed.

For our example, the first step, as in the forward selection procedure, is to add the variable HGT to the model, since it has the highest significant correlation with WGT. Next, as before, we add AGE to the model, since it has a higher significant partial correlation with WGT than does

$(AGE)^2$, controlling for HGT. Now, before testing to see whether $(AGE)^2$ should also be added to the model, we look at the partial F of HGT, given that AGE is already in the model, to see whether HGT should now be removed. This partial is given by $F(X_1 \mid X_2) = 7.665$ (see Table 16-1), which exceeds $F_{1,9,0.90} = 3.36$. Thus, we do not remove HGT from the model. Next we check to see whether we should add $(AGE)^2$; of course, the answer is no, since we have dealt with this situation before.

The analysis-of-variance table that best summarizes the results obtained for our example is as follows:

Source		d.f.	SS	MS	F	R^2
Regression	$\begin{cases} X_1 \\ X_2 \mid X_1 \end{cases}$	1 1	588.92 103.90	588.92 103.90	19.67** 4.79 $(P < .10)$	.7800
Residual		9	195.43	21.71		
Total		11	888.25			

The ANOVA table that considers all variables is

Source		d.f.	SS	MS	F	R^2
Regression	$\begin{cases} X_1 \\ X_2 \mid X_1 \\ X_3 \mid X_1, X_2 \end{cases}$	1 1 1	588.92 103.90 0.24	588.92 103.90 0.24	19.67** 4.79 $(P < .10)$ 0.01	.7802
Error		8	195.19	24.40		
Total		11	888.25			

16-5-5 Using Computer Programs

So far we have discussed decisions in stepwise variable selection in terms of F_p and its p-value, R_p^2, C_p, and MSE(p). In backward elimination, the F statistic is often called an "F-to-leave"; while in forward selection, the F statistic is often called an "F-to-enter." Stepwise computer programs require specifying comparison or critical values for these F's or specifying their associated significance levels. These values must not be interpreted as they are in a single-hypothesis test. Their limitation derives from the fact that the probability of finding at least one significant independent variable when there are actually none increases rapidly as the number k of candidate independent variables increases—an approximate upper bound on this overall significance level being $1 - (1 - \alpha)^k$, where α is the significance level of any one test.

To prevent this overall significance level from exceeding α in value, a conservative but easily implemented approach is to conduct any one test at level α/k (see Pope and Webster 1972, and Kupper, Stewart, and Williams 1976, for further discussion). In section 16-7 we discuss other techniques for helping to ensure the reliability of our conclusions.

Since almost all model selection methods involve using the data to generate hypotheses for further study (often called "exploratory data analysis"), p-values and other variable selection criteria must be utilized cleverly. For example, model selection programs often allow you to specify a p-value to help decide whether a variable is to be considered significant enough to be included in a model. It is very helpful to specify a "p-to-enter" of 1.00 for forward selection.

This guarantees that the process will go as far as possible, thereby providing the maximum information for choosing the best model. Similarly, specifying a value of 0.00 for the *p*-to-leave guarantees that a backward elimination process will remove all variables.

16-5-6 Chunkwise Methods

The methods just described for selecting single variables can be generalized in a very useful way. The basic idea is that any selection method in which a single variable is added or deleted can be generalized to apply to adding or deleting a group of variables. Consider, for example, using backward elimination to build a model for which the response variable is blood cholesterol level. Assume that three groups of predictors are available: demographic (gender, race, age, and their pairwise interactions), anthropometric (height, weight, and their interaction), and diet recall (amounts of five food types); hence, a total of $6 + 3 + 5 = 14$ predictors exists. The three groups of variables constitute *chunks*—sets of predictors that are logically related and equally important (within a chunk) as candidate predictors.

Several possible chunkwise testing methods are available. Choosing among them depends on two factors: the analyst's preference for backward, forward, or other selection strategies; and the extent to which the analyst can logically group variables (i.e., form chunks) and then order the groups in importance. In many applications, an a priori order exists among the chunks. For this example, the researcher may wish to consider diet variables *only* after controlling for important demographic and anthropometric variables. Imposing order in this fashion simplifies the analysis, typically increases reliability, and increases the chance of finding a scientifically plausible model. Hence, whenever possible, we recommend imposing an order on chunk selection.

We illustrate the use of chunkwise testing methods by describing a backward elimination strategy. Other strategies may be preferred, depending on the situation. One approach to a backward elimination method for chunkwise testing involves requiring all but one specified chunk of variables to stay in the model, the variables in that specified chunk being candidates for deletion.[3] For our example, assume that the set of diet variables is the first chunk considered for deletion. (If an a priori order among chunks exists, that order determines which chunk is considered first for deletion. Otherwise, we choose first the chunk that makes the least important predictive contribution (e.g., because it has the smallest multiple partial *F* statistic).)

Thus, in our example, all of the demographic and anthropometric variables are forced to remain in the model. If, for example, the chunkwise multiple partial *F* test for the set of diet variables is not significant, the entire chunk can be deleted. If this test shows that the chunk is significant, at least one of the variables in the diet chunk should be retained.

Of course, the simplest chunkwise method adds or deletes all variables in a chunk together. A more sensitive approach is to manipulate single variables within a significant chunk while keeping the other chunks in place. If we assume that the diet chunk is important, we must decide which of the diet variables to retain as important predictors. A reasonable second step here is then to require all demographic variables and the important individual diet variables to be retained, while considering the (second) chunk of anthropometric variables for deletion. The final step in this three-chunk example requires that the individual variables selected from the first two chunks remain in the model, while variables in the third (demographic) chunk become

---

[3]Stepwise regression computer packages permit this approach by letting the user specify variables to be forced into the model. The same feature can be used for subsequent chunkwise steps.

candidates for deletion. Forward and stepwise single-variable selection methods can also be generalized for use in chunkwise testing.

Chunkwise methods for selecting variables can have substantial advantages over single-variable selection methods. First, chunkwise methods effectively incorporate into the analysis prior scientific knowledge and preferences about sets of variables. Second, the number of possible models to be evaluated is reduced. If a chunk test is not significant and the entire chunk of variables is deleted, no tests on individual variables in that chunk need be carried out. In many situations, such testing for group (or chunk) significance is more effective and reliable than testing variables one at a time.

16-6 Step 4: Conducting the Analysis

Having specified the maximum model, the criterion for selecting a model, and the strategy for applying the criterion, we must conduct the analysis as planned. Obviously this requires some type of computer program. The goodness of fit of the model chosen should certainly be examined. And the regression diagnostic methods of Chapter 12, such as residual analysis, are needed to demonstrate that the model chosen is reasonable for the data at hand. In the next section, we discuss methods for evaluating whether the model chosen has a good chance of fitting new samples of data from the same (or similar) populations.

16-7 Step 5: Evaluating Reliability with Split Samples

Having chosen a model that is best for a particular sample of data, we have no assurance that the model can reliably be applied to other samples. In effect, we are now asking the question "Will our conclusions generalize?" If the chosen model predicts well for subsequent samples from the population of interest, we say that the model is *reliable*. In this section we discuss methods for evaluating the reliability of a model. Most generally accepted methods for assessing model reliability involve some form of a *split-sample* approach. We discuss three approaches to assessing model reliability here: the follow-up study, the split-sample analysis, and the holdout sample.

The most compelling way to assess the reliability of a chosen model is to conduct a new study and test the fit of the chosen model to the new data. However, this approach is usually expensive and sometimes intractable. Can a single study achieve the two goals of finding the best model and assessing its reliability?

A split-sample analysis attempts to do so. For the cholesterol example, the simplest split-sample analysis would proceed as follows. First, we randomly assign all observations to one or the other of two groups, the training group or the holdout group. This must be done before any analysis is conducted. Subjects may be grouped in strata based on one or more important categorical variables. For the cholesterol example, appropriate strata might be defined by various gender–race combinations. If such strata are important, we may use a stratum-specific split-sample scheme. With this method, we randomly assign all subjects within a stratum to the training group or the holdout group; such assignment is done separately for each stratum. The goal of stratum-specific random assignment is to ensure that the two groups of observations (training and holdout) are equally representative of the parent population.

An alternative to stratified random splitting is a pair-matching assignment scheme. With pair-matching assignment, we finds pairs of subjects that are as similar as possible and then randomly assign one member of the pair to the training sample and the other to the holdout sample. Unfortunately, the differences between the resulting training and holdout groups tend to be fewer than corresponding differences among subsequent randomly chosen samples. This tends to create an unrealistically optimistic opinion about model reliability, which we wish to avoid.

Either of two alternatives is usually recommended as the second step in a split-sample analysis. The first is to conduct model selection separately for each of the two groups of data. Typically, any difference in predictor variables chosen by the two selection processes is taken as an indication of unreliability. In practice, the two models obtained *almost always* differ somewhat, which is the primary reason model selection methods have a reputation for being unreliable. This approach for assessing reliability is far too stringent when prediction is the goal. More realistically, we suggest that a *very good* predictive model should accomplish two things: predict as well in any new sample as it does in the sample at hand; and pass all regression diagnostic tests for model adequacy applied to any new sample. These comments lead us to consider a second approach to split-sample analysis for assessing reliability.

The second approach attempts to answer the question "Will the chosen model predict well in a new sample?" This approach thus addresses a relatively modest, yet very desirable goal. With this second approach, we begin by conducting a model-building process using the data for the training group. Suppose that the fitted prediction equation obtained for the training group data is denoted as

$$\hat{Y}_1 = \hat{\beta}_0 + \hat{\beta}_1 X_1 + \hat{\beta}_2 X_2 + \cdots + \hat{\beta}_p X_p$$

For the children's weight example, this equation takes the form

$$\widehat{\text{WGT}} = 6.553 + 0.722\text{HGT} + 2.050\text{AGE}$$

Next, we use the estimated prediction equation, which is based only on the training group data, to compute predicted values for this group. We denote this set of predicted values as $\{\hat{Y}_{i1}\}$, $i = 1, 2,\ldots, n_1$, with the subscript i indexing the subject and the subscript 1 denoting the training group. For this training sample, we let

$$R^2(1) = R^2(Y_1 \mid X_1, X_2,\ldots, X_p) = r^2(Y_1, \hat{Y}_1)$$

denote the sample squared *multiple* correlation, which equals the sample squared *univariate* correlation between the observed and predicted response values.

Next, we use the prediction equation for the training group to compute predicted values for the holdout (or "validation") sample. We denote this set of predicted values as

$$\{\hat{Y}_{i2}^*\}, \qquad i = n_1 + 1, n_1 + 2,\ldots, n_1 + n_2$$

where n_2 is the number of observations in the holdout group and $n = n_1 + n_2$.

Finally, we compute the univariate correlation between these predicted values and the observed responses in the holdout sample (group 2)—namely,

$$R_*^2(2) = r^2(Y_2, \hat{Y}_2^*)$$

The quantity $R_*^2(2)$ is called the *cross-validation correlation,* and the quantity

$$R^2(1) - R_*^2(2)$$

is called the *shrinkage on cross-validation.* Typically, the cross-validation correlation $R_*^2(2)$ is a less biased estimator of the population squared multiple correlation than is the (positively) biased $R^2(1)$. Hence, the shrinkage statistic is almost always positive. How large must shrinkage be to cast doubt on model reliability? No firm rules can be given. Certainly the fitted model is unreliable if shrinkage is .90 or more. In contrast, shrinkage values less than .10 indicate a reliable model.

Using only half the data to estimate the prediction equation appears rather wasteful. If the shrinkage is small enough to indicate reliability of the model, it seems reasonable to pool the data and calculate pooled estimates of the regression coefficients.

Depending on the situation, using only half of the data for the training sample analysis may be inadvisable. A useful rule of thumb is to increase the relative size of the training sample as the total sample size decreases and to decrease the relative size of the training sample as the total sample size increases. To illustrate this, consider two situations. In the first situation, data from over 3,000 subjects were available for regression analysis. The maximum model included fewer than 20 variables, consisting of a chunk of demographic control variables and a chunk of pollution exposure variables. The primary goal of the study was to test for the presence of a relationship between the response variable (a measure of pulmonary function) and the pollution variables, controlling for demographic effects. As a framework for choosing the best set of demographic control variables, a 10% stratified random sample ($n \approx 300$) was used as a training sample.

The second situation involved a study of the relationship between measurements of amount of body fat by X-ray methods (CAT scans) and by traditional parameters such as skin-fold thickness. The primary goal was to provide an equation to predict the X-ray-measured body fat from demographic information (gender and age) and simple anthropometric measures (body weight, height, and three readings of skinfold thickness). The maximum model contained fewer than 20 predictors. Data from approximately 200 subjects were available. These data constituted nearly one year's collection from the hospital X-ray laboratory and demanded substantial human effort and computer processing to extract. Consequently, approximately 75% of the data were used as a training sample. These two examples illustrate the general concept that the splitting proportion should be tailored to the problem under study.

16-8 Example Analysis of Actual Data

We are now ready to apply the methods discussed in this chapter to a set of actual data. In Chapter 12 (on regression diagnostics), we introduced data on more than 200 children, reported by Lewis and Taylor (1967). The categories reported for all subjects include body weight (WGT), standing height (HGT), AGE, and GENDER. Our goal here is to provide a reliable equation for predicting weight.

The first step is to choose a maximum model. The number of subjects is large enough that we can consider a fairly wide range of models. It is natural to want to include the *linear* effects of AGE, HGT, and GENDER in the model. Quadratic and cubic terms for both AGE and HGT will also be included, since WGT is expected to increase in a nonlinear way as a function of AGE and HGT. Because the same model is unlikely to hold for both males and females, the interaction of the dichotomous variable GENDER with each power of AGE and HGT will be included. Terms involving cross-products between AGE and HGT powers will not be included, since no biological fact suggests that such interactions are important. Hence, 13 variables will be included in the maximum model as predictors of WGT: AGE; AGE2 $= (AGE)^2$; AGE3 $= (AGE)^3$; HGT; HT2 $= (HGT)^2$; HT3 $= (HGT)^3$; MALE $= 1$ for a boy, and 0 for a girl; MALE $\times$ AGE; MALE $\times$ AGE2; MALE $\times$ AGE3; MALE $\times$ HGT; MALE $\times$ HT2; and MALE $\times$ HT3. (In the analysis to follow, AGE and HGT denote centered variables.)

The second step is to choose a selection criterion. We choose to emphasize C_p, while also looking at R_p^2. The former criterion is helpful in deciding the size of the best model, while the latter provides an easily interpretable measure of predictive ability.

The third step is to choose a strategy for selecting variables. We prefer backward elimination methods, with as much structure imposed on the process as possible. As mentioned earlier, one procedure is to group the variables into chunks and then to order the chunks by degree of importance. However, we shall not consider a chunkwise strategy here.

After the (seemingly) best model is chosen, the fourth step is to evaluate the reliability of the model. Since the goal here is to produce a reliable prediction equation, this step is especially important. To implement a split-sample approach to assessing reliability, we first randomly assigned the subjects to either data set ONE (the training group) or data set TWO (the holdout group). Random splitting was sex-specific. The following table summarizes the observed split-sample subject distribution.

Sample	Female	Male	Total
ONE	55	63	118
TWO	56	63	119
Total	111	126	237

Data set ONE will be used to build the best model. Then data set TWO will be used to compute cross-validation correlation and shrinkage statistics.

16-8-1 Analysis of Data Set ONE

Table 16-2 provides descriptive statistics for data set ONE, and Table 16-3 provides the matrix of intercorrelations. To avoid computational problems, the predictor variables were centered (see Table 16-2 and Chapter 12).

Table 16-4 summarizes the backward elimination analysis based on data set ONE. For data set ONE, SSY $= 35,700.6949$ and MSE $= 120.0027676$ for the full model. Table 16-4 specifies the order of variable elimination, the variables in the best model for each possible model size, and C_p and R_p^2 for each such model. Since the maximum model includes $k = 13$ variables, the top row corresponds to $p = 13$, for which $R_p^2 = .65042$. Since the maximum model always has

TABLE 16-2 Descriptive statistics for data set ONE ($n = 118$).

Variable	Mean	S.D.	Minimum	Maximum
WGT (lb)	100.05	17.47	63.5	150.0
AGE[a] (yr)	0	1.50	−2.00	3.91
HGT[b] (in.)	0	3.7	−8.47	10.73

[a] Equals (Age $-\overline{AGE}_1$), where $\overline{AGE}_1 \doteq 13.59$.

[b] Equals (Height $-\overline{HGT}_1$), where $\overline{HGT}_1 \doteq 61.27$.

TABLE 16-3   Correlations ($\times 100$) for data set ONE with centered age and height variables ($n = 118$).

WGT													
AGE	59												
AGE2	25	44											
AGE3	48	78	80										
HGT	77	67	16	43									
HT2	05	10	23	11	15								
HT3	63	46	14	32	79	35							
MALE	03	−01	−04	−03	18	18	24						
MALE × AGE	46	72	28	50	62	29	46	00					
MALE × AGE2	18	26	55	38	27	35	32	51	37				
MALE × AGE3	38	58	49	60	46	30	40	12	81	70			
MALE × HGT	57	55	16	34	81	41	68	10	77	27	57		
MALE × HT2	26	22	19	17	43	78	70	50	31	53	36	47	
MALE × HT3	48	38	17	27	64	65	88	16	54	32	46	78	75

TABLE 16-4   Backward elimination based on data set ONE ($n = 118$).

p	C_p	R_p^2	Variables Used
13	14.00	.65042	HT3
12	12.00	.65042	MALE × AGE
11	10.00	.65042	AGE2
10	8.00	.65040	MALE × AGE2
9	6.00	.65040	MALE × AGE3
8	4.03	.65032	HT2
7	2.08	.65015	AGE
6	0.38	.64914	MALE
5	−0.82	.64644	MALE × HT3
4	1.15	.63312	MALE × HT2
3	−0.12	.63066	MALE × HGT
2	0.52	.62177	AGE3
1	6.72	.59422	HGT

$C_p = k + 1$, the C_p-value of 14 provides no information regarding the best model size. The second row corresponds to a $p = 12$ variable model. The variable HT3 (listed in the preceding row) has been deleted, leaving $C_p = 12$. Similarly, the bottom row tells us that the best single-variable model includes HGT ($C_p = 6.72$, $R_p^2 = .59422$), the best two-variable model includes HGT and AGE3 ($C_p = 0.52$, $R_p^2 = .62177$), and so on. In general, the variable identified in row p in the table and all variables listed in rows below it are included in the best p-variable model; while all variables listed in rows above row p are excluded.

Table 16-4 does not suggest any obvious choice for a best model. The maximum R^2 is about .650, and the minimum is around .594, which is a small operating range, indicating that many different-size models predict about equally well. This is not unusual with moderate to strong predictor intercorrelations—say, above .50 in absolute value (see Table 16-3). Since

$R_7^2 \doteq .650$ and $R_{13}^2 \doteq .650$, it seems unreasonable to use more than seven predictors. However, on the basis of R_p^2 alone, it is difficult to decide whether a model containing fewer than seven predictors is appropriate.

In contrast, the C_p statistic suggests that only two variables are needed. (Recall that C_p should be compared with the value $p + 1$.) C_p is greater than $p + 1$ only when $p = 1$, which calls into question only the one-variable model in Table 16-4. Unfortunately, the $p = 2$ model with HGT and AGE3 as predictors is not appealing, since the linear and quadratic terms AGE and AGE2 are not included. As discussed in detail in Chapter 13, we strongly recommend including such lower-order terms in polynomial-type regression models.

An apparently nonlinear relationship may indicate the need to transform the response and/or predictor variables. Since quadratic and cubic terms are included as candidate predictors, log(WGT) and $(\text{WGT})^{1/2}$ can be tried as alternative response variables. Such transformations can sometimes linearize a relationship (see Chapter 12). Based on all 13 predictors, $R^2 = .668$ for the log transformation and .659 for the square root transformation. Because neither transformation produces a substantial gain in R^2, they will not be considered further.

In additional exploratory analyses of the data in data set ONE, models can be fitted in two fixed orders: an interaction ordering (Table 16-5), and a power ordering (Table 16-6). We can interpret Tables 16-5 and 16-6 similarly to Table 16-4; each row gives p, the number of variables in the model, and the associated C_p- and R_p^2-values. All variables on or below line p are included in the fitted model, and all those above line p are excluded.

In both tables, certain values of p are natural break points for model evaluation. For example, in Table 16-5, $p = 3$ corresponds to including just the linear terms of the predictors HGT, AGE, and MALE. The fact that $4.98 > 3 + 1$ leads us to consider larger models. Including the pairwise interactions MALE $\times$ HGT and MALE $\times$ AGE gives $p = 5$ and $C_5 = 5.36 < 5 + 1$, which suggests that this five-variable model is reasonable. Notice that $R_5^2 = .626$ and that each subsequent variable addition increases R^2 very little. In fact, the best model might be of size $p = 4$, since $C_4 = 3.60$ and R^2 is not appreciably reduced. Taken together, these comments suggest choosing a four-variable model containing HGT, AGE, MALE, and MALE $\times$ HGT, with

TABLE 16-5   Interaction-ordered fitting based on data set ONE ($n = 118$).

p	C_p	R_p^2	Variables Used
13	14.00	.65042	MALE $\times$ AGE3
12	12.01	.65040	MALE $\times$ HT3
11	10.39	.64909	MALE $\times$ AGE2
10	8.39	.64909	MALE $\times$ HT2
9	6.94	.64727	AGE3
8	6.97	.64045	HT3
7	6.37	.63575	AGE2
6	7.35	.62571	HT2
5	5.36	.62567	MALE $\times$ AGE
4	3.60	.62486	MALE $\times$ HGT
3	4.98	.61352	MALE
2	5.86	.60383	AGE
1	6.72	.59422	HGT

TABLE 16-6   Power-ordered fitting based on data
set ONE ($n = 118$).

p	C_p	R_p^2	Variables Used
13	14.00	.65042	MALE × AGE3
12	12.01	.65040	MALE × HT3
11	10.39	.64909	MALE × AGE2
10	8.39	.64909	MALE × HT2
9	6.94	.64727	MALE × AGE
8	8.00	.64698	MALE × HGT
7	4.77	.64112	AGE3
6	5.05	.63343	HT3
5	4.83	.62747	AGE2
4	6.27	.61589	HT2
3	4.98	.61352	MALE
2	5.86	.60383	AGE
1	6.72	.59422	HGT

$R_4^2 = .625$ and $C_4 = 3.60$.

Table 16-6 summarizes a similar analysis in which the powers of the continuous predictors are entered into the model in a logical ordering. Here, as in Table 16-5, $p = 3$ gives $C_3 = 4.98$, encouraging us to consider larger models. The model with all three original variables and the two quadratic terms gives $p = 5$, $C_p = 4.83$, and $R_p^2 = .62747$. Adding higher-order powers or any interactions does not improve R^2 appreciably, nor does it lead to C_p-values noticeably greater than $p + 1$. All models smaller than $p = 5$, on the other hand, do have C_p-values noticeably greater than $p + 1$. The results in Table 16-6 lead to choosing a five-variable model containing HGT, AGE, MALE, HT2, and AGE2, with $R_5^2 = .627$ and $C_5 = 4.83$.

In this example, then, two possible models have been suggested, both with roughly the same R^2-value. Personal preference may be exercised within the constraints of parsimony and scientific knowledge. In this case, we prefer the five-variable model with HGT, AGE, MALE, HT2, and AGE2 as predictors. We choose this model as best because we expect growth to be nonlinear and because we prefer to avoid using interaction terms. The chosen model is thus of the general form

$$\widehat{WGT} = \hat{\beta}_0 + \hat{\beta}_1 MALE + \hat{\beta}_2 HGT + \hat{\beta}_3 AGE + \hat{\beta}_4 HT2 + \hat{\beta}_5 AGE2$$

A predicted weight for the ith subject may be computed with the formula

$$\widehat{Weight}_i = 100.9 - 3.153 MALE_i + 3.53(Height_i - 61.27) + .4199(Age_i - 13.59)$$
$$- .0731(Height_i - 61.27)^2 + .8067(Age_i - 13.59)^2$$

Recall that MALE has a value of 1 for boys and 0 for girls. For males, the estimated intercept is $100.9 - 3.153(1) = 97.747$; while for females, it is $100.9 - 3.153(0) = 100.9$. For these particular data, then, the fitted model predicts that, on average, girls outweigh boys of the same height and age by about 3 pounds. After selecting a model, we must consider residual analysis and other regression diagnostic procedures (see Chapter 12). Some large residuals are present, but these do not justify a more complex model. Since they turn out not to be influential, they need not be deleted to improve estimation accuracy.

16-8-2 Analysis of Sample TWO

Since we used exploratory analysis of data set ONE to choose the best (most appealing) model, we will use analysis of the holdout data set TWO to evaluate the reliability of the chosen prediction equation. Table 16-7 provides descriptive statistics for data set TWO. These appear very similar to those of data set ONE, as one would hope. In particular, the AGE and HGT variables are centered on the sample ONE means. This continuity is necessary to maintain the definitions of these variables as used in the fitted model. Cross-validation analysis would look spuriously bad if this were not done.

In comparison to Table 16-4, Table 16-8 illustrates the instability of stepwise methods for selecting variables. Despite using the same backward elimination strategy, we find that almost no variable appears in the same place as it did in the analysis for data set ONE.

Our recommended cross-validation analysis begins by using the regression equation estimated from data set ONE (with predictors HGT, AGE, MALE, HT2, and AGE2) to predict WGT values for data set TWO. The squared multiple correlation between these predicted values and the observed WGT values in data set TWO is .621, and this is the cross-validation correlation. Since $R_2(1) = .627$ for sample ONE, shrinkage is $.627 - .621 = .006$, which is quite small and indicates excellent reliability of estimation.

TABLE 16-7   Descriptive statistics for data set TWO ($n = 119$).

Variable	Mean	S.D.	Minimum	Maximum
WGT (lb)	102.6	21.22	50.5	171.5
AGE[a] (yr)	0.20	1.46	−1.92	3.58
HGT[b] (in.)	0.18	4.16	−10.77	9.73

[a] Equals (Age − $\overline{AGE}_1$), where $\overline{AGE}_1 \doteq 13.59$.

[b] Equals (Height − $\overline{HGT}_1$), where $\overline{HGT}_1 \doteq 61.27$.

TABLE 16-8   Backward elimination based on data set TWO ($n = 119$).

p	C_p	R_p^2	Variables Used
13	14.00	.72328	AGE2
12	12.00	.72328	MALE
11	10.02	.72323	AGE3
10	8.31	.72245	MALE × AGE2
9	7.22	.72006	HT2
8	6.06	.71781	MALE × HT2
7	4.15	.71763	MALE × HGT
6	3.16	.71496	HT3
5	1.52	.71402	MALE × HT3
4	2.58	.70595	MALE × AGE
3	7.99	.68642	AGE
2	9.60	.67690	MALE × AGE3
1	33.90	.60762	HGT

An important aspect of assessing the reliability of a model involves considering difference scores of the form

$$y_i - \hat{y}_i$$

or (in our example)

$$\text{WGT}_i - \widehat{\text{WGT}}_i$$

where only data set TWO subjects are used and where the data set ONE equation is used to compute the predicted values. These "unstandardized residuals" can be subjected to various residual analyses. The most helpful such analysis entails univariate descriptive statistics, a box-and-whisker plot, and a plot of the difference scores $(y_i - \hat{y}_i)$ versus the predicted values $(\hat{y}_i)$ (such plots are described in Chapter 12). In our example, a few large positive residuals are present, but they are not sufficiently implausible or influential to require further investigation. Their presence does hint at why cubic terms and interactions keep trying to creep into the model. These residuals could be reduced in size, but only by adding many more predictors.

For prediction purposes, if the model is finally deemed acceptable, we should pool all the data and report the coefficient estimates based on the combined data. If the model is not acceptable, we must review the model-building process, paying particular attention to variable selection and large differences between training and holdout groups of data.

16-9 Issues in Selecting the Most Valid Model

In this chapter, we have considered variable selection strategies for situations where the primary study goal is prediction. In contrast, validity-oriented strategies are directed toward providing valid estimates of one or more regression coefficients in a model. Essentially, a valid estimate accurately reflects the true relationship(s) in the target population being studied. As described in Chapter 11, in a validity-based strategy for selecting variables, both confounding and interaction must be considered. Moreover, the sample of data under consideration must accurately reflect the population of interest; and the assumptions attending the model and analysis selected must be reasonably well satisfied.

More specifically, a variable selection strategy with a validity goal first involves determining important interaction effects, followed by evaluating the effect of confounding, which is contingent on the results of assessing interaction in the analysis. For examples of this type of strategy, see the epidemiologic-research-oriented text of Kleinbaum, Kupper, and Morgenstern (1982, Chapters 21–24), and Kleinbaum (1994).

Problems

1. Add the variables $(\text{HGT})^2$ and $(\text{AGE} \times \text{HGT})$ to the data set for WGT, HGT, AGE, and $(\text{AGE})^2$ given in section 16-3. Then, using an available regression program and the accompanying computer output, determine the best regression model relating WGT to the five independent variables as follows.
   a. Use the forward approach.
   b. Use the backward approach.

    c. Use an approach that first determines the best model, employing HGT and AGE as the only candidate independent variables, and then determines whether any second-order terms should be added to the model. (Use $\alpha = .10$.)

    d. Compare and discuss the models obtained for each of the three approaches.

Edited SAS Output (PROC REG) for Problem 1

```
SSY = 888.25000
N = 12      Regression Models for Dependent Variable: WGT
Number in   R-square      C(p)          MSE Variables in Model
    Model
        1   0.75353959  -0.44530   21.891845 AGEHGT
        1   0.66751456   2.19160   29.533019 HGT2
        1   0.66301438   2.32955   29.932748 HGT
        1   0.59261791   4.48739   36.185714 AGE
        1   0.58759589   4.64133   36.631795 AGE2
     ----------------------------------------------------------
        2   0.77998605   0.74404   21.714155 HGT AGE
        2   0.77641424   0.85352   22.066672 HGT AGE2
        2   0.77551977   0.88094   22.154952 HGT AGEHGT
        2   0.77313977   0.95390   22.389844 AGE2 AGEHGT
        2   0.77240359   0.97646   22.462502 AGE HGT2
        2   0.77113982   1.01520   22.587228 HGT2 AGEHGT
        2   0.76770289   1.12055   22.926434 AGE2 HGT2
        2   0.76687655   1.14588   23.007989 AGE AGEHGT
        2   0.66812487   4.17289   32.754231 HGT HGT2
        2   0.59271434   6.48444   40.196832 AGE AGE2
     ----------------------------------------------------------
        3   0.80277917   2.04537   21.897676 HGT HGT2 AGEHGT
        3   0.80018558   2.12487   22.185645 HGT AGE2 HGT2
        3   0.79586296   2.25737   22.665591 HGT AGE HGT2
        3   0.78025383   2.73583   24.398692 HGT AGE AGE2
        3   0.78022600   2.73668   24.401782 HGT AGE AGEHGT
        3   0.77693268   2.83763   24.767444 AGE AGE2 AGEHGT
        3   0.77641992   2.85335   24.824376 HGT AGE2 AGEHGT
        3   0.77388224   2.93114   25.106138 AGE2 HGT2 AGEHGT
        3   0.77338486   2.94638   25.161362 AGE AGE2 HGT2
        3   0.77247774   2.97419   25.262081 AGE HGT2 AGEHGT
     ----------------------------------------------------------
        4   0.80425435   4.00015   24.838725 HGT AGE HGT2 AGEHGT
        4   0.80313839   4.03436   24.980332 HGT AGE2 HGT2 AGEHGT
        4   0.80058226   4.11271   25.304687 HGT AGE AGE2 HGT2
        4   0.78026281   4.73556   27.883080 HGT AGE AGE2 AGEHGT
        4   0.77705760   4.83380   28.289798 AGE AGE2 HGT2 AGEHGT
     ----------------------------------------------------------
        5   0.80425917   6.00000   28.977798 HGT AGE AGE2 HGT2 AGEHGT
     ----------------------------------------------------------
```

  2. Using the data given in Problem 2 in Chapter 5 (with SBP as the dependent variable) and the accompanying computer output, find the best regression model, using $\alpha = .05$ and the independent variables AGE, QUET, and SMK as follows.

    a. Use the forward approach.

b. Use the backward approach.

c. Use the all possible regressions approach.

d. Based on your results in parts (a) through (c), select a model for further analysis to determine which, *if any*, of the following interaction (i.e., product) terms should be added to the model: AQ = AGE × QUET, AS = AGE × SMK, QS = QUET × SMK, and AQS = AGE × QUET × SMK.

Edited SAS Output (PROC REG) for Problem 2

```
SSY = 6425.96875
N = 32        Regression Models for Dependent Variable: SBP
Number in    R-square        C(p)          MSE    Variables in Model
   Model
       1     0.60094136    18.75391     85.47795   AGE
       1     0.55057002    24.65545     96.26743   QUET
       1     0.06117337    81.99341    201.09569   SMK
    ------------------------------------------------------------------------
       2     0.72980190     5.65655     59.87188   AGE SMK
       2     0.64124063    16.03243     79.49574   AGE QUET
       2     0.64120550    16.03655     79.50353   SMK QUET
    ------------------------------------------------------------------------
       3     0.76094763     4.00750     54.86225   AGE SMK QUET
       3     0.74049903     6.40327     59.55518   AGE SMK AS
       3     0.67763692    13.76822     73.98197   AGE QUET AQ
       3     0.65112480    16.87440     80.06647   SMK QUET QS
    ------------------------------------------------------------------------
       4     0.79009614     2.59244     49.95688   AGE SMK QUET AQ
       4     0.76481454     5.55445     55.97387   AGE SMK QUET AQS
       4     0.76389199     5.66254     56.19343   AGE SMK QUET AS
       4     0.76156044     5.93570     56.74834   AGE SMK QUET QS
    ------------------------------------------------------------------------
       5     0.79222668     4.34283     51.35172   AGE SMK QUET AQ AS
       5     0.79184514     4.38753     51.44602   AGE SMK QUET AQ AQS
       5     0.79083289     4.50613     51.69621   AGE SMK QUET AQ QS
       5     0.77498209     6.36321     55.61377   AGE SMK QUET QS AQS
       5     0.76501101     7.53143     58.07815   AGE SMK QUET AS AQS
       5     0.76496332     7.53702     58.08994   AGE SMK QUET AS QS
    ------------------------------------------------------------------------
       6     0.79350600     6.19294     53.07696   AGE SMK QUET AQ QS AQS
       6     0.79270441     6.28686     53.28300   AGE SMK QUET AS QS AQS
       6     0.79253885     6.30625     53.32556   AGE SMK QUET AQ AS QS
       6     0.79226386     6.33847     53.39624   AGE SMK QUET AQ AS AQS
    ------------------------------------------------------------------------
       7     0.79515282     8.00000     54.84757   AGE SMK QUET AQ AS QS AQS
    ------------------------------------------------------------------------
```

3. For the same data set you considered in Problem 2 (and employing the accompanying computer output), use the stepwise regression approach to find the best regression models of SBP on QUET, AGE, and QUET × AGE for smokers and nonsmokers separately. (Use $\alpha = .05$.) Compare the results for each group here with those you got for Problem 2(d).

Edited SAS Output (PROC REG) for Problem 3

```
----------------------------SMK=0----------------------------
           Stepwise Procedure for Dependent Variable SBP
                 Statistics for Entry: Step 1
                          DF = 1,13
                                     Model
           Variable      Tolerance     R**2            F        Prob > F
           AGE           1.000000     0.8029       52.9435       0.0001
           QUET          1.000000     0.7307       35.2746       0.0001
           AQ            1.000000     0.8381       67.3098       0.0001
Step 1   Variable AQ Entered   R-square = 0.83812692
         C(p) = 0.28924556
                             Sum of              Mean             Prob
             DF            Squares              Square         F    > F
Regression    1      1953.17096690      1953.17096690      67.31  0.0001
Error        13       377.22903310        29.01761793
Total        14      2330.40000000
                                                 Type II
              Parameter        Standard          Sum of              Prob
Variable       Estimate         Error           Squares        F     > F
INTERCEP     93.07315336      5.98128777     7026.21151957   242.14  0.0001
AQ            0.24951048      0.03041232     1953.17096690    67.31  0.0001
Bounds on condition number:        1,        1
----------------------------------------------------------------
                 Statistics for Entry:   Step 2
                          DF = 1,12
                                     Model
           Variable      Tolerance     R**2            F        Prob > F
           AGE           0.070103     0.8406        0.1873        0.6729
           QUET          0.089970     0.8419        0.2896        0.6003
All variables left in the model are significant at the 0.0500 level.
No other variable met the 0.0500 significance level for entry into
the model.
       Summary of Stepwise Procedure for Dependent Variable SBP
           Variable
           Entered   Number  Partial   Model                         Prob
Step       Removed     In      R**2     R**2      C(p)        F        > F
 1   AQ                 1     0.8381   0.8381   0.2892   67.3098    0.0001
----------------------------SMK=1----------------------------
           Stepwise Procedure for Dependent Variable SBP
                 Statistics for Entry: Step 1
                          DF = 1,15
                                     Model
           Variable      Tolerance     R**2            F        Prob > F
           AGE           1.000000     0.6737       30.9693       0.0001
           QUET          1.000000     0.5640       19.4032       0.0005
           AQ            1.000000     0.6978       34.6296       0.0001
Step 1   Variable AQ Entered   R-square = 0.69776125
         C(p) = 2.33277164
                             Sum of              Mean             Prob
             DF            Squares              Square         F    > F
Regression    1      2583.44048925      2583.44048925      34.63  0.0001
Error        15      1119.03009898        74.60200660
Total        16      3702.47058824
```

```
Step 1  (Continued)
                                              Type II
                   Parameter        Standard   Sum of                  Prob
Variable           Estimate          Error     Squares        F        > F
INTERCEP     102.21983310      8.02768730  12095.94906067   162.14    0.0001
AQ             0.25167268      0.04276733   2583.44048925    34.63    0.0001

Bounds on condition number:          1,         1
-------------------------------------------------------------------------
                  Statistics for Entry: Step 2
                         DF = 1,14

                                     Model
       Variable      Tolerance       R**2              F       Prob > F
       AGE           0.126133        0.7104        0.6109      0.4475
       QUET          0.078094        0.7311        1.7374      0.2086

All variables left in the model are significant at the 0.0500 level.
No other variable met the 0.0500 significance level for entry into
the model.
      Summary of Stepwise Procedure for Dependent Variable SBP
         Variable
         Entered   Number  Partial   Model                          Prob
Step     Removed     In     R**2     R**2      C(p)        F        > F
  1      AQ           1     0.6978   0.6978   2.3328    34.6296    0.0001
```

4. For the data given in Problem 4 in Chapter 8 (plus the accompanying computer output), find (using $\alpha = .10$) the best regression model relating homicide rate (Y) to population size (X_1), percentage of families with income less than \$5,000 (X_2), and unemployment rate (X_3).
   a. Use the stepwise approach.
   b. Use the backward approach.
   c. Use the all possible regressions approach.

Edited SAS Output (PROC REG) for Problem 4

```
SSY = 1855.20200
N = 20     Regression Models for Dependent Variable: Y
Number in   R-square         C(p)          MSE     Variables in Model
  Model
    1       0.74795075      6.19694    25.977905    X3
    1       0.70522751      9.95940    30.381251    X2
    1       0.00450219     71.66941   102.602751    X1
-------------------------------------------------------------------------
    2       0.80199321      3.43765    21.608388    X2 X3
    2       0.76715744      6.50549    25.409999    X1 X3
    2       0.71032611     11.51039    31.611976    X1 X2
-------------------------------------------------------------------------
    3       0.81831787      4.00000    21.066066    X1 X2 X3
-------------------------------------------------------------------------
```

5. The data set listed in the following table contains information on AGE, SEX (1 = male, 2 = female), work problems index (WP), marital conflict index (MC), and depression index (DEP) for a sample of 39 new admissions to a psychiatric clinic at a large university hospital. For each sex *separately,* determine (using $\alpha = .10$) the best regression model relating DEP to MC and WP, controlling for AGE, using the following sequential procedure: (1) force AGE into the model first; (2) use the all possible regressions approach on the remaining two independent variables WP and MC; (3) determine whether the interaction

ID No.	AGE	SEX	WP	MC	DEP
1	45	2	90	70	69
2	35	1	90	75	75
3	32	2	70	32	35
4	32	2	80	30	73
5	39	2	85	55	86
6	25	2	85	6	161
7	22	1	75	20	202
8	30	2	70	63	91
9	49	2	75	4	113
10	47	1	84	12	68
11	48	1	64	11	109
12	49	2	85	7	92
13	45	2	80	8	80
14	41	2	80	15	82
15	45	2	82	6	156
16	59	2	72	5	198
17	42	2	70	17	170
18	35	1	70	29	188
19	31	2	70	80	82
20	45	1	70	126	37
21	28	1	85	30	194
22	37	1	90	9	294
23	29	1	80	14	94
24	29	1	70	24	126
25	31	1	80	21	192
26	29	1	60	11	232
27	29	1	70	10	184
28	23	2	80	10	238
29	44	2	78	19	112
30	28	1	70	22	141
31	32	2	70	21	108
32	36	2	74	77	87
33	22	2	78	67	33
34	46	2	70	25	73
35	21	1	70	14	168
36	34	1	80	17	218
37	27	2	80	18	175
38	31	2	80	42	126
39	19	2	75	36	135

term (MC × WP) should be added to the model. Compare and discuss the results obtained for each sex.

6. For the data in Problem 15 in Chapter 5, use LN_BRNTL as the response variable and LN_PPMTL, LN_BLDTL, AGE, and WEIGHT as predictors. (Use $\alpha = .05$.)

   a. Indicate a plausible fixed order for testing predictors, based on the nature of the data. Briefly defend your choice.

   b. Use a computer program to fit the full model. Provide a test of whether the multiple correlation is 0.

   c. Using the *fixed* order, choose a best model, adding variables in a forward fashion. Do not adhere to a particular α, but use your judgment.

   d. Repeat part (c) but delete the variables in the fixed order (a backward method).

   e. Using a computer program employing a stepwise procedure, use a backward algorithm to find the best model.

   f. Repeat part (e), using a forward algorithm.

   g. Compare and contrast your conclusions for parts (c), (d), (e), and (f). Indicate your preferred model.

   h. Using the ideas presented in Chapter 12, indicate how you could evaluate the adequacy of the best model and the validity of the underlying assumptions.

   i. Suggest a practical split-sample approach for this particular set of data. Include recommended sample sizes, any stratification variables, and variable selection strategy.

7. Use the data of Bethel and others (1985), discussed in Problem 19 in Chapter 12. Delete the three female subjects, leaving 16 observations. Use FEV_1 as the response, and AGE, WEIGHT, and HEIGHT as predictors.

   a. Use the all possible regressions procedure to suggest a best model.

   b. Consider a model with centered AGE, WEIGHT, HEIGHT, and their squares as predictors. Suggest a plausible forward chunkwise strategy for choosing a model, and implement it.

   c. Use the all possible regressions procedure for the expanded model to choose a best model.

   d. Compare results from parts (a), (b), and (c). What model seems most plausible? How do the data limit your conclusions?

8. Use the data from Freund (1979), presented in Problem 22 in Chapter 12. Taking the model discussed there as the maximum model, repeat parts (a) through (h) of Problem 6. In part (h), note the possible role of collinearity.

9. A random sample of data were collected on residential sales in a large city. The accompanying table shows the selling price (Y, in \$1,000s), area (X_1, in hundreds of square feet), number of bedrooms (X_2), total number of rooms (X_3), house age (X_4, in years), and location ($Z = 0$ for in-town and inner suburbs, $Z = 1$ for outer suburbs).

   In parts (a) through (c), use variables X_1, X_2, X_3, X_4, and Z as the predictor variables. Use the accompanying computer output to answer parts (a)–(d).

   a. Use the all possible regressions procedure to suggest a best model.

   b. Use the stepwise regression algorithm to suggest a best model.

   c. Use the backward elimination algorithm to suggest a best model.

   d. Which of the models selected in parts (a), (b), and (c) seems to be the best model, and why?

House	Y	X_1	X_2	X_3	X_4	Z
1	84.0	13.8	3	7	10	0
2	93.0	19.0	2	7	22	1
3	83.1	10.0	2	7	15	1
4	85.2	15.0	3	7	12	1
5	85.2	12.0	3	7	8	1
6	85.2	15.0	3	7	12	1
7	85.2	12.0	3	7	8	1
8	63.3	9.1	3	6	2	1
9	84.3	12.5	3	7	11	1
10	84.3	12.5	3	7	11	1
11	77.4	12.0	3	7	5	0
12	92.4	17.9	3	7	18	0
13	92.4	17.9	3	7	18	0
14	61.5	9.5	2	5	8	0
15	88.5	16.0	3	7	11	0
16	88.5	16.0	3	7	11	0
17	40.6	8.0	2	5	5	0
18	81.6	11.8	3	7	8	1
19	86.7	16.0	3	7	9	0
20	89.7	16.8	2	7	12	0
21	86.7	16.0	3	7	9	0
22	89.7	16.8	2	7	12	0
23	75.9	9.5	3	6	6	1
24	78.9	10.0	3	6	11	0
25	87.9	16.5	3	7	15	0
26	91.0	15.1	3	7	8	1
27	92.0	17.9	3	8	13	1
28	87.9	16.5	3	7	15	0
29	90.9	15.0	3	7	8	1
30	91.9	17.8	3	8	13	1

Edited SAS Output (PROC REG) for Problem 9

```
N = 30       Regression models for Dependent Variable: Y
Number in    R-square          C(p)          MSE      Variables in
Model                                                 Model
   1        0.74168607      11.59122       32.78751    X3
   1        0.63919733      26.50593       45.79629    X1
   1        0.37277539      65.27708       79.61294    X4
   1        0.11031111     103.47228      112.92725    X2
   1        0.01448589     117.41728      125.09024    Z
- - - - - - - - - - - - - - - - - - - - - - - - - - - - - - - - - - -
   2        0.80692546       4.09724       25.41440    X1 X3
   2        0.80634720       4.18139       25.49052    X3 X4
   2        0.75352898      11.86778       32.44298    X3 Z
   2        0.74172797      13.58513       33.99634    X2 X3
```

```
N = 30       Regression models for Dependent Variable: Y (Continued)
Number in    R-square          C(p)            MSE    Variables in
Model                                                 Model
     2       0.70553756     18.85175        38.76009    X1 Z
     2       0.69355902     20.59493        40.33682    X1 X2
     2       0.64446398     27.73950        46.79920    X1 X4
     2       0.57233554     38.23602        56.29347    X2 X4
     2       0.40513240     62.56831        78.30242    X4 Z
     2       0.11465711    104.83983       116.53767    X2 Z
-------------------------------------------------------------------
     3       0.82237293      3.84924        24.28032    X1 X3 X4
     3       0.81899611      4.34065        24.74191    X2 X3 X4
     3       0.81039947      5.59168        25.91700    X1 X2 X3
     3       0.80863284      5.84877        26.15849    X3 X4 Z
     3       0.80801358      5.93889        26.24314    X1 X3 Z
     3       0.75353085     13.86751        33.69053    X2 X3 Z
     3       0.74058026     15.75215        35.46078    X1 X2 Z
     3       0.72307049     18.30026        37.85424    X1 X2 X4
     3       0.70977138     20.23562        39.67213    X1 X4 Z
     3       0.58509132     38.37973        56.71498    X2 X4 Z
-------------------------------------------------------------------
     4       0.83459939      4.06998        23.51341    X1 X2 X3 X4
     4       0.82270177      5.80139        25.20478    X1 X3 X4 Z
     4       0.82090242      6.06324        25.46058    X2 X3 X4 Z
     4       0.81180550      7.38707        26.75380    X1 X2 X3 Z
     4       0.76269650     14.53367        33.73516    X1 X2 X4 Z
-------------------------------------------------------------------
     5       0.83508027      6.00000        24.42193    X1 X2 X3 X4 Z
-------------------------------------------------------------------

               Stepwise Procedure for Dependent Variable Y

Step 1   Variable X3 Entered   R-square = 0.74168607
         C(p) = 11.59122300
                         Sum of            Mean                  Prob
              DF        Squares           Square        F        > F
Regression    1     2635.95947489     2635.95947489    80.40    0.0001
Error        28      918.05019178       32.78750685
Total        29     3554.00966667
                                                 Type II
              Parameter        Standard          Sum of         Prob
Variable      Estimate          Error            Squares    F    > F
INTERCEP  -17.08438356      11.26623604       75.39617765   2.30  0.1406
X3         14.71917808       1.64160396     2635.95947489  80.40  0.0001
Bounds on condition number:      1,       1
-------------------------------------------------------------------
Step 2   Variable X1 Entered   R-square = 0.80692546
         C(p) = 4.09723891
                         Sum of            Mean                  Prob
              DF        Squares           Square        F        > F
Regression    2     2867.82088345     1433.91044172    56.42    0.0001
Error        27      686.18878322       25.41439938
Total        29     3554.00966667
```

(continued)

```
Step 2  (Continued)

                                          Type II
           Parameter      Standard        Sum of                 Prob
Variable   Estimate       Error           Squares        F       > F
INTERCEP   -4.21355813    10.79550361     3.87161368     0.15    0.6994
X1          1.30267900     0.43128376   231.86140856     9.12    0.0055
X3         10.14195666     2.09410969   596.10738063    23.46    0.0001

Bounds on condition number:    2.099378,    8.397511

- - - - - - - - - - - - - - - - - - - - - - - - - - - - - - - - - - - - - - - -

Step 3   Variable X4 Entered   R-square = 0.82237293
         C(p) = 3.84924057

                          Sum of          Mean                   Prob
           DF             Squares         Square         F       > F
Regression  3          2922.72134013   974.24044671    40.12    0.0001
Error      26           631.28832653    24.28032025
Total      29          3554.00966667

                                          Type II
           Parameter      Standard        Sum of                 Prob
Variable   Estimate       Error           Squares        F       > F
INTERCEP   -4.92269366    10.56242154     5.27391395     0.22    0.6451
X1          0.81475833     0.53197084    56.95559271     2.35    0.1377
X3         10.52638005     2.06275692   632.29012250    26.04    0.0001
X4          0.45796577     0.30455957    54.90045668     2.26    0.1447

Bounds on condition number:          3.343224,    22.37951

- - - - - - - - - - - - - - - - - - - - - - - - - - - - - - - - - - - - - - - -

All variables left in the model are significant at the 0.1500 level.

No other variable met the 0.1500 significance level for entry into
the model.

                Summary of Stepwise Procedure for
                    Dependent Variable Y

        Variable
        Entered  Number  Partial   Model                            Prob
Step    Removed   In     R**2      R**2      C(p)        F          > F
  1     X3        1      0.7417    0.7417   11.5912    80.3952     0.0001
  2     X1        2      0.0652    0.8069    4.0972     9.1232     0.0055
  3     X4        3      0.0154    0.8224    3.8492     2.2611     0.1447
```

10. In Problem 9, the first-order interactions between Z and the predictor variables X_1, X_3, and X_4 were not included in the model selection process. Investigate whether this was appropriate, as follows (using the accompanying computer output).

   a. Conduct a partial test to investigate the importance of the interactions, given that X_1, X_2, X_3, X_4, and Z are in the model.

   b. In view of your answer in part (a), and in light of the model selected in Problem 9, was it appropriate to exclude the interactions?

   c. Why is it necessary to ignore the interaction term $X_2 \times Z$?

Edited SAS Output (PROC REG) for Problem 10

```
                        Analysis of Variance
                         Sum of            Mean
Source        DF        Squares           Square      F Value     Prob > F
Model          8     3100.73976        387.59247      17.957       0.0001
Error         21      453.26991         21.58428
C Total       29     3554.00967

          Root MSE        4.64589      R-square      0.8725
          Dep Mean       83.49667      Adj R-sq      0.8239
          C.V.            5.56416

                        Parameter Estimates
                      Parameter        Standard      T for H0:        Prob
Variable      DF       Estimate           Error     Parameter=0      > |T|
INTERCEP       1     -24.707976      16.10563434        -1.534      0.1399
X1             1       0.277195       1.30211784         0.213      0.8335
X2             1       0.283568       3.48718098         0.081      0.9360
X3             1      13.607861       4.98736294         2.728      0.0126
X4             1       1.006553       0.57142756         1.761      0.0927
Z              1      53.467221      26.89803697         1.988      0.0600
X1Z            1       0.706832       1.55572552         0.454      0.6542
X3Z            1      -8.124224       5.76509590        -1.409      0.1734
X4Z            1      -0.676632       0.83800926        -0.807      0.4285

Variable      DF      Type I SS
INTERCEP       1        209151
X1             1   2271.713503
X2             1    193.201955
X3             1    415.252079
X4             1     86.006770
Z              1      1.709055
X1Z            1     89.289861
X3Z            1     29.494884
X4Z            1     14.071648
```

11. In 1990, *Business Week* magazine compiled financial data on the 1,000 companies that had the biggest impact on the U.S. economy.[4] Data from a sample of the top 500 companies in *Business Week*'s report were presented in Problem 13 in Chapter 8. In addition to the company name, the following variables were shown:

1990 Rank (X_1):   Based on company's market value (share price on March 16, 1990, multiplied by available common shares outstanding).

1989 Rank (X_2):   Rank in 1989 compilation.

P-E Ratio (X_3):   Price-to-earning ratio based on 1989 earnings and March 16, 1990, share price.

Yield (Y):   Annual dividend rate as a percentage of March 16, 1990, share price.

---

[4] "The 1990 Business Week 1000," in *Business Week* magazine, April 13, 1990.

In parts (a) and (b), use variables X_1, X_2, and X_3 as the predictor variables.

a. Use the all possible regressions procedure to suggest a best model.

b. Use the stepwise regression algorithm to suggest a best model.

c. Which model seems to be the better one? Why?

Edited SAS Output (PROC REG) for Problem 11

```
                 ALL POSSIBLE REGRESSIONS OUTPUT
N = 20    Regression Models for Dependent Variable: Y
Number in      R-square          C(p)            MSE       Variables in
Model

    1        0.38695207       10.05134       1.6375292       X3
    1        0.03803947       24.87831       2.5695192       X1
    1        0.03670921       24.93484       2.5730725       X2
    -----------------------------------------------------------------
    2        0.59531344        3.19707       1.1445558       X1 X3
    2        0.55288183        5.00019       1.2645632       X2 X3
    2        0.03807258       26.87690       2.7205738       X1 X2
    -----------------------------------------------------------------
    3        0.62348324        4.00000       1.1314398       X1 X2 X3
    -----------------------------------------------------------------

                   STEPWISE REGRESSION OUTPUT
         Stepwise Procedure for Dependent Variable Y
Step 1   Variable X3 Entered   R-square = 0.38695207
         C(p) = 10.05134226
                           Sum of          Mean              Prob
               DF          Squares         Square        F    > F
Regression      1       18.60476972     18.60476972    11.36  0.0034
Error          18       29.47552528      1.63752918
Total          19       48.08029500

                                              Type II
               Parameter      Standard        Sum of          Prob
Variable       Estimate       Error           Squares     F    > F
INTERCEP       5.49230032     0.93681453     56.28470078  34.37  0.0001
X3            -0.19274361     0.05718240     18.60476972  11.36  0.0034
Bounds on condition number:       1,             1
-----------------------------------------------------------------
Step 2   Variable X1 Entered   R-square = 0.59531344
         C(p) = 3.19706996
                           Sum of          Mean              Prob
               DF          Squares         Square        F    > F
Regression      2       28.62284574     14.31142287    12.50  0.0005
Error          17       19.45744926      1.14455584
Total          19       48.08029500
```

```
Step 2 (Continued)
                                           Type II
                Parameter     Standard     Sum of                Prob
Variable        Estimate      Error        Squares       F       > F
INTERCEP        7.21134513    0.97521036   62.58538586   54.68   0.0001
X1             -0.00513171    0.00173456   10.01807602    8.75   0.0088
X3             -0.24882557    0.05142752   26.79389702   23.41   0.0002
Bounds on condition number: 1.157227,   4.628907
---------------------------------------------------------------------
All variables left in the model are significant at the 0.0500 level.
No other variables met the 0.0500 significance level for entry into
the model.
        Summary of Stepwise Procedure for Dependent Variable Y
        Variable
        Entered   Number   Partial    Model                          Prob
Step    Removed   In       R**2       R**2       C(p)       F        > F
 1      X3        1        0.3870     0.3870     10.0513    11.3615   0.0034
 2      X1        2        0.2084     0.5953      3.1971     8.7528   0.0088
```

12. Radial keratotomy is a type of refractive surgery in which radial incisions are made in a myopic (nearsighted) patient's cornea to reduce the person's myopia. Theoretically, the incisions alow the curvature of the cornea to become less steep, thereby reducing the patient's refractive error. This and other vision-correction surgery techniques have been growing in popularity in the 1980s and 1990s, both among the public and among ophthalmologists.

The Prospective Evaluation of Radial Keratotomy (PERK) clinical trial was begun in 1983 to evaluate the effects of radial keratotomy. As part of the study, Lynn et al. (1987) examined the variables associated with the sample patients' five-year postsurgical change in refractive error (Y, measured in diopters, D). Several independent variables were under consideration:

Baseline refractive error (X_1, diopters).

Patient age (X_2, in years).

Patient's sex (X_3).

Baseline average central keratometric power (X_4, a measure of corneal curvature, in diopters).

Depth of incision scars (X_5, in mm).

Baseline horizontal corneal diameter (X_6, in mm).

Baseline intraocular pressure (X_7, in mm Hg).

Baseline central corneal thickness (X_8, in mm).

Diameter of clear zone (X_9, in mm). (The clear zone is the circular central portion of the cornea that is left uncut during the surgery; the surgical incisions are made radially from the periphery of the cornea to the edge of the clear zone. Smaller clear zones are used for more myopic patients, the thinking being that "more" surgery, in the form of longer incisions, is probably needed for such patients.)

Some of the PERK study results from the all possible subsets analysis that was performed are shown in the following table.

Variable Added to Model	Model R^2 as Variables Are Added
Diameter of clear zone (X_9)	0.28
Patient age (X_2)	0.40
Depth of incision scars (X_5)	0.44
Baseline refractive error (X_1)	0.45
Baseline horizontal corneal diameter (X_6)	0.47
Baseline avg. central keratometric power (X_4)	0.48
Baseline intraocular pressure (X_7)	0.49
Patient's sex (X_3)	0.49
Baseline central corneal thickness (X_8)	0.49

a. Using this table of R^2 values, perform an all possible regressions analysis to suggest a best model.

b. The PERK study researchers concluded that "The regression analysis of the factors affecting the outcome of radial keratotomy [i.e., the change in refractive error] showed that the diameter of the clear zone, patient age, and the average depth of the incision scars were the most important factors." Do you agree with this assessment? Explain.

References

Bethel, R. A.; Sheppard, D.; Geffroy, B.; Tam, E.; Nadel, J. A.; and Boushey, J. A. 1985. "Effect of 0.25 ppm Sulfur Dioxide on Airway Resistance in Freely Breathing, Heavily Exercising, Asthmatic Subjects." *American Review of Respiratory Diseases* 131: 659–61.

Freund, R. J. 1979. "Multicollinearity etc., Some 'New' Examples." *Proceedings of the Statistical Computing Section,* American Statistical Association, pp. 111–12.

Hocking, R. R. 1976. "The Analysis and Selection of Variables in Linear Regression." *Biometrics* 32: 1–49.

Kleinbaum, D. G. 1994. *Logistic Regression: A Self-Learning Text.* New York: Springer-Verlag.

Kleinbaum, D. G.; Kupper, L. L.; and Morgenstern, H. 1982. *Epidemiologic Research: Principles and Quantitative Methods.* Belmont, Calif.: Lifetime Learning Publications.

Kupper, L. L.; Stewart, J. R.; and Williams, K. A. 1976. "A Note on Controlling Significance Levels in Stepwise Regression." *American Journal of Epidemiology* 103(1): 13–15.

Lewis, T., and Taylor, L. R. 1967. *Introduction to Experimental Ecology.* New York: Academic Press.

Lynn, M. J.; Waring, G. O., III; Sperduto, R. D.; et al. 1987. "Factors Affecting Outcome and Predictability of Radial Keratotomy in the PERK Study." *Archives of Ophthalmology* 105: 42–51.

Marquardt, D. W., and Snee, R. D. 1975. "Ridge Regression in Practice." *The American Statistician* 29(1): 3–20.

Pope, P. T., and Webster, J. T. 1972. "The Use of an F-statistic in Stepwise Regression Procedures." *Technometrics* 14(2): 327–40.

17

One-way Analysis of Variance

17-1 Preview

This chapter is the first of four that focus on *analysis of variance* (ANOVA). Earlier we described ANOVA as a technique for assessing how several *nominal* independent variables affect a *continuous* dependent variable. An example given in Table 2-2 (the Ponape study) involved describing the effects on blood pressure of two cultural incongruity indices, each dichotomized into "high" and "low" categories (see Example 1.5). In this case, the dependent variable (blood pressure) was continuous, and the two independent variables (the cultural incongruity indices) were both nominal.

The fact that ANOVA is generally restricted to use with nominal independent variables suggests an interesting interpretation of the purpose of the technique. Loosely speaking, *ANOVA is usually employed in comparisons involving several population means*. In fact, in the simplest special case (involving a comparison of two population means), the ANOVA comparison procedure is equivalent to the usual two-sample t test, which requires the assumption of equal population variances.

The population means to be compared can generally be easily specified by cross-classifying the nominal independent variables under consideration to form different combinations of categories.[1] In the example dealing with the Ponape study, we need only cross-classify the HI and LO categories of incongruity index 1 with the HI and LO categories of index 2. This yields the four population means corresponding to the four combinations HI–HI, HI–LO, LO–HI, and LO–LO, as indicated in the following configuration:

---

[1]Such specification is not possible if the categories of any nominal variable are viewed as being only a sample from a much larger population of categories of interest. We consider such situations later when discussing *random-effects models*.

INDEX 2

HI   LO

		HI	LO
	HI	μ_1	μ_2
INDEX 1			
	LO	μ_3	μ_4

Assessing whether the two indices have some effect on the dependent variable "blood pressure" is equivalent to determining what kind of differences, if any, exist among the four means.

17-1-1 Why the Name ANOVA?

If the ANOVA technique usually involves comparing means, it seems somewhat inappropriate to call it analysis of *variance*. Why not instead use the acronym ANOME, where *ME* stands for *means*? Actually, the designation *ANOVA* is quite justifiable: although typically means are compared, the comparisons are made using estimates of variance. As with regression analysis, the ANOVA test statistics are F statistics and are actually ratios of estimates of variance. In fact, it is even possible, and in some cases appropriate, to specify the null hypotheses of interest in terms of population variances.

17-1-2 ANOVA versus Regression

Another general distinction relates to the difference between an "ANOVA problem" and a "regression problem." For ANOVA, *all* independent variables must be treated as nominal; for regression analysis, any mixture of measurement scales (nominal, ordinal, or interval) is permitted for the independent variables. In fact, ANOVA is often viewed as a special case of regression analysis, and almost any ANOVA model can be represented by a regression model whose parameters can be estimated and inferred about in the usual manner. The same may be said for certain other multivariable techniques, such as analysis of covariance. Hence, we may view the various names given to these techniques as indicators of different (linear) models having the same general form

$$Y = \beta_0 + \beta_1 X_1 + \beta_2 X_2 + \cdots + \beta_p X_p + E$$

yet involving different types of variables and perhaps different assumptions about these variables. The choice of method can thus be regarded as equivalent to the choice of an appropriate linear model.

17-1-3 Factors and Levels

Some additional terms must be introduced at this point. In Chapter 14, in connection with using dummy variables in regression, we saw that a nominal variable with k categories can generally be incorporated into a regression model if we define $k - 1$ dummy variables. These $k - 1$ variables collectively describe the *basic* nominal variable under consideration. To refer to a basic variable without having to identify the specific dummy variables used to define it in the regression model, we can follow the approach of calling the basic nominal variable a *factor*. The different categories of the factor are often referred to as its *levels*.

For example, if we wanted to compare the effects of several drugs on some human health response, we would consider the nominal variable "drugs" as a single factor and the specific drug categories as the levels. If we were comparing k drugs, we would incorporate them into a regression model by defining $k - 1$ dummy variables. If, in addition to comparing the drugs, we wanted to consider whether males and females responded differently, we would consider the nominal variable "sex" as a second factor and the specific categories (male and female) as the levels of this dichotomous factor.

17-1-4 Fixed versus Random Factors

A *random factor* is a factor whose levels may be regarded as a sample from some large population of levels.[2] A *fixed factor* is a factor whose levels are the only ones of interest. The distinction is important in any ANOVA, since different tests of significance are required for different configurations of random and fixed factors. We will see this more specifically when considering two-way ANOVA. For now, we can simply look at some examples of random and fixed factors (summarized in Table 17-1):

1. "Subjects" or "litters" is usually considered a random factor, since we ordinarily wish to make inferences about a large population of potential subjects on the basis of the subjects sampled.

2. "Observers" is a random factor we often consider when examining the effect of different observers on the response variable of interest.

3. "Days," "weeks," and so on are usually considered random factors in investigations of the effect of time on a response variable observed during different time periods. We normally use many levels for such temporal factors, to represent a large number of time periods.

4. "Sex" is always a fixed factor, since its two levels include all possible levels of interest.

5. "Locations" (for example, cities, plants, or states) may be fixed or random, depending on whether a set of specific sites or a larger geographical universe is to be considered.

6. "Age" is usually treated as a fixed factor regardless of how the different age groups are defined.

TABLE 17-1 Examples of random and fixed factors.

Random	Fixed	Random or Fixed
Subjects	Sex	Locations
Litters	Age	Treatments
Observers	Marital status	Drugs
Days	Education	Tests
Weeks		

---

[2]In practice, the experimental levels of a random factor need not be selected at random as long as they are reasonably representative of the larger population of levels of interest.

7. "Treatments," "drugs," "tests," and so on are usually considered fixed factors, but they may be considered random if their levels represent a much larger group of possible levels.

8. "Marital status" is treated as a fixed factor.

9. "Education" is treated as a fixed factor.

17-2 One-way ANOVA: The Problem, Assumptions, and Data Configuration

One-way ANOVA deals with the effect of a single factor on a single response variable. When that one factor is a fixed factor, one-way ANOVA (often referred to as *fixed-effects one-way ANOVA*) involves a comparison of several (two or more) population means.[3] The different populations effectively correspond to the different levels of the single-factor "populations."

17-2-1 The Problem

The main analysis problem in fixed-effects one-way ANOVA is to determine whether the population means are all equal or not. Thus, given k means (denoted as $\mu_1, \mu_2, \ldots, \mu_k$), the basic null hypothesis of interest is

$$H_0: \mu_1 = \mu_2 = \cdots = \mu_k \tag{17.1}$$

The alternative hypothesis is given by

H_A: "The k population means are not all equal."

If the null hypothesis (17.1) is rejected, the next problem is to find out where the differences are. For example, if $k = 3$ and $H_0: \mu_1 = \mu_2 = \mu_3$ is rejected, we might wish to determine whether the main differences are between μ_1 and μ_2, between μ_1 and μ_3, between μ_1 and the average of the other two means, or the like. Such questions fall under the general statistical subject of multiple-comparison procedures, which are discussed in section 17-7.

17-2-2 The Assumptions

Four assumptions must be made for fixed-effects one-way ANOVA:

1. Random samples (individuals, animals, etc.) have been selected from each of k populations or groups.

2. A value of a specified dependent variable has been recorded for each experimental unit (individual, animal, etc.) sampled.

---

[3]In this section, we focus on situations involving fixed factors. Random factors are discussed in section 17-6.

3. The dependent variable is normally distributed in each population.

4. The variance of the dependent variable is the same in each population (this common variance is denoted as σ^2).

Although these assumptions provide the theoretical justification for applying this method, it is sometimes necessary to use fixed-effects one-way ANOVA to compare several means even when the necessary assumptions are not clearly satisfied. Indeed, these assumptions rarely hold exactly. It is therefore important to consider the consequences of applying fixed-effects one-way ANOVA when the assumptions are in question.

In general, fixed-effects one-way ANOVA can be applied as long as none of the assumptions is badly violated. This is true for more complex ANOVA situations as well as for fixed-effects one-way ANOVA. The term generally used to denote this property of broad applicability is *robustness*: a procedure is robust if moderate departures from the basic assumptions do not adversely affect its performance in any meaningful way.

We must nevertheless avoid asserting robustness as an automatic justification for carelessly applying the ANOVA method. Certain facts should be kept in mind when considering the use of ANOVA in a given situation. It is true that the normality assumption does not have to be exactly satisfied as long as we are dealing with relatively large samples (e.g., 20 or more observations from each population), although the consequences of large deviations from normality are somewhat more severe for random factors than for fixed factors. Similarly, the assumption of variance homogeneity can be mildly violated without serious risk, provided that the numbers of observations selected from each population are more or less the same (again, the consequences are more severe for random factors).

On the other hand, an inappropriate assumption of independence of the observations can lead to serious errors in inference for both fixed and random cases. In general, great care should be taken to ensure that the observations are independent. This concern arises primarily in studies where repeated observations are recorded on the same experimental subjects: the level of response of a subject on one occasion commonly has a decided effect on subsequent responses.

What should we do when one or more of these assumptions are in serious question? One option is to transform the data (e.g., by means of a log, square root, or other transformation) so that they more closely satisfy the assumptions. Another thing is to select a more appropriate method of analysis (e.g., nonparametric ANOVA methods or growth curve analysis or longitudinal data analysis procedures).[4] An introduction to the analysis of repeated measures data is provided in Chapter 21.

17-2-3 Data Configuration for One-way ANOVA

Computations necessary for one-way ANOVA can easily be performed even with an ordinary calculator when the data are conveniently arranged. Table 17-2 illustrates a useful way of presenting the data for the general one-way situation. Clearly, the number of observations

---

[4]Descriptions of nonparametric methods that can be used when these assumptions are clearly and strongly violated can be found in Siegel (1956), Lehmann (1975), and Hollander and Wolfe (1973). A discussion of growth curve analysis can be found in Allen and Grizzle (1969). An excellent discussion of longitudinal data analysis methods can be found in Diggle, Liang, and Zeger (1994).

TABLE 17-2   General data configuration for one-way ANOVA.

Population	Sample Size	Observations	Total	Sample Mean
1	n_1	$Y_{11}, Y_{12}, Y_{13}, \ldots, Y_{1n_1}$	$T_1 = Y_1.$	$\bar{Y}_1. = T_1/n_1$
2	n_2	$Y_{21}, Y_{22}, Y_{23}, \ldots, Y_{2n_2}$	$T_2 = Y_2.$	$\bar{Y}_2. = T_2/n_2$
3	n_3	$Y_{31}, Y_{32}, Y_{33}, \ldots, Y_{3n_3}$	$T_3 = Y_3.$	$\bar{Y}_3. = T_3/n_3$
$\vdots$	$\vdots$	$\vdots$	$\vdots$	$\vdots$
k	n_4	$Y_{k1}, Y_{k2}, Y_{k3}, \ldots, Y_{kn_k}$	$T_k = Y_k.$	$\bar{Y}_k. = T_k/n_k$

$$n = \sum_{i=1}^{k} n_i \qquad\qquad G = Y.. \qquad \bar{Y} = G/n$$

selected from each population does *not* have to be the same; that is, there are n_i observations from the ith population, and n_i need not equal n_j if $i \neq j$. Double-subscript notation (Y_{ij}) is used to distinguish one observation from another. The first subscript for a given observation denotes the population number, and the second distinguishes that observation from the others in that sample. Thus, Y_{23} denotes the third observation from the second population, Y_{62} denotes the second observation from the sixth population, and Y_{kn_k} denotes the last observation from the kth population. The totals for each sample (from each population) are denoted alternatively by T_i or $Y_i.$ for the ith sample, where the · denotes that we are summing over all values of j (i.e., we are adding together all observations within the given sample). The grand total over all samples is denoted as $Y.. = G$. The sample means are alternatively denoted by $\bar{Y}_i.$ or T_i/n_i for the ith sample; these statistics are particularly important because they represent the estimates of the population means of interest. Finally, the grand mean over all samples is $\bar{Y} = G/n$.

■ **Example**   In a study by Daly (1973) of the effects of neighborhood characteristics on health, a stratified random sample of 100 households was selected—25 from each of four turnkey neighborhoods included in the study. The data configuration of Cornell Medical Index (CMI) scores for female heads of household is given in Table 17-3. Such scores are measures (derived from questionnaires) of the overall (self-perceived) health of individuals; the higher the score, the poorer the health. Each of the turnkey neighborhoods differed in total number of residents and in percentage of blacks in the surrounding neighborhoods. The racial composition of the turnkey neighborhoods themselves was over 95% black. Daly's main thesis was that the health of persons living in similar federal or state housing projects varied according to the racial composition of the surrounding neighborhoods: the "friendlier" (in terms of similar racial composition) the surrounding neighborhood was, the better would be the health of the residents in the project. According to Daly, federal housing planners had never considered information about overall neighborhood racial composition and its relationship to health as criteria for selecting areas for such projects. This study, it was hoped, might provide some concrete recommendations for improved federal planning.

The sample means of the data in Table 17-3 vary. To determine whether the observed differences in these sample means are attributable solely to chance, we can perform a one-way fixed-effects ANOVA. The possibility of violations of the assumptions underlying this methodology are not of great concern for this data set, since the sample sizes are equal and reasonably large and since observations on women from different households may be treated as independent.   ■

TABLE 17-3   Cornell Medical Index scores for a sample of women from different households in four turnkey housing neighborhoods.

Neighborhood	No. of Households	% Blacks in Surrounding Neighborhoods	Sample Size (n_i)	Observations (Y_{ij})	Total (T_i)	Sample Mean $(\overline{Y}_{i.})$
Cherryview	98	17	25	49, 12, 28, 24, 16, 28, 21, 48, 30, 18, 10, 10, 15, 7, 6, 11, 13, 17, 43, 18, 6, 10, 9, 12, 12	$T_1 = 473$	$\overline{Y}_{1.} = 18.92$
Morningside	211	100	25	5, 1, 44, 11, 4, 3, 14, 2, 13, 68, 34, 40, 36, 40, 22, 25, 14, 23, 26, 11, 20, 4, 16, 25, 17	$T_2 = 518$	$\overline{Y}_{2.} = 20.72$
Northhills	212	36	25	20, 31, 19, 9, 7, 16, 11, 17, 9, 14, 10, 5, 15, 19, 29, 23, 70, 25, 6, 62, 2, 14, 26, 7, 55	$T_3 = 521$	$\overline{Y}_{3.} = 20.84$
Easton	40	65	25	13, 10, 20, 20, 22, 14, 10, 8, 21, 35, 17, 23, 17, 23, 83, 21, 17, 41, 20, 25, 49, 41, 27, 37, 57	$T_4 = 671$	$\overline{Y}_{4.} = 26.84$
			$\sum_{i=1}^{4} n_i = 100$		$G = 2,183$	$\overline{Y} = 21.83$

17-3 Methodology for One-way Fixed-effects ANOVA

The null hypothesis of equal population means (H_0: $\mu_1 = \mu_2 = \cdots = \mu_k$) is tested by using an F test. The test statistic is calculated as follows:

$$F = \frac{\text{MST}}{\text{MSE}} \qquad F = \frac{\sum_{i=1}^{k} n_i(\overline{x}_i - \overline{x})^2 / k - 1}{S_p^2} \tag{17.2}$$

where

$$\text{MST} = \frac{\sum_{i=1}^{k} (T_i^2/n_i) - G^2/n}{k - 1} \tag{17.3}$$

and

$$\text{MSE} = \frac{\sum_{i=1}^{k} \sum_{j=1}^{n_i} Y_{ij}^2 - \sum_{i=1}^{k} (T_i^2/n_i)}{n - k} = S_p^2 \tag{17.4}$$

When H_0 is true (i.e., when the population means are all equal), the F statistic of (17.2) has an F distribution with $k - 1$ numerator and $n - k$ denominator degrees of freedom. Thus, for a given α, we would reject H_0 and conclude that some (i.e., at least two) of the population means differ from one another if

$$F \geq F_{k-1, n-k, 1-\alpha}$$

where $F_{k-1, n-k, 1-\alpha}$ is the $100(1 - \alpha)\%$ point of the F distribution with $k - 1$ and $n - k$ degrees of freedom. The critical region for this test involves only upper percentage points of the F distribution, since only large values of the F statistic (usually values much greater than 1) will provide significant evidence for rejecting H_0.

17-3-1 Numerical Illustration

For the data given in Table 17-3, the calculations needed to perform the F test proceed as follows:

$$\underbrace{\sum_{i=1}^{4} \sum_{j=1}^{25} Y_{ij}^2 = (49)^2 + (12)^2 + \cdots + (37)^2 + (57)^2}_{\text{Sum of 100 squared observations}} = 72{,}851.00$$

$$\sum_{i=1}^{4} \frac{T_i^2}{n_i} = \frac{(473)^2}{25} + \frac{(518)^2}{25} + \frac{(521)^2}{25} + \frac{(671)^2}{25} = 48{,}549.40$$

$$\frac{G^2}{n} = \frac{(2183)^2}{100} = 47{,}654.89$$

$$
\begin{aligned}
\text{MST} &= \frac{\sum_{i=1}^{4} (T_i^2/n_i) - G^2/n}{4 - 1} \\
&= \frac{48{,}549.40 - 47{,}654.82}{3} \\
&= 298.17
\end{aligned}
$$

$$
\begin{aligned}
\text{MSE} &= \frac{\sum_{i=1}^{4} \sum_{j=1}^{25} Y_{ij}^2 - \sum_{i=1}^{4} (T_i^2/n_i)}{100 - 4} \\
&= \frac{72{,}851.00 - 48{,}549.40}{96} \\
&= 253.14
\end{aligned}
$$

$$
\begin{aligned}
F &= \frac{\text{MST}}{\text{MSE}} \\
&= \frac{298.17}{253.14} \\
&= 1.178
\end{aligned}
$$

The preceding calculations can be conveniently performed by a computer program. An example of computer output for these data (using SAS's GLM procedure) is shown next.

SAS Output for ANOVA of CMI Scores

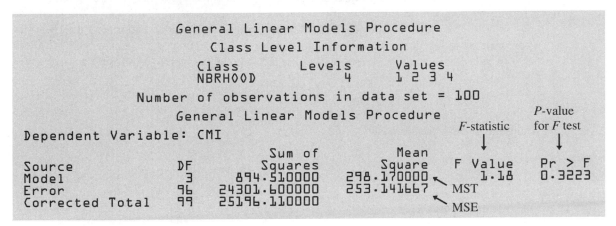

```
                    General Linear Models Procedure
                       Class Level Information
                    Class        Levels      Values
                    NBRHOOD         4         1 2 3 4
            Number of observations in data set = 100
                    General Linear Models Procedure
Dependent Variable: CMI
                                 Sum of          Mean
Source              DF           Squares         Square     F Value    Pr > F
Model                3          894.510000     298.170000     1.18     0.3223
Error               96        24301.600000     253.141667
Corrected Total     99        25196.110000
```

Using these calculations, we may test H_0: $\mu_1 = \mu_2 = \mu_3 = \mu_4$ (i.e., the hypothesis that there are *no* differences among the true mean CMI scores for the four neighborhoods) against H_A: "There are differences among the true mean CMI scores." For example, if $\alpha = .10$, we would find from the F tables that $F_{3,96,0.90} = 2.15$, which is greater than the computed F of 1.178. Thus, we would not reject H_0 at $\alpha = .10$.

To find the P-value for this test, we first note that $F_{3,96,0.75} = 1.41$, which also exceeds the computed F. Thus, we know that $P > .25$. From the preceding SAS output, we see that the P-value is in fact .3223. Therefore, we conclude (as did Daly) that the observed mean CMI scores for the four neighborhoods do not significantly differ.

If a significant difference among the sample means had been found, it would still have been up to the investigator to determine whether the actual magnitude of the difference(s) was meaningful in a practical sense and whether the pattern of the difference(s) was as hypothesized. In our example, the distribution of percentages of blacks in the surrounding neighborhoods (see Table 17-3) indicates that the observed differences among the sample means clearly does not match the pattern hypothesized. Under Daly's conjecture, Cherryview (with 17% black in the surrounding neighborhood) would have been expected to register the highest observed mean CMI score, followed by Northhills (36%), Easton (65%), and finally Morningside (100%). This was not the order actually obtained. Daly also examined whether her conjecture was supported when she controlled for other possibly relevant factors, such as "months lived in the neighborhood," "number of children," and "marital status." However, no significant results were obtained from these analyses either.

17-3-2   Rationale for the *F* Test in One-way ANOVA

The use of the F test described in the preceding subsection may be motivated by various considerations. We therefore offer an intuitive theoretical appreciation of its purpose and also provide some insight into the rationale behind more complex ANOVA testing procedures.

1. The F test in one-way fixed-effects ANOVA is a generalization of the two-sample t test.
We can easily show with a little algebra that the numerator and denominator components in the
F statistic for one-way ANOVA (17.2) are simple generalizations of the corresponding components in the square of the ordinary two-sample t test statistic. In fact, when $k = 2$, the F statistic
for one-way ANOVA is exactly equal to the square of the corresponding t statistic. Such a result
is intuitively reasonable, since the numerator degrees of freedom of F when $k = 2$ is 1, and we
have previously noted that the square of a t statistic with v degrees of freedom has the F distribution with 1 and v degrees of freedom in numerator and denominator, respectively.

In particular, recall that the two-sample t test statistic is given by the formula

$$T = \frac{(\overline{Y}_{1.} - \overline{Y}_{2.})/\sqrt{1/n_1 + 1/n_2}}{S_p}$$

where the pooled sample variance S_p^2 is given by

$$S_p^2 = \frac{1}{n_1 + n_2 - 2} \sum_{i=1}^{2} \sum_{j=1}^{n_i} (Y_{ij} - \overline{Y}_{i.})^2$$

or, equivalently, by

$$S_p^2 = \frac{(n_1 - 1)S_1^2 + (n_2 - 1)S_2^2}{n_1 + n_2 - 2}$$

where S_1^2 and S_2^2 are the sample variances for groups 1 and 2, respectively. Focusing first on the
denominator (MSE) of the F statistic (17.2), we can show with some algebra that

$$
\begin{aligned}
\text{MSE} &= \frac{\sum_{i=1}^{k} \sum_{j=1}^{n_i} Y_{ij}^2 - \sum_{i=1}^{k} (T_i^2/n_i)}{n - k} \\
&= \frac{(n_1 - 1)S_1^2 + (n_2 - 1)S_2^2 + \cdots + (n_k - 1)S_k^2}{n_1 + n_2 + \cdots + n_k - k}
\end{aligned}
$$

Thus, MSE is a pooled estimate of the common population variance σ^2, since it is a weighted
sum of the k estimates of σ^2 obtained by using the k different sets of observations. Furthermore,
when $k = 2$, MSE is equal to S_p^2.

Looking at the numerator (MST) of the F statistic, we can show that

$$
\begin{aligned}
\text{MST} &= \frac{\sum_{i=1}^{k} \left(T_i^2/n_i\right) - G^2/n}{k - 1} \\
&= \frac{1}{k - 1} \sum_{i=1}^{k} n_i(\overline{Y}_{i.} - \overline{Y})^2
\end{aligned}
$$

which simplifies to $(\overline{Y}_{1.} - \overline{Y}_{2.})^2/(1/n_1 + 1/n_2)$ when $k = 2$. Thus, the equivalence is established.

2. The F statistic is the ratio of two variance estimates.   We have already seen that MSE is
a pooled estimate of the common population variance σ^2; that is, the true average (or mean)

value (μ_{MSE}, say) of MSE is σ^2. It turns out, however, that MST estimates σ^2 *only* when H_0 is true—that is, *only* when the population means $\mu_1, \mu_2, \ldots, \mu_k$ are all equal. In fact, the true mean value (μ_{MST}, say) of MST has the general form

$$\mu_{MST} = \sigma^2 + \frac{1}{k-1}\sum_{i=1}^{k} n_i(\mu_i' - \overline{\mu})^2 \tag{17.5}$$

where $\overline{\mu} = \sum_{i=1}^{k} n_i\mu_i/n$. By inspection of expression (17.5), we can see that MST estimates σ^2 *only* when all the μ_i are equal, in which case $\mu_i = \overline{\mu}$ for every i, and so $\sum_{i=1}^{k} n_i(\mu_i - \overline{\mu})^2 = 0$. Otherwise, both terms on the right-hand side of (17.5) are positive and MST estimates something greater in value than σ^2. In other words,

$$\mu_{MST} = \sigma^2 \quad \text{when } H_0 \text{ is true}$$

and

$$\mu_{MST} > \sigma^2 \quad \text{when } H_0 \text{ is not true.}$$

Loosely speaking, then, the F statistic MST/MSE may be viewed as approximating in some sense the ratio of population means

$$\frac{\mu_{MST}}{\mu_{MSE}} = \frac{\sigma^2 + [1/(k-1)]\sum_{i=1}^{k} n_i(\mu_i - \overline{\mu})^2}{\sigma^2} \tag{17.6}$$

When H_0 is true, the numerator and denominator of (17.6) both equal σ^2, and the F statistic is the ratio of two estimates of the same variance. Furthermore, F can be expected to give different values depending on whether H_0 is true; that is, F should take a value close to 1 if H_0 is true (since in that case it approximates $\sigma^2/\sigma^2 = 1$), whereas F should be larger than 1 if H_0 is false (since the numerator of (17.6) is greater than the denominator).

3. *The F statistic compares the variability between groups to the variability within groups.*   As with regression analysis, the total variability in the observations in a one-way ANOVA situation is measured by a total sum of squares:

$$\text{TSS} = \sum_{i=1}^{k}\sum_{j=1}^{n_i}(Y_{ij} - \overline{Y})^2 \tag{17.7}$$

Furthermore, it can be shown that

$$\text{TSS} = \text{SST} + \text{SSE} \tag{17.8}$$

where

$$\text{SST} = (k-1)\text{MST} = \sum_{i=1}^{k} n_i(\overline{Y}_{i\cdot} - \overline{Y})^2$$

and

$$\text{SSE} = (n-k)\text{MSE} = \sum_{i=1}^{k}\sum_{j=1}^{n_i}(Y_{ij} - \overline{Y}_{i\cdot})^2$$

SST can be considered a measure of the variability *between* (or *across*) populations. (The designation SST is read "sum of squares due to treatments," since the populations often represent treatment groups.) It involves components of the general form $\bar{Y}_{i\cdot} - \bar{Y}$, which is the difference between the *i*th group mean and the overall mean.

SSE is a measure of the variability *within* populations and gives no information about variability between populations. It involves components of the general form $Y_{ij} - \bar{Y}_{i\cdot}$, which is the difference between the *j*th observation in the *i*th group and the mean for the *i*th group.

If SST is quite large in comparison to SSE, we know that most of the total variability is due to differences *between* populations rather than to differences *within* populations. Thus, it is natural in such a case to suspect that the population means are not all equal.

By writing the *F* statistic (17.2) in the form

$$F = \frac{\text{SST}}{\text{SSE}} \left(\frac{n-k}{k-1} \right)$$

we can see that *F* will be large whenever SST accounts for a much larger proportion of the total sum of squares than does SSE.

17-3-3 ANOVA Table for One-way ANOVA

As in regression analysis, the results of any ANOVA procedure can be summarized in an ANOVA table. The ANOVA table for one-way ANOVA is given in general form in Table 17-4; Table 17-5 is the ANOVA table for our example involving the CMI data. This ANOVA table was also previously shown in the SAS output on page 431.

The "Source" and "SS" columns in Table 17-4 display the components of the fundamental equation of one-way ANOVA:

$$\text{TSS} = \text{SST} + \text{SSE}$$

The "MS" column contains the sums of squares divided by their corresponding degrees of freedom. The two mean squares are then used to form the numerator and denominator for the *F* test.

TABLE 17-4 General ANOVA table for one-way ANOVA (*k* populations).

Source	d.f.	SS	MS	F
Between	$k-1$	SST	$\text{MST} = \dfrac{\text{SST}}{k-1}$	$\dfrac{\text{MST}}{\text{MSE}}$
Within	$n-k$	SSE	$\text{MSE} = \dfrac{\text{SSE}}{n-k}$	
Total	$n-1$	TSS		

TABLE 17-5 ANOVA table for CMI data (*k* = 4).

Source	d.f.	SS	MS	F
Between (neighborhoods)	3	894.51	298.17	1.18
Within (error)	96	24,301.60	253.14	
Total	99	25,196.11		

17-4 Regression Model for Fixed-effects One-way ANOVA

We observed earlier that most ANOVA procedures can also be considered in a regression analysis setting; this can be done by defining appropriate dummy variables in a regression model.[5] The ANOVA F tests are then formulated in terms of hypotheses concerning the coefficients of the dummy variables in the regression model.[6]

▪ **Example**   For the example involving the CMI data of Daly's (1973) study (see Table 17-3), a number of alternative regression models could be used to describe the situation, depending on the coding schemes used for the dummy variables. One such model is

$$Y = \mu + \alpha_1 X_1 + \alpha_2 X_2 + \alpha_3 X_3 + E \tag{17.9}$$

where the regression coefficients are denoted as μ, α_1, α_2, and α_3, and the independent variables are defined as

$$X_1 = \begin{cases} 1 & \text{if neighborhood 1} \\ -1 & \text{if neighborhood 4} \\ 0 & \text{if otherwise} \end{cases} \qquad X_2 = \begin{cases} 1 & \text{if neighborhood 2} \\ -1 & \text{if neighborhood 4} \\ 0 & \text{if otherwise} \end{cases}$$

$$X_3 = \begin{cases} 1 & \text{if neighborhood 3} \\ -1 & \text{if neighborhood 4} \\ 0 & \text{if otherwise} \end{cases}$$

Although we previously used the Greek letter β with subscripts to denote regression coefficients, we have changed the notation for our ANOVA regression model so that these coefficients correspond to the parameters in the classical fixed-effects ANOVA model described in section 17-5.

The coding scheme used here to define the dummy variables is called an *effect* coding scheme. The coefficients μ, α_1, α_2, and α_3 for this (dummy variable) model can each be expressed in terms of the underlying population means (μ_1, μ_2, μ_3, and μ_4), as follows:

$$\mu = \frac{\mu_1 + \mu_2 + \mu_3 + \mu_4}{4} \qquad (= \bar{\mu}^*, \text{ say})$$

$$\alpha_1 = \mu_1 - \bar{\mu}^* \tag{17.10}$$
$$\alpha_2 = \mu_2 - \bar{\mu}^*$$
$$\alpha_3 = \mu_3 - \bar{\mu}^*$$

The coefficients can be expressed as in (17.10) as follows: $\mu_{Y|X_1,X_2,X_3} = \mu + \alpha_1 X_1 + \alpha_2 X_2 + \alpha_3 X_3$.

---

[5]As mentioned earlier, we are restricting our attention here entirely to models with fixed factors. Models involving random factors will be treated in section 17-6.

[6]We will see later that a regression formulation is often desirable, if not mandatory, for dealing with certain nonorthogonal ANOVA problems involving two or more factors. We will discuss such problems in Chapter 20.

Thus,

$$\mu_1 = \mu_{Y|1,0,0} = \mu + \alpha_1 \qquad \text{since } X_1 = 1, X_2 = 0, X_3 = 0 \text{ for neighborhood 1}$$

$$\mu_2 = \mu_{Y|0,1,0} = \mu + \alpha_2 \qquad \text{since } X_1 = 0, X_2 = 1, X_3 = 0 \text{ for neighborhood 2}$$

$$\mu_3 = \mu_{Y|0,0,1} = \mu + \alpha_3 \qquad \text{since } X_1 = 0, X_2 = 0, X_3 = 1 \text{ for neighborhood 3}$$

$$\mu_4 = \mu_{Y|-1,-1,-1}$$
$$= \mu - \alpha_1 - \alpha_2 - \alpha_3 \qquad \text{since } X_1 = X_2 = X_3 = -1 \text{ for neighborhood 4}$$

Adding the left-hand sides and right-hand sides of these equations yields

$$\mu_1 + \mu_2 + \mu_3 + \mu_4 = 4\mu$$

or

$$\mu = \frac{1}{4} \sum_{i=1}^{4} \mu_i \qquad (= \bar{\mu}^*)$$

Solution (17.10) is obtained by replacing μ with $\bar{\mu}^*$ in the preceding equations and then solving for the regression coefficients α_i in terms of $\mu_1, \mu_2, \mu_3,$ and μ_4.

Model (17.9) involves coefficients that describe separate comparisons of the first three population means with the overall unweighted mean $\bar{\mu}^*$. In this model, $\mu_4 - \bar{\mu}^*$ can be expressed as the negative sum of $\alpha_1, \alpha_2,$ and α_3. Moreover, model (17.9) can be fitted to provide *exactly* the same F statistic as is required in one-way ANOVA for the test of $H_0: \mu_1 = \mu_2 = \mu_3 = \mu_4$. The equivalent regression null hypothesis is $H_0: \alpha_1 = \alpha_2 = \alpha_3 = 0,$[7] the regression F statistic will have the same degrees of freedom (i.e., $k - 1 = 3$ and $n - k = 96$) as given previously, and the ANOVA table will be exactly the same as the one given in the last section (where we pooled the dummy variable effects into one source of variation with 3 degrees of freedom).

Other coding schemes for the independent variables yield exactly the same ANOVA table and F test as model (17.9), although the regression coefficients themselves represent different parameters and have different least-squares estimators. One frequently used coding scheme defines the independent variables as

$$X_i = \begin{cases} 1 & \text{if neighborhood } i \\ 0 & \text{otherwise} \end{cases} \qquad i = 1, 2, 3$$

This coding scheme is an example of *reference cell* coding. The referent group in this case is group 4, and the regression coefficients describe separate comparisons of the first three population means with μ_4:

$$\mu = \mu_4$$
$$\alpha_1 = \mu_1 - \mu_4$$
$$\alpha_2 = \mu_2 - \mu_4$$
$$\alpha_3 = \mu_3 - \mu_4$$

---

[7]When $\alpha_1 = \alpha_2 = \alpha_3 = 0$, it follows from simple algebra based on (17.10) that $\mu_1 = \mu_2 = \mu_3 = \mu_4$ (e.g., $\alpha_1 = \mu_1 - \bar{\mu}^* = 0$ implies that $\mu_1 = \bar{\mu}^*$; $\alpha_2 = \mu_2 - \bar{\mu}^* = 0$ implies that $\mu_2 = \bar{\mu}^* = \mu_1$; etc.).

17-4-1 Effect Coding Model

For the general situation involving k populations, the following model using effect coding is analogous to that of the previous section:

$$Y = \mu + \alpha_1 X_1 + \alpha_2 X_2 + \cdots + \alpha_{k-1} X_{k-1} + E \qquad (17.11)$$

in which

$$X_i = \begin{cases} 1 & \text{for population } i \\ -1 & \text{for population } k \\ 0 & \text{otherwise} \end{cases} \qquad i = 1, 2, \ldots, k-1$$

The coefficients of this model can be expressed in terms of the k population means $\mu_1, \mu_2, \ldots, \mu_k$ as

$$\mu = \frac{\mu_1 + \mu_2 + \cdots + \mu_k}{k} = \bar{\mu}^*$$

$$\alpha_1 = \mu_1 - \bar{\mu}^*$$

$$\alpha_2 = \mu_2 - \bar{\mu}^* \qquad (17.12)$$

$$\vdots$$

$$\alpha_{k-1} = \mu_{k-1} - \bar{\mu}^*$$

$$-(\alpha_1 + \alpha_2 + \cdots + \alpha_{k-1}) = \mu_k - \bar{\mu}^*$$

for model (17.11). The F statistic for one-way ANOVA with k populations can be obtained equivalently by testing the null hypothesis H_0: $\alpha_1 = \alpha_2 = \cdots = \alpha_{k-1} = 0$ in model (17.11).

17-4-2 Reference Cell Coding Model

Another coding scheme for one-way ANOVA of k populations uses the following reference cell coding:

$$X_i = \begin{cases} 1 & \text{for population } i \\ 0 & \text{otherwise} \end{cases} \qquad i = 1, 2, \ldots, k-1$$

Again, only $k - 1$ such dummy variables are needed, and the population "left out" becomes the reference population (group or cell). Thus, given X_i as defined here, group k is the reference group. (If X_1 had been left out instead of X_k, then group 1 would have been the reference group.)

With the specified reference cell coding, the responses in each group under the model $Y = \mu + \alpha_1 X_1 + \alpha_2 X_2 + \cdots + \alpha_{k-1} X_{k-1} + E$ are as follows:

Group 1: $Y = \mu + \alpha_1 \quad + E$
Group 2: $Y = \mu + \alpha_2 \quad + E$
$\qquad \vdots$
Group $k - 1$: $Y = \mu + \alpha_{k-1} + E$
Group k: $Y = \mu \qquad \quad + E$

In turn, the regression coefficients can be written in terms of group means as follows:

$$\alpha_1 = \mu_1 - \mu_k$$
$$\alpha_2 = \mu_2 - \mu_k$$
$$\vdots$$
$$\alpha_{k-1} = \mu_{k-1} - \mu_k$$
$$\mu = \mu_k$$

Thus, different coding schemes (e.g., an effect or some kind of reference cell scheme) yield regression coefficients representing different parameters (for example, $\alpha_1 = \mu_1 - \overline{\mu}^*$ for the effect coding, in this subsection, but $\alpha_1 = \mu_1 - \mu_k$ for the reference cell coding). Regardless of the coding scheme used, the test of the hypothesis $H_0: \mu_1 = \mu_2 = \cdots = \mu_k$ can be obtained equivalently by testing $H_0: \alpha_1 = \alpha_2 = \cdots = \alpha_{k-1} = 0$ in the regression model (17.11). In other words, the correct SST, SSE, and $F_{k-1, n-k}$-values are obtained from the regression analysis regardless of the (legitimate) coding scheme chosen.

17-5 Fixed-effects Model for One-way ANOVA

Many textbooks and articles that deal strictly with ANOVA procedures use a more classical type of model than the regression model given earlier to describe the fixed-effects one-way ANOVA situation. The more classical type of model is often referred to as a *fixed-effects ANOVA model;* in it, all factors under consideration are fixed (i.e., the levels of each factor are the only levels of interest). The *effects* referred to in this type of model represent measures of the influence (i.e., the effect) that different levels of the factor have on the dependent variable.[8] Such measures are often expressed in the form of differences between a given mean and an overall mean; that is, the effect of the ith population is often measured as the amount by which the ith population mean differs from an overall mean.

▨ **Example**   For the CMI data ($k = 4$), the fixed-effects ANOVA model is

$$Y_{ij} = \mu + \alpha_i + E_{ij} \qquad i = 1, 2, 3, 4; \quad j = 1, 2, \ldots, 25 \tag{17.13}$$

where

$Y_{ij} = j$th observation from the ith population

$\mu = \dfrac{\mu_1 + \mu_2 + \mu_3 + \mu_4}{4} = \overline{\mu}^*$ (the overall unweighted mean), since $n_i = 25$ for all i

$\alpha_1 = \mu_1 - \mu =$ Differential effect of neighborhood 1

$\alpha_2 = \mu_2 - \mu =$ Differential effect of neighborhood 2

$\alpha_3 = \mu_3 - \mu =$ Differential effect of neighborhood 3

$\alpha_4 = \mu_4 - \mu =$ Differential effect of neighborhood 4

$E_{ij} = Y_{ij} - \mu - \alpha_i =$ Error component associated with the jth observation from the ith population

---

[8]In situations involving models with two or more factors, *effects* can also refer to measures of the influence of combinations of levels of the different factors on the dependent variable.

One important property of this model is that the sum of the four α effects is 0; that is, $\alpha_1 + \alpha_2 + \alpha_3 + \alpha_4 = 0$. Thus, these effects represent differentials from the overall population mean μ that average out to 0. Nevertheless, the effect of one level (i.e., a neighborhood) may differ considerably from the effect of another. If this proved to be the case, we would probably find that our F test leads to rejection of the null hypothesis of equal population mean CMI scores for the four neighborhoods.

Another important property of this model is that the effects α_1, α_2, α_3, and α_4, which are population parameters defined in terms of population means, can each be estimated from the data by appropriately substituting the usual estimates of the means into the expression for the effect. For our example, the estimated effects are given by

$$\hat{\alpha}_1 = \overline{Y}_1. - \overline{Y} = \text{Sample mean CMI score for neighborhood 1} - \text{Overall sample mean}$$
$$\text{CMI score for all neighborhoods}$$

$$\hat{\alpha}_2 = \overline{Y}_2. - \overline{Y} = \text{Sample mean CMI score for neighborhood 2} - \text{Overall sample mean}$$
$$\text{CMI score for all neighborhoods}$$

$$\hat{\alpha}_3 = \overline{Y}_3. - \overline{Y} = \text{Sample mean CMI score for neighborhood 3} - \text{Overall sample mean}$$
$$\text{CMI score for all neighborhoods}$$

$$\hat{\alpha}_4 = \overline{Y}_4. - \overline{Y} = \text{Sample mean CMI score for neighborhood 4} - \text{Overall sample mean}$$
$$\text{CMI score for all neighborhoods}$$

The actual numerical values obtained from these formulas are as follows:

$$\hat{\alpha}_1 = 18.92 - 21.83 = -2.91$$

$$\hat{\alpha}_2 = 20.72 - 21.83 = -1.11$$

$$\hat{\alpha}_3 = 20.84 - 21.83 = -0.99$$

$$\hat{\alpha}_4 = 26.84 - 21.83 = 5.01$$

Like the population effects, the estimated effects sum to 0; that is, $\sum_{i=1}^{4} \hat{\alpha}_i = 0$.

If we consider the general one-way ANOVA situation (with k populations and n_i observations from the ith population), the fixed-effects one-way ANOVA model may be written as follows:

$$Y_{ij} = \mu + \alpha_i + E_{ij} \qquad i = 1, 2,..., k; \quad j = 1, 2,..., n_i \qquad (17.14)$$

where

$Y_{ij} = j$th observation from the ith population

$$\mu = \frac{\mu_1 + \mu_2 + \cdots + \mu_k}{k} \quad (= \overline{\mu}^*)$$

$\alpha_i = \mu_i - \mu = $ Differential effect of population i

$E_{ij} = Y_{ij} - \mu - \alpha_i = $ Error component associated with the jth observation from the ith population

Here, it is easy to show that the sum of the α effects is 0; that is, $\sum_{i=1}^{k} \alpha_1 = 0$. Similarly, the estimated effects, $\hat{\alpha}_i = \overline{Y}_i. - \overline{Y}^*$, where $\overline{Y}^* = \sum_{i=1}^{k} \overline{Y}_i./k$, satisfy the constraint $\sum_{i=1}^{k} \hat{\alpha}_i = 0$.

An alternative definition of μ is $\overline{\mu} = \sum_{i=1}^{k} n_i \mu_i / n$, the overall weighted mean of the means. In this case, the weighted sum $\sum_{i=1}^{k} n_i \alpha_i = 0$, and the weighted sum of the estimated effects $\hat{\alpha}_i = \overline{Y}_{i\cdot} - \overline{Y}$, where $\overline{Y} = \sum_{i=1}^{k} n_i \overline{Y}_{i\cdot} / n$ satisfies $\sum_{i=1}^{k} n_i \hat{\alpha}_i = 0$.

Model (17.14) corresponds in structure to the regression model given by (17.9): the regression coefficients $\alpha_1, \alpha_2, \ldots, \alpha_{k-1}$ are precisely the effects $\alpha_1 = \mu_1 - \overline{\mu}^*$, $\alpha_2 = \mu_2 - \overline{\mu}^*$,..., $\alpha_{k-1} = \mu_{k-1} - \overline{\mu}^*$; the regression constant μ represents the overall (unweighted) mean $\overline{\mu}^*$; and the negative sum of the regression coefficients $(-\sum_{i=1}^{k-1} \alpha_i)$ represents the effect $\alpha_k = \mu_k - \overline{\mu}^*$. This is why we have defined each of these models using the same notation for the unknown parameters:

$$Y = \mu + \sum_{i=1}^{k-1} \alpha_i X_i + E \qquad \text{(dummy variable regression model)}$$

$$Y_{ij} = \mu + \alpha_i + E_{ij} \qquad \text{(fixed-effects ANOVA model)}$$

Notice that μ represents the unweighted average of the k population means, $\overline{\mu}^*$, rather than the weighted average $\overline{\mu}$, even though the sample sizes can be different in the different populations. Correspondingly, the least-squares estimate of μ is $\sum_{i=1}^{k} \overline{Y}_{i\cdot} / k$, the unweighted average of the k sample means, rather than $\overline{Y}$. Nevertheless, the dummy variables in the regression model can be redefined to obtain a least-squares solution yielding $\overline{Y}$ as the estimate of μ. The following dummy variable definitions

$$X_i = \begin{cases} -n_i & \text{if population } k \\ n_k & \text{if population } i \qquad i = 1, 2, \ldots, k-1 \\ 0 & \text{otherwise} \end{cases}$$

are required.   ▪

17-6 Random-effects Model for One-way ANOVA

In section 17-1 we distinguished between fixed and random factors. We have also observed that, in ANOVA situations involving two or more factors, the F tests required for making inferences differ depending on whether all factors are fixed, all factors are random, or some of both are present. The null hypotheses to be tested must be stated in different terms when random factors are involved than when only fixed factors are involved.

▮ **Example**   To get some insight into the structure of random-effects models, reconsider Daly's (1973) study (see Table 17-3). It might be argued that the four different turnkey neighborhoods form a representative sample of a larger population of similar types of neighborhoods (some of which might even be predominantly white with differing percentages of blacks in the surrounding neighborhoods). If so, the neighborhood factor would have to be considered random, and the appropriate ANOVA model would be a *random-effects* one-way ANOVA model.[9]

---

[9]This type of model is also referred to as a *variance-components model*.

Its form would be essentially the same as that given in (17.13), except that the α components would be treated differently; that is, the random-effects model would be of the form[10]

$$Y_{ij} = \mu + A_i + E_{ij} \qquad i = 1, 2, 3, 4; \quad j = 1, 2, ..., 25 \qquad (17.15)$$

In this model, the A_i's can be viewed as constituting a random sample of random variables that have a common distribution—one that represents the entire population of possible effects (in our example, neighborhoods).

To perform the appropriate analysis, we must assume that the distribution of A_i is normal with zero mean:

$$A_i \frown N(0, \sigma_A^2) \qquad i = 1, 2, 3, 4 \qquad (17.16)$$

where σ_A^2 denotes the variance of A_i. We must also assume that the A_i's are independent of the E_{ij}'s and of each other.[11]

The requirement of zero mean in (17.16) is similar in philosophy to the requirement that $\sum_{i=1}^{k} \alpha_i = 0$ for the fixed-effects model. When the random model (17.15) applies, we assume that the average (i.e., mean) effect of neighborhoods is 0 over the entire population of neighborhoods; that is, we assume that $\mu_{A_i} = 0$, $i = 1, 2, 3, 4$.

How do we state our null hypothesis? Because we have specified that the neighborhood effects average out to 0 over the entire population of possible effects, the only one way to assess whether any significant neighborhood effects are present at all involves considering σ_A^2. If there is no variability (i.e., $\sigma_A^2 = 0$), all neighborhood effects must be 0. If there is variability (i.e., $\sigma_A^2 > 0$), some nonzero effects must exist in the population of neighborhood effects.

Thus, our null hypothesis of no neighborhood effects should be stated as follows:

$$H_0: \sigma_A^2 = 0 \qquad (17.17)$$

This hypothesis is analogous to the null hypothesis (17.1) used in the fixed-effects case, although it happens to be stated in terms of a population variance rather than in terms of population means.

We must still explain why the F test given by (17.2) for the fixed-effects model is exactly the same as that used for the random-effects model.[12] Such an explanation is best made by considering the properties of the mean squares MST and MSE. In section 17-3-2, in connection with the fixed-effects model, we saw that the F statistic, MST/MSE, could be considered a rough approximation to the ratio of the means of these mean squares,

$$\frac{\mu_{\text{MST}}}{\mu_{\text{MSE}}} = \frac{\sigma^2 + [1/(k-1)] \sum_{i=1}^{k} n_i (\mu_i - \overline{\mu})^2}{\sigma^2}$$

---

[10]Our usual convention has been to use Latin letters (X, Y, Z, etc.) to denote random variables and to use Greek letters (β, μ, σ, τ) to denote parameters. This requires using A_i's rather than α_i's to denote random effects.

[11]Since the *same* random variable A_i defined by (17.16) appears in (17.15) for each of the observations from population i, it follows that corr$(Y_{ij}, Y_{ij'}) = \sigma_A^2/(\sigma_A^2 + \sigma^2)$, the so-called intraclass (or intrapopulation) correlation. Thus, in contrast to the one-way *fixed-effects* ANOVA model, the one-way *random-effects* ANOVA model introduces a dependency among the set of observations from the same population.

[12]Again, the F tests are computationally equivalent for fixed-effects and random-effects models only in one-way ANOVA. When dealing with two-way or higher-way ANOVA, the testing procedures may be different.

The parameters μ_{MST} and μ_{MSE} are often called *expected mean squares*.

A similar argument can be made with regard to the F statistic for the random-effects model. In particular, for the random-effects model as well as for the fixed-effects model, the denominator MSE estimates σ^2; that is,

$$\mu_{MSE} = \sigma^2$$

Furthermore, for the random-effects model applied to our example ($k = 4$, $n_i = 25$), it can be shown that MST estimates

$$\mu_{MST(random)} = \sigma^2 + 25\sigma_A^2 \tag{17.18}$$

Thus, for the random-effects model, F approximates the ratio

$$\frac{\mu_{MST(random)}}{\mu_{MSE}} = \frac{\sigma^2 + 25\sigma_A^2}{\sigma^2} \tag{17.19}$$

Since the null hypothesis in this case is H_0: $\sigma_A^2 = 0$, the ratio (17.19) simplifies to $\sigma^2/\sigma^2 = 1$ when H_0 is true. Thus, the F statistic under H_0 again consists of the ratio of two estimates of the same variance σ^2. Furthermore, because $\sigma_A^2 > 0$ when H_0 is not true, the greater the variability among neighborhood effects is, the larger should be the observed value of F.

In general, the random-effects model for one-way ANOVA is given by

$$Y_{ij} = \mu + A_i + E_{ij} \qquad i = 1, 2,\dots, k; \qquad j = 1, 2,\dots, n_i \tag{17.20}$$

where A_i and E_{ij} are independent random variables satisfying $A_i \frown N(0, \sigma_A^2)$ and $E_{ij} \frown N(0, \sigma^2)$.[13]

For this model, F approximates the following ratio of expected mean squares:

$$\frac{\mu_{MST(random)}}{\mu_{MSE}} = \frac{\sigma^2 + n_0 \sigma_A^2}{\sigma^2}$$

where

$$n_0 = \frac{\sum_{i=1}^{k} n_i - \left(\sum_{i=1}^{k} n_i^2 / \sum_{i=1}^{k} n_i \right)}{k - 1}$$

functions as an average of the n_i observations selected from each population.[14] The F statistic for the random-effects model is therefore the ratio of two estimates of σ^2 when H_0: $\sigma_A^2 = 0$ is true.

Table 17-6 summarizes the similarities and the differences between the fixed- and random-effects models. Tables with similar formats will be used in subsequent chapters to highlight distinctions for ANOVA situations with more than two factors. ∎

---

[13]Under these assumptions, $Y_{ij} \sim N(\mu, \sigma_A^2 + \sigma^2)$ for all (i, j). And, $\text{corr}(Y_{ij}, Y_{ij'}) = \sigma_A^2/(\sigma_A^2 + \sigma^2)$ for all $j \neq j'$ with fixed i. In other words, observations from *different* populations are independent, but observations from the *same* population are correlated.

[14]When all the n_i are equal, as in the Daly (1973) example (i.e., $n_i = n^*$), then n_0 is equal to n^*, since

$$n_0 = \frac{kn^* - (kn^{*2}/kn^*)}{k - 1} = n^*$$

In the Daly (1973) example, $n^* = 25$.

TABLE 17-6   Combined one-way ANOVA table for fixed- and random-effects models.

Source	d.f.	MS	F	Expected Mean Square (EMS)	
				Fixed Effects	Random Effects
Between	$k-1$	MST	$\dfrac{\text{MST}}{\text{MSE}}$	$\sigma^2 + \dfrac{1}{k-1}\sum_{i=1}^{k} n_i(\mu_i - \bar{\mu})^2$	$\sigma^2 + n_0\sigma_A^2$
Within	$n-k$	MSE		σ^2	σ^2
Total	$n-1$				
				$H_0: \mu_1 = \mu_2 = \cdots = \mu_k$	$H_0: \sigma_A^2 = 0$

17-7 Multiple-comparison Procedures for Fixed-effects One-way ANOVA

When we find that an ANOVA F test for simultaneously comparing several population means is statistically significant, our next step customarily is to determine which *specific* differences exist among the population means. For example, if we are comparing four means (fixed-effects case) and the null hypothesis $H_0: \mu_1 = \mu_2 = \mu_3 = \mu_4$ is rejected,[15] we next try to determine which subgroups of means are different, by considering more specific hypotheses such as $H_{01}: \mu_1 = \mu_2$, $H_{02}: \mu_2 = \mu_3$, $H_{03}: \mu_3 = \mu_4$, or even H_{04} $(\mu_1 + \mu_2)/2 = (\mu_3 + \mu_4)/2$, which compares the average effect of populations 1 and 2 with the average effect of populations 3 and 4. Such specific comparisons may have been of interest to us before (a priori) the data were collected, or they may arise in completely exploratory studies only after (a posteriori) the data have been examined. In either event, a seemingly reasonable first approach to drawing inferences about differences among the population means would be to conduct several t tests and to focus on all the tests found significant. For example, if all pairwise comparisons among the means are desired, then $_4C_2 = 6$ such tests must be performed with regard to 4 means (or in general, $_kC_2 = k(k-1)/2$ tests with regard to k means). Thus, in testing $H_0: \mu_i = \mu_j$ at the α level of significance, we could reject this H_0 when

$$|T| \geq t_{n-k, 1-\alpha/2}$$

where

$$T = \frac{(\bar{Y}_i - \bar{Y}_j) - 0}{\sqrt{\text{MSE}(1/n_i + 1/n_j)}}$$

and where n is the total number of observations; k is the number of means under consideration; n_i and n_j are the sizes of the samples selected from the ith and jth populations, respectively; $\bar{Y}_i$ and $\bar{Y}_j$ are the corresponding sample means; and MSE is the mean-square-error term, with $n - k$ degrees of freedom, that estimates the (homoscedastic) variance σ^2. MSE is used instead of a

---

[15]This section deals only with fixed-effects ANOVA problems. The random-effects model treats the observed factor levels as a sample from a larger population of levels of interest and therefore is not directed exclusively at inferences of the sampled levels.

$$S_p = \sqrt{MSE\left(\frac{1}{n_i} - \frac{1}{n_j}\right)}$$

simple two-sample estimate of σ^2 based entirely on data from groups i and j; this is because MSE is a better estimate of σ^2 (in terms of degrees of freedom) under the assumption of variance homogeneity over all k populations.

Equivalently, one could reject H_0 if the $100(1 - \alpha)\%$ confidence interval

$$(\overline{Y}_i - \overline{Y}_j) \pm t_{n-k,\,1-\alpha/2}\sqrt{\text{MSE}\left(\frac{1}{n_i} + \frac{1}{n_j}\right)} \qquad = \overline{Y}_i - \overline{Y}_j \pm t\, S_p\sqrt{\frac{1}{n_i} + \frac{1}{n_j}}$$

does not include 0.

Unfortunately, performing several such t tests has a serious drawback: the more null hypotheses there are to be tested, the more likely one of them is to be rejected even if all the null hypotheses are actually true. In other words, if several such tests are made, each at the α level, the probability of incorrectly rejecting *at least one* H_0 is much larger than α and continues to increase with each additional test made. Moreover, if in an exploratory study the investigator decides to compare only the sample means that are most discrepant (e.g., the largest versus the smallest), the testing procedure becomes biased in favor of rejecting H_0 because only the comparisons most likely to be significant are made. This bias will be reflected in the fact that the actual probability of falsely rejecting a given null hypothesis exceeds the α level specified for the test.

17-7-1 The LSD Approach   *Bonferoni*

An approximate way to circumvent the problem of distorted significance levels when making several tests involves reducing the significance level used for each individual test sufficiently to fix the *overall significance level* (i.e., the probability of falsely rejecting at least one of the null hypotheses being tested) at some desired value (say, α). If we make l such tests, the maximum possible value for this overall significance level is $l\alpha$. Thus, one simple way to ensure an overall significance level of at most α is to use α/l as the significance level for each separate test. This approach, which is often referred to as the *least-significant-difference* (LSD) method, is an application of the Bonferroni correction for multiple testing. For example, if all $l = {}_k C_2 = k(k - 1)/2$ pairwise comparisons of k population means are to be made, each test can be performed at the $\alpha/{}_k C_2$ significance level to ensure that the overall significance level does not exceed α.

One disadvantage of the LSD method is that the *true* overall significance level may be so much less than the value α that none of the individual tests are likely to be rejected (i.e., the overall power of the method is low). Consequently, several more powerful procedures have been devised to provide an overall significance level of α. All of these procedures are grouped under the heading "multiple-comparison procedures." We shall focus here on two such methods—one due to Tukey, and the other to Scheffé. Discussions of these and other multiple-comparison methods can be found in Miller (1966), Guenther (1964), Lindman (1974), and Neter and Wasserman (1974).

Example   Consider the set of data given in Table 17-7, which was collected from an experiment designed to compare the relative potencies of four cardiac substances. In the experiment, a suitable dilution of one of the substances was slowly infused into an anesthetized guinea pig, and the dosage at which the pig died was recorded. Ten guinea pigs were used for each substance, and the laboratory environment and the measurement procedures were assumed to be identical for each guinea pig. The main research goal was to determine whether any differences

overall
$$\alpha = (\alpha)\neq k$$

TABLE 17-7 Potencies (dosages at death) of four cardiac substances.

Substance	Sample Size (n_i)	Dosage at Death (Y_{ij})	Total	Sample Mean ($\bar{Y}_i$)	Sample Variance (S_i^2)
1	10	29, 28, 23, 26, 26, 19, 25, 29, 26, 28	259	25.9	9.4333
2	10	17, 25, 24, 19, 28 21, 20, 25, 19, 24	222	22.2	12.1778
3	10	17, 16, 21, 22, 23 18, 20, 17, 25, 21	200	20.0	8.6667
4	10	18, 20, 25, 24, 16 20, 20, 17, 19, 17	196	19.6	8.7111

existed among the potencies of the four substances and, if so, to quantify those differences. The overall ANOVA table for comparing the mean potencies of the four cardiac substances is given in Table 17-8.

The global F test strongly rejects ($P < .001$) the null hypothesis of equality of the four population means. Therefore, the multiple-comparison question arises: What is the best way to account for the differences found? As a crude first step, we can examine the nature of the differences with the help of a schematic diagram of ordered sample means (Figure 17-1). In the diagram, an overbar has been drawn over the labels for substances 3 and 4 to indicate that the sample mean potencies for these two substances are quite similar. On the other hand, no overbar has been drawn connecting 1 and 2 with each other or with 3 and 4, suggesting that substances 1 and 2 differ from each other as well as from both 3 and 4.

Such an overall quantification of the differences among the population means is desired from a multiple-comparison analysis. Nevertheless, the purely descriptive approach taken does not account for the sampling variability associated with any estimated comparison of interest. As a result, two sample means that seem practically different may not, in fact, be statistically different. Since the only multiple-comparison method we have discussed so far is the LSD method, let us consider how to apply this method to the data of Table 17-7, using an overall significance level of $\alpha = .05$ for all pairwise comparisons of the mean potencies of the four cardiac

TABLE 17-8 ANOVA table for data of Table 17-7.

Source	d.f.	SS	MS	F
Substances	3	249.875	83.292	8.545 ($P < .001$)
Error	36	350.900	9.747	
Total	39	600.775		

FIGURE 17-1 Crude comparison of sample means for potency data.

$l = k^* = \dfrac{4!}{2 \cdot 2!} = 6$

substances. This approach requires computing $_4C_2 = 6$ confidence intervals, each associated with a significance level of $\alpha/6 = .05/6 = .0083$, utilizing the formula

$$(\bar{Y}_i - \bar{Y}_j) \pm t_{36,\, 1-0.0083/2}\sqrt{\text{MSE}\left(\frac{1}{10} + \frac{1}{10}\right)}$$

The right-hand side of this expression is calculated as

$$t_{36,\, 0.99585}\sqrt{9.747\left(\frac{1}{5}\right)} = 2.79(1.396) = 3.895$$

Thus, the pairwise confidence intervals are given as

$\mu_1 - \mu_4$: 6.3 ± 3.895; i.e., $(2.405,\ 10.195)$*

$\mu_1 - \mu_3$: 5.9 ± 3.895; i.e., $(2.005,\ 9.795)$*

$\mu_1 - \mu_2$: 3.7 ± 3.895; i.e., $(-0.195,\ 7.595)$

$\mu_2 - \mu_4$: 2.6 ± 3.895; i.e., $(-1.295,\ 6.495)$

$\mu_2 - \mu_3$: 2.2 ± 3.895; i.e., $(-1.695,\ 6.095)$

$\mu_3 - \mu_4$: 0.4 ± 3.895; i.e., $(-3.495,\ 4.295)$

In this example, the term *least significant difference* refers to the fact that *any two* sample means are considered significantly different if their absolute difference exceeds 3.895. (No such blanket statement can be made in instances where the n_i's are not all equal.)

The preceding intervals reveal only two significant comparisons (the ones starred) and translate into the diagrammatic overall ranking shown in Figure 17-2. These results are somewhat ambiguous due to overlapping "sets of similarities," which indicate that substances 2, 3, and 4 have essentially the same potency; that 1 and 2 have about the same potency; but also that 1 differs from both 3 and 4. In other words, one possible conclusion is that 2, 3, and 4 are to be grouped together and that 1 and 2 are to be grouped together—which is difficult to reconcile because substance 2 is common to both groups. Having to confront this ambiguity is quite fortuitous from a pedagogical standpoint, since such ambiguous results are not infrequently encountered in connection with multiple-comparison procedures. In our case the results indicate that the procedure used was not sensitive enough to permit an adequate evaluation of substance 2. Repeating the analysis with a larger data set would help clear up the ambiguity. Alternatively, since the LSD approach tends to be conservative (i.e., the confidence intervals tend to be wider than necessary to achieve the overall significance level desired), other multiple-comparison methods—such as those of Tukey or Scheffé—may provide more precise results (i.e., narrower confidence intervals) and so may reduce or eliminate any ambiguity.

FIGURE 17-2   LSD comparison of sample means for potency data.

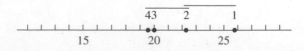

The LSD approach can also be implemented by using SAS's GLM procedure. Although SAS does not show the confidence intervals, it does display a graphic that indicates which pairs of means significantly differ. The accompanying SAS output shows this graphic; means that are similar (i.e., not significantly different) at $\alpha = 0.05$ are labeled with the same letter.

SAS's GLM procedure does not control for the number of pairwise comparisons when implementing the LSD approach; instead, the value $\alpha = .05$ is used for each pairwise comparison. As a result, the SAS output given below shows an LSD of 2.8317 instead of 3.798, since SAS uses a critical value of $t = 2.03, 36, 0.95$. As mentioned earlier, it is desirable to control for the number of comparisons.

SAS Output for Data of Table 17-7

```
               General Linear Models Procedure
                       .
                       . [Portion of output omitted]
                       .
            T tests (LSD) for variable: DOSAGE
NOTE: This test controls the type I comparisonwise error rate not
      the experimentwise error rate.
         Alpha= 0.05        df= 36      MSE= 9.747222
                    Critical Value of T= 2.03
             Least Significant Difference= 2.8317
Means with the same letter are not significantly different.
         T Grouping        Mean        N    SUBSTANC
                   A      25.900       10    1

                   B      22.200       10    2
                   B
                   B      20.000       10    3
                   B
                   B      19.600       10    4
```

17-7-2 Tukey's Method

Tukey's method is applicable when two conditions are met: *If $n_i = n_j$ best is Tukey*

1. The sizes of the samples selected from each population are equal; that is, $n_i \equiv n^*$, say, for all $i = 1, 2,..., k$, where k represents the number of means that are being compared.

2. Pairwise comparisons of the means are of primary interest; that is, null hypotheses of the form $H_0: \mu_i = \mu_j$ are to be considered.

A generalized version of Tukey's procedure is available for considering more complex comparisons than simple pairwise differences between means. However, since this procedure is most powerful for comparing simple differences between means and not for making more complex comparisons, Tukey's method should generally be used only in the former situation; the latter situation is best handled by using Scheffé's method (discussed in section 17-7-3).

To use Tukey's method, we compute the following confidence interval for the pairwise comparison of population means μ_i and μ_j:

$$(\overline{Y}_i - \overline{Y}_j) \pm T\sqrt{\text{MSE}}$$

(17.21)

where   *Table* A-6

$$T = \frac{1}{\sqrt{n^*}} q_{k,n-k,1-\alpha}$$

and where $q_{k,n-k,1-\alpha}$ is the $100(1-\alpha)\%$ point of the distribution of the studentized range with k and $n-k$ degrees of freedom (see Table A-6 in Appendix A); n^* is the common sample size (i.e., $n_i \equiv n^*$); k is the number of populations or groups, and $n(= kn^*)$ is the total number of observations.

The studentized range distribution with k and r degrees of freedom is the distribution of a statistic of the form R/S, where $R = \{\max_i (Y_i) - \min_i (Y_i)\}$ is the range of a set of k independent observations $Y_1, Y_2, \ldots, Y_k$ from a normal distribution with mean μ and variance σ^2, and where S^2 is an estimate of σ^2 based on $(n-k)$ degrees of freedom (which is independent of the Y's). In particular, when k means are being compared in fixed-effects one-way ANOVA, the statistic $\{\max_i (\overline{Y}_i) - \min_i (\overline{Y}_i)\}/\sqrt{\text{MSE}/n^*}$ has the studentized range distribution with k and $n-k$ degrees of freedom under $H_0: \mu_1 = \mu_2 = \cdots = \mu_k$, where $n_i = n^*$ for each i and where $n = kn^*$.

An alternative to Tukey's procedure, called the *Student–Newman–Keuls (SNK)* method, uses the studentized range distribution but with a modified number of numerator degrees of freedom. The SNK procedure replaces the first k in $q_{k,n-k,1-\alpha}$ by $k^* = $ Number of means in the range of means being tested. Thus, we would have $k^* = 3$ when comparing the second largest with the smallest of four means, whereas we would have $k^* = 2$ when comparing the second largest with the third largest.

For unequal sample sizes, Steele and Torrie (1960) recommended a slight modification of the Tukey procedure. For each comparison involving unequal sample sizes, they replace the term $T\sqrt{\text{MSE}}$ in (17.21) with $q_{k,n-k,1-\alpha}\sqrt{(\text{MSE}/2)(1/n_i + 1/n_j)}$, where n_i and n_j are the sample sizes associated with the ith and jth groups, respectively.

In the set of all $_kC_2$ Tukey pairwise confidence intervals of the form (17.21), the probability is $1-\alpha$ that these intervals simultaneously contain the associated population mean differences that are being estimated; that is, $1-\alpha$ is the *overall confidence coefficient* for all pairwise comparisons taken together. In particular, if each confidence interval is used to test the appropriate pairwise hypothesis of the general form $H_0: \mu_i = \mu_j$ by determining whether the value 0 is contained in the calculated interval, the probability of falsely rejecting the null hypothesis for *at least one* of the $_kC_2$ tests is equal to α.

The procedure for applying Tukey's method to a set of data is usually carried out stepwise, as follows:

1. Rank-order the sample means $\overline{Y}_i$ from largest to smallest (e.g., in our example, the order is $\overline{Y}_1 > \overline{Y}_2 > \overline{Y}_3 > \overline{Y}_4$).

2. Compare the largest sample mean with the smallest, using (17.21); then compare the largest with the next smallest; and so on, until either the largest has been compared with the second largest or a nonsignificant result has been obtained, whichever comes first (e.g., in our example, we would first look at 1 versus 4, then at 1 versus 3, and finally at 1 versus 2).

3. Continue by comparing the second largest mean with the smallest, the second largest with the next smallest, and so on, but make no further comparisons with the second largest mean once any nonsignificant result is obtained.

4. Continue making such comparisons involving the third largest mean, then the fourth largest, and so on. At each stage, once a nonsignificant comparison is obtained, conclude that no difference exists between any means enclosed by the first nonsignificant pair.

5. Represent the overall conclusion about similarities and differences among the population means by using a schematic diagram of the ordered sample means, drawing overbars to connect means that do not significantly differ.

Let us now illustrate using Tukey's method with an overall significance level of $\alpha = .05$ for the potency data of Table 17-7. The ordering of sample means from largest to smallest indicates that the following sequence of pairwise comparisons should be made:

1 vs. 4,   1 vs. 3,   1 vs. 2,   2 vs. 4,   2 vs. 3,   3 vs. 4

Since the value of $T\sqrt{\text{MSE}}$ in (17.21) is needed for any such comparison, we compute it first, as follows: for $n^* = 10$, $k = 4$, $n = 40$, and $\text{MSE} = 9.747$,

$$T = \frac{1}{\sqrt{n^*}} q_{k,n-k,1-\alpha} = \frac{1}{\sqrt{10}} q_{4,36,0.95} = \frac{1}{\sqrt{10}} (3.81) = 1.205$$

(The value $q_{4,36,0.95} = 3.81$ was obtained from Table A-6 by interpolation.) Therefore,

$$T\sqrt{\text{MSE}} = 1.205 \sqrt{9.747} = 3.762$$

Now, using (17.21), we compare 1 and 4 as follows:

$$(\overline{Y}_1 - \overline{Y}_4) \pm T\sqrt{\text{MSE}}$$
$$(25.9 - 19.6) \pm 3.762$$
$$6.3 \pm 3.762$$

or

$$(2.538, 10.062)$$

Because this confidence interval does not contain the value 0, we can conclude (based on an overall significance level of $\alpha = .05$) that $\mu_1 \neq \mu_4$.

Next, we look at 1 versus 3, as follows:

$$(\overline{Y}_1 - \overline{Y}_3) \pm T\sqrt{\text{MSE}}$$
$$(25.9 - 20.0) \pm 3.762$$

or

$$(2.138, 9.662)$$

Because this confidence interval also does not contain the value 0, we can conclude (at an overall significance level of $\alpha = .05$) that $\mu_1 \neq \mu_3$.

Next, we compare 1 and 2, obtaining

$$3.7 \pm 3.762$$

or

$$(-0.062, 7.462)$$

which contains the value 0. We thus conclude that there is insufficient evidence to permit us to reject H_0: $\mu_1 = \mu_2$ and also that all remaining comparisons, which involve smaller (in absolute value) pairwise mean differences, are nonsignificant. This conclusion supports the hypothesis that $\mu_2 = \mu_3 = \mu_4$. We may schematically represent the results based on applying Tukey's method as shown in Figure 17-3.

In general, a pairwise difference between, say, the largest mean and the second largest may not be significant, even though another pairwise difference (say, between the third- and fourth-largest means) is significant. Thus, in general, one should not stop making *all* remaining comparisons when the first nonsignificant pairwise difference is encountered unless all remaining pairwise differences are smaller than the one involved in the first nonsignificant pair. As an example, if we found that $\overline{Y}_1 = 100$, $\overline{Y}_2 = 99$, $\overline{Y}_3 = 80$, and $\overline{Y}_4 = 20$, then $\overline{Y}_1 - \overline{Y}_2 = 1$ would be quite small (and possibly nonsignificant) while $\overline{Y}_3 - \overline{Y}_4 = 60$ would be large.

Like the LSD method, the Tukey method leaves some ambiguity as to results: substance 2 has again been associated with substance 1 and also with substances 3 and 4. Such ambiguity is not uncommon. In this instance, it suggests that the amount of data collected was insufficient to permit a clear categorization of substance 2.

The Tukey approach can also be implemented by using SAS's GLM procedure. Once again SAS does not show the confidence intervals, but it does display a graphic indicating which pairs of means significantly differ. The accompanying SAS output shows this graphic; means that are similar (i.e., not significantly different) at an overall level of $\alpha = .05$ are labeled with the same letter.

SAS Output for Data of Table 17-7

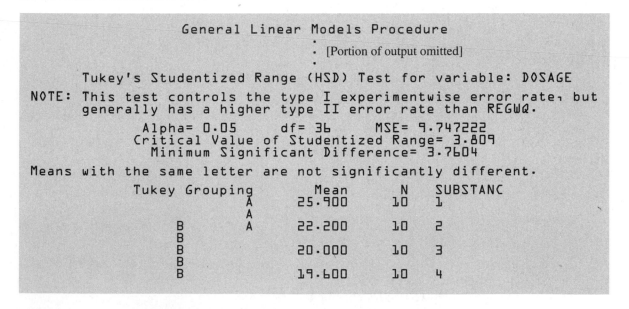

```
            General Linear Models Procedure
                            .
                            . [Portion of output omitted]
                            .

     Tukey's Studentized Range (HSD) Test for variable: DOSAGE
NOTE: This test controls the type I experimentwise error rate, but
      generally has a higher type II error rate than REGWQ.

            Alpha= 0.05      df= 36      MSE= 9.747222
         Critical Value of Studentized Range= 3.809
            Minimum Significant Difference= 3.7604
Means with the same letter are not significantly different.

            Tukey Grouping         Mean       N    SUBSTANC
                         A         25.900      10   1
                         A
              B          A         22.200      10   2
              B
              B                    20.000      10   3
              B
              B                    19.600      10   4
```

FIGURE 17-3 Tukey comparison of sample means for potency data.

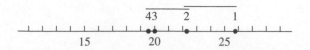

17-7-3 Scheffé's Method

Can compare pairwise, $c_1 = -c_2 = 1$, $c_n = 0$

Scheffé's method is generally recommended when either of two conditions are met:

1. The sizes of the samples selected from the different populations are not all equal.

2. Comparisons other than simple pairwise comparisons between two means are of interest. (These more general types of comparisons are referred to as *contrasts*.)

To illustrate the meaning of *contrast*, suppose that the investigator who collected the potency data of Table 17-7 suspected that substances 1 and 3 had similar potencies, that substances 2 and 4 had similar potencies, and that the potencies of 1 and 3, on the average, differed significantly from those of 2 and 4. Then it would be of interest to compare the average results obtained for 1 and 3 with the average results for 2 and 4, to assess whether $(\mu_1 + \mu_3)/2$ really differs from $(\mu_2 + \mu_4)/2$. In other words, we could consider the contrast

$$L_1 = \frac{\mu_1 + \mu_3}{2} - \frac{\mu_2 + \mu_4}{2}$$

which would be 0 if the null hypothesis H_0: $(\mu_1 + \mu_3)/2 = (\mu_2 + \mu_4)/2$ were true. We can rewrite L_1 as follows:

$$L_1 = \frac{\mu_1 + \mu_3}{2} - \frac{\mu_2 + \mu_4}{2} = \frac{1}{2}\mu_1 - \frac{1}{2}\mu_2 + \frac{1}{2}\mu_3 - \frac{1}{2}\mu_4$$

or

$$L_1 = \sum_{i=1}^{4} c_{1i}\mu_i$$

so that L_1 is a *linear* function of the population means, with $c_{11} = \frac{1}{2}$, $c_{12} = -\frac{1}{2}$, $c_{13} = \frac{1}{2}$, and $c_{14} = -\frac{1}{2}$. Further,

$$c_{11} + c_{12} + c_{13} + c_{14} = \tfrac{1}{2} - \tfrac{1}{2} + \tfrac{1}{2} - \tfrac{1}{2} = 0$$

In general, a *contrast* is defined as any linear function of the population means, say,

$$L = \sum_{i=1}^{k} c_i\mu_i$$

such that

$$\sum_{i=1}^{k} c_i = 0$$

The associated null hypothesis is

$$H_0: \quad \sum_{i=1}^{k} c_i\mu_i = 0$$

and the two-tailed alternative hypothesis is

$$H_A: \quad \sum_{i=1}^{k} c_i\mu_i \neq 0$$

The data in Table 17-7 suggest that the mean potency of substance 1 is definitely higher than the mean potencies of the other three substances. Such an observation invites a comparison of the mean potency of substance 1 with the average potency of substances 2, 3, and 4. In this case, the appropriate contrast to consider is

$$L_2 = \mu_1 - \frac{\mu_2 + \mu_3 + \mu_4}{3}$$

or

$$L_2 = \sum_{i=1}^{4} c_{2i} \mu_i$$

where $c_{21} = 1$, $c_{22} = c_{23} = c_{24} = -\frac{1}{3}$. (Again, $\sum_{i=1}^{4} c_{2i} = 0$.)

A third possible contrast of interest involves a comparison of the average results for substances 1 and 2 with those for 3 and 4. The contrast here is

$$L_3 = \frac{\mu_1 + \mu_2}{2} - \frac{\mu_3 + \mu_4}{2} = \sum_{i=1}^{4} c_{3i} \mu_i$$

where $c_{31} = c_{32} = \frac{1}{2}$ and $c_{33} = c_{34} = -\frac{1}{2}$.

Finally, any pairwise comparison is also a contrast. For example, a comparison of 1 with 4 takes the form

$$L_4 = \mu_1 - \mu_4 = \sum_{i=1}^{4} c_{4i} \mu_i$$

where $c_{41} = 1$, $c_{42} = c_{43} = 0$, and $c_{44} = -1$.

Scheffé's method provides a family of confidence intervals for evaluating *all possible contrasts* that can be defined, given a fixed number k of population means, such that the overall confidence coefficient associated with the entire family is $1 - \alpha$, where α is specified by the investigator. In other words, the probability is $1 - \alpha$ that these confidence intervals simultaneously contain the true values of all the contrasts being considered. The overall significance level is α for testing hypotheses of the general form $H_0: L = \sum_{i=1}^{k} c_i \mu_i = 0$ concerning all possible contrasts; that is, the probability is α that at least one such null hypothesis will falsely be rejected.

The general form of a Scheffé-type confidence interval is as follows. Let $L = \sum_{i=1}^{k} c_i \mu_i$ be some contrast of interest. Then the appropriate confidence interval concerning L is given by

$$\sum_{i=1}^{k} c_i \overline{Y}_i \pm S \sqrt{\text{MSE} \left(\sum_{i=1}^{k} \frac{c_i^2}{n_i} \right)} \tag{17.22}$$

where $\hat{L} = \sum_{i=1}^{k} c_i \overline{Y}_i$ is the unbiased point estimator of L, and where $S^2 = (k-1)F_{k-1, n-k, 1-\alpha}$ with $n = \sum_{i=1}^{k} n_i$.

As a special case, when only pairwise comparisons are of interest, this formula simplifies to

$$(\overline{Y}_i - \overline{Y}_j) \pm S \sqrt{\text{MSE} \left(\frac{1}{n_i} + \frac{1}{n_j} \right)} \tag{17.23}$$

when inferences regarding $\mu_i = \mu_j$ are being considered.

If the investigator is interested only in pairwise comparisons, Scheffé's method using (17.23) is not recommended; in such cases, Tukey's method will always provide narrower confidence intervals (i.e., will give more precise estimates of the true pairwise differences). However, if the sample sizes are unequal and/or if contrasts other than pairwise comparisons are of interest, Scheffé's method is preferable. Furthermore, Scheffé's method has another desirable property: whenever the overall F test of the null hypothesis that all k population means are equal is rejected, at least one contrast will be found that differs significantly from 0. In contrast, Tukey's method may not turn up any significant pairwise comparisons, even when the overall F statistic is significant.

To illustrate Scheffé's method, let us first consider all pairwise comparisons for the data of Table 17-7 and follow the same procedure as is used with Tukey's method; that is, let us consider, in order, 1 versus 4, 1 versus 3, 1 versus 2, 2 versus 4, 2 versus 3, and 3 versus 4. We begin by first computing the quantity

$$S\sqrt{\text{MSE}\left(\frac{1}{n_i} + \frac{1}{n_j}\right)}$$

which will have the same value for all pairwise comparisons, since all the sample sizes are equal to 10. So

$$\sqrt{\text{MSE}\left(\frac{1}{n_i} + \frac{1}{n_j}\right)} = \sqrt{9.747\left(\frac{1}{10} + \frac{1}{10}\right)} = 1.3962$$

And with $k = 4$, $n = 40$, and $\alpha = .05$, we have

$$S = \sqrt{(k-1)F_{k-1,n-k,1-\alpha}} = \sqrt{3F_{3,36,0.95}} = \sqrt{3(2.886)} = 2.9424$$

Thus, $S\sqrt{\text{MSE}(1/n_i + 1/n_j)} = 2.9424(1.3962) = 4.1082$. Now, to compare substances 1 and 4, we obtain the following confidence interval:

$$(\overline{Y}_1 - \overline{Y}_4) \pm S\sqrt{\text{MSE}\left(\frac{1}{n_1} + \frac{1}{n_4}\right)}$$

$$(25.9 - 19.6) \pm 4.1082$$

or

$$(2.192, 10.408)$$

Since this interval does not contain the value 0 (which is also the case for the corresponding Tukey interval), we support the contention that $\mu_1 \neq \mu_4$. Thus we proceed to the 1 versus 3 comparison:

$$(\overline{Y}_1 - \overline{Y}_3) \pm S\sqrt{\text{MSE}\left(\frac{1}{n_1} + \frac{1}{n_3}\right)}$$

$$(25.9 - 20.0) \pm 4.1082$$

or

$$(1.792, 10.008)$$

Again, like the corresponding interval obtained with Tukey's method, this interval does not contain the value 0, which supports rejection of H_0: $\mu_1 = \mu_3$.

The next comparison, between substances 1 and 2, yields the interval

$$(-0.408, 7.808)$$

which contains the value 0. Thus, we cannot reject H_0: $\mu_1 = \mu_2$; and observing that the remaining pairwise comparisons are nonsignificant, we conclude that $\mu_2 = \mu_3 = \mu_4$ and that $\mu_1 = \mu_2$. Thus, when applied to pairwise comparisons for this data set, Tukey's and Scheffé's methods yield the same general conclusions about the relative potencies of the four substances, including the ambiguity associated with substance 2. On closer inspection of the pairwise intervals derived (see Table 17-9), however, we see that the Tukey intervals are narrower and so support more precise inferences.

Scheffé's multiple comparisons approach can also be implemented by using SAS's GLM procedure. Once again, the SAS output does not show the confidence intervals, but it does display a graphic showing which pairs of means significantly differ. The accompanying SAS output shows this graphic; means that are similar (i.e., not significantly different) at an overall level of $\alpha = .05$ are labeled with the same letter.

Edited SAS Output (PROC GLM) for Data of Table 17-7

```
                General Linear Models Procedure
                                 .
                             .   [Portion of output omitted]
                                 .
                Scheffe's test for variable: DOSAGE
 NOTE: This test controls the type I experimentwise error rate but
       generally has a higher type II error rate than REGWF for all
       pairwise comparisons
             Alpha= 0.05        df= 36      MSE= 9.747222
                   Critical Value of F= 2.86627
             Minimum Significant Difference= 4.0942
    Means with the same letter are not significantly different.
          Scheffe Grouping          Mean        N    SUBSTANC
                           A        25.900      10     1
                           A
                  B        A        22.200      10     2
                  B
                  B        20.000      10     3
                  B
                  B        19.600      10     4
```

Let us now use Scheffé's method to make inferences regarding two contrasts of interest. In particular, let us consider

$$L_2 = \mu_1 - \frac{\mu_2 + \mu_3 + \mu_4}{3} = \sum_{i=1}^{4} c_{2i} \mu_i$$

TABLE 17-9   Comparison of some Tukey and Scheffé confidence intervals for the potency data of Table 17-7.

Pairwise Comparison	Tukey Lower Limit	Tukey Upper Limit	Scheffé Lower Limit	Scheffé Upper Limit
$\mu_1 - \mu_4$	2.535	10.065	2.192	10.408
$\mu_1 - \mu_3$	2.135	9.665	1.792	10.008
$\mu_1 - \mu_2$	−0.065	7.465	−0.408	7.808
$\mu_2 - \mu_4$	−1.165	6.365	−1.508	6.708
$\mu_2 - \mu_3$	−1.565	5.965	−1.908	6.308
$\mu_3 - \mu_4$	−3.365	4.165	−3.708	4.508

where $c_{21} = 1$ and $c_{22} = c_{23} = c_{24} = -\frac{1}{3}$, and

$$L_3 = \frac{\mu_1 + \mu_2}{2} - \frac{\mu_3 + \mu_4}{2} = \sum_{i=1}^{4} c_{3i} \mu_i$$

where $c_{31} = c_{32} = \frac{1}{2}$ and $c_{33} = c_{34} = -\frac{1}{2}$.

Using our previously computed value $S = \sqrt{(k-1)F_{k-1,n-k,1-\alpha}} = 2.9424$, we calculate from (17.22) as follows:

$$\left(\overline{Y}_1 - \frac{\overline{Y}_2 + \overline{Y}_3 + \overline{Y}_4}{3} \right) \pm S \sqrt{\text{MSE}\left[\frac{(1)^2}{10} + \frac{(-1/3)^2}{10} + \frac{(-1/3)^2}{10} + \frac{(-1/3)^2}{10} \right]}$$

$$\left(25.9 - \frac{22.2 + 20.0 + 19.6}{3} \right) \pm 2.9424 \sqrt{9.747\left(\frac{12}{90} \right)}$$

$$(25.9 - 20.6) \pm 2.9424(1.1400)$$

$$5.3 \pm 3.354$$

or

$$1.946 \leq L_2 \leq 8.654$$

Since this interval does not contain the value 0, we have evidence that the potency of substance 1 differs from the average potency of substances 2, 3, and 4.

Next we calculate the following Scheffé interval regarding L_3:

$$\left(\frac{\overline{Y}_1 + \overline{Y}_2}{2} - \frac{\overline{Y}_3 + \overline{Y}_4}{2} \right) \pm S \sqrt{\text{MSE}\left[\frac{(1/2)^2}{10} + \frac{(1/2)^2}{10} + \frac{(-1/2)^2}{10} + \frac{(-1/2)^2}{10} \right]}$$

$$\left(\frac{25.9 + 22.2}{2} - \frac{20.0 + 19.6}{2} \right) \pm 2.9424 \sqrt{9.747\left(\frac{1}{10} \right)}$$

$$(24.05 - 19.80) \pm 2.9424(0.9873)$$

$$4.25 \pm 2.9049$$

or

$$(1.345, 7.155)$$

FIGURE 17-4   Conclusion regarding sample means for potency data.

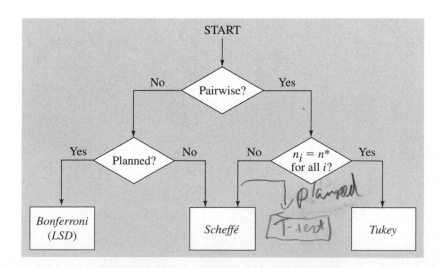

Because this interval also does not contain the value 0, we conclude that the average potency of substances 1 and 2 differs from the average potency of substances 3 and 4.

How do these results about the contrasts L_2 and L_3 help clear up the ambiguity created by considering all pairwise comparisons? First, uncertainty regarding substance 2 remains. Nevertheless, the confidence interval for comparing 1 with the average of 2, 3, and 4 (i.e., 1.946 to 8.654) is farther away from 0 than is the confidence interval for comparing the average of 1 and 2 with the average of 3 and 4 (i.e., 1.345 to 7.155); this suggests that substance 2 is closer to 3 and 4 in potency than it is to substance 1. The pairwise comparisons support this contention, too, since the confidence interval for $\mu_1 - \mu_2$ is farther from 0 than is the interval for $\mu_2 - \mu_3$, again indicating that 2 is closer to 3 than to 1. Thus, if a definite decision regarding substance 2 had to be made on the basis of this set of data, the most logical thing to do would be to consider the potency of 1 as distinct from the potencies of 2, 3, and 4, which, as a threesome, are too similar to separate. Schematically, this conclusion is represented in Figure 17-4.

17-8 Choosing a Multiple-comparison Technique

Figure 17-5 summarizes a strategy for choosing a multiple-comparison technique for ANOVA. The first choice is between pairwise and nonpairwise comparisons. If only pairwise comparisons are being considered and if all cell-specific sample sizes are equal, the Tukey method is preferable. Otherwise, the less-sensitive Scheffé method must be used.

FIGURE 17-5   Recommended procedure for choosing a multiple-comparison technique
in ANOVA.

If any nonpairwise comparisons are to be considered, a choice must be made between planned (a priori) and unplanned comparisons. The Bonferroni method should be used for planned comparisons; unplanned comparisons should be evaluated using the Scheffé method.

Although we have discussed only the Bonferroni (LSD), Tukey, and Scheffé methods, a large number of other multiple-comparison techniques have been suggested in the statistical literature. Miller (1981) provides the most comprehensive review of these procedures. The methods differ with regard to the target set of contrasts (e.g., pairwise versus nonpairwise), the cell-specific sample sizes, and other properties. The developers of these methods have sought to minimize the Type II error rate (i.e., to maximize power; see Chapter 3) for a particular set of comparisons while controlling the Type I error rate.

The reasons for considering other multiple-comparison methods are to improve power and to achieve robustness. A multiple-comparison procedure is robust if a violation of its underlying assumptions, such as due to nonhomogeneity of variance, does not seriously affect the validity of an analysis. If a robust procedure is used, even when some assumptions are violated, Type I and Type II error rates from statistical testing are unlikely to be compromised. Both the Bonferroni and Scheffé methods tend to be fairly robust; the Tukey method is less so. Scheffé's method is always the least powerful, since all possible comparisons are considered. The LSD method is generally presumed to be insensitive (i.e., to have low power), but this presumption can be discounted if the sets of comparisons are well planned.

Although many acceptable multiple-comparison procedures are available, most are limited to a rather narrow range of application (e.g., Tukey's method requires equal cell-specific sample sizes). Both the Bonferroni and Scheffé procedures are completely general methods—the former for planned (a priori) and the latter for unplanned (a posteriori) multiple comparisons. The tasks of covering all multiple-comparison methods completely is beyond the scope of this book.

17-9 Orthogonal Contrasts and Partitioning an ANOVA Sum of Squares

Our previous discussions of multiple regression have touched on the notion of a partitioned sum of squares in regression analysis, where the sum of squares due to regression (SSR) is broken down into various components reflecting the relative contributions of various terms in the fitted model. In an ANOVA framework, it is possible (via the use of orthogonal contrasts) to partition the treatment sum of squares SST into meaningful components associated with certain specific comparisons of particular interest. To illustrate how such a partitioning can be accomplished, we must discuss two new concepts: *orthogonal contrasts* and the *sum of squares associated with a contrast*.

In the notation of section 17-7, two estimated contrasts

$$\hat{L}_A = \sum_{i=1}^{k} c_{Ai}\overline{Y}_i \qquad \text{and} \qquad \hat{L}_B = \sum_{i=1}^{k} c_{Bi}\overline{Y}_i$$

are *orthogonal* to one another (i.e., are orthogonal contrasts) if

$$\sum_{i=1}^{k} \frac{c_{Ai}c_{Bi}}{n_i} = 0 \qquad\qquad\qquad (17.24)$$

In the special case where the n_i's are equal, the (17.24) equation reduces to the condition

$$\sum_{i=1}^{k} c_{Ai}c_{Bi} = 0 \tag{17.25}$$

Consider the three contrasts discussed earlier with regard to the potency data of Table 17.7:

$$\hat{L}_1 = \frac{\overline{Y}_1 + \overline{Y}_3}{2} - \frac{\overline{Y}_2 + \overline{Y}_4}{2} = \frac{1}{2}\overline{Y}_1 - \frac{1}{2}\overline{Y}_2 + \frac{1}{2}\overline{Y}_3 - \frac{1}{2}\overline{Y}_4$$

where $c_{11} = c_{13} = \frac{1}{2}$ and $c_{12} = c_{14} = -\frac{1}{2}$;

$$\hat{L}_2 = \overline{Y}_1 - \frac{1}{3}(\overline{Y}_2 + \overline{Y}_3 + \overline{Y}_4) = \overline{Y}_1 - \frac{1}{3}\overline{Y}_2 - \frac{1}{3}\overline{Y}_3 - \frac{1}{3}\overline{Y}_4$$

where $c_{21} = 1$ and $c_{22} = c_{23} = c_{24} = -\frac{1}{3}$; and

$$\hat{L}_3 = \frac{\overline{Y}_1 + \overline{Y}_2}{2} - \frac{\overline{Y}_3 + \overline{Y}_4}{2} = \frac{1}{2}\overline{Y}_1 + \frac{1}{2}\overline{Y}_2 - \frac{1}{2}\overline{Y}_3 - \frac{1}{2}\overline{Y}_4$$

where $c_{31} = c_{32} = \frac{1}{2}$ and $c_{33} = c_{34} = -\frac{1}{2}$. Since $n_i = 10$ for every i, we need only verify that condition (17.25) holds to demonstrate orthogonality. In particular, for $\hat{L}_1$ and $\hat{L}_2$, we have

$$\sum_{i=1}^{4} c_{1i}c_{2i} = \left(\frac{1}{2}\right)(1) + \left(-\frac{1}{2}\right)\left(-\frac{1}{3}\right) + \left(\frac{1}{2}\right)\left(-\frac{1}{3}\right) + \left(-\frac{1}{2}\right)\left(-\frac{1}{3}\right) = \frac{2}{3} \neq 0$$

For $\hat{L}_1$ and $\hat{L}_3$, we have

$$\sum_{i=1}^{4} c_{1i}c_{3i} = \left(\frac{1}{2}\right)\left(\frac{1}{2}\right) + \left(-\frac{1}{2}\right)\left(\frac{1}{2}\right) + \left(\frac{1}{2}\right)\left(-\frac{1}{2}\right) + \left(-\frac{1}{2}\right)\left(-\frac{1}{2}\right) = 0$$

And, for $\hat{L}_2$ and $\hat{L}_3$, we have

$$\sum_{i=1}^{4} c_{2i}c_{3i} = (1)\left(\frac{1}{2}\right) + \left(-\frac{1}{3}\right)\left(\frac{1}{2}\right) + \left(-\frac{1}{3}\right)\left(-\frac{1}{2}\right) + \left(-\frac{1}{3}\right)\left(-\frac{1}{2}\right) = \frac{2}{3} \neq 0$$

Thus, we conclude that $\hat{L}_1$ and $\hat{L}_3$ are orthogonal to one another but that neither is orthogonal to $\hat{L}_2$.

Orthogonality is a desirable property for several reasons. Suppose that SST denotes the treatment sum of squares with $k - 1$ degrees of freedom for a fixed-effects one-way ANOVA, and suppose that $\hat{L}_1, \hat{L}_2,\ldots, \hat{L}_t$ are a set of $t(\leq k - 1)$ mutually orthogonal contrasts of the k sample means (*mutually orthogonal* means that any two contrasts selected from the set of t contrasts are orthogonal to one another). Then SST can be partitioned into $t + 1$ statistically independent sums of squares, with t of these sums of squares having 1 degree of freedom each and being associated with the t orthogonal contrasts, and with the remaining sum of squares having $k - 1 - t$ degrees of freedom and being associated with what remains after the t orthogonal contrast sums of squares have been accounted for. In other words, we can write

$$\text{SST} = \text{SS}(\hat{L}_1) + \text{SS}(\hat{L}_2) + \cdots + \text{SS}(\hat{L}_t) + \text{SS(remainder)}$$

where $\text{SS}(\hat{L})$ is the notation for the sum of squares (with 1 degree of freedom) associated with the contrast $\hat{L}$. In particular, it can be shown that

$$\text{SS}(\hat{L}) = \frac{(\hat{L})^2}{\sum\limits_{i=1}^{k} (c_i^2/n_i)} \qquad \text{when} \qquad \hat{L} = \sum_{i=1}^{k} c_i \overline{Y}_i \qquad (17.26)$$

and that

$$\frac{\text{SS}(\hat{L})}{\text{MSE}} \frown F_{1,n-k}$$

under H_0: $L = \sum_{i=1}^{k} c_i \mu_i = 0$; and, in general, that

$$\frac{[\text{SS}(\hat{L}_1) + \text{SS}(\hat{L}_2) + \cdots + \text{SS}(\hat{L}_t)]/t}{\text{MSE}} \frown F_{t,n-k}$$

under H_0: $L_1 = L_2 = \cdots = L_t = 0$ when $\hat{L}_1, \hat{L}_2, \ldots, \hat{L}_t$ are mutually orthogonal. Thus, by partitioning SST as described, we can test hypotheses concerning sets of orthogonal contrasts that are of more specific interest than the global hypothesis H_0: $\mu_1 = \mu_2 = \cdots = \mu_k$.

For example, to test H_0: $L_2 = \mu_1 - \frac{1}{3}(\mu_2 + \mu_3 + \mu_4) = 0$ for the potency data in Table 17-7, we first use (17.26) to calculate

$$\text{SS}(\hat{L}_2) = \frac{(5.30)^2}{[(1)^2 + (-1/3)^2 + (-1/3)^2 + (-1/3)^2]/10} = 210.675$$

and then we form the ratio

$$\frac{\text{SS}(\hat{L}_2)}{\text{MSE}} = \frac{210.675}{9.747} = 21.614$$

which is highly significant ($P < .001$ based on the $F_{1,36}$ distribution). Modifying Table 17-8 to reflect this partitioning of SST (the sum of squares for "substances") yields

Source		d.f.	SS	MS	F
Substances	1 vs. (2, 3, 4)	1	210.675	210.675	21.614 ($P < .001$)
	Remainder	2	39.200	irrelevant	—
Error		36	350.900	9.747	
Total		39	600.775		

Similarly, the partitioned ANOVA table for testing

$$H_0: L_3 = \frac{\mu_1 + \mu_2}{2} - \frac{\mu_3 + \mu_4}{2} = 0$$

is as follows:

Source		d.f.	SS	MS	F
Substances	(1, 2) vs. (3, 4)	1	179.776	179.776	18.444 ($P < .001$)
	Remainder	2	70.099	irrelevant	—
Error		36	350.900	9.747	
Total		39	600.775		

This partition follows from the fact that

$$SS(\hat{L}_3) = \frac{(4.25)^2}{[(1/2)^2 + (1/2)^2 + (-1/2)^2 + (-1/2)^2]/10} = 180.625$$

Finally, since $\hat{L}_1$ and $\hat{L}_3$ are orthogonal, we can represent the independent contributions of these two contrasts to SST in one ANOVA table:

Source		d.f.	SS	MS	F
	(1, 3) vs. (2, 4)	1	42.025	42.025	4.312 (n.s.)
Substances	(1, 2) vs. (3, 4)	1	180.625	180.625	18.444 ($P < .001$)
	Remainder	1	27.225	irrelevant	—
Error		36	350.900	9.747	
Total		39	600.775		

Here,

$$SS(\hat{L}_1) = \frac{\left(\dfrac{25.9 + 20.0}{2} - \dfrac{22.2 + 19.6}{2}\right)^2}{[(1/2)^2 + (-1/2)^2 + (1/2)^2 + (-1/2)^2]/10} = 42.025$$

It would not be valid to present a partitioned ANOVA table that simultaneously included partitions due to $\hat{L}_2$ and to $\hat{L}_1$ or $\hat{L}_3$ or both. This is because $\hat{L}_2$ is not orthogonal to these contrasts, and so its sum of squares does not represent a separate and independent contribution to SST; this can easily be confirmed by observing from the preceding tables that

$$SST \neq SS(\hat{L}_1) + SS(\hat{L}_2) + SS(\hat{L}_3)$$

A final example of using orthogonal contrasts involves the need to assess whether the sample means exhibit a trend of some sort; this need often arises when the treatments (or populations) being studied represent, for example, different levels of the same factor (e.g., different concentrations of the same material, or different temperature or pressure settings). In such situations it is of interest to quantify how the sample means vary with changes in the level of the factor—that is, to clarify whether the change in mean response takes place in a linear, quadratic, or other way as the level of the factor increases or decreases.

A qualitative first step in assessing such a trend is to plot the observed treatment means as a function of the factor levels. This may yield some general idea of the pattern (if any) present. Standard regression techniques can then be used to quantify any trends suggested by such plots, but our goal here is not to fit a regression model; rather, it is to evaluate a possible general trend in the means statistically, instead of forming an opinion on the basis of a simple examination of a plot of the means.

In a standard regression approach, the independent variable ("treatments") would be treated as an interval variable, and the actual value of the variable at each treatment setting would be used. To test for a linear trend by using regression, therefore, we would apply the model $Y = \beta_0 + \beta_1 X + E$, where X denotes the (interval) treatment variable. With this model, the test for linear trend is the usual test for zero slope. In general, this test is not equivalent to the test for a linear trend in *mean* response using the orthogonal polynomials described in this section, except when the pure error mean square is used in place of the residual mean square in the denominator of the F statistic to test for zero slope; this is because the regression pure error mean square and the one-way ANOVA error mean square are identical. The usual test for lack of fit of the straight-line regression model is equivalent to the test for a nonlinear trend in mean response discussed in this section. (See Problem 10 at the end of this chapter for more on this.)

A statistical trend analysis may be carried out by determining how much of the sum of squares due to treatments (SST) is associated with each of the terms (linear, quadratic, cubic, etc.) in a polynomial regression. If the various levels of the treatment or factor being studied are equally spaced, this determination is best carried out by using the method of orthogonal polynomials. (For a discussion, see Armitage (1971) and also Chapter 13.)

To illustrate the use of orthogonal polynomials, let us again turn to the potency data of Table 17-7. Further, let us suppose that the four substances actually represent four equally spaced concentrations of some toxic material, with substance 1 the least concentrated solution and substance 4 the most concentrated. For example, substance 1 might represent a 10% solution of the toxic material; substance 2 a 20% solution; substance 3 a 30% solution; and substance 4 a 40% solution. In this case, a plot of the four sample means versus concentration takes the form shown in Figure 17-6. This plot suggests at least a linear and possibly a quadratic relationship between concentration and response, and this general impression can be quantified via the use of orthogonal polynomials. In particular, given $k = 4$ sample means, it is possible to fit up to a third-order ($k - 1 = 3$) polynomial to these means; the cubic model (with four terms) would pass through all four points on the Figure 17-6 graph, thus explaining all the variation in the four sample means (or equivalently, in SST). Because the concentrations are equally spaced, it is possible via the use of orthogonal polynomials to define three orthogonal contrasts of the four sample means—one (say, $\hat{L}_l$) measuring the strength of the linear component of the third-degree polynomial, one (say, $\hat{L}_q$) the quadratic component contribution, and one (say, $\hat{L}_c$) the cubic component effect. The sums of squares associated with these three

FIGURE 17-6   Plot of the sample means versus concentration.

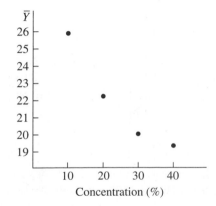

orthogonal contrasts each have 1 degree of freedom, are statistically independent, and satisfy the relationship

$$SST = SS(\hat{L}_l) + SS(\hat{L}_q) + SS(\hat{L}_c)$$

It can be shown that these three orthogonal contrasts have the following coefficients:

Contrast	Coefficient of				Calculated Value of Contrast
	$\bar{Y}_1 = 25.9$	$\bar{Y}_2 = 22.2$	$\bar{Y}_3 = 20.0$	$\bar{Y}_4 = 19.6$	
$\hat{L}_l$	-3	-1	1	3	-21.1
$\hat{L}_q$	1	-1	-1	1	3.3
$\hat{L}_c$	-1	3	-3	1	0.3

Condition (17.25) clearly holds for these three sets of coefficients, which is sufficient to establish mutual orthogonality here, since the n_i's are all equal. These coefficients were taken from Table A-7 in Appendix A. The table may be used only for equally spaced treatment values and equal cell-specific sample sizes. Kirk (1969) summarizes an algorithm for the general case that does not require these two restrictions. More conveniently, computer software can be used to obtain orthogonal contrast coefficients.

Now, from (17.26), the sums of squares for these three particular contrasts are

$$SS(\hat{L}_l) = \frac{(-21.1)^2}{[(-3)^2 + (-1)^2 + (1)^2 + (3)^2]/10} = 222.605$$

$$SS(\hat{L}_q) = \frac{(3.3)^2}{[(1)^2 + (-1)^2 + (-1)^2 + (1)^2]/10} = 27.225$$

$$SS(\hat{L}_c) = \frac{(0.3)^2}{[(-1)^2 + (3)^2 + (-3)^2 + (1)^2]/10} = 0.045$$

Notice that

$$SST = 249.875 = 222.605 + 27.225 + 0.045$$

Finally, to assess the significance of these sums of squares, we form the following partitioned ANOVA table:

Source		d.f.	SS	MS	F
Substances	$\hat{L}_l$	1	222.605	222.605	22.838 ($P < .001$)
	$\hat{L}_q$	1	27.225	27.225	2.793 ($.1 < P < .25$)
	$\hat{L}_c$	1	0.045	0.045	< 1 (n.s.)
Error		36	350.900	9.747	
Total		39	600.775		

It is clear from the preceding F tests that the relationship between potency (as measured by the amount of injected material needed to cause death) and concentration is strongly linear, with no evidence of higher-order effects.

Problems

1. Five treatments for fever blisters, including a placebo, were randomly assigned to 30 patients. The data in the accompanying table identify the number of days from initial appearance of the blisters until healing is complete, for each treatment.

Treatment	No. of Days
Placebo (1)	5, 8, 7, 7, 10, 8
2	4, 6, 6, 3, 5, 6
3	6, 4, 4, 5, 4, 3
4	7, 4, 6, 6, 3, 5
5	9, 3, 5, 7, 7, 6

   a. Compute the sample means and the sample standard deviations for each treatment.
   b. Complete the ANOVA table in the accompanying computer output for the data given.
   c. Do the effects of the five treatments differ significantly with regard to healing fever blisters? In other words, test H_0: $\mu_1 = \mu_2 = \mu_3 = \mu_4 = \mu_5$ against H_A: "At least two treatments have different population means."
   d. What are the estimates of the true effects $(\mu_i - \mu)$ of the treatments? Verify that the sum of these estimated effects is 0. [*Note*: μ_i is the population mean for the ith treatment, and $\mu = \frac{1}{5}\sum_{i=1}^{5} \mu_i$ is the overall population mean.]
   e. Using dummy variables, create an appropriate regression model that describes this experiment. Give two possible ways of defining these dummy variables (one using 0's and 1's, and the other using 1's and −1's), and describe for each coding scheme how the regression coefficients are related to the population means $\mu_1, \mu_2, \mu_3, \mu_4, \mu_5$, and μ.
   f. For the data of this problem, carry out the Scheffé, Tukey, and LSD multiple-comparison procedures for pairwise differences between means, as described in section 17.7. Also, compare the widths of the confidence intervals obtained by the three procedures.

Edited SAS Output (PROC GLM) for Problem 1

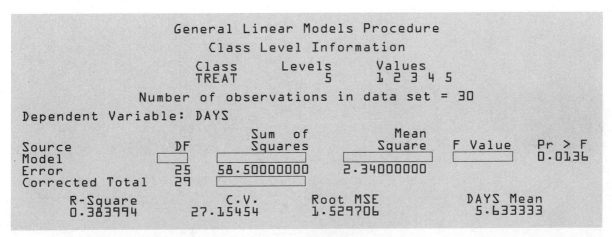

```
                     General Linear Models Procedure
                         Class Level Information
                     Class       Levels      Values
                     TREAT          5         1 2 3 4 5
              Number of observations in data set = 30
Dependent Variable: DAYS
                             Sum of              Mean
Source                DF     Squares            Square      F Value     Pr > F
Model            [      ]  [          ]      [          ]  [       ]     0.0136
Error              25      58.50000000       2.34000000
Corrected Total    29    [          ]
         R-Square              C.V.         Root MSE          DAYS Mean
         0.383994           27.15454        1.529706          5.633333
```

(continued)

```
                General Linear Models Procedure (Continued)
Source     DF       Type I SS      Mean Square    F Value    Pr > F
TREAT       4    36.46666667       9.11666667       3.90     0.0136

Source     DF      Type III SS     Mean Square    F Value    Pr > F
TREAT       4    36.46666667       9.11666667       3.90     0.0136
```

Scheffe's test for variable: DAYS

NOTE: This test controls the type I experimentwise error rate, but
generally has a higher type II error rate than REGWF for all
pairwise comparisons

Alpha= 0.05   df= 25   MSE= 2.34
Critical Value of F= 2.75871
Minimum Significant Difference= 2.9338

Means with the same letter are not significantly different.

Scheffe Grouping		Mean	N	TREAT
	A	7.5000	6	1
	A			
B	A	6.1667	6	5
B	A			
B	A	5.1667	6	4
B	A			
B	A	5.0000	6	2
B				
B		4.3333	6	3

General Linear Models Procedure

Tukey's studentized Range (HSD) Test for variable: DAYS

NOTE: This test controls the type I experimentwise error rate, but
generally has a higher type II error rate than REGWQ.

Alpha= 0.05   df= 25   MSE= 2.34
Critical Value of Studentized Range= 4.153
Minimum Significant Difference= 2.5938

Means with the same letter are not significantly different.

Tukey Grouping		Mean	N	TREAT
	A	7.5000	6	1
	A			
B	A	6.1667	6	5
B	A			
B	A	5.1667	6	4
B	A			
B	A	5.0000	6	2
B				
B		4.3333	6	3

T tests (LSD) for variable: DAYS

NOTE: This test controls the type I comparisonwise error rate not
the experimentwise error rate.

Alpha= 0.05   df= 25   MSE= 2.34
Critical Value of T= 2.06
Least Significant Difference= 1.8189

Means with the same letter are not significantly different.

```
            General Linear Models Procedure (Continued)
        T Grouping        Mean        N   TREAT
                  A       7.5000       6   1
                  A
           B      A       6.1667       6   5
           B
           B      C       5.1667       6   4
           B      C
           B      C       5.0000       6   2
                  C
                  C       4.3333       6   3
```

2. The following data are replicate measurements of the sulfur dioxide concentration in each of three cities:

   City I: 2, 1, 3

   City II: 4, 6, 8

   City III: 2, 5, 2

a. Complete the ANOVA table in the accompanying computer output for simultaneously comparing the mean sulfur dioxide concentrations in the three cities.

b. Test whether the three cities differ significantly in mean sulfur dioxide concentration levels.

c. What is the estimated effect associated with each city?

d. State precisely the appropriate ANOVA fixed-effects model for these data.

e. Using dummy variables, state precisely the regression model that corresponds to the fixed-effects model in part (d). What is the relationship between the coefficients in the regression model and the effects in the ANOVA model?

f. Using the t distribution, find a 90% confidence interval for the true difference between the effects of cities I and II (making sure to use the best estimate of σ^2 provided by the data).

Edited SAS Output (PROC GLM) for Problem 2

```
                  General Linear Models Procedure
                           .
                           . [Portion of output omitted]
                           .

Dependent Variable: SO2
                              Sum of              Mean
Source            DF          Squares            Square      F Value    Pr > F
Model              2       26.0000000        13.0000000     [        ]  0.0553
Error          [    ]      [          ]
Corrected Total    8       [          ]

                R-Square          C.V.        Root MSE            SO2 Mean
                0.619048       44.53618       1.63299             3.66667

Source            DF        Type I SS     Mean Square      F Value    Pr > F
CITY               2       26.0000000      13.0000000        4.87     0.0553

Source            DF      Type III SS     Mean Square      F Value    Pr > F
CITY               2       26.0000000      13.0000000        4.87     0.0553
```

3. Each of three chemical laboratories performed four replicate determinations of the concentration of suspended particulate matter in a certain area using the "Hi-Vol" method of analysis. The resulting data are presented next.

Lab 1	Lab II	Lab III
4	2	5
4	2	2
6	5	5
10	3	8

a. Identify the appropriate ANOVA table for these data.
b. Test the null hypothesis of no differences among the laboratories.
c. With large-scale interlaboratory studies, analysts usually make inferences about a large population of laboratories of which only a random sample (e.g., laboratories I, II, and III) can be investigated. In such a case, describe the appropriate random-effects model for the data.
d. Two quantities of particular interest in a large-scale interlaboratory study are repeatability (i.e., a measure of the variability among replicate measurements within a single laboratory) and reproducibility (i.e., a measure of the variability between results from different laboratories). Using the random-effects model defined in part (c), define what you think are reasonable measures of repeatability and reproducibility, and obtain estimates of the quantities you have defined, using the data in the accompanying computer output.

Edited SAS Output (PROC GLM) for Problem 3

```
                  General Linear Models Procedure
                      .
                      .  [Portion of output omitted]
                      .

Dependent Variable: CONC
                          Sum of            Mean
Source          DF        Squares          Square     F Value    Pr > F
Model            2       18.6666667      9.3333333      1.75      0.2280
Error            9       48.0000000      5.3333333
Corrected Total 11       66.6666667

       R-Square             C.V.          Root MSE           CONC Mean
       0.280000           49.48717        2.30940            4.66667

Source          DF        Type I SS      Mean Square    F Value    Pr > F
LAB              2       18.6666667      9.3333333       1.75      0.2280

Source          DF        Type III SS    Mean Square    F Value    Pr > F
LAB              2       18.6666667      9.3333333       1.75      0.2280
```

4. Ten randomly selected mental institutions were examined to determine the effects of three different antipsychotic drugs on patients with the same types of symptoms. Each institution

used one and only one of the three drugs exclusively for a one-year period. The proportion of treated patients in each institution who were discharged after one year of treatment is as follows for each drug used:

Drug 1: 0.10, 0.12, 0.08, 0.14 $(\overline{Y}_1 = 0.11, S_1 = 0.0192)$

Drug 2: 0.12, 0.14, 0.19 $(\overline{Y}_2 = 0.15, S_2 = 0.0361)$

Drug 3: 0.20, 0.25, 0.15 $(\overline{Y}_3 = 0.20, S_3 = 0.0500)$

a. Determine the appropriate ANOVA table for this data set, using the accompanying computer output.

b. Test to see whether significant differences exist among drugs with regard to the average proportion of patients discharged.

c. What other factors should be considered in comparing the effects of the three drugs?

d. What basic ANOVA assumptions might be violated here?

Edited SAS Output (PROC GLM) for Problem 4

```
                    General Linear Models Procedure
                               .
                               .  [Portion of output omitted]
                               .

Dependent Variable: DISCH
                                Sum of              Mean
Source              DF          Squares            Square      F Value    Pr > F
Model                2        0.01389000        0.00694500      5.06      0.0436
Error                7        0.00960000        0.00137143
Corrected Total      9        0.02349000

          R-Square              C.V.          Root MSE          DISCH Mean
          0.591315           24.85423         0.03703            0.14900

Source              DF        Type I SS      Mean Square     F Value    Pr > F
DRUG                 2        0.01389000      0.00694500       5.06      0.0436

Source              DF        Type III SS    Mean Square     F Value    Pr > F
DRUG                 2        0.01389000      0.00694500       5.06      0.0436
```

5. Suppose that a random sample of five active members in each of four political parties in a certain western European country was given a questionnaire purported to measure (on a 100-point scale) the extent of "general authoritarian attitude toward interpersonal relationships." The means and standard deviations of the authoritarianism scores for each party are given in the following table.

	Party 1	Party 2	Party 3	Party 4
$\overline{Y}_i$	85	80	95	50
S_i	6	7	4	10
n_i	5	5	5	5

a. Determine the appropriate ANOVA table for this data set.

    b. Test to see whether significant differences exist among parties with respect to mean authoritarianism scores.

    c. Using dummy variables, state an appropriate regression model for this experimental situation.

    d. Apply Tukey's method of multiple comparisons to identify the pairs in which the means significantly differ from one another. (Use $\alpha = .05$.)

6. A psychosociological questionnaire was administered to a random sample of 200 persons on an island in the South Pacific that has become increasingly westernized over the past 30 years. From the questionnaire data, each of the 200 persons was classified into one of three groups—HI-POS, NO-DIF, and HI-NEG—according to the discrepancy between the amount of prestige in that person's traditional culture and the amount of prestige in the modern (westernized) culture. On the basis of the questionnaire data, a measure of "anomie" (i.e., social disorientation), denoted as Y, was determined on a 100-point scale, with the results summarized in the following table.

Group	n_i	$\overline{Y}_i$	S_i
HI-POS	50	65	9
NO-DIF	75	50	11
HI-NEG	75	55	10

    a. Determine the appropriate ANOVA table.

    b. Test whether the three different categories of prestige discrepancy have significantly different sample mean anomie scores.

    c. How would you test whether a significant difference exists between the NO-DIF category and the other two categories combined?

7. To determine whether reading skills of the average high school graduate have declined over the past 10 years, the average verbal college aptitude scores (VSAT) were compared for a random sample of five big-city high schools for the years 1965, 1970, and 1975. The results are shown in the following table and accompanying computer output.

Year	School 1	School 2	School 3	School 4	School 5
1965	550	560	535	545	555
1970	545	560	528	532	541
1975	536	552	526	527	530

    a. Determine the sample means for each year.

    b. Comment on the independence assumption required for using one-way ANOVA on the data given.

    c. Determine the one-way ANOVA table for these data.

    d. Test by means of one-way ANOVA whether any significant differences exist among the three VSAT average scores.

    e. Use Scheffé's method to locate any significant differences between pairs of means. (Use $\alpha = .05$.)

Edited SAS Output (PROC GLM) for Problem 7

```
                    General Linear Models Procedure
                          ·  [Portion of output omitted]
                          ·

Dependent Variable: VSAT
                            Sum of              Mean
Source              DF      Squares             Square      F Value    Pr > F
Model                2     548.133333         274.066667      2.26     0.1466
Error               12    1453.600000         121.133333
Corrected Total     14    2001.733333

        R-Square              C.V.          Root MSE            VSAT Mean
        0.273829           2.032638         11.0061              541.467

Source              DF     Type I SS      Mean Square     F Value    Pr > F
YEAR                 2     548.133333      274.066667       2.26     0.1466

Source              DF     Type III SS    Mean Square     F Value    Pr > F
YEAR                 2     548.133333      274.066667       2.26     0.1466
```

8. Three persons (denoted A, B, and C) claiming to have unusual psychic ability underwent ESP tests at an eastern U.S. psychic research institute. On each of the five randomly selected days, each person was asked to specify for 26 pairs of cards whether both cards in a given pair were of the same color or not. The numbers of correct answers are given in the accompanying table.

Person	Day 1	Day 2	Day 3	Day 4	Day 5
A	20	22	20	21	18
B	24	21	18	22	20
C	16	18	14	13	16

a. Determine the mean score for each person, and interpret the results.
b. Test whether the three persons have significantly different ESP ability.
c. Carry out Scheffé's multiple-comparison procedure to determine which pairs of persons, if any, significantly differ in ESP ability.
d. On the basis of the results in parts (b) and (c), can one conclude that any of these persons has statistically significant ESP ability? Explain.

9. The average generation times for four different strains of influenza virus were determined using six cultures for each strain. The data are summarized in the following table.

Statistic	Strain A	Strain B	Strain C	Strain D
$\bar{Y}$	420.3	330.7	540.4	450.8
S	30.22	28.90	31.08	33.29

a. Test whether the true mean generation time differs among the four strains.
b. What is the appropriate ANOVA table for these data?
c. Use Tukey's multiple-comparison procedure to identify where any differences occur among the means. (Use $\alpha = .05$.)

10. Three replicate water samples were taken at each of four locations in a river to determine whether the quantity of dissolved oxygen—a measure of water pollution—varied from one location to another (the higher the level of pollution, the lower the dissolved oxygen reading). Location 1 was adjacent to the wastewater discharge point for a certain industrial plant, and locations 2, 3, and 4 were selected at points 10, 20, and 30 miles downstream from this discharge point. The resulting data appear in the accompanying table. The quantity Y_{ij} denotes the value of the dissolved oxygen content for the jth replicate at location i ($j = 1, 2, 3$; $i = 1, 2, 3, 4$), and $\overline{Y}_i$ denotes the mean of the three replicates taken at location i.

Location	Dissolved Oxygen Content (Y_{ij})	Mean ($\overline{Y}_i$)
1	4, 5, 6	5
2	6, 6, 6	6
3	7, 8, 9	8
4	8, 9, 10	9

a. Do the data provide sufficient evidence to suggest that values for mean dissolved oxygen content differ significantly among the four locations? (Use $\alpha = .05$.) Make sure to construct the appropriate ANOVA table.

b. Given that μ_i represents the true mean level of dissolved oxygen at location i ($i = 1, 2, 3, 4$), test the null hypothesis

$$H_0: -3\mu_1 - \mu_2 + \mu_3 + 3\mu_4 = 0$$

versus

$$H_A: -3\mu_1 - \mu_2 + \mu_3 + 3\mu_4 \neq 0$$

at the 2% level. The quantity $(-3\mu_1 - \mu_2 + \mu_3 + 3\mu_4)$ is a contrast based on orthogonal polynomials (see section 17-9), which can be shown to be a measure of the linear relationship between "location" (the four equally spaced distances 0, 10, 20, and 30 miles downstream from the plant) and "dissolved oxygen content."

c. Another way to quantify the strength of this linear relationship is to fit by least squares the model

$$Y = \beta_0 + \beta_1 X + E$$

where

$$X = \begin{cases} 0 & \text{for location 1} \\ 10 & \text{for location 2} \\ 20 & \text{for location 3} \\ 30 & \text{for location 4} \end{cases}$$

Fitting such a regression model to the $n = 12$ data points yields the accompanying ANOVA table. Use this table to perform a test of $H_0: \beta_1 = 0$ at the 2% significance level.

Source		d.f.		SS	
Regression		1		29.40	
Residual	Lack of fit	10 $\begin{cases} 2 \\ 8 \end{cases}$		6.60 $\begin{cases} 0.60 \\ 6.00 \end{cases}$	
	Pure error				

d. The regression model in part (c) amounts to saying that $\mu_i = \beta_0 + \beta_1 X_i$. Show that the hypothesis tested in part (b) is equivalent to the hypothesis tested in part (c).

e. Why do the two test statistics calculated in parts (b) and (c) *not* have the same numerical value? What reasonable modification of the test in part (c) would yield the same F-value as that obtained in part (b)?

f. Given the results of part (b), a test for a nonlinear trend in mean response can be obtained by subtracting the sum of squares for the contrast

$$\hat{L} = -3\overline{Y}_1 - \overline{Y}_2 + \overline{Y}_3 + 3\overline{Y}_4$$

from the sum of squares for treatments, and then dividing this difference by the appropriate degrees of freedom to yield an F statistic of the form

$$F(\text{Nonlinear trend}) = \frac{[\text{SST} - \text{SS}(\hat{L})]/\text{df}}{\text{MSE}}$$

Carry out this test based on the results obtained in parts (a) and (b). (Use $\alpha = .05$.)

g. Carry out the usual regression lack-of-fit test for adequacy of the straight-line model fit in part (c), using $\alpha = .05$. Does the value of the F statistic equal the value obtained in part (f)?

11. Consider the data from Problem 15 in Chapter 5. The dependent variable of interest is LN_BRNTL. Use $\alpha = .05$.

a. Conduct a one-way ANOVA, with dosage level (PPM_TOLU) as a categorical predictor.

b. Using the Bonferroni (LSD) technique, compute all pairwise comparisons.

c. Repeat part (b) using Tukey's method.

d. Repeat part (b) using Scheffé's method.

e. The ANOVA overall test corresponds to fitting a polynomial model of order k. What is k?

f. Examine the ANOVA assumptions by computing the estimates of the within-cell variances. Provide frequency histograms for each cell to aid in this.

12. Repeat Problem 11, skipping part (e), but use LN_BLDTL as the dependent variable.

13. The accompanying source table relates to a study involving the effects of trimethylin doses of 0, 3, 6, and 9 mg/kg on a sample of 48 rats. Each rat received only one dose. The response variable was the log of the activity counts in 1 hour in a residential maze. Compute the unknown values for $a, b, c, d, e, f,$ and g. Show your work, even if it seems obvious and trivial to you.

Source	SS	d.f.	MS	F
Dosage	a	d	g	14.71
Within dosage	b	e	12.84	
Total	c	f		

14. For each of the following contrasts, indicate the null hypothesis that is being tested.

	$\overline{X}_1$	$\overline{X}_2$	$\overline{X}_3$	$\overline{X}_4$	$\overline{X}_5$
c_1	1	0	-1	0	0
c_2	0	$\frac{1}{2}$	0	$\frac{1}{2}$	-1
c_3	$\frac{1}{3}$	$\frac{1}{3}$	$\frac{1}{3}$	0	-1

15. Persistent patent ductus arteriosus (PDA) is a medical condition that affects more than 40% of very low-birthweight infants (see Cotton et al. 1979). Infants with PDA have an increased risk of ailments such as cerebral hemorrhage, bronchopulmonary dysplasia, as well as of death. In a recent study, Varvarigou et al. (1996) determined that early ibuprofen administration is effective in preventing PDA in preterm neonates. In a sample of 34 such infants, 11 were treated with one dose of ibuprofen, 12 with three doses, and 11 with a saline solution, beginning within three hours of birth.

    As part of the study, clinical factors such as birth weight, gestational age, and one-minute Apgar scores were compared for the three groups. Summary statistics for these variables are presented in the following table.

Treatment Group	Mean Birth Weight (SD in parentheses)	Mean Gestational Age (weeks) (SD in parentheses)	Mean One-minute Apgar Score (SD in parentheses)
Saline ($n = 11$)	843.0 (290.5)	27.3 (2.4)	5.5 (1.9)
1 dose of ibuprofen ($n = 11$)	908.6 (334.0)	26.6 (2.9)	3.7 (2.3)
3 doses of ibuprofen ($n = 12$)	947.5 (245.3)	26.5 (2.4)	5.0 (2.1)

   a. Suppose that an ANOVA is to be performed to compare the average birth weight for the different treatment groups. State precisely the ANOVA model. Is Treatment Group a fixed-effects factor or a random-effects factor?
   b. Using the summary statistics given in the table, determine the ANOVA table for the model in (a).
   c. Test whether the three treatment groups differ significantly in mean birth weights.
   d–f. Repeat parts (a)–(c) for gestational age.
   g–i. Repeat parts (a)–(c) for Apgar score.

16. In the PDA research described in Problem 15, one of the outcome variables was mean airway pressure (measured in cm H_2O) seven days after birth—a measure of respiratory status. The data on this outcome are summarized in the accompanying table.

Treatment Group	Mean Airway Pressure (SD in parentheses)
Saline ($n = 11$)	4.7 (1.8)
1 dose of ibuprofen ($n = 11$)	3.4 (0.7)
3 doses of ibuprofen ($n = 12$)	2.6 (0.6)

   a.–c. Repeat parts (a)–(c) of Problem 15, using mean airway pressure as the dependent variable.

    d. Use Tukey's method to locate any significant differences between pairs of means. (Use $\alpha = .05$.)

17. A survey was conducted to examine the effects of pharmaceutical drug advertising on physician practices. Self-administered questionnaires were delivered to randomly selected physicians in various practice types: general practice, family practice, and internal medicine. Physicians reported on their attitude toward drug advertising, and on the influence of drug advertising on their prescription-writing habits. Likert Scale measurements were used in the survey (1 = Strongly disagree, 5 = Strongly agree; the higher the score, the more favorably disposed the respondent is to advertising). The sample data are presented in the following table and in the accompanying computer output.

Practice Type	Attitude Toward Advertising	Influence on Prescription-writing Habits
General Practice	3, 4, 4, 4, 4, 3	4, 5, 5, 3, 5, 4
Family Practice	3, 4, 2, 2, 2, 1	3, 2, 2, 2, 1, 3
Internal Medicine	2, 4, 2, 1, 1, 2	1, 2, 3, 1, 1, 1

    a. Suppose that an ANOVA is to be performed to compare the average attitude toward advertising for the different practice types. State precisely the ANOVA model for these data.

    b. Complete the ANOVA table in the computer output for the model in part (a).

    c. Test whether the physician types differ significantly in average attitude toward advertising.

    d. Use Scheffé's method to locate differences between pairs of means.

    e.–h. Repeat parts (a)–(d) with regard to influence in prescription-writing habits.

Edited SAS Output (PROC GLM) for Problem 17

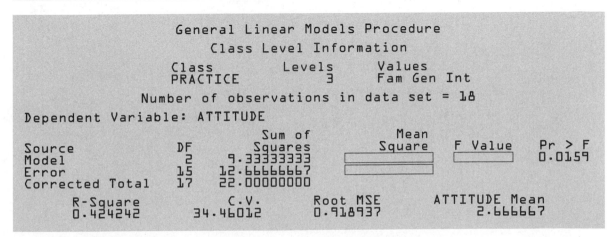

(continued)

```
              General Linear Models Procedure (Continued)
Source           DF        Type I SS      Mean Square     F Value    Pr > F
PRACTICE          2       9.33333333       4.66666667        5.53    0.0159

Source           DF       Type III SS     Mean Square     F Value    Pr > F
PRACTICE          2       9.33333333       4.66666667        5.53    0.0159
              Scheffe's test for variable: ATTITUDE
```

NOTE: This test controls the type I experimentwise error rate but
      generally has a higher type II error rate than REGWF for all
      pairwise comparisons

```
          Alpha= 0.05      df= 15      MSE= 0.844444
               Critical Value of F= 3.68232
            Minimum Significant Difference= 1.4398
```

Means with the same letter are not significantly different.

```
              Scheffe Grouping        Mean        N    PRACTICE
                       A            3.6667         6    Gen
                       A
               B       A            2.3333         6    Fam
               B
               B                    2.0000         6    Int
```

```
              General Linear Models Procedure
                   Class Level Information
               Class        Levels       Values
               PRACTICE        3         Fam Gen Int
           Number of observations in data set = 18
```

Dependent Variable: HABITS

Source	DF	Sum of Squares	Mean Square	F Value	Pr > F
Model	2				0.0001
Error		9.66666667	0.64444444		
Corrected Total	17				

```
          R-Square          C.V.         Root MSE      HABITS Mean
          0.731481        30.10399       0.802773       2.666667

Source           DF       Type I SS      Mean Square     F Value    Pr > F
PRACTICE          2      26.33333333     13.16666667       20.43    0.0001

Source           DF       Type III SS    Mean Square     F Value    Pr > F
PRACTICE          2      26.33333333     13.16666667       20.43    0.0001
              Scheffe's test for variable: HABITS
```

NOTE: This test controls the type I experimentwise error rate but
      generally has a higher type II error rate than REGWF for all
      pairwise comparisons

```
          Alpha= 0.05      df= 15      MSE= 0.644444
               Critical Value of F= 3.68232
            Minimum Significant Difference= 1.2578
```

Means with the same letter are not significantly different.

```
              Scheffe Grouping        Mean        N    PRACTICE
                       A            4.3333         6    Gen

               B                    2.1667         6    Fam
               B
               B                    1.5000         6    Int
```

18. Customer portfolio analyses provide retailers with data on customer characteristics, including expenditure patterns. Based on this information retailers can make marketing and operational adjustments to improve their market position. ABC Foods, a large supermarket chain, conducted a survey of the average weekly expenditures of 542 randomly selected residents in a large metropolitan market. The customers were classified as Loyal to ABC, New to ABC, Defectors from ABC, Loyal to a competitor, and Unaffiliated. The average weekly expenditure data are summarized in the following table.

Customer Type	n	Avg. Weekly Expenditure (SD)
Loyal to ABC	84	$75 ($13)
New to ABC	25	$65 ($10)
Defectors from ABC	27	$82 ($14)
Loyal to competitors	173	$93 ($17)
Unaffiliated	233	$82 ($15)

a. Suppose that an ANOVA is to be performed to compare the average weekly expenditures for the different customer types. State precisely the ANOVA model. Is customer type a fixed-effects factor or a random-effects factor?

b. Using the summary statistics given in the table, determine the ANOVA table for the model in part (a).

c. Test whether the five customer types differ significantly in average weekly expenditures.

d. Use Scheffé's method to locate any significant differences between pairs of means (use $\alpha = .05$).

19. The data in the accompanying table were sampled from *U.S. News & World Report*'s 1996 story on the "Best Mutual Funds."[16] The following variables are shown for each fund:

CAT (fund category): 1 = Aggressive growth; 2 = Long-term growth; 3 = Growth and income; 4 = Income.

LOAD (load status): N = No load; L = Load.

VOL (volatility): A letter grade from A+ to F indicating how much the month-to-month return varied from the fund's three-year total return: A+ = Least variability; F = Most variability.

OPI (Overall Performance Index): An overall measure of the relative performance of each fund over the past 1, 3, 5, and 10 years. The higher the OPI, the better the performance.

---

[16]"1996 Mutual Funds Guide, Best Mutual Funds," *U.S. News & World Report* (January 29, 1996), pp. 88–100.

Fund	CAT	LOAD	VOL	OPI
20th Century Giftrust Investors	1	N	F	89.9
AIM Aggressive Growth	1	L	F+	92.4
Stein Roe Capital Opportunity	1	N	F+	88.1
FPA Capital	1	L	D	92.2
Third Avenue Value	1	N	B	88.2
Phoenix U.S. Govt. A	1	L	D	88.0
Vanguard Primecap	2	N	D+	93.9
MFS Research A	2	L	D+	94.5
Fidelity Value	2	N	B	91.4
Mairs & Power Growth	2	N	C	99.1
Guardian Park Avenue	2	L	C	92.2
AIM Value A	2	L	D+	90.6
Lexington Corp. Leaders	3	N	C	91.3
Fundamental Investors	3	L	B	92.7
Oppenheimer Main St. Inc. & Growth A	3	L	D+	90.8
MAS Value Portfolio	3	N	B	99.8
Vanguard Index Value	3	N	B	88.9
Putnam Growth and Income	3	L	B+	90.8
Federated Liberty Equity Income A	4	L	B	89.5
United Income A	4	L	C+	76.8
Benham Inc. & Growth	4	N	B+	89.8
Manager's Income Equity	4	N	B	84.7
Pioneer Equity Income	4	L	B+	84.0
One Group Income Equity A	4	N	B+	83.3

a. Suppose that an ANOVA is to be performed to compare the average OPI values for the different fund categories. State precisely the ANOVA model. Is fund category a fixed- or random-effects factor?

b. Using the SAS output that follows, complete the ANOVA table.

c. Test whether the average overall performance indices differ significantly across the four fund categories.

d. Use Tukey's method to locate any significant differences between pairs of means. (Use $\alpha = .05$.)

Edited SAS Output (PROC GLM) for Problem 19

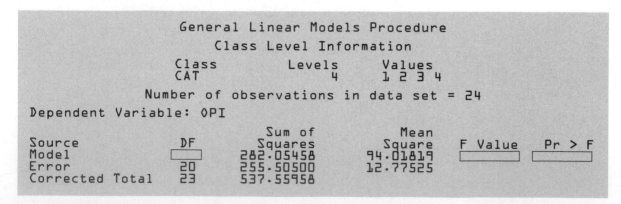

```
                    General Linear Models Procedure
                      Class Level Information
                 Class          Levels        Values
                  CAT              4           1 2 3 4
           Number of observations in data set = 24
Dependent Variable: OPI
                                Sum of              Mean
Source            DF           Squares             Square      F Value    Pr > F
Model          [      ]        282.05458          94.01819   [        ]  [        ]
Error             20           255.50500          12.77525
Corrected Total   23           537.55958
```

```
              General Linear Models Procedure (Continued)
      R-Square            C.V.         Root MSE              OPI Mean
      0.524695         3.966062         3.5742                90.121

Source          DF       Type I SS    Mean Square    F Value    Pr > F
CAT              3       282.05458       94.01819       7.36     0.0016

Source          DF      Type III SS    Mean Square    F Value    Pr > F
CAT              3       282.05458       94.01819       7.36     0.0016

Tukey's Studentized Range (HSD) Test for variable: OPI

NOTE: This test controls the type I experimentwise error rate, but
      generally has a higher type II error rate than REGWQ.

          Alpha= 0.05      df= 20       MSE= 12.77525
          Critical Value of Studentized Range= 3.958
              Minimum Significant Difference= 5.7759

Means with the same letter are not significantly different.

              Tukey Grouping          Mean      N    CAT
                            A        93.617      6     2
                            A
                            A        92.383      6     3
                            A
                      B     A        89.800      6     1
                      B
                      B              84.683      6     4
```

20. This problem refers to the data of Problem 19.

    a. Suppose that an ANOVA is to be performed to compare the average OPI values for funds with different volatilities. State precisely the ANOVA model. Is fund volatility a fixed- or random-effects factor?

    b. In the SAS output that follows, complete the ANOVA table.

    c. Test whether the average overall performance indices differ significantly by volatility rating. Interpret your results.

Edited SAS Output (Proc GLM) for Problem 20

```
                   General Linear Models Procedure
                      Class Level Information
                  Class        Levels       Values
                  VOL            8         B B+ C C+ D D+ F F+
             Number of observations in data set = 24

Dependent Variable: OPI
                            Sum of          Mean
Source           DF        Squares         Square     F Value    Pr > F
Model          [    ]     291.43994       41.63428    [      ]   [      ]
Error           16        246.11964       15.38248
Corrected Total 23        537.55958

      R-Square            C.V.         Root MSE              OPI Mean
      0.542154         4.351991         3.9221                90.121

Source          DF       Type I SS    Mean Square    F Value    Pr > F
VOL              7       291.43994       41.63428       2.71     0.0470

Source          DF      Type III SS    Mean Square    F Value    Pr > F
VOL              7       291.43994       41.63428       2.71     0.0470
```

21. Radial keratotomy is a type of refractive surgery in which radial incisions are made in a myopic (nearsighted) patient's cornea to reduce the person's myopia. Theoretically, the incisions allow the curvature of the cornea to become less steep, thereby reducing the patient's refractive error. [*Note*: Myopic patients have negative refractive errors. Patients who are farsighted have positive refractive errors. Patients who are neither near- nor far-sighted have zero refractive error.]

The incisions extend radially from the periphery toward the center of the cornea. A circular central portion of the cornea, known as the clear zone, remains uncut. The diameter of the clear zone is determined by the baseline refraction of the patient. Patients with a greater degree of myopia may receive longer incisions, leaving them with smaller clear zones. The thinking here is that "more surgery" is needed to correct the worse initial vision of these patients.

Radial keratotomy and other vision-correction surgery techniques have been growing in popularity in the 1980s and 1990s, both among the public and among ophthalmologists. The Prospective Evaluation of Radial Keratotomy (PERK) study was begun in 1983 to evaluate the effects of radial keratotomy. Lynn et al. (1987) examined the variables associated with the five-year postsurgical change in refractive error (Y, measured in diopters, D). One of the independent variables under consideration was diameter of the clear zone (X). In the PERK study, three clear zone sizes were used: 3.0 mm, 3.5 mm, and 4.0 mm.

The following computer output is based on data adapted from the PERK study.

a. Suppose that an ANOVA is to be performed to compare the average five-year change in refraction for patients with different clear zones. State precisely the ANOVA model. Is clear zone size a fixed- or random-effects factor?

b. In the SAS output that follows, complete the ANOVA table.

c. Test whether the average five-year change in refraction differs significantly by clear zone size. Interpret your results.

d. Use Tukey's method to locate any significant pairwise differences between clear zones. (Use $\alpha = .05$.) Interpret your results.

Edited SAS Output (Proc GLM) for Problem 21

```
                    General Linear Models Procedure
                       Class Level Information
                   Class          Levels      Values
                   CLRZONE          3         3.0 3.5 4.0
             Number of observations in data set = 54
  NOTE: Due to missing values, only 51 observations can be used in
        this analysis
  Dependent Variable: Y
  Source              DF       Sum of Squares      F Value    Pr > F
  Model            [      ]       14.70441590    [       ]  [        ]
  Error              48          64.59163802
  Corrected Total    50          79.29605392
         R-Square              C.V.              Y Mean
         0.185437           30.43590           3.81137255
  Source           DF         Type I SS       F Value      Pr > F
  CLRZONE           2        14.70441590        5.46       0.0073
  Source           DF        Type III SS      F Value      Pr > F
  CLRZONE           2        14.70441590        5.46       0.0073
```

```
            Tukey's Studentized Range (HSD) Test for variable: Y
 NOTE: This test controls the type I experimentwise error rate.
    Alpha= 0.05    Confidence= 0.95    df= 48    MSE= 1.345659
                Critical Value of Studentized Range= 3.420
 Comparisons significant at the 0.05 level are indicated by  *** .
                     Simultaneous                 Simultaneous
                        Lower        Difference       Upper
        CLRZONE       Confidence      Between      Confidence
       Comparison       Limit          Means          Limit
       3.0 - 3.5       -0.2562        0.7021        1.6603
       3.0 - 4.0        0.3368        1.2778        2.2188      ***
       3.5 - 3.0       -1.6603       -0.7021        0.2562
       3.5 - 4.0       -0.4326        0.5757        1.5840
       4.0 - 3.0       -2.2188       -1.2778       -0.3368      ***
       4.0 - 3.5       -1.5840       -0.5757        0.4326
```

22. Data on law and business schools were sampled from *U.S. News & World Report*'s 1996 report on the "America's Best Graduate Schools 1996 Annual Guide."[17] The school's reputation rank among academics and the 1995 median starting salary for graduates are shown in the accompanying table for 12 randomly sampled law and 12 randomly sampled business schools.

University	School	Reputation Rank by Academics	1995 Median Starting Salaries (in $1000s)
Vanderbilt University	Law	17	62
University of Chicago	Law	2	70
Brigham Young University	Law	45	45.5
George Washington University	Law	24	60
Cornell University	Law	11	70
Rutgers University	Law	45	62
University of Pennsylvania	Law	6	70
University of Illinois (Urbana-Champaign)	Law	23	53
Villanova University	Law	65	62.2
University of Florida	Law	37	43.4
Tulane University	Law	49	53
Case Western Reserve University	Law	41	47
University of Georgia	Business	45	45
Tulane University	Business	40	50
University of Southern California	Business	24	55
Brigham Young University	Business	57	68
Emory University	Business	33	58
University of Illinois (Urbana-Champaign)	Business	24	45.6
Cornell University	Business	10	60
University of North Carolina (Chapel Hill)	Business	16	59
Massachusetts Institute of Technology	Business	1	75
University of Texas (Austin)	Business	19	55
College of William and Mary	Business	67	48
Penn State University	Business	33	49.7

---

[17]"America's Best Graduate Schools," *U.S. News & World Report* (March 18, 1996), pp. 79–91.

a. Suppose that an ANOVA is to be performed to compare the average 1995 starting salaries (SAL) for business and law schools (let the variable SCHOOL = Law or Bus, depending on the type of school). State precisely the ANOVA model. Is SCHOOL a fixed- or random-effects factor?

b. In the SAS output that follows, complete the ANOVA table.

c. Test whether the average 1995 starting salaries differ significantly for graduates of law and of business schools.

d. Suppose that an ANOVA is to be performed to compare the average 1995 starting salaries (SAL) for top 25 schools and schools not in the top 25 in terms of reputation rank (let the variable REP = 1 if a school's reputation rank is 25 or less, REP = 2 if the rank is 26 or more). State precisely the ANOVA model.

e. In the second part of the SAS output that follows, complete the ANOVA table.

f. Test whether the average 1995 starting salaries differ significantly for top 25 schools and schools not in the top 25 in terms of reputation rank.

Edited SAS Output (Proc GLM) for Problem 22

```
                    General Linear Models Procedure
                      Class Level Information
                Class          Levels       Values
                SCHOOL           2          Law Bus
            Number of observations in data set = 24
Dependent Variable: SAL
                             Sum of           Mean
Source              DF       Squares          Square    F Value    Pr > F
Model           [      ]    37.001667        37.001667  [       ]  [       ]
Error             22       1929.391667       87.699621
Corrected Total   23       1966.393333
        R-Square           C.V.         Root MSE          SAL Mean
        0.018817         16.44873        9.3648            56.933
Source         DF      Type I SS      Mean Square    F Value    Pr > F
SCHOOL          1      37.001667       37.001667      0.42      0.5227

Source         DF      Type III SS    Mean Square    F Value    Pr > F
SCHOOL          1       37.001667      37.001667      0.42      0.5227
                                 .
                          .  [Portion of output omitted]
                                 .
                    General Linear Models Procedure
                      Class Level Information
                Class          Levels       Values
                REP              2           1 2
            Number of observations in data set = 24
Dependent Variable: SAL
                             Sum of           Mean
Source              DF       Squares          Square    F Value    Pr > F
Model           [      ]    440.32667        440.32667  [       ]  [       ]
Error             22       1526.06667        69.36667
Corrected Total   23       1966.393333
```

```
          General Linear Models Procedure (Continued)
    R-Square              C.V.        Root MSE          SAL Mean
    0.223926           14.62880        8.3287            56.933

Source       DF      Type I SS     Mean Square    F Value    Pr > F
REP           1      440.32667      440.32667        6.35    0.0195

Source       DF     Type III SS    Mean Square    F Value    Pr > F
REP           1      440.32667      440.32667        6.35    0.0195
```

23. In September 1996, *U.S. News & World Report* published a report on America's health maintenance organizations (HMOs).[18] The report was intended to serve as a consumer guide to HMO quality. For each HMO included in the report, data were provided on several variables, including the following:

PHYSTURN: Physician turnover rate (%).

PHYSCERT: Percentage of doctors who were board certified.

PREV: Prevention score, indicating how well the HMO meets Public Health Service goals in various measures of preventive care (including immunizations and prenatal care). The results can be negative (indicating that the HMO falls short of the goals), or positive (indicating that goals are exceeded).

HMO	PREV	PHYSTURN	PHYSCERT
HMO Kentucky	−114	1	64
Kaiser Foundation (HI Region)	−6	7	92
CIGNA HealthCare of LA	−97	6	80
CIGNA HealthCare of S. California	−82	8	81
CIGNA HealthCare of N. California	−42	21	82
HIP Health Plan of Florida	−8	3	80
CIGNA HealthCare of San Diego	−36	5	76
NYLCare of the Mid-Atlantic	−9	6	80
Personalcare Insurance of Illinois	−26	13	74
Prudential Healthcare Tri-state	−78	9	71
CIGNA HealthCare of S. Florida	−26	4	77
Health New England	11	5	79
Group Health Northwest	4	3	90
Kaiser Foundation—Mid-Atlantic	29	4	92
Health Alliance Medical Plans	−28	6	82
Pilgrim Health Care	22	2	83
Partners National Health, NC	−1	6	84
Healthsource of New Hampshire	11	5	81

a. In the *U.S. News & World Report* story, the average physician turnover rate for HMOs is reported to be 6%. Suppose that an ANOVA is to be performed to compare the average prevention scores for HMOs with low physician turnover rates ($\leq 6\%$) and HMOs with higher rates. State precisely the ANOVA model.

---

[18]"Rating the HMOs," *U.S. News & World Report* (September 2, 1996), pp. 52–63.

b. In the SAS output that follows, complete the ANOVA table. To produce the output, the following coding scheme was followed: **TURN** = 1 if PHYSTURN $\leq 6\%$; 2 otherwise.

c. Test whether the average prevention scores differ significantly for the two types of HMOs mentioned in part (a).

d. Suppose that an ANOVA is to be performed to compare the average prevention scores for HMOs with low percentages of board-certified primary care physicians ($\leq 75\%$) and HMOs with higher percentages of board-certified physicians. State precisely the ANOVA model.

e. In the second part of the SAS output that follows, complete the ANOVA table. To produce the output, the following coding scheme was followed: **CERT** = 1 if PHYSCERT $\leq 75\%$; 2 otherwise.

f. Test whether the average prevention scores differ significantly for the two types of HMOs mentioned in part (d).

Edited SAS Output (Proc GLM) for Problem 23

```
                    General Linear Models Procedure
                       Class Level Information
                  Class        Levels       Values
                  TURN           2           1 2
              Number of observations in data set = 18
Dependent Variable: PREV
                           Sum of          Mean
Source            DF       Squares        Square     F Value    Pr > F
Model          [     ]    2868.5675      2868.5675   [      ]   [      ]
Error            16       26717.8769     1669.8673
Corrected Total  17       29586.4444

        R-Square         C.V.         Root MSE        PREV Mean
        0.096955       -154.5278       40.864         -26.444
Source        DF        Type I SS     Mean Square    F Value    Pr > F
TURN           1        2868.5675     2868.5675       1.72      0.2085
Source        DF        Type III SS   Mean Square    F Value    Pr > F
TURN           1        2868.5675     2868.5675       1.72      0.2085
                    General Linear Models Procedure
                       Class Level Information
                  Class        Levels       Values
                  BORDCERT       2           1 2
              Number of observations in data set = 18
Dependent Variable: PREV
                           Sum of          Mean
Source            DF       Squares        Square     F Value    Pr > F
Model          [     ]    7691.3778      7691.3778   [      ]   [      ]
Error            16       21895.0667     1368.4417
Corrected Total  17       29586.4444

        R-Square         C.V.         Root MSE        PREV Mean
        0.259963       -139.8874       36.992         -26.444
Source        DF        Type I SS     Mean Square    F Value    Pr > F
BORDCERT       1        7691.3778     7691.3778       5.62      0.0306
Source        DF        Type III SS   Mean Square    F Value    Pr > F
BORDCERT       1        7691.3778     7691.3778       5.62      0.0306
```

References

Allen, D. M., and Grizzle, J. E. 1969. "Analysis of Growth and Dose Response Curves." *Biometrics* 25: 357–82.

Armitage, P. 1971. *Statistical Methods in Medical Research.* Oxford: Blackwell Scientific.

Cotton, R. B.; Stahlman, M. T.; Kovar, I.; and Catterton, W. Z. 1979. "Medical Management of Small Preterm Infants with Symptomatic Patent Ductus Arteriosus." *Journal of Pediatrics* 2: 467–73.

Daly, M. B. 1973. "The Effect of Neighborhood Racial Characteristics on the Attitudes, Social Behavior, and Health of Low Income Housing Residents." Ph.D. dissertation, Department of Epidemiology, University of North Carolina, Chapel Hill, N.C.

Diggle, P. J.; Liang, K. Y.; and Zeger, S. L. 1994. *Analysis of Longitudinal Data.* Oxford: Oxford University Press.

Guenther, W. C. 1964. *Analysis of Variance.* Englewood Cliffs, N.J.: Prentice-Hall.

Hollander, M., and Wolfe, D. A. 1973. *Nonparametric Statistical Methods.* New York: John Wiley & Sons.

Kirk, R. E. 1969. *Experimental Design: Procedures for the Behavioral Sciences.* Belmont, Calif.: Wadsworth.

Lehmann, E. L. 1975. *Non-parametrics: Statistical Methods Based on Ranks.* San Francisco: Holden-Day.

Lindman, H. R. 1974. *Analysis of Variance in Complex Experimental Designs.* San Francisco: W. H. Freeman.

Lynn, M. J.; Waring, G. O., III; Sperduto, R. D.; et al. 1987. "Factors Affecting Outcome and Predictability of Radial Keratotomy in the PERK Study." *Archives of Ophthalmology* 105: 42–51.

Miller, R. G., Jr. 1966. *Simultaneous Statistical Inference.* New York: McGraw-Hill.

———. 1981. *Simultaneous Statistical Inference,* 2nd ed. New York: Springer-Verlag.

Neter, J., and Wasserman, W. 1974. *Applied Linear Statistical Models.* Homewood, Ill.: Richard D. Irwin.

Siegel, S. 1956. *Nonparametric Statistics for the Behavioral Sciences.* New York: McGraw-Hill.

Steele, R. G. D., and Torrie, J. H. 1960. *Principles and Procedures of Statistics, with Special Reference to Biological Sciences.* New York: McGraw-Hill.

U.S. News & World Report. 1996a. "1996 Mutual Funds Guide, Best Mutual Funds," *U.S. News & World Report* (January 29, 1996), pp. 88–100.

———. 1996b. "America's Best Graduate Schools," *U.S. News & World Report* (March 18, 1996), pp. 79–91.

———. 1996c. "Rating the HMOs," *U.S. News & World Report* (September 2, 1996), pp. 52–63.

Varvarigou, A.; Bardin, C. L.; Beharry, K.; Chemtob, S.; Papageorgiou, A.; and Aranda, J. 1996. "Early Ibuprofen Administration to Prevent Patent Ductus Arteriosus in Premature Newborn Infants." *Journal of American Medical Association* 275: 539–44.

Randomized Blocks: Special Case of Two-way ANOVA

18-1 Preview

In Chapter 17 we considered the simplest kind of ANOVA problem—one involving a single factor (or independent variable). We now focus on the two-factor case, which is generally referred to as two-way ANOVA. This extension is by no means trivial. In fact, we will devote three chapters (Chapters 18, 19, and 20) to different aspects of the two-factor case. In this first chapter we first examine how a two-factor situation may be classified according to the data pattern. We then restrict our attention to a specific type of pattern, which will lead us to consider the randomized-blocks design, the main topic of this chapter.

18-1-1 Two-way Data Patterns

Several different types of data patterns for two-way ANOVA are illustrated in Figure 18-1. Each of these tables describes a two-factor study with four levels of factor 1 (the "row" factor) and three levels of factor 2 (the "column" factor). The Y's in each table correspond to individual observations on a single dependent variable Y. The number of Y's in a given cell is denoted by n_{ij} for the ith level of factor 1 and the jth level of factor 2. The marginal total for the ith row is denoted by $n_{i.}$, and the marginal total for the jth column is denoted by $n_{.j}$. The total number of observations is denoted by $n...$

The simplest two-factor pattern, which is illustrated in Figure 18-1(a), arises when each cell holds a single observation (i.e., when $n_{ij} = 1$ for all i and j). This pattern incorporates the randomized-blocks design to be discussed in this chapter, although such single-observation-per-cell data may arise in other ways.

A second type of pattern, illustrated in Figure 18-1(b), occurs when equal numbers of observations occur in each cell. Here, $n_{ij} = 4$ for all i and j. Chapter 19 will focus on this equal-replications situation.

FIGURE 18-1   Some two-way data patterns for a 4 × 3 table.

(a) Single Observation
per Cell ($n_{ij} = 1$)

(b) Equal Replications per Cell
($n_{ij} = 4$)

(c) Equal Replications by Column
and Proportionate Replications by
Row ($n_{ij} = n_{.j}/4$)

(d) Proportionate Row and Column
Replications ($n_{ij} = n_{i.}n_{.j}/n_{..}$)

(e) Nonsystematic Replications

The patterns in parts (c) through (e) of Figure 18-1 present different problems in statistical analysis,[1] which we will discuss in Chapter 20. The common property of these three patterns is that not all cells have the same number of observations. Unequal cell numbers often arise in

---

[1]In Chapters 18, 19, and 20 we assume that each cell in a table contains *at least* one observation. When one or more cells contain no observations, the analysis required is considerably more complicated. Further discussion of this situation is provided in texts by Ostle (1963), Peng (1967), and Armitage (1971).

observational studies in which the levels of certain factors are determined after, rather than before, the data are collected.

For the pattern in Figure 18-1(c), cells in the same column have the same number of observations, whereas cells in the same row are in the ratio $4:2:3$. For this table, each of the four cell frequencies in the jth column is equal to the same fraction of the corresponding total column frequency (i.e., $n_{ij} = n_{\cdot j}/4$ in this case). Note, for example, that $n_{\cdot 1}/4 = 16/4 = 4$, which is the number of observations in any cell in column 1.

For Figure 18-1(d) the cells in a given column are in the ratio $1:2:3:2$, whereas the cells in a given row are in the ratio $4:2:3$. This pattern results because n_{ij} is determined as

$$n_{ij} = \frac{n_i . n_{\cdot j}}{n_{\cdot \cdot}}$$

which means that any cell frequency can be obtained by multiplying the corresponding row and column marginal frequencies together and then dividing by the total number of observations. Thus, for cell (1, 2) in Figure 18-1(d), we have $n_1 . n_{\cdot 2}/n_{\cdot \cdot} = 9(16)/72 = 2$, which equals n_{12}. Similarly, for cell (4, 3), $n_4 . n_{\cdot 3}/n_{\cdot \cdot} = 18(24)/72 = 6$, which equals n_{43}.

There is no mathematical rule for describing the pattern of cell frequencies in Figure 18-1(e), so we say that such a pattern is nonsystematic. As we will see in Chapter 20, the ANOVA procedures required for the patterns in Figure 18-1(c) and 18-1(d) differ from those required for the irregular pattern in Figure 18-1(e). For the former two patterns, the same computational procedure may be used as when an equal number of observations exists per cell. For the nonsystematic case, a different procedure is required.

18-1-2 The Case of a Single Observation per Cell

A two-way table with a single observation in each cell can arise in a number of different experimental situations. Consider the following three examples:

1. Six hypertensive individuals, matched pairwise by age and sex, are randomly assigned (within each pair) to either a treatment or a control group. For each individual, a measure of change in self-perception of health is determined after 1 year. The main question of interest is whether the true mean change in self-perception for the treatment group differs from that for the control group.

2. Six growth-inducing treatment combinations are randomly assigned to six mice from the same litter. The treatment combinations are defined by the cross-classification of the levels of two factors: factor A (drug A1 or placebo A0) and factor B (drug B2 [high dose], drug B1 [low dose], or placebo B0). The dependent variable of interest is weight gain measured 1 week after treatment is initiated. The questions to be considered include (a) whether the effect of drug A1 differs from that of placebo A0; (b) whether differences exist among the effects of drugs B1 and B2 and placebo B0; and (c) whether the drug A1 effect differs from that of placebo A0 in the same way at each level of factor B.

3. Scores of satisfaction with medical care are recorded for six hypertensive patients assigned to one of six categories depending on whether the nurse practitioner assigned to the patient was measured to have high or low autonomy (factor A) and high, medium, or low knowledge of hypertension (factor B). The main questions of interest

are (a) whether mean satisfaction scores differ between patients with high-autonomy nurses and those with low-autonomy nurses, and (b) whether these differences in mean satisfaction scores differ in the same way at each level of knowledge grouping.

Each of these experiments may be represented by a 2×3 two-way table, as shown in Figure 18-2.
In the figure, the third example involves two factors whose levels (or categories) were determined after the data were gathered. Such a study is often referred to as an *observational study* rather than as an *experiment,* since the latter term is usually reserved for studies involving factors whose levels are decided beforehand. Epidemiologic, sociological, and psychological studies are usually observational rather than experimental. For such studies, the levels of the various factors of interest are determined after the frequency distributions of these factors have been considered.[2] For this example, the autonomy groupings were determined after the frequency distribution of autonomy scores on nurses was considered. Similarly, the knowledge categories were determined using the observed knowledge scores. Thus, one patient was in each of the six groups because of a posteriori (rather than a priori) considerations. In actual practice it is often impossible to arrange things so nicely, especially when large samples are involved. That is why most such observational studies have unequal numbers of observations in various cells.

FIGURE 18-2   Different experimental situations resulting in two-way tables with a single observation per cell.

	Pair 1	Pair 2	Pair 3
Treatment	Y	Y	Y
Control	Y	Y	Y

Y = Change in self-perception of health after 1 year

(a)

	Placebo B0	Drug B1	Drug B2
Placebo A0	Y	Y	Y
Drug A1	Y	Y	Y

Y = Weight gain after 1 week

(b)

	Low Knowledge	Medium Knowledge	High Knowledge
Low Autonomy	Y	Y	Y
High Autonomy	Y	Y	Y

Y = Patient satisfaction with medical care

(c)

[2] In this chapter we focus on the case where the levels of each factor are considered fixed; nevertheless, even if one or both factors were considered random, the tests of hypotheses of interest, as with one-way ANOVA, would be computed in exactly the same way as in the fixed-factor case. This is not so when more than one observation occurs per cell, as discussed in Chapter 19.

The second example, Figure 18-2(b), involves two factors whose levels were determined before the data were collected; the resulting six *treatment combinations* can be viewed, in one sense, as representing the different levels of a single factor that have been randomly assigned to the six individuals. Although it may be necessary because of limited experimental material or prohibitive cost to apply or assign each treatment combination to only a single experimental unit, considerably more information is obtained if each treatment combination has several replications.

This brings us to Figure 18-2(a), which is of the general type to be focused on in this chapter. Like Figure 18-2(b), this example represents a designed rather than an observational study. It differs from the other tables in Figure 18-2, however, in its allocation of individuals to cells (i.e., treatment combinations). The allocation here was done by randomization within each pair rather than, say, by randomization among all six individuals, ignoring any pairing. Another feature unique to this table is that the effect of only one of the two factors involved (in this case "treatment") is of primary interest. The other factor, "pair" (with three levels corresponding to the three pairs), is used only to help increase the precision of the comparison between the effects of the treatment and the behavior of control groups. Thus, if pair matching (on age and sex) is used and significant differences are found between the change scores for treatment and control groups, such differences cannot solely be attributed to one group's being older, say, or having a different sex composition than the other. The pairing, therefore, serves to eliminate or block out noise that otherwise would affect the comparison of treatment and control groups due to the confounding effects of age and sex. Such pairs are often referred to as *blocks,* and the associated experimental design is called a *randomized-blocks design.*

The analysis required for data arranged as in Figure 18-2(a) is described in most introductory statistics texts. Since two groups are involved (treatment and control) and since matching has been done, the generally recommended method of analysis involves using the paired-difference t test, which focuses on differences (changes) in scores within pairs. Hence, the key test statistic involved is of the form $T = \bar{d}\sqrt{n}/S_d$, where $\bar{d}$ is the difference between the treatment group mean and the control group mean, S_d is the standard deviation of the difference scores for all pairs, and n is the number of pairs (3, in Figure 18-2(a)). This statistic has the t distribution with $n - 1$ degrees of freedom under H_0 so the critical region is of the form $|T| \geq t_{n-1, 1-\alpha/2}$ for a two-sided test of the null hypothesis that no difference in true average change score exists between treatment and control groups.

The paired-difference t test can be viewed as a special case of the general F test used in a randomized-blocks ANOVA, involving more than two treatments per block. In fact, this randomized-blocks F test represents a generalization of the paired-difference t test just as the one-way ANOVA F test represents a generalization of the two-sample t test.

18-2 Equivalent Analysis of a Matched-pairs Experiment

Let us make the matched-pairs example of the previous section more realistic by considering the data given in Table 18-1, which involves 15 pairs of individuals matched on age and sex. The main inferential question for these data involves whether the mean change score for the treatment group significantly differs from that for the control group. Stated in terms of population means, the null hypothesis is

$$H_0: \mu_T = \mu_C$$

TABLE 18-1   Matched-pairs design involving change scores in self-perception of health (Y) among hypertensives.

Group	1	2	3	4	5	6	7	8	9	10	11	12	13	14	15	Total	Mean
								Pair									
Treatment	10	12	8	8	13	11	15	16	4	13	2	15	5	6	8	146	9.73
Control	6	5	7	9	10	12	9	8	3	14	6	10	1	2	1	103	6.87
Total	16	17	15	17	23	23	24	24	7	27	8	25	6	8	9	249	8.30
Difference	4	7	1	−1	3	−1	6	8	1	−1	−4	5	4	4	7	43	2.86

where μ_T and μ_C denote the population means of the treatment and control groups, respectively. The alternative hypothesis is then

$$H_A: \mu_T \neq \mu_C$$

assuming that the analyst did not theorize in advance that one particular group would have a higher or lower population mean than the other. (If, however, the analyst thought a priori that the treatment group would have a significantly higher mean change score than the control group, the alternative would be one-sided: $H_A: \mu_T > \mu_C$.)

18-2-1 Paired-difference t Test

One method for testing the null hypothesis H_0 was described in the previous section—the paired-difference t test. To perform this test, we first determine from Table 18-1 that $\bar{d} = \bar{Y}_T - \bar{Y}_C = 2.86$ and

$$S_d^2 = \frac{1}{14}\left[\sum_{i=1}^{15} d_i^2 - \frac{\left(\sum_{i=1}^{15} d_i\right)^2}{15}\right] = 12.695$$

where d_i is the observed difference between the scores for the treatment and control groups for the ith pair. Then we compute the test statistic as

$$T = \frac{\bar{d}\sqrt{n}}{S_d} = \frac{2.86\sqrt{15}}{\sqrt{12.695}} = 3.109$$

Since $t_{14,0.995} = 2.976$ and $t_{14,0.9995} = 4.140$, the P-value for this two-sided test is given by

$$.001 < P < .01$$

We therefore reject H_0 and conclude that the mean change score for the treatment group significantly differs from that for the control group.

18-2-2 Randomized-blocks F Test

Another way to test the null hypothesis $H_0: \mu_T = \mu_C$ for the data of Table 18-1 is to use an F test, based on the ANOVA table given in Table 18-2.

This ANOVA table differs in form from the one used for one-way ANOVA in that the total sum of squares is partitioned into three components (treatments, pairs, and errors) instead of just

TABLE 18-2   ANOVA table for matched-pairs data of Table 18-1.

Source	SS	d.f.	MS	F
Treatment	61.63	1	61.63	9.71 (.005 < P < .01)
Pairs (blocks)	391.80	14	27.99	4.40 (.001 < P < .005)
Error	88.87	14	6.35	
Total	542.30	29		

two (between and within). The treatment component in Table 18-2 is similar to the between (i.e., treatment) source in the one-way ANOVA case; and the error component here is analogous to the within component in the one-way ANOVA case, because this source is used as an estimate of the population variance σ^2. Finally, we have a pairs (or blocks) component in Table 18-2 that has no corresponding component in the one-way ANOVA case.

To recap, whereas the total sum of squares in one-way ANOVA may be split up into two components, as indicated by the equation

$$SS(Total) = SS(Between) + SS(Within),$$

the total sum of squares in a randomized-blocks ANOVA can be partitioned into three meaningful components:

$$SS(Total) = SS(Treatments) + SS(Blocks) + SS(Error) \qquad (18.1)$$

The last two components on the right-hand side of (18.1) can be seen as representing a partition of the experimental error (or within) sum of squares associated with one-way ANOVA (i.e., with the blocking ignored). By separating out the blocking effect, we obtain a more precise estimate of experimental error—one not contaminated by any noise due to the effects of the blocking variables (in our case, age and sex).

The computated sums of squares in expression (18.1) are shown in the ANOVA table in Table 18-2. They are

$$SS(Treatments) \quad = \quad 61.63$$
$$SS(Blocks) \quad = 391.80$$
$$SS(Error) \quad = \quad 88.87, \text{ and}$$
$$SS(Total) \quad = 542.30.$$

The degress of freedom (d.f.) corresponding to the sums of squares are shown in the second column of Table 18-2. The d.f. for "treatments" is always equal to 1 less than the number of treatments (in our example, $2 - 1 = 1$). The d.f. for "blocks" is equal to 1 less than the number of blocks (in our example, $15 - 1 = 14$). The d.f. for "error" is obtained as the product of the treatment degrees of freedom and the block degrees of freedom:

Error d.f. = [Treatment d.f.][Block d.f.] = 1(14) = 14

Finally, the d.f. for the total sum of squares is equal to 1 less than the number of observations (in our example, $30 - 1 = 29$).

The mean squares, as usual, are obtained by dividing the sums of squares by their corresponding degrees of freedom. Finally, the F statistic for the test of the null hypothesis that no differences exist among the treatments is given by the formula

$$F = \frac{MS(Treatments)}{MS(Error)} \qquad (18.2)$$

In particular, to test H_0: $\mu_T = \mu_C$ for the data of Table 18-1, we calculate

$$F = \frac{61.63/1}{88.87/14} = \frac{61.63}{6.35} = 9.71$$

which is the observed value for F with 1 and 14 degrees of freedom under H_0. Thus, H_0 is rejected at significance level α if the observed value of F is greater than $F_{1,14,1-\alpha}$. Since $F_{1,14,0.99} = 8.86$ and $F_{1,14,0.995} = 11.06$, the P-value for this test satisfies the inequality $.005 < P < .01$. This is quite small, so it is reasonable to reject H_0 and to conclude that the treatment group has a significantly different mean change score from the control group.

Thus, the conclusions reached via the paired-difference t test and the randomized-blocks F test are exactly the same: reject H_0. This happens because the two tests are completely equivalent. It can be shown mathematically that the square of a paired-difference T statistic with v degrees of freedom is exactly equal to a randomized-blocks F statistic with 1 and v degrees of freedom (i.e., $F_{1,v} = T_v^2$), and that $F \geq F_{1,v,1-\alpha}$ whenever $|T| \geq t_{v,1-\alpha/2}$; and vice versa.

In our example,

$$T_{14}^2 = (3.109)^2 = 9.67 = F_{1,14}$$

and

$$t_{14,0.995}^2 = (2.977)^2 = 8.86 = F_{1,14,0.99}$$

An F test of the null hypothesis H_0: "No significant differences exist among the blocks" may also be performed. This test is not of primary interest, since the very use of a randomized-blocks design is based on the a priori assumption that significant block-to-block variation exists. Nevertheless, an a posteriori F test may be used to check the reasonableness of this assumption. The test statistic to be used in this case is

$$F = \frac{\text{MS(Blocks)}}{\text{MS(Error)}}$$

which, for the example of Table 18-1, has the F distribution under H_0 with 14 degrees of freedom in the numerator and 14 degrees of freedom in the denominator. From Table 18-2, we find that this F statistic is computed to be 4.39, with a P-value satisfying $.001 < P < .005$. The conclusion for this test, as expected, is to reject the null hypothesis that no block differences exist.

18-3 Principle of Blocking

For the matched-pairs example, we indicated that the primary reason for "pairing up" the data was to prevent the confounding factors age and sex from blurring the comparison between the treatment and control groups. In other words, using the matched-pairs design represented an attempt to account for the fact that the experimental units (i.e., the subjects) were not homogeneous with regard to factors (other than experimental group membership status) that were likely to affect the response variable. The key point here is not simply that subjects of different age and sex are different, but more precisely that age and sex are likely to affect the response variable. In another experimental situation, age and sex might not be important covariables and so would not have to be controlled or adjusted for.

In a case where the experimental units under study are heterogeneous relative to certain concomitant variables that affect the response variable (but are not of primary interest), using a randomized-blocks design requires the following two steps:

1. The experimental units (e.g., people, animals) that are homogeneous are collected together to form a block.

2. The various treatments are assigned at random to the experimental units within each block.

These steps are illustrated in Figure 18-3, where six blocks are formed, each consisting of three homogeneous experimental units. Three treatments (labeled A, B, and C) are then assigned at random to the three units within each block.

Figure 18-4, on the other hand, provides an example of incorrect blocking. In this case, one type of experimental unit might predominantly receive one kind of treatment, and another type might not get that treatment at all. For example, the experimental unit type ☆ was assigned

FIGURE 18-3   Steps used in forming randomized blocks.

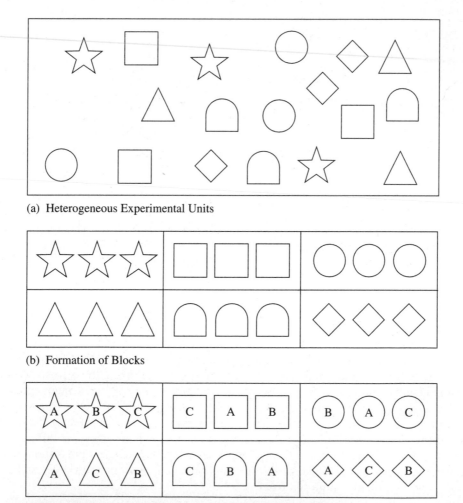

(a) Heterogeneous Experimental Units

(b) Formation of Blocks

(c) Randomization of Treatments A, B, and C Within Each Block

FIGURE 18-4   Example of incorrect blocking.

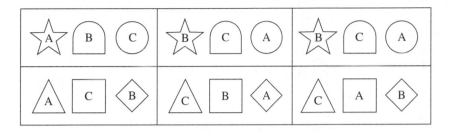

treatment B twice, whereas experimental unit types ○ and △ were not assigned treatment B at all. If the blocking had been done correctly, every distinct type of experimental unit would have been assigned each treatment exactly once.

With regard to our matched-pairs design in Table 18-1, if we had not blocked on age, the treatment might have been assigned mostly to older subjects, and the control group might have consisted mostly of younger subjects. Any differences found subsequently might very well have been due entirely to age differences between the two groups and not to the effect of the treatment itself.

18-4 Analysis of a Randomized-blocks Experiment

In this section we discuss the appropriate analysis for a randomized-blocks experiment, focusing on the case where both factors (treatments and blocks) are considered fixed, although the tests of hypotheses of interest are computed in exactly the same way even if one or both factors are considered random; that is, practically speaking, it does not matter how the factors are defined.[3]

TABLE 18-3   Data layout for a randomized-block design.

Treatment	Block 1	2	3	...	b	Total	Mean
1	Y_{11}	Y_{12}	Y_{13}	...	Y_{1b}	T_1	$\overline{Y}_{1\cdot}$
2	Y_{21}	Y_{22}	Y_{23}	...	Y_{2b}	T_2	$\overline{Y}_{2\cdot}$
3	Y_{31}	Y_{32}	Y_{33}	...	Y_{3b}	T_3	$\overline{Y}_{3\cdot}$
⋮	⋮	⋮	⋮		⋮	⋮	⋮
k	Y_{k1}	Y_{k2}	Y_{k3}	...	Y_{kb}	T_k	$\overline{Y}_{k\cdot}$
Total	B_1	B_2	B_3	...	B_b	G	—
Mean	$\overline{Y}_{\cdot 1}$	$\overline{Y}_{\cdot 2}$	$\overline{Y}_{\cdot 3}$	...	$\overline{Y}_{\cdot b}$	—	$\overline{Y}_{\cdot\cdot}$

[3]In Chapter 19, which discusses the situation of equal replications (at least two) per cell, the required *F* tests differ depending on how the factors are defined.

18-4-1 Data Configuration

Table 18-3 on the prior page gives the general layout of the data for a randomized-blocks design involving k treatments and b blocks. In this table Y_{ij} denotes the value of the observation on the dependent variable Y corresponding to the ith treatment in the jth block (e.g., Y_{23} denotes the value of the observation associated with treatment 2 in block 3). The (row) total for treatment i is denoted by T_i; that is, $T_i = \sum_{j=1}^{b} Y_{ij}$. The (column) total for block j is denoted by B_j; that is, $B_j = \sum_{i=1}^{k} Y_{ij}$. The grand total of all bk observations is $G = \sum_{i=1}^{k} \sum_{j=1}^{b} Y_{ij}$. Finally, the treatment (row) means are denoted by $\bar{Y}_{i.}$ for the ith treatment; the block (column) means are denoted by $\bar{Y}_{.j}$ for the jth block; and $\bar{Y}_{...}$ denotes the grand mean.

A special case of this general format was given in Table 18-1, where $k = 2$ and $b = 15$. Another example (with $k = 4$ and $b = 8$) appears in Table 18-4. Although based on artificial data, this example illustrates the type of information being considered in several ongoing, large-scale U.S. intervention studies dealing with risk factors associated with heart disease. Table 18-4 presents the data for a small experiment designed to assess the effects of four different cholesterol-reducing diets on persons who have hypercholesterolemia. In such a study it would seem logical to try to take into account (or adjust for) the effects of age, sex, and body size on the dependent variable Y, which represents reduction in cholesterol level after one year. One way to do this is by using a randomized-blocks design, where the blocks are chosen to represent combinations of the various age–sex–body size categories of interest. Thus, in Table 18-4, block 1 consists of males over 50 years of age with a quetelet index $(= 100[\text{weight}]/[\text{height}]^2)$ above 3.5; block 3 consists of males under 50 with a quetelet index above 3.5; and so on. For each block of subjects, the four diets are randomly assigned to the sample of four persons in the block. Each subject is then followed for one year, after which the change in cholesterol level (Y) is recorded.

The primary research question of interest in this study concerns whether any of the average reductions in cholesterol level achieved by the four diets significantly differ. By inspecting Table 18-4, we can see that diet 1 appears to be the best, since it is associated with the largest average reduction (12.57 units). Nevertheless, this conclusion needs to be appraised statistically, which involves determining whether the observed differences among the mean reductions

TABLE 18-4   Randomized-blocks experiment for comparing the effects of four cholesterol-reducing diets on persons with hypercholesterolemia (Y = reduction in cholesterol level after one year).

	Block									
	1 (Male, Age > 50, QUET > 3.5)	2 (Male, Age > 50, QUET < 3.5)	3 (Male, Age < 50, QUET > 3.5)	4 (Male, Age < 50, QUET < 3.5)	5 (Female, Age > 50, QUET > 3.5)	6 (Female, Age > 50, QUET < 3.5)	7 (Female, Age < 50, QUET > 3.5)	8 (Female, Age < 50, QUET < 3.5)		
Treatment (Diet)									Total	Mean
1	11.2	6.2	16.5	8.4	14.1	9.5	21.5	13.2	100.6	12.58
2	9.3	4.1	14.2	6.9	14.2	8.9	15.2	10.1	82.9	10.36
3	10.4	5.1	14.0	6.2	11.1	8.4	17.3	11.2	83.7	10.46
4	9.0	4.9	13.7	6.1	11.8	8.4	15.9	9.7	79.5	9.94
Total	39.9	20.3	58.4	27.6	51.2	35.2	69.9	44.2	346.7	—
Mean	9.98	5.08	14.60	6.90	12.80	8.80	17.48	11.05	—	10.83

for the four diets can be attributed solely to chance. The randomized-blocks F test provides a method for making such a statistical evaluation.

Further inspection of Table 18-4 indicates that considerable differences exist among the block means. This is to be expected, since the reason for blocking was that different blocks (or equivalently, different categories of the covariates age, sex, and body size) were expected to have different effects on the response. We can nevertheless perform a statistical test to satisfy ourselves that such block differences are statistically significant. In fact, if this test yields a finding of nonsignificance, we probably should not have used these blocks in the first place, since they cost us degrees of freedom for estimating σ^2 and consequently produce a less sensitive comparison of treatment effects.

18-4-2 Hypotheses Tested in a Randomized-blocks Analysis

The primary null hypothesis of interest in a randomized-blocks analysis is the equality of treatment means:[4]

$$H_0: \mu_1. = \mu_2. = \cdots = \mu_k. \tag{18.3}$$

where $\mu_i.$ denotes the population mean response associated with the ith treatment. The alternative hypothesis may be stated as

$$H_A: \text{"Not all the } \mu_i.\text{'s are equal."}$$

If performing this test leads to rejection of H_0, it becomes of interest to determine where the important differences are among the treatment effects. One qualitative way to make such a determination is simply to look at the observed treatment means and to make visual comparisons. Thus, from Table 18-4, rejecting the null hypothesis that no differences exist among the diets leads to the conclusion that diet 1 is best in reducing cholesterol level, that diets 2 and 3 are next in line and are about equally effective, and that diet 4 is the worst of the four. To verify such conclusions statistically, we can use multiple-comparison techniques to conduct appropriate tests, as described in Chapter 17.

As mentioned earlier, another null hypothesis sometimes of interest is that no significant differences exist among the block means. Since use of a randomized-blocks design is based on the a priori suspicion that the block means are different, such a hypothesis is usually tested only to check whether the blocking was justified.

18-5 ANOVA Table for a Randomized-blocks Experiment

Table 18-5 summarizes the procedures necessary to test the equality of the treatment and block means. Since computation of the ANOVA table sums of squares is routinely carried out

---

[4]Again, we are considering the null hypothesis for the fixed-effects case. In one-way ANOVA, this hypothesis would be stated differently if the treatment effects were considered random, but the test of hypothesis would be computed in exactly the same way.

TABLE 18-5   ANOVA table for a randomized-blocks experiment with k treatments and b blocks.

Source	d.f.	SS	MS	F
Treatments	$k-1$	SST	$MST = \dfrac{SST}{k-1}$	$\dfrac{MST}{MSE}$
Blocks	$b-1$	SSB	$MSB = \dfrac{SSB}{b-1}$	$\dfrac{MSB}{MSE}$
Error	$(k-1)(b-1)$	SSE	$MSE = \dfrac{SSE}{(k-1)(b-1)}$	
Total	$kb-1$	TSS		

by a computer program, (e.g., by SAS's GLM procedure), specific computational formulas are not included in the SS column of Table 18-5. The ANOVA table quantities for the data given in Table 18-4 are shown in the accompanying SAS computer output.

From Table 18-5 and the SAS output, it appears that the total (corrected) sum of squares, TSS, is split up into the three components SST (treatments), SSB (blocks), and SSE (error) using the fundamental equation

$$\text{TSS(Total)} = \text{SST(Treatments)} + \text{SSB(Blocks)} + \text{SSE(Error)} \qquad (18.4)$$

Edited SAS Output (PROC GLM) for Data of Table 18-4

```
                      Class Level Information
              Class        Levels        Values
              TRT            4           1 2 3 4
              BLK            8           1 2 3 4 5 6 7 8
           Number of observations in data set = 32
Dependent Variable: Y
                              Sum of  SSE    Mean
Source              DF        Squares        Square      F Value    Pr > F
Model               10       496.420625 /   49.642063     52.89     0.0001
Error               21        19.711563 /    0.938646  ← MSE
Corrected Total     31       516.132188  ← TSS

              R-Square          C.V.        Root MSE       Y Mean
              0.961809        8.942254      0.96884       10.8344      F Statistics

Source         DF SST Type I SS    Mean Square MST F Value    Pr > F
TRT            3     33.560938       11.186979       11.92     0.0001
BLK            7    462.859688       66.122813 ← MSB 70.44     0.0001
Source         DF SSB Type III SS   Mean Square     F Value    Pr > F
TRT            3     33.560938       11.186979       11.92     0.0001
BLK            7    462.859688       66.122813       70.44     0.0001
```

The formula for the total sum of squares is:

$$\text{TSS(Total)} = \sum_{i=1}^{k} \sum_{j=1}^{b} (Y_{ij} - \bar{Y}..)^2 \qquad (18.5)$$

As is the case in any ANOVA situation, TSS measures the total unexplained variation in the data, corrected for the grand mean. The degrees of freedom associated with TSS is $kb - 1$. From the SAS output, we see that TSS is 516.132.

The treatment sum of squares SST is defined as

$$\text{SST} = b \sum_{i=1}^{k} (\overline{Y}_{i\cdot} - \overline{Y}_{\cdot\cdot})^2 \qquad (18.6)$$

SST reflects the variation among the treatment means. Its basic components are of the form $(\overline{Y}_{i\cdot} - \overline{Y}_{\cdot\cdot})$, which is the difference between the ith treatment mean and the grand mean. SST has $k - 1$ degrees of freedom. For the SAS output, SST is computed as 33.561.

The block sum of squares SSB is defined as

$$\text{SSB} = k \sum_{j=1}^{b} (\overline{Y}_{\cdot j} - \overline{Y}_{\cdot\cdot})^2 \qquad , \; SSM = SST + SSB \qquad (18.7)$$

with $b - 1$ degrees of freedom. The basic components of SSB take the form $(\overline{Y}_{\cdot j} - \overline{Y}_{\cdot\cdot})$, which is the difference between the jth block mean and the grand mean. From the SAS output, we see that SSB is computed as 462.860.

Finally, the residual sum of squares SSE is given by

$$\text{SSE} = \sum_{i=1}^{k} \sum_{j=1}^{b} (Y_{ij} - \overline{Y}_{i\cdot} - \overline{Y}_{\cdot j} + \overline{Y}_{\cdot\cdot})^2 \qquad (18.8)$$

with degrees of freedom equal to $(k - 1)(b - 1)$. For the SAS output, SSE is computed as 19.712.

The complexity of expression (18.8) indicates that SSE is not easily recognizable as an estimate of σ^2. In fact, its basic component, $(Y_{ij} - \overline{Y}_{i\cdot} - \overline{Y}_{\cdot j} + \overline{Y}_{\cdot\cdot})$, can be written in the form

$$(Y_{ij} - \overline{Y}_{\cdot j}) - (\overline{Y}_{i\cdot} - \overline{Y}_{\cdot\cdot})$$

which is an estimate of the difference between the effect of the ith treatment relative to the average effect of all treatments in the jth block (i.e., $Y_{ij} - \overline{Y}_{\cdot j}$) and the overall effect of the ith treatment relative to the overall mean (i.e., $\overline{Y}_{i\cdot} - \overline{Y}_{\cdot\cdot}$). Actually, SSE measures block–treatment interaction here, in the sense that SSE is large when the treatment effects vary from block to block. Although we discuss the concept of interaction more thoroughly in Chapter 19, using a randomized-blocks design requires us to assume that no such block-treatment interaction exists; as a result, the residual variation reflected in SSE can be attributed solely to experimental error.[5] The expected effect of violating the no-interaction assumption is a reduction in the power of the tests of treatment and block effects. This assumption must not be violated; otherwise, we would have no way of obtaining an unbiased estimate of σ^2 to use in the denominator of the F statistic. After all, for each block in a randomized-blocks design, no treatment is applied more than once, which makes it impossible to obtain a pure error estimate of σ^2 associated with a particular treatment in a given block. In the case of two-way ANOVA with more than one observation per

---

[5]A method for testing the validity of the assumption of no block–treatment interaction, developed by Tukey (1949), is described in Problem 3(f) at the end of this chapter.

cell, a pure error estimate of σ^2 can be developed by utilizing the available information on within-cell variability.

Each mean-square term in Table 18-5 is, as usual, obtained by dividing the corresponding sum-of-squares term by its degrees of freedom. The F statistics are formed as the ratios of mean-square terms, with MSE in the denominator in each case (regardless of whether each factor is treated as fixed or random).

18-5-1 F Test for the Equality of Treatment Means

The test of H_0: $\mu_1. = \mu_2. = \cdots = \mu_k.$ is performed by using the following F statistic (defined in terms of the notation of Table 18-3):

$$F = \frac{\text{MST}}{\text{MSE}} \tag{18.9}$$

where

$$\text{MST} = \frac{1}{k-1}(\text{SST})$$

and

$$\text{MSE} = \frac{1}{(k-1)(b-1)}(\text{SSE})$$

The assumptions required for this test are as follows:

1. The observations are statistically independent of one another.

2. Each observation is selected from a normally distributed population.

3. Each observation is selected from a population with variance σ^2 (i.e., variance homogeneity is assumed).

4. There is no block–treatment interaction effect (i.e., the true extent to which treatments differ is the same, regardless of the block considered).

If H_0 is true, the F statistic in (18.9) has the F distribution with $k-1$ degrees of freedom in the numerator and $(k-1)(b-1)$ degrees of freedom in the denominator. Thus, for a given α, we would reject H_0 and conclude that not all treatments have the same effect on the response, provided that

$$F \geq F_{k-1,\,(k-1)(b-1),\,1-\alpha}$$

From the SAS output shown in section 18-5, we see that MST is equal to 11.187 and MSE is equal to 0.939, so

$$F = \frac{11.187}{0.939} = 11.9$$

Since $F_{3,21,0.999} = 7.94$, the P-value for this test satisfies $P < .001$. Thus, we reject the null hypothesis H_0 and conclude that significant differences exist among the four diets.

18-5-2 *F* Test for the Equality of Block Means

As previously mentioned, this test is rarely used except as an a posteriori check to confirm that blocking was effective. To perform this test, we calculate the following F statistic:

$$F = \frac{\text{MSB}}{\text{MSE}} \qquad (18.10)$$

where

$$\text{MSB} = \frac{1}{b-1}\left(\frac{1}{k}\sum_{j=1}^{b} B_j^2 - \frac{G^2}{bk}\right)$$

and where MSE is calculated as before.

Under the null hypothesis H_0: "There are no differences among the true block means," the F statistic (18.10) has the F distribution with $b-1$ and $(k-1)(b-1)$ degrees of freedom. Thus, H_0 is rejected at significance level α when F exceeds $F_{b-1,(k-1)(b-1),1-\alpha}$. As shown in the SAS output in section 18-5, this test yields the following F statistic:

$$F = \frac{66.123}{0.939} = 70.4$$

The P-value for this test satisfies $P < .001$, so H_0 is rejected, as expected.

18-6 Regression Models for a Randomized-blocks Experiment

The randomized-blocks experiment, like the one-way ANOVA situation, can be described by either a regression model or a classical ANOVA effects model. The regression formulation, which we consider in this section, is technically equivalent to the fixed-effects ANOVA approach in terms of estimating the unknown parameters in each model. Nevertheless, for testing purposes, mean-square terms in the ANOVA table obtained from the fit of a proper regression model can be used to compute appropriate F statistics, regardless of whether the actual ANOVA model is fixed, random, or mixed.

An appropriate regression model for the randomized-blocks experiment should contain $k-1$ dummy variables for the k treatments and $b-1$ dummy variables for the b blocks. One such model formulation is

$$Y = \mu + \sum_{i=1}^{k-1} \alpha_i X_i + \sum_{j=1}^{b-1} \beta_j Z_j + E \qquad (18.11)$$

where

$$X_i = \begin{cases} -1 & \text{if treatment } k \\ 1 & \text{if treatment } i \\ 0 & \text{otherwise} \end{cases} \quad \text{and} \quad Z_j = \begin{cases} -1 & \text{if block } b \\ 1 & \text{if block } j \\ 0 & \text{otherwise} \end{cases}$$

(for $i = 1, 2,..., k-1$; and $j = 1, 2,..., b-1$).

As in one-way ANOVA, other coding schemes for the independent variables are possible. For example, we may let

$$X_i = \begin{cases} 1 & \text{if treatment } i \\ 0 & \text{otherwise} \end{cases} \quad \text{and} \quad Z_j = \begin{cases} 1 & \text{if block } j \\ 0 & \text{otherwise} \end{cases}$$

(for $i = 1, 2,\ldots, k - 1$; and $j = 1, 2,\ldots, b - 1$). For any such coding scheme, the F tests obtained from fitting the regression model are exactly equivalent to the randomized-blocks ANOVA F tests described previously. The regression coefficients, however, represent different functions of the cell population means.

As in the one-way ANOVA situation considered in Chapter 17, the regression coefficients in the preceding model for Y may be expressed in terms of underlying cell (i.e., block–treatment combination) means. To see this, consider the matrix of population cell means associated with the general randomized-blocks layout presented in Table 18-6. In this table, the (cell) mean for the ith treatment in the jth block is denoted by μ_{ij}; the mean for the ith treatment (averaged over the b blocks) is denoted by $\mu_{i\cdot}$; the mean for the jth block (averaged over the k treatments) is denoted by $\mu_{\cdot j}$; and $\mu_{\cdot\cdot}$ denotes the overall mean. Hence, $\mu_{i\cdot}$, $\mu_{\cdot j}$, and $\mu_{\cdot\cdot}$ satisfy

$$\mu_{i\cdot} = \frac{1}{b}\sum_{j=1}^{b} \mu_{ij}, \qquad \mu_{\cdot j} = \frac{1}{k}\sum_{i=1}^{k} \mu_{ij}, \qquad \mu_{\cdot\cdot} = \frac{1}{bk}\sum_{i=1}^{k}\sum_{j=1}^{b} \mu_{ij}$$

For model (18.11), the coefficients α_i and β_i can be expressed as follows:

$$
\begin{aligned}
\mu &= \mu_{\cdot\cdot} \\
\alpha_i &= \mu_{i\cdot} - \mu_{\cdot\cdot} \qquad i = 1, 2,\ldots, k - 1 \\
\beta_j &= \mu_{\cdot j} - \mu_{\cdot\cdot} \qquad j = 1, 2,\ldots, b - 1 \\
-\sum_{i=1}^{k-1} \alpha_i &= \mu_{k\cdot} - \mu_{\cdot\cdot} \\
-\sum_{j=1}^{b-1} \beta_j &= \mu_{\cdot b} - \mu_{\cdot\cdot}
\end{aligned}
$$

       (18.12)

TABLE 18-6   Matrix of cell means for a randomized-blocks layout.

Treatment (i)	Block (j) 1	2	3	...	b	Mean
1	μ_{11}	μ_{12}	μ_{13}	...	μ_{1b}	$\mu_{1\cdot}$
2	μ_{21}	μ_{22}	μ_{23}	...	μ_{2b}	$\mu_{2\cdot}$
3	μ_{31}	μ_{32}	μ_{33}	...	μ_{3b}	$\mu_{3\cdot}$
$\vdots$	$\vdots$	$\vdots$	$\vdots$		$\vdots$	$\vdots$
k	μ_{k1}	μ_{k2}	μ_{k3}	...	μ_{kb}	$\mu_{k\cdot}$
Mean	$\mu_{\cdot 1}$	$\mu_{\cdot 2}$	$\mu_{\cdot 3}$	...	$\mu_{\cdot b}$	$\mu_{\cdot\cdot}$

Thus, the coefficient μ represents the overall mean, the coefficient α_i represents the difference between the ith treatment mean and the overall mean, and the coefficient β_j represents the difference between the jth block mean and the overall mean. Furthermore, the negative sum of the α_i (i.e., $-\sum_{i=1}^{k-1}\alpha_i$) gives the difference between the kth treatment mean and the overall mean; and the negative sum of the β_j (i.e., $-\sum_{j=1}^{b-1}\beta_j$) gives the difference between the bth block mean and the overall mean.

These results can also be used to express the cell, treatment, and block means as a function of the regression parameters. The cell means are

$$\mu_{ij} = \mu + \alpha_i + \beta_j \qquad i = 1, 2,\dots, k-1; j = 1, 2,\dots, b-1$$

$$\mu_{kj} = \mu - \sum_{i=1}^{k-1}\alpha_i + \beta_j \qquad j = 1, 2,\dots, b-1$$

$$\mu_{ib} = \mu + \alpha_i - \sum_{j=1}^{b-1}\beta_j \qquad i = 1, 2,\dots, k-1$$

$$\mu_{kb} = \mu - \sum_{i=1}^{k-1}\alpha_i - \sum_{j=1}^{b-1}\beta_j$$

In turn, the treatment means are

$$\mu_{i\cdot} = \mu + \alpha_i \qquad i = 1, 2,\dots, k-1$$

$$\mu_{k\cdot} = \mu - \sum_{i=1}^{k-1}\alpha_i$$

and the block means are

$$\mu_{\cdot j} = \mu + \beta_j \qquad j = 1, 2,\dots, b-1$$

$$\mu_{\cdot b} = \mu - \sum_{j=1}^{b-1}\beta_j$$

Similar results can be obtained by using reference cell coding. The model statement remains the same, but the dummy variables change as follows:

$$X_i = \begin{cases} 1 & \text{if treatment } i \quad i = 1, 2,\dots, k-1 \\ 0 & \text{otherwise} \end{cases}$$

$$Z_j = \begin{cases} 1 & \text{if block } j \quad j = 1, 2,\dots, b-1 \\ 0 & \text{otherwise} \end{cases}$$

Although the same parameter symbols are used, their meanings change. In particular, the cell means (for reference cell coding) are

$$\mu_{ij} = \mu + \alpha_i + \beta_j \qquad i = 1, 2,\dots, k-1; j = 1, 2,\dots, b-1$$
$$\mu_{kj} = \mu + \beta_j \qquad j = 1, 2,\dots, b-1$$
$$\mu_{ib} = \mu + \alpha_i \qquad i = 1, 2,\dots, k-1$$
$$\mu_{kb} = \mu$$

The treatment means are

$$\mu_{i.} = \mu + \alpha_i + \sum_{j=1}^{b-1} \frac{\beta_j}{b} \quad i = 1, 2, \ldots, k-1$$

$$\mu_{k.} = \mu + \sum_{j=1}^{b-1} \frac{\beta_j}{b}$$

and the block means are

$$\mu_{.j} = \mu + \beta_j + \sum_{i=1}^{k-1} \frac{\alpha_i}{k} \quad j = 1, 2, \ldots, b-1$$

$$\mu_{.b} = \mu + \sum_{i=1}^{k-1} \frac{\alpha_i}{k}$$

These results demonstrate that estimates of means and contrasts between means can be expressed as linear combinations of estimated regression coefficients. [Caution: When using a computer package to estimate effects in ANOVA, make sure you understand exactly what parameters the program is estimating.]

As with one-way ANOVA, the F tests resulting from the fit of regression model (18.11) or from any other properly coded regression model used for testing the hypotheses of equality of treatment means and of equality of block means are exactly the same as those obtained for the ANOVA procedures presented earlier. The multiple-partial F test of $H_0: \alpha_1 = \alpha_2 = \cdots = \alpha_{k-1} = 0$ under model (18.11) provides exactly the same F-value as that given by

$$F = \frac{\text{MST}}{\text{MSE}}$$

under (18.9). Similarly, the multiple-partial F test of $H_0: \beta_1 = \beta_2 = \cdots = \beta_{b-1} = 0$ in model (18.11) yields exactly the same F-value as is given by

$$F = \frac{\text{MSB}}{\text{MSE}}$$

under (18.10).

Thus, it does not matter which, if any, regression model (i.e., coding scheme) we use if we are interested only in performing the preceding global F tests. Furthermore, if we want to make certain specific comparisons of means, we can always calculate these comparisons directly without using a regression model.

18-7 Fixed-effects ANOVA Model for a Randomized-blocks Experiment

If both the block effects and the treatment effects are considered fixed,[6] a classical ANOVA model may be written in terms of these effects as was done in the one-way ANOVA case. The

---

[6]Random-effects models for two-way ANOVA are discussed in Chapter 19.

effects for a randomized-blocks experiment are defined as differences between a given treatment mean and the overall mean (i.e., $\mu_{i.} - \mu_{..}$) and as differences between a given block mean and the overall mean (i.e., $\mu_{.j} - \mu_{..}$). The fixed-effects model may be written in the form

$$Y_{ij} = \mu + \alpha_i + \beta_j + E_{ij} \qquad i = 1, 2,\ldots, k; j = 1, 2,\ldots, b \qquad (18.13)$$

where

Y_{ij} = Observed response associated with the ith treatment in the jth block

$\mu = \mu_{..}$ = Grand (overall) mean

$\alpha_i = \mu_{i.} - \mu_{..}$ = Effect of treatment i

$\beta_j = \mu_{.j} - \mu_{..}$ = Effect of block j

$E_{ij} = Y_{ij} - \mu - \alpha_i - \beta_j$ = Error component associated with the ith treatment in the jth block

In model (18.13), the effect of any given treatment is the same, regardless of the block; and similarly, the effect of any given block is the same, regardless of the treatment. In other words, there is no block–treatment interaction. Thus, to determine the mean response for a given cell, we need only know the treatment effect and block effect associated with that cell—and not the particular contribution of the cell itself. A model incorporating block–treatment interaction would be of the form

$$Y_{ij} = \mu + \alpha_i + \beta_j + \gamma_{ij} + E_{ij}$$

which contains the (interaction) term γ_{ij} specific to cell (i, j).

A few additional properties of the fixed-effects model (18.13) are worth noting. First, it can be shown that

$$\sum_{i=1}^{k} \alpha_i = 0 \qquad \text{and} \qquad \sum_{j=1}^{b} \beta_j = 0$$

since $\alpha_i = \mu_{i.} - \mu_{..}$ and $\beta_j = \mu_{.j} - \mu_{..}$.

Another property of interest is that the various treatment and block effects can be estimated from the data, as follows:

$$\hat{\alpha}_i = \overline{Y}_{i.} - \overline{Y}_{..} \qquad \text{and} \qquad \hat{\beta}_j = \overline{Y}_{.j} - \overline{Y}_{..}$$

A final property of the fixed-effects model (18.13) is that it corresponds in structure to the regression model given by (18.11); that is, the coefficients α_1 through α_{k-1} of model (18.11) correspond to the first $k - 1$ treatment effects in model (18.13). Similarly, the coefficients β_1 through β_{b-1} correspond to the first $b - 1$ block effects in model (18.13). Finally, the negative sums $-\sum_{i=1}^{k-1}\alpha_i$ and $-\sum_{j=1}^{b-1}\beta_j$ represent the effects of the kth treatment and the bth block, respectively.

Problems

1. A private research corporation conducted an experiment to investigate the toxic effects of three chemicals (I, II, and III) used in the tire-manufacturing industry. In this experiment, three

adjacent 1-inch squares were marked on the back of each of eight rats, and the three chemicals were applied separately to the three squares on each rat. The squares were then rated from 0 to 10, depending on the degree of irritation. The data are as shown in the following table.

Chemical	\multicolumn{8}{c}{Rat}	Total
	1	2	3	4	5	6	7	8	
I	6	9	6	5	7	5	6	6	50
II	5	9	9	8	8	7	7	7	60
III	3	4	3	6	8	5	5	6	40
Total	14	22	18	19	23	17	18	19	150

a. What are the blocks and what are the treatments in this randomized-blocks design?
b. Complete the ANOVA table in the accompanying computer output for the data set given.
c. Do the data provide sufficient evidence to indicate a significant difference in the toxic effects of the three chemicals?
d. Using a confidence interval of the form

$$(\overline{Y}_{1\cdot} - \overline{Y}_{2\cdot}) \pm t_{v, 1-\alpha/2} \sqrt{MSE\left(\frac{1}{n_1} + \frac{1}{n_2}\right)}$$

where v is the degrees of freedom, find a 98% confidence interval for the true difference in the toxic effects of chemicals I and II.
e. Provide a reasonable measure of the proportion of total variation that is explained by the particular statistical model used in analyzing this data set.
f. State the fixed-effects ANOVA model and the corresponding regression model for this analysis.
g. State the assumptions on which the validity of the analysis depends.

Edited SAS Output (PROC GLM) for Problem 1

```
Dependent Variable: SCORE
                              Sum of             Mean
Source            DF         Squares           Square    F Value    Pr > F
Model              9      43.5000000        4.8333333       2.71    0.0463
Error             14      [          ]      [          ]
Corrected Total   23      68.5000000

        R-Square              C.V.         Root MSE            SCORE Mean
        0.635036           21.38090         1.33631              6.25000

Source            DF       Type I SS     Mean Square    F Value    Pr > F
CHEM               2      [          ]   [          ]      7.00    0.0078
RAT                7      18.5000000       2.6428571      1.48    0.2518

Source            DF     Type III SS     Mean Square    F Value    Pr > F
CHEM               2      25.0000000      12.5000000      7.00    0.0078
RAT                7      18.5000000       2.6428571      1.48    0.2518
```

2. In a study of the psychosocial changes in individuals who participated in a community-based intervention program, individuals who were clinically identified as hypertensives were randomly assigned to one of two treatment groups (with n individuals in each group). Group 1 was given the usual care that existing community facilities provide for hypertensives; group 2 was given the special care that the intervention study team provided. Among the variables measured on each individual were SP1, an index of the individual's self-perception of health immediately after being identified as hypertensive but before being assigned to one of the two groups; SP2, an index of the individual's self-perception of health one year after assignment to one of the two groups; AGE; and SEX. One main research question of interest was whether the change in self-perception of health after one year would be greater for individuals in group 2 than for those in group 1. Several different analytical approaches to answering this question are possible, depending on the dependent variable chosen and on the treatment of the variables SP1, AGE, and SEX in the analysis. Among these approaches are the following six:

- Matching pairwise on AGE and SEX (which we assume is possible) and then performing a paired-difference t test to determine whether the mean of group 1 change scores significantly differs from the mean of group 2 change scores. [*Note:* This is equivalent to performing a randomized-blocks analysis, where the blocks are the pairs of individuals.]
- Matching pairwise on AGE and SEX and then performing a regression analysis, with the change score $Y = SP2 - SP1$ as the dependent variable and with SP1 as one of the independent variables.
- Matching pairwise on AGE and SEX and then performing a regression analysis, with SP2 as the dependent variable and with SP1 as one of the independent variables.
- Controlling for AGE and SEX (without any prior matching) via analysis of covariance, with the change score $Y = SP2 - SP1$ as the dependent variable.
- Controlling for AGE and SEX via analysis of covariance, with the change score $Y = SP2 - SP1$ as the dependent variable and with SP1 as one of the independent variables.
- Controlling for AGE and SEX via analysis of covariance, with SP2 as the dependent variable and with SP1 as one of the independent variables.

   a. What regression model is associated with each of the preceding six approaches? Make sure to define your variables carefully.

   b. For each of the preceding regression models, state the appropriate null hypothesis (in terms of regression coefficients) for testing for group differences with respect to self-perception scores. For each regression model, indicate how to set up the appropriate ANOVA table to carry out the desired test.

   c. Assuming that you have decided to match pairwise on AGE and SEX, which of the preceding regression models would you prefer to use, and why? [*Note:* Actually, you have only two models to choose from, since the second and third models described in the preceding list produce exactly the same test statistic for comparing the two groups.]

3. An experiment was conducted at the University of North Carolina to see whether the BOD test for water pollution is biased by the presence of copper. In this test, the amount of dissolved oxygen in a sample of water is measured at the beginning and at the end of a five-day period; the difference in dissolved oxygen content is ascribed to the action of bacteria on the impurities in the sample and is called the *biochemical oxygen demand* (BOD). The question

is whether dissolved copper retards the bacterial action and produces artificially low responses for the test.

The data in the following table are partial results from this experiment. The three samples (which are from different sources) are split into five subsamples, and the concentration of copper ions in each subsample is given. The BOD measurements are given for each subsample–copper ion concentration combination.

Sample	Copper Ion Concentration (ppm)					Mean
	0	0.1	0.3	0.5	0.75	
1	210	195	150	148	140	168.60
2	194	183	135	125	130	153.40
3	138	98	89	90	85	100.00
Mean	180.67	158.67	124.67	121.00	118.33	140.67

a. Using dummy variables and treating the copper ion concentration as a categorical (or nominal) variable, provide an appropriate regression model for this experiment. Is this a randomized-blocks experiment?

b. If copper ion concentration is treated as an interval (continuous) variable, one appropriate regression model would be

$$Y = \beta_0 + \beta_1 Z_1 + \beta_2 Z_2 + \beta_3 X + \beta_4 Z_1 X + \beta_5 Z_2 X + E$$

where

$$Z_1 = \begin{cases} 1 & \text{for sample 1} \\ 0 & \text{for sample 2} \\ -1 & \text{for sample 3} \end{cases} \qquad Z_2 = \begin{cases} 0 & \text{for sample 1} \\ 1 & \text{for sample 2} \\ -1 & \text{for sample 3} \end{cases}$$

and $X = $ Copper ion concentration. What are the advantages and disadvantages of using the models in parts (a) and (b)? Which model would you prefer to use, and why?

c. Compare (without doing any statistical tests) the average BOD responses at the various levels of copper ion concentration.

d. Use the following table, which is based on a randomized-blocks analysis, to test (at $\alpha = .05$) the null hypothesis that copper ion concentration has no effect on the BOD test.

Source	d.f.	SS	MS	F
Samples	2	12,980.9333	6,490.4667	56.83
Concentrations	4	9,196.6667	2,299.1667	20.13
Error	8	913.7333	114.2167	

e. Judging from the ANOVA table and the observed block means, does blocking appear to be justified?

f. The randomized-blocks analysis assumes that the relative differences in BOD responses at different levels of copper ion concentration are the same regardless of the sample used; in other words, there is no copper ion concentration–sample interaction. One method (see Tukey 1949) for testing whether such an interaction effect actually exists is *Tukey's test for additivity.* It addresses the null hypothesis

H_0: No interaction exists (i.e., the model is additive in the block and treatment effects).

versus the alternative hypothesis

H_A: The model is not additive, and a transformation $f(Y)$ exists that removes the nonadditivity in the model for Y.

Tukey's test statistic is given by

$$F = \frac{\text{SSN}}{(\text{SSE} - \text{SSN})/[(k-1)(b-1)-1]}$$

(which is distributed as $F_{1,(k-1)(b-1)-1}$ under H_0), where

$$\text{SSN} = \frac{\left[\displaystyle\sum_{i=1}^{k}\sum_{j=1}^{b} Y_{ij}(\overline{Y}_{i.} - \overline{Y}_{..})(\overline{Y}_{.j} - \overline{Y}_{..})\right]^2}{\displaystyle\sum_{i=1}^{k}(\overline{Y}_{i.} - \overline{Y}_{..})^2 \sum_{j=1}^{b}(\overline{Y}_{.j} - \overline{Y}_{..})^2}$$

using the notation in this chapter. (Since the numerator degrees of freedom for this test statistic is 1, the square root of F has the t distribution with $(k-1)(b-1)-1$ degrees of freedom.)

Given that the computed t statistic is $T = 2.090$ in Tukey's test for additivity, is there significant evidence of nonadditivity? (Use $\alpha = .05$.)

g. The following tables are based on fitting the multiple regression model given in part (b). Use these to test whether evidence exists of a significant effect due to copper ion concentration (i.e., test $H_0: \beta_3 = \beta_4 = \beta_5 = 0$). Use the result $R^2 = .888$.

Variable	Regression Coefficient	Beta Coefficient
(intercept)	167.195	
Z_1	32.513	0.67659
Z_2	16.972	0.35319
X	−80.389	−0.55585
$Z_1 X$	−13.877	−0.12337
$Z_2 X$	−12.844	−0.11419

Source	d.f.	MS
Z_1	1	11,764.90
$Z_2 \mid Z_1$	1	1,216.03
$X \mid Z_1, Z_2$	1	7,134.57
$Z_1 X \mid Z_1, Z_2, X$	1	303.27
$Z_2 X \mid Z_1, Z_2, X, Z_1 X$	1	91.07
Error	9	286.83

4. Using the accompanying computer output based on the data in Problem 7 in Chapter 17, conduct a randomized-blocks analysis, treating the high schools as blocks, to test whether significant differences exist among the mean VSAT scores for the years 1965, 1970, and 1975. Do the results obtained from this randomized-blocks analysis differ from those obtained earlier from the one-way ANOVA?

Edited SAS Output (PROC GLM) for Problem 4

```
Dependent Variable: VSAT
                           Sum of           Mean
Source             DF      Squares          Square     F Value     Pr > F
Model              6     1875.20000       312.53333      19.76     0.0002
Error              8      126.53333        15.81667
Corrected Total   14     2001.73333

        R-Square              C.V.        Root MSE           VSAT Mean
        0.936788           0.734490       3.97702             541.467

Source             DF     Type I SS      Mean Square     F Value     Pr > F
YEAR               2      548.13333       274.06667       17.33      0.0012
SCHOOL             4     1327.06667       331.76667       20.98      0.0003

Source             DF    Type III SS     Mean Square     F Value     Pr > F
YEAR               2      548.13333       274.06667       17.33      0.0012
SCHOOL             4     1327.06667       331.76667       20.98      0.0003
```

5. Using the accompanying computer output based on the data in Problem 8 in Chapter 17, conduct a randomized-blocks analysis, treating the days as blocks, to test whether the three persons have significantly different ESP ability. Does blocking on days seem appropriate?

Edited SAS Output (PROC GLM) for Problem 5

```
Dependent Variable: SCORE
                           Sum of           Mean
Source             DF      Squares          Square     F Value     Pr > F
Model              6     111.466667       18.577778       6.12     0.0113
Error              8      24.266667        3.033333
Corrected Total   14     135.733333

        R-Square              C.V.        Root MSE           SCORE Mean
        0.821218           9.231343       1.74165             18.8667

Source             DF     Type I SS      Mean Square     F Value     Pr > F
PERSON             2      91.7333333      45.8666667      15.12      0.0019
DAY                4      19.7333333       4.9333333       1.63      0.2585

Source             DF    Type III SS     Mean Square     F Value     Pr > F
PERSON             2      91.7333333      45.8666667      15.12      0.0019
DAY                4      19.7333333       4.9333333       1.63      0.2585
```

6. The promotional policies of four companies in a certain industry were compared to determine whether their rates for promoting blacks and whites differed. Data on the variable "rate discrepancy" were obtained for the four companies in each of three different two-year periods. This variable was defined as

$$d = \hat{p}_W - \hat{p}_B$$

where

$$\hat{p}_W = \frac{[\text{Number of whites promoted}](100)}{[\text{Number of whites eligible for promotion}]}$$

and

$$\hat{p}_B = \frac{[\text{Number of blacks promoted}](100)}{[\text{Number of blacks eligible for promotion}]}$$

The resulting data are reproduced in the following table.

		Company		
Period	1	2	3	4
1	3	5	5	4
2	4	4	3	5
3	8	12	10	9

a. Using dummy variables, create an appropriate regression model for this data set. Is this a randomized-blocks design?

b. Use the following table to test whether any significant differences exist among the average rate discrepancies for the four companies. Tukey's T for testing additivity equals 1.893 with 5 degrees of freedom.

Source	d.f.	SS	MS	F
Periods	2	84.5000	42.2500	33.800
Companies	3	6.0000	2.0000	1.600
Error	6	7.5000	1.2500	

c. Does Tukey's test for these data indicate that a removable interaction effect is present?

d. If no significant differences are found among the rate discrepancies for the four companies, would this support the contention that none of the companies has a discriminatory promotional policy?

e. Comment on the suitability of this analysis in view of the fact that the response variable is a difference in proportions.

7. Suppose that, in a study to compare body sizes of three genotypes of fourth-instar silkworm, the mean lengths (in millimeters) for separately reared cocoons of heterozygous (HET), homozygous (HOM), and wild (WLD) silkworms were determined at five laboratory sites, as detailed in the following table and in the accompanying computer output.[7]

---

[7]Adapted from a study by Sokal and Karlen (1964).

Variable	Site				
	1	2	3	4	5
HOM	29.87	28.16	32.08	30.84	29.44
HET	32.51	30.82	34.17	33.46	32.99
WLD	35.76	33.14	36.29	34.95	35.89

a. Assuming that this is a randomized-blocks experiment, what are the blocks and what are the treatments?
b. Why do you think a randomized-blocks analysis is appropriate (or inappropriate) for this experiment?
c. Carry out an appropriate analysis of the data for this experiment, and state your conclusions. Be sure to state the null hypothesis for each test performed.

Edited SAS Output (PROC GLM) for Problem 7

```
Dependent Variable: LENGTH

                           Sum of            Mean
Source              DF     Squares           Square      F Value      Pr > F
Model                6     84.7616800        14.1269467   46.97        0.0001
Error                8      2.4062933         0.3007867
Corrected Total     14     87.1679733

      R-Square             C.V.          Root MSE            LENGTH Mean
      0.972395           1.677632        0.54844                32.6913

Source              DF     Type I SS         Mean Square  F Value      Pr > F
VARIABLE             2     65.8139733        32.9069867   109.40       0.0001
SITE                 4     18.9477067         4.7369267    15.75       0.0007

Source              DF     Type III SS       Mean Square  F Value      Pr > F
VARIABLE             2     65.8139733        32.9069867   109.40       0.0001
SITE                 4     18.9477067         4.7369267    15.75       0.0007
```

8. In Problem 3, the independent variable of interest, copper ion concentration, is an interval-scale variable.
   a. Treating copper ion concentration as a five-level categorical predictor (as in the randomized-blocks analysis) corresponds to including a collection of polynomial terms for copper ion concentration (e.g., X, X^2,...) in the model. What is the highest order of such implicit terms?
   b. State, in terms of k levels of a predictor, a general principle based on the example in part (a).
   c. State the expansion of the regression model given in Problem 3(b) that encompasses the polynomial terms discussed in part (a).
   d. Use a computer program to fit the model stated in part (c). Center X by subtracting 0.330. (Why?) Provide estimated regression coefficients.
   e. Provide a multiple partial F test comparing the model in parts (b) and (g) of Problem 3 to the one fitted here; the latter is equivalent to the model fitted in part (d) of Problem 3.

9. Assume that the following table came from the analysis of a randomized-blocks design ANOVA.

Source	d.f.	SS	MS	F
Treatments	4	b	e	5.00
Blocks	a	c	48.00	6.00
Error	20	d	f	

a.–f. Complete the source table by determining the values for *a* through *f*. Show your work.
g. Provide tests of treatments and blocks with $\alpha = .05$. What are your conclusions?

10. An educator has obtained class average scores on a standardized test for classes from each grade from first through seventh at each of four schools in a city. These are shown in the following table. To assess differences between schools, the educator decides to conduct a randomized-blocks ANOVA, treating grade as a blocking factor.

Observation	School	Grade	Score
1	1	1	70.4
2	1	2	74.5
3	1	3	49.6
4	1	4	72.9
5	1	5	60.3
6	1	6	59.5
7	1	7	77.0
8	2	1	73.8
9	2	2	71.0
10	2	3	65.0
11	2	4	73.2
12	2	5	79.7
13	2	6	82.5
14	2	7	60.9
15	3	1	71.3
16	3	2	62.7
17	3	3	71.9
18	3	4	99.4
19	3	5	74.4
20	3	6	82.8
21	3	7	74.4
22	4	1	68.5
23	4	2	79.5
24	4	3	72.1
25	4	4	77.0
26	4	5	86.5
27	4	6	101.8
28	4	7	100.7

a. Conduct the appropriate analysis, using the accompanying computer output.
b. Provide *F* tests of school and grade differences, using $\alpha = .05$. What do you conclude?
c. Why may using the children's original scores be better than using means?

Edited SAS Output (PROC GLM) for Problem 10

```
Dependent Variable: SCORE
                             Sum of          Mean
Source              DF       Squares        Square      F Value      Pr > F
Model                9     2005.67821     222.85313        2.07      0.0905
Error               18     1939.26857     107.73714
Corrected Total     27     3944.94679

           R-Square            C.V.       Root MSE      SCORE Mean
           0.508417        13.88383        10.3797        74.7607

Source              DF     Type I SS     Mean Square    F Value      Pr > F
SCHOOL               3    1131.06393      377.02131        3.50      0.0370
GRADE                6     874.61429      145.76905        1.35      0.2857

Source              DF    Type III SS    Mean Square    F Value      Pr > F
SCHOOL               3    1131.06393      377.02131        3.50      0.0370
GRADE                6     874.61429      145.76905        1.35      0.2857
```

11. A company evaluated three new production-line technologies, one involving high automation, the second involving moderate automation, and the third involving low automation. All three technologies were intended to improve productivity through increased automation, but they continued to require operator intervention. Three groups were identified, each consisting of three line operators with similar years of production-line experience. Within each group, the operators were randomly assigned to one of the three new production-line technologies. The output (in units per hour) was recorded for each operator, and is shown in the accompanying table.

Operator Experience	Technology		
	High Automation (H)	Moderate Automation (M)	Low Automation (L)
<1 Year	10	8	5
1–2 Years	18	12	10
>2 Years	19	13	12

a. What are the blocks, and what are the treatments in this randomized-blocks design?
b. Complete the ANOVA table in the accompanying computer output for the data set given.
c. Do the data provide sufficient evidence to indicate that any significant difference exists in the outputs for the three technologies?
d. Using a confidence interval of the form

$$(\overline{Y}_{1.} - \overline{Y}_{2.}) \pm t_{v, 1-\alpha/2} \sqrt{MSE\left(\frac{1}{n_1} + \frac{1}{n_2}\right)}$$

where v represents the degrees of freedom, find a 99% confidence interval for the true difference in average output between the high and moderate automation lines.

Edited SAS Output (PROC GLM) for Problem 11

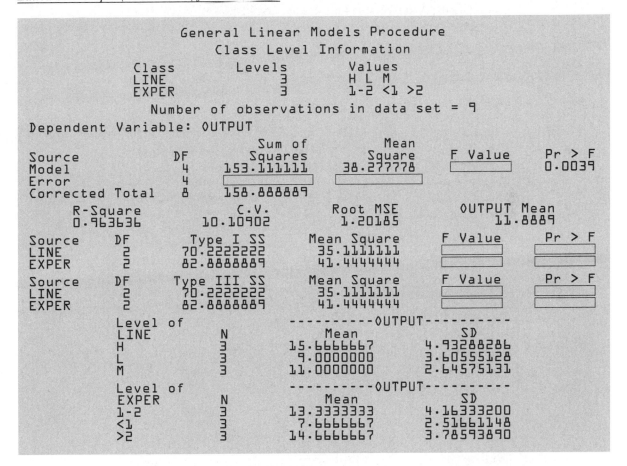

```
                    General Linear Models Procedure
                      Class Level Information
            Class       Levels        Values
            LINE          3           H L M
            EXPER         3           1-2 <1 >2
              Number of observations in data set = 9
Dependent Variable: OUTPUT
                              Sum of         Mean
Source              DF       Squares        Square      F Value    Pr > F
Model                4     153.111111     38.277778                0.0039
Error                4                  [            ]
Corrected Total      8     158.888889
        R-Square            C.V.        Root MSE       OUTPUT Mean
        0.963636         10.10902        1.20185         11.8889
Source       DF        Type I SS      Mean Square    F Value    Pr > F
LINE          2       70.2222222      35.1111111
EXPER         2       82.8888889      41.4444444
Source       DF      Type III SS     Mean Square    F Value    Pr > F
LINE          2       70.2222222      35.1111111
EXPER         2       82.8888889      41.4444444
        Level of                     ----------OUTPUT----------
        LINE         N                  Mean              SD
        H            3              15.6666667       4.93288286
        L            3               9.0000000       3.60555128
        M            3              11.0000000       2.64575131
        Level of                     ----------OUTPUT----------
        EXPER        N                  Mean              SD
        1-2          3              13.3333333       4.16333200
        <1           3               7.6666667       2.51661148
        >2           3              14.6666667       3.78593890
```

12. An advertising company evaluated three types of television ads for a new low-cost subcompact automobile: visual-appeal ads, budget-appeal ads, and feature-appeal ads. To control for age differences, the company randomly selected viewers from four age groups to evaluate the persuasiveness of the ads (measured on a scale ranging from 1 to 10, with 1 being the lowest level of persuasion, and 10 the highest). Within each age group, subgroups of three viewers were randomly assigned to view one or another of the three types of ads. The sample persuasion scores are reproduced in the following table and processed as shown in the accompanying computer output.

Viewer Age	Type of Ad		
	Visual Appeal (V)	Budget Appeal (B)	Feature Appeal (F)
18–25 Years	7	8	5
26–35 Years	6	9	4
36–45 Years	8	9	4
46 and Older	10	10	5

a. Identify the blocks and treatments in this randomized-blocks design.

b. Complete the ANOVA table for the data set given.

c. Do the data provide sufficient evidence to indicate a significant difference in the persuasion scores for the three types of ads?

d. Using a confidence interval of the form

$$(\overline{Y}_{i\cdot} - \overline{Y}_{j\cdot}) \pm t_{v,1-\alpha/2}\sqrt{\text{MSE}\left(\frac{1}{n_i} + \frac{1}{n_j}\right)}$$

where v is the degrees of freedom, find a 99% confidence interval for the true difference in average output between the visual-appeal and budget-appeal ads.

e. Repeat part (d) for visual-appeal and feature-appeal ads.

f. Repeat part (d) for budget-appeal and feature-appeal ads.

g. Based on your results from parts (b)–(f), what do you conclude about the average persuasion scores for the different types of ads?

Edited SAS Output (PROC GLM) for Problem 12

```
                    General Linear Models Procedure
                       Class Level Information
                 Class      Levels            Values
                  AD           3              B F V
                  AGE          4              18-25 26-35 36-45 >45
              Number of observations in data set = 12
Dependent Variable: SCORE
                             Sum of           Mean
Source            DF         Squares         Square    F Value      Pr > F
Model              5       50.0833333     10.0166667   [        ]    0.0040
Error              6
Corrected Total   11       54.9166667

      R-Square              C.V.          Root MSE          SCORE Mean
      0.911988           12.67098         0.89753            7.08333

Source            DF      Type I SS      Mean Square    F Value     Pr > F
AD                 2     43.1666667      21.5833333    [        ]   [       ]
AGE                3      6.9166667       2.3055556    [        ]   [       ]

Source            DF      Type III SS    Mean Square    F Value     Pr > F
AD                 2     43.1666667      21.5833333    [        ]   [       ]
AGE                3      6.9166667       2.3055556    [        ]   [       ]

       Level of                       -----------SCORE-----------
         AD              N            Mean                  SD
          B             4         9.00000000          0.81649658
          F             4         4.50000000          0.57735027
          V             4         7.75000000          1.70782513

       Level of                       -----------SCORE-----------
        AGE             N            Mean                  SD
        18-25           3         6.66666667          1.52752523
        26-35           3         6.33333333          2.51661148
        36-45           3         7.00000000          2.64575131
         >45            3         8.33333333          2.88675135
```

References

Armitage, P. 1971. *Statistical Methods in Medical Research.* Oxford: Blackwell Scientific.

Ostle, B. 1963. *Statistics in Research,* 2d ed. Ames, Iowa: Iowa State University Press.

Peng, K. C. 1967. *Design and Analysis of Scientific Experiments.* Reading, Mass.: Addison-Wesley.

Sokal, R. R., and Karlen, I. 1964. "Competition Among Genotypes in *Tibolium castaneum* at Varying Densities and Gene Frequencies (the Black Locus)." *Genetics* 49: 195–211.

Tukey, J. W. 1949. "One Degree of Freedom for Nonadditivity." *Biometrics* 5: 232.

Two-way ANOVA with Equal Cell Numbers

$$SSR_{ow} = \sum_{c}^{r} \left(\bar{y}_{i..} - \bar{y}_{...} \right)^2 \qquad F = \frac{df(r-1)}{df \cdot c(m-1)}$$

$$SSE = \sum_{ijk} \left(y_{ijk} - \bar{y}_{ij} \right)^2$$

19-1 Preview

In this chapter we shall consider the analysis of the simplest two-way data layout involving more than one observation per cell; the layout for which the number of observations in any given cell is at least two and is exactly the same as in any other cell, as previously illustrated in Figure 18-1(b). As we will see in this chapter, having equal cell numbers makes for a straightforward analysis requiring only slightly more involved calculations than those used for a randomized-blocks experiment. On the other hand, having unequal cell numbers leads to a much more complicated analysis (see Chapter 20).

Two-way layouts with equal cell numbers are rarely seen in observational studies, but they are often generated intentionally in experimental studies, where the investigator chooses the levels of the factors and the allocation of subjects to the various factor combinations. Such two-way layouts with equal cell numbers can be obtained in any of three ways:

1. *Blocking* whereby several (but equal numbers of) observations on each treatment occur in each block.

2. *Stratifying* according to the levels of the two factors of interest and then sampling within each stratum.

3. *Forming treatment combinations* (i.e., cells) and then allocating these combinations to individuals.

The design chosen depends, of course, on the study characteristics. For example, if we want to eliminate the effects of a confounding factor, we can use blocking. On the other hand, if we are interested in measuring the respiratory function of industrial workers in different plants (factor 1) subject to different levels of exposure to some substance (factor 2), a stratified sampling procedure is appropriate. Or if we are interested in the effects of combinations of different dosages of two different drugs, we may randomly assign the different drug combinations to different groups of subjects.

Regardless of the experimental design, having more than one observation at each combination of factor levels is essential. In the respiratory function example, if only one person subject to a certain level of exposure was examined from a given plant, we would have no direct way to determine how the responses of other persons in the same circumstances differed from the response of that individual. Similarly, for the drug example, if no more than one individual received a specific treatment combination of drugs, we could not assess the variation in response among persons receiving that same treatment combination. Thus, a major reason for having more than one observation at each combination of factor levels (i.e., in each cell) is to be able to compute a pure estimate of experimental error (σ^2).

Pure in this context means within-cell. Using a randomized-blocks design (with a single observation per cell) precludes the possibility of obtaining a within-cell estimate of σ^2. However, if the blocking does *only* what it is assumed to do (i.e., eliminate the effects of confounding factors), we can still obtain an estimate of σ^2 (although not a pure one) from a test that would ordinarily measure block–treatment interaction (which is assumed not to exist).

The detection of an interaction effect between two factors, although not of interest for the randomized-blocks design (in which the blocks are not considered as the levels of a factor or an independent variable), is an important reason to have multiple observations in each cell in two-way layouts. We will elaborate on this notion of interaction later in the chapter.

Example   In Table 19-1 the Cornell Medical Index (CMI) data of Table 17-3 from Daly's (1973) study has been categorized according to the levels of two factors:

Factor 1:   Percentage black in the surrounding neighborhood (PSN);
Level 1 = Low ($\leq 50\%$), Level 2 = High ($> 50\%$)

Factor 2:   Number of households in turnkey neighborhood (NHT);
Level 1 = Low (≤ 100), Level 2 = High (> 100)

TABLE 19-1   Cornell Medical Index scores for a sample of women from four turnkey neighborhoods.

Number of Households (NHT)	Percentage Black in Surrounding Neighborhoods (PSN)		Total
	Low ($\leq 50\%$)	High ($> 50\%$)	
Low (≤ 100)	Cherryview: 49, 12, 28, 24, 16, 28, 21, 48, 30, 18, 10, 10, 15, 7, 6, 11, 13, 17, 43, 18, 6, 10, 9, 12, 12 ($n_{11} = 25$, $\bar{Y}_{11.} = 18.92$)	Easton: 13, 10, 20, 20, 22, 14, 10, 8, 21, 35, 17, 23, 17, 23, 83, 21, 17, 41, 20, 25, 49, 41, 27, 37, 57 ($n_{12} = 25$, $\bar{Y}_{12.} = 26.84$)	$n_1. = 50$ $\bar{Y}_{1..} = 22.88$
High (> 100)	Northhills: 20, 31, 19, 9, 7, 16, 11, 17, 9, 14, 10, 5, 15, 19, 29, 23, 70, 25, 6, 62, 2, 14, 26, 7, 55 ($n_{21} = 25$, $\bar{Y}_{21.} = 20.84$)	Morningside: 5, 1, 44, 11, 4, 3, 14, 2, 13, 68, 34, 40, 36, 40, 22, 25, 14, 23, 26, 11, 20, 4, 16, 25, 17 ($n_{22} = 25$, $\bar{Y}_{22.} = 20.72$)	$n_2. = 50$ $\bar{Y}_{2..} = 20.78$
Total	$n_{.1} = 50$, $\bar{Y}_{.1.} = 19.88$	$n_{.2} = 50$, $\bar{Y}_{.2.} = 23.78$	$n_{..} = 100$ $\bar{Y}_{...} = 21.83$

Each of the four cells in Table 19-1 represents a combination of a level of factor 1 and a level of factor 2. The 25 observations in each cell constitute random samples of 25 women who were heads of household selected from the four turnkey neighborhoods, as defined by the stratification of the two factors NHT and PSN.

It may be argued that the categorization scheme in Table 19-1 is inappropriate, since the neighborhood with 98 households (Cherryview) almost qualifies as large ("high"). Thus, perhaps the cut point for dichotomizing the variable NHT should be less than 100. Such a decision is often subjective and is by no means easy to make. Recognizing this problem, we nevertheless have proceeded on the assumption that this categorization scheme is reasonable.

We have already seen that the F test for one-way ANOVA, which treats the four cells of Table 19-1 as the levels of a single variable "neighborhood," was nonsignificant for these data. Thus, we concluded that the mean CMI scores for the four neighborhoods, when compared simultaneously, do not significantly differ from one another.

Consequently, a researcher would probably not expect much from further analysis, especially if no additional variables were taken into account. On the other hand, if the one-way ANOVA F test had led to the conclusion that the four neighborhoods have significantly different mean CMI scores, the nature of these differences would have been of considerable interest. For example, do neighborhoods with a high percentage of blacks in the surrounding environs have significantly smaller mean CMI scores than those with a low percentage of blacks in surrounding environs? Do neighborhoods with a large number of households have significantly smaller mean CMI scores than neighborhoods with a small number of households? Do the amount and direction of the difference in CMI scores between neighborhoods of different size depend significantly on the racial makeup of the surrounding environs (i.e., is there an interaction effect)?

These questions can be answered by using two-way ANOVA. In fact, in spite of the nonsignificance of the one-way ANOVA F test, we will perform a two-way ANOVA to quantify the separate effects of the factors PSN and NHT and (even more important) to examine the possibility of an interaction between these two factors.

19-2 Using a Table of Cell Means

Our first step in examining a two-way layout should always be to construct a table of cell means. For our CMI data, a table of cell means is presented in Table 19-2. From this table, we can make three significant observations:

1. The mean CMI score for low NHT is larger than that for high NHT:

$$\hat{\mu}_{1.} - \hat{\mu}_{2.} = \overline{Y}_{1..} - \overline{Y}_{2..} = 22.88 - 20.78 = 2.10$$

This comparison measures the *main effect* of NHT.

TABLE 19-2   Cell means for CMI data.

NHT	PSN		Row Mean
	Low	High	
Low	$\hat{\mu}_{11} = \overline{Y}_{11.} = 18.92$	$\hat{\mu}_{12} = \overline{Y}_{12.} = 26.84$	$\hat{\mu}_{1.} = \overline{Y}_{1..} = 22.88$
High	$\hat{\mu}_{21} = \overline{Y}_{21.} = 20.84$	$\hat{\mu}_{22} = \overline{Y}_{22.} = 20.72$	$\hat{\mu}_{2.} = \overline{Y}_{2..} = 20.78$
Column Mean	$\hat{\mu}_{.1} = \overline{Y}_{.1.} = 19.88$	$\hat{\mu}_{.2} = \overline{Y}_{.2.} = 23.78$	$\hat{\mu}_{..} = \overline{Y}_{...} = 21.83$

2. The mean CMI score for low PSN is smaller than that for high PSN:

$$\hat{\mu}_{.1} - \hat{\mu}_{.2} = \overline{Y}_{.1.} - \overline{Y}_{.2.} = 19.88 - 23.78 = -3.90$$

This comparison measures the *main effect* of PSN.

3. There is little difference between high PSN and low PSN when NHT is high:

$$\hat{\mu}_{22} - \hat{\mu}_{21} = \overline{Y}_{22.} - \overline{Y}_{21.} = 20.72 - 20.84 = -0.12$$

but there is considerable difference between high PSN and low PSN when NHT is low:

$$\hat{\mu}_{12} - \hat{\mu}_{11} = \overline{Y}_{12.} - \overline{Y}_{11.} = 26.84 - 18.92 = 7.92$$

These two comparisons measure the *interaction* between NHT and PSN.

Observation 1 suggests that persons from small turnkey neighborhoods might not be as healthy as persons from large turnkey neighborhoods (the lower the CMI score, the healthier), which is consistent with what Daly theorized. Observation 2 suggests that persons from turnkey neighborhoods containing a high percentage of blacks in the surrounding neighborhood might not be as healthy as persons from turnkey neighborhoods with a low percentage of blacks in the surrounding neighborhood. This observation runs counter to Daly's theory. Finally, observation 3 suggests that there is little difference between neighborhoods containing high and low black percentages in the surroundings when the turnkey neighborhood size is large, whereas there is considerable difference between neighborhoods with high and low black percentages in the surroundings when the size of the turnkey neighborhood is small.

Another way of describing the interaction effect pointed out in observation 3 is to say that, when PSN is low, persons from neighborhoods with low NHT seem to be healthier than persons from neighborhoods with high NHT; but that, when PSN is high, persons from neighborhoods with high NHT seem to be healthier than persons from neighborhoods with low NHT.

Clearly the most important of these observations is observation 3, which suggests some kind of interaction between NHT and PSN; that is, any difference in the health of persons from different PSN categories seems to depend on what NHT category is being considered. Equivalently, any difference between persons from different NHT categories appears to depend on the PSN category.

Nevertheless, we must remember that the differences found in this study were obtained from a sample, so the differences could have occurred solely by chance. In other words, we must determine whether the differences found are statistically significant. This can be done by using two-way ANOVA.

19-2-1 Fixed, Random, or Mixed Model

To determine the appropriate significance tests for a two-way ANOVA, we first have to specify whether each of the two factors is fixed or random. Although such a specification in the one-way ANOVA situation alters only the statement of the null and alternative hypotheses (and not the form of the F test used), the way the factors are classified in the two-way case affects the F test as well.

In fact, three different cases must be considered, depending on the classification of the two factors:

1. The fixed-effects case, where both factors are fixed.

2. The random-effects case, where both factors are random.

3. The mixed-effects case, where one factor is fixed and the other is random.

In our example (i.e., Table 19-1), the classification of the factors NHT and PSN depends on the researcher's point of view. The fixed-effects case would apply if the researcher were interested only in the particular turnkey neighborhoods selected for study or (in terms of the two factors NHT and PSN) did not wish to make inferences to neighborhoods of different sizes or to different black percentages for surrounding neighborhoods. The random-effects case would apply if the turnkey neighborhood sizes chosen were considered representative of a larger population of sizes of interest and if the black percentages were considered representative of a larger population of black percentages of interest. The mixed-effects case would be applicable if one of the factors were considered fixed and the other random. Of these three cases, the random-effects case probably best represents the true situation. Nevertheless, we will discuss the appropriate analysis for each case.

19-2-2 Two-way ANOVA Results for the Data of Table 19-1

The computer output that follows gives the two-way ANOVA results for the CMI data of Table 19-1. The four sources of variation shown in the output correspond to the two main effects (for NHT and PSN), the interaction effect (NHT $\times$ PSN), and the error variation. Corresponding to these four sources are three null hypotheses that may be tested:

1. H_0: No main effect of NHT.

2. H_0: No main effect of PSN.

3. H_0: No interaction effect between NHT and PSN.

Edited SAS Output (PROC GLM) for the Data of Table 19-1

```
                    Class Level Information
                  Class    Levels      Values
                  NHT         2          1 2
                  PSN         2          1 2
            Number of observations in data set = 100
Dependent Variable: CMI
                              Sum of           Mean
Source              DF       Squares         Square    F Value    Pr > F
Model                3    894.510000    298.170000       1.18    0.3223
Error               96  24301.600000    253.141667
Corrected Total     99  25196.110000
        R-Square            C.V.        Root MSE            CMI Mean
        0.035502        72.88331        15.9104            21.8300
Source        DF     Type I SS    Mean Square    F Value    Pr > F
NHT            1    110.250000    110.250000       0.44    0.5109
PSN            1    380.250000    380.250000       1.50    0.2233
NHT*PSN        1    404.010000    404.010000       1.60    0.2095
Source        DF   Type III SS    Mean Square    F Value    Pr > F
NHT            1    110.250000    110.250000       0.44    0.5109
PSN            1    380.250000    380.250000       1.50    0.2233
NHT*PSN        1    404.010000    404.010000       1.60    0.2095
Source          Type III Expected Mean Square
NHT             Var(Error) + 25 Var(NHT*PSN) + 50 Var(NHT)
PSN             Var(Error) + 25 Var(NHT*PSN) + 50 Var(PSN)
NHT*PSN         Var(Error) + 25 Var(NHT*PSN)
```

For random-effects model ←

Fixed-effects tests

```
                  Class Level Information (Continued)
                                              Random- and
Source: NHT                                   mixed-effects tests
Error: MS(NHT*PSN)
                                                  ↓
        Type III    Denominator   Denominator
DF        MS            DF             MS        F Value      Pr > F
1       110.25          1           404.01      0.2729       0.6935

Source: PSN
Error: MS(NHT*PSN)

        Type III    Denominator   Denominator
DF        MS            DF             MS        F Value      Pr > F
1       380.25          1           404.01      0.9412       0.5096

Source: NHT*PSN
Error: MS(Error)

        Type III    Denominator   Denominator
DF        MS            DF             MS        F Value      Pr > F
1       404.01          96        253.14166667  1.5960       0.2095
```

Each of these hypotheses can be stated more precisely in terms of population cell means and/or variances, depending on whether the fixed-, random-, or mixed-effects case applies. For example, in the fixed-effects case, the null hypotheses may be given in terms of cell means (see Table 19-2) as follows:

1. H_0: $\mu_{1.} = \mu_{2.}$ (no main effect of NHT).

2. H_0: $\mu_{.1} = \mu_{.2}$ (no main effect of PSN).

3. H_0: $\mu_{22} - \mu_{21} - \mu_{12} + \mu_{11} = 0$ (no interaction effect between NHT and PSN).

More will be said in sections 19-4 and 19-7 about testing null hypotheses in the fixed-, random-, and mixed-effects cases. We shall describe in section 19-3 how the sum-of-squares and degrees-of-freedom terms are determined for the general two-way ANOVA case. At present, we focus entirely on the F statistics given in the output, which differ according to how the factors are classified. The two numbers in parentheses next to each F statistic below indicate the appropriate degrees of freedom to be used for that F test. None of the tests turns out to be significant, as we might have expected from the previous results for one-way ANOVA.

Tests for Fixed Effects

F tests for the fixed-effects case *always* involve dividing the mean square for the particular source being considered by the error mean square. The degrees of freedom correspond to the particular mean squares used. Thus,

$$F(\text{NHT}) = \frac{\text{MS(NHT)}}{\text{MS(Error)}} = \frac{110.25}{253.14} = 0.44_{(1,96)}$$

$$F(\text{PSN}) = \frac{\text{MS(PSN)}}{\text{MS(Error)}} = \frac{380.25}{253.14} = 1.50_{(1,96)}$$

$$F(\text{NHT} \times \text{PSN}) = \frac{\text{MS(NHT} \times \text{PSN)}}{\text{MS(Error)}} = \frac{404.01}{253.14} = 1.60_{(1,96)}$$

Tests for Random or Mixed Effects[1]

In either the random- or mixed-effects case, the F test for each main effect consists of dividing the mean square for the particular main effect under consideration by the interaction mean square. Again, the degrees of freedom correspond to the particular mean squares used. Thus,

$$F(\text{NHT}) = \frac{\text{MS(NHT)}}{\text{MS(NHT} \times \text{PSN)}} = \frac{110.25}{404.01} = 0.27_{(1,1)}$$

$$F(\text{PSN}) = \frac{\text{MS(PSN)}}{\text{MS(NHT} \times \text{PSN)}} = \frac{380.25}{404.01} = 0.94_{(1,1)}$$

The F test for interaction is the same for the random- and mixed-effects cases as for the fixed-effects case.

19-3 General Methodology

In this section we describe the data configuration, computational formulas, and ANOVA table for the general balanced (i.e., having equal cell numbers) two-way situation consisting of r levels of one factor (which we call the *row factor*), c levels of the other factor (which we call the *column factor*), and n observations in each of the rc cells.

19-3-1 General Layout of Data for Two-way ANOVA

Table 19-3 offers a convenient format for presenting the data for the general two-way situation when all cells contain an equal number of observations. Table 19-4 gives the corresponding table of (sample) cell means.

Table 19-3 uses three subscripts to differentiate the individual observations. The first two subscripts index the row and column (i.e., the cell), and the third subscript denotes the observa-

TABLE 19-3 General layout of data for two-way ANOVA with equal numbers of observations per cell.

Row Factor	Column Factor 1	Column Factor 2	$\cdots$	Column Factor c	Row Total
1	$(Y_{111}, Y_{112}, \ldots, Y_{11n})$ T_{11}	$(Y_{121}, Y_{122}, \ldots, Y_{12n})$ T_{12}	$\cdots$	$(Y_{1c1}, Y_{1c2}, \ldots, Y_{1cn})$ T_{1c}	R_1
2	$(Y_{211}, Y_{212}, \ldots, Y_{21n})$ T_{21}	$(Y_{221}, Y_{222}, \ldots, Y_{22n})$ T_{22}	$\cdots$	$(Y_{2c1}, Y_{2c2}, \ldots, Y_{2cn})$ T_{2c}	R_2
$\vdots$	$\vdots$	$\vdots$	$\vdots$	$\vdots$	$\vdots$
r	$(Y_{r11}, Y_{r12}, \ldots, Y_{r1n})$ T_{r1}	$(Y_{r21}, Y_{r22}, \ldots, Y_{r2n})$ T_{r2}	$\cdots$	$(Y_{rc1}, Y_{rc2}, \ldots, Y_{rcn})$ T_{rc}	R_r
Column Total	C_1	C_2	$\cdots$	C_c	G

---

[1]See section 19-7 (and footnote 8 on page 542) for a discussion of the rationale for these F tests in terms of "expected mean squares."

TABLE 19-4   Sample cell means for two-way ANOVA.

Row Factor	Column Factor				Row Mean
	1	2	...	c	
1	$\overline{Y}_{11\cdot}$	$\overline{Y}_{12\cdot}$	...	$\overline{Y}_{1c\cdot}$	$\overline{Y}_{1\cdot\cdot}$
2	$\overline{Y}_{21\cdot}$	$\overline{Y}_{22\cdot}$	...	$\overline{Y}_{2c\cdot}$	$\overline{Y}_{2\cdot\cdot}$
$\vdots$	$\vdots$	$\vdots$		$\vdots$	$\vdots$
r	$\overline{Y}_{r1\cdot}$	$\overline{Y}_{r2\cdot}$	...	$\overline{Y}_{rc\cdot}$	$\overline{Y}_{r\cdot\cdot}$
Column Mean	$\overline{Y}_{\cdot1\cdot}$	$\overline{Y}_{\cdot2\cdot}$	...	$\overline{Y}_{\cdot c\cdot}$	$\overline{Y}_{\cdots}$

tion number within that cell. For example, Y_{122} denotes the second observation in cell $(1, 2)$, which corresponds to row 1 and column 2. In general, Y_{ijk} denotes the kth observation in the (i, j)th cell of the table. The cell total for the (i, j)th cell is denoted by T_{ij}; the ith row total is R_i; the jth column total is C_j; and the grand total is G. In other words,

$$R_i = \sum_{j=1}^{c} \sum_{k=1}^{n} Y_{ijk}, \qquad C_j = \sum_{i=1}^{r} \sum_{k=1}^{n} Y_{ijk}, \qquad G = \sum_{i=1}^{r} \sum_{j=1}^{c} \sum_{k=1}^{n} Y_{ijk}$$

In Table 19-4, the mean of the n observations in cell (i, j) is denoted by $\overline{Y}_{ij\cdot}$; this sample mean estimates the population cell mean μ_{ij}. The ith row mean is $\overline{Y}_{i\cdot\cdot}$; the jth column mean is $\overline{Y}_{\cdot j\cdot}$; and $\overline{Y}_{\cdots}$ is the grand (overall) mean. Thus, we have

$$\overline{Y}_{ij\cdot} = \frac{1}{n} \sum_{k=1}^{n} Y_{ijk}, \qquad \overline{Y}_{i\cdot\cdot} = \frac{R_i}{cn}, \qquad \overline{Y}_{\cdot j\cdot} = \frac{C_j}{rn}, \qquad \overline{Y}_{\cdots} = \frac{G}{rcn}$$

■ **Example**   In our earlier example (Table 19-1), $r = c = 2$ and $n = 25$. Tables 19-5 and 19-6 give the data layout and the table of cell means for an example in which $r = 3$, $c = 3$, and $n = 12$. This example (although artificial) deals with one kind of data set used in occupational health studies for evaluating the health status of industrial workers. The dependent variable here is forced expiratory volume (FEV), which is a measure of respiratory function. Very low FEV indicates possible respiratory dysfunction, whereas high FEV indicates good respiratory function. In this example, observations are taken on $n = 12$ individuals in each of three plants in a given industry where workers are exposed primarily to one of three toxic substances. Thus, we have two factors, each with three levels. The categories of the row factor (Plant) are labeled 1, 2, and 3 in Table 19-5. The categories of the column factor (Toxsub) are labeled A, B, and C. Three questions are of particular interest here:

1. Does the mean FEV level differ among plants (i.e., is there a main effect due to Plant)?

2. Does the mean FEV level differ among exposure categories (i.e., is there a main effect due to Toxsub)?

3. Do the differences in mean FEV levels among plants depend on the exposure category, and vice versa (i.e., is there an interaction effect between Plant and Toxsub)?

We can evaluate these questions preliminarily by examining the cell means in Table 19-6. Of the three plants, plant 1 has the lowest mean FEV (3.70), followed by plant 2 (3.90), and then plant 3 (4.13). This suggests that the workers in plant 1 have poorer respiratory health than those in plant 2,

TABLE 19-5 Forced expiratory volume classified by plant and toxic exposure.

Plant	Toxic Substance A	B	C	Row Total
1	4.64, 5.92, 5.25, 6.17, 4.20, 5.90, 5.07, 4.13, 4.07, 5.30, 4.37, 3.76 ($T_{11} = 58.78$)	3.21, 3.17, 3.88, 3.50, 2.47, 4.12, 3.51, 3.85, 4.22, 3.07, 3.62, 2.95 ($T_{12} = 41.57$)	3.75, 2.50, 2.65, 2.84, 3.09, 2.90, 2.62, 2.75, 3.10, 1.99, 2.42, 2.37 ($T_{13} = 32.98$)	$R_1 = 133.33$
2	5.12, 6.10, 4.85, 4.72, 5.36, 5.41, 5.31, 4.78, 5.08, 4.97, 5.85, 5.26 ($T_{21} = 62.81$)	3.92, 3.75, 4.01, 4.64, 3.63, 3.46, 4.01, 3.39, 3.78, 3.51, 3.19, 4.04 ($T_{22} = 45.33$)	2.95, 3.21, 3.15, 3.25, 2.30, 2.76, 3.01, 2.31, 2.50, 2.02, 2.64, 2.27 ($T_{23} = 32.37$)	$R_2 = 140.51$
3	4.64, 4.32, 4.13, 5.17, 3.77, 3.85, 4.12, 5.07, 3.25, 3.49, 3.65, 4.10 ($T_{31} = 49.56$)	4.95, 5.22, 5.16, 5.35, 4.35, 4.89, 5.61, 4.98, 5.77, 5.23, 4.86, 5.15 ($T_{32} = 61.52$)	2.95, 2.80, 3.63, 3.85, 2.19, 3.32, 2.68, 3.35, 3.12, 4.11, 2.90, 2.75 ($T_{33} = 37.65$)	$R_3 = 148.73$
Column Total	$C_1 = 171.15$	$C_2 = 148.42$	$C_3 = 103.00$	$G = 422.57$

TABLE 19-6 Cell means for data of Table 19-6.

Plant	Toxic Substance A	B	C	Row Mean
1	4.90	3.46	2.75	3.70
2	5.23	3.78	2.70	3.90
3	4.13	5.13	3.14	4.13
Column Mean	4.75	4.12	2.86	3.91

and so on. Nevertheless, if the 3.70 value for plant 1 is considered clinically normal, then—despite the differences observed—all plants will be given a "clean bill of health." Furthermore, these differences might have occurred solely by chance (i.e., might not be statistically significant).

In addition, it can be seen from Table 19-6 that exposure to substance C is associated with the poorest respiratory health (2.86), whereas exposure to substance A (4.75) and exposure to substance B (4.12) are associated with considerably better respiratory health. Again, we must decide whether to view the 2.86 value as meaningfully low and determine whether the differences among substances A, B, and C are statistically significant.

Finally, Table 19-6 indicates that the differences among plants depend somewhat on the toxic exposure being considered. For example, with respect to toxic substance A, plant 3 has the lowest mean FEV (4.13). With respect to toxic substance B, however, plant 1 has the lowest mean (3.46); and with respect to toxic substance C, plant 2 has the lowest (2.70). Furthermore, the magnitude of the differences among plants varies with the toxic substance. For toxic substance B, the difference between the highest and lowest plant means is $5.13 - 3.46 = 1.67$, whereas for toxic substances A and C the maximum differences are smaller ($5.23 - 4.13 = 1.10$ and $3.14 - 2.70 = 0.44$, respectively). Such fluctuations suggest that a significant interaction effect may be present, although this must be verified statistically.

19-3-2 ANOVA Table for Two-way ANOVA

Table 19-7 gives the general form of the two-way ANOVA table with r levels of the row factor and c levels of the column factor. The computer output given after that shows the corresponding ANOVA results associated with the FEV data of Table 19-5. From Table 19-7 we see that the total (corrected) sum of squares (TSS) has been divided into the four components—SSR (rows), SSC (columns), SSRC ($r \times c$ interaction), and SSE (error)—based on the following fundamental equation:

$$TSS = SSR + SSC + SSRC + SSE, \tag{19.1}$$

or equivalently

$$\sum_{i=1}^{r}\sum_{j=1}^{c}\sum_{k=1}^{n}(Y_{ijk} - \overline{Y}...)^2 = cn\sum_{i=1}^{r}(\overline{Y}_{i..} - \overline{Y}...)^2 + rn\sum_{j=1}^{n}(\overline{Y}_{.j.} - \overline{Y}...)^2$$

$$+ n\sum_{i=1}^{r}\sum_{j=1}^{c}(\overline{Y}_{ij.} - \overline{Y}_{i..} - \overline{Y}_{.j.} + \overline{Y}...)^2 + \sum_{i=1}^{r}\sum_{j=1}^{c}\sum_{k=1}^{n}(Y_{ijk} - \overline{Y}_{ij.})^2$$

TABLE 19-7   General (balanced) two-way ANOVA table.[2]

Source	d.f.	SS	MS	F Fixed	F Mixed or Random
Row (main effect)	$r - 1$	SSR	$MSR = \dfrac{SSR}{r-1}$	$\dfrac{MSR}{MSE}$	$\dfrac{MSR}{MSRC}$
Column (main effect)	$c - 1$	SSC	$MSC = \dfrac{SSC}{c-1}$	$\dfrac{MSC}{MSE}$	$\dfrac{MSC}{MSRC}$
Row $\times$ column (interaction)	$(r-1)(c-1)$	SSRC	$MSRC = \dfrac{SSRC}{(r-1)(c-1)}$	$\dfrac{MSRC}{MSE}$	$\dfrac{MSRC}{MSE}$
Error	$rc(n-1)$	SSE	$MSE = \dfrac{SSE}{rc(n-1)}$		
Total	$rcn - 1$	TSS			

[2]See section 19-7 for a discussion of the rationale for these F tests in terms of "expected mean squares."

Edited SAS Output (PROC GLM) for the Data of Table 19-5

```
                       Class Level Information
                  Class    Levels    Values
                  PLANT       3       1 2 3
                  SUBSTNCE    3       A B C
            Number of observations in data set = 108
Dependent Variable: FEV
                              Sum of         Mean
Source            DF          Squares        Square    F Value    Pr > F
Model              8        94.6984630    11.8373079     44.10    0.0001
Error             99        26.5758583     0.2684430
Corrected Total  107       121.2743213
        R-Square             C.V.        Root MSE           FEV Mean
        0.780862          13.24193        0.51811           3.91269
Source           DF       Type I SS      Mean Square   F Value    Pr > F
PLANT             2       3.2988963       1.6494481       6.14    0.0031
SUBSTNCE          2      66.8893685      33.4446843     124.59    0.0001
PLANT*SUBSTNCE    4      24.5101981       6.1275495      22.83    0.0001
Source           DF      Type III SS     Mean Square   F Value    Pr > F
PLANT             2       3.2988963       1.6494481       6.14    0.0031
SUBSTNCE          2      66.8893685      33.4446843     124.59    0.0001
PLANT*SUBSTNCE    4      24.5101981       6.1275495      22.83    0.0001
Source          Type III Expected Mean Square
PLANT           Var(Error)+12 Var(PLANT*SUBSTNCE)+36 Var(PLANT)
SUBSTNCE        Var(Error)+12 Var(PLANT*SUBSTNCE)+36 Var(SUBSTNCE)
PLANT*SUBSTNCE  Var(Error)+12 Var(PLANT*SUBSTNCE)
Source: PLANT
Error: MS(PLANT*SUBSTNCE)
                 Denominator    Denominator
  DF    Type III MS      DF         MS         F Value    Pr > F
   2  1.6494481481        4    6.127549537     0.2692     0.7768
Source: SUBSTNCE
Error: MS(PLANT*SUBSTNCE)
                 Denominator    Denominator
  DF    Type III MS      DF         MS         F Value    Pr > F
   2  33.444684259        4    6.127549537     5.4581     0.0719
Source: PLANT*SUBSTNCE
Error: MS(Error)
                 Denominator    Denominator
  DF    Type III MS      DF         MS         F Value    Pr > F
   4  6.127549537        99    0.2684430135   22.8263     0.0001
```

Fixed-effects tests

For random-effects model

Random- and mixed-effects tests

Since the sums of squares for each of these sources are routinely calculated by computer programs for ANOVA (e.g., SAS's GLM procedure), we do not provide specific computation formulas in the SS column of Table 19-7.

For the FEV data of Table 19-5, the sums of squares obtained (see the above SAS output) are:

SSR(Plant) = 3.299

SSC(Toxsub) = 66.889

$$SSRC(\text{Plant} \times \text{Toxsub}) = 24.510$$
$$SSE = 26.576$$
$$TSS = 121.274$$

The degrees of freedom associated with these sums of squares are

SSR has $r - 1$ d.f.

SSC has $c - 1$ d.f.

SSRC has $(r - 1)(c - 1)$ d.f.   *(19.2)*

SSE has $rc(n - 1)$ d.f.

TSS has $rcn - 1$ d.f.

Each mean-square term is obtained (as usual) by dividing the corresponding sum of squares by its associated degrees of freedom. Again, the appropriate *F* statistics to use depend on whether the row and column factors are classified as fixed or as random. These *F* tests are described in section 19-4.

19-4 *F* Tests for Two-way ANOVA

The null hypotheses of interest for two-way ANOVA—as well as the basic statistical assumptions required for validly testing them—can be stated quite generally to encompass the four possible schemes of factor classification.[3]

Null Hypotheses

1. $H_0(R)$: *There is no row factor (main) effect* (i.e., there are no differences among the effects of the levels of the row factor).

2. $H_0(C)$: *There is no column factor (main) effect* (i.e., there are no differences among the effects of the levels of the column factor).

3. $H_0(RC)$: *There is no interaction effect between rows and columns* (i.e., the row level effects within any one column are the same as those within any other column; and the column level effects within any one row are the same as those within any other row).

Assumptions

1. All observations are *statistically independent* of one another for fixed-effects models.[4]

2. Each observation comes from a *normally distributed population*.

3. Each observation has the same population variance (i.e., the usual assumption of *variance homogeneity* applies).

---

[3]However, the null hypotheses are quite different when stated more precisely in terms of population cell means and/or variances.

[4]The responses are *not* mutually independent when random effects are included in the regression model for two-way ANOVA with equal cell numbers.

As previously stated, the appropriate F statistics depend on whether the row and column factors are classified as fixed or random. The mean-square term associated with a given source of variation will estimate different quantities, depending on whether the row and column factors are fixed or random; this explains why different denominators are used in the various F tests of two-way ANOVA.

Regardless of how the factors are classified, however, the F statistic used to test $H_0(RC)$ of no row–column interaction always has the form

$$F(RC) = \frac{MSRC}{MSE}$$

with $(r - 1)(c - 1)$ and $rc(n - 1)$ degrees of freedom.[5]

The tests for main effects differ as follows with respect to the factor classification scheme:

1. *Rows and columns fixed.* Divide the mean squares for rows and for columns by the mean square for error:

$$F(R) = \frac{MSR}{MSE}$$

with $r - 1$ and $rc(n - 1)$ degrees of freedom, or

$$F(C) = \frac{MSC}{MSE}$$

with $c - 1$ and $rc(n - 1)$ degrees of freedom.

2. *Rows and columns random, or one factor fixed and the other factor random.* Divide the mean squares for rows and columns by the mean square for interaction:

$$F(R) = \frac{MSR}{MSRC}$$

with $r - 1$ and $(r - 1)(c - 1)$ degrees of freedom, or

$$F(C) = \frac{MSC}{MSRC}$$

with $c - 1$ and $(r - 1)(c - 1)$ degrees of freedom.

For the FEV data, the classification of the factors depends on the point of view of the researcher. If, for example, the plants and toxic substances were the only ones of interest, both factors would be considered fixed. However, if the plants were considered to represent a sample from a large population of plants of interest and if the toxic substances likewise were viewed as representing a population of toxic agents of interest, both factors would be treated as random. Of course, the classification would be mixed if one of these factors were considered fixed and the other random.

We will not argue for any particular classification scheme for the factors in the FEV example, since the example is artificial. The decisions regarding certain null hypotheses differ, how-

---

[5]$F(RC)$ denotes the F test of $H_0(RC)$. Similarly, $F(R)$ and $F(C)$ denote the F tests of $H_0(R)$ and $H_0(C)$, respectively.

ever, depending on how the factors are classified. This can be seen from the earlier SAS output, as follows:

1. *Both factors fixed.* Both main effects are significant, since $F(\text{Plant}) = 6.14^{**}$ (with 2 and 99 degrees of freedom) and $F(\text{Toxsub}) = 124.60^{**}$ (with 2 and 99 degrees of freedom).

2. *Both factors random, or one factor fixed and the other factor random.* Neither main effect is significant, since $F(\text{Plant}) = 0.27$ (with 2 and 4 degrees of freedom) and $F(\text{Toxsub}) = 5.46$ (with 2 and 4 degrees of freedom).

Despite these differences among the main-effect test results, the most important finding from this analysis is that the interaction effect is significant:

$$F(\text{Plant} \times \text{Toxsub}) = 22.83^{**} \text{ (with 4 and 99 d.f.)}$$

In section 19-6 we discuss the interpretation of such interaction effects. Certainly, it does not make much sense to talk about the separate or independent effects (i.e., main effects) of Plant and Toxsub on FEV, since there is strong evidence that these factors do not affect FEV independently of one another.

Regardless of whether the factors are fixed or random, eight distinct patterns of significance may result, depending on the significance (or nonsignificance) of the three tests involved (two main-effect tests and the interaction test). Suppose that one factor is labeled A and the second factor B. Then Table 19-8 summarizes the possible outcomes (with significant results indicated by asterisks). In pattern I, the *F* tests for the A effect, the B effect, and the A × B interaction are all nonsignificant. In pattern VIII, all three are significant.

Each pattern deserves some comment. Pattern I leads us to conclude that there is no evidence of any relationship between the response variable and the predictors (considered either separately or together). Pattern II implies that only factor A is related to the response variable. Similarly, Pattern III implies that only factor B is related to the response variable. In Pattern IV, we conclude that both A and B affect the response. Furthermore, the nature of the relationship between factor A and the response remains the same at all levels of factor B studied, and conversely. Patterns V, VI, VII, and VIII all involve significant interaction. Most statisticians recommend discussing only this interaction in such cases. A significant interaction implies that both factors are important and that the level of the other factor must be known to characterize the effect of one designated factor on the response. Section 19-6 discusses interpreting interactions.

TABLE 19-8 Patterns of significance possible in two-way ANOVA with equal cell numbers.

Source	Pattern							
	I	II	III	IV	V	VI	VII	VIII
A		*		*		*		*
B			*	*			*	*
A × B					*	*	*	*

Asterisk denotes significance for the effect in that row.

19-5 Regression Model for Fixed-effects Two-way ANOVA

In this section we describe a particular regression model[6] and a related classical fixed-effects ANOVA model for two-way ANOVA when both factors are considered fixed. Random-effects and mixed-effects models are discussed in detail in section 19-7. As in the cases of one-way ANOVA and randomized-blocks ANOVA, a regression model for two-way ANOVA can be interpreted in terms of the cell, marginal, and overall means associated with the two-way layout (see Table 19-9). In the table,

$$\mu_{i\cdot} = \frac{1}{c} \sum_{j=1}^{c} \mu_{ij} \qquad i = 1, 2, \ldots, r$$

$$\mu_{\cdot j} = \frac{1}{r} \sum_{i=1}^{r} \mu_{ij} \qquad j = 1, 2, \ldots, c$$

$$\mu_{\cdot\cdot} = \frac{1}{rc} \sum_{i=1}^{r} \sum_{j=1}^{c} \mu_{ij}$$

19-5-1 Regression Model

When there are r rows and c columns, a regression model can be formulated involving $r - 1$ dummy variables for the row factor, $c - 1$ dummy variables for the column factor, and $(r - 1)(c - 1)$ interaction dummy variables, which are constructed by forming products of each of the row dummy variables with each of the column dummy variables. Such a model can be expressed as

$$Y = \mu + \sum_{i=1}^{r-1} \alpha_i X_i + \sum_{j=1}^{c-1} \beta_j Z_j + \sum_{i=1}^{r-1} \sum_{j=1}^{c-1} \gamma_{ij} X_i Z_j + E \tag{19.3}$$

TABLE 19-9 Table of population cell means for two-way layout.

Row	Column 1	2	...	c	Row Mean
1	μ_{11}	μ_{12}	...	μ_{1c}	$\mu_{1\cdot}$
2	μ_{21}	μ_{22}	...	μ_{2c}	$\mu_{2\cdot}$
⋮	⋮	⋮		⋮	⋮
r	μ_{r1}	μ_{r2}	...	μ_{rc}	$\mu_{r\cdot}$
Column Mean	$\mu_{\cdot 1}$	$\mu_{\cdot 2}$	...	$\mu_{\cdot c}$	$\mu_{\cdot\cdot}$

---

[6]Several other regression models can be defined, of course, depending on the coding scheme for the dummy variables. The regression model given here is the one most commonly used, because of its natural connection with the classical fixed-effects two-way ANOVA model.

in which

$$X_i = \begin{cases} -1 & \text{for level } r \text{ of the row factor} \\ 1 & \text{for level } i \text{ of the row factor} \\ 0 & \text{otherwise} \end{cases}$$

and

$$Z_j = \begin{cases} -1 & \text{for level } c \text{ of the column factor} \\ 1 & \text{for level } j \text{ of the column factor} \\ 0 & \text{otherwise} \end{cases}$$

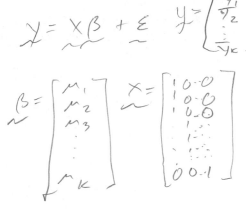

(for $i = 1, 2, \ldots, r-1$; and $j = 1, 2, \ldots, c-1$).

The formulas relating the coefficients α_i, β_j, and γ_{ij} to the various means of Table 19-9 are

$$\mu = \mu_{..}$$

$$\alpha_i = \mu_{i.} - \mu_{..} \qquad\qquad i = 1, 2, \ldots, r-1$$

$$\beta_j = \mu_{.j} - \mu_{..} \qquad\qquad j = 1, 2, \ldots, c-1$$

$$\gamma_{ij} = \mu_{ij} - \mu_{i.} - \mu_{.j} + \mu_{..} \qquad i = 1, 2, \ldots, r-1; j = 1, 2, \ldots, c-1$$

$$-\sum_{i=1}^{r-1} \alpha_i = \mu_{r.} - \mu_{..}$$

$$-\sum_{j=1}^{c-1} \beta_j = \mu_{.c} - \mu_{..} \tag{19.4}$$

$$-\sum_{i=1}^{r-1} \gamma_{ij} = \mu_{rj} - \mu_{r.} - \mu_{.j} + \mu_{..} \qquad j = 1, 2, \ldots, c-1$$

$$-\sum_{j=1}^{c-1} \gamma_{ij} = \mu_{ic} - \mu_{i.} - \mu_{.c} + \mu_{..} \qquad i = 1, 2, \ldots, r-1$$

As with the ANOVA regression analogies made in earlier chapters, the same F tests given in Table 19-7 (for the case where both factors are fixed) can be obtained by using the appropriate multiple partial F tests concerning subsets of the coefficients in the regression model (19.3). Specifically, the multiple partial F test of $H_0: \alpha_1 = \alpha_2 = \cdots = \alpha_{r-1} = 0$ for model (19.3) yields exactly the same F statistic as that used in standard (balanced) two-way fixed-effects ANOVA for testing the significance of the main effect of the row factor (i.e., $F = \text{MSR}/\text{MSE}$). Similarly, the multiple partial F test of $H_0: \beta_1 = \beta_2 = \cdots = \beta_{c-1} = 0$ in model (19.3) yields exactly the same F statistic as is used in standard (balanced) two-way fixed-effects ANOVA to test the significance of the main effect of the column factor (i.e., $F = \text{MSC}/\text{MSE}$). Finally, the multiple partial F test of $H_0: \gamma_{ij} = 0$ (for $i = 1, 2, \ldots, r-1$; and $j = 1, 2, \ldots, c-1$) is identical to the (balanced) two-way fixed-effects ANOVA F test for interaction (i.e., $F = \text{MSRC}/\text{MSE}$).

The preceding formulas may also be used to express the cell, row marginal, and column marginal means as functions of the regression coefficients, as follows:

$$\mu_{ij} = \mu + \alpha_i + \beta_j + \gamma_{ij} \qquad\qquad i = 1, 2,\ldots, r - 1; j = 1, 2,\ldots, c - 1$$

$$\mu_{rj} = \mu - \sum_{i=1}^{r-1} \gamma_{ij} - \sum_{i=1}^{r-1} \alpha_i + \beta_j \qquad j = 1, 2,\ldots, c - 1$$

$$\mu_{ic} = \mu - \sum_{j=1}^{c-1} \gamma_{ij} - \sum_{j=1}^{c-1} \beta_j + \alpha_i \qquad i = 1, 2,\ldots, r - 1$$

$$\mu_{rc} = \mu - \sum_{i=1}^{r-1} \alpha_i - \sum_{j=1}^{c-1} \beta_j + \sum_{i=1}^{r-1}\sum_{j=1}^{c-1} \gamma_{ij}$$

The row marginal means are

$$\mu_{i\cdot} = \mu + \alpha_i \qquad i = 1, 2,\ldots, r - 1$$

$$\mu_{r\cdot} = \mu - \sum_{i=1}^{r-1} \alpha_i$$

and the column marginal means are

$$\mu_{\cdot j} = \mu + \beta_j \qquad j = 1, 2,\ldots, c - 1$$

$$\mu_{\cdot c} = \mu - \sum_{j=1}^{c-1} \beta_j$$

If reference cell coding is used, the general form of model (19.3) stays the same, but the definitions of X_i and Z_j change. For example, we can define

$$X_i = \begin{cases} 1 & \text{for level } i \text{ of the row factor} \\ 0 & \text{otherwise} \end{cases} \qquad i = 1, 2,\ldots, r - 1$$

and

$$Z_j = \begin{cases} 1 & \text{for level } j \text{ of the column factor} \\ 0 & \text{otherwise} \end{cases} \qquad j = 1, 2,\ldots, c - 1$$

With these definitions in place, the parameters in model (19.3) have different interpretations from those given earlier. In particular, the cell means (for the above reference cell coding) can be expressed as the following functions of the parameters in model (19.3):

$$\mu_{ij} = \mu + \alpha_i + \beta_j + \gamma_{ij} \qquad i = 1, 2,\ldots, r - 1; j = 1, 2,\ldots, c - 1$$
$$\mu_{rj} = \mu + \beta_j \qquad\qquad j = 1, 2,\ldots, c - 1$$
$$\mu_{ic} = \mu + \alpha_i \qquad\qquad i = 1, 2,\ldots, r - 1$$
$$\mu_{rc} = \mu$$

The row marginal means are

$$\mu_{i.} = \mu + \alpha_i + \sum_{j=1}^{c-1} \frac{\beta_j + \gamma_{ij}}{c} \quad i = 1, 2,..., r-1$$

$$\mu_{r.} = \mu + \sum_{j=1}^{c-1} \frac{\beta_j}{c}$$

and the column marginal means are

$$\mu_{.j} = \mu + \beta_j + \sum_{i=1}^{r-1} \frac{\alpha_i + \gamma_{ij}}{r} \quad j = 1, 2,..., c-1$$

$$\mu_{.c} = \mu + \sum_{i=1}^{r-1} \frac{\alpha_i}{r}$$

These expressions must be modified when cell-specific sample sizes are not all equal. Chapter 20 addresses two-way ANOVA with unequal cell-specific sample sizes.

19-5-2 Classical Two-way Fixed-effects ANOVA Model

When both factors are considered fixed, there are three types of effects to consider:

1. *Main effects of the row factor:* The differences between the various row means and the overall mean (i.e., $\mu_{i.} - \mu_{..}, i = 1, 2,..., r$).

2. *Main effects of the column factor:* The differences between the various column means and the overall mean (i.e., $\mu_{.j} - \mu_{..}, j = 1, 2,..., c$).

3. *Interaction effects:* The differences between differences, of the form $(\mu_{ij} - \mu_{i.}) - (\mu_{.j} - \mu_{..})$ or $(\mu_{ij} - \mu_{.j}) - (\mu_{i.} - \mu_{..})$, for $i = 1, 2,..., r$ and $j = 1, 2,..., c$.

The classical two-way fixed-effects ANOVA model involving such effects is of the following form:

$$Y_{ijk} = \mu + \alpha_i + \beta_j + \gamma_{ij} + E_{ijk} \tag{19.5}$$

where

$\mu = \mu_{..} = $ Overall mean

$\alpha_i = \mu_{i.} - \mu_{..} = $ Effect of row i

$\beta_j = \mu_{.j} - \mu_{..} = $ Effect of column j

$\gamma_{ij} = \mu_{ij} - \mu_{i.} - \mu_{.j} + \mu_{..} = $ Interaction effect associated with cell (i, j)

$E_{ijk} = Y_{ijk} - \mu - \alpha_i - \beta_j - \gamma_{ij} = $ Error (or residual) associated with the kth observation in cell (i, j)

(for $i = 1, 2,..., r; j = 1, 2,..., c;$ and $k = 1, 2,..., n$).

The following relationships are clearly satisfied by the effects in the preceding model:

$$\sum_{i=1}^{r} \alpha_i = 0, \qquad \sum_{j=1}^{c} \beta_j = 0, \qquad \sum_{i=1}^{r} \gamma_{ij} = 0, \qquad \sum_{j=1}^{c} \gamma_{ij} = 0 \qquad (19.6)$$

A comparison of the regression coefficients in model (19.3) with the ANOVA effects in model (19.5) makes clear that the models are completely equivalent.

Finally, each of the effects in model (19.5) can be simply estimated by using sample means, as follows:

$$\hat{\mu} = \overline{Y}_{...}$$

$$\hat{\alpha}_i = \overline{Y}_{i..} - \overline{Y}_{...} \qquad\qquad i = 1, 2, \dots, r$$

$$\hat{\beta}_{j.} = \overline{Y}_{.j.} - \overline{Y}_{...} \qquad\qquad j = 1, 2, \dots, c$$

$$\hat{\gamma}_{ij} = \overline{Y}_{ij.} - \overline{Y}_{i..} - \overline{Y}_{.j.} + \overline{Y}_{...} \qquad i = 1, 2, \dots, r; j = 1, 2, \dots, c$$

19-6 Interactions in Two-way ANOVA

In this section we investigate several ways to evaluate interaction in two-way ANOVA. For convenience, we focus on the fixed-effects case. Nevertheless, even though the parameters involved and the test statistics used for making inferences in the fixed-effects case differ from those used in the random- and mixed-effects cases, the interpretations of interactions are generally the same.

19-6-1 Concept of Interaction

An interaction exists between two factors if the relationship among the effects associated with the levels of one factor differs according to the levels of the second factor. In other words, an interaction represents an effect due to the joint influence of two factors, over and above the effects of each factor considered separately.

More specifically, if we consider the two-way table of cell means and the various ways of writing the statistical model in the fixed-effects case, there are three equivalent ways of describing or representing an interaction in statistical terms.

Method 1: Interaction as a Difference in Differences of Means   In two-way ANOVA, an interaction exists between the row and column factors if any of the following equivalent statements are true:

1. For some pair of columns, the difference between the means in these columns for a given row is not equal to the difference between the corresponding means for some other row. For example, for rows 1 and 2 and columns 1 and 2, $\mu_{11} - \mu_{12} \neq \mu_{21} - \mu_{22}$.

2. For some pair of rows, the difference between the means in these rows for a given column is not equal to the difference between the corresponding means for some other column. For example, for rows 1 and 2 and columns 1 and 2, $\mu_{11} - \mu_{21} \neq \mu_{12} - \mu_{22}$.

3. For some cell in the table, the difference between that cell's mean and its associated marginal row mean is not equal to the difference between its associated marginal column mean and the overall mean. For example, for the (i, j)th cell, $\mu_{ij} - \mu_{i.} \neq \mu_{.j} - \mu_{..}$, or $\mu_{ij} - \mu_{i.} - \mu_{.j} + \mu_{..} \neq 0$.

4. For some cell in the table, the difference between that cell's mean and its associated marginal column mean is not equal to the difference between its associated marginal row mean and the overall mean. For example, $\mu_{ij} - \mu_{\cdot j} \neq \mu_{i\cdot} - \mu_{\cdot\cdot}$, or $\mu_{ij} - \mu_{i\cdot} - \mu_{\cdot j} + \mu_{\cdot\cdot} \neq 0$.

Thus, when there is no interaction, the relationship among the column effects (β_j's) is the same regardless of the row being considered, and vice versa. And, since, from statements 3 and 4,

$$\mu_{ij} - \mu_{i\cdot} - \mu_{\cdot j} + \mu_{\cdot\cdot} = 0$$

when there is no interaction, it follows that

$$\text{MSRC} = \frac{n}{(r-1)(c-1)} \sum_{i=1}^{r} \sum_{j=1}^{c} (\overline{Y}_{ij} - \overline{Y}_{i\cdot\cdot} - \overline{Y}_{\cdot j\cdot} + \overline{Y}_{\cdot\cdot\cdot})^2$$

(which estimates

$$\frac{n}{(r-1)(c-1)} \sum_{i=1}^{r} \sum_{j=1}^{c} (\mu_{ij} - \mu_{i\cdot} - \mu_{\cdot j} + \mu_{\cdot\cdot})^2$$

for the population) is small when there is no interaction and large when there is interaction.

Method 2: Interaction as an Effect in the Fixed-effects Model   An interaction exists if the appropriate fixed-effects ANOVA model has the form

$$Y_{ijk} = \mu + \alpha_i + \beta_j + \gamma_{ij} + E_{ijk}$$

where $\gamma_{ij} \neq 0$ for at least one (i, j) pair.

Representations 1 and 2 are completely equivalent. For example, if there is no interaction, then $\mu_{ij} = \mu + \alpha_i + \beta_j$, so $\mu_{1j} - \mu_{2j} = (\mu + \alpha_1 + \beta_j) - (\mu + \alpha_2 + \beta_j) = \alpha_1 - \alpha_2$, which is independent of j. Thus, $\mu_{11} - \mu_{21} = \mu_{12} - \mu_{22} = \cdots = \mu_{1c} - \mu_{2c}$.

Method 3: Interaction as a Term in a Regression Model   An interaction exists if the appropriate regression model (using dummy variables) contains a term that involves the product (or in general, any function) of variables from different factors, for example, if the appropriate model is of the form

$$Y = \mu + \sum_{i=1}^{r-1} \alpha_i X_i + \sum_{j=1}^{c-1} \beta_j Z_j + \sum_{i=1}^{r-1}\sum_{j=1}^{c-1} \gamma_{ij} X_i Z_j + E$$

where at least one of the γ_{ij} is not 0.

When $r = c = 2$, the model simplifies to

$$Y = \mu + \alpha_1 X_1 + \beta_1 Z_1 + \gamma_{11} X_1 Z_1 + E$$

where

$$X_1 = \begin{cases} -1 & \text{if level 2 of the row factor} \\ 1 & \text{if level 1 of the row factor} \end{cases}$$

$$Z_1 = \begin{cases} -1 & \text{if level 2 of the column factor} \\ 1 & \text{if level 1 of the column factor} \end{cases}$$

Then, $\mu = \mu_{\cdot\cdot}$, $\alpha_1 = \mu_{1\cdot} - \mu_{\cdot\cdot}$, $\beta_1 = \mu_{\cdot 1} - \mu_{\cdot\cdot}$, and $\gamma_{11} = \mu_{11} - \mu_{1\cdot} - \mu_{\cdot 1} + \mu_{\cdot\cdot}$.

Representation 3 is equivalent to the other two representations, provided that both independent variables (i.e., factors) are considered nominal and so are represented by dummy variables. If, however, both independent variables are continuous, a regression model with any product term will exhibit a somewhat different interaction effect, not necessarily characterized by a nonzero difference of mean differences.

19-6-2 Some Hypothetical Examples

We now consider some hypothetical two-way tables of population cell means that illustrate different patterns of interaction. These tables pertain to the example in section 19-2, for which the factors are NHT and PSN and the dependent variable is CMI score. Subsequently, we will examine the table of sample cell means actually obtained (Table 19-2), keeping in mind that the statistical test for interaction may negate any tentative trends suggested by the sample means. We will also examine the example of section 19-3 in this light.

Row and Column Main Effects but No Interaction Effect

Table 19-10 presents three alternative layouts, each representing the general situation involving *both* a row main effect and a column main effect but no interaction effect. Each of these tables gives *population* (and not sample) mean values, so there is no sampling variation to consider.

The main effects are reflected in the differences between marginal row means and between marginal column means in each table. The lack of an interaction effect can be established by comparing the differences among the cell means, as discussed in section 19-6-1. From Table 19-10(a), for example, we have $\mu_{11} - \mu_{21} = \mu_{21} - \mu_{22}$, since $26 - 23 = 3 = 20 - 17$. For this same table, $\mu_{11} - \mu_{1.} - \mu_{.1} + \mu_{..} = 26 - 24.5 - 23 + 21.5 = 0$; and similar terms associated with the other three cells in the table are also 0. Furthermore, for this table, the model (19.5) can be shown to have the following specific structure:

$$\mu_{ij} = 21.5 + \alpha_i + \beta_j$$

where

$$\alpha_i = \begin{cases} 3 & \text{if } i = 1 \\ -3 & \text{if } i = 2 \end{cases} \quad \text{and} \quad \beta_j = \begin{cases} 1.5 & \text{if } j = 1 \\ -1.5 & \text{if } j = 2 \end{cases}$$

TABLE 19-10 Alternative layouts for main effects but no interaction.

(a) NHT	PSN Low	High	Row Mean	(b) NHT	PSN Low	High	Row Mean	(c) NHT	PSN Low	High	Row Mean
Low	26	23	24.5	Low	18	26	22	Low	18	26	22
High	20	17	18.5	High	20	28	24	High	16	24	20
Column Mean	23	20	21.5	Column Mean	19	27	23	Column Mean	17	25	21

This model does not involve any γ_{ij} term (i.e., there is no interaction term in the model). Thus, we have

$$\mu_{11} = 21.5 + 3 + 1.5 = 26$$

$$\mu_{12} = 21.5 + 3 - 1.5 = 23$$

$$\mu_{21} = 21.5 - 3 + 1.5 = 20$$

$$\mu_{22} = 21.5 - 3 - 1.5 = 17$$

The layouts for tables (b) and (c) also represent no-interaction models; they have the particular forms, respectively, of

$$\mu_{ij} = 23 + \alpha_i + \beta_j$$

where

$$\alpha_i = \begin{cases} -1 & \text{if } i = 1 \\ 1 & \text{if } i = 2 \end{cases} \quad \text{and} \quad \beta_j = \begin{cases} -4 & \text{if } j = 1 \\ 4 & \text{if } j = 2 \end{cases}$$

and

$$\mu_{ij} = 21 + \alpha_i + \beta_j$$

where

$$\alpha_i = \begin{cases} 1 & \text{if } i = 1 \\ -1 & \text{if } i = 2 \end{cases} \quad \text{and} \quad \beta_j = \begin{cases} -4 & \text{if } j = 1 \\ 4 & \text{if } j = 2 \end{cases}$$

Exactly One Main Effect and No Interaction Effect

This situation is depicted in Table 19-11. Table 19-11(a) contains a main effect due to PSN but no main effect due to NHT. Table 19-11(b) contains a main effect due to NHT but no main effect due to PSN. There is no PSN $\times$ NHT interaction.

Same-direction Interaction

Three examples of same-direction interaction are given in Table 19-12. In Table 19-12(a) we see that $\mu_{11} - \mu_{12} = 26 - 23 = 3$, whereas $\mu_{21} - \mu_{22} = 20 - 13 = 7$. Also, $\mu_{11} - \mu_{1.} =$

TABLE 19-11 Layouts for one main effect and no interaction.

(a) Main Effect Due to PSN

NHT	PSN Low	PSN High	Row Mean
Low	18	26	22
High	18	26	22
Column Mean	18	26	22

(b) Main Effect Due to NHT

NHT	PSN Low	PSN High	Row Mean
Low	19	19	19
High	24	24	24
Column Mean	21.5	21.5	21.5

TABLE 19-12 Layouts for same-direction interaction.

(a) NHT	PSN Low	High	Row Mean	(b) NHT	PSN Low	High	Row Mean	(c) NHT	PSN Low	High	Row Mean
Low	26	23	24.5	Low	18	26	22	Low	18	26	22
High	20	13	16.5	High	20	36	28	High	12	24	18
Column Mean	23	18	20.5	Column Mean	19	31	25	Column Mean	15	25	20

$26 - 24.5 = 1.5$ and $\mu_{.1} - \mu_{..} = 23 - 20.5 = 2.5$, so $\mu_{11} - \mu_{1.} - \mu_{.1} + \mu_{..} = -1.0$. The other interactions of the general form $(\mu_{ij} - \mu_{i.} - \mu_{.j} + \mu_{..})$ are similarly determined to be either 1.0 (for $i = 1, j = 2$; or for $i = 2, j = 1$) or -1.0 (for $i = 2, j = 2$).

These hypothetical results indicate that in _both_ low- and high-NHT neighborhoods, persons in friendly surroundings (i.e., high PSN) are healthier (i.e., have a lower CMI score) than persons in unfriendly surroundings (i.e., low PSN), but that the extent of this difference is greater when there is a large number of households (high NHT) than when there is a small number (low NHT). In other words, at each level of NHT, the difference between the PSN-level effects lies in the same direction (i.e., high PSN is associated with a lower mean CMI score than is low PSN), but the magnitude of the difference depends on the NHT level. This is the meaning of _same-direction interaction._

The model for Table 19-12(a) may be expressed as

$$\mu_{ij} = 20.5 + \alpha_i + \beta_j + \gamma_{ij}$$

where

$$\alpha_i = \begin{cases} 4.0 & \text{if } i = 1 \\ -4.0 & \text{if } i = 2 \end{cases} \qquad \beta_j = \begin{cases} 2.5 & \text{if } j = 1 \\ -2.5 & \text{if } j = 2 \end{cases}$$

$$\gamma_{ij} = \begin{cases} -1.0 & \text{if } i = 1, j = 1 \\ 1.0 & \text{if } i = 1, j = 2 \\ 1.0 & \text{if } i = 2, j = 1 \\ -1.0 & \text{if } i = 2, j = 2 \end{cases}$$

Thus,

$$\mu_{11} = 20.5 + 4.0 + 2.5 - 1.0 = 26$$

$$\mu_{12} = 20.5 + 4.0 - 2.5 + 1.0 = 23$$

$$\mu_{21} = 20.5 - 4.0 + 2.5 + 1.0 = 20$$

$$\mu_{22} = 20.5 - 4.0 - 2.5 - 1.0 = 13$$

The same type of model holds for tables (b) and (c).

TABLE 19-13   Layouts for reverse interaction.

(a) NHT	PSN Low	PSN High	Row Mean		(b) NHT	PSN Low	PSN High	Row Mean
Low	18	26	22		Low	26	22	24
High	22	20	21		High	18	24	21
Column Mean	20	23	21.5		Column Mean	22	23	22.5

Reverse Interaction

Two examples of reverse interaction are given in Table 19-13. In these instances, the direction of the difference between two cell means for one row (column) is opposite to, or reversed from, the direction of the difference between the corresponding cell means for some other row (column).

In Table 19-13(a), we see that $\mu_{11} - \mu_{12} = 18 - 26 = -8$, whereas $\mu_{21} - \mu_{22} = 22 - 20 = 2$. Also, $\mu_{21} - \mu_{2.} = 22 - 21 = 1$ and $\mu_{.1} - \mu_{..} = 20 - 21.5 = -1.5$, so $\mu_{21} - \mu_{2.} - \mu_{.1} + \mu_{..} = 2.5$.

These hypothetical results for this table indicate that, for neighborhoods with a small number of households (low NHT), persons in unfriendly surroundings (low PSN) are healthier than are persons in friendly surroundings; but for neighborhoods with a large number of households, the situation is reversed. In other words, the difference between the effects of the high and low PSN levels is positive for low NHT but negative for high NHT (i.e., there is a reversal in sign). This is the meaning of *reverse interaction*.

The model in this case is given as

$$\mu_{ij} = 21.5 + \alpha_i + \beta_j + \gamma_{ij}$$

where

$$\alpha_i = \begin{cases} 0.5 & \text{if } i = 1 \\ -0.5 & \text{if } i = 2 \end{cases} \qquad \beta_j = \begin{cases} -1.5 & \text{if } j = 1 \\ 1.5 & \text{if } j = 2 \end{cases}$$

$$\gamma_{ij} = \begin{cases} -2.5 & \text{if } i = 1, j = 1 \\ 2.5 & \text{if } i = 1, j = 2 \\ 2.5 & \text{if } i = 2, j = 1 \\ -2.5 & \text{if } i = 2, j = 2 \end{cases}$$

Thus,

$$\mu_{11} = 21.5 + 0.5 - 1.5 - 2.5 = 18$$

$$\mu_{12} = 21.5 + 0.5 + 1.5 + 2.5 = 26$$

$$\mu_{21} = 21.5 - 0.5 - 1.5 + 2.5 = 22$$

$$\mu_{22} = 21.5 - 0.5 + 1.5 - 2.5 = 20$$

It can be shown that a reverse interaction is indicated in Table 19-13(b), as well.

TABLE 19-14   Cell means for data of Table 19-1.

NHT	PSN Low	PSN High	Row Mean
Low	18.92	26.84	22.88
High	20.84	20.72	20.78
Column Mean	19.88	23.78	21.83

19-6-3 Interaction Effects for Data of Table 19-1

The table of sample cell means actually obtained for the CMI example of section 19-2 is given in Table 19-14. From this table, the following differences of means can be determined:

$$\overline{Y}_{11.} - \overline{Y}_{12.} = 18.92 - 26.84 = -7.92 \quad \text{whereas} \quad \overline{Y}_{21.} - \overline{Y}_{22.} = 20.84 - 20.72 = 0.12$$

$$\overline{Y}_{11.} - \overline{Y}_{21.} = 18.92 - 20.84 = -1.92 \quad \text{whereas} \quad \overline{Y}_{12.} - \overline{Y}_{22.} = 26.84 - 20.72 = 6.12$$

$$\overline{Y}_{11.} - \overline{Y}_{1..} - \overline{Y}_{.1.} + \overline{Y}_{...} = 18.92 - 22.88 - 19.88 + 21.83 = -2.01$$

$$\overline{Y}_{12.} - \overline{Y}_{1..} - \overline{Y}_{.2.} + \overline{Y}_{...} = 26.84 - 22.88 - 23.78 + 21.83 = 2.01$$

$$\overline{Y}_{21.} - \overline{Y}_{2..} - \overline{Y}_{.1.} + \overline{Y}_{...} = 20.84 - 20.78 - 19.88 + 21.83 = 2.01$$

$$\overline{Y}_{22.} - \overline{Y}_{2..} - \overline{Y}_{.2.} + \overline{Y}_{...} = 20.72 - 20.78 - 23.78 + 21.83 = -2.01$$

These comparisons suggest a possible reverse interaction. Specifically, for small turnkey neighborhoods, persons from friendly surroundings (high PSN) appear to have worse health (higher mean CMI scores) than do persons from unfriendly surroundings; but little difference in mean CMI scores is found on the friendliness variable for large turnkey neighborhoods. As was mentioned in section 19-2, this pattern runs counter to what Daly (1973) expected to find. However, these observed differences are subject to sampling variation (i.e., they are sample values and not population values); and the test for interaction for these data does not reveal significant interaction.

19-6-4 Interaction Effects for Data of Table 19-5

The table of sample cell means for the FEV example (Table 19-5) is as shown in Table 19-15. This table of means is slightly more difficult to interpret than the one for the CMI data, because it contains three rows and columns instead of two. Nevertheless, we can immediately observe that the relative magnitudes of the means vary from column to column. For example, for Toxsub A, the order of Plants by increasing mean FEV is 3, 1, 2. For Toxsub B, on the other hand, the order is 1, 2, 3, and for Toxsub C the order is 2, 1, 3. These differences in ordering indicate some interaction, the significance of which was established earlier. The following comparisons of cell means should help in interpreting the nature of this significant interaction effect:

$$\overline{Y}_{11.} - \overline{Y}_{12.} = 4.90 - 3.46 = 1.44 \quad \text{whereas} \quad \overline{Y}_{31.} - \overline{Y}_{32.} = 4.13 - 5.13 = -1.00$$

$$\overline{Y}_{21.} - \overline{Y}_{31.} = 5.23 - 4.13 = 1.10 \quad \text{whereas} \quad \overline{Y}_{22.} - \overline{Y}_{32.} = 3.78 - 5.13 = -1.35$$

$$\overline{Y}_{21.} - \overline{Y}_{.1.} = 5.23 - 4.75 = 0.48 \quad \text{whereas} \quad \overline{Y}_{2..} - \overline{Y}_{...} = 3.90 - 3.91 = -0.01$$

TABLE 19-15   Cell means for data of Table 19-5.

| | Toxsub | | | Row |
Plant	A	B	C	Mean
1	4.90	3.46	2.75	3.70
2	5.23	3.78	2.70	3.90
3	4.13	5.13	3.14	4.13
Column Mean	4.75	4.12	2.86	3.91

TABLE 19-16   Interaction effects ($\hat{\gamma}_{ij}$ values) for data of Table 19-5.

| | Toxsub | | | Row |
Plant	A	B	C	Mean
1	0.35	−0.45	0.10	0.00
2	0.49	−0.34	−0.15	0.00
3	−0.84	0.78	0.06	0.00
Column Mean	0.00	0.00	0.00	0.00

The set of interaction effects of the form $\hat{\gamma}_{ij} = \overline{Y}_{ij.} - \overline{Y}_{i..} - \overline{Y}_{.j.} + \overline{Y}_{...}$ is given in Table 19-16. These patterns demonstrate that some plants are associated with better respiratory health than others for one kind of toxic exposure but are worse for other kinds of such exposure. Therefore, we cannot conclude that one plant was better overall than another. Rather, the differences in respiratory health among plants depend on which toxic substance is being considered.

19-7 Random- and Mixed-effects Two-way ANOVA Models

In this section we examine the classical two-way ANOVA statistical models appropriate when both factors are random or when one factor is fixed and the other is random. We will specify the appropriate null hypotheses of interest and the expected mean squares associated with each model. The expected mean square for a particular source is the true average (population) value of the mean-square term in the ANOVA table.

19-7-1 Random-effects Model[7]

When both factors are random, the two-way ANOVA model is given as

$$Y_{ijk} = \mu + A_i + B_j + C_{ij} + E_{ijk} \tag{19.7}$$

---

[7]As previously footnoted in Chapter 17 for one-way ANOVA models, whenever random effects occur in a two-way ANOVA model, we can no longer assume that the Y_{ijk} are mutually independent.

where A_i, B_j, C_{ij}, and E_{ijk} are mutually independent random variables satisfying

$$A_i \frown N\left(0, \sigma_R^2\right)$$

$$B_j \frown N\left(0, \sigma_C^2\right)$$

$$C_{ij} \frown N\left(0, \sigma_{RC}^2\right)$$

$$E_{ijk} \frown N\left(0, \sigma^2\right) \qquad i = 1, 2, \ldots, r; \quad j = 1, 2, \ldots, c; \quad k = 1, 2, \ldots, n$$

19-7-2 Mixed-effects Model with Fixed Row Factor and Random Column Factor

One particular model[8] is

$$Y_{ijk} = \mu + \alpha_i + B_j + C_{ij} + E_{ijk} \tag{19.8}$$

where each α_i is a constant such that $\sum_{i=1}^{r} \alpha_i = 0$, and where B_j, C_{ij}, and E_{ijk} are random variables satisfying

$$B_j \frown N(0, \sigma_C^2)$$

$$C_{ij} \frown N(0, \sigma_{RC}^2) \qquad i = 1, 2, \ldots, r; \quad j = 1, 2, \ldots, c; \quad k = 1, 2, \ldots, n$$

$$E_{ijk} \frown N(0, \sigma^2)$$

19-7-3 Mixed-effects Model with Random Row Factor and Fixed Column Factor

One particular model is

$$Y_{ijk} = \mu + A_i + \beta_j + C_{ij} + E_{ijk} \tag{19.9}$$

---

[8]There are several ways to define the assumptions for a two-way mixed model. In particular, one must choose between two alternative forms of summation restrictions for the random components of a two-way mixed model (see Searle et al. 1992). The assumptions we have made (in the body of the text) for mixed models (19.9) and (19.10) include summation restrictions for the fixed effects—e.g., $\sum_{i=1}^{r} \alpha_i = 0$ for model (19.9)—but no such summation restrictions for random effects, particularly the (random) interaction effects C_{ij}. Alternatively, we can choose to assume (for either model 19.9 or model 19.10) that the sum of the interaction factors C_{ij} over the levels of the fixed factor are zero; i.e., $\sum_{i=1}^{r} C_{ij} = 0$ for each j. The latter summation restriction is in fact what we assumed in previous editions of this text. Moreover, this restriction implied that C_{ij} and $C_{i'j}$, $i \neq i'$, are correlated, which in turn, leads to a different formula for the F statistic for the test of hypothesis about the random main effect in the mixed model; specifically, the denominator mean square is MSE instead of MSRC. The question of which form of mixed model to use—that without the restrictions of the C_{ij} or that with them—remains open (Searle et al. 1992). Nevertheless, we have chosen to assume (in the main body of the text) no restrictions on the C_{ij}, because the F tests implied by no restrictions correspond to those calculated by standard computer programs for ANOVA models with mixed effects, including SAS's GLM, whose output we illustrate in the text. Also, the assumptions without summation restrictions correspond to assumptions statisticians typically make when considering the analysis of repeated measures data, which we discuss in Chapter 21.

where each β_j is a constant such that $\sum_{j=1}^{c} \beta_j = 0$, and where A_j, C_{ij}, and E_{ijk} are random variables satisfying

$$A_i \sim N(0, \sigma_R^2)$$
$$C_{ij} \sim N(0, \sigma_{RC}^2) \qquad i = 1, 2, ..., r; \quad j = 1, 2, ..., c; \quad k = 1, 2, ..., n$$
$$E_{ijk} \sim N(0, \sigma^2)$$

19-7-4 Null Hypotheses and Expected Mean Squares for Two-way ANOVA Models

For fixed-, random-, and mixed-effects models, Table 19-17 gives the specific null hypotheses being tested, with regard to row main effects, column main effects, and interaction effects. Table 19-18 gives the expected mean square for each factor in each of the models. These two tables demonstrate why different F statistics are required for testing the various hypotheses of interest. In this regard, the primary consideration is the choice of the appropriate denominator mean squares to use in the various F statistics. The numerator mean square always corresponds to the factor being considered; for example, if the factor is "rows," the numerator mean square is MSR, regardless of the type of model. Similarly, if the factor is "columns" or "interaction," the numerator mean square is MSC or MSRC, respectively. The denominator mean square, however, is chosen to correspond to the expected mean square to which the numerator expected mean square reduces under the null hypothesis of interest. For example, in a test for significant row effects in a random-effects model, the numerator expected mean square, $\sigma^2 + n\sigma_{RC}^2 + cn\sigma_R^2$ from Table 19-18, reduces to $\sigma^2 + n\sigma_{RC}^2$ under $H_0: \sigma_R^2 = 0$. This requires that the denominator mean square be MSRC, since the expected mean square of MSRC under the random-effects model is exactly $\sigma^2 + n\sigma_{RC}^2$.

Thus, the ratio of expected mean squares

$$\frac{\text{EMS(R)}}{\text{EMS(RC)}} = \frac{\sigma^2 + n\sigma_{RC}^2 + cn\sigma_R^2}{\sigma^2 + n\sigma_{RC}^2}$$

reduces to $(\sigma^2 + n\sigma_{RC})/(\sigma^2 + n\sigma_{RC}^2) = 1$ under $H_0: \sigma_R^2 = 0$, so the F statistic MSR/MSRC is the ratio of two estimates of the same variance under H_0.

TABLE 19-17 Null hypotheses for two-way ANOVA.

			Model Type	
			Mixed Effects	
Source	Fixed Effects	Random Effects	Rows Fixed, Columns Random	Rows Random, Columns Fixed
Rows	$\alpha_1 = \alpha_2 = ... = \alpha_r = 0$	$\sigma_R^2 = 0$	$\alpha_1 = \alpha_2 = ... = \alpha_r = 0$	$\sigma_R^2 = 0$
Columns	$\beta_1 = \beta_2 = ... = \beta_c = 0$	$\sigma_C^2 = 0$	$\sigma_C^2 = 0$	$\beta_1 = \beta_2 = ... = \beta_c = 0$
Interactions	$\gamma_{ij} = 0$ for all i, j	$\sigma_{RC}^2 = 0$	$\sigma_{RC}^2 = 0$	$\sigma_{RC}^2 = 0$

TABLE 19-18   Expected mean squares for two-way ANOVA (r = rows, c = columns, n = replications per cell).

			Mixed Effects	
Source	Fixed Effects	Random Effects	Rows Fixed, Columns Random	Rows Random, Columns Fixed
Rows	$\sigma^2 + cn\sum_{i=1}^{r}\dfrac{\alpha_i^2}{r-1}$	$\sigma^2 + n\sigma_{RC}^2 + cn\sigma_R^2$	$\sigma^2 + n\sigma_{RC}^2 + cn\sum_{i=1}^{r}\dfrac{\alpha_i^2}{r-1}$	$\sigma^2 + n\sigma_{RC}^2 + cn\sigma_R^2$
Columns	$\sigma^2 + rn\sum_{j=1}^{c}\dfrac{\beta_j^2}{c-1}$	$\sigma^2 + n\sigma_{RC}^2 + rn\sigma_C^2$	$\sigma^2 + n\sigma_{RC}^2 + rn\sigma_C^2$	$\sigma^2 + n\sigma_{RC}^2 + rn\sum_{j=1}^{c}\dfrac{\beta_j^2}{c-1}$
Interactions	$\sigma^2 + n\sum_{i=1}^{r}\sum_{j=1}^{c}\dfrac{\gamma_{ij}^2}{(r-1)(c-1)}$	$\sigma^2 + n\sigma_{RC}^2$	$\sigma^2 + n\sigma_{RC}^2$	$\sigma^2 + n\sigma_{RC}^2$
Error	σ^2	σ^2	σ^2	σ^2

As another example, let us consider the F test for significant row effects based on the mixed-effects model with the row factor fixed and column factor random. The test statistic in this case, $F = \text{MSR}/\text{MSRC}$, involves the following ratio of expected mean squares (see Table 19-18):

$$\frac{\text{EMS(R)}}{\text{EMS(RC)}} = \frac{\sigma^2 + n\sigma_{RC}^2 + cn\sum_{i=1}^{r}\alpha_i^2/(r-1)}{\sigma^2 + n\sigma_{RC}^2}$$

Under $H_0: \alpha_1 = \alpha_2 = \cdots = \alpha_r = 0$, this ratio simplifies to $(\sigma^2 + n\sigma_{RC}^2)/(\sigma^2 + n\sigma_{RC}^2) = 1$. Thus, the F statistic is the ratio of two estimates of the same variance under H_0.

As a final example, we consider the F test for significant row effects based on the mixed-effects model with the row factor random and the column factor fixed. The test statistic is $F = \text{MSR}/\text{MSRC}$, which involves

$$\frac{\text{EMS(R)}}{\text{EMS(RC)}} = \frac{\sigma^2 + n\sigma_{RC}^2 + cn\sigma_R^2}{\sigma^2 + n\sigma_{RC}^2}$$

Under $H_0: \sigma_R^2 = 0$, this ratio simplifies to

$$\frac{\sigma^2 + n\sigma_{RC}^2}{\sigma^2 + n\sigma_{RC}^2} = 1$$

as desired.

Problems

1. The data in the following table and accompanying computer output come from an animal experiment designed to investigate whether levorphanol reduces stress as reflected in the cortical sterone level. The four treatment groups contained five animals each.

Control	Levorphanol Only	Epinephrine Only	Levorphanol and Epinephrine
1.90	0.82	5.33	3.08
1.80	3.36	4.84	1.42
1.54	1.64	5.26	4.54
4.10	1.74	4.92	1.25
1.89	1.21	6.07	2.57

a. These data may be analyzed by means of two-way ANOVA. What are the two factors?

b. Classify each factor as either fixed or random.

c. Rearrange the data into a two-way table appropriate for two-way ANOVA.

d. Form the table of sample means, and comment on its content.

e. Determine the appropriate ANOVA table for this data set.

f. Analyze the data to determine whether significant main effects exist due to levorphanol and epinephrine, and whether a significant interaction effect exists between epinephrine and levorphanol.

Edited SAS Output (PROC GLM) for Problem 1

```
Dependent Variable: LEVEL
                          Sum of          Mean
Source           DF       Squares         Square    F Value    Pr > F
Model             3     37.5784400     12.5261467     12.30     0.0002
Error            16     16.2978400      1.0186150
Corrected Total  19     53.8762800

        R-Square              C.V.        Root MSE        LEVEL Mean
        0.697495           34.05076       1.00926          2.96400

Source    DF    Type I SS     Mean Square    F Value    Pr > F
L          1   12.8320200     12.8320200      12.60     0.0027
E          1   18.5859200     18.5859200      18.25     0.0006
L*E        1    6.1605000      6.1605000       6.05     0.0257

Source    DF    Type III SS   Mean Square    F Value    Pr > F
L          1   12.8320200     12.8320200      12.60     0.0027
E          1   18.5859200     18.5859200      18.25     0.0006
L*E        1    6.1605000      6.1605000       6.05     0.0257
```

2. The following table gives the performance competency scores for a random sample of family nurse practitioners (FNPs) with different specialties, from hospitals in three cities.

Specialty	City 1	City 2	City 3
Pediatrics	91.7, 74.9, 88.2, 79.5	86.3, 88.1, 92.0, 69.5	82.3, 78.7, 89.8, 84.5
Obstetrics and gynecology	80.1, 76.2, 70.3, 89.5	71.3, 73.4, 76.9, 87.2	90.1, 65.6, 74.6, 79.1
Diabetes and hypertension	71.5, 49.8, 55.1, 75.4	80.2, 76.1, 44.2, 50.5	48.7, 54.4, 60.1, 70.8

    a. Classify each factor as either fixed or random, and justify your classification.

    b. Form the table of sample means (you may use the accompanying computer output to do so), and then comment.

    c. Using the computer output, compute the appropriate F statistic for each of the four possible factor classification schemes (i.e., both factors fixed, both random, and one factor of each type).

    d. Analyze the data based on each possible factor classification scheme. How do the results compare?

    e. Using Scheffé's method, as described in Chapter 17, find a 95% confidence interval for the true difference in mean scores between pediatric FNPs and ob-gyn FNPs.

    f. Using the sample means obtained in part (b), and assuming each factor to be fixed, state a regression model appropriate for the two-way ANOVA table, and provide estimates of the regression coefficients associated with the factor main effects.

Edited SAS Output (PROC GLM) for Problem 2

```
                        Class Level Information
                   Class    Levels    Values
                   SPEC        3      D&H OBGYN PED
                   CITY        3      1 2 3
              Number of observations in data set = 36
Dependent Variable: SCORE
                            Sum of          Mean
Source            DF       Squares        Square     F Value    Pr > F
Model              8    3288.95000     411.11875        3.82    0.0040
Error             27    2902.18000     107.48815
Corrected Total   35    6191.13000
         R-Square          C.V.       Root MSE        SCORE Mean
         0.531236      13.94438        10.3676           74.3500
Source            DF    Type I SS    Mean Square    F Value    Pr > F
SPEC               2   3229.87167    1614.93583       15.02    0.0001
CITY               2     24.54167      12.27083        0.11    0.8925
SPEC*CITY          4     34.53667       8.63417        0.08    0.9877
Source            DF   Type III SS    Mean Square    F Value    Pr > F
SPEC               2   3229.87167    1614.93583       15.02    0.0001
CITY               2     24.54167      12.27083        0.11    0.8925
SPEC*CITY          4     34.53667       8.63417        0.08    0.9877
Source             Type III Expected Mean Square ←[For random-effects model]
SPEC               Var(Error) + 4 Var(SPEC*CITY) + 12 Var(SPEC)
CITY               Var(Error) + 4 Var(SPEC*CITY) + 12 Var(CITY)
SPEC*CITY          Var(Error) + 4 Var(SPEC*CITY)
Source: CITY                                  [Random- or
Error: MS(SPEC*CITY)                           mixed-effects tests]
                                                        ↓
               Denominator     Denominator
DF    Type III MS        DF              MS    F Value    Pr > F
 2   1614.9358333         4    8.6341666667   187.0402    0.0001
```

```
Source: CITY
Error: MS(SPEC*CITY)

                    Denominator      Denominator
DF      Type III MS      DF              MS      F Value    Pr > F
 2     12.270833333       4     8.6341666667     1.4212    0.3417

Source: SPEC*CITY
Error: MS(Error)

                    Denominator      Denominator
DF      Type III MS      DF              MS      F Value    Pr > F
 4     8.6341666667      27     107.48814815     0.0803    0.9877

    Level of     Level of             ----------SCORE----------
    SPEC         CITY        N          Mean            SD
    D&H           1          4       62.9500000     12.4184003
    D&H           2          4       62.7500000     18.0452210
    D&H           3          4       58.5000000      9.4286797
    OBGYN         1          4       79.0250000      8.0619993
    OBGYN         2          4       77.2000000      7.0554943
    OBGYN         3          4       77.3500000     10.1857744
    PED           1          4       83.5750000      7.301897
    PED           2          4       83.9750000      9.9389386
    PED           3          4       83.8250000      4.6456969
```

3. The following table gives the average patient waiting time in minutes for patients of a sample of 16 physicians, classified by type of practice and type of physician.

Physician Type	Type of Practice	
	Group	Solo
General practitioner	15, 20, 25, 20	20, 25, 30, 25
Specialist	30, 25, 30, 35	25, 20, 30, 30

a. Classify each factor as either fixed or random, and justify your classification scheme.
b. Using the accompanying computer output, compute the F statistic corresponding to each of the four possible factor classification schemes.
c. Discuss the analysis of the data when both factors are considered fixed.
d. What is the estimate of the (fixed) effect due to "general practitioner," the (fixed) effect due to "group practice," and the interaction effect $\mu_{11} - \mu_{12} - \mu_{21} + \mu_{22}$, where μ_{ij} denotes the cell mean in the ith row and jth column of the table of cell means?
e. Interpret the interaction effect observed.
f. What is an appropriate regression model for this two-way ANOVA?
g. How might you modify the model in part (f) to reflect the conclusions made in part (c)?

Edited SAS Output (PROC GLM) for Problem 3

```
                    Class Level Information
                Class     Levels     Values
                PHYS         2       GEN SPEC
                PRACTICE     2       GROUP SOLO
          Number of observations in data set = 16
Dependent Variable: TIME

                              Sum of          Mean
Source              DF        Squares         Square     F Value    Pr > F
Model                3      204.687500      68.229167      3.74     0.0415
Error               12      218.750000      18.229167
Corrected Total     15      423.437500

          R-Square            C.V.        Root MSE         TIME Mean
          0.483395          16.86741      4.26956          25.3125

Source              DF      Type I SS     Mean Square    F Value    Pr > F
PHYS                 1      126.562500    126.562500       6.94     0.0218
PRACTICE             1        1.562500      1.562500       0.09     0.7778
PHYS*PRACTICE        1       76.562500     76.562500       4.20     0.0629

Source              DF     Type III SS    Mean Square    F Value    Pr > F
PHYS                 1      126.562500    126.562500       6.94     0.0218
PRACTICE             1        1.562500      1.562500       0.09     0.7747
PHYS*PRACTICE        1       76.562500     76.562500       4.20     0.0629

Source             Type III Expected Mean Square  ◄── For random-effects model
PHYS               Var(Error) + 4 Var(PHYS*PRACTICE) + 8 Var(PHYS)
PRACTICE           Var(Error) + 4 Var(PHYS*PRACTICE) + 8 Var(PRACTICE)
PHYS*PRACTICE      Var(Error) + 4 Var(PHYS*PRACTICE)

Source: PHYS
Error: MS(PHYS*PRACTICE)

                  Denominator    Denominator
DF    Type III MS     DF             MS        F Value    Pr > F
 1      126.5625       1          76.5625      1.6531     0.4208

Source: PRACTICE
Error: MS(PHYS*PRACTICE)

                  Denominator    Denominator
DF    Type III MS     DF             MS        F Value    Pr > F
 1        1.5625       1          76.5625      0.0204     0.9097

Source: PHYS*PRACTICE
Error: MS(Error)

                  Denominator    Denominator
DF    Type III MS     DF             MS        F Value    Pr > F
 1       76.5625      12       18.229166667    4.2000     0.0629

     Level of      Level of                ----------TIME-----------
     PHYS          PRACTICE      N          Mean              SD
     GEN           GROUP         4       20.0000000       4.08248290
     GEN           SOLO          4       25.0000000       4.08248290
     SPEC          GROUP         4       30.0000000       4.08248290
     SPEC          SOLO          4       26.2500000       4.78713554
```

4. A study was undertaken to measure and compare sexist attitudes of students at various types of colleges. Random samples of 10 undergraduate seniors of each sex were selected from each of three types of colleges. A questionnaire was then administered to each student, from which a score for "degree of sexism"—defined as the extent to which a student considered males and females to have different life roles—was determined (the higher the score, the more sexist the attitude). The resulting data are given in the following table.

College Type	Male	Female
Coed with 75% or more males	50, 35, 37, 32, 46, 38, 36, 40, 38, 41	38, 27, 34, 30, 22, 32, 26, 24, 31, 33
Coed with less than 75% males	30, 29, 31, 27, 22, 20, 31, 22, 25, 30	28, 31, 28, 26, 20, 24, 31, 24, 31, 26
Not coed	45, 40, 32, 31, 26, 28, 39, 27, 37, 35	40, 35, 32, 29, 24, 26, 36, 25, 35, 35

a. Form the table of cell means, and interpret the results obtained (see the accompanying computer printout).

b. Using the computer output, calculate the F statistics corresponding to a model with both factors fixed.

c. Discuss the analysis of the data for this fixed-effects model case.

Edited SAS Output (PROC GLM) for Problem 4

```
                                    ⋮  [Portion of output omitted]
                                    ⋮
Dependent Variable: SCORE
                              Sum of          Mean
Source            DF         Squares        Square     F Value    Pr > F
Model              5      1144.88333     228.97667       9.06     0.0001
Error             54      1365.30000      25.28333
Corrected Total   59      2510.18333

R-Square                    C.V.         Root MSE            SCORE Mean
0.456096                16.02205          5.02825              31.3833

Source            DF       Type I SS   Mean Square    F Value    Pr > F
COLLEGE            2      657.433333    328.716667      13.00     0.0001
SEX               1      228.150000    228.150000       9.02     0.0040
COLLEGE*SEX       2      259.300000    129.650000       5.13     0.0091
                                    ⋮  [Portion of output omitted]
                                    ⋮
Level of     Level of                 ----------SCORE----------
COLLEGE      SEX            N          Mean               SD
<75%         FEMALE        10      26.9000000        3.63470922
<75%         MALE          10      26.7000000        4.16466619
>75%         FEMALE        10      29.7000000        4.92273637
>75%         MALE          10      39.3000000        5.31350481
NOT          FEMALE        10      31.7000000        5.41705127
NOT          MALE          10      34.0000000        6.27162924
```

5. Random samples of 100 persons awaiting trial on felony charges were selected from rural, urban, and suburban court locations in each of two states, one (state 1) in the Northeast and the other (state 2) in the South. The following table summarizes the data on the time $\overline{Y}$ (in months) between arrest and beginning of trial for these random samples.

	Court Location		
State	Rural	Suburban	Urban
1	$\overline{Y} = 3.4, S = 1.3$	$\overline{Y} = 5.8, S = 1.2$	$\overline{Y} = 6.8, S = 1.5$
2	$\overline{Y} = 2.4, S = 1.5$	$\overline{Y} = 3.5, S = 1.7$	$\overline{Y} = 4.7, S = 1.7$

a. Do the sample means in the table suggest the average waiting times for state 1 vary by court location differently from how they vary for state 2? Is there an interaction effect?

b. Analyze these data. Use the following ANOVA table. Assume that both factors are fixed.

Source	d.f.	SS	MS
States	1	486.00	486.00
Court locations	2	826.33	413.17
Interaction	2	49.00	24.50
Error	594	1,327.591	2.235

c. Define an appropriate regression model for this two-way ANOVA.

d. How might you revise the model in part (c) and the associated ANOVA table in order to investigate whether a linear trend exists in waiting time measured against degree of urbanization (as determined by treating the categories rural, suburban, and urban on an ordinal scale)? What difficulty does one encounter when considering such a model?

6. An experiment was conducted at a large state university to determine whether two different instructional methods for teaching a beginning statistics course would yield different levels of cognitive achievement. One instructional method involved using a self-instructional format, including a sequence of slide-tape presentations; the other method utilized the standard lecture format. The 100 students who registered for the course were randomly assigned to one of four sections, 25 per section, corresponding to the combinations of one of the two methods with one of two instructors. The results obtained from identical final exams given to each section are summarized in the following table.

	Method	
Instructor	Lecture	Self-instruction
A	$\overline{Y} = 71.2, S = 13.8$	$\overline{Y} = 80.2, S = 12.1$
B	$\overline{Y} = 73.8, S = 11.7$	$\overline{Y} = 77.5, S = 14.1$

a. What do the results suggest about the comparative effects of the two instructional methods?

b. Classify each factor as either fixed or random, and explain your classification.

Source	d.f.	SS	MS
INSTRUC	1	$6.2500E - 02$	$6.2500E - 02$
METHOD	1	$6.0081E + 03$	$1.0081E + 03$
INSTRUC $\times$ METHOD	1	$1.7556E + 02$	$1.7556E + 02$
Error	96	$1.6141E + 04$	$1.6814E + 02$

c. Using the ANOVA table, perform the appropriate F tests for each of the four types of factor classification schemes. Compare the conclusions reached under each scheme.

d. What factors should be controlled for in this experiment?

e. Given a continuous variable C to be controlled for, write an appropriate regression model for this data set that takes C into account. What general method of analysis is characterized by such a model?

7. The following table and accompanying computer output present data on the uric acid level found in the bloodstreams of persons with Down's syndrome, and in the bloodstreams of non–Down's syndrome subjects. All subjects were between the ages of 21 and 25. Analyze these data, using the ANOVA table to determine whether evidence of a higher uric acid level exists in the Down's syndrome group, making sure to characterize any sex relationships that exist.

	Sex	
Group	Male	Female
Down's syndrome	5.84, 6.30, 6.95, 5.92, 7.94	4.90, 6.95, 6.73, 5.32, 4.81
Others	5.50, 6.08, 5.12, 7.58, 6.78	4.94, 7.20, 5.22, 4.60, 3.88

Edited SAS Output (PROC GLM) for Problem 7

```
                                    ⋮
                                    ⋮  [Portion of output omitted]
                                    ⋮

Dependent Variable: ACID
                             Sum of           Mean
Source            DF        Squares         Square    F Value    Pr > F
Model              3      5.65548000     1.88516000       1.74    0.1987
Error             16     17.31484000     1.08217750
Corrected Total   19     22.97032000

        R-Square            C.V.       Root MSE           ACID Mean
        0.246208        17.54854        1.04028             5.92800

Source            DF     Type I SS    Mean Square    F Value    Pr > F
GROUP              1    1.13288000     1.13288000       1.05    0.3215
SEX                1    4.47458000     4.47458000       4.13    0.0589
GROUP*SEX          1    0.04802000     0.04802000       0.04    0.8380

                                    ⋮
                                    ⋮  [Portion of output omitted]
                                    ⋮

        Level of     Level of                -----------ACID-----------
        GROUP        SEX          N           Mean                 SD
        DOWNS-S      FEMALE       5      5.74200000         1.02360637
        DOWNS-S      MALE         5      6.59000000         0.87286883
        OTHERS       FEMALE       5      5.16800000         1.24149909
        OTHERS       MALE         5      6.21200000         0.98879725
```

8. An experiment was conducted to investigate the survival of diplococcus pneumonia bacteria in chick embryos under relative humidities (RH) of 0%, 25%, 50%, and 100% and under temperatures (Temp) of 10°C, 20°C, 30°C, and 40°C, using 10 chicks for each RH–Temp combination.[9] The partially completed ANOVA table is as given next.

Source	d.f.	MS
RH		2.010
Temp		7.816
Interaction		1.642
Error		0.775
Total		

a. Should the two factors RH and Temp be considered as fixed or random? Explain.
b. Carry out the analysis of variance for both the fixed-effects case and the random-effects case. Do your conclusions differ in the two cases?
c. Write the fixed-effects and the random-effects models that could describe this experiment.
d. Using dummy variables, provide a regression model that can be used to obtain the results in the ANOVA table.
e. What regression model would be appropriate for describing the relationship of RH and Temp to survival time (Y) if the data for the independent variables are to be treated as interval rather than as nominal?

9. The diameters (Y) of three species of pine trees were compared at each of four locations, using samples of five trees per species at each location. The resulting data are given in the following table.

Species	Location 1	Location 2	Location 3	Location 4
A	23	25	21	14
	15	20	17	17
	26	21	16	19
	13	16	24	20
	21	18	27	24
B	28	30	19	17
	22	26	24	21
	25	26	19	18
	19	20	25	26
	26	28	29	23
C	18	15	23	18
	10	21	25	12
	12	22	19	23
	22	14	13	22
	13	12	22	19

---

[9]Adapted from a study by Price (1954).

a. Comment on whether each of the two factors should be considered fixed or random.
b. Use the following partially completed ANOVA table to carry out your analysis, first considering both factors as fixed and then considering a mixed model with "locations" treated as random. Compare your conclusions.

Source	d.f.	SS
Species		344.9333
Locations		46.0500
Interaction		113.6000
Error		875.6000

10. Consider an ANOVA table of the following form.

Source	d.f.	SS	MS	F	P
A					P_1
B					P_2
A $\times$ B					P_3
Error					

Use $\alpha = .05$ for all parts of this problem. In each case, decide what effects (if any) are significant, and what conclusions to draw, based on a two-way ANOVA table with the following P-values:
a. $P_1 = .03$, $P_2 = .51$, $P_3 = .31$
b. $P_1 = .001$, $P_2 = .63$, $P_3 = .007$
c. $P_1 = .093$, $P_2 = .79$, $P_3 = .02$
d. $P_1 = .56$, $P_2 = .38$, $P_3 = .24$

11. Assume that a total of 75 subjects were tested in a balanced two-way fixed-effects factorial. A plot of the means from the study is shown next. The dependent variable is Y, and the factors are A and B.

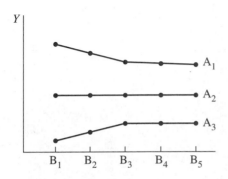

a. Give the left two columns (the source and degrees-of-freedom columns) of the source table for the data plotted.
b. Assume that the plot shows the population means. Indicate which significance tests should yield significant results in the ANOVA table.

12. Assume that the following ANOVA table came from a balanced two-way fixed-effects ANOVA. Show the formula used, the numbers filled in, and the value for each letter in the table.

Source	d.f.	SS	MS	F
A	a	5.12	g	6.40*
B	b	e	0.76	3.80*
A × B	c	4.32	0.36	i
Error	20	4.00	h	
Total	d	f		

a. Indicate which (if any) family or families of means should be evaluated with multiple comparisons.

13. The data in the following table and accompanying computer output are from a hypothetical study of human body temperature as affected by air temperature and a dietary supplement that is hoped to increase heat tolerance. Body temperatures (in degrees Celsius) were measured for 36 athletes immediately following a standard exercise routine in a room controlled to a fixed air temperature (in degrees Celsius). Each subject had been receiving a steady, fixed dose (in milligrams per kilogram of body weight) of the dietary supplement.

a. Provide an appropriate two-way ANOVA source table.

b. Provide F tests of the two main effects and the interaction. Use $\alpha = .05$. What do you conclude?

c. Define dummy variables, and specify an appropriate multiple regression model corresponding to the analysis done in part (a). [*Hint:* Use dummy variables that have the value -1 for Airtemp = 21 and for Dose = 0. Why is this a scientifically sensible choice?]

d. Since both factors are interval-scale variables, specify a corresponding natural-polynomial multiple regression model.

Edited SAS Output (PROC GLM) for Problem 13

```
                              ⋮   [Portion of output omitted]
                              ⋮
Dependent Variable: BODYTEMP

                          Sum of           Mean
Source            DF      Squares          Square      F Value    Pr > F
Model             11      0.13888889       0.01262626     0.43     0.9255
Error             24      0.70000000       0.02916667
Corrected Total   35      0.83888889

        R-Square            C.V.         Root MSE       BODYTEMP Mean
        0.165563          0.461644        0.17078           36.9944

Source            DF      Type I SS     Mean Square   F Value    Pr > F
AIRTEMP            3      0.02777778     0.00925926     0.32     0.8126
DOSE              2      0.06055556     0.03027778     1.04     0.3695
AIRTEMP*DOSE      6      0.05055556     0.00842593     0.29     0.9364

Source            DF      Type III SS   Mean Square   F Value    Pr > F
AIRTEMP            3      0.02777778     0.00925926     0.32     0.8126
DOSE              2      0.06055556     0.03027778     1.04     0.3695
AIRTEMP*DOSE      6      0.05055556     0.00842593     0.29     0.9364
```

Observation	Airtemp	Dose	Bodytemp
1	21	0.00	37.2
2	21	0.00	37.2
3	21	0.00	36.8
4	21	0.05	37.1
5	21	0.05	36.9
6	21	0.05	36.8
7	21	0.10	37.1
8	21	0.10	37.1
9	21	0.10	37.1
10	25	0.00	36.9
11	25	0.00	37.0
12	25	0.00	37.1
13	25	0.05	37.1
14	25	0.05	36.7
15	25	0.05	37.0
16	25	0.10	36.9
17	25	0.10	37.0
18	25	0.10	37.3
19	29	0.00	36.9
20	29	0.00	37.0
21	29	0.00	36.8
22	29	0.05	36.9
23	29	0.05	37.0
24	29	0.05	36.9
25	29	0.10	36.9
26	29	0.10	37.0
27	29	0.10	37.2
28	33	0.00	37.1
29	33	0.00	37.3
30	33	0.00	36.7
31	33	0.05	36.9
32	33	0.05	37.0
33	33	0.05	37.0
34	33	0.10	36.9
35	33	0.10	36.8
36	33	0.10	37.2

14. The experiment in Problem 14 in Chapter 18 was repeated—this time using three workers per technology/experience combination rather than just one. For each cell in the following table, the outputs (in number of units per hour) are listed for three randomly chosen workers at the given experience level.

	Technology		
Operator Experience	High Automation (H)	Moderate Automation (M)	Low Automation (L)
< 1 Year	16, 14, 12	8, 12, 10	5, 4, 8
1–2 Years	15, 18, 17	10, 10, 12	8, 10, 10
> 2 Years	20, 18, 19	13, 13, 14	12, 11, 13

    a. Are the two factors in this problem—Technology and Operator Experience—fixed or random factors?

    b. Form the table of sample means for these data, and comment on its content.

    c. The data may be analyzed by using a two-way ANOVA. Determine the ANOVA table for these data.

    d. Analyze the accompanying computer output to determine whether significant main effects exist due to technology type and worker experience, and whether these factors significantly interact.

Edited SAS Output (PROC GLM) for Problem 14

```
                   General Linear Models Procedure
                      Class Level Information
                   Class    Levels    Values
                   LINE        3      H L M
                   EXPER       3      1-2 <1 >2
             Number of observations in data set = 27
Dependent Variable: OUTPUT
                         Sum of           Mean
Source            DF     Squares          Square     F Value    Pr > F
Model              8    386.296296      48.287037     22.10     0.0001
Error             18     39.333333       2.185185
Corrected Total   26    425.629630

        R-Square          C.V.       Root MSE         OUTPUT Mean
        0.907588       12.02181      1.47824            12.2963

Source            DF   Type I SS    Mean Square    F Value    Pr > F
LINE               2  269.407407    134.703704      61.64     0.0001
EXPER              2  107.629630     53.814815      24.63     0.0001
LINE*EXPER         4    9.259259      2.314815       1.06     0.4050

Source            DF   Type III SS  Mean Square    F Value    Pr > F
LINE               2  269.407407    134.703704      61.64     0.0001
EXPER              2  107.629630     53.814815      24.63     0.0001
LINE*EXPER         4    9.259259      2.314815       1.06     0.4050

        Level of                 ----------OUTPUT----------
        LINE           N            Mean              SD
        H              9        16.5555556        2.55495162
        L              9         9.0000000        3.04138127
        M              9        11.3333333        1.93649167

        Level of                 ----------OUTPUT----------
        EXPER          N            Mean              SD
        1-2            9        12.2222222        3.56292639
        <1             9         9.8888889        4.01386486
        >2             9        14.7777778        3.30823887

Level of        Level of                ----------OUTPUT----------
LINE            EXPER        N            Mean              SD
H               1-2          3        16.6666667        1.52752523
H               <1           3        14.0000000        2.00000000
H               >2           3        19.0000000        1.00000000
L               1-2          3         9.3333333        1.15470054
L               <1           3         5.6666667        2.08166600
L               >2           3        12.0000000        1.00000000
M               1-2          3        10.6666667        1.15470054
M               <1           3        10.0000000        2.00000000
M               >2           3        13.3333333        0.57735027
```

15. An advertising company evaluated three types of television ads for a new, low-cost, sub-compact automobile: visual appeal ads, budget appeal ads, and feature appeal ads. To control for age differences, viewers from four age groups were chosen to evaluate the persuasiveness of the ads (as measured on a scale from 1 to 10, where 1 represented the lowest level of persuasion, and 10 the highest). Within each age group were six viewers; two each were randomly assigned to view one of the three types of ads. The sample persuasion scores are presented in the following table.

| Viewer Age | Type of Ad | | |
	Visual Appeal (V)	Budget Appeal (B)	Feature Appeal (F)
18–25 years	6, 5	8, 7	5, 4
26–35 years	7, 6	9, 10	4, 8
36–45 years	8, 9	9, 8	4, 2
46 and older	10, 9	10, 8	5, 4

a. Form the table of sample means for these data, and comment on its content.
b. The data may be analyzed by using a two-way ANOVA. Determine the ANOVA table for these data.
c. Analyze the accompanying computer output to determine whether significant main effects exist due to ad type and viewer age, and whether these factors significantly interact.

Edited SAS Output (PROC GLM) for Problem 15

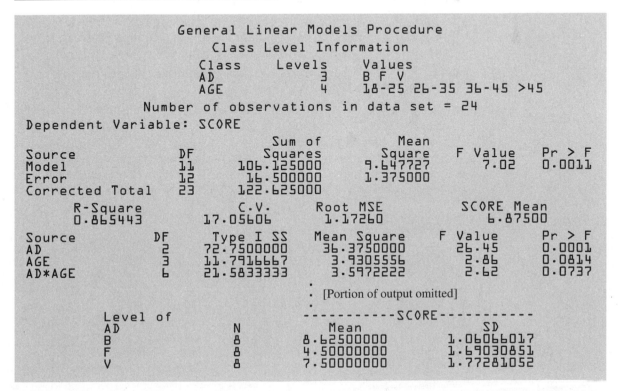

```
                    General Linear Models Procedure
                       Class Level Information
                     Class     Levels    Values
                     AD           3      B F V
                     AGE          4      18-25 26-35 36-45 >45
            Number of observations in data set = 24
Dependent Variable: SCORE
                                Sum of          Mean
Source               DF         Squares        Square    F Value    Pr > F
Model                11      106.125000       9.647727       7.02    0.0011
Error                12       16.500000       1.375000
Corrected Total      23      122.625000

       R-Square             C.V.        Root MSE         SCORE Mean
       0.865443          17.05606        1.17260           6.87500

Source               DF      Type I SS    Mean Square    F Value    Pr > F
AD                    2      72.7500000     36.3750000      26.45    0.0001
AGE                   3      11.7916667      3.9305556       2.86    0.0814
AD*AGE                6      21.5833333      3.5972222       2.62    0.0737
                                       .
                                       .   [Portion of output omitted]
                                       .
       Level of                      ----------SCORE----------
         AD            N            Mean                  SD
         B             8        8.62500000          1.06066017
         F             8        4.50000000          1.69030851
         V             8        7.50000000          1.77281052
```

```
          Level of                      ------------SCORE----------
          AGE              N             Mean                SD
          18-25            6         5.83333333          1.47196014
          26-35            6         7.33333333          2.16024690
          36-45            6         6.66666667          2.94392029
          >45              6         7.66666667          2.58198890
  Level of       Level of                   ----------SCORE----------
  AD             AGE            N             Mean                SD
  B              18-25          2         7.50000000          0.70710678
  B              26-35          2         9.50000000          0.70710678
  B              36-45          2         8.50000000          0.70710678
  B              >45            2         9.00000000          1.41421356
  F              18-25          2         4.50000000          0.70710678
  F              26-35          2         6.00000000          2.82842712
  F              36-45          2         3.00000000          1.41421356
  F              >45            2         4.50000000          0.70710678
  V              18-25          2         5.50000000          0.70710678
  V              26-35          2         6.50000000          0.70710678
  V              36-45          2         8.50000000          0.70710678
  V              >45            2         9.50000000          0.70710678
```

16. This question refers to the *U.S. News & World Report* mutual fund data presented in Problem 19 in Chapter 17. The variables described in that question were:

CAT (fund category): 1 = Aggessive growth; 2 = Long-term growth; 3 = Growth and income; 4 = Income.

LOAD (load status): N = No load; L = Load.

VOL (volatility): A letter grade from A+ to F indicating how much the month-to-month return varied from the fund's three-year total return: A+ = Least variability; F = Most variability.

OPI (Overall Performance Index): A measure of the relative performance of each fund over the past 1, 3, 5, and 10 years.

a. Suppose that a two-way ANOVA is to be performed, with OPI as the dependent variable and fund category (CAT) and load status (LOAD) as the two factors. State precisely the ANOVA model. Are the factors fixed or random?

b. In the SAS output that follows, complete the ANOVA table.

c. Analyze the data to determine whether there are significant main effects due to fund category and load status, and whether these factors significantly interact.

Edited SAS Output (PROC GLM) for Problem 16

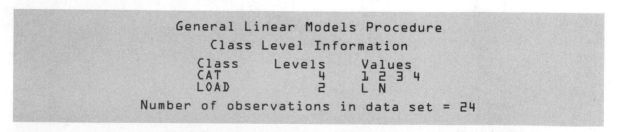

```
              General Linear Models Procedure
                  Class Level Information
                 Class     Levels    Values
                 CAT          4      1 2 3 4
                 LOAD         2      L N
          Number of observations in data set = 24
```

```
           General Linear Models Procedure (Continued)
              Class Level Information (Continued)
Dependent Variable: OPI
                              Sum of         Mean
Source              DF        Squares       Square    F Value    Pr > F
Model               ☐       312.07292     44.58185   ☐          ☐
Error               16      225.48667     14.09292
Corrected Total     23      537.55958

      R-Square            C.V.        Root MSE          OPI Mean
      0.580536         4.165578        3.7541           90.121

Source       DF     Type I SS    Mean Square   F Value    Pr > F
CAT          3      282.05458     94.01819      6.67      0.0039
LOAD         1        8.05042      8.05042      0.57      0.4608
CAT*LOAD     3       21.96792      7.32264      0.52      0.6748

Source       DF     Type III SS  Mean Square   F Value    Pr > F
CAT          3      282.05458     94.01819      6.67      0.0039
LOAD         1        8.05042      8.05042      0.57      0.4608
CAT*LOAD     3       21.96792      7.32264      0.52      0.6748

         Level of                  ------------OPI------------
         CAT          N               Mean               SD
          1           6           89.8000000        2.06009709
          2           6           93.6166667        3.06425630
          3           6           92.3833333        3.83218823
          4           6           84.6833333        4.77301442

         Level of                  ------------OPI------------
         LOAD         N               Mean               SD
          L          12           89.5416667        4.82972959
          N          12           90.7000000        4.98105502

   Level of      Level of          -----------OPI----------
   CAT           LOAD        N         Mean              SD
    1             L          3      90.8666667        2.48461935
    1             N          3      88.7333333        1.01159939
    2             L          3      92.4333333        1.96044213
    2             N          3      94.8000000        3.92810387
    3             L          3      91.4333333        1.09696551
    3             N          3      93.3333333        5.72741943
    4             L          3      83.4333333        6.36893502
    4             N          3      85.9333333        3.42101350
```

17. This question refers back to the *U.S. News & World Report* graduate school data presented in Problem 22 in Chapter 17.

    a. Suppose that a two-way ANOVA is to be performed, with 1995 starting salary as the dependent variable and school type (SCHOOL) and reputation rank (REP = 1 if in top 25; 2 if not) as the two factors. State precisely the ANOVA model. Are the factors fixed or random?

    b. In the SAS output that follows, complete the ANOVA table.

    c. Analyze the data to determine whether there are significant main effects due to school type and reputation rank, and whether these factors significantly interact. Use $\alpha = .10$.

Edited SAS Output (PROC GLM) for Problem 17

```
                    General Linear Models Procedure
                      Class Level Information
                    Class    Levels    Values
                    SCHOOL      2       1 2
                    REP         2       1 2

              Number of observations in data set = 24
Dependent Variable: SAL
                           Sum of          Mean
Source             DF      Squares         Square      F Value    Pr > F
Model              [ ]    547.37000      182.45667     [      ]   [      ]
Error               20    1419.02333      70.95117
Corrected Total     23    1966.39333

      R-Square             C.V.        Root MSE           SAL Mean
      0.278362          14.79494        8.4233            56.933

Source             DF    Type I SS     Mean Square    F Value    Pr > F
SCHOOL              1     37.00167       37.00167        0.52     0.4862
REP                 1    440.32667      440.32667        6.21     0.0216
SCHOOL*REP          1     70.04167       70.04167        0.99     0.3428

Source             DF    Type III SS   Mean Square    F Value    Pr > F
SCHOOL              1     37.00167       37.00167        0.52     0.4862
REP                 1    440.32667      440.32667        6.21     0.0216
SCHOOL*REP          1     70.04167       70.04167        0.99     0.3428

        Level of                      ----------SAL------------
        SCHOOL           N              Mean              SD
          1             12           58.1750000       9.65628622
          2             12           55.6916667       9.06396044

        Level of                      ----------SAL------------
        REP              N              Mean              SD
          1             12           61.2166667       8.62657979
          2             12           52.6500000       8.01969167

      Level of     Level of           ----------SAL-----------
      SCHOOL       REP         N         Mean              SD
        1           1          6      64.1666667       7.05454936
        1           2          6      52.1833333       8.31923474
        2           1          6      58.2666667       9.64710665
        2           2          6      53.1166667       8.47122581
```

References

Daly, M. B. 1973. "The Effect of Neighborhood Racial Characteristics on the Attitudes, Social Behavior, and Health of Low Income Housing Residents." Ph.D. dissertation, Department of Epidemiology, University of North Carolina, Chapel Hill, N.C.

Hocking, R. R. 1973. "A Discussion of the Two-Way Mixed Model." *American Statistician* 27(4): 148–52.

Price, R. D. 1954. "The Survival of Bacterium Tularense in Lice and Louse Feces." *American Journal of Tropical Medicine Hygiene* 3: 179–86.

Searle, S. R.; Casella, G.; and McCulloch, C. E. 1992. *Variance Components.* New York: John Wiley & Sons, pp. 118–27.

20

Two-way ANOVA with Unequal Cell Numbers

20-1 Preview

When we first began discussing two-way ANOVA in Chapter 18, we saw in Figure 18-1 several ways to classify a two-factor problem according to the observed data pattern. We have already covered methods for handling the single-observation-per-cell case (Chapter 18) and the equal-cell-number case (Chapter 19). In this chapter we examine procedures for analyzing two-factor patterns that have unequal cell numbers. This situation presents special problems in statistical analysis—both in computation and interpretation.

In treating these problems, we must make several distinctions among patterns of cell frequency. The first distinction is between *balanced* and *unbalanced* patterns. A balanced design has an equal number of observations in each cell, whereas an unbalanced design does not. Similarly, a *complete* design has at least one observation per cell, whereas an *incomplete* design has zero observations in one or more cells. All incomplete designs are unbalanced. Finally, some unbalanced designs exhibit *proportional cell frequencies*.

The general data configuration for the unequal-cell-number case in two-way ANOVA is presented in Table 20-1. A numerical example is given in Table 20-2, to which we will refer throughout this chapter. In the table we have

c = Number of columns (i.e., number of levels of the column factor)

r = Number of rows (i.e., number of levels of the row factor)

Y_{ijk} = kth observation in the cell associated with the ith row and jth column

n_{ij} = Number of observations in the cell associated with the ith row and jth column

Also,

$$\overline{Y}_{ij\cdot} = \frac{1}{n_{ij}} \sum_{k=1}^{n_{ij}} Y_{ijk}$$

$$\overline{Y}_{i..} = \frac{1}{n_{i.}} \sum_{j=1}^{c} \sum_{k=1}^{n_{ij}} Y_{ijk} \qquad \text{where} \qquad n_{i.} = \sum_{j=1}^{c} n_{ij}$$

$$\overline{Y}_{.j.} = \frac{1}{n_{.j}} \sum_{i=1}^{r} \sum_{k=1}^{n_{ij}} Y_{ijk} \qquad \text{where} \qquad n_{.j} = \sum_{i=1}^{r} n_{ij}$$

$$\overline{Y}_{...} = \frac{1}{n_{..}} \sum_{i=1}^{r} \sum_{j=1}^{c} \sum_{k=1}^{n_{ij}} Y_{ijk} \qquad \text{where} \qquad n_{..} = \sum_{i=1}^{r} \sum_{j=1}^{c} n_{ij}$$

TABLE 20-1   Data layout for the unequal-cell-number case (two-way ANOVA).

Row Factor	Column Factor 1	Column Factor 2	...	Column Factor c	Row Marginals
1	$Y_{111}, Y_{112},\ldots, Y_{11n_{11}}$ (Sample size $= n_{11}$) (Cell mean $= \overline{Y}_{11.}$)	$Y_{121}, Y_{122},\ldots, Y_{12n_{12}}$ (Sample size $= n_{12}$) (Cell mean $= \overline{Y}_{12.}$)	...	$Y_{1c1}, Y_{1c2},\ldots, Y_{1cn_{1c}}$ (Sample size $= n_{1c}$) (Cell mean $= \overline{Y}_{1c.}$)	$n_{1.}, \overline{Y}_{1..}$
2	$Y_{211}, Y_{212},\ldots, Y_{21n_{21}}$ (Sample size $= n_{21}$) (Cell mean $= \overline{Y}_{21.}$)	$Y_{221}, Y_{222},\ldots, Y_{22n_{22}}$ (Sample size $= n_{22}$) (Cell mean $= \overline{Y}_{22.}$)	...	$Y_{2c1}, Y_{2c2},\ldots, Y_{2cn_{2c}}$ (Sample size $= n_{2c}$) (Cell mean $= \overline{Y}_{2c.}$)	$n_{2.}, \overline{Y}_{2..}$
$\vdots$	$\vdots$	$\vdots$	...	$\vdots$	$\vdots$
r	$Y_{r11}, Y_{r12},\ldots, Y_{r1n_{r1}}$ (Sample size $= n_{r1}$) (Cell mean $= \overline{Y}_{r1.}$)	$Y_{r21}, Y_{r22},\ldots, Y_{r2n_{r2}}$ (Sample size $= n_{r2}$) (Cell mean $= \overline{Y}_{r2.}$)	...	$Y_{rc1}, Y_{rc2},\ldots, Y_{rcn_{rc}}$ (Sample size $= n_{rc}$) (Cell mean $= \overline{Y}_{rc.}$)	$n_{r.}, \overline{Y}_{r..}$
Column Marginals	$n_{.1}, \overline{Y}_{.1.}$	$n_{.2}, \overline{Y}_{.2.}$	...	$n_{.c}, \overline{Y}_{.c.}$	$n_{..}, \overline{Y}_{...}$

TABLE 20-2   Satisfaction with medical care (Y), classified by patient worry and affective communication between patient and physician.

Affective Communication	Worry Negative	Worry Positive	Row Marginals
High	2, 5, 8, 6, 2, 4, 3, 10 ($n_{11} = 8$) ($\overline{Y}_{11.} = 5$)	7, 5, 8, 6, 3, 5, 6, 4, 5, 6, 8, 9 ($n_{12} = 12$) ($\overline{Y}_{12.} = 6$)	$n_{1.} = 20$ $\overline{Y}_{1..} = 5.6$
Medium	4, 6, 3, 3 ($n_{21} = 4$) ($\overline{Y}_{21.} = 4$)	7, 7, 8, 6, 4, 9, 8, 7 ($n_{22} = 8$) ($\overline{Y}_{22.} = 7$)	$n_{2.} = 12$ $\overline{Y}_{2..} = 6$
Low	8, 7, 5, 9, 9, 10, 8, 6, 8, 10 ($n_{31} = 10$) ($\overline{Y}_{31.} = 8$)	5, 8, 6, 6, 9, 7, 7, 8 ($n_{32} = 8$) ($\overline{Y}_{32.} = 7$)	$n_{3.} = 18$ $\overline{Y}_{3..} = 7.56$
Column Marginals	$n_{.1} = 22$ $\overline{Y}_{.1.} = 6.18$	$n_{.2} = 28$ $\overline{Y}_{.2.} = 6.57$	$n_{..} = 50$ $\overline{Y}_{...} = 6.40$

The unequal-cell-number case arises quite frequently in observational studies. In such studies, one or more of the following statements are typically true:

1. All the variables of interest are not categorized before the data are collected.

2. New variables are often considered after the data are collected.

3. When all the variables are separately categorized, it is often impractical or even impossible to control in advance how the various categories will combine to form combinations of interest.

The unequal-cell-number case can also arise in experimental studies when a posteriori consideration is given to variables other than those of primary interest, even if the design based on the primary variables calls for equal cell numbers. Furthermore, unequal cell numbers generally result whenever data points are missing, which may occur (for example) because of study dropouts or incomplete records.

The example presented in Table 20-2 is derived from a study by Thompson (1972) of the relationship of two factors—patient perception of pregnancy and physician–patient communication—to patient satisfaction with medical care. Two main variables of interest were the patient's WORRY and a measure of affective communication (AFFCOM). These variables were developed from scales based on questionnaires administered to patients and their physicians. Based on the distribution of scores, the WORRY variable was grouped into the categories "positive" and "negative"; and the AFFCOM variable was grouped into the categories "high," "medium," and "low." Table 20-2 presents artificial data of this type, showing scores for satisfaction with medical care (Y = TOTSAT), classified according to these six combinations of levels of the factors WORRY and AFFCOM.

As the table indicates, the categorization scheme used leads to a two-way table with unequal cell numbers. For WORRY, there are 22 negatives and 28 positives; for AFFCOM, there are 20 high, 12 medium, and 18 low scores. When the separate categories for the two variables are considered together, the resulting six categories have different cell sample sizes, ranging from 4 (for medium AFFCOM, negative WORRY) to 12 (for high AFFCOM, positive WORRY).

20-2 Problem with Unequal Cell Numbers: Nonorthogonality

The key statistical concept associated with the special analytical problems encountered in the unequal-cell-number case pertains to the *nonorthogonality* of the sums of squares usually used to describe the sources of variation in a two-way ANOVA table. To clarify what *orthogonality* means, we first state the general formulas for these sums of squares (given in section 19-3 for the equal-cell-number case) in terms of unequal cell numbers:

$$\text{SSR} = \sum_{i=1}^{r} \sum_{j=1}^{c} \sum_{k=1}^{n_{ij}} (\overline{Y}_{i..} - \overline{Y}...)^2, \qquad \text{SSC} = \sum_{i=1}^{r} \sum_{j=1}^{c} \sum_{k=1}^{n_{ij}} (\overline{Y}_{.j.} - \overline{Y}...)^2,$$

$$\text{SSRC} = \sum_{i=1}^{r} \sum_{j=1}^{c} \sum_{k=1}^{n_{ij}} (\overline{Y}_{ij.} - \overline{Y}_{i..} - \overline{Y}_{.j.} + \overline{Y}...)^2, \qquad (20.1)$$

$$\text{SSE} = \sum_{i=1}^{r} \sum_{j=1}^{c} \sum_{k=1}^{n_{ij}} (Y_{ijk} - \overline{Y}_{ij.})^2, \qquad \text{TSS} = \sum_{i=1}^{r} \sum_{j=1}^{c} \sum_{k=1}^{n_{ij}} (Y_{ijk} - \overline{Y}...)^2$$

These formulas for SSR, SSC, and SSRC are often referred to as the *unconditional* sums of squares for rows, columns, and interaction, respectively; here, *unconditional* means that each sum of squares may be separately defined from basic principles to describe the variability associated with the estimated effects $(\overline{Y}_{i..} - \overline{Y}_{...})$, $(\overline{Y}_{.j.} - \overline{Y}_{...})$, and $(\overline{Y}_{ij.} - \overline{Y}_{i..} - \overline{Y}_{.j.} + \overline{Y}_{...})$ for rows, columns, and interaction, respectively. (Later we will explore an equivalent way to illustrate the meaning of the term *unconditional,* using regression analysis methodology.)

If the collection of sums of squares in (20.1) are orthogonal, the following fundamental equation holds:

$$\text{SSR} + \text{SSC} + \text{SSRC} + \text{SSE} = \text{TSS}$$

That is, the terms on the left-hand side must partition the total sum of squares into nonoverlapping sources of variation.

We have already seen that this fundamental equation holds true for the equal-cell-number case (Chapter 19). Unfortunately, when unequal cell numbers exist, the unconditional sums of squares no longer represent completely separate (i.e., orthogonal) sources of variation; thus,

$$\text{Unequal cell numbers} \Rightarrow \text{SSR} + \text{SSC} + \text{SSRC} + \text{SSE} \neq \text{TSS}$$

To see why this is so, consider the general regression formulation for two-way ANOVA, which incorporates the unequal-cell-number case as well as the equal-cell-number case. The general regression equation is

$$Y = \mu + \sum_{i=1}^{r-1} \alpha_i X_i + \sum_{j=1}^{c-1} \beta_j Z_j + \sum_{i=1}^{r-1}\sum_{j=1}^{c-1} \gamma_{ij} X_i Z_j + E \tag{20.2}$$

where μ, α_i, β_j, and γ_{ij} are regression coefficients and X_i and Z_j are appropriately defined dummy variables. The general form of the fundamental regression equation for this model may be written

$$\text{TSS} = \text{SSReg} + \text{SSE}$$

where

$$\text{TSS} = \sum (Y_l - \overline{Y})^2$$

$$\text{SSReg} = \sum (\hat{Y}_l - \overline{Y})^2$$
$$= \text{Regression SS}(X_1, X_2,\ldots, X_{r-1};\, Z_1, Z_2,\ldots, Z_{c-1};\, X_1 Z_1, X_1 Z_2,\ldots, X_{r-1}Z_{c-1})$$

$$\text{SSE} = \sum (Y_l - \hat{Y}_l)^2$$

and where the summation is over all $n_{..}$ observations.

Now, using the *extra-sum-of-squares principle* (see Chapter 9), we can partition the regression sum of squares in various ways to emphasize the contribution due to adding sets of variables to a regression model that already contains other sets of variables. In particular, we can partition the fundamental regression equation as follows with regard to model (20.2):

$$\text{TSS} = \text{Regression SS}(\overbrace{X_1, X_2,\ldots, X_{r-1}}^{R})$$

$$+ \text{Regression SS}(\overbrace{Z_1, Z_2,\ldots, Z_{c-1}}^{C} \mid \overbrace{X_1, X_2,\ldots, X_{r-1}}^{R}) \tag{20.3}$$

$$+ \text{Regression SS}(\overbrace{X_1 Z_1, X_1 Z_2,\ldots, X_{r-1}Z_{c-1}}^{RC} \mid \overbrace{X_1, X_2,\ldots, X_{r-1}, Z_1, Z_2,\ldots, Z_{c-1}}^{R,\,C})$$

$$+ \text{SSE}$$

On the other hand, if we wish to enter the column effects into the model first, the appropriate equation becomes

$$\text{TSS} = \text{Regression SS}(\overbrace{Z_1, Z_2, \ldots, Z_{c-1}}^{C})$$

$$+ \text{Regression SS}(\overbrace{X_1, X_2, \ldots, X_{r-1}}^{R} \mid \overbrace{Z_1, Z_2, \ldots, Z_{c-1}}^{C}) \qquad (20.4)$$

$$+ \text{Regression SS}(\overbrace{X_1Z_1, X_1Z_2, \ldots, X_{r-1}Z_{c-1}}^{RC} \mid \overbrace{X_1, X_2, \ldots, X_{r-1}, Z_1, Z_2, \ldots, Z_{c-1}}^{R, C})$$

$$+ \text{SSE}$$

As suggested by (20.3) and (20.4), it can be shown that

$$\text{Regression SS}(X_1, X_2, \ldots, X_{r-1}) \equiv \text{SSR}$$
$$\text{Regression SS}(Z_1, Z_2, \ldots, Z_{c-1}) \equiv \text{SSC} \qquad (20.5)$$
$$\text{Regression SS}(X_1Z_1, X_1Z_2, \ldots, X_{r-1}Z_{c-1}) \equiv \text{SSRC}$$

where SSR, SSC, and SSRC are the unconditional sums of squares given by (20.1). For example, we can express (20.3) and (20.4) as

$$\text{SSR} + \text{SS}(C \mid R) + \text{SS}(RC \mid R, C) + \text{SSE} = \text{TSS}$$

and

$$\text{SSC} + \text{SS}(R \mid C) + \text{SS}(RC \mid R, C) + \text{SSE} = \text{TSS}$$

respectively. Both of these equations involve conditional sums of squares.

When all the cell sample sizes are equal, however, it is also true that

$$\text{Equal cell numbers} \Rightarrow \begin{cases} \text{SSR} = \text{SS}(R \mid C) \\ \text{SSC} = \text{SS}(C \mid R) \\ \text{SSRC} = \text{SS}(RC \mid R, C) \end{cases}$$

Consequently, when all the cell sample sizes are equal, the extra sums of squares are not affected by variables in the model, and the following relationship holds:

$$\text{Equal cell numbers} \Rightarrow \text{SSR} + \text{SSC} + \text{SSRC} + \text{SSE} = \text{TSS} \qquad (20.6)$$

In the unequal-cell-number case, we have

$$\text{Unequal cell numbers} \Rightarrow \begin{cases} \text{SSR} \neq \text{SS}(R \mid C) \\ \text{SSC} \neq \text{SS}(C \mid R) \\ \text{SSRC} \neq \text{SS}(RC \mid R, C) \end{cases}$$

Therefore, in the unequal-cell-number case, (20.6) does not hold, and we must consider such expressions as (20.3) and (20.4), which reflect the importance of the order in which the effects are entered into the model. As discussed in section 20-3, the unequal-cell-number case is best handled by using regression analysis to carry out the two-way ANOVA calculations.

One exception to this occurs, when proportional cell frequencies satisfy

$$n_{ij} = \frac{n_{i.}n_{.j}}{n_{..}} \qquad (20.7)$$

When (20.7) holds, the following statement can be made:

$$n_{ij} = \frac{n_i.n_{.j}}{n_{..}} \Rightarrow \begin{cases} \text{SSR} = \text{SS}(\text{R} \mid \text{C}) \\ \text{SSC} = \text{SS}(\text{C} \mid \text{R}) \\ \text{SSRC} \neq \text{SS}(\text{RC} \mid \text{R, C}) \end{cases}$$

Thus, although (20.6) still does not hold in this case, (20.3) and (20.4) simplify to the single equation

$$\text{SSR} + \text{SSC} + \text{SS}(\text{RC} \mid \text{R, C}) + \text{SSE} = \text{TSS} \qquad (20.8)$$

Hence, (20.8) contains only one term—SS(RC | R, C)—that differs from the terms in (20.6). This sum of squares, however, can easily be obtained by subtraction, as is clear from (20.8). Thus, *when the proportional cell frequency allocation of (20.7) is used, the standard equal-cell-number ANOVA calculations can be performed,* without any need to resort to regression analysis methods.

To summarize, we have the flow diagram for two-way ANOVA shown in Figure 20-1.

FIGURE 20-1   Flow diagram for two-way ANOVA.

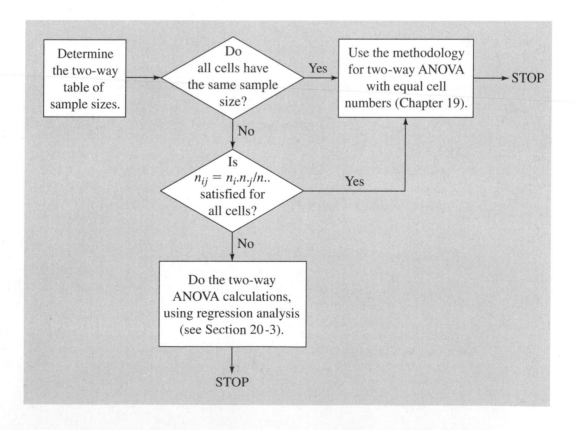

20-3 Regression Approach for Unequal Cell Sample Sizes

As described in section 20-2, the general regression model applicable to both the equal- and unequal-cell-number cases is

$$Y = \mu + \sum_{i=1}^{r-1} \alpha_i X_i + \sum_{j=1}^{c-1} \beta_j Z_j + \sum_{i=1}^{r-1}\sum_{j=1}^{c-1} \gamma_{ij} X_i Z_j + E \qquad (20.9)$$

where the $\{X_i\}$ and $\{Z_j\}$ are sets of dummy variables representing the r levels of the row factor and the c levels of the column factor, respectively. In general, *any* two-way ANOVA problem can be analyzed through a regression approach utilizing such a model. When there are unequal cell numbers, however, the order in which effects (row, column, or interaction) are tested becomes important, and a careless decision can yield inappropriate conclusions. The procedure we recommend is a backward-type algorithm in which interaction is considered before main effects (see Appelbaum and Cramer 1974). This algorithm involves the following steps:

Step 1. After fitting the full model, perform a chunk test for interaction (i.e., test H_0: $\gamma_{ij} = 0$ for all i and j in the preceding model). The F statistic is given by

$$F(X_1Z_1, X_1Z_2,\ldots, X_{r-1}Z_{c-1} \mid X_1, X_2,\ldots, X_{r-1}, Z_1, Z_2,\ldots, Z_{c-1})$$

and its numerator and denominator degrees of freedom are $(r-1)(c-1)$ and $n_{..} - rc$, respectively.

Step 2.

a.   If the step 1 test is significant, two primary options are available:
   i. Do no further testing and use the above full model as the final model.
   ii. Do individual testing to eliminate any nonsignificant interaction terms. The final model will then contain all main effects and all significant product terms.[1]

b.   If the step 1 test is not significant, reduce the model by eliminating all interaction terms. This reduced model is of the form

$$Y = \mu + \sum_{i=1}^{r-1} \alpha_i X_i + \sum_{j=1}^{c-1} \beta_j Z_j + E \qquad (20.10)$$

Then conduct two main-effect chunk tests of H_0: $\alpha_i = 0$ for all i and of H_0: $\beta_j = 0$ for all j, using the F statistics $F(X_1, X_2,\ldots, X_{r-1} \mid Z_1, Z_2,\ldots, Z_{c-1})$ and $F(Z_1, Z_2,\ldots, Z_{c-1} \mid X_1, X_2,\ldots, X_{r-1})$, respectively.[2] These tests consider the significance of the row effects given the column effects and the significance of the column effects given the row effects, respectively.

---

[1]Another possible element of option **ii** is to allow for the possible removal of main-effect terms that are not components of significant interaction terms.

[2]As mentioned elsewhere (e.g., Chapter 9), some statisticians prefer to use the mean-square residual for the full (interaction) model in these main-effect F tests, rather than using the mean-square residual for the reduced model (20.10).

Step 3.

a. If step 2(b) yields nonsignificant results for both tests, reduce the model further by eliminating the chunk of variables (the set of row or column main effects) corresponding to the least significant chunk test (the test having the larger P-value). Thus, if the test of H_0: $\alpha_i = 0$ for all i (the test for "rows" given "columns") has the larger P-value, the new reduced model is

$$Y = \mu + \sum_{j=1}^{c-1} \beta_j Z_j + E \qquad (20.11)$$

Alternatively, if H_0: $\beta_j = 0$ for all j (the test for "columns" given "rows") is less significant, the new reduced model is

$$Y = \mu + \sum_{i=1}^{r-1} \alpha_i X_i + E \qquad (20.12)$$

After reducing the model, conduct a chunk test for the main effects in this new reduced model, using either $F(X_1, X_2, \ldots, X_{r-1})$ or $F(Z_1, Z_2, \ldots, Z_{c-1})$, depending on which set of main effects remains.[3] If this final test is nonsignificant, the overall conclusion is that the row, column, and interaction effects are all unimportant. If the test is significant, the final model contains only the significant (row or column) main effects, and the conclusion is that only these effects are important.

b. If step 2(b) yields significant results for both tests, the reduced model (20.10) is the final model, and the conclusion is that both row and column effects are important but that there are no important interaction terms.

c. If step 2(b) produces exactly one significant test, there is no need to reduce the model further; the conclusion is that one of the two sets of main effects is important and that there is no significant interaction.[4]

Figure 20-2 provides a flow diagram for the preceding strategy.

20-3-1 Example of Regression Approach to Analyzing Unbalanced Two-way ANOVA Data

For the data on satisfaction with medical care from Table 20-2, two regression-model-based ANOVA tables (Tables 20-3 and 20-4) can be produced depending on whether rows precede columns or columns precede rows into the model. If we follow the strategy for regression analysis previously outlined, our first step is to conduct a chunk test for interaction. The regression model we are using for this test is

$$Y = \mu + \alpha_1 X_1 + \alpha_2 X_2 + \beta_1 Z_1 + \gamma_{11} X_1 Z_1 + \gamma_{21} X_2 Z_1 + E$$

---

[3]As in step 2, some statisticians prefer to use the mean-square residual for the full interaction model, rather than the mean-square residual for the reduced model, for the denominator of these tests.

[4]An alternative here is to reduce the model further by eliminating the nonsignificant set of main-effect variables. However, the unconditional test for the remaining set of main-effect variables may then be nonsignificant, even though the corresponding conditional test under model (20.10) is significant. In this situation, we believe that the conclusions based on model (20.10) are more appropriate.

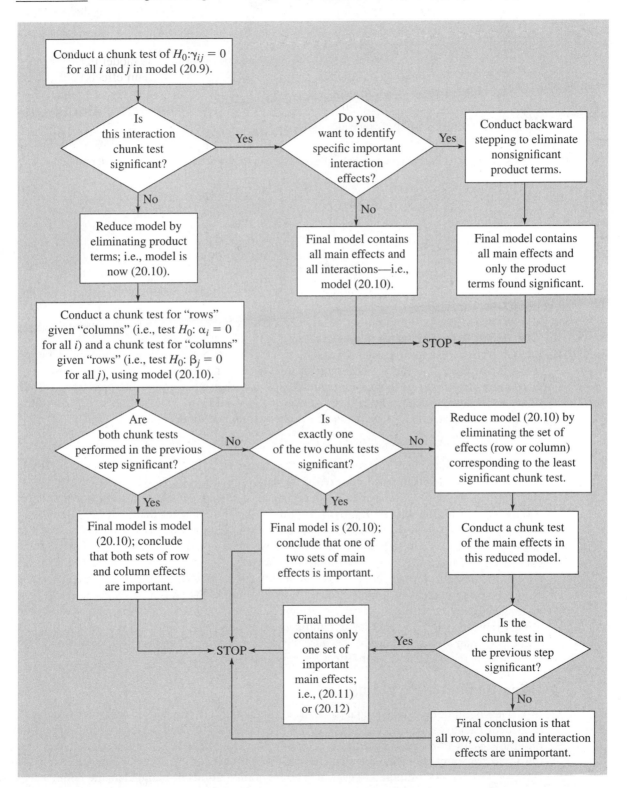

TABLE 20-3 ANOVA table when rows precede columns for regression analysis of data in Table 20-2.

Source	d.f.	SS	MS	F	P
X_1, X_2	2	38.756	19.378	$4.97_{2,47}$	$.01 < P < .025$
$Z_1 \mid X_1, X_2$	1	5.861	5.861	$1.52_{1,46}$	$.10 < P < .25$
$X_1Z_1, X_2Z_1 \mid X_1, X_2, Z_1$	2	27.383	13.692	$4.02_{2,44}$	$.01 < P < .025$
Residual	44	150.000	3.409		

TABLE 20-4 ANOVA table when columns precede rows for regression analysis of data in Table 20-2.

Source	d.f.	SS	MS	F	P
Z_1	1	1.870	1.870	$0.41_{1,48}$	$P > .25$
$X_1, X_2 \mid Z_1$	2	42.747	21.373	$5.54_{2,46}$	$.005 < P < .01$
$X_1Z_1, X_2Z_1 \mid X_1, X_2, Z_1$	2	27.383	13.692	$4.02_{2,44}$	$.01 < P < .025$
Residual	44	150.000	3.409		

where X_1 and X_2 are dummy variables for the row effects corresponding to the variable AFFCOM, and where Z_1 is a dummy variable for the column effects corresponding to the variable WORRY. The null hypothesis of no interaction is $H_0: \gamma_{11} = \gamma_{21} = 0$. As expected, Tables 20-3 and 20-4 provide the same numerical results for this conditional test. The P-value for this interaction test indicates significance at the .05 level but not at the .01 level. Under our strategy, if the investigator is using a .05 significance level, the analysis would stop at this point; we would conclude that significant interaction exists and that main-effect interpretations are not relevant. If the significance level is .01, however, we would conclude that no significant interaction effects exist, and then we would consider the reduced model

$$Y = \mu + \alpha_1 X_1 + \alpha_2 X_2 + \beta_1 Z_1 + E$$

The conditional test for "rows" given "columns" (i.e., $H_0: \alpha_1 = \alpha_2 = 0$) based on fitting the preceding model leads to an F ratio of 5.54, as given in the "$X_1, X_2 \mid Z_1$" row of Table 20-4. The corresponding P-value indicates significance at the .01 level. The conditional test for "columns" given "rows" (i.e., $H_0: \beta_1 = 0$) in the reduced model results in an F-value of 1.52 (see the "$Z_1 \mid X_1, X_2$" row of Table 20-3). The corresponding P-value ($> .10$) clearly indicates nonsignificance. Thus, based on our strategy, we would conclude (using $\alpha = .01$) that a significant AFFCOM main effect and a nonsignificant WORRY main effect are present.

If the model is further reduced to contain only the AFFCOM main effects (X_1 and X_2), the corresponding unconditional main-effect F test, as found in the "X_1, X_2" row of Table 20-3, is no longer significant at the .01 level. Nevertheless, since the model containing both sets of main effects indicates the existence of a significant AFFCOM main effect, we would argue that AFFCOM is an important variable because the latter model is taking the WORRY variable into account (i.e., the AFFCOM main-effect F test is conditional).

20-3-2 Using Computer Programs

Nearly all complex data analysis is now conducted with the aid of computer programs. Statistical packages usually contain ANOVA programs or ANOVA options within regression programs. With unbalanced data, and especially with incomplete data, the user of such packages must make an extra effort to understand the dummy variable coding schemes being used and the partial F tests being computed (e.g., which linear combinations of parameters are being tested in which model). Without this understanding, the user may inadvertently conduct tests quite different from those desired.

As mentioned earlier, ANOVA may be thought of as an analysis of means. We have found it helpful to consider such means as linear combinations of regression parameters. This approach can be particularly helpful with unbalanced data and is a lifesaver with incomplete data.

20-4 Higher-way ANOVA

As a special case of regression analysis, ANOVA can be generalized to any number of factors (i.e., independent variables). Nevertheless, extreme emphasis on complex ANOVA models and the associated testing procedures is likely to be unwarranted, especially for researchers in the health, medical, social, and behavioral sciences. Two reasons underlie this contention:

1. As more independent variables are considered, the researcher's wish to treat them all as nominal variables becomes less likely.

2. Even if all independent variables are treated as nominal, sufficient numbers of observations may not be available in all cells (e.g., there are some empty cells) or the researcher may not be able to place equal sample sizes in each cell.

Methods are available, however, for designing and analyzing experimental studies in which only a *fraction* of the total number of possible cells need be used; these methods permit the researcher to estimate the effects of primary interest. Refer to texts by Ostle (1963), Snedecor and Cochran (1967), and Peng (1967) for applications of such methods.

In general, however, regression analysis should predominate in higher-way ANOVA situations, especially since so much research in the health, medical, social, and behavioral sciences is observational. Consequently, we will not extend our discussion to three-way or higher ANOVA situations. For reference purposes, however, we present Table 20-5, the general three-way ANOVA table for the equal-cell-number situation when all factors are assumed to be fixed. An exercise at the end of the chapter deals with the three-way case.

TABLE 20-5   General three-way ANOVA table for equal cell sample sizes (all factors fixed).

Source	d.f.	MS	F
A (a levels)	$a - 1$	MSA	MSA/MSE
B (b levels)	$b - 1$	MSB	MSB/MSE
C (c levels)	$c - 1$	MSC	MSC/MSE
AB	$(a-1)(b-1)$	MSAB	MSAB/MSE
AC	$(a-1)(c-1)$	MSAC	MSAC/MSE
BC	$(b-1)(c-1)$	MSBC	MSBC/MSE
ABC	$(a-1)(b-1)(c-1)$	MSABC	MSABC/MSE
Error	$n - abc$	MSE	
Total	$n - 1$		

Note: See Ostle (1963) for information about F statistics when one or more factors are assumed to be random.

Problems

1. Consider hypothetical data based on a study concerning the effects of rapid cultural change on blood pressure levels for native citizens of an island in Micronesia. Blood pressures were taken on a random sample of 30 males over age 40 from a certain province. These persons, who commuted to work in the nearby westernized capital city, were also given a sociological questionnaire from which their social rankings in both their traditional and modern (i.e., westernized) cultures were determined. The results are summarized in the following table.

Modern Rank (Factor A)	Traditional Rank (Factor B)		
	HI	MED	LO
HI	130, 140, 135	150, 145	175, 160, 170, 165, 155
MED	145, 140, 150	150, 160, 155	165, 155, 165, 170, 160
LO	180, 160, 145	155, 140, 135	125, 130, 110

a. Discuss the table of sample means for this data set.
b. Give an appropriate regression model for this data set, treating the two factors as nominal variables.
c. Using the regression ANOVA tables that follow (where X pertains to factor A, and Z to factor B), carry out two different main-effect tests for each factor, and also test for interaction. Compare the results of the two main-effect tests for each factor.

Source	d.f.	MS
X_1	1	469.17985
$X_2 \mid X_1$	1	508.52217
$Z_1 \mid X_1, X_2$	1	187.97673
$Z_2 \mid X_1, X_2, Z_1$	1	7.54570
$X_1 Z_1 \mid X_1, X_2, Z_1, Z_2$	1	3,925.29395
$X_1 Z_2 \mid X_1, X_2, Z_1, Z_2, X_1 Z_1$	1	9.70621
$X_2 Z_1 \mid X_1, X_2, Z_1, Z_2, X_1 Z_1, X_1 Z_2$	1	633.17613
$X_2 Z_2 \mid X_1, X_2, Z_1, Z_2, X_1 Z_1, X_1 Z_2, X_2 Z_1$	1	2.67593
Residual	21	75.83333

Source	d.f.	MS
Z_1	1	278.59213
$Z_2 \mid Z_1$	1	22.71129
$X_1 \mid Z_1, Z_2$	1	391.77041
$X_2 \mid Z_1, Z_2, X_1$	1	480.15062
$X_1 Z_1 \mid Z_1, Z_2, X_1, X_2$	1	3,925.29395
$X_1 Z_2 \mid Z_1, Z_2, X_1, X_2, X_1 Z_1$	1	9.70621
$X_2 Z_1 \mid Z_1, Z_2, X_1, X_2, X_1 Z_1, X_1 Z_2$	1	633.17613
$X_2 Z_2 \mid Z_1, Z_2, X_1, X_2, X_1 Z_1, X_1 Z_2, X_2 Z_1$	1	2.67593
Residual	21	75.83333

d. How might you modify the regression model given in part (b) so that any trends in blood pressure levels could be quantified in terms of increasing social rankings for the two factors? (This requires assigning numerical values to the categories of each factor.) What difficulty do you encounter in defining such a model?

2. A study was conducted to assess the combined effects of patient attitude and patient–physician communication on patient satisfaction with medical care during pregnancy. A random sample of 110 pregnant women under the care of private physicians was followed from the first visit with the physician until delivery. On the basis of specially devised questionnaires, the following variables were measured for each patient: Y = Satisfaction score; X_1 = Attitude score; and X_2 = Communication score. Each score was developed as an interval variable, but some question remains as to whether the analysis should treat the attitude and/or communication scores as nominal variables.

   a. What would be an appropriate regression model for describing the joint effect of X_1 and X_2 on Y if an interaction between communication and attitude is possible and if all variables are treated as interval variables?

   b. What would be an appropriate regression model (using dummy variables) if the analyst wished to allow for an interaction effect but desired only to compare high values versus low values (i.e., to make group comparisons) for both the communication and attitude variables? What kind of ANOVA model would this regression model correspond to?

   c. When would the model in part (a) be preferable to that in (b), and vice versa?

   d. If both independent variables are treated nominally, as in part (b), would you expect the associated 2 × 2 table to have equal numbers in each of the four cells?

3. The data listed in the table on the following page are from an unpublished study by Harbin and others (1985). Subjects were young (ages 18 through 29) or old (ages 60 through 86) men. Each subject was exposed to 0 or 100 parts per million (PPM) of carbon monoxide (CO) for a period before and during testing. Median reaction times for 30 trials are reported for two different tasks: (1) simple reaction time (no choice), REACTIM1; (2) two-choice reaction time, REACTIM2. Pilot data analysis from an earlier study established that this dependent variable followed an appropriate Gaussian distribution. For this problem, consider only REACTIM1 as a dependent variable. Use $\alpha = .01$.

   a. Observe the cross-tabulation of AGEGROUP and PPM_CO in the accompanying SAS Proc FREQ computer output. Why might this be called a nearly orthogonal two-way design? Any missing data were due to technical problems unrelated to the response variable or treatments.

   b. Using dummy variables coded -1 (young, PPM_CO = 0) and 1, define dummy variables and a corresponding multiple regression model for a two-way ANOVA.

   c. Use an appropriate computer program to fit the regression model in part (b). Report appropriate tests of the AGEGROUP × CO interaction and tests of their main effects in an appropriate summary source table.

   d. SAS's Proc GLM procedure automatically codes dummy variables (and properly treats unequal sample sizes). The accompanying Proc GLM computer output shows the two-way ANOVA results for this problem. Complete the F-statistics in the ANOVA table, and report the results of the same tests as in part (c). Compare the results to those in part (c), and explain any differences.

Observation	AGEGROUP	PPM_CO	REACTIM1	REACTIM2
1	Young	0	291.5	632.0
2	Young	0	471.0	607.5
3	Young	0	692.0	859.0
4	Young	0	376.0	484.0
5	Young	0	372.5	501.0
6	Young	0	307.0	381.0
7	Young	0	501.0	559.0
8	Young	0	466.0	632.0
9	Young	0	375.0	434.0
10	Young	0	425.0	454.0
11	Young	0	343.0	542.0
12	Young	0	348.0	471.0
13	Young	0	503.0	521.0
14	Young	0	382.5	519.0
15	Young	100	472.5	515.0
16	Young	100	354.0	521.0
17	Young	100	350.0	456.0
18	Young	100	486.0	522.0
19	Young	100	402.0	472.0
20	Young	100	347.0	414.0
21	Young	100	320.0	363.0
22	Young	100	446.0	591.0
23	Young	100	410.0	539.5
24	Young	100	302.0	472.5
25	Young	100	692.5	656.0
26	Young	100	447.5	548.0
27	Young	100	525.5	527.0
28	Young	100	322.5	574.0
29	Young	100	468.5	559.5
30	Young	100	378.0	499.5
31	Young	100	497.5	529.5
32	Old	0	542.0	595.0
33	Old	0	599.0	606.0
34	Old	0	562.0	598.0
35	Old	0	586.0	744.0
36	Old	0	674.0	724.0
37	Old	0	762.0	836.5
38	Old	0	697.0	834.0
39	Old	0	583.0	698.5
40	Old	0	533.5	668.0
41	Old	0	524.5	670.0
42	Old	0	500.0	587.0
43	Old	0	680.0	912.5
44	Old	0	563.5	619.0
45	Old	100	523.5	646.5
46	Old	100	770.0	862.5
47	Old	100	712.0	829.0
48	Old	100	653.0	697.0
49	Old	100	699.5	818.0
50	Old	100	561.0	819.5
51	Old	100	751.0	872.0
52	Old	100	520.5	889.0
53	Old	100	523.0	601.0

Edited SAS Output (PROC FREQ and PROC GLM) for Problem 3

```
              TABLE OF AGEGROUP BY PPM_CO

        AGEGROUP |    PPM_CO
        Frequency|
        Percent  |
        Row Pct  |
        Col Pct  |        0|      100|    Total
        ---------+---------+---------+
        Old      |       13|        9|       22
                 |    24.53|    16.98|    41.51
                 |    59.09|    40.91|
                 |    48.15|    34.62|
        ---------+---------+---------+
        Young    |       14|       17|       31
                 |    26.42|    32.08|    58.49
                 |    45.16|    54.84|
                 |    51.85|    65.38|
        ---------+---------+---------+
        Total           27        26        53
                     50.94     49.06    100.00

           General Linear Models Procedure
              Class Level Information

           Class       Levels      Values
           AGEGROUP          2      Old Young
           PPM_CO            2      0 100

      Number of observations in data set = 53
```

Dependent Variable: REACTIM1

Source	DF	Sum of Squares	Mean Square	F Value	Pr > F
Model	3	484794.098	161598.033	17.51	0.0001
Error	49	452176.619	9228.094		
Corrected Total	52	936970.717			

R-Square	C.V.	Root MSE	REACTIM1 Mean
0.517406	19.14396	96.0630	501.792

SS(AGEGROUP)

Source	DF	Type I SS	Mean Square	F Value	Pr > F
AGEGROUP	1	478181.843	478181.843	☐	☐
PPM_CO	1	4210.892	4210.892	☐	☐
AGEGROUP*PPM_CO	1	2401.364	2401.364	☐	☐

SS(PPM_CO | AGEGROUP)   SS(PPM_CO | AGEGROUP, AGEGROUP * PPM_CO)

Source	DF	Type II SS	Mean Square	F Value	Pr > F
AGEGROUP	1	481452.369	481452.369	☐	☐
PPM_CO	1	4210.892	4210.892	☐	☐
AGEGROUP*PPM_CO	1	2401.364	2401.364	☐	☐

SS(AGEGROUP | PPM_CO), AGEGROUP * PPM_CO)   SS(AGEGROUP * PPM_CO | AGEGROUP, PPM_CO)

4. a.–d. Repeat Problem 3, using REACTIM2 as the dependent variable and referring to the accompanying computer output.

Edited SAS Output (PROC GLM) for Problem 4

```
                    General Linear Models Procedure
                       Class Level Information
                  Class        Levels       Values
                  AGEGROUP        2          Old Young
                  PPM_CO          2          0    100
Number of observations in data set = 53
Dependent Variable: REACTIM2
                              Sum of              Mean
Source              DF        Squares            Square      F Value    Pr > F
Model                3       584722.361        194907.454     20.04     0.0001
Error               49       476633.960          9727.224
Corrected Total     52      1061356.321
        R-Square            C.V.         Root MSE       REACTIM2 Mean
        0.550920          16.09215        98.6267          612.887
Source              DF      Type I SS      Mean Square     F Value    Pr > F
AGEGROUP             1     543059.011      543059.011      ┌──────┐  ┌──────┐
PPM_CO               1       3971.106        3971.106      └──────┘  └──────┘
AGEGROUP*PPM_CO      1      37692.243       37692.243      ┌──────┐  ┌──────┐
                                                           └──────┘  └──────┘
Source              DF     Type II SS      Mean Square     F Value    Pr > F
AGEGROUP             1     545527.925      545527.925      ┌──────┐  ┌──────┐
PPM_CO               1       3971.106        3971.106      └──────┘  └──────┘
AGEGROUP*PPM_CO      1      37692.243       37692.243      ┌──────┐  ┌──────┐
                                                           └──────┘  └──────┘
```

5. A crime victimization study was undertaken in a medium-size southern city. The main purpose was to determine the effects of being a crime victim on confidence in law enforcement authority and in the legal system itself. A questionnaire was administered to a stratified random sample of 40 city residents; among the information elicited were data on the number of times victimized, a measure of social class status (SCLS), and a measure of the respondent's confidence in law enforcement and in the legal system. The data are reproduced in the following table.

No. of Times Victimized	Social Class Status		
	LO	MED	HI
0	4, 14, 15, 19, 17, 17, 16	7, 10, 12, 15, 16	8, 19, 10, 17
1	2, 7, 18	6, 19, 12, 12	7, 6, 5, 3, 16
2+	7, 8, 2, 11, 12	1, 2, 4	4, 2, 8, 9

a. Determine the table of sample means, and comment on any patterns noted.
b. Analyze this data set using the following ANOVA table.
c. How would you analyze this data set by using the two tables of regression results that follow?
d. What ANOVA assumption(s) might not hold for these data?

Source	d.f.	SS	MS
VICTIM	2	4.0000E + 02	2.0000E + 02
SCLS	2	2.2739E + 01	1.1370E + 01
VICTIM × SCLS	4	1.0993E + 02	2.7483E + 01
Error	31	7.0408E + 02	2.2712E + 01

Source	d.f.	MS
Z_1	1	44.03235
$Z_2 \mid Z_1$	1	1.03496
$X_1 \mid Z_1, Z_2$	1	395.75734
$X_2 \mid Z_1, Z_2, X_1$	1	0.06778
$X_1 Z_1 \mid Z_1, Z_2, X_1, X_2$	1	1.68985
$X_1 Z_2 \mid Z_1, Z_2, X_1, X_2, X_1 Z_1$	1	3.31635
$X_2 Z_1 \mid Z_1, Z_2, X_1, X_2, X_1 Z_1, X_1 Z_2$	1	0.40190
$X_2 Z_2 \mid Z_1, Z_2, X_1, X_2, X_1 Z_1, X_1 Z_2, X_2 Z_1$	1	94.59353
Residual	31	22.71229

Source	d.f.	MS
X_1	1	407.86993
$X_2 \mid X_1$	1	0.52174
$Z_1 \mid X_1, X_2$	1	27.98766
$Z_2 \mid X_1, X_2, Z_1$	1	4.51309
$X_1 Z_1 \mid X_1, X_2, Z_1, Z_2$	1	1.68985
$X_1 Z_2 \mid X_1, X_2, Z_1, Z_2, X_1 Z_1$	1	3.31635
$X_2 Z_1 \mid X_1, X_2, Z_1, Z_2, X_1 Z_1, X_1 Z_2$	1	0.40190
$X_2 Z_2 \mid X_1, X_2, Z_1, Z_2, X_1 Z_1, X_1 Z_2, X_2 Z_1$	1	94.59353
Residual	31	22.71229

Note: X pertains to number of times victimized; *Z* pertains to social class status.

6. The effect of a new antidepressant drug on reducing the severity of depression was studied in manic–depressive patients at two state mental hospitals. In each hospital all such patients were randomly assigned to either a treatment (new drug) or a control (old drug) group. The results of this experiment are summarized in the following table; a high mean score indicates more of a lowering in depression level than does a low mean score.

	Group	
Hospital	Treatment	Control
A	$n = 25, \bar{Y} = 8.5, S = 1.3$	$n = 31, \bar{Y} = 4.6, S = 1.8$
B	$n = 25, \bar{Y} = 2.3, S = 0.9$	$n = 31, \bar{Y} = -1.7, S = 1.1$

a. Without performing any statistical tests, interpret the means in the table.
b. What regression model is appropriate for analyzing the data? For this model, describe how to test whether the new drug has a significant effect.

7. A study was conducted by a television network in a certain state to evaluate the viewing characteristics of adult females. Each individual in a stratified random sample of 480 women was sent a questionnaire; the strata were formed on the basis of the following three factors: season (winter or summer), region (eastern, central, or western), and residence (rural or urban). The averages of the total time reported watching TV (hours per day) are summarized in the accompanying table of sample means and standard deviations.

Residence and Region	Summer			Winter			Marginals	
	n	$\overline{Y}$	S	n	$\overline{Y}$	S	n	$\overline{Y}$
Rural								
East	40	2.75	1.340	40	4.80	0.851	80	3.78
Central	40	2.75	1.380	40	4.85	0.935	80	3.80
West	40	2.65	1.180	40	4.78	0.843	80	3.71
Marginals	120	2.72		120	4.81		240	3.76
Urban								
East	40	3.38	0.958	40	3.65	0.947	80	3.52
Central	40	3.15	1.130	40	4.50	0.743	80	3.83
West	40	3.65	0.779	40	4.05	0.781	80	3.85
Marginals	120	3.39		120	4.07		240	3.73
Marginals	240	3.06		240	4.44		480	3.75

a. Suppose that the questionnaire contained items concerning additional factors such as occupation (categorized as housewife, blue-collar worker, white-collar worker, or professional), age (categorized as 20 to 34, 35 to 50, and over 50), and number of children (categorized as 0, 1–2, and 3+). What is the likelihood of obtaining equal cell numbers when carrying out an ANOVA to consider these additional variables?

b. Examine the table of sample means for main effects and interactions. (You may want to form two-factor summary tables for looking at two-factor interactions.)

c. Using the following table of ANOVA results, carry out appropriate F tests, and discuss your results.

d. State a regression model appropriate for obtaining information equivalent to the ANOVA results presented next.

Source	d.f.	SS	MS
RESID	1	1.3333E − 01	1.3333E − 01
REGION	2	2.5527E + 00	1.2763E + 00
SEASON	1	2.2963E + 02	2.2963E + 02
RESID × REGION	2	3.3247E + 00	1.6623E + 00
RESID × SEASON	1	6.0492E + 01	6.0492E + 01
REGION × SEASON	2	7.2247E + 00	3.6123E + 00
RESID × REGION × SEASON	2	6.7460E + 00	3.3730E + 00
Error	468	4.7821E + 02	1.0218E + 00

8. Suppose that the following data were obtained by an investigator studying the influence of estrogen injections on change in the pulse rate of adolescent chimpanzees:

Male
- Control: 5.1, −2.3, 4.2, 3.8, 3.2, −1.5, 6.1, −2.5, 1.9, −3.0, −2.8, 1.7
- Estrogen: 15.0, 6.2, 4.1, 2.3, 7.6, 14.8, 12.3, 13.1, 3.4, 8.5, 11.2, 6.9

Female
- Control: −2.3, −5.8, −1.5, 3.8, 5.5, 1.6, −2.4, 1.9
- Estrogen: 7.3, 2.4, 6.5, 8.1, 10.3, 2.2, 12.7, 6.3

a. What are the factors in this experiment? Should they be designated as fixed or random?

b. Demonstrate that the cell frequency for two-way ANOVA in this problem is proportional; that is, $n_{ij} = n_{i.}n_{.j}/n_{..}$ for each of the four cells.

c. Use the following table of sample means

Sex	Control	Estrogen
Male	1.158333	8.783333
Female	0.100000	6.975000

and the general formulas

$$\text{SS(rows)} = \sum_{i=1}^{r} \sum_{j=1}^{c} \sum_{k=1}^{n_{ij}} (\overline{Y}_{i..} - \overline{Y}_{...})^2$$

$$\text{SS(columns)} = \sum_{i=1}^{r} \sum_{j=1}^{c} \sum_{k=1}^{n_{ij}} (\overline{Y}_{.j.} - \overline{Y}_{...})^2$$

$$\text{SS(cells)} = \sum_{i=1}^{r} \sum_{j=1}^{c} \sum_{k=1}^{n_{ij}} (\overline{Y}_{ij.} - \overline{Y}_{...})^2$$

to analyze the data for this problem, employing the usual methodology for equal-cell-number two-way ANOVA and the fact that SSE = 530.24078 (i.e., compute the sums of squares for rows, columns, and cells directly, and then obtain the sum of squares for interaction by subtraction).

d. What regression model is appropriate for analyzing this data set?

e. Using the regression analysis results given in the following table, check whether SS(Rows) = Regression SS(SEX) = Regression SS(SEX | TREATMENT) and SS(Columns) = Regression SS(TREATMENT) = Regression SS(TREATMENT | SEX), where SS(Rows) and SS(Columns) are as obtained in part (c). What has been demonstrated here?

Source	d.f.	SS
SEX	1	19.72667
TREATMENT | SEX	1	536.55619
SEX × TREATMENT | SEX, TREATMENT	1	1.35000
Residual	36	530.24078

Source	d.f.	SS
TREATMENT	1	536.55619
SEX | TREATMENT	1	19.72667
SEX × TREATMENT | SEX, TREATMENT	1	1.35000
Residual	36	530.24078

Note: $\text{SEX} = \begin{cases} -1 & \text{if male} \\ 1 & \text{if female} \end{cases}$ and $\text{TREATMENT} = \begin{cases} -1 & \text{if control} \\ 1 & \text{if estrogen} \end{cases}$

9. The data listed in the following table relate to a study by Reiter and others (1981) of the effects of injecting triethyl-tin (TET) into rats once at age 5 days. The animals were injected with 0, 3, or 6 mg per kilogram of body weight. The response was the log of the activity count for 1 hour, counting at 21 days of age. The rat was left to move about freely in a figure 8 maze. Analysis of other studies with this type of activity count confirms that log counts should yield Gaussian errors if the model is correct.

a. Tabulate the DOSAGE $\times$ SEX cell sample sizes. Explain why this might be called a nearly orthogonal design.

b. Using the accompanying SAS computer output, conduct a two-way ANOVA with SEX and DOSAGE as factors. (First complete the F-statistics and p-values in the ANOVA table.)

c. Using $\alpha = .05$, report your conclusions based on the ANOVA.

d. Which, if any, families of means should be followed up with multiple-comparison tests? What type of comparisons would you recommend?

Observation	LOGACT21	DOSAGE	SEX	CAGE
1	2.636	0	Male	5
2	2.736	0	Male	6
3	2.775	0	Male	7
4	2.672	0	Male	9
5	2.653	0	Male	11
6	2.569	0	Male	12
7	2.737	0	Male	15
8	2.588	0	Male	16
9	2.735	0	Male	17
10	2.444	3	Male	3
11	2.744	3	Male	5
12	2.207	3	Male	6
13	2.851	3	Male	7
14	2.533	3	Male	9
15	2.630	3	Male	11
16	2.688	3	Male	12
17	2.665	3	Male	15
18	2.517	3	Male	16
19	2.769	3	Male	17
20	2.694	6	Male	3
21	2.845	6	Male	5
22	2.865	6	Male	6
23	3.001	6	Male	7
24	3.043	6	Male	9
25	3.066	6	Male	11
26	2.747	6	Male	12
27	2.894	6	Male	15
28	1.851	6	Male	16
29	2.489	6	Male	17
30	2.494	0	Female	3
31	2.723	0	Female	5
32	2.841	0	Female	6
33	2.620	0	Female	7
34	2.682	0	Female	9
35	2.644	0	Female	11
36	2.684	0	Female	12

Observation	LOGACT21	DOSAGE	SEX	CAGE
37	2.607	0	Female	15
38	2.591	0	Female	16
39	2.737	0	Female	17
40	2.220	3	Female	3
41	2.371	3	Female	5
42	2.679	3	Female	6
43	2.591	3	Female	7
44	2.942	3	Female	9
45	2.473	3	Female	11
46	2.814	3	Female	12
47	2.622	3	Female	15
48	2.730	3	Female	16
49	2.955	3	Female	17
50	2.540	6	Female	3
51	3.113	6	Female	5
52	2.468	6	Female	6
53	2.606	6	Female	7
54	2.764	6	Female	9
55	2.859	6	Female	11
56	2.763	6	Female	12
57	3.000	6	Female	15
58	3.111	6	Female	16
59	2.858	6	Female	17

Edited SAS Output (PROC FREQ and PROC GLM) for Problem 9

```
                    TABLE OF DOSAGE BY SEX
          DOSAGE    |   SEX
          Frequency |
          Percent   |
          Row Pct   |
          Col Pct   |Female   |Male     |    Total
          ----------+---------+---------+
              0     |    10   |     9   |      19
                    | 16.95   | 15.25   |   32.20
                    | 52.63   | 47.37   |
                    | 33.33   | 31.03   |
          ----------+---------+---------+
              3     |    10   |    10   |      20
                    | 16.95   | 16.95   |   33.90
                    | 50.00   | 50.00   |
                    | 33.33   | 34.48   |
          ----------+---------+---------+
              6     |    10   |    10   |      20
                    | 16.95   | 16.95   |   33.90
                    | 50.00   | 50.00   |
                    | 33.33   | 34.48   |
          ----------+---------+---------+
          Total          30        29          59
                       50.85     49.15      100.00
```

Continued

```
                    General Linear Models Procedure
                      Class Level Information
                  Class      Levels      Values
                  DOSAGE       3          0 3 6
                  SEX          2          Female Male
            Number of observations in data set = 59
Dependent Variable: LOGACT21
                            Sum of          Mean
Source           DF         Squares         Square    F Value    Pr > F
Model             5        0.28197725     0.05639545    1.16      0.3412
Error            53        2.57750079     0.04863209
Corrected Total  58        2.85947803
          R-Square            C.V.        Root MSE     LOGACT21 Mean
          0.098611          8.196165      0.22053         2.69061
Source           DF        Type I SS     Mean Square   F Value    Pr > F
DOSAGE            2        0.25750763     0.12875381   ┌──────┐  ┌──────┐
SEX               1        0.01054232     0.01054232   └──────┘  └──────┘
DOSAGE*SEX        2        0.01392730     0.00696365   ┌──────┐  ┌──────┐
                                                       └──────┘  └──────┘
                                                       ┌──────┐  ┌──────┐
                                                       └──────┘  └──────┘
Source           DF        Type II SS    Mean Square   F Value    Pr > F
DOSAGE            2        0.25806594     0.12903297   ┌──────┐  ┌──────┐
SEX               1        0.01054232     0.01054232   └──────┘  └──────┘
DOSAGE*SEX        2        0.01392730     0.00696365   ┌──────┐  ┌──────┐
                                                       └──────┘  └──────┘
                                                       ┌──────┐  ┌──────┐
                                                       └──────┘  └──────┘
```

10. The experimenters described in Problem 9 hoped that home cage would not affect activity level in any systematic fashion. Explore this question by repeating Problem 9, but replacing SEX with CAGE in your analysis.

Edited SAS Output (PROC FREQ and PROC GLM) for Problem 10

```
                              TABLE OF DOSAGE BY CAGE
DOSAGE   | CAGE
Frequency|
Percent  |
Row Pct  |
Col Pct  |      3|       5|       6|       7|       9|    Total
---------+--------+--------+--------+--------+--------+
    0    |     1  |     2  |     2  |     2  |     2  |      19
         |  1.69  |  3.39  |  3.39  |  3.39  |  3.39  |   32.20
         |  5.26  | 10.53  | 10.53  | 10.53  | 10.53  |
         | 20.00  | 33.33  | 33.33  | 33.33  | 33.33  |
---------+--------+--------+--------+--------+--------+
    3    |     2  |     2  |     2  |     2  |     2  |      20
         |  3.39  |  3.39  |  3.39  |  3.39  |  3.39  |   33.90
         | 10.00  | 10.00  | 10.00  | 10.00  | 10.00  |
         | 40.00  | 33.33  | 33.33  | 33.33  | 33.33  |
---------+--------+--------+--------+--------+--------+
    6    |     2  |     2  |     2  |     2  |     2  |      20
         |  3.39  |  3.39  |  3.39  |  3.39  |  3.39  |   33.90
         | 10.00  | 10.00  | 10.00  | 10.00  | 10.00  |
         | 40.00  | 33.33  | 33.33  | 33.33  | 33.33  |
---------+--------+--------+--------+--------+--------+
Total          5        6        6        6        6        59
            8.47    10.17    10.17    10.17    10.17    100.00
```

```
                    TABLE OF DOSAGE BY CAGE (Continued)
  DOSAGE   |  CAGE
  Frequency|
  Percent  |
  Row Pct  |
  Col Pct  |      11|        12|        15|        16|        17|     Total
  ---------+---------+---------+---------+---------+---------+
        0  |       2 |       2 |       2 |       2 |       2 |       19
           |    3.39 |    3.39 |    3.39 |    3.39 |    3.39 |    32.20
           |   10.53 |   10.53 |   10.53 |   10.53 |   10.53 |
           |   33.33 |   33.33 |   33.33 |   33.33 |   33.33 |
  ---------+---------+---------+---------+---------+---------+
        3  |       2 |       2 |       2 |       2 |       2 |       20
           |    3.39 |    3.39 |    3.39 |    3.39 |    3.39 |    33.90
           |   10.00 |   10.00 |   10.00 |   10.00 |   10.00 |
           |   33.33 |   33.33 |   33.33 |   33.33 |   33.33 |
  ---------+---------+---------+---------+---------+---------+
        6  |       2 |       2 |       2 |       2 |       2 |       20
           |    3.39 |    3.39 |    3.39 |    3.39 |    3.39 |    33.90
           |   10.00 |   10.00 |   10.00 |   10.00 |   10.00 |
           |   33.33 |   33.33 |   33.33 |   33.33 |   33.33 |
  ---------+---------+---------+---------+---------+---------+
  Total          6         6         6         6         6         59
             10.17     10.17     10.17     10.17     10.17    100.00
```

```
                    General Linear Models Procedure
                       Class Level Information
              Class        Levels        Values
              DOSAGE          3           0 3 6
              CAGE           10           3 5 6 7 9 11 12 15 16 17
          Number of observations in data set = 59
```

Dependent Variable: LOGACT21

Source	DF	Sum of Squares	Mean Square	F Value	Pr > F
Model	29	1.30579603	0.04502745	0.84	0.6786
Error	29	1.55368200	0.05357524		
Corrected Total	58	2.85947803			

R-Square	C.V.	Root MSE	LOGACT21 Mean
0.456655	8.602631	0.23146	2.69061

Source	DF	Type I SS	Mean Square	F Value	Pr > F
DOSAGE	2	0.25750763	0.12875381		
CAGE	9	0.47895137	0.05321682		
DOSAGE*CAGE	18	0.56933703	0.03162984		

Source	DF	Type II SS	Mean Square	F Value	Pr > F
DOSAGE	2	0.26793817	0.13396908		
CAGE	9	0.47895137	0.05321682		
DOSAGE*CAGE	18	0.56933703	0.03162984		

11. A manufacturer conducted a pricing experiment to explore the effects of price decreases on sales of one of its breakfast cereals. The two largest supermarket chains in a particular market participated in the experiment. Ten stores from each chain were randomly selected, and each store was assigned a price level for the cereal (either the original price or a 10% reduced price). If the competing chain had a store in the same vicinity, the two stores both were assigned the same price level. Some stores failed to complete the experiment due to competition from other supermarkets chains. Sales volumes (in hundreds of units) over the period of the study were noted for each of the remaining 17 stores, and are shown in the following table.

| | Price Level | |
Supermarket Chain	Original Price (Price = 0)	10% Reduced Price (Price = R)
1	14, 14, 15	14, 14, 17, 19, 13
2	9, 7, 12, 8	10, 12, 14, 15, 13

a. Using the accompanying SAS computer output, conduct a two-way ANOVA with Chain and Price as factors. (First complete the F-statistics and p-values in the ANOVA table, as part of your analysis.)

b. Do the effects of a price decrease on sales volume significantly differ at the different chains? If not, does it appear that a price decrease will significantly increase sales volume? How do the sales volumes compare at the two chains? Report your conclusions at the $\alpha = .05$ level. Also, do the conclusions differ for the different approaches taken in parts (a) and (b)?

Edited SAS Output (PROC GLM) for Problem 11

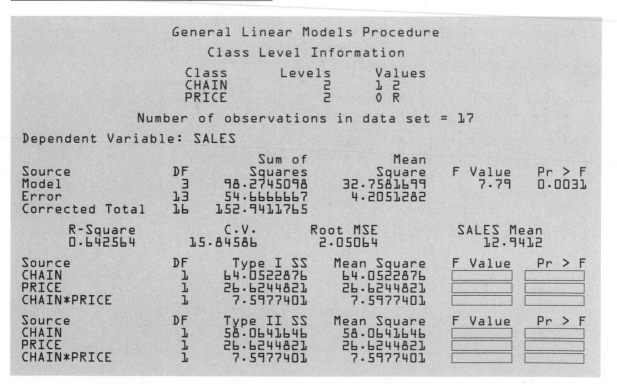

```
                    General Linear Models Procedure
                       Class Level Information
                    Class        Levels       Values
                    CHAIN           2          1 2
                    PRICE           2          0 R
              Number of observations in data set = 17
Dependent Variable: SALES
                            Sum of            Mean
Source              DF      Squares           Square     F Value    Pr > F
Model                3    98.2745098       32.7581699      7.79     0.0031
Error               13    54.6666667        4.2051282
Corrected Total     16   152.9411765

       R-Square            C.V.         Root MSE       SALES Mean
       0.642564          15.84586        2.05064         12.9412

Source              DF     Type I SS     Mean Square    F Value    Pr > F
CHAIN                1    64.0522876     64.0522876    [      ]   [      ]
PRICE                1    26.6244821     26.6244821    [      ]   [      ]
CHAIN*PRICE          1     7.5977401      7.5977401    [      ]   [      ]

Source              DF    Type II SS     Mean Square    F Value    Pr > F
CHAIN                1    58.0641646     58.0641646    [      ]   [      ]
PRICE                1    26.6244821     26.6244821    [      ]   [      ]
CHAIN*PRICE          1     7.5977401      7.5977401    [      ]   [      ]
```

12. The manager of a market research company conducted an experiment to investigate the productivity of three employees on each of two computerized data-entry systems. The employees conducted phone surveys, entering the survey data into the computer during the phone call. Productivity was measured as the time (in minutes) taken to complete a call in which the respondent agreed to complete the survey. Each employee used each system for one hour, and the order of use was randomized. The productivity data are recorded in the following table.

	System	
Employee	1	2
1	8, 7, 8, 9, 8	6, 4, 7, 4, 3, 3, 4, 5, 6
2	5, 4, 6, 3, 8, 8, 7	6, 2, 3, 3, 4, 4, 5, 6, 7, 4
3	3, 4, 5, 4, 3, 5, 5, 6	3, 3, 4, 5, 4, 3, 2, 4, 3, 3, 3, 4

a. Using the accompanying SAS computer output, conduct a two-way ANOVA. (First complete the F-statistics and p-values in the ANOVA table, as part of your analysis.)

b. Report your conclusions at the $\alpha = .01$ level. Is there an interaction between Employee and System? If not, does it appear that one system is significantly more productive than the other? Do the employees differ in terms of productivity?

Edited SAS Output (PROC GLM) for Problem 12

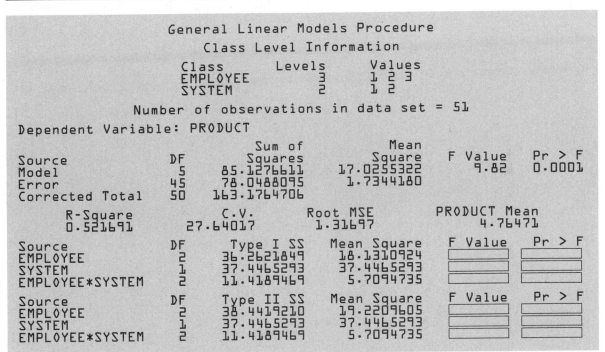

13. This question refers to the radial keratotomy data of Problem 21 in Chapter 17.

a. Suppose that a two-way ANOVA is to be performed, with five-year postoperative change in refractive error as the dependent variable (Y) and diameter of clear zone (CLRZONE) and baseline average corneal curvature (BASECURV = 1 if curvature is < 43 diopters; = 2 if curvature is between 43 and 44 diopters; and = 3 if curvature is > 44 diopters). State precisely the ANOVA model. Are the factors fixed or random?

b. In the SAS output that follows, complete the ANOVA table.

c. Analyze the data to determine whether there are significant main effects due to the clear zone and baseline curvature, and whether these factors significantly interact. Use $\alpha = .10$.

Edited SAS Output (PROC GLM) for Problem 13

```
                    General Linear Models Procedure
                      Class Level Information
                  Class     Levels    Values
                  CLRZONE      3       3.0 3.5 4.0
                  BASECURV     3       1 2 3
              Number of observations in data set = 54
```

NOTE: Due to missing values, only 51 observations can be used in this analysis.

Dependent Variable: Y

Source	DF	Sum of Squares	F Value	Pr > F
Model	☐	32.22787390	☐	☐
Error	42	47.06818002		
Corrected Total	50	79.29605392		

R-Square	C.V.	Y Mean
0.406425	27.77523	3.81137255

Source	DF	Type I SS	F Value	Pr > F
CLRZONE	2	14.70441590	6.56	0.0033
BASECURV	2	5.37753783	2.40	0.1031
CLRZONE*BASECURV	4	12.14592017	2.71	0.0428

Source	DF	Type III SS	F Value	Pr > F
CLRZONE	2	12.92686249	5.77	0.0061
BASECURV	2	5.97019625	2.66	0.0814
CLRZONE*BASECURV	4	12.14592017	2.71	0.0428

Level of CLRZONE	N	Mean	SD
3.0	20	4.41875000	1.15228249
3.5	15	3.71666667	1.33836476
4.0	16	3.14093750	0.97595119

Level of BASECURV	N	Mean	SD
1	11	4.30681818	1.20592759
2	15	3.60833333	1.03818466
3	25	3.71520000	1.38615361

Level of CLRZONE	Level of BASECURV	N	Mean	SD
3.0	1	3	5.00000000	0.99215674
3.0	2	6	3.66666667	0.79320027
3.0	3	11	4.67045455	1.22509276
3.5	1	4	4.65625000	0.71716310
3.5	2	5	4.22500000	1.26676261
3.5	3	6	2.66666667	1.06555932
4.0	1	4	3.43750000	1.42339090
4.0	2	4	2.75000000	0.46770717
4.0	3	8	3.18812500	0.96893881

14. This question refers to the *U.S. News & World Report* graduate school data presented in Problem 22 in Chapter 17.

a. Suppose that a two-way ANOVA is to be performed, with 1995 starting salary as the dependent variable and school type (SCHOOL) and reputation rank among academics (REP1 = 1 if reputation rank is in the top 10; = 2 if rank is 11 to 20; = 3 if rank is 21 or worse) as the two factors. State precisely the ANOVA model. Are the factors fixed or random?

b. In the SAS output that follows, complete the ANOVA table.

c. Analyze the data to determine whether there are significant main effects due to the school type and reputation rank, and whether these factors significantly interact.

Edited SAS Output (PROC GLM) for Problem 14

```
                    General Linear Models Procedure
                      Class Level Information
                    Class    Levels    Values
                    SCHOOL       2      1 2
                    REP          3      1 2 3
             Number of observations in data set = 24
Dependent Variable: SAL
                             Sum of          Mean
Source              DF       Squares         Square    F Value    Pr > F
Model              [    ]    1000.5058      200.1012   [     ]    [     ]
Error               18        965.8875       53.6604
Corrected Total     23       1966.3933

       R-Square            C.V.        Root MSE           SAL Mean
       0.508802          12.86650       7.3253             56.933

Source         DF     Type I SS      Mean Square   F Value    Pr > F
SCHOOL          1       37.00167       37.00167      0.69     0.4172
REP             2      910.36583      455.18292      8.48     0.0025
SCHOOL*REP      2       53.13833       26.56917      0.50     0.6176

Source         DF     Type III SS    Mean Square   F Value    Pr > F
SCHOOL          1       67.78778       67.78778      1.26     0.2758
REP             2      910.36583      455.18292      8.48     0.0025
SCHOOL*REP      2       53.13833       26.56917      0.50     0.6176

       Level of                  ------------SAL-----------
       SCHOOL          N            Mean              SD
         1            12         58.1750000        9.65628622
         2            12         55.6916667        9.06396044

       Level of                  ------------SAL-----------
       REP             N            Mean              SD
         1             4         68.7500000        6.29152870
         2             4         61.5000000        6.35085296
         3            16         52.8375000        7.37688959

  Level of     Level of                  -----------SAL-----------
  SCHOOL         REP          N            Mean              SD
    1             1           2         70.0000000        0.0000000
    1             2           2         66.0000000        5.6568542
    1             3           8         53.2625000        7.5450906
    2             1           2         67.5000000       10.6066017
    2             2           2         57.0000000        2.8284271
    2             3           8         52.4125000        7.6986896
```

References

Appelbaum, M. I., and Cramer, E. M. 1974. "Some Problems in the Non-Orthogonal Analysis of Variance." *Psychology Bulletin* 81(6): 335–43.

Harbin, T. J.; Benignus, V. A.; Muller, K. E.; and Barton, C. N. 1985. "Effects of Low-Level Carbon Monoxide Exposure upon Evoked Cortical Potentials in Young and Elderly Men." Manuscript in review, U.S. Environmental Protection Agency, Washington, D.C.

Ostle, B. 1963. *Statistics in Research,* 2d ed. Ames: Iowa State University Press.

Peng, K. C. 1967. *Design and Analysis of Scientific Experiments.* Reading, Mass.: Addison-Wesley.

Reiter, L. W.; Heavner, G. G.; Dean, K. F.; and Ruppert, P. H. 1981. "Developmental and Behavioral Effects of Early Postnatal Exposure to Triethyltin in Rats." *Neurobehavioral Toxicolology and Teratology* 3: 285–93.

Searle, S. R. 1971. *Linear Models.* New York: John Wiley & Sons.

————. 1987. *Linear Models for Unbalanced Data.* New York: John Wiley & Sons.

Snedecor, G. W., and Cochran, W. G. 1967. *Statistical Methods,* 6th ed. Ames: Iowa State University Press.

Thompson, S. J. 1972. "The Doctor–Patient Relationship and Outcomes of Pregnancy." Ph.D. dissertation, Department of Epidemiology, University of North Carolina, Chapel Hill, N.C.

Analysis of Repeated Measures Data

21-1 Preview

Chapters 17 through 20 describe ANOVA methods for analyzing continuous response data for which a single response is measured on each observational unit (e.g., subject). In many studies, however, two or more responses are observed on each unit. The responses on each unit generally are correlated, necessitating an analysis that accounts for such correlation.

For example, several blood pressure measurements may be taken over time on each of several subjects. Since the same response variable (blood pressure) is being measured at different times, we refer to such data as *repeated measures data,* and the collection of responses on the same subject is often called a *cluster* of responses. When such repeated measures are taken over *time,* the design is called a *longitudinal study.*

It is possible for clustered data not to be longitudinal. For example, suppose that the responses in each cluster are measurements of blood pressure taken over a short time period (i.e., during a single household visit) on a set of persons living in the same household. Then the data are not longitudinal, because the responses are effectively measured at the same point in time. Nevertheless, responses on persons from the same household are not likely to be independent, since members of a household typically share certain environmental and genetic factors.

This chapter describes ANOVA methods for analyzing repeated measures (i.e., clustered) data when the response variable is continuous. We focus on an approach that involves partitioning sums of squares according to various sources of variation, analogously to the classical ANOVA techniques described in Chapters 17–20. This approach requires a regression (ANOVA) model that contains both fixed and random effects, where the factor *Subjects* is considered a random factor. Inferences about main effects and interactions will be carried out using F tests based on appropriate ratios of mean square terms. In this chapter, we primarily consider balanced data sets (i.e., data sets for which the number of measurements per cluster is the same for all clusters), although we briefly discuss the analysis of unbalanced data as well. We also look at some other "recent" approaches to analyzing repeated measures data, although the details of these methods are beyond the scope of this text.

21-2 Examples

21-2-1 A Study of the Posture of Computer Operators

Over the past two decades, the use of computers in business, government, and academic settings has grown considerably. Accompanying this growth, public health specialists have become increasingly concerned about the possible health consequences of spending long working hours operating a computer. In particular, persons whose daily work assignments require intensive computer use may be at increased risk for developing chronic musculoskeletal disorders associated with degraded posture. Proper study of such health consequences requires having a reliable measure of posture.

In recent years, ergonomists have proposed various approaches to measuring posture. In a study by Ortiz and associates (1997), the subjects were computer operators recruited from a major utility company and from a large hospital in Atlanta. A primary goal of this study was to evaluate the effects of selected factors on the measurement of posture. Two such factors were *day of the work week* (early, middle, or late), denoted as Days, and time of day (A.M. or P.M.), denoted as Time. The investigators were interested in determining whether posture measurements taken on a given subject exhibited little within-subject variability over the days and time periods of measurement.

The Ortiz study considered several different measurements (responses) of posture on 19 subjects. Without going into full detail here, one of the measurements was labeled as Shoulder Flexion (SF), measured in degrees; the higher the SF score, the worse the posture. The study was designed so that each of the 19 subjects was measured sequentially in the A.M. and in the P.M. on Monday, Wednesday, and Friday of the same week. Thus, a total of 6 repeated measurements were made on each subject for this response (SF), involving all combinations of the two factors (Days and Time). In ANOVA terminology, there were three levels of Days (Monday, Wednesday, and Friday) and two levels of Time (A.M. and P.M.).

The study just described is an example of a *repeated measures design involving two factors,* in which each subject is observed at each level of each of the two factors.[1] This study design is *balanced* in that each subject has the same number of repeated measurements—namely, 6. Table 21-1 lists the data for the posture measurement SF. In this table, the 6 repeated measures on subject 1 appear as the first row of observations in the table; thus, the cluster of possibly correlated measurements for subject 1 is given by

$$\{17, 5, 10, 1, 5, 1\}$$

The sample means given at the bottom of the table suggest that the SF scores are lower on Friday than on the other two days; that is, there appears to be a possible main effect of the factor Days. There is also some suggestion of interaction between Days and Time, since the Wednesday A.M. mean score is clearly higher than the Wednesday P.M. score, whereas the Friday A.M. mean score is lower than the Friday P.M. mean score. The (repeated measures) ANOVA method

---

[1]The Ortiz study used two different ergonomists to take measurements on each subject, thus introducing a third factor (Raters) that can also be considered in the analysis of these data. To simplify the discussion of the repeated measures ANOVA approach, we have ignored the factor Raters in our discussion of these data. However, Problem 10 at the end of this chapter considers an analysis involving the three factors Days, Time, and Raters.

TABLE 21-1   Shoulder flexion (SF) data from repeated measures study of posture among computer operators (Ortiz et al. 1997).

Subject #	Sample Size	Monday A.M.	Monday P.M.	Wednesday A.M.	Wednesday P.M.	Friday A.M.	Friday P.M.
1	6	17	5	10	1	5	1
2	6	1	7	7	4	19	12
3	6	36	22	31	30	27	25
4	6	21	24	22	30	20	24
5	6	10	14	7	10	8	5
6	6	15	18	22	26	2	17
7	6	19	10	12	8	12	14
8	6	46	48	46	52	41	52
9	6	9	11	7	15	5	8
10	6	31	31	39	28	38	40
11	6	17	26	27	24	19	16
12	6	25	24	29	5	24	31
13	6	8	14	7	9	6	9
14	6	23	18	21	24	27	25
15	6	7	7	7	12	0	0
16	6	22	19	7	13	6	14
17	6	24	20	32	27	22	17
18	6	28	28	42	34	9	7
19	6	1	5	0	0	6	2
Sample Means:		18.95	18.47	19.74	18.53	15.58	16.79

used to assess whether there are significant main effects or interaction involving the two factors Days and Time is described in section 21-4.

21-2-2 A Study of Treatments for Heartburn

In this example, we consider a (fictitious) study to compare two treatments for relieving heartburn. Suppose that 30 subjects are given two symptom-provoking meals, spaced 3 days apart; upon completing each meal, each subject is given either an active treatment (*A*) for heartburn or a placebo (*P*). The subjects have been randomly allocated to one of two groups of 15 subjects, and each subject gets the same treatment for both meals. The response of interest is an index of physical discomfort measured by a questionnaire administered to each subject two hours after receiving a given treatment. This index provides a numerical value between 0 and 100, where 0 denotes no sign of any discomfort and 100 denotes the extreme end of discomfort. A fictitious set of data for this study is presented in Table 21-2.

The study design considered here is a repeated measures design because two measurements are made on each subject. This design differs from the previous (posture measurement) design in that each subject in this design receives only one of the two treatments; in the previous study, each subject was observed over all three days and at both time periods. In the next section, we distinguish the factors of such repeated measures designs by calling the Treatments factor in the heartburn study a *nest factor* and by referring to both the Days and Time factors in the posture measurement study as *crossover factors*.

TABLE 21-2   Data layout for repeated measures study to compare treatments for heartburn.

	Subject Number	Meal			Subject Number	Meal	
		1	2			1	2
	1	65	55		16	85	75
	2	60	70		17	60	60
	3	70	70		18	80	75
	4	35	30		19	55	50
	5	50	50		20	50	45
Active	6	40	40		21	70	65
Treatment	7	50	65	Placebo	22	70	80
	8	55	35		23	65	50
	9	20	50		24	60	40
	10	30	70		25	90	35
	11	65	80		26	85	50
	12	45	70		27	65	55
	13	30	45		28	60	50
	14	55	65		29	75	70
	15	25	45		30	55	40
Average		46.33	56.00	Average		68.33	56.00

Overall averages: $\overline{Y}_A = 51.17$, $\overline{Y}_P = 62.17$

The overall averages provided at the bottom of Table 21-2 indicate that the active treatment ($\overline{Y}_A$) produces, on average, lower discomfort scores than the placebo ($\overline{Y}_P$). The (repeated measures) ANOVA method used to assess whether this difference between the treatments is significant is described in section 21-4.

21-3 General Approach for Repeated Measures ANOVA

This section outlines the steps involved in carrying out a repeated measures ANOVA. These steps apply to many different repeated measures designs, and they allow for both balanced and unbalanced data. In a _balanced_ repeated measures design, each subject is observed the same number of times; in _unbalanced_ designs, the number of responses per subject may differ. Later sections provide details of repeated measures ANOVA methods, as well as numerical examples for some relatively common repeated measures designs involving balanced data. Unbalanced designs are briefly discussed in section 21-5. Appendix C at the end of the text presents ANOVA table formats and other pertinent information for the study designs described in the main body of the chapter.

21-3-1 Crossover versus Nest Factors

Before looking at the steps involved in the ANOVA approach to data analysis, let us distinguish between two types of factors often considered in a repeated measures study: crossover factors and nest factors.

A *crossover factor* is a factor for which each subject is observed at two or more levels of that factor. In a balanced design, each subject is observed at every level of a crossover factor. For example, consider a balanced design involving only one factor—say, Treatments—and suppose that each subject in the study is given every treatment over the time period of the study. Then Treatments is a crossover factor in such a study.

As another example, suppose that the study design involves the two factors, Factor A (e.g., drug types), with levels $A_1, A_2,\ldots, A_a$; and Factor B (e.g., doses), with levels $B_1, B_2,\ldots, B_b$. Further, suppose that each subject is observed at each of the $a \times b$ combinations of the two factors. Then both Factor A and Factor B are crossover factors in this study. The posture measurements study on computer operators that we discussed earlier involves two crossover factors (namely, Days, and Time), since each subject is observed at each of the six combinations of these two factors.

In contrast to a crossover factor, a *nest factor* is one for which each subject is observed at only one level of the factor. For example, in the heartburn study described earlier, each subject is allocated to only one of the two treatments, so Treatments is a nest factor in this study.

As another example, suppose each subject is observed at several doses ($B_1, B_2,\ldots, B_b$) of Factor B, but for only one drug type (say, A_j) out of several drug types ($A_1, A_2,\ldots, A_a$) of Factor A being studied. Then Factor A is a nest factor and Factor B is a crossover factor.[2]

As a final example, consider a study of student grade point averages (GPAs) from two different major departments (say, economics and mathematics) at the same university. Suppose that the GPAs for the students are observed at the end of their first year and at the end of their fourth year. This study involves a balanced repeated measures design because exactly two responses (i.e., two GPAs) are observed on each student. The factor Departments is a nest factor because economics majors are a different collection of students (assuming no "double" majors) from mathematics majors. The factor Years (first versus fourth) is a crossover factor, since each student is observed at both years.

The reason we have made the above distinction between crossover and nest factors is that the process of partitioning sources of variability in the repeated measures ANOVA depends on which factors are crossover factors and which factors are nest factors. In particular, the variability attributable to nest factors is derived from the between-subjects partition, whereas variability attributable to crossover factors is derived from the within-subjects partition.

In longitudinal studies where repeated measures are taken over time, a crossover factor is typically referred to as a *time-dependent variable,* whereas a nest factor is typically called a *time-independent variable.* A time-dependent variable takes on values that may change for different observations on the same subject. For example, in the posture measurement study, the factors Days and Time are both time-dependent variables, because the day and time period vary among the six observations on each subject. In contrast, a time-independent variable has the same value for all observations on the same subject. In the heartburn study, the Treatments factor is a time-independent variable: for a subject receiving the active treatment, Treatments has the same value (say, 1) at both meals; and for a subject receiving the placebo, Treatments has the same value (say, 0) at both meals.

---

[2]The factor Subjects, which is a component of any repeated measures design and describes the observational units under study, cannot be characterized as either a crossover or a nest factor. Nevertheless, in repeated measures designs, we typically say that the factor Subjects is *nested within* any crossover factor. We also say that (the crossover) Factor B is *nested within* (the nest) Factor A. For any repeated measures design containing a nest factor and a crossover factor, the crossover factor is always nested within the nest factor.

21-3-2 Overview of the Steps for Repeated Measures ANOVA

The following steps can serve as guidelines for carrying out an ANOVA-based analysis of repeated measures data:

1. *Describe the data layout* in terms of nest and crossover factors.

2. *Specify the ANOVA model,* making sure to distinguish fixed factors from random factors and to state the distributional (i.e., normality) assumptions for the random factors (including defining the relevant variance components).

3. *Partition the total sums of squares* according to the different sources of variation, including subject-to-subject effects.

4. *Form the ANOVA table* corresponding to the ANOVA model specified in step 2 and incorporating the partition of sums of squares delineated in step 3.

5. *Determine the formulas for expected mean squares* in terms of fixed effects and variance component parameters.

6. *Determine the appropriate F statistic* for each test of hypothesis of interest, as the ratio of the mean square for the factor being evaluated to the mean square for the source of variation that yields the same expected mean square as does the numerator expected mean square under the null hypothesis.

7. *Carry out F tests of interest, using a convenient computer program for ANOVA,* and then *summarize your results,* using (as appropriate) the hypothesis test results, sample means, and standard errors for subgroups of interest, and the confidence intervals for differences of means of interest.

SAS has two computational procedures—GLM and MIXED—for carrying out these analyses (for continuous response data only). MIXED uses a more general model than the (ANOVA) model used by GLM. For most of the examples discussed in this chapter, including the problems at the end of the chapter, we present program statements and computer output based on the GLM procedure. This procedure is particularly well-suited for carrying out balanced (and unbalanced) repeated measures ANOVA. Appendix 21-A at the end of this chapter presents program statements and computer output based on applying both GLM and MIXED procedures to the same (posture measurement study) data set. Users of other packages should check the list of procedures available in their packages to determine whether a repeated measures ANOVA option is available.

21-4 Overview of Selected Repeated Measures Designs and ANOVA-based Analyses

In this section, we examine the data layout, ANOVA model, and corresponding partitioning of variability for several relatively common balanced repeated measures designs.

The ANOVA tables, expected mean squares formulas, and appropriate F statistics for each of these designs are detailed in Appendix C at the end of the text. For each design, we assume that the response is continuous and approximately normally distributed, and that the design is balanced.

We consider the following designs in the subsections of section 21-4:

1. A balanced repeated measures design with one crossover factor.

2. A balanced repeated measures design with two crossover factors.

3. A balanced repeated measures design with one nest factor.

4. A balanced repeated measures design with one crossover factor and one nest factor.

In the first two cases, all factors[3] are crossover factors; in the latter two cases, at least one factor is a nest factor. The process of partitioning the variability differs depending on whether some of the factors are nest factors.

21-4-1 Balanced Repeated Measures Design with One Crossover Factor

The data layout for this design is given in Table 21-3. A single crossover factor, Treatments, is being considered. The design is balanced because each subject is observed the same number of times t, where t is the number of treatments. The factor Treatments is a crossover factor because each subject is observed at different (in this case, all) levels of this factor; that is, each subject receives every treatment. This design is typically called a *crossover design*. When using a crossover design, most investigators leave enough time between successive responses on each subject to ensure that the effects of each treatment do not "carry over" to the next response (involving a different treatment); in this case, we say that the study does not exhibit a *carryover*

TABLE 21-3 Data layout for a balanced repeated measures design with one crossover factor.

		Treatments			
		1	2	$\cdots$	t
	1	Y_{11}	Y_{12}		Y_{1t}
	2	Y_{21}	Y_{22}		Y_{2t}
Subject Number	3	Y_{31}	Y_{32}		Y_{3t}
	$\vdots$	$\vdots$	$\vdots$	$\cdots$	$\vdots$
	s	Y_{s1}	Y_{s2}		Y_{st}

[3]We are ignoring the factor Subjects in describing the factors of each design, since Subjects cannot obviously be characterized as either a crossover or a nest factor.

effect. Even if there is no carryover effect, we must treat the responses on the same subject as possibly correlated because they constitute repeated measurements on the same subject.

The ANOVA model for the crossover design in Table 21-3 is given as

$$Y_{ij} = \mu + S_i + \tau_j + E_{ij} \tag{21.1}$$

where $i = 1,\ldots, s$ and $j = 1,\ldots, t$. Here s denotes the number of subjects, t denotes the number of treatments, S_i is the random effect of subject i, τ_j is the fixed effect[4] of treatment j, and E_{ij} is the random error for treatment j within subject i. In this model, typically the $\{S_i\}$ and $\{E_{ij}\}$ are assumed to be mutually independent, S_i is assumed to be distributed as $N(0, \sigma_s^2)$, and E_{ij} is assumed to be distributed as $N(0, \sigma^2)$. Also, if Treatments is a fixed factor, we require the summation restriction $\sum_{j=1}^{t} \tau_j = 0$.[5]

In Table 21-3, each subject is observed only once at each treatment, so a "pure" estimate of the error for each subject is impossible to compute. In fact, the model given by (21.1) is a *randomized blocks ANOVA* model, where the blocks are the subjects (treated as a random factor in the model). As we saw in Chapter 18, a randomized block design does not permit direct estimation of the error variance σ^2; thus, in order to estimate σ^2, we have to assume that there is no block-by-treatment interaction. In model (21.1), therefore, we estimate σ^2 by using the mean square estimate of the Treatments-by-Subjects (TS) interaction; that is, we assume that $MS_{TS} = MS_{Error}$. (See Table C-1 in Appendix C for the general ANOVA table for model (21.1).)

In Chapter 17 (see footnotes on pp. 440–442), we noted that the responses $\{Y_{ij}\}$ are not mutually independent in a random effects model such as (21.1). In particular, for model (21.1), it can be shown that the correlation between repeated observations on the same subject is given by the formula

$$\rho_{ijj'} = \frac{\sigma_s^2}{\sigma_s^2 + \sigma^2} \tag{21.2}$$

where j and j' indicate different responses on the ith subject. Thus, even though model (21.1) assumes that the random effect terms $\{S_i\}$ and the error terms $\{E_{ij}\}$ are mutually independent, the responses on the same subject are not independent in this model, provided that $\sigma_s^2 > 0$. The correlation given by (21.2) is always nonnegative. Moreover, since the right-hand side of (21.2) does not depend on i or on the (j, j') pair chosen, $\rho_{ijj'}$ has the same value regardless of the sub-

---

[4]Treatments could be treated as a random factor instead of as a fixed factor; if so, model (21.1) must be modified by replacing τ_j with a random effect—say, T_j. Nevertheless, for a design involving a single crossover factor, the F test for significance of the treatment factor (see Appendix C) is computed in the same way whether the crossover factor is considered fixed or random. For repeated measures designs involving two or more factors, however, the F tests differ depending on whether the factors are designated as fixed or random.

[5]The (fixed-effects) regression model corresponding to model (21.1) is given by

$$Y = \mu + \sum_{i=1}^{s-1} S_i X_i + \sum_{j=1}^{t-1} \tau_j Z_j + E$$

where the $\{X_i\}$ are dummy variables for the s subjects, the $\{Z_j\}$ are dummy variables for the t treatments, and the dummy variables are coded using effect coding, as described in section 17-4 of Chapter 17.

ject and regardless of the pair of responses measured on that subject (i.e., $\rho_{ijj'} \equiv \rho$, a constant). In the parlance of repeated measures/longitudinal data literature, we say that model (21.1) has an *exchangeable* (or *compound symmetric*) *correlation structure.*

For the design described in Table 21-3, the variability associated with model (21.1) can be partitioned as shown in Figure 21-1. In Figure 21-1, we first partition the total variation into Between Subjects variation and Within Subjects variation. We then partition the Within Subjects variation into Variation Between Treatments and Residual Variation. Since model (21.1) assumes that there is no interaction effect, we must assume that the Residual Variation estimates within-subject error (especially since there is no pure error estimate available). The reason this partitioning comes from the Within Subjects box and not from the Between Subjects box relates to the crossover characteristic of the Treatments factor: since each subject is observed on every treatment, the Variation between Treatments must be derived from differences among treatment responses within subjects. The ANOVA table, expected mean square terms, and F test formulas for this design are provided in Appendix C.

As an example of this design, consider a two-period crossover design to compare two treatments for relieving heartburn. This is a different version of the heartburn study previously described in section 21-3. The data for this design are given in Table 21-4. In this (fictitious) study, 30 subjects are each given two symptom-provoking meals. Upon completing each meal, each subject is given either an active treatment (A) for heartburn or a placebo (P), with a different treatment being used after each meal. To avoid possible carryover effects, the two meals are spaced three days apart for each subject. Also, 15 of the 30 subjects are randomly assigned to the treatment sequence $A:P$, whereas the remaining 15 are assigned to the treatment sequence $P:A$; here, $A:P$ means that the subject received the active treatment after the first meal and the placebo after the second, and $P:A$ means that the subject received the placebo first and the active treatment second. As in the previously described heartburn study, the response of interest is an index of physical discomfort measured by a questionnaire administered to each subject two hours after receiving a given treatment. This index provides a numerical value between 0 and 100, where 0 denotes no sign of discomfort and 100 denotes the extreme end of discomfort.

FIGURE 21-1   Partitioning the sums of squares:
Balanced repeated measures ANOVA
with one crossover factor.

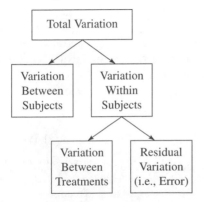

TABLE 21-4   Data layout for crossover study comparing treatments for heartburn.

	A : P				P : A	
	A	P			P	A
1	65	75		16	85	55
2	60	60		17	60	70
3	70	75		18	80	70
4	35	50		19	55	30
5	50	45		20	50	50
Subject 6	40	65	Subject 21		70	40
Number 7	50	80	Number 22		70	65
8	55	50		23	65	35
9	20	40		24	60	50
10	30	35		25	90	70
11	65	50		26	85	80
12	45	55		27	65	70
13	30	50		28	60	45
14	55	70		29	75	65
15	25	40		30	55	45
Average	46.33	56.00		Average	68.33	56.00

Overall averages: $\overline{Y}_A = 51.17, \quad \overline{Y}_P = 62.17$

Assuming that there are no carryover effects and that the only two factors contributing to the variability of the responses are Subjects and Treatments,[6] we can use model (21.1) to analyze these repeated measures data, where $t = 2$ and $s = 30$. Table 21-5 gives the ANOVA table summarizing the repeated measures analysis performed using model (21.1) for the heartburn data in Table 21-4. In this table, the F statistic for testing the hypothesis of no effect of Treatments (i.e., $H_0: \tau_1 = \tau_2 = 0$) is given by the formula

$$F = \mathrm{MS}_T / \mathrm{MS}_{\mathrm{Error}}$$

where the $\mathrm{MS}_{\mathrm{Error}}$ is given by the mean square for the Treatments-by-Subjects interaction, which is assumed to estimate σ^2. This F statistic has 1 and 29 degrees of freedom under H_0. From Table 21-5, the computed value of this F statistic is 23.82, which is highly significant ($P = .0001$). We can therefore conclude that the active treatment for heartburn yielded significantly lower discomfort scores than did the placebo.

As discussed in Chapter 19 with regard to models containing random-effect terms, the reason the preceding F statistic is appropriate for testing for no effect of Treatments is that both MS_T and $\mathrm{MS}_{\mathrm{Error}}$ terms estimate the same variance under H_0; that is, both mean square terms have the same _expected mean square_ under H_0. The expected mean square terms for

---

[6]We can take the possibility of carryover effects into account by modifying model (21.1) to allow for the effect of the sequence used in applying treatments to each subject. A dichotomous variable indicating the sequence used (i.e., either $A:P$ or $P:A$) can be added to the model. This variable, which we denote as Seq, is a nest variable. Thus, if Seq is added to the model, we are considering a repeated measures design involving a nest factor (Seq) and a crossover factor (Treatments). We discuss the appropriate model and corresponding analysis for this design in section 21-4-4.

TABLE 21-5   ANOVA table for analysis of crossover study comparing treatments for heartburn[7]

Source	d.f.	MS	F
Between Subjects	29	$MS_S = 388.22$	
Within Subjects	30	134.17	
⎧ Treatments	1	$MS_T = 1815.00$	$\frac{1815.00}{76.21} = 23.82\ (P = .0001)$
⎨ Treatments × Subjects	29	$MS_{Error} = 76.21$	
⎩ (i.e., Error)			
Total (corrected)	59		

model (21.1) are provided in Table C-2 in Appendix C. In our example, whether the null hypothesis is true or not, the expected mean square for Error equals σ^2, whereas the expected mean square for Treatments equals $\sigma^2 + 30(\tau_1^2 + \tau_2^2)$. Thus, under H_0: $\tau_1 = \tau_2 = 0$, both the numerator and the denominator are equal to σ^2.

As mentioned earlier, we are performing a randomized blocks analysis here, with the Treatments factor being considered fixed and the Subjects factor (which is the "blocking" factor) being random. Also, as described in Chapter 18, the F test for the effect of Treatments is equivalent to a paired difference test that considers the difference score between responses for each subject as the dependent variable. Although the format of Table 21-5 differs from that of Table 18-5 in Chapter 18, the two tables provide essentially identical information. The Within Subjects source is not provided in Table 18-5, because it is not used in the analysis. The Within Subjects information is included in Table 21-5 because of its role in the partitioning process, even though the MS Within Subjects is not used in either F test presented in the table.

Edited computer output using SAS's GLM procedure in analyzing these data is provided next. We distinguish different sections of this output by using circled numbers ① through ⑤. The information in section ① gives the program statements used. The data set is called "hrtbrn," the Subjects factor is called "subj," and the Treatments factor is called "trt." In the last line of these program statements, the Subjects factor is identified as a random factor, and the Treatments factor (by default) is fixed. The "test" statement on the last line tells the computer to perform F tests for all independent variables listed in the model statement, based on the specification of fixed and random factors. The information in section ② gives an overall ANOVA table that separates the variation (involving 30 degrees of freedom) due to the combined effects of the Subjects factor and the Treatments factor from the variation due to Error (involving 29 degrees of freedom). Section ③ presents Type I and Type III ANOVA information, assuming that all independent variables are treated as fixed factors (which is not the case here, since Subjects is a random factor); the Type I and Type III sets of information are identical in the output because the design is balanced. Section ④ describes the expected mean squares (EMS) formula for each factor in the model. This information is used by the program to determine the appropriate denominator degrees of freedom

---

[7]By using Table 21-5 and the expected mean square formulas given in Table C-2 of Appendix C, we can compute the estimate of the exchangeable correlation ($\hat{\rho}$) given by formula (21.2). This calculation involves substituting estimates for σ_s^2 and σ^2 into (21.2) to obtain the formula

$$\hat{\rho} = (MS_s - MS_{Error})/(MS_s + MS_{Error})$$

where $MS_{Error} \equiv MS_{TS}$. The resulting estimate is $\hat{\rho} = 0.67$, which is a relatively high correlation.

to use for the F tests for the significance of each factor. The EMS formulas given here are identical to the formulas provided in Table C-2 of Appendix C, although the notation used is different; that is, Var(Error) denotes σ^2, Var(SUBJ) denotes σ_s^2, and Q(TRT) denotes $\frac{s}{t-1}\sum_{j=1}^{t}\tau_j^2$.

The EMS output allows the investigator to verify the formulas used for the F tests, which are provided in section ⑤. For example, the F test for the effect of treatments involves dividing the MS(TRT) by the MS(Error). The MS(Error) term is the correct denominator for this test because, under the null hypothesis that all τ_j are zero, the Q(TRT) term becomes zero, causing the EMS for TRT to reduce to Var(Error). Thus, both the MS for TRT and the MS for Error estimate the same quantity—namely, Var(Error)—under H_0.

Edited SAS Output (PROC GLM) for Analysis of Crossover Study Comparing Treatments for Heartburn

```
①  proc glm data = hrtbrn;
      class subj trt;
      model discom = trt subj;   ← Program statements
      random subj / test;
--------------------------------------------------------------------
②  Source            DF   Sum of Squares    Mean Square   F Value   Pr > F
   Model             30   13073.33333333    435.77777778    5.72    0.0001
   Error             29    2210.00000000     76.20689655
   Corrected Total   59   15283.33333333

③  Source    DF      Type I SS      Mean Square   F Value   Pr > F
   TRT        1    1815.00000000   1815.00000000    23.82   0.0001
   SUBJ      29   11258.33333333    388.21839080     5.09   0.0001

   Source    DF     Type III SS     Mean Square   F Value   Pr > F
   TRT        1    1815.00000000   1815.00000000    23.82   0.0001
   SUBJ      29   11258.33333333    388.21839080     5.09   0.0001

④  Source:      Type III Expected Mean Square
   TRT          Var(Error) + Q(TRT)
   SUBJ         Var(Error) + 2 Var(SUBJ)

        Tests of Hypotheses for Mixed Model Analysis of Variance
⑤  Source: TRT
   Error: MS(Error)

                   Denominator   Denominator
   DF   Type III MS        DF            MS   F Value   Pr > F
    1          1815        29   76.206896552   23.8167   0.0001

   Source: SUBJ
   Error: MS(Error)

                   Denominator   Denominator
   DF   Type III MS        DF            MS   F Value   Pr > F
   29   388.2183908        29   76.206896552    5.0943   0.0001
```

21-4-2 Balanced Repeated Measures Design with Two Crossover Factors

The data layout for this design is given in Table 21-6. Two factors, denoted as Factor A (with a levels) and Factor B (with b levels), are under consideration here. Both are crossover factors because each subject is observed at more than one level (actually, all levels) of each factor;

TABLE 21-6 Data layout for balanced repeated measures design with two crossover factors.

		Subject Number*	Factor B: 1	2	…	b
Factor A	1	1	Y_{111}	Y_{112}		Y_{11b}
		2	Y_{211}	Y_{212}		Y_{21b}
		⋮	⋮	⋮		⋮
		s	Y_{s11}	Y_{s12}		Y_{s1b}
	2	1	Y_{121}	Y_{122}		Y_{12b}
		2	Y_{221}	Y_{222}		Y_{22b}
		⋮	⋮	⋮		⋮
		s	Y_{s21}	Y_{s22}		Y_{s2b}
	⋮		⋮	⋮	…	⋮
	a	1	Y_{1a1}	Y_{1a2}		Y_{1ab}
		2	Y_{2a1}	Y_{2a2}		Y_{2ab}
		⋮	⋮	⋮		⋮
		s	Y_{sa1}	Y_{sa2}		Y_{sab}

* The same s subjects are observed at every combination of Factors A and B.

that is, each subject is observed once at every possible combination of both factors. Since this design involves crossover factors, investigators typically leave enough time between administering successive treatment combinations to each subject to prevent carryover effects. The $a \times b$ observations on each subject are potentially correlated because they constitute repeated measurements on the same subject.

The ANOVA model for the preceding two-factor crossover design, which assumes that both Factor A and Factor B are fixed factors,[8] is given as

$$Y_{ijk} = \mu + S_i + \alpha_j + \beta_k + \delta_{jk} + S_{ij} + S_{ik} + E_{ijk} \qquad (21.3)$$

where $i = 1,\ldots, s$; $j = 1,\ldots, a$; $k = 1,\ldots, b$; s denotes the number of subjects; a denotes the number of levels of Factor A; b denotes the number of levels of Factor B; μ represents the overall mean; S_i is the random effect of subject i; α_j is the fixed effect of level j of Factor A, β_k is the fixed effect of level k of Factor B; δ_{jk} is the fixed interaction effect of level j of Factor A with level k of Factor B; S_{ij} is the random effect of level j of Factor A for subject i; S_{ik} is the random effect of level k of Factor B for subject i; and E_{ijk} is the random error for level j of Factor A and

[8]If either or both of Factors A and B are considered to be random (rather than fixed) factors, model (21.2) must be modified so that the fixed effects in the model are replaced with random effects, as appropriate; the interaction effect must be changed to a random effect if either of the two factors is random. Also, summation restrictions usually are not imposed on random effects terms (see footnote 8 in Chapter 19 for further details).

level k of Factor B for subject i. In addition, the usual summation restrictions are assumed for the fixed-effects terms:

$$\sum_{j=1}^{a} \alpha_j = 0 \qquad \sum_{k=1}^{b} \beta_k = 0 \qquad \sum_{j=1}^{a} \delta_{jk} = 0 \text{ for all } k \qquad \sum_{k=1}^{b} \delta_{jk} = 0 \text{ for all } j$$

For model (21.3), it is typically assumed that $\{S_i\}$, $\{S_{ij}\}$, $\{S_{ik}\}$, and $\{E_{ijk}\}$ are mutually independent, that S_i is distributed as $N(0, \sigma_S^2)$, that S_{ij} is distributed as $N(0, \sigma_{SA}^2)$, that S_{ik} is distributed as $N(0, \sigma_{SB}^2)$, and that E_{ijk} is distributed as $N(0, \sigma^2)$. Because each subject is observed only once at each combination of levels of Factors A and B, a pure estimate of the error for each subject is not obtainable. Therefore, to obtain an error term for model (21.3), we will assume that there is no three-way (random) interaction involving Factor A, Factor B, and Subjects, so that the mean square for this three-way interaction estimates σ^2.

For the design described in Table 21-6 and for the corresponding model (21.3), the variability can be partitioned as shown in Figure 21-2. As Figure 21-2 indicates, we first partition the total variation into between-subjects and within-subjects variation, just as we did in Figure 21-1. We then partition the within-subjects variation into variations contributed separately by Factor A, by Factor B, and by their interaction. As in the case of a repeated measures design with one crossover factor, this partitioning comes from the within-subjects box because both Factor A and Factor B are crossover factors: since each subject is observed at every combination of the levels of Factors A and B, the variation explained by the two factors and their interaction is derived from responses "within" subjects. A final partitioning is carried out by separately dividing the total variation of each factor and their interaction into two sources: the variation in the factor (or interaction) "ignoring" subjects, and the variation in the factor (or interaction)

FIGURE 21-2 Partitioning the total sums of squares for balanced repeated measures design with two crossover factors.

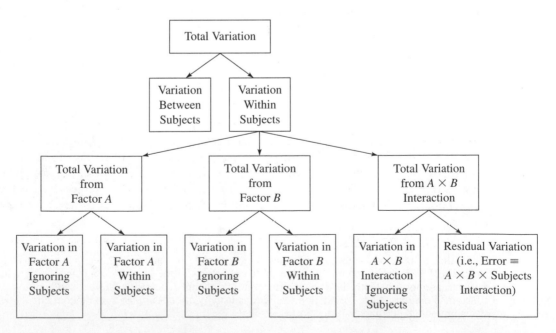

"within" subjects. To explain the latter partitioning, let us consider the regression model that corresponds to the ANOVA model (21.3):

$$Y = \mu + \sum_{i=1}^{s-1} S_i X_i + \sum_{j=1}^{a-1} \alpha_j U_j + \sum_{k=1}^{b-1} \beta_k V_k + \sum_{j=1}^{a-1}\sum_{k=1}^{b-1} \delta_{jk} U_j V_k$$
$$+ \sum_{i=1}^{s-1}\sum_{j=1}^{a-1} S_{ij} X_i U_j + \sum_{i=1}^{s-1}\sum_{k=1}^{b-1} S_{ik} X_i V_k + E$$

(21.4)

The regression model given by (21.4) contains $s - 1$ dummy variables X_i for the s subjects, $(a - 1)$ dummy variables U_j for the a levels of Factor A, and $(b - 1)$ dummy variables V_k for the b levels of Factor B. The dummy variables are usually defined by effect coding, as described in section 17-4 of Chapter 17 (see also Appendix C). The terms from this model corresponding to the various sources of variation due to each factor and their interactions are shown in Figure 21-3.

The boxes at the bottom of Figure 21-3 indicate that the "Variation in Factor A Ignoring Subjects" describes the main effect of Factor A, whereas the "Variation in Factor A Within Subjects" describes the Subjects-by-Factor A Interaction; a similar interpretation holds for Factor B. The residual variation (σ^2) is actually the variation due to the interaction of Factor A with Factor B within Subjects (i.e., $A \times B \times$ Subjects); but because there is only one observation per subject at each combination of levels of Factor A and Factor B, we must assume that these three factor-interaction effects represent random error.

The ANOVA table, expected mean squares terms, and F test formulas for this design are provided in Appendix C. These tables consider the cases when both factors are fixed and when one or both factors are random.

The posture measurement study (introduced in section 21-2) offers an example of this design, with Table 21-1 providing data from measurements of Shoulder Flexion (SF). These data describe 6 observations on each of 19 subjects, where each subject was observed at the six combi-

FIGURE 21-3   Sources of variation for partitioning sums of squares for balanced repeated measures design with two crossover factors.

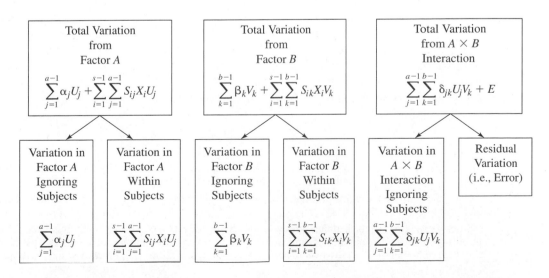

nations of three days (Monday, Wednesday and Friday) and two times (A.M. and P.M.). If we consider Days and Time to be fixed factors,[9] model (21.3) is appropriate for this design, where $s = 19$, $a = 3$, $b = 2$, $\{\alpha_j\}$ are the fixed effects of the Days factor, $\{\beta_k\}$ are the fixed effects of the Time factor, $\{\delta_{jk}\}$ are the fixed interaction effects of Days with Time, and $\{S_i\}$, $\{S_{ij}\}$, $\{S_{ik}\}$ are the appropriate random effects due to Subjects and their interactions with Days and Time, respectively.

Table 21-7 gives the ANOVA table that summarizes the repeated measures analysis for these data,[10] using model (21.3). This table indicates that all three F tests are nonsignificant. We can therefore conclude that neither Days nor Time nor their interaction is a significant factor in predicting Shoulder Flexion response for these data.

TABLE 21-7 ANOVA table for repeated measures ANOVA on shoulder flexion (SF) responses in Table 21-1.

Source	d.f.	MS	F
Between Subjects	18	781.51	
Within Subjects	95		
Days	2	96.56	1.78 $(P = .18)$
Subjects × Days	36	54.39	
Time	1	0.71	0.02 $(P = .88)$
Subjects × Time	18	30.51	
Days × Time	2	14.63	0.79 $(P = .50)$
Subjects × Days × Time (i.e., Error)	36	18.43	
Total (corrected)	113		

21-4-3 Balanced Repeated Measures Design with One Nest Factor

The data layout for this design is given in Table 21-8. A single factor (Treatments) with t levels is under consideration. There is a total of $t \times s$ different subjects, with each treatment level being given to a different subset of s subjects. Because each subject is given only a single treatment, the Treatments factor is a nest factor. Also, since each subject is observed several (r) times on the same treatment (level), there are repeated measures for each subject. The repeated observations on the same subject are therefore almost surely correlated to some extent. The total number of observations in this design is the product $s \times t \times r$. This is a balanced design because every subject is observed the same (r) number of times and the same number of subjects (s) receive each treatment.[11]

---

[9]To justify treating Days and Time as fixed factors, we must be specifically interested in the three days and two time periods of the day considered in the study.

[10]Computer output, including program statements, for this analysis is provided in Appendix C at the end of the text.

[11]Strictly speaking, a repeated measures design is *balanced* if two conditions hold: first, each subject is observed the same number of times; and second, every combination of levels of the factors involved (whether they be crossover or nest factors) has observations on the same number of subjects.

TABLE 21-8   Data layout for balanced repeated measures design with one nest factor.

		Repeats			
	Subject Number	1	2	...	r
1	(1, 1)	Y_{111}	Y_{112}	...	Y_{11r}
	(2, 1)	Y_{211}	Y_{212}	...	Y_{21r}
	$\vdots$	$\vdots$	$\vdots$		$\vdots$
	(s, 1)	Y_{s11}	Y_{s12}	...	Y_{s1r}
2	(1, 2)	Y_{121}	Y_{122}	...	Y_{12r}
	(2, 2)	Y_{221}	Y_{222}	...	Y_{22r}
	$\vdots$	$\vdots$	$\vdots$		$\vdots$
	(s, 2)	Y_{s21}	Y_{s22}	...	Y_{s2r}
			...		
t	(1, t)	Y_{1t1}	Y_{1t2}	...	Y_{1tr}
	(2, t)	Y_{2t1}	Y_{2t2}	...	Y_{2tr}
	$\vdots$	$\vdots$	$\vdots$		$\vdots$
	(s, t)	Y_{st1}	Y_{st2}	...	Y_{str}

(left margin label: Treatments)

The ANOVA model in which Treatments is assumed to be a fixed factor[12] in this nested design is

$$Y_{ijk} = \mu + S_{i(j)} + \tau_j + E_{k(ij)} \qquad (21.5)$$

where $i = 1,\ldots, s$; $j = 1,\ldots, t$; $k = 1,\ldots, r$; μ is the overall mean; $S_{i(j)}$ is the random effect of subject i within treatment j; τ_j is the fixed effect of treatment j; and $E_{k(ij)}$ is the random error for repeat k on subject i within treatment j. We typically assume that $\{S_{i(j)}\}$ and $\{E_{k(ij)}\}$ are mutually independent, that $S_{i(j)}$ is normally distributed as $N(0, \sigma_S^2)$, and that $E_{k(ij)}$ is distributed as $N(0, \sigma^2)$. We also assume that the summation restriction $\sum_{j=1}^{t} \tau_j = 0$ applies for the fixed effects.

For the design described in Table 21-8 and the corresponding model (21.5), the Total Variation can be partitioned as shown in Figure 21-4. Because Treatments is a nest factor rather than a crossover factor, the Variation Between Treatments and the Variation Due to Subjects Within Treatments are obtained by partitioning the Variation Between Subjects. The Variation Within Subjects is assumed to estimate pure error.

The ANOVA table, expected mean squares terms, and F test formulas for this design are provided in Appendix C.

Let us return to the heartburn study first described in section 21-2. The (fictitious) data for this study were presented in Table 21-2. The study uses a different design from that used in the

---

[12]If Treatments is viewed as a random factor instead of as a fixed factor, model (21.5) must be modified by replacing τ_j with a random effect—say, T_j. Nevertheless, the F test for significance of the treatment factor (see Appendix C) is computed the same way whether the nest factor is considered fixed or random.

FIGURE 21-4 Partitioning total sums of squares for balanced
repeated measures design with one nest factor.

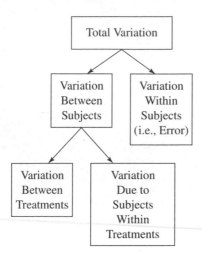

crossover study described in section 21-4-1. Since each of the 30 subjects received only one of the two treatments, Treatments is a nest factor (rather than a crossover factor). The response of interest is once again an index of physical discomfort measured by a questionnaire administered to each subject two hours after the subject received a given treatment. In this example, $s = 15$, $t = 2$, and $r = 2$.

Table 21-9 gives the ANOVA table summarizing the repeated measures analysis based on model (21.5) for the (nested) heartburn study data presented in Table 21-2. Also, software analysis of these data produced by using SAS's GLM program is presented in the block of computer output that follows. From these results, we find that the F statistic of 6.20 for testing the significance of the Treatments factor is significant at the .05 level. We can therefore conclude that the rated effects of the active treatment and the placebo significantly differ, with the active treatment producing meaningfully lower discomfort scores than the placebo. These results are similar to those obtained in the crossover version of the heartburn study (see Table 21-5); however, the F test for Treatments in Table 21-9 ($P = .02$) is not nearly as significant as the corresponding F test when a crossover design (where each subject serves as its own control) is used ($P = .0001$).

TABLE 21-9 ANOVA table based on model (21.5) for data in Table 21-2
comparing two treatments for heartburn.

Source	d.f.	MS	F
Between Subjects	29	345.11	
⎧ Treatments	1	1815.00	$\frac{1815.00}{292.62} = 6.20\ (P = .02)$
⎩ Subjects Within Treatments	28	292.62	$\frac{292.62}{175.83} = 1.66\ (P = .09)$
Within Subjects (i.e., Error)	30	175.83	
Total (corrected)	59		

Edited SAS Output (PROC GLM) for Repeated Measures ANOVA
of Data in Table 21-2 to Compare Treatments for Heartburn

```
proc glm data = hrtbrn2;
   class subj trt;                    ← Program statements
   model discom = trt subj(trt);
   random subj(trt) / test;
Dependent Variable: DISCOM
Source            DF    Sum of Squares    Mean Square    F Value   Pr > F
Model             29    10008.33333333    345.11494253      1.96   0.0356
Error             30     5275.00000000    175.83333333
Corrected Total   59    15283.33333333
Source            DF         Type I SS      Mean Square    F Value   Pr > F
TRT                1     1815.00000000    1815.00000000     10.32   0.0031
SUBJ (TRT)        28     8193.33333333     292.61904762      1.66   0.0869
Source            DF       Type III SS      Mean Square    F Value   Pr > F
TRT                1     1815.00000000    1815.00000000     10.32   0.0031
SUBJ (TRT)        28     8193.33333333     292.61904762      1.66   0.0869
Source:          Type III Expected Mean Square
TRT              Var(Error) + 2 Var(SUBJ(TRT)) + Q(TRT)
SUBJ (TRT)       Var(Error) + 2 Var(SUBJ(TRT))
       Tests of Hypotheses for Mixed Model Analysis of Variance
Source: TRT
Error: MS(SUBJ(TRT))
                      Denominator      Denominator
DF      Type III MS          DF               MS     F Value   Pr > F
 1             1815           28     292.61904762      6.2026   0.0190
Source: SUBJ(TRT)
Error: MS(Error)
                      Denominator      Denominator
DF      Type III MS          DF               MS     F Value   Pr > F
28     292.61904762          30     175.83333333      1.6642   0.0869
```

21-4-4 Balanced Repeated Measures Design with One Crossover Factor and One Nest Factor

The data layout for this design is given in Table 21-10. Here, two factors, denoted as Factor A (with a levels) and Factor B (with b levels), are under consideration. In this design, each subject is allocated to only one of the a levels of Factor A but is observed at each of the b levels of Factor B. Thus, Factor A is a nest factor and Factor B is a crossover factor, with Factor B being nested within Factor A. This is a balanced design because each subject is observed the same (b) number of times, and each level of Factor A has observations on the same number (s) of subjects. The total number of observations in the study is $s \times a \times b$. Since each subject is observed several (b) times (over the b levels of Factor B), these repeated observations on the same subject are most probably correlated.

The ANOVA model for the preceding design (assuming that Factor A and Factor B are fixed factors) is

$$Y_{ijk} = \mu + S_{i(j)} + \alpha_j + \beta_k + \delta_{jk} + E_{k(ij)} \tag{21.6}$$

TABLE 21-10 Data layout for balanced repeated measures design with one crossover factor and one nest factor.

		Subject Number	Factor B 1	Factor B 2	$\cdots$	Factor B b
Factor A	1	$(1, 1)$	Y_{111}	Y_{112}	$\cdots$	Y_{11b}
		$(2, 1)$	Y_{211}	Y_{212}	$\cdots$	Y_{21b}
		$\vdots$	$\vdots$	$\vdots$		$\vdots$
		$(s, 1)$	Y_{s11}	Y_{s12}	$\cdots$	Y_{s1b}
	2	$(1, 2)$	Y_{121}	Y_{122}	$\cdots$	Y_{12b}
		$(2, 2)$	Y_{221}	Y_{222}	$\cdots$	Y_{22b}
		$\vdots$	$\vdots$	$\vdots$		$\vdots$
		$(s, 2)$	Y_{s21}	Y_{s22}	$\cdots$	Y_{s2b}
	$\vdots$	$\vdots$	$\vdots$	$\vdots$	$\cdots$	$\vdots$
	a	$(1, a)$	Y_{1a1}	Y_{1a2}	$\cdots$	Y_{1ab}
		$(2, a)$	Y_{2a1}	Y_{2a2}	$\cdots$	Y_{2ab}
		$\vdots$	$\vdots$	$\vdots$	$\vdots$	$\vdots$
		(s, a)	Y_{sa1}	Y_{sa2}	$\cdots$	Y_{sab}

where $i = 1,\ldots, s$; $j = 1,\ldots, a$; $k = 1,\ldots, b$; μ is the overall mean; $S_{i(j)}$ is the random effect of subject i within level j of Factor A; α_j is the fixed effect of level j of Factor A; β_k is the fixed effect of level k of Factor B; δ_{jk} is the fixed interaction effect of level j of Factor A with level k of Factor B; and $E_{k(ij)}$ is the random error for the kth level of Factor B within level j of Factor A on subject i.

Model (21.6) differs from model (21.5) in that the β_k and δ_{jk} terms for the effects of Factor B and its interaction with Factor A are included. In model (21.5), the Repeats factor is considered as contributing only to error.

For model (21.6), we typically assume that $\{S_{i(j)}\}$ and $\{E_{k(ij)}\}$ are mutually independent, that $S_{i(j)}$ is distributed as $N(0, \sigma_s^2)$, and that $E_{k(ij)}$ is distributed as $N(0, \sigma^2)$. We also assume that the usual summation restrictions for the fixed-effects terms apply.

An important special case of this two-factor design with repeated measures on one of the two factors is given by a *two-group pre/posttest design*. The data layout for this design is shown in Table 21-11. In this special case, one of the factors is Group and the other factor is Test. In essence, the study design involves comparing responses on two (independent) groups of subjects, where each subject is measured at two times. The responses from the first time are typically called the *pretest scores*, and the responses from the second time are called the *posttest scores*. The objective of the study is to determine whether the change in response from pretest to posttest differs in some overall (i.e., average) way between the two groups. The Group factor (A), which is of primary interest, is a nest factor, whereas the Test factor (B) is a crossover factor; and Test is nested within Group. This is a repeated measures design because responses are measured at two times for each subject. The ANOVA model given by (21.6) applies here with $a = b = 2$. The parameters of primary interest for this special case are α_j and δ_{jk}, which relate to the main effect of the Group factor and the interaction effect of Group with Test. In particular, if the δ_{jk} terms are found

TABLE 21-11   Data layout for a balanced two-group pre/posttest design.

	Subject Number	Pretest	Posttest
Group 1	$(1, 1)$	Y_{111}	Y_{112}
	$(2, 1)$	Y_{211}	Y_{212}
	$\vdots$	$\vdots$	$\vdots$
	$(s, 1)$	Y_{s11}	Y_{s12}
Group 2	$(1, 2)$	Y_{121}	Y_{122}
	$(2, 2)$	Y_{221}	Y_{222}
	$\vdots$	$\vdots$	$\vdots$
	$(s, 2)$	Y_{s21}	Y_{s22}

to be nonsignificant, they may be dropped from the model; and a test of H_0: $\alpha_j = 0$ in a reduced model is carried out to assess whether the two groups significantly differ.

Two alternative approaches to using model (21.6) deserve mention. In the first of these, for each subject in each group, we define the difference scores $d_{ij} = Y_{ij2} - Y_{ij1}$ and use a two-sample sample t statistic to compare the mean difference scores $\bar{d}_1$ and $\bar{d}_2$ for each group. In the second approach, we use an analysis of covariance model in which the dependent variable is the posttest score (i.e., Y_{ij2}) and the independent variables are the pretest score (Y_{ij1}) and the group status (defined, say, by a 0–1 dummy variable). Brogan and Kutner (1980) have shown that the first approach is mathematically equivalent to using the repeated measures model (21.6), but that it differs from the second (analysis of covariance) approach.

For the general data layout described in Table 21-10, with corresponding model (21.6), the variability can be partitioned as shown in Figure 21-5. For the two-group pre/posttest design described by Table 21-11, Factor A denotes the Group factor and Factor B denotes the Test factor.

FIGURE 21-5   Partitioning total sums of squares for balanced repeated measures design with one crossover and one nest factor.

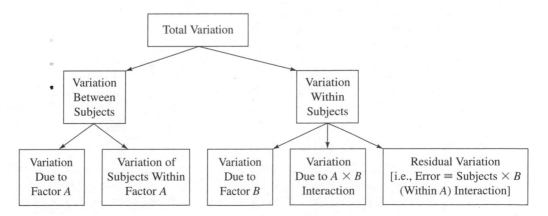

As Figure 21-5 indicates, the variation due to the nest factor (A) is obtained from the Variation Between Subjects. The variation associated with the crossover factor (B), due to the main effect of Factor B or to its interaction with Factor A, is obtained from the Variation Within Subjects. This partitioning again illustrates the general rule that variation associated with a nest factor derives from the Between Subjects variation; in contrast, the variation associated with a crossover factor, with the interaction of two or more crossover factors, or with the interaction of a crossover factor and a nest factor, derives from the Within Subjects variation.

The ANOVA table, expected mean squares terms, and F test formulas for this design are provided in Appendix C.

To see an example of this design, we return to the two-period crossover design for comparing two treatments for the relief of heartburn (see section 21-4-1). The data for this design are given in Table 21-4. In our previous analysis of these data, we viewed the crossover factor (Treatments) as the only factor of interest. If we wish to consider the possibility that the sequence used (either $A:P$ or $P:A$) may have an effect on the response, however, we can modify the model by including a dichotomous variable to indicate the sequence. This latter variable, which we denote as Seq, is a nest variable; thus, if we add Seq to the model, we create a repeated measures design involving a nest factor (i.e., Seq) and a crossover factor (i.e., Treatments). Model (21.6) is appropriate for the analysis of this study design, where Factor A is Seq, Factor B is Treatments, $a = 2$, and $b = 2$.

Table 21-12 gives the ANOVA table that summarizes the repeated measures analysis based on model (21.6) for these data. The results of a software analysis are provided in the block of computer output that follows. In these analyses, we have assumed that both Treatments and Seq are fixed factors. From the results, we see that the F statistic of 0.34 for testing the interaction of Treatments with Seq is nonsignificant ($P = .56$). The F statistic of 23.28 for testing the significance of the Treatments factor is highly significant ($P \ll .01$). And the Seq factor is significant ($F = 5.38$, $P = .03$) at the .05 level. We can therefore conclude that the active treatment for heartburn produced significantly lower discomfort scores than did the placebo, when we control for the Seq variable.

TABLE 21-12   ANOVA table for data in Table 21-4 from heartburn study with one crossover (Treatments) and one nest (Seq) factor.

Source	d.f.	MS	F
Between Subjects	29	388.22	
⎰ Seq	1	1815.00	$\frac{1815}{337.26} = 5.38$ $(P = .03)$
⎱ Subjects (within Seq)	28	337.26	
Within Subjects	30	134.17	
⎰ Treatments	1	1815.00	$\frac{1815}{77.98} = 23.28$ $(P = .0001)$
⎨ Seq × Treatments	1	26.67	$\frac{26.67}{77.98} = 0.34$ $(P = .56)$
⎩ Subjects × Treatments (within Seq) (i.e., Error)	28	77.98	
Total (corrected)	59		

Edited SAS Output (PROC GLM) for Data in Table 21-4
from Heartburn Study with One Nest Factor and One Crossover Factor

```
proc glm data = hrtbrn;                          ← Program statements
    class subj trt seq;
    model discom = trt seq trt*seq subj(seq);
    random subj(seq) / test;
```

Source	DF	Type III SS	Mean Square	F Value	Pr > F
TRT	1	1815.00000000	1815.00000000	23.28	0.0001
SEQ	1	1815.00000000	1815.00000000	23.28	0.0001
TRT*SEQ	1	26.66666667	26.66666667	0.34	0.5634
SUBJ(SEQ)	28	9443.33333333	337.26190476	4.33	0.0001

Source:	Type III Expected Mean Square
TRT	Var(Error) + Q(TRT,TRT*SEQ)
SEQ	Var(Error) + 2 Var(SUBJ(SEQ)) + Q(SEQ,TRT*SEQ)
TRT*SEQ	Var(Error) + Q(TRT*SEQ)
SUBJ (SEQ)	Var(Error) + 2 Var(SUBJ(SEQ))

```
        Tests of Hypotheses for Mixed Model Analysis of Variance
Source: TRT *
Error: MS(Error)
```

		Denominator	Denominator		
DF	Type III MS	DF	MS	F Value	Pr > F
1	1815	28	77.976190476	23.2763	0.0001

```
Source: SEQ *
Error: MS(SUBJ(SEQ))
```

		Denominator	Denominator		
DF	Type III MS	DF	MS	F Value	Pr > F
1	1815	28	337.26190476	5.3816	0.0279

```
Source: TRT*SEQ
Error: MS(Error)
```

		Denominator	Denominator		
DF	Type III MS	DF	MS	F Value	Pr > F
1	26.66666667	28	77.976190476	0.3420	0.5634

```
Source: SUBJ(SEQ)
Error: MS(Error)
```

		Denominator	Denominator		
DF	Type III MS	DF	MS	F Value	Pr > F
28	337.26190476	28	77.976190476	4.3252	0.0001

21-5   Repeated Measures ANOVA for Unbalanced Data

So far, we have focused on balanced repeated measures designs—namely, designs for which each subject is observed the same number of times. More generally, however, the number of observations per subject may differ. For example, the repeated part of the design may be unbalanced, either by study plan or because of missing data. In an observational study of the growth of children, for example, subjects may have different numbers of height measurements recorded because of different numbers of visits to a clinic. In this section, we briefly consider methods available for dealing with such unbalanced repeated measures data.

Analyzing unbalanced repeated measures designs by using ANOVA methods is complicated because the partitioned sums of squares for different factors are not independent of one another. As described in Chapter 20 for two-way single-response ANOVA designs with unequal cell numbers, one way of handling unbalanced repeated measures data is to perform regression modeling that considers partial (e.g., "extra") sums of squares for carrying out inference-making procedures. This is the approach used by SAS's GLM procedure.

To understand the approach used by GLM, suppose that the posture measurement study described in section 21-2 had produced unbalanced data in which not all subjects were measured six times (e.g., some subjects were measured on fewer than three days and/or on fewer than both time periods). In such a case, model (21.3) and its corresponding regression model (21.4) would still be appropriate, although the number of repeated observations per subject would vary. For convenience, we restate model (21.4) here as

$$Y = \mu + \sum_{i=1}^{s-1} S_i X_i + \sum_{j=1}^{a-1} \alpha_j U_j + \sum_{k=1}^{b-1} \beta_k V_k + \sum_{j=1}^{a-1}\sum_{k=1}^{b-1} \delta_{jk} U_j V_k$$
$$+ \sum_{i=1}^{s-1}\sum_{j=1}^{a-1} S_{ij} X_i U_j + \sum_{i=1}^{s-1}\sum_{k=1}^{b-1} S_{ik} X_i V_k + E$$

We will refer to the preceding model as the *full model*. Now, based on this model, suppose that we want to test whether there is significant Days × Time interaction; that is, we want to test the null hypothesis H_0: $\delta_{jk} = 0$ for all j, k. If the design were balanced, the appropriate F statistic, as illustrated in ANOVA Table 21-7, would be given by the formula

$$F = MS_{DT}/MS_{Error}$$

where MS_{DT} denotes the mean square for the Days × Time interaction effect. This same formula is used with unbalanced data. For unbalanced data (as well as for balanced data), the MS_{Error} term is obtained by fitting the full model, treating all factors (including the S_i, S_{ij}, and S_{ik} terms) as fixed, and then using the Residual MS as the MS_{Error} term. To obtain MS_{DT} from unbalanced data, we fit the reduced model (again treating all effects as fixed) that omits all terms involving δ_{jk} from the full model to obtain the Residual SS for this reduced model. We then subtract the Residual SS for the full model from the Residual SS for the reduced model to get the Extra SS for the addition of the Days × Time interaction terms to the reduced model. Finally, we divide this Extra SS term (called a *Type III SS term* by the GLM procedure) by the appropriate degrees of freedom (e.g., 2 in our posture measurement study example) to yield the MS_{DT} value.

21-6 Other Approaches to Analyzing Repeated Measures Data

The ANOVA approach we have been describing in this chapter has a number of limitations. First, it requires that all independent variables in the model—including interval variables—be categorized. Such categorization generally forfeits information if the interval variables are reliably measured, so researchers may be disinclined to categorize interval variables. Also, the assumption of normality may not be appropriate, particularly when the response of interest is a binary (0 or 1) variable or a count variable (such as the number of cases of a disease).

In light of these limitations, investigators have developed a number of other approaches to analyzing repeated measures data, including use of a more general form of linear regression model with both fixed and random effects, called the *general mixed model* (see Ware 1985). The general mixed model allows for interval-independent variables and uses the method of maximum likelihood (described in Chapter 22) to estimate parameters. A recently developed SAS procedure called MIXED can carry out the computations required to fit such a model. For more detailed information on the general mixed linear model, see Searle 1987; Searle et al. 1992; and SAS user manuals for the MIXED procedure. Appendix 21-A illustrates the results of using the MIXED procedure on the posture measurement data, and compares the results from using MIXED with those obtained from using GLM.

To address situations where the (repeated measures) response is binary, a count variable, or continuous but not normally distributed, a method involving the use of *generalized estimating equations* (GEE) has been developed (Zeger and Liang 1986). The GEE approach uses a procedure called *quasi-likelihood estimation,* which is a generalization of maximum likelihood estimation. In recent years, the GEE approach has become very popular among researchers who need to consider nonlinear regression models for correlated binary or count response data (e.g., the logistic model discussed in Chapter 23 and the Poisson regression model discussed in Chapter 24). Some currently available software can apply the GEE method—in particular, SAS's GENMOD procedure in Release 6.2. Also, a recently developed SAS procedure (currently in macro form) called GLIMMIX permits the use of model to analyze correlated responses that can be either continuous or discrete (see Wolfinger and O'Connell, 1993). A thorough discussion of GEE and related methods is beyond the scope of this text. For further information, consult Diggle, Liang, and Zeger (1994).

Appendix 21-A
Example of SAS's GLM and MIXED Procedures

This appendix provides program statements and edited output obtained from applying SAS's GLM and MIXED procedures (release 8.08) to the posture measurement data (Ortiz et al. 1996) described in section 21-2 and analyzed in section 21-4 of Chapter 21. The study design is a balanced repeated measures ANOVA with two crossover factors. The data analyzed involve 19 subjects whose shoulder flexion (SF) posture was measured on each of three days (Monday, Wednesday, and Friday) and at each of two times (A.M. and P.M.) on each day. The analyses reported in this appendix assume that both the Days and Time factors are fixed factors, and use model (21.3) to fit the data.

SAS Program Statements for PROC GLM

```
PROC GLM DATA = SF;
   CLASS subj day time;
   MODEL sf = day time day*time subj subj*day, subj*time;
   RANDOM subj subj*day subj*time / test;
```

SAS Program Statements for PROC MIXED

```
PROC MIXED DATA = SF ASYCOV,
    CLASS subj day time,
    MODEL sf = day time day*time / E3 s,
    RANDOM INTERCEPT day time / SUBJECT = subj s g,
```

Edited SAS Output for PROC GLM

```
              SHOULDER FLEXION DATA BY DAYS AND TIME
                 General Linear Models Procedure
Dependent Variable: SF

Source          DF    Sum of Squares    Mean Square   F Value   Pr > F
Model           77    16797.58771930   218.15048986    11.84    0.0001
Error           36      663.40350877    18.42787524            Incorrect
Corrected Total 113   17460.99122807                       ╱ F Tests

Source          DF         Type I SS     Mean Square   F Value   Pr > F
DAY              2      193.12280702    96.56140351     5.24    0.0101
TIME             1        0.71052632     0.71052632     0.04    0.8454
DAY*TIME         2       29.26315789    14.63157895     0.79    0.4598
SUBJ            18    14067.15789474   781.50877193    42.41    0.0001
SUBJ*DAY        36     1958.21052632    54.39473684     2.95    0.0008
SUBJ*TIME       18      549.12280702    30.50682261     1.66    0.0973
                                                              Incorrect
                                                           ╱ F Tests

Source          DF       Type III SS     Mean Square   F Value   Pr > F
DAY              2      193.12280702    96.56140351     5.24    0.0101
TIME             1        0.71052632     0.71052632     0.04    0.8454
DAY*TIME         2       29.26315789    14.63157895     0.79    0.4598
SUBJ            18    14067.15789474   781.50877193    42.41    0.0001
SUBJ*DAY        36     1958.21052632    54.39473684     2.95    0.0008
SUBJ*TIME       18      549.12280702    30.50682261     1.66    0.0973

Source:     Type III Expected Mean Square
DAY         Var(Error) + 2 Var(SUBJ*DAY) + Q(DAY,DAY*TIME)
TIME        Var(Error) + 3 Var(SUBJ*TIME) + Q(TIME,DAY*TIME)
DAY*TIME    Var(Error) + Q(DAY*TIME)
SUBJ        Var(Error) + 3 Var(SUBJ*TIME) + 2 Var(SUBJ*DAY)
              + 6 Var(SUBJ)
SUBJ*DAY    Var(Error) + 2 Var(SUBJ*DAY)
SUBJ*TIME   Var(Error) + 3 Var(SUBJ*TIME)

        Tests of Hypotheses for Mixed Model Analysis of Variance
Dependent Variable: SF

Source: DAY *
Error: MS(SUBJ*DAY)                                         Correct
                                                         ╱ F Tests
            Denominator    Denominator
DF    Type III MS      DF           MS    F Value   Pr > F
 2   96.561403509      36   54.394736842   1.7752    0.1840

Source: DAY*TIME
Error: MS(Error)

            Denominator    Denominator
DF    Type III MS      DF           MS    F Value   Pr > F
 2   14.631578947      36   18.427875244   0.7940    0.4598
```

```
Source: TIME *
Error: MS(SUBJ*TIME)

                   Denominator        Denominator
DF     Type III MS         DF                MS    F Value    Pr > F
 1     0.7105263158        18       30.50682612     0.0233    0.8804
```

Edited SAS Output from PROC MIXED

```
                      The MIXED Procedure
              Model Fitting Information for SF
            Description                            Value
            Observations                       114.0000
            Variance Estimate                   18.4279
            Standard Deviation Estimate          4.2928
            REML Log Likelihood               -377.174
            Akaike's Information Criterion     -381.174
            Schwartz's Bayesian Criterion     -386.538
            -2 REML Log Likelihood             754.3473
            Null Model LRT Chi-Square          117.6654
            Null Model LRT DF                    3.0000
            Null Model LRT P-Value               0.0000
                      The MIXED Procedure
                 Solution for Fixed Effects
Parameter            Estimate     Std Error    DDF        T    Pr > |T|
INTERCEPT         16.78947368    2.89836738     36     5.79      0.0000
DAY 1              1.68421053    1.95774591     36     0.86      0.3953
DAY 2              1.73684211    1.95774591     36     0.89      0.3809
DAY 3              0.00000000         .          .      .         .
TIME 1            -1.21052632    1.53740010     36    -0.79      0.4362
TIME 2             0.00000000         .          .      .         .
DAY*TIME 1 1       1.68421053    1.96965801     36     0.86      0.3982
DAY*TIME 1 2       0.00000000         .          .      .         .
DAY*TIME 2 1       2.42105263    1.96965801     36     1.23      0.2270
DAY*TIME 2 2       0.00000000         .          .      .         .
DAY*TIME 3 1       0.00000000         .          .      .         .
DAY*TIME 3 2       0.00000000         .          .      .         .
             Solution for Random Effects ← Not provided in Proc GLM
Parameter   Subject      Estimate     Std Error    DDF        T    Pr > |T|
INTERCEPT   SUBJ 1    -10.52985717    3.98438520     36    -2.64      0.0121
DAY 1       SUBJ 1      1.98181353    3.19561709     36     0.62      0.5391
DAY 2       SUBJ 1     -1.93330203    3.19561709     36    -0.60      0.5490
DAY 3       SUBJ 1     -1.63749330    3.19561709     36    -0.51      0.6115
TIME 1      SUBJ 1      1.44062296    1.80047099     36     0.80      0.4289
TIME 2      SUBJ 1     -1.79638058    1.80047099     36    -1.00      0.3251
                          .
                        . [Portion of output omitted]
                          .
Parameter   Subject      Estimate     Std Error    DDF        T    Pr > |T|
INTERCEPT   SUBJ 19   -14.34211491    3.98438520     36    -3.60      0.0010
DAY 1       SUBJ 19    -0.74462103    3.19561709     36    -0.23      0.8171
DAY 2       SUBJ 19    -3.00668780    3.19561709     36    -0.94      0.3530
DAY 3       SUBJ 19     1.58704779    3.19561709     36     0.50      0.6225
TIME 1      SUBJ 19    -0.27353715    1.80047099     36    -0.15      0.8801
TIME 2      SUBJ 19    -0.21101991    1.80047099     36    -0.12      0.9074
```

(continued)

```
                              G Matrix  ← Gives variance components estimates
 Row          COL1         COL2         COL3         COL4         COL5         COL6
  1   119.17251462
  2                  17.98343080
  3        σ̂²_S                  17.98343080
  4                      σ̂²_SD               17.98343080
  5                                               4.02631579
  6                                                            4.02631579
              Covariance Parameter Estimates (REML) σ̂²_ST
σ²_S  Cov Parm         Ratio        Estimate      Std Error        Z    Pr > |Z|
      INTERCEPT     6.46696991   119.17251462   43.50875511    2.74   0.0062
σ²_SD  DAY           0.97588195    17.98343080    6.76836389    2.66   0.0079
      TIME          0.21849051     4.02631579    3.68590830    1.09   0.2747
σ²_ST  Residual      1.00000000    18.42787524    4.34349185    4.24   0.0000
σ²           Asymptotic Covariance Matrix of Estimates
 Cov Parm        INTERCEPT            DAY            TIME         Residual
 INTERCEPT     1893.0117711    -15.27024992    -6.79295999     3.14432024
 DAY            -15.27024992     45.81074975     3.14432024    -9.43296072
 TIME            -6.79295999      3.14432024    13.58591999    -6.28864048
 Residual         3.14432024     -9.43296072    -6.28864048    18.86592144
              Tests of Fixed Effects  ← Correct F Tests
 Source          NDF       DDF     Type III F     Pr > F
 DAY               2        36        1.78         0.1840
 TIME              1        18        0.02         0.8804
 DAY*TIME          2        36        0.79         0.4598
```

Problems

1. The data set for this problem derives from the posture measurement study described in the main body of the chapter. Here, we consider the data on shoulder flexion (SF) for 19 subjects who were each observed by two different raters on each of the three days (Monday, Wednesday, and Friday) and at two time periods (A.M. and P.M.) each day. Thus, 12 observations were made on each subject, with each observation corresponding to one of the 12 combinations of three days, two times, and two raters. In this problem, we assume that the investigator is interested only in assessing whether the SF measurements vary significantly by day of measurement. Consequently, the accompanying table contains the average of the four SF scores taken on each subject (at two times and by two raters) for each of the three days.

   a. For the data in the table, is the factor Days a crossover factor or a nest factor? Explain.

   b. Assuming that the only important effect is that of Days (i.e., the factors Time and Raters are assumed not to have important effects), state the ANOVA model for analyzing the tabulated data and identify the assumptions made about the random effects (including the error term) in the model.

   c. State the null hypothesis that the factor Days has no significant effect in terms of a statement about parameters in your ANOVA model given in part (b).

   d. Based on a comparison of averages at the bottom of the table, does the factor Days appear to have a meaningful effect? Explain.

Average Shoulder Flexion Score by Days

Subject	Monday	Wednesday	Friday
1	10.50	7.78	6.00
2	4.00	2.50	16.75
3	25.75	28.00	22.75
4	20.00	22.50	20.00
5	4.75	8.25	6.50
6	17.25	22.00	18.00
7	24.00	9.25	12.50
8	45.50	50.00	41.75
9	10.00	10.00	5.50
10	28.25	31.25	41.25
11	21.75	22.25	15.50
12	23.00	29.00	26.00
13	10.50	7.00	7.00
14	22.00	19.50	22.50
15	7.25	5.00	0.00
16	19.50	9.00	9.00
17	23.25	26.50	17.50
18	28.50	35.25	8.75
19	3.00	1.75	2.50
Average of Averages	18.36	18.25	15.78

e. Computer results are given next for a repeated measures ANOVA of these data. Based on these results, does the factor Days have a significant effect? Explain by specifying the F statistic, its degrees of freedom, and its P-value appropriate for these data.

f. State the expression for the expected mean square of the factor Days (see Appendix C). Under the null hypothesis that the factor Days has no significant effect, what does this expression reduce to?

g. Use footnote 7 in section 21-4-1 to calculate an estimate of the exchangeable correlation (ρ) given by expression (21.2). Interpret your result.

Edited SAS Output (PROC GLM) for Problem 1

```
TITLE 'MEAN SF BY DAYS';
DATA mnsf;
INPUT subj days mnsf;              Program statements
CARDS;
(the data is inserted here)
;
PROC GLM DATA = mnsf;
    CLASS subj day;
    MODEL mnsf = day subj;
    RANDOM subj / test;
```

(continued)

```
                        MEAN SF BY DAYS
                 General Linear Models Procedure
Dependent Variable: MNSF

Source            DF  Sum of Squares   Mean Square   F Value  Pr > F
Model             20    6697.69777193  334.88488860    12.94  0.0001
Error             36     931.81916491   25.88386569
Corrected Total   56    7629.51693684

       R-Square           C.V.        Root MSE        MNSF Mean
       0.877867        29.13695      5.08761886      17.46105263

Source       DF         Type I SS    Mean Square   F Value  Pr > F
DAYS          2       80.99476842   40.49738421      1.56   0.2231
SUBJ         18     6616.70300351  367.59461131     14.20   0.0001

Source       Type III Expected Mean Square
DAYS         Var(Error) + Q(DAYS)
SUBJ         Var(Error) + 3 Var(SUBJ)

      Tests of Hypotheses for Mixed Model Analysis of Variance
Dependent Variable: MNSF

Source: DAYS
Error: MS(Error)

                    Denominator    Denominator
DF    Type III MS       DF             MS       F Value  Pr > F
 2   40.497384211       36       25.883865692    1.5646  0.2231

Source: SUBJ
Error: MS(Error)

                    Denominator    Denominator
DF    Type III MS       DF             MS       F Value  Pr > F
18  367.59461131        36       25.883865692   14.2017  0.0001
```

2. The analysis described in Problem 1 for the posture measurement data on shoulder flexion (SF) may be criticized because information is lost when the 4 observations for a given subject on a given day are combined into an average score, rather than being treated individually in the analysis. The data set shown in the following table allows for an analysis that considers all 12 observations per subject. In analyzing this data set, we assume that the 4 observations for a given subject are true *replicates* (i.e., we assume that neither the time of day observed nor the rater used—which combine to provide the 4 observations each day for a given subject—are important factors in predicting SF response).

Shoulder Flexion Score by Days

Subject	Sample Size	Monday				Wednesday				Friday			
1	12	15	5	17	5	11	9	10	1	8	10	5	1
2	12	0	8	1	7	−2	1	7	4	18	18	19	12
3	12	29	16	36	22	25	26	31	30	13	26	27	25
4	12	17	18	21	24	14	24	22	30	16	20	20	24

Shoulder Flexion Score by Days (Continued)

Subject	Sample Size	Monday				Wednesday				Friday			
5	12	6	17	10	14	8	8	7	10	9	4	8	5
6	12	18	18	15	18	16	24	22	26	20	1	2	17
7	12	12	7	19	10	10	7	12	8	7	17	12	14
8	12	41	49	46	48	50	52	46	52	30	44	41	52
9	12	8	12	9	11	7	11	7	15	1	8	5	8
10	12	27	34	31	31	36	22	39	28	42	45	38	40
11	12	17	27	17	26	20	18	27	24	17	10	19	16
12	12	23	20	25	24	20	32	29	5	19	30	24	31
13	12	5	15	8	14	6	6	7	9	9	4	6	9
14	12	25	18	23	18	15	18	21	24	16	22	27	25
15	12	5	10	7	7	0	1	7	12	2	-2	0	0
16	12	20	17	22	19	5	11	7	13	3	13	6	14
17	12	28	21	24	20	27	20	32	27	16	15	22	17
18	12	33	25	28	28	38	27	42	34	5	14	9	7
19	12	2	4	1	5	2	5	0	0	2	0	6	2
Averages		18.36				18.25				15.78			

a. How should the ANOVA model for Problem 1 be modified to take the data layout fully into account? [*Hint:* You need to replace Y_{ij} with Y_{ijk} where k denotes the kth of r replicates on the jth day; you also need to add an interaction term to the model.]

b. Use the computer printout provided at the end of this problem to test whether the factor Days has a significant main effect. [*Hint:* See the ANOVA table provided next.] What do you conclude?

ANOVA table for balanced repeated measures ANOVA with one crossover factor and replications for each subject[†]

Source	d.f.	MS	F
Between Subjects	$s-1$	MS_S	MS_S/MS_{TS}
Within Subjects	$s(tr-1)$	MS_w	
Treatments	$t-1$	MS_T	MS_T/MS_{TS}
Treatments × Subjects	$(s-1)(t-1)$	MS_{TS}	MS_{TS}/MS_E
Error	$st(r-1)$		MS_E
Total (corrected)	$str-1$		

[†] s = Number of subjects, t = Number of treatments, and r = Number of replicates per subject. In applying this ANOVA table to the data of Problem 2, we have Treatments = Days, $t = 3$, $s = 19$, $r = 4$.

c. Use the same printout to test whether a significant interaction effect exists between Subjects and Days. Based on the results of this test, what aspect(s) of the test you carried out in part (b) might be troubling?

Edited SAS Output (PROC GLM) for Problem 2

```
TITLE  SF BY DAYS ;
DATA sf;
INPUT subj days sf;
CARDS;
(the data is inserted here)      ← Program statements
;
PROC GLM DATA = sf;
   CLASS subj day;
   MODEL sf = day subj subj*day;
   RANDOM subj subj*day / test;
```

General Linear Models Procedure

Dependent Variable: SF

Source	DF	Sum of Squares	Mean Square	F Value	Pr > F
Model	56	30273.81578947	540.60385338	28.97	0.0001
Error	171	3191.25000000	18.66228070		
Corrected Total	227	33465.06578947			

R-Square	C.V.	Root MSE	SF Mean
0.904639	24.81625	4.31998619	17.40789474

Source	DF	Type I SS	Mean Square	F Value	Pr > F
DAY	2	303.57894737	151.78947368	8.13	0.0004
SUBJ	18	26738.81578947	1485.48976608	79.60	0.0001
SUBJ*DAY	36	3231.42105263	89.76169591	4.81	0.0001

Incorrect F Tests

Source:	Type III Expected Mean Square
DAY	Var(Error) + 4 Var(SUBJ*DAY) + Q(DAY)
SUBJ	Var(Error) + 4 Var(SUBJ*DAY) + 12 Var(SUBJ)
SUBJ*DAY	Var(Error) + 4 Var(SUBJ*DAY)

Tests of Hypotheses for Mixed Model Analysis of Variance

Dependent Variable: SF

Correct F Tests

Source: DAY
Error: MS(SUBJ*DAY)

DF	Type III MS	Denominator DF	Denominator MS	F Value	Pr > F
2	151.78947368	36	89.76169590б	1.6910	0.1986

Source: SUBJ
Error: MS(SUBJ*DAY)

DF	Type III MS	Denominator DF	Denominator MS	F Value	Pr > F
18	1485.4897661	36	89.76169590б	16.5493	0.0001

Source: SUBJ*DAY
Error: MS(Error)

DF	Type III MS	Denominator DF	Denominator MS	F Value	Pr > F
36	89.76169590б	171	18.66228070б	4.8098	0.0001

3. A study by Holder, Plikaytis, and Carlone (1996) compared two laboratory protocols designed to measure antibody levels in an enzyme-linked immunosorbent assay (ELISA) for *Streptococcus pneumoniae*. One protocol incorporates a "blocking step" that is thought to increase the specificity of the assay, maximizing the yield of the specific antibody; the other protocol does not use the blocking step. The data shown in the following table provide the ELISA results obtained from using each protocol on six specimens (i.e., samples), each with *Streptococcus pneumoniae* serogroup 4. This is a Balanced Repeated Measures ANOVA design with six measurements on each sample, three of which use the protocol with the blocking step and three of which do not use the blocking step.

Antibody yields for two protocols measuring *Streptococcus pneumoniae* serogroup 4

| | | Sample # | | | | | | |
		1	2	3	4	5	6	Sample Mean
Blocking Step	Yes	7.3	1.0	2.8	2.3	28.4	0.4	
		9.3	1.0	3.1	3.2	29.2	0.5	7.66
		6.4	1.0	2.8	2.8	35.9	0.4	
	No	9.3	1.5	3.6	3.7	27.2	0.6	
		9.1	1.4	3.4	3.6	25.5	0.6	7.67
		9.3	2.0	4.5	3.6	28.3	0.8	

a. Is the factor Blocking Step a crossover factor or a nest factor? Should this factor be considered a fixed or a random factor? Explain.

b. Since this study involves a balanced repeated measures design with replications at each level of the single treatment factor (Blocking Step), the ANOVA model for analyzing the data is the same one used for the design in Problem 2. State the corresponding regression model for this analysis.

c. State the null hypothesis that Blocking Step has no significant effect in terms of a statement about the parameters in the ANOVA and/or corresponding regression model given in part (b).

d. Based on a comparison of sample means, does the Blocking Step factor appear to have a meaningful effect? Explain.

e. Computer results for a repeated measures ANOVA of these data are given next. Based on these results, does the Blocking Step factor have a significant effect? Explain by specifying the F statistic, its degrees of freedom, and its P-value appropriate for these data. [*Hint:* The ANOVA table provided in Problem 2 applies to the study design considered here.]

Edited SAS Output (PROC GLM) for Problem 3

```
TITLE 'SEROGROUP4 DATA';
DATA serogroup4;
INPUT block sample result;          Program statements
CARDS;
(the data is inserted here)
;
PROC GLM DATA = serogroup4;
   CLASS block sample;
   MODEL result = block sample block*sample;
   RANDOM sample block*sample / test;
```

(continued)

```
                  General Linear Models Procedure
Dependent Variable: RESULT
Source              DF   Sum of Squares   Mean Square   F Value   Pr > F
Model               11   3565.02555556    324.09323232   177.83   0.0001
Error               24     43.74000000      1.82250000
Corrected Total     35   3608.76555556

        R-Square             C.V.         Root MSE       RESULT Mean
        0.987880           17.62146      1.35000000      7.66111111

Source         DF        Type I SS     Mean Square    F Value   Pr > F
BLOCK           1       0.00111111      0.00111111       0.00    0.9805
SAMPLE          5    3532.18555556    706.43711111     387.62    0.0001
BLOCK*SAMPLE    5      32.83888889      6.56777778       3.60    0.0143
```

Incorrect F Tests

```
Source:        Type I Expected Mean Square
BLOCK          Var(Error) + 3 Var(BLOCK*SAMPLE) + Q(BLOCK)
SAMPLE         Var(Error) + 3 Var(BLOCK*SAMPLE) + 6 Var(SAMPLE)
BLOCK*SAMPLE   Var(Error) + 3 Var(BLOCK*SAMPLE)

          Tests of Hypotheses for Mixed Model Analysis of Variance
Dependent Variable: RESULT

Source: BLOCK
Error: MS(BLOCK*SAMPLE)
```

Correct F Tests ↓

```
              Denominator    Denominator
DF   Type I MS       DF            MS      F Value    Pr > F
 1   0.00111111       5    6.5677777778   0.000169    0.9901

Source: SAMPLE
Error: MS(BLOCK*SAMPLE)

              Denominator    Denominator
DF    Type I MS      DF            MS      F Value    Pr > F
 5   706.43711111     5    6.5677777778   107.5611    0.0001

Source: BLOCK*SAMPLE
Error: MS(Error)

              Denominator    Denominator
DF    Type I MS      DF            MS      F Value    Pr > F
 5   6.5677777778    24       1.8225      3.6037     0.0143
```

4. A study by Heffner, Drawbaugh, and Zigmond (1974) investigated the effects of an amphetamine on the behavior of rats. Before the study began, 24 "thirsty" rats were trained to press a lever to obtain water. The rats were categorized into three groups (slow, medium, and fast) of equal size according to their initial press rates. Each rat received three doses of the drug (i.e., the amphetamine under study) and one dose of a placebo on separate occasions and in random order. One hour after the drug injection, an experimental session began in which the rat received water after pressing the lever a prespecified number of times. Half of the rats received water after two presses of the lever, and the other half received water after five presses.

　　The response measured was the lever press rate (LPR, the total number of lever presses divided by the elapsed time in seconds) achieved by each thirsty rat in attempting to obtain water. The primary research question was whether the drug affected the LPR. Also of interest were whether the number of presses (PRS) required to obtain water (two versus five) and/or the initial press rate (IPR)—slow, medium, or fast—affected the response, and whether any interaction effects existed. The resulting data are listed in the accompanying table.

Lever press rate for 24 thirsty rats

IPR	PRS	Rat #	Drug 1 (Placebo)	Drug 2	Drug 3	Drug 4	Mean
Slow	2	1	.81	.80	.82	.50	.76
		2	.77	.78	.79	.51	
		3	.80	.82	.83	.52	
		4	.95	.95	.91	.60	
	5	5	2.18	2.44	1.92	.92	1.80
		6	2.02	2.20	1.75	.82	
		7	2.06	2.28	1.86	.80	
		8	2.28	2.46	1.90	.90	
Medium	2	9	1.03	1.13	1.04	.82	.98
		10	.96	.93	1.02	.63	
		11	.98	1.00	.98	.74	
		12	1.17	1.20	1.18	.91	
	5	13	2.62	2.58	2.21	1.03	2.09
		14	2.60	2.60	2.34	1.14	
		15	2.39	2.41	2.09	.90	
		16	2.70	2.64	2.23	1.02	
Fast	2	17	1.20	1.24	1.27	.96	1.20
		18	1.25	1.23	1.30	1.01	
		19	1.23	1.20	1.18	.95	
		20	1.31	1.42	1.41	1.08	
	5	21	2.98	2.64	2.34	1.28	2.34
		22	3.10	2.85	2.40	1.35	
		23	2.80	2.48	2.16	1.01	
		24	3.21	2.92	2.56	1.40	
Mean			1.81	1.80	1.60	.91	

IPR = Initial press rate; PRS = Number of presses

a. State whether each of the factors Drug, IPR, and PRS is a crossover or a nest factor. Which of these factors should be treated as fixed factors, and which as random factors? Explain.

b. Consider an analysis of these data that focuses only on the effect of the factor Drug (i.e., an analysis that ignores the factors IPR and PRS in the analysis). Then for each of the 24 rats we have four responses, corresponding to the administration of each of the four levels of Drug. What kind of repeated measures study design is this? (Choose from any of the study designs described in the chapter.)

c. Identify the ANOVA model for the design described in part (b), and state the assumptions typically made about the random effects (including the error term) in the model.

d. Based on a comparison of sample means, does the factor Drug appear to have a meaningful effect? Explain.

e. Use the computer output provided next to test for the significance of the factor Drug. Make sure to describe the null hypothesis being tested, the F statistic used, its degrees of freedom, and the resulting P-value for this test. What do you conclude?

f. Use the same computer output to test for the significance of the subject (i.e., Rat) factor. Why might you expect the results for this test to be significant?

g. State the expression for the expected mean square of the factor Drug. Under the null hypothesis that the factor Drug has no significant effect, what does this expression reduce to?

h. Why is it not possible to test whether a significant interaction exists between Drug and Rat?

Edited SAS Output (PROC GLM) for Problem 4

```
DATA rats;
   SET DIR1.RATSDAT;
TITLE 'GLM.SAS:DRUG ONLY';
PROC GLM DATA = rats;                    Program statements
   CLASS rat drug;
   MODEL lpr = drug rat;
   RANDOM rat / test;
------------------------------------------------------------------
Dependent Variable: LPR

Source              DF   Sum of Squares   Mean Square   F Value   Pr > F
Model               26      46.64734375    1.79412861     18.91   0.0001
Error               69       6.54735521    0.09488921
Corrected Total     95      53.19469896

Source              DF   Sum of Squares   Mean Square   F Value   Pr > F
DRUG                 3      13.01466979    4.33822326     45.72   0.0001
RAT                 23      33.63267396    1.46229017     15.41   0.0001

Source         Type III Expected Mean Square
DRUG           Var(Error) + Q(DRUG)
RAT            Var(Error) + 4 Var(RAT)

      Tests of Hypotheses for Mixed Model Analysis of Variance

Dependent Variable: LPR

Source: DRUG
Error: MS(Error)

                      Denominator     Denominator
DF      Type III MS          DF              MS   F Value   Pr > F
 3      4.33822326           69      0.09488921     45.72   0.0001

Source: RAT
Error: MS(Error)

                      Denominator     Denominator
DF      Type III MS          DF              MS   F Value   Pr > F
 3      1.46229017           69      0.09488921     15.41   0.0001
```

5. Consider the same study by Heffner, Drawbaugh, and Zigmond (1974) and the same data set described in Problem 4, where the response measured was the lever press rate (LPR) achieved by a thirsty rat in attempting to obtain water. The three factors identified as possible predictors were Drug (three active doses plus a placebo), IPR (initial press rate, categorized as slow, medium, or fast), and PRS (number of presses required to obtain water: two or five). Another factor is the random factor Rat (with 24 levels). In this problem, suppose that the combination of the factors IPR and PRS is treated as a single Factor A with six levels, according to the following categorization:

IPR	PRS	Level of Factor A
Slow	2	1
Slow	5	2
Medium	2	3
Medium	5	4
Fast	2	5
Fast	5	6

A repeated measures ANOVA was carried out to evaluate whether Factor A significantly affected the response. (This analysis ignores the effects of the factor Drug, so each of the four responses per rat is treated as a repeated observation at the same level of Factor A.) The ANOVA model here pertains to a balanced repeated measures ANOVA design with one nest factor.

a. State whether Factor A is a crossover or a nest factor. Also, should Factor A be treated as a fixed or a random factor? Explain briefly.

b. Based on a comparison of sample means, does Factor A appear to have a meaningful effect? Explain.

c. Identify the formula for the ANOVA model for this situation, and specify the assumptions typically made about the random effects (and error term) for this model.

d. The ANOVA table produced by using SAS's PROC GLM for analyzing the effect of Factor A on the response is given in the following computer output. For this ANOVA, carry out a test of whether the effect of Factor A is significant. What do you conclude?

Edited SAS Output (PROC GLM) for Problem 5

```
DATA rats;
   SET DIR1.RATSDAT;
INPUT id prs rat ipr drug lpr;
IF ipr=1 and prs=1 THEN factora=1;
IF ipr=1 and prs=2 THEN factora=2;
IF ipr=2 and prs=1 THEN factora=3;      Program statements
IF ipr=2 and prs=2 THEN factora=4;
IF ipr=3 and prs=1 THEN factora=5;
IF ipr=3 and prs=2 THEN factora=6;
TITLE 'GLM.SAS:FACTOR A ONLY';
PROC GLM DATA = rats;
   CLASS rat factora;
   MODEL lpr = factora rat(factora);
   RANDOM rat(factora) / test;
```

(continued)

```
Dependent Variable: LPR
Source              DF   Sum of Squares   Mean Square   F Value   Pr > F
Model               23      33.63267396    1.46229017      5.38   0.0001
Error               72      19.56202500    0.27169479
Corrected Total     95      53.19469896

Source              DF   Sum of Squares   Mean Square   F Value   Pr > F
FACTORA              5      32.80713021    6.56142604     24.15   0.0001
RAT(FACTORA)        18       0.82554375    0.04586354      0.17   0.9999

Source:             Type III Expected Mean Square
FACTORA             Var(Error) + 4 Var(RAT(FACTORA)) + Q(FACTORA)
RAT(FACTORA)        Var(Error) + 4 Var(RAT(FACTORA))

      Tests of Hypotheses for Mixed Model Analysis of Variance
Dependent Variable: LPR

Source: FACTORA
Error: MS(RAT(FACTORA))

                    Denominator     Denominator
DF     Type III MS      DF              MS        F Value   Pr > F
 5    6.5614260417      18        0.0458635417   143.0641   0.0001
```

e. Using Table C-8 in Appendix C, give the expression for the expected mean square of Factor A (i.e., the one corresponding to the Treatments source in the table). Under the null hypothesis that Factor A has no significant effect, what does this expression reduce to, and what is the corresponding (reduced) expected mean square term?

f. In this analysis, the exchangeable correlation (ρ) between two responses on the same subject is given by the expression

$$\rho = \frac{\sigma_S^2}{\sigma_S^2 + \sigma^2}$$

where σ_S^2 denotes the variance component for subjects (i.e., Rat), and σ^2 denotes pure error. Use the computer output to develop an estimate of this correlation coefficient, and interpret your result. [*Hint:* In computing your estimate of ρ, you need to estimate σ_S^2 by solving the equation

$$E(\mathrm{MS}_{S(T)}) = \sigma_S^2 + r\sigma^2$$

for σ_S^2, where r is the number of repeats measured on the ith rat for level j of Factor A, and then substituting the estimates $\mathrm{MS}_{S(T)}$ for $E(\mathrm{MS}_{S(T)})$ and $\mathrm{MS}_{\mathrm{Error}}$ for σ^2.]

6. Consider the same study by Heffner, Drawbaugh, and Zigmond (1974) and the same data set described in Problems 4 and 5, where the response measured was the lever press rate (LPR) achieved by a thirsty rat in attempting to obtain water. The three factors identified as possible predictors were Drug (three doses plus a placebo), IPR (initial press rate, categorized as slow, medium, or fast), and PRS (number of presses required to obtain water: two or five). Another factor of importance is the random factor Rat (with 24 levels). Furthermore, in Problem 5, we combined the factors IPR and PRS into a single Factor A with six levels, according to the following categorization:

IPR	PRS	Level of Factor A
Slow	2	1
Slow	5	2
Medium	2	3
Medium	5	4
Fast	2	5
Fast	5	6

Now we consider both Factor A and Drug together in a repeated measures ANOVA to evaluate whether either or both factors and their interaction have an effect on the response.

a. Describe the data layout for this analysis. [*Hint:* Consider Table 21-10 in the text.]

b. Using the data layout in part (a) and the data provided in Problem 4, form a table of sample means. Based on this table, describe whether there appears to be a meaningful main effect of Factor A, a meaningful main effect of Drug, and/or a meaningful interaction effect of Factor A with Drug. Explain.

c. What kind of repeated measures study design is this? (Choose from any of the study designs described in this chapter.)

d. Identify the formula for the ANOVA model for this design, and specify the assumptions typically made about the random effects (and error term) for this model.

e. For the model stated in part (d), use a flow diagram to describe how the variation (sums of squares) is partitioned into contributions from various sources of predictors. [*Hint:* Consider Figure 21-4 in the text.]

f. The ANOVA table produced by using SAS's PROC GLM for analyzing the effect of Factor A on the response is given in the following computer output. For this ANOVA, carry out tests of whether the main effect of Factor A, the main effect of Drug, and the Factor A-by-Drug interaction are significant. For each test, state the null hypothesis being tested, the form of the F statistic, its degrees of freedom under the null hypothesis, and the resulting P-value. What do you conclude?

g. Using Tables C-11 and C-12 in Appendix C, give expressions for the expected mean squares of Factor A, Drug, and their interaction. Under the null hypothesis that these factors have no significant effect, what does each expression reduce to, and what is the corresponding (reduced) mean square term?

Edited SAS Output (PROC GLM) for Problem 6

```
DATA rats;
   SET DIR1.RATSDAT;
TITLE 'GLM.SAS:DRUG AND FACTOR A';          Program statements
PROC GLM DATA = rats;
   CLASS rat factora drug;
   MODEL lpr = factora drug factora*drug rat(factora);
   RANDOM rat(factora) / test;
-----------------------------------------------------------------
```

(continued)

```
Dependent Variable: LPR

Source              DF    Sum of Squares    Mean Square    F Value   Pr > F
Model               41       46.64734375    1.29540409      841.46   0.0001
Error               54        0.08313125    0.00153947
Corrected Total     95       53.19469896

Source              DF    Sum of Squares    Mean Square    F Value   Pr > F
FACTORA              5       32.80713021    6.56142604     4262.14   0.0001
DRUG                 3       13.01466979    4.33822326     2818.00   0.0001
FACTORA*DRUG        15        6.46422396    0.43094826      279.93   0.0001
RAT(FACTORA)        18        0.82554375    0.04586354       29.29   0.0001

Source:              Type III Expected Mean Square
FACTORA              Var(Error) + 4 Var(RAT(FACTORA))
                        + Q(FACTORA, FACTORA*DRUG)
DRUG                 Var(Error) + Q(DRUG, FACTORA, FACTORA*DRUG)
FACTORA*DRUG         Var(Error) + Q(FACTORA*DRUG)
RAT(FACTORA)         Var(Error) + 4 Var(RAT(FACTORA))

      Tests of Hypotheses for Mixed Model Analysis of Variance

Dependent Variable: LPR

Source: FACTORA
Error: MS(RAT(FACTORA))

                   Denominator      Denominator
DF    Type III MS          DF               MS    F Value   Pr > F
 5     6.56142604          18       0.04586354     143.06   0.0001

Source: DRUG
Error: MS(ERROR)

                   Denominator      Denominator
DF    Type III MS          DF               MS    F Value   Pr > F
 3     4.33822326          54       0.00153947    2818.00   0.0001

Source: FACTORA*DRUG
Error: MS(ERROR)

                   Denominator      Denominator
DF    Type III MS          DF               MS    F Value   Pr > F
15     0.43094826          54       0.00153947     279.93   0.0001
```

7. Consider the same study by Heffner, Drawbaugh, and Zigmond (1974) and the same data set described in Problems 4 through 6, where the response measured was the lever press rate (LPR) achieved by a thirsty rat in attempting to obtain water.

In this problem, we consider the effects of IPR and PRS individually, as well as the effects of Drug and Rat and various interactions among these factors, in a repeated measures ANOVA. The ANOVA model for this situation is a modification of the model used in Problem 6, except that the effect of Factor A is split into individual components. The formula for the ANOVA model for this design is

$$Y_{ijkl} = \mu + S_{i(jk)} + \alpha_j + \beta_k + \gamma_l + (\alpha\beta)_{jk} + (\alpha\gamma)_{jl} + (\beta\gamma)_{kl} + (\alpha\beta\gamma)_{jkl} + E_{l(ijk)}$$

where

μ = Overall mean

α_j = jth fixed effect of IPR, j = 1, 2, 3

β_k = kth fixed effect of PRS, k = 1, 2

γ_l = lth fixed effect of Drug, l = 1, 2, 3, 4

$(\alpha\beta)_{jk}$ = Fixed interaction effect of the jth level of IPR with the kth level of PRS

$(\alpha\gamma)_{jl}$ = Fixed interaction effect of the jth level of IPR with the lth level of Drug

$(\beta\gamma)_{kl}$ = Fixed interaction effect of the kth level of PRS with the lth level of Drug

$(\alpha\beta\gamma)_{jkl}$ = Three-way fixed interaction for the j-k-l combination of IPR, PRS, and Drug

$S_{i(jk)}$ = Random effect of rat i within levels j and k of IPR and PRS, respectively

$E_{l(ijk)}$ = Random error of the lth level of Drug within levels j and k of IPR and PRS, respectively, for rat i

a. For the preceding model, produce a flow diagram to describe how the total variation (i.e., total sums of squares) is partitioned into contributions from various sources of predictors. [*Hint:* Use Figure 21-4 in the text as a basis for starting the flow diagram.]

b. Using the following computer output produced by fitting the preceding ANOVA model, carry out tests for main effects and interactions of each predictor in the model. What do you conclude about whether the Drug factor has a significant effect on the response?

Edited SAS Output (PROC GLM) for Problem 7

```
DATA rats;
   SET DIR1.RATSDAT;                              Program statements
TITLE 'GLM.SAS:INDIVIDUAL FACTORS MODEL';
PROC GLM DATA = rats;
   CLASS ipr prs drug rat;
   MODEL lpr = ipr prs drug ipr*prs ipr*drug prs*drug ipr*prs*drug
               rat(ipr*prs);
   RANDOM rat(ipr*prs) / test;
```

Dependent Variable: LPR

Source	DF	Sum of Squares	Mean Square	F Value	Pr > F
Model	41	46.64734375	1.29540409	841.46	0.0001
Error	54	0.08313125	0.00153947		
Corrected Total	95	53.19469896			

Source	DF	Sum of Squares	Mean Square	F Value	Pr > F
IPR	2	3.88891458	1.94445729	1263.07	0.0001
PRS	1	28.87523438	28.87523438	18756.64	0.0001
DRUG	3	13.01466979	4.33822326	2818.00	0.0001
IPR*PRS	2	0.04298125	0.02149063	13.96	0.0001
IPR*DRUG	6	0.15725208	0.02620868	17.02	0.0001
PRS*DRUG	3	6.09505312	2.03168437	1319.73	0.0001
IPR*PRS*DRUG	6	0.21191875	0.03531979	22.94	0.0001
RAT(IPR*PRS)	18	0.82554375	0.04586354	29.29	0.0001

(continued)

```
      Tests of Hypotheses for Mixed Model Analysis of Variance
Dependent Variable: LPR
Source: IPR
Error: MS(RAT(IPR*PRS))
                    Denominator      Denominator
DF      Type III MS         DF              MS     F Value    Pr > F
 2    1.9444572917          18      0.04586354       42.40    0.0001
Source: PRS
Error: MS(RAT(IPR*PRS))
                    Denominator      Denominator
DF      Type III MS         DF              MS     F Value    Pr > F
 1    28.875234375          18      0.04586354      629.59    0.0001
Source: IPR*PRS
Error: MS(ERROR)

                    Denominator      Denominator
DF      Type III MS         DF              MS     F Value    Pr > F
15    0.021490625          18      0.04586354        0.47    0.6333
```

Note: *F* statistics in ANOVA table for DRUG (i.e., 2818.00), IPR*DRUG (i.e., 17.02), PRS*DRUG (i.e., 1319.73), and IPR*PRS*DRUG (i.e., 22.94) are correct.

8. A study by Rikkers et al. (1978) involved a prospective randomized surgical trial that compared cirrhotic patients who had bled from use of either a nonselective shunt (a standard operation) or a selective shunt (a new operation). The response calibrates the maximal rate of urea synthesis (MRUS), a measure of liver function; low values of MRUS indicate poor liver function. The study sample consisted of eight selective shunt patients and thirteen nonselective shunt patients. MRUS was measured both preoperatively and early postoperatively on each of the 21 patients. The purpose of the study was to compare the change in liver function in each of the two groups. The data are described in the accompanying table.

   a. Decide whether each of the factors Group and Time should be a nest or a crossover factor. Also, state whether each of these factors should be considered as a fixed or as a random factor. Explain.

   b. What kind of repeated measures study design is this? (Choose from any of the study designs described in this chapter.)

   c. Identify the ANOVA model for analyzing these data, and specify the assumptions typically made about the random effects (including the error term) in the model.

   d. For the model stated in part (c), use a flow diagram to describe how the total variation (i.e., total sums of squares) is partitioned into contributions from various sources of predictors [*Hint:* Consider Figure 21-4 in the text.]

   e. State the null hypothesis that the factor Group has no significant effect in terms of a statement about parameters in your ANOVA model given in part (c).

   f. Based on a comparison of sample means, does the change in liver function appear to differ significantly between the two groups? Explain.

   g. Computer results for a repeated measures ANOVA of these data are given next. Based on these results, does the factor Group have a significant effect? Explain by describing the *F* statistic, its degrees of freedom, and its *P*-value appropriate for these data.

   h. State the expression for the expected mean square of the factor Group. Under the null hypothesis that the factor Group has no significant effect, what does this expression reduce to?

Group	Subject #	Pre	Post
Selective Shunt	1	51	48
	2	35	55
	3	66	60
	4	40	35
	5	39	36
	6	46	43
	7	52	46
	8	42	54
	Means	46.375	47.125
Nonselective Shunt	9	34	16
	10	40	36
	11	34	16
	12	36	18
	13	38	32
	14	32	14
	15	44	20
	16	50	43
	17	60	45
	18	63	67
	19	50	36
	20	42	34
	21	43	32
	Means	43.538	31.462

Edited SAS Output (PROC GLM) for Problem 8

```
DATA mrus;
   SET DIR1.MRUSDAT;
TITLE 'GLM.SAS:MRUS DATA';              Program statements
PROC GLM DATA = mrus;
   CLASS subj group time;
   MODEL mrus = group time group*time subj(group);
   RANDOM subj(group) / test;
------------------------------------------------------------------
                   MRUS BY GROUP AND TIME
                General Linear Models Procedure
Dependent Variable: MRUS

Source            DF    Sum of Squares    Mean Square    F Value   Pr > F
Model             22    6237.76465201    283.53475691      7.91    0.0001
Error             19     681.21153846     35.85323887
Corrected Total   41    6918.97619048

     R-Square              C.V.        Root MSE         MRUS Mean
     0.901544           14.59581     5.98775742       41.02380952
```

(continued)

```
          General Linear Models Procedure (Continued)
Source          DF        Type I SS       Mean Square  F Value   Pr > F
GROUP            1     847.47619048     847.47619048     23.64   0.0001
TIME             1     542.88095238     542.88095238     15.14   0.0010
GROUP*TIME       1     407.40750916     407.40750916     11.36   0.0032
SUBJ(GROUP)     19    4440.00000000     233.68421053      6.52   0.0001

Source          Type III Expected Mean Square
GROUP           Var(Error) + 2 Var(SUBJ(GROUP)) + Q(GROUP, GROUP*TIME)
TIME            Var(Error) + Q(TIME, GROUP*TIME)
GROUP*TIME      Var(Error) + Q(GROUP*TIME)
SUBJ(GROUP)     Var(Error) + 2 Var(SUBJ(GROUP))

      Tests of Hypotheses for Mixed Model Analysis of Variance
Dependent Variable: MRUS
Source: GROUP *
Error: MS(SUBJ(GROUP))
                        Denominator     Denominator
DF     Type III MS          DF              MS     F Value     Pr > F
 1     847.47619048         19     233.68421053     3.6266     0.0721
* This test assumes one or more other fixed effects are zero.
Source: TIME *
Error: MS(Error)
                        Denominator     Denominator
DF     Type III MS          DF              MS     F Value     Pr > F
 1     317.69322344         19      35.853238866    8.8609     0.0078
* This test assumes one or more other fixed effects are zero.
Source: GROUP*TIME
Error: MS(Error)
                        Denominator     Denominator
DF     Type III MS          DF              MS     F Value     Pr > F
 1     407.40750916         19      35.853238866   11.3632     0.0032
Source: SUBJ(GROUP)
Error: MS(Error)
                        Denominator     Denominator
DF     Type III MS          DF              MS     F Value     Pr > F
19     233.68421053         19      35.853238866    6.5178     0.0001
```

9. For the posture measurement study considered in the main body of this chapter, we previously described in Table 21-1 the data from measuring shoulder flexion (SF) for one of two raters (i.e., rater 2); and we described the analysis of these data in section 21-4-2. The following table lists the corresponding data (on SF) for rater 1.

   a. Identify the ANOVA model appropriate for analyzing the tabulated data for rater 1. Also, state the assumptions typically made about the random effects (including the error term) in the model.

   b. Computer output from an analysis of the data for rater 1 is provided next. Based on this output, answer the following questions.

   i. Does the factor Days have a significant main effect? Explain by describing the F statistic, its degrees of freedom, and the P-value for this test.

   ii. Does the factor Time have a significant main effect? Explain by describing the F statistic, its degrees of freedom, and the P-value for this test.

   iii. Do the factors Days and Time have a significant main effect? Explain by describing the F statistic, its degrees of freedom, and the P-value for this test.

Shoulder flexion (SF) scores for rater 1 from study of posture

Subject #	Sample Size	Monday		Wednesday		Friday	
		A.M.	P.M.	A.M.	P.M.	A.M.	P.M.
1	6	15	5	11	9	8	10
2	6	0	8	−2	1	18	18
3	6	29	16	25	26	13	26
4	6	17	18	14	24	16	20
5	6	6	17	8	8	9	4
6	6	18	18	16	24	20	1
7	6	12	7	10	7	7	17
8	6	41	49	50	52	30	44
9	6	8	12	7	11	1	8
10	6	27	34	36	22	42	45
11	6	17	27	20	18	17	10
12	6	23	20	20	32	19	30
13	6	5	15	6	6	9	4
14	6	25	18	15	18	16	22
15	6	5	10	0	1	2	−2
16	6	20	17	5	11	3	13
17	6	28	21	27	20	16	15
18	6	33	25	38	27	5	14
19	6	2	4	2	5	2	0
Sample Means:		17.42	17.95	16.21	16.95	13.32	15.74

Edited SAS Output (PROC GLM) for Problem 9

```
TITLE 'SF BY DAYS AND TIME, RATER 1';
    DATA sfr1;
    INPUT subj day time sf;
    CARDS;
    .
    . (the data)
    .
    ;                                        Program statements
PROC GLM DATA = sfr1;
    CLASS subj day time;
    MODEL sf = day time day*time subj subj*day subj*time;
    RANDOM subj subj*day subj*time / test;
------------------------------------------------------------
                General Linear Models Procedure
Dependent Variable: SF

Source            DF   Sum of Squares   Mean Square   F Value   Pr > F
Model             77   14961.27192982   194.30223285     5.95   0.0001
Error             36    1175.29824561    32.64717349
Corrected Total  113   16136.57017544
```

(continued)

```
              General Linear Models Procedure (Continued)
Source          DF          Type III SS        Mean Square   F Value   Pr > F
DAY             2         191.70175439         95.85087719     2.94    0.0659
TIME            1          44.21929825         44.21929825     1.35    0.2522
DAY*TIME        2          21.70175439         10.85087719     0.33    0.7194
SUBJ           18       12464.07017544        692.44834308    21.21    0.0001
SUBJ*DAY       36        1893.29824561         52.59161793     1.61    0.0787
SUBJ*TIME      18         346.28070175         19.23781676     0.59    0.8840

Source       Type III Expected Mean Square
DAY          Var(Error) + 2 Var(SUBJ*DAY) + Q(DAY, DAY*TIME)
TIME         Var(Error) + 3 Var(SUBJ*TIME) + Q(TIME, DAY*TIME)
DAY*TIME     Var(Error) + Q(DAY*TIME)
SUBJ         Var(Error) + 3 Var(SUBJ*TIME) + 2 Var(SUBJ*DAY)
               + 6 Var(SUBJ)
SUBJ*DAY     Var(Error) + 2 Var(SUBJ*DAY)
SUBJ*TIME    Var(Error) + 3 Var(SUBJ*TIME)

      Tests of Hypotheses for Mixed Model Analysis of Variance
Dependent Variable: SF

Source: DAY
Error: MS(SUBJ*DAY)

                   Denominator     Denominator
DF      Type III MS        DF            MS      F Value    Pr > F
 2     95.850877193        36   52.591617934    1.8226     0.1762

Source: TIME
Error: MS(SUBJ*TIME)

                   Denominator     Denominator
DF      Type III MS        DF            MS      F Value    Pr > F
 1     44.219298246        18   19.237816764    2.2986     0.1469

Source: DAY*TIME
Error: MS(Error)

                   Denominator     Denominator
DF      Type III MS        DF            MS      F Value    Pr > F
 2     10.850877193        36   32.647173489    0.3324     0.7194
```

c. Based on the preceding analyses and using Table C-5 in Appendix C, specify the expressions for the expected mean squares of the factor Days, the factor Time, and the interaction of Days with Time. State the appropriate null hypothesis corresponding to each of these expected mean squares in terms of parameters in the ANOVA model. What do each of these expected means squares reduce to under appropriate null hypotheses?

d. Based on the ANOVA model from part (a), the formula for the correlation between any two responses on the same subject is given by the expression

$$\text{Corr}(Y_{ijk}, Y_{ij'k'}) = (\sigma_S^2 + p_{jj'}\sigma_{SA}^2 + q_{kk'}\sigma_{SB}^2 + r_{jj'kk'}\sigma^2)/(\sigma_S^2 + \sigma_{SA}^2 + \sigma_{SB}^2 + \sigma^2)$$

where

$$p_{jj'} = \begin{cases} 1 \text{ if } j=j' \\ 0 \text{ if } j \neq j' \end{cases} \qquad q_{kk'} = \begin{cases} 1 \text{ if } k=k' \\ 0 \text{ if } k \neq k' \end{cases} \qquad r_{jj'kk'} = \begin{cases} 1 \text{ if } j=j' \text{ and } k=k' \\ 0 \text{ otherwise} \end{cases}$$

for days j and j' and times k and k' on the ith subject; here, σ_S^2 denotes the variance component for the random effect of subjects, σ_{SA}^2 denotes the variance component for the (random) subject-by-days interaction effect, and σ_{SB}^2 denotes the variance component for the (random) subject-by-time interaction effect.

From the preceding formula, construct the (6×6) correlation matrix for the six observations on the same subject. Based on the correlation matrix obtained, is the correlation structure for this model exchangeable? Explain.

10. Consider the combined information from both raters on the shoulder flexion scores in the posture measurement study, using the data for rater 2 given in Table 21-1 of the main text and the data for rater 1 given in Problem 9. The questions below thus concern the data on 19 subjects, with 12 observations on each subject, and with each observation corresponding to one of the 12 combinations of three days, two times, and two raters. The design is a balanced repeated measures design with three crossover factors.

a. Using the information provided in Table 21-1 for rater 2 and in Problem 9 for rater 1, fill in the sample means in the following tables.

Rater 1

	Monday	Wednesday	Friday	
A.M.	_____	_____	_____	_____
P.M.	_____	_____	_____	_____
	_____	_____	_____	

Rater 2

	Monday	Wednesday	Friday	
A.M.	_____	_____	_____	_____
P.M.	_____	_____	_____	_____
	_____	_____	_____	

b. Based on the results you obtained for the tables in part (a), does Days and/or Time appear to have a meaningful main effect? Explain.

c. Based on the results you obtained for the tables in part (a), does there appear to be a meaningful interaction effect between Days and Time? Explain.

d. Based on the results you obtained for the tables in part (a), does there appear to be a meaningful three-way interaction effect among Days, Time, and Raters? Explain.

e. Edited computer output from SAS's GLM procedure for the analysis of the SF data appears next. In it, the Days and Time factors are treated as fixed, and the Raters factor as random. Based on this output, answer the following questions.

   i. Identify the ANOVA model that was used for this analysis.

   ii. Construct an ANOVA table for this analysis. In this table, provide appropriate degrees of freedom, mean square terms, and F statistics that consider tests for main effects and interactions involving the three factors Days, Time, and Raters.

   iii. Summarize the results of the F tests for the main effects and interactions from the ANOVA table you constructed in the previous part. What are your conclusions about the effects of the three factors involved?

f. Answer questions i, ii, and iii of part (e) assuming this time that all three factors are fixed. [Hint: You can answer this part by using appropriate information provided in the computer output.]

g. Compare your results from parts (e) and (f).

h. Based on the ANOVA model used in either part (f) or (g) of this problem, is the correlation structure for a given individual an exchangeable correlation structure? Explain.

Edited SAS Output (PROC GLM) for Problem 10

```
PROC glm DATA = sf;                    Program statements
    CLASS subj rater time day;
    MODEL sf = rater time day rater*time rater*day time*day
rater*time*day subj subj*rater subj*time subj*day subj*rater*time
subj*rater*day subj*time*day;
    RANDOM rater rater*time rater*day rater*time*day subj
subj*rater subj*time subj*day subj*rater*time subj*rater*day
subj*time*day / test;
--------------------------------------------------------------------
                    General Linear Models Procedure
Dependent Variable: SF
```

Source	DF	Sum of Squares	Mean Square	F Value	Pr > F
Model	191	33260.82017544	174.14041977	30.69	0.0001
Error	36	204.24561404	5.67348928		
Corrected Total	227	33465.06578947			

Source	DF	Type III SS	Mean Square	F Value	Pr > F
RATER	1	242.21491228	242.21491228	42.69	0.0001
TIME	1	31.68859649	31.68859649	5.59	0.0236
DAY	2	303.57894737	151.78947368	26.75	0.0001
RATER*TIME	1	28.77631579	28.77631579	5.07	0.0305
RATER*DAY	2	52.66666667	26.33333333	4.64	0.0161
TIME*DAY	2	26.35087719	13.17543860	2.32	0.1126
RATER*TIME*DAY	2	16.42105263	8.21052632	1.45	0.2486
SUBJ	18	26738.81578947	1485.48976608	261.83	0.0001
SUBJ*RATER	18	239.20175439	13.28898635	2.34	0.0146
SUBJ*TIME	18	686.06140351	38.11452242	6.72	0.0001
SUBJ*DAY	36	3231.42105263	89.76169591	15.82	0.0001
SUBJ*RATER*TIME	18	68.30701754	3.79483431	0.67	0.8175
SUBJ*RATER*DAY	36	373.66666667	10.37962963	1.83	0.0370
SUBJ*TIME*DAY	36	1221.64912281	33.93469786	5.98	0.0001

Source	Type III Expected Mean Square
RATER	Var(Error) + 2 Var(SUBJ*RATER*DAY) + 3 Var(SUBJ*RATER*TIME) + 6 Var(SUBJ*RATER) + 19 Var(RATER*TIME*DAY) + 38 Var(RATER*DAY) + 57 Var(RATER*TIME) + 114 Var(RATER)
TIME	Var(Error) + 2 Var(SUBJ*TIME*DAY) + 3 Var(SUBJ*RATER*TIME) + 6 Var(SUBJ*TIME) + 19 Var(RATER*TIME*DAY) + 57 Var(RATER*TIME) + Q(TIME TIME*DAY)
DAY	Var(Error) + 2 Var(SUBJ*TIME*DAY) + 2 Var(SUBJ*RATER*DAY) + 4 Var(SUBJ*DAY) + 19 Var(RATER*TIME*DAY) + 38 Var(RATER*DAY) + Q(DAY TIME*DAY)
RATER*TIME	Var(Error) + 3 Var(SUBJ*RATER*TIME) + 19 Var(RATER*TIME*DAY) + 57 Var(RATER*TIME)
RATER*DAY	Var(Error) + 2 Var(SUBJ*RATER*DAY) + 19 Var(RATER*TIME*DAY) + 38 Var(RATER*DAY)
TIME*DAY	Var(Error) + 2 Var(SUBJ*TIME*DAY) + 19 Var(RATER*TIME*DAY) + Q(TIME*DAY)
RATER*TIME*DAY	Var(Error) + 19 Var(RATER*TIME*DAY)

```
SUBJ                Var(Error) + 2 Var(SUBJ*TIME*DAY)
                    + 2 Var(SUBJ*RATER*DAY) + 3 Var(SUBJ*RATER*TIME)
                    + 4 Var(SUBJ*DAY) + 6 Var(SUBJ*TIME)
                    + 6 Var(SUBJ*RATER) + 12 Var(SUBJ)

SUBJ*RATER          Var(Error) + 2 Var(SUBJ*RATER*DAY)
                    + 3 Var(SUBJ*RATER*TIME) + 6 Var(SUBJ*RATER)

SUBJ*TIME           Var(Error) + 2 Var(SUBJ*TIME*DAY)
                    + 3 Var(SUBJ*RATER*TIME) + 6 Var(SUBJ*TIME)

SUBJ*DAY            Var(Error) + 2 Var(SUBJ*TIME*DAY)
                    + 2 Var(SUBJ*RATER*DAY) + 4 Var(SUBJ*DAY)

SUBJ*RATER*TIME Var(Error) + 3 Var(SUBJ*RATER*TIME)

SUBJ*RATER*DAY  Var(Error) + 2 Var(SUBJ*RATER*DAY)

SUBJ*TIME*DAY   Var(Error) + 2 Var(SUBJ*TIME*DAY)
```

```
        Tests of Hypotheses for Mixed Model Analysis of Variance
Source: RATER
Error: MS(RATER*TIME) + MS(RATER*DAY) - MS(RATER*TIME*DAY)
     + MS(SUBJ*RATER) - MS(SUBJ*RATER*TIME) - MS(SUBJ*RATER*DAY)
     + MS(Error)

                    Denominator   Denominator
DF    Type III MS       DF            MS       F Value   Pr > F
 1   242.21491228      2.18      51.687134503   4.6862   0.1520

Source: TIME
Error: MS(RATER*TIME) + MS(SUBJ*TIME) - MS(SUBJ*RATER*TIME)

                    Denominator   Denominator
DF    Type III MS       DF            MS       F Value   Pr > F
 1   31.688596491      4.38      63.096003899   0.5022   0.5145

Source: DAY
Error: MS(RATER*DAY) + MS(SUBJ*DAY) - MS(SUBJ*RATER*DAY)

                    Denominator   Denominator
DF    Type III MS       DF            MS       F Value   Pr > F
 2   151.78947368     19.49     105.71539961    1.4358   0.2620

Source: RATER*TIME
Error: MS(RATER*TIME*DAY) + MS(SUBJ*RATER*TIME) - MS(Error)

                    Denominator   Denominator
DF    Type III MS       DF            MS       F Value   Pr > F
 1   28.776315789      1.13      6.331871345    4.5447   0.2555

Source: RATER*DAY
Error: MS(RATER*TIME*DAY) + MS(SUBJ*RATER*DAY) - MS(Error)

                    Denominator   Denominator
DF    Type III MS       DF            MS       F Value   Pr > F
 2   26.333333333      4.44      12.916666667   2.0387   0.2355

Source: TIME*DAY
Error: MS(RATER*TIME*DAY) + MS(SUBJ*TIME*DAY) - MS(Error)

                    Denominator   Denominator
DF    Type III MS       DF            MS       F Value   Pr > F
 2   13.175438596     19.98      36.471734893   0.3613   0.7013

Source: RATER*TIME*DAY
Error: MS(Error)

                    Denominator   Denominator
DF    Type III MS       DF            MS       F Value   Pr > F
 2   8.2105263158      36       5.6734892788    1.4472   0.2486
```

References

Brogan, D. R., and Kutner, M. H. 1980. "Comparative Analysis of Pretest-Posttest Research Designs." *American Statistician* 34: 229–32.

Diggle, P. J.; Liang, K. Y.; and Zeger, S. L. 1994. *Analysis of Longitudinal Data.* Oxford: Oxford University Press.

Dixon, W. J. 1990. *BMDP, Statistical Software Manual.* Berkeley: University of California Press.

Heffner, T. G.; Drawbaugh, R. B.; and Zigmond, M. J. 1974. "Amphetamine and Operant Behavior in Rats: Relationship Between Drug Effect and Control Response Rate." *Journal of Comparative and Physiological Psychology* 86: 1031–43.

Holder, P.; Plikaytis, B. D.; and Carlone, G. 1996. "Need for Blocking to Reduce Non-specific Binding." Paper presented at WHO Workshop on Pneumococcal ELISA Standardization, Atlanta, May 15–16, 1996.

Karim, and Zeger, S. L. 1988. *A SAS Macro for Longitudinal Data Analysis,* Johns Hopkins University, Department of Biostatistics Technical Report #674.

Neter, J.; Wasserman, W.; and Kutner, M. H. 1985. *Applied Linear Statistical Models,* Homewood, Ill.: Charles C. Irvin.

Ortiz, D. J.; Marcus, M.; Gerr, F.; Jones, W.; and Cohen, S. 1997. "Estimation of Within Subject Variability in Upper Extremity Posture Measured With Manual Goniometry Among Video Display Terminal Users." *Applied Ergonomics* 28(2) [in press].

Rikkers, L. F.; Rudman, D.; Galambos, J. T.; Fulenwidr, J. T.; Millikan, W. J.; Kutner, M. H.; Smith, R. B.; Salam, A. A.; Sones, P. J.; and Warren, W. D. 1978. "A Randomized Controlled Trial of Distal Spenorenal Shunt." *Annals of Surgery* 188: 271–82.

SAS Institute, Inc. 1992. *SAS/STAT Software: Changes and Enhancements, Release 6.07.* SAS Technical Report P-229. Cary, N.C.

———. 1993. *SAS/STAT User's Guide,* 4th ed., Version 6, Cary, N.C.

Searle, S. R. 1971. *Linear Models.* New York: John Wiley & Sons.

———. 1987. *Linear Models for Unbalanced Data.* New York: John Wiley & Sons.

Searle, S. R.; Casella, G.; McCulloch, C. E. 1992. *Variance Components.* New York: John Wiley & Sons.

Ware, J. 1985. "Linear Models for the Analysis of Several Measurements in Longitudinal Studies," *American Statistician* 39: 95–101.

Wolfinger, R., and O'Connell, S. S. 1993. "Generalized Linear Models: A Pseudo-Likelihood Approach." *Journal of Statistical Computation and Simulation* 48:233–243.

Zeger, S. L., and Liang, K. Y. 1986. "Longitudinal Data Analysis for Discrete and Continuous Outcomes." *Biometrics* 42: 121–30.

The Method of Maximum Likelihood

22-1 Preview

This chapter describes the methodology of maximum likelihood (ML) estimation and its associated inference-making procedures. The term *maximum likelihood* refers to a very general algorithm for obtaining estimators of population parameters; such estimators have excellent (large-sample) statistical properties. One major advantage of the ML method of estimating parameters is its applicability to a wide variety of situations. In particular, when a multiple linear regression model is fitted to normally distributed data, the least squares estimators of the regression coefficients are identical to the ML estimators. Furthermore, ML estimation is the method of choice for estimating the parameters in nonlinear models such as the logistic regression model (discussed in Chapter 23) and the Poisson regression model (discussed in Chapter 24).

This chapter begins with a general discussion of the basic theoretical principles underlying the method of maximum likelihood. We then describe and illustrate inference-making procedures based on the use of ML estimators.

22-2 The Principle of Maximum Likelihood

In introducing the underlying principle of maximum likelihood estimation, let us focus on a relatively simple problem of statistical estimation. Suppose that a large population contains a certain unknown proportion θ ($0 \leq \theta \leq 1$) of individuals with a particular genetic disorder. In epidemiology, the parameter θ is referred to as the *prevalence* of the disorder of interest in the population under study. Further, suppose that a random sample of m individuals is selected from this population, and that Y represents the random variable denoting the number of individuals in the random sample of size m who have this genetic disorder. The possible values of Y are the $m + 1$ integer values $0, 1, 2,\ldots, m$. The statistical estimation problem concerns how to use m

and Y to obtain an estimate of θ with good statistical properties. By "good," we mean that the chosen estimator of θ (which is itself a random variable, since it is a function of the random variable Y) has little or no bias (i.e., its expected value is equal to θ) and has a small variance.

Given the above scenario, it is reasonable to assume that the underlying probability distribution of the discrete random variable Y is binomial; in particular, we have

$$\text{pr}(Y; \theta) = {}_mC_Y \theta^Y (1 - \theta)^{m-Y} \quad Y = 0, 1, 2, \ldots, m \tag{22.1}$$

where ${}_mC_Y = m!/Y!(m - Y)!$ denotes the number of combinations of m distinct objects chosen Y at a time. To simplify our discussion for now, let us assume (unrealistically, of course) that only two values of θ are possible; specifically, let us assume that θ is equal either to .2 or to .6. Then the true underlying probability distribution of Y is, from (22.1), either

$$\text{pr}(Y; .2) = {}_mC_Y (.2)^Y (.8)^{m-Y} \quad Y = 0, 1, 2, \ldots, m \tag{22.2}$$

or

$$\text{pr}(Y; .6) = {}_mC_Y (.6)^Y (.4)^{m-Y} \quad Y = 0, 1, 2, \ldots, m \tag{22.3}$$

Given this situation, the method of maximum likelihood will choose the point estimate of θ (namely, either .2 or .6, since these are the only two permissible values) that, loosely speaking, agrees more closely with the observed data (i,e., with the observed value y of Y or the observed sample proportion y/m).

To illustrate the above concept numerically, assume that we have a random sample of $m = 5$ individuals. Then the possible observed values y of Y and their associated probabilities based on (22.2) and (22.3) are given in the following table:

θ	0	1	2	3	4	5
.2	.328	.409	.205	.051	.007	.000
.6	.010	.077	.230	.346	.259	.078

Thus, if $\theta = .2$, the probability that $Y = 4$ is, from (22.2),

$$\text{pr}(Y = 4; .2) = {}_5C_4(.2)^4(.8)^{5-4} = 5(.2)^4(.8)^1 \approx .007$$

In contrast, if $\theta = .6$, the probability that $Y = 4$ is, from (22.3),

$$\text{pr}(Y = 4; .6) = {}_5C_4(.6)^4(.4)^{5-4} \approx .259$$

Based on these computations, if the observed value of Y is actually 4, the method of maximum likelihood will choose $\hat{\theta} = .6$ as the estimate of θ, since the probability of observing $Y = 4$ is higher when $\theta = .6$ than when $\theta = .2$ (i.e., the observed data agree better, or are more consistent, with the value .6 than with the value .2). More generally, then, it follows by inspection of the entries in the preceding table that we will estimate θ by .2 when $Y = 0$ or 1 and by .6 when $Y = 2, 3, 4,$ or 5; in compact notation, the ML estimator $\hat{\theta}$ of θ in this simple example is defined as

$$\hat{\theta} = \begin{cases} .2 & Y = 0, 1 \\ .6 & Y = 2, 3, 4, 5 \end{cases} \tag{22.4}$$

Clearly, from (22.4), the principle of ML estimation dictates selecting for each value of Y the value of θ, denoted as $\hat{\theta}$, that satisfies the inequality

$$\text{pr}(Y; \hat{\theta}) > \text{pr}(Y; \theta^*)$$

where θ^* is the alternative value of θ.

In general, in the realistic situation in which all possible values of θ satisfying $0 \le \theta \le 1$ are possible, it makes sense to do the following: For an observed value of Y, the ML estimator of θ is the value of θ, denoted $\hat{\theta}$, for which the expression

$$\text{pr}(Y; \theta) = {}_mC_Y\theta^Y(1-\theta)^{m-Y} \tag{22.5}$$

attains its *maximum value* as a function of θ. We can find the specific value $\hat{\theta}$ of θ that maximizes the function (22.5), using calculus, by setting the derivative of (22.5) with respect to θ equal to 0 and then solving the resulting equation for $\hat{\theta}$. In particular,

$$\begin{aligned}
\frac{d}{d\theta}[\text{pr}(Y; \theta)] &= {}_mC_Y[Y\theta^{Y-1}(1-\theta)^{m-Y} - (m-Y)\theta^Y(1-\theta)^{m-Y-1}]\\
&= {}_mC_Y\theta^{Y-1}(1-\theta)^{m-Y-1}[Y(1-\theta) - (m-Y)\theta]\\
&= {}_mC_Y\theta^{Y-1}(1-\theta)^{m-Y-1}(Y-m\theta)
\end{aligned} \tag{22.6}$$

Equating (22.6) to 0 yields the three solutions 0, 1, and Y/m. The first two solutions minimize (22.5). However, the solution Y/m maximizes (22.5), as can be verified by confirming that the second derivative of (22.5) with respect to θ is negative when θ is replaced by the value Y/m. Thus, the ML estimator of θ is $\hat{\theta} = Y/m$ when Y has the binomial distribution (22.1); here, $\hat{\theta}$ is the *sample proportion* of subjects with the genetic disorder in question. Figure 22-1 illustrates graphically that the ML estimator $\hat{\theta} = Y/m$ maximizes the function $\text{pr}(Y; \theta)$ given by (22.1).

The ML estimator $\hat{\theta} = Y/m$ has the property that

$$\text{pr}(Y; \hat{\theta}) > \text{pr}(Y; \theta^*) \tag{22.7}$$

where θ^* is any other value of θ satisfying $0 \le \theta^* \le 1$. In other words, since $\text{pr}(Y; \theta)$ gives the likelihood (or probability) of the observed Y, the estimator $\hat{\theta} = Y/m$, having been chosen to maximize this likelihood function, is naturally called the ML estimator of θ.

FIGURE 22-1   Illustration of the principle of maximum likelihood, using the function (22.1).

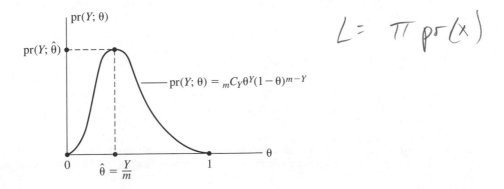

In the example that we have been considering, the data set consists of the observed value of Y, so $\text{pr}(Y; \theta)$ is the likelihood function for (or equivalently, the probability distribution of) the observed data. Thus, as mentioned earlier, the ML estimator $\hat{\theta}$ of θ is the estimator that agrees most closely with the observed data in the sense of (22.7). With this concept in mind, we are now in a position to make a more general definition of the principle of maximum likelihood.

Let $L(\mathbf{Y}; \boldsymbol{\theta})$ denote the *likelihood function* for a data set $\mathbf{Y} = (Y_1, Y_2, \ldots, Y_n)$ of n observations from some population characterized by the parameter set $\boldsymbol{\theta} = (\theta_1, \theta_2, \ldots, \theta_p)$.[1] The likelihood function $L(\mathbf{Y}; \boldsymbol{\theta})$ can informally be treated as the probability distribution of the multivariate variable $\mathbf{Y} = (Y_1, Y_2, \ldots, Y_n)$ involving the n random variables $(Y_1, Y_2, \ldots, Y_n)$. By a straightforward extension of the general principles of maximum likelihood discussed earlier, the ML estimator of $\boldsymbol{\theta}$ is the set $\hat{\boldsymbol{\theta}} = (\hat{\theta}_1, \hat{\theta}_2, \ldots, \hat{\theta}_p)$ of estimators for which

$$L(\mathbf{Y}; \hat{\boldsymbol{\theta}}) > L(\mathbf{Y}; \boldsymbol{\theta}^*) \tag{22.8}$$

where $\boldsymbol{\theta}^*$ denotes any other set of estimators of the elements of $\boldsymbol{\theta}$. The similarity between expressions (22.7) and (22.8) is clear.

In practice, finding the set $\hat{\boldsymbol{\theta}}$ of numerical functions of the observed data for which $L(\mathbf{Y}; \hat{\boldsymbol{\theta}})$ is a maximum requires solving a system of p equations in p unknowns. Specifically, since maximizing $L(\mathbf{Y}; \boldsymbol{\theta})$ is equivalent to the computationally easier problem of maximizing the natural logarithm of $L(\mathbf{Y}; \boldsymbol{\theta})$, the elements of $\hat{\boldsymbol{\theta}}$ are typically found as the solutions of the p equations obtained by setting the partial derivative of $\ln L(\mathbf{Y}; \boldsymbol{\theta})$ with respect to each $\theta_j (j = 1, 2, \ldots, p)$ equal to 0. In particular, this set of ML equations can be written in the form

$$\frac{\partial}{\partial \theta_j}[\ln L(\mathbf{Y}; \boldsymbol{\theta})] = 0 \quad j = 1, 2, \ldots, p \tag{22.9}$$

Except for some special cases, the system of ML equations (22.9) does not lead to closed-form expressions for the ML estimators; consequently, these equations usually must be solved by using sophisticated computer algorithms. This complexity results because the set of equations (22.9) typically involves nonlinear (as opposed to linear) functions of the elements of $\boldsymbol{\theta}$, thus requiring the use of so-called *iterative* computational procedures. Such calculations are not a major problem, however, since sophisticated computer algorithms are available that have been designed specifically to perform such numerical manipulations.

Chapters 23 (on logistic regression) and 24 (on Poisson regression) introduce some particular forms of $L(\mathbf{Y}; \boldsymbol{\theta})$ that have important practical applications. For now, it is sufficient to appreciate that ML estimation requires the specification of a likelihood function $L(\mathbf{Y}; \boldsymbol{\theta})$, which is then used to produce the estimator set $\hat{\boldsymbol{\theta}}$ of ML estimators of the elements of $\boldsymbol{\theta}$ via the system of equations (22.9).

22-3 Statistical Inference via Maximum Likelihood

Once the ML estimates have been obtained, the next step is to use the elements of $\hat{\boldsymbol{\theta}}$ to make statistical inferences about the corresponding elements of $\boldsymbol{\theta}$. The computations involved

---

[1]Note that, in our earlier example, our data set consisted of the single observation $\mathbf{Y} = Y_1 = Y$, the number of individuals with a certain genetic disorder in a sample of size m; the parameter set $\boldsymbol{\theta}$ thus consisted of the single parameter $\theta_1 = \theta$. Now we are generalizing the situation so that the data set can consist of several observations and we can consider more than one parameter of interest.

in making such inferences are based on two quantities that are generally included in the output provided by standard ML computer programs. The first of these two quantities is the *maximized likelihood value*, which is the numerical value of the likelihood function $L(\mathbf{Y}; \boldsymbol{\theta})$ when the numerical values of the ML estimates (i.e., the elements of $\hat{\boldsymbol{\theta}}$) are substituted for the corresponding elements of $\boldsymbol{\theta}$ in the expression for $L(\mathbf{Y}; \boldsymbol{\theta})$. In particular, this ML value is simply $L(\mathbf{Y}; \hat{\boldsymbol{\theta}})$, the quantity that appears in expression (22.8). The second quantity of interest is the *estimated covariance matrix* $\hat{\mathbf{V}}(\hat{\boldsymbol{\theta}})$ of the ML estimators. This $p \times p$ symmetric matrix has as its elements the p estimated variances of the ML estimators (appearing on the diagonal) and the $p(p - 1)/2$ estimated covariances between pairs of estimators (appearing off the diagonal).

To see how these two quantities are used in making inferences about the elements of $\boldsymbol{\theta}$, let us consider the following simple regression analysis setting. Suppose that we are interested in the following three models:

Model 0: $E(Y) = \beta_0$

Model 1: $E(Y) = \beta_0 + \beta_1 X$

Model 2: $E(Y) = \beta_0 + \beta_1 X + \beta_2 X^2$

Here, $E(Y)$ denotes the expectation (i.e., the true mean) of the random variable Y, which is assumed to be at most a second-degree polynomial function of a single regressor variable X. If we are given n pairs of data points (X_i, Y_i), where $i = 1, 2,..., n$, the objective of our analysis is to decide which of the preceding three models best fits the data, using statistical estimation and inference-making methodology based on the principle of maximum likelihood.

To fit any of the three models by the method of maximum likelihood, we must first specify a likelihood function. To illustrate, suppose that we make the typical regression analysis assumptions regarding the preceding models. Specifically, let us assume that $Y_i (i = 1, 2,..., n)$ is normally distributed with mean $\mu_i = E(Y_i)$ possibly depending on X_i and with variance Var $(Y_i) = \sigma^2$ not varying with i. Further, let us assume that X_i is measured without error (i.e., that X_i is nonstochastic) and that the n random variables $Y_1, Y_2,..., Y_n$ are mutually independent. In more compact notation, this set of assumptions implies, for model 1, say, that

$$Y_i \frown N(\beta_0 + \beta_1 X_i, \sigma^2), \quad i = 1, 2,..., n \tag{22.10}$$

and that the $\{Y_i\}_{i=1}^{n}$ are mutually independent. (As elsewhere in this book, the symbol $\frown$ means "is distributed as," and the notation "$N(\beta_0 + \beta_1 X_i, \sigma^2)$" refers to a normal distribution with mean $\beta_0 + \beta_1 X_i$ and variance σ^2.)

Employing the expression for the distribution (density function) of a normally distributed random variable, we find from (22.10) that the distribution of Y_i is

$$f\left(Y_i; \beta_0, \beta_1, \sigma^2\right) = \frac{1}{\sqrt{2\pi\sigma^2}} \exp\left\{-\frac{1}{2\sigma^2}[Y_i - (\beta_0 + \beta_1 X_i)]^2\right\} \tag{22.11}$$

where $-\infty < Y_i < +\infty$. Note that (22.11) is a function of three parameters—namely, β_0, β_1, and σ^2. Under the assumption that the $\{Y_i\}_{i=1}^{n}$ are mutually independent,[2] it can be shown (from

---

[2]The assumption of mutual independence for a set of random variables is the strongest assumption that can be made about the joint behavior of this set. Such an assumption allows precise description of the joint distribution of the variables (i.e., the likelihood function) *solely* on the basis of knowledge of the *separate* behavior (i.e., the so-called marginal distribution) of each variable in the set. In particular, the joint distribution under mutual independence is simply the product of the marginal distributions.

certain principles of statistical theory) that the joint distribution of $Y_1, Y_2,\ldots, Y_n$ (i.e., the likelihood function) is, from (22.11),

$$L\left(\mathbf{Y}; \beta_0, \beta_1, \sigma^2\right) = \prod_{i=1}^{n} f\left(Y_i; \beta_0, \beta_1, \sigma^2\right)$$

$$= \frac{1}{(2\pi\sigma^2)^{n/2}} \exp\left\{ -\frac{1}{2\sigma^2}\sum_{i=1}^{n} [Y_i - (\beta_0 + \beta_1 X_i)]^2 \right\} \qquad (22.12)$$

where $-\infty < Y_i < +\infty$, $i = 1, 2,\ldots, n$.

By solving simultaneously the three ML equations

$$\frac{\partial}{\partial \beta_0}\left[\ln L\left(\mathbf{Y}; \beta_0, \beta_1, \sigma^2\right)\right] = 0 \qquad \frac{\partial}{\partial \beta_1}\left[\ln L\left(\mathbf{Y}; \beta_0, \beta_1, \sigma^2\right)\right] = 0$$

and

$$\frac{\partial}{\partial (\sigma^2)}\left[\ln L\left(\mathbf{Y}; \beta_0, \beta_1, \sigma^2\right)\right] = 0$$

we can show that the ML estimators of β_0, β_1, and σ^2 under model 1 are, respectively,

$$\hat{\beta}_0 = \overline{Y} - \hat{\beta}_1 \overline{X}_1 \qquad \hat{\beta}_1 = \frac{\displaystyle\sum_{i=1}^{n} (X_i - \overline{X})(Y_i - \overline{Y})}{\displaystyle\sum_{i=1}^{n} (X_i - \overline{X})^2} \qquad (22.13)$$

and

$$\hat{\sigma}_1^2 = \frac{1}{n}\sum_{i=1}^{n} [Y_i - (\hat{\beta}_0 + \hat{\beta}_1 X_i)]^2 = \frac{\text{SSE}_1}{n} \qquad (22.14)$$

for normal distribution

$\hat{\mu} = \overline{X}$

where SSE_1 is the sum of squares of residuals about the fitted straight line.

The ML estimators $\hat{\beta}_0$ and $\hat{\beta}_1$ given by (22.13) are the same estimators of the intercept and slope of a straight-line model that would be obtained by the method of unweighted least squares.

This equivalence between maximum likelihood and unweighted least-squares estimation carries over to the more general situation of multiple *linear* regression analysis. In particular, assume (for $i = 1, 2,\ldots, n$) that Y_i is normal with mean $E(Y_i) = \beta_0 + \sum_{j=1}^{k} \beta_j X_{ij}$; further, assume that $\text{Var}(Y_i) = \sigma^2$ and that the $\{Y_i\}_{i=1}^{n}$ are mutually independent. Then it can be shown that the set $\hat{\boldsymbol{\beta}} = (\hat{\beta}_0, \hat{\beta}_1,\ldots, \hat{\beta}_k)$ of ML estimators of the elements of $\boldsymbol{\beta} = (\beta_0, \beta_1,\ldots,\beta_k)$ is identical to the set of unweighted least-squares estimators of $\boldsymbol{\beta}$, which are chosen to minimize the quantity $\sum_{i=1}^{n}(Y_i - \beta_0 - \sum_{j=1}^{k}\beta_j X_{ij})^2$ with respect to $\beta_0, \beta_1,\ldots, \beta_k$. More generally, under the same assumptions as previously, except that $\text{Var } Y_i = \sigma^2$, it can be shown that ML estimation and weighted least-squares estimation using the function $\sum_{i=1}^{n}(Y_i - \beta_0 - \sum_{j=1}^{k}\beta_j X_{ij})^2/\sigma_i^2$ both lead to the same set of weighted least-squares estimators.

The ML estimator $\hat{\sigma}_1^2$ of σ^2, given by (22.14), is a biased estimator of σ^2; the unbiased estimator is

$$\left(\frac{n}{n-2}\right)\hat{\sigma}_1^2 = \frac{\text{SSE}_1}{n-2}$$

under model 1. Thus, the method of maximum likelihood does not always produce unbiased estimators of the parameters of interest,[3] although the extent of such bias generally decreases as the sample size increases.

Once the ML estimators (22.13) and (22.14) of the parameters in the likelihood function (22.12) have been obtained, the next step is to calculate the value of the maximized likelihood function $L(\mathbf{Y}; \hat{\beta}_0, \hat{\beta}_1, \hat{\sigma}_1^2)$ and to determine the 3×3 estimated covariance matrix $\hat{\mathbf{V}}(\hat{\beta}_0, \hat{\beta}_1, \hat{\sigma}_1^2)$ of the three estimators $\hat{\beta}_0$, $\hat{\beta}_1$, and $\hat{\sigma}_1^2$. The maximized likelihood $L(\mathbf{Y}; \hat{\beta}_0, \hat{\beta}_1, \hat{\sigma}_1^2)$ and the estimated covariance matrix $\hat{\mathbf{V}}(\hat{\beta}_0, \hat{\beta}_1, \hat{\sigma}_1^2)$ are needed to make statistical inferences.

In general, the numerical value of the maximized likelihood function and the numerical values of the entries in the estimated covariance matrix must be calculated by an ML computer program, since the form of likelihood function typically under consideration does not lead (as we will later illustrate) to explicit algebraic expressions either for the ML estimators or for their variances and covariances. For our particular example, however, the likelihood function (22.12) is sufficiently well-behaved to allow us to obtain the explicit algebraic expressions (22.13) and (22.14) for the ML estimators, from which exact algebraic expressions for $L(\mathbf{Y}; \hat{\beta}_0, \hat{\beta}_1, \hat{\sigma}_1^2)$ and $\hat{\mathbf{V}}(\hat{\beta}_0, \hat{\beta}_1, \hat{\sigma}_1^2)$ can be obtained.

A specific algebraic expression for the maximized likelihood $L(\mathbf{Y}; \hat{\beta}_0, \hat{\beta}_1, \hat{\sigma}_1^2)$ can be specified by substituting (22.13) and (22.14) into (22.12) and then simplifying; the resulting maximized likelihood function can be written in the form

$$L(\mathbf{Y}; \hat{\beta}_0, \hat{\beta}_1, \hat{\sigma}_1^2) = \left(2\pi\hat{\sigma}_1^2 e\right)^{-n/2} \tag{22.15}$$

For a numerical example, let us refer to the data in Table 13-2 of Chapter 13; for convenience, these data are reproduced next:

Dosage level (X)	1	2	3	4	5	6	7	8
Weight gain (Y)	1.0	1.2	1.8	2.5	3.6	4.7	6.6	9.1

For this particular set of $n = 8$ data points (which admittedly is too small to justify the use of asymptotic theory, but is appropriate for illustrative purposes), formulas (22.13) and (22.14) give $\hat{\beta}_0 = -1.20$, $\hat{\beta}_1 = 1.11$, $\text{SSE}_1 = 5.03$, and $\hat{\sigma}_1^2 = \text{SSE}_1/n = 5.03/8 = 0.6288$. (Note that, from Table 13-2, $\text{MSE} = [n/(n-2)]\hat{\sigma}_1^2 = \text{SSE}_1/6 = 5.03/6 = 0.84$.) Hence, from (22.15),

$$L(\mathbf{Y}; \hat{\beta}_0, \hat{\beta}_1, \hat{\sigma}_1^2) = L(\mathbf{Y}; -1.20, 1.11, 0.6288)$$
$$= [2(3.1416)(0.6288)(2.7183)]^{-8/2} = (10.7397)^{-4}$$

This is the number that would appear in the computer output from an ML estimation program using the likelihood function (22.12) and the data set under consideration.

---

[3]Formal mathematical arguments aside, the ML method produces estimators whose properties are optimal for *large* samples (under certain conditions of mathematical regularity) when the assumed likelihood function is correct. ML estimators are said to be *asymptotically optimal* in the sense that desirable properties such as unbiasedness, minimum variance, and normality hold exactly in the limit only as the amount of data becomes infinitely large. In practice, this upholds the reasonableness of assuming *for large data sets* that an ML estimator is essentially unbiased, has a small variance, and is approximately normally distributed when the appropriate likelihood function is being used.

The estimated covariance matrix for the ML estimators $\hat{\beta}_0$, $\hat{\beta}_1$, and $\hat{\sigma}_1^2$ has the following general form:

$$\hat{\mathbf{V}}(\hat{\beta}_0, \hat{\beta}_1, \hat{\sigma}_1^2) = \begin{bmatrix} \widehat{\mathrm{Var}}\hat{\beta}_0 & \widehat{\mathrm{Cov}}(\hat{\beta}_0, \hat{\beta}_1) & \widehat{\mathrm{Cov}}(\hat{\beta}_0, \hat{\sigma}_1^2) \\ \widehat{\mathrm{Cov}}(\hat{\beta}_0, \hat{\beta}_1) & \widehat{\mathrm{Var}}\hat{\beta}_1 & \widehat{\mathrm{Cov}}(\hat{\beta}_1, \hat{\sigma}_1^2) \\ \widehat{\mathrm{Cov}}(\hat{\beta}_0, \hat{\sigma}_1^2) & \widehat{\mathrm{Cov}}(\hat{\beta}_1, \hat{\sigma}_1^2) & \widehat{\mathrm{Var}}\hat{\sigma}_1^2 \end{bmatrix}$$

The estimated variances of the ML estimators appear on the diagonal of this symmetric matrix, and the estimated covariances are given by the off-diagonal elements of the matrix.

Without going into the mathematical development, it can be shown that the elements of $\hat{\mathbf{V}}(\hat{\beta}_0, \hat{\beta}_1, \hat{\sigma}_1^2)$ satisfy the following relationships:

$$\widehat{\mathrm{Var}}\,\hat{\beta}_0 = \hat{\sigma}_1^2 \left[\frac{1}{n} + \frac{\overline{X}^2}{\sum_{i=1}^{n}(X_i - \overline{X})^2} \right]$$

$$\widehat{\mathrm{Var}}\,\hat{\beta}_1 = \frac{\hat{\sigma}_1^2}{\sum_{i=1}^{n}(X_i - \overline{X})^2}$$

$$\widehat{\mathrm{Var}}\,\hat{\sigma}_1^2 = \frac{2\hat{\sigma}_1^4}{n}$$

$$\widehat{\mathrm{Cov}}(\hat{\beta}_0, \hat{\beta}_1) = \frac{-\overline{X}\hat{\sigma}_1^2}{\sum_{i=1}^{n}(X_i - \overline{X})^2}$$

$$\widehat{\mathrm{Cov}}(\hat{\beta}_0, \hat{\sigma}_1^2) = \widehat{\mathrm{Cov}}(\hat{\beta}_1, \hat{\sigma}_1^2) = 0$$

For the data under consideration, $\overline{X} = 4.50$ and $\sum_{i=1}^{n}(X_i - \overline{X})^2 = 42$. Thus, we can verify that

$$\hat{\mathbf{V}}(\hat{\beta}_0, \hat{\beta}_1, \hat{\sigma}_1^2) = \begin{bmatrix} 0.3818 & -0.0674 & 0 \\ -0.0674 & 0.0150 & 0 \\ 0 & 0 & 0.0988 \end{bmatrix} \tag{22.16}$$

This is the matrix that a computer program would print out based on the use of the likelihood function (22.12) for the data set under consideration.

Later, we will use the maximized likelihood value of $(10.7397)^{-4}$ to carry out *likelihood ratio tests,* which involve comparing the ratios of maximized likelihoods for different models. For now, let us focus on the estimated covariance matrix (22.16) and use it to perform certain exercises in statistical inference making.

22-3-1 Hypothesis Testing and Interval Estimation

Hypothesis Testing

Based on the large-sample properties of ML estimators, it can be shown that (under model 1) the quantity

$$\frac{\hat{\beta}_1 - \beta_1}{\sqrt{\widehat{\text{Var}}\ \hat{\beta}_1}}$$

will behave approximately as a standard normal random variable (i.e., a Z variate) when the sample size is large. Hence, a test of H_0: $\beta_1 = 0$ versus H_1: $\beta_1 \neq 0$ can be based on the Z statistic $\hat{\beta}_1/\sqrt{\widehat{\text{Var}}\ \hat{\beta}_1}$, which (for large n) has approximately a standard normal distribution under H_0: $\beta_1 = 0$. This test statistic is called a *Wald statistic*.

For the data under analysis, the (Wald) test statistic $\hat{\beta}_1/\sqrt{\widehat{\text{Var}}\hat{\beta}_1}$ has the numerical value $1.11/\sqrt{0.0150} = 9.063$, which is highly significant. The value 0.0150—the second diagonal element of the matrix (22.16)—represents the ML-based estimate of Var $\hat{\beta}_1$ under the likelihood function (22.12). An equivalent test can be based on the chi-square distribution; in particular, since $Z^2 \sim \chi_1^2$ when $Z \sim N(0, 1)$, it follows that $\hat{\beta}_1^2/\widehat{\text{Var}}\hat{\beta}_1$ will have approximately a χ_1^2 distribution under H_0: $\beta_1 = 0$ when the sample size is sufficiently large. In our numerical example, then, this χ_1^2 statistic has the highly significant value[4] of $(9.063)^2 = 82.14$.

ML-based test statistics are usually assumed to have *large-sample* chi-square distributions (with appropriately specified degrees of freedom) under the null hypotheses of interest. The large-sample requirement is crucial to the validity of such *asymptotic* procedures. In large-sample situations, the P-values associated with the use of ML-based chi-square statistics are comparable to those based on other methods of parameter estimation and statistical inference. In small-sample situations (such as the present example, with $n = 8$), discrepancies among different analysis procedures can be large.[5]

Interval Estimation

To obtain confidence intervals for parameters by the method of maximum likelihood, we can again appeal to the approximate normality of ML estimators when the sample size is large. In our example, since

$$\frac{\hat{\beta}_1 - \beta_1}{\sqrt{\widehat{\text{Var}}\hat{\beta}_1}}$$

---

[4]A reader who inspects Table 13-3 may wonder about the difference between the ML-based $\chi_1^2(= Z^2)$ statistic just discussed and the $F_{1,6}(= t_6^2)$ statistic in Table 13-3 (with the value 61.95), both of which have been used to test H_0: $\beta_1 = 0$ versus H_1: $\beta_1 \neq 0$ under model 1. For this particular example, the difference lies in the fact that the estimated variance of $\hat{\beta}_1$ for the $t_6^2 = F_{1,6}$ statistic is $n/(n-2)$ times that for the $Z^2 = \chi_1^2$ statistic, as noted earlier. Thus, $[n/(n-2)](61.95) = [8/(8-2)](61.95) = 82.60$, which agrees closely with the χ_1^2-value of 82.14.

[5]So far, we have been dealing with so-called *unconditional* ML methods. *Conditional* ML procedures offer a viable alternative for dealing with small-sample problems. We discuss conditional ML procedures at the end of Chapter 23.

is approximately $N(0, 1)$ for large samples under model 1, an approximate $100(1 - a)\%$ large-sample ML-based confidence interval for β_1 has the general form

$$\hat{\beta}_1 \pm Z_{1-(\alpha/2)}\sqrt{\widehat{\text{Var}}\hat{\beta}_1}$$

where $\text{pr}[Z > Z_{1-(\alpha/2)}] = \alpha/2$ when $Z \sim N(0, 1)$. For instance, a 95% ML-based confidence interval for β_1 using our data set of size $n = 8$ is

$$1.11 \pm 1.96\sqrt{0.0150} = 1.11 \pm 0.24$$

which yields the interval $(0.87, 1.35)$.

As another example, let us obtain a $100(1 - \alpha)\%$ large-sample confidence interval for the true mean of Y when X is set at some value X_0 within the range of X-values for the data under consideration (i.e., interpolation, but not extrapolation, is permitted). If we denote this parameter as $E(Y \mid X = X_0)$, the ML estimator of $E(Y \mid X = X_0)$ based on fitting model 1 is

$$\hat{E}(Y \mid X = X_0) = \hat{Y}_0 = \hat{\beta}_0 + \hat{\beta}_1 X_0$$

Since $\hat{Y}_0$ is a linear combination of random variables (namely, $\hat{\beta}_0$ and $\hat{\beta}_1$), it follows that $\hat{Y}_0 = \hat{\beta}_0 + \hat{\beta}_1 X_0$ has estimated variance[6]

$$\widehat{\text{Var}}\hat{Y}_0 = \widehat{\text{Var}}\hat{\beta}_0 + X_0^2\widehat{\text{Var}}\hat{\beta}_1 + 2X_0\,\widehat{\text{Cov}}(\hat{\beta}_0, \hat{\beta}_1)$$

Under the assumption that the quantity

$$\frac{\hat{Y}_0 - E(Y \mid X = X_0)}{\sqrt{\widehat{\text{Var}}\hat{Y}_0}}$$

is approximately $N(0, 1)$ in large samples, it follows that a $100(1 - \alpha)\%$ confidence interval for $E(Y \mid X = X_0)$ is

$$\hat{Y}_0 \pm Z_{1-(\alpha/2)}\sqrt{\widehat{\text{Var}}\hat{Y}_0}$$

For our particular numerical example, suppose that $X_0 = \overline{X} = 4.50$. Then, from (22.16), it follows that

$$\begin{aligned} \text{Var}\,\hat{Y}_0 &= 0.3818 + (4.50)^2(0.0150) + 2(4.50)(-0.0674) \\ &= 0.3818 + 0.3038 - 0.6066 = 0.0790 \end{aligned}$$

Since $\hat{Y}_0 = -1.20 + 1.11(4.50) = -1.20 + 5.00 = 3.80$, it follows that the 95% ML-based large-sample confidence interval for $E(Y \mid X = 4.50)$ is

$$3.80 \pm 1.96\sqrt{0.0790} = 3.80 \pm 0.55$$

which yields the interval $(3.25, 4.35)$.

---

[6]In general, if $\hat{L} = \sum_{j=1}^{k} a_j X_j$ is a linear function of the k random variables $X_1, X_2, \ldots, X_k$, and if $a_1, a_2, \ldots, a_k$ are known constants, then

$$\text{Var}\,\hat{L} = \sum_{j=1}^{k} a_j^2 \,\text{Var}\,X_j + 2\sum_{\text{all } j<j'} \sum a_j a_{j'} \,\text{Cov}(X_j, X_{j'})$$

For example, when $k = 2$, we have

$$\text{Var}\,\hat{L} = a_1^2 \,\text{Var}\,X_1 + a_2^2 \,\text{Var}\,X_2 + 2a_1 a_2 \,\text{Cov}(X_1, X_2)$$

22-3-2 Hypothesis Testing Using Likelihood Ratio Tests

As its name suggests, a *likelihood ratio test* involves a *ratio* comparison of (maximized) likelihood values. To see how such a comparison can be used in making statistical inferences, let us again focus on models 0, 1, and 2 introduced at the beginning of section 22-3. For models 0 and 2, we can obtain the value of the maximized likelihood function in the same way we did for model 1. For model 1, maximization of the likelihood function (22.12) led to the estimators (22.13) and (22.14) and then to the maximized likelihood expression (22.15); for the data under consideration, (22.15) had a specific numerical value of $(10.7397)^{-4}$.

Dispensing with the mathematical details,[7] it can be shown that $L(\mathbf{Y}; \hat{\beta}_0, \hat{\sigma}_0^2)$ and $L(\mathbf{Y}; \hat{\beta}_0, \hat{\beta}_1, \hat{\beta}_2, \hat{\sigma}_2^2)$, the maximized likelihood values for models 0 and 2 under the same general assumptions used for the ML-fitting of model 1, are equal to $(121.8426)^{-4}$ and $(0.4270)^{-4}$, respectively.

Note that

$$L(\mathbf{Y}; \hat{\beta}_0, \hat{\sigma}_0^2) < L(\mathbf{Y}; \hat{\beta}_0, \hat{\beta}_1, \hat{\sigma}_1^2) < L(\mathbf{Y}; \hat{\beta}_0, \hat{\beta}_1, \hat{\beta}_2, \hat{\sigma}_2^2) \qquad (22.17)$$

for the particular data set under consideration. This result reflects the principle of multiple regression analysis that the squared multiple correlation coefficient R^2 will always increase somewhat (for any set of data) whenever another regression parameter is included in the model under consideration. Analogously, since model 0 is a special case of model 1 when $\beta_1 = 0$, and since model 1 in turn is a special case of model 2 when $\beta_2 = 0$, (22.17) must hold for any set of data.

The fact that models 0, 1, and 2 constitute a *hierarchical class* of models leads to an important application of likelihood ratio tests.[8] This is because the magnitude of the ratio of two maximized likelihood values reflects how much the maximized likelihood value for one specific model has changed based on the addition (or deletion) of one or more parameters to (or from)

---

[7]Under model 0, for example,

$$L(\mathbf{Y}; \beta_0, \sigma^2) = \frac{1}{(2\pi\sigma^2)^{n/2}} \exp\left[-\frac{1}{2\sigma^2} \sum_{i=1}^{n} (Y_i - \beta_0)^2 \right]$$

The ML estimators of β_0 and σ^2 using $L(\mathbf{Y}; \beta_0, \sigma^2)$ are $\hat{\beta}_0 = \overline{Y}$ and $\hat{\sigma}_0^2 = (1/n)\sum_{i=1}^{n}(Y_i - \overline{Y})^2$; so

$$L(\mathbf{Y}; \hat{\beta}_0, \hat{\sigma}_0^2) = (2\pi\hat{\sigma}_0^2 e)^{-n/2}$$

Similarly, if $\hat{\beta}_0$, $\hat{\beta}_1$, and $\hat{\beta}_2$ are the ML estimators of β_0, β_1, and β_2 under model 2, then

$$L(\mathbf{Y}; \hat{\beta}_0, \hat{\beta}_1, \hat{\beta}_2, \hat{\sigma}_2^2) = (2\pi\hat{\sigma}_2^2 e)^{-n/2}$$

where

$$\hat{\sigma}_2^2 = \frac{1}{n} \sum_{i=1}^{n} [Y_i - (\hat{\beta}_0 + \hat{\beta}_1 X_i + \hat{\beta}_2 X_i^2)]^2$$

[8]For our purposes, a *hierarchical class* of models is a group of models each of which, except for the most complex (i.e., that one containing the largest number of regression coefficients), can be obtained as a special case of another more complex model in the class by setting one or more regression coefficients equal to 0 in the more complex model. For example, the three models $E_1(Y) = \beta_0 + \beta_1 X_1$, $E_2(Y) = \beta_0 + \beta_1 X_1 + \beta_2 X_2$, and $E_3(Y) = \beta_0 + \beta_1 X_1 + \beta_2 X_2 + \beta_3 X_3 + \beta_{12} X_1 X_2$ constitute a hierarchical class, since $E_2(Y)$ becomes $E_1(Y)$ when $\beta_2 = 0$, and $E_3(Y)$ reduces to $E_2(Y)$ when $\beta_3 = \beta_{12} = 0$.

the given model. A decision to reject some null hypothesis based on a likelihood ratio test depends on whether some appropriate function of the ratio of maximized likelihoods for the two models being compared (i.e., the test statistic) is large enough to indicate a statistically significant disparity between the two maximized likelihood values. This philosophy of assessing the significance of a change in maximized likelihood values (via a test statistic that is a function of the likelihood ratio) is completely analogous to the philosophy in standard multiple regression analysis of assessing the significance of a change in R^2 based on the addition of one or more parameters to a given model.

In earlier chapters, we discussed how to use a partial F test (or, equivalently, a t test) to assess whether an increase in R^2 is statistically significant when a parameter is added to a given regression model, and how to use a multiple-partial F test to make the same assessment when more than one new parameter is introduced into that model.

A likelihood ratio test is performed analogously. Instead of having an F distribution, however, the likelihood ratio statistic to be used will have, in large samples, approximately a chi-square distribution under the null hypothesis. The degrees of freedom for this distribution will equal the number of parameters in the more complex model that must be set equal to 0 to obtain the less complex model as a special case.[9]

The specific test statistic to be used in performing a likelihood ratio test is of the general form $-2\ln(\hat{L}_1/\hat{L}_2)$, where $\hat{L}_1$ is the maximized likelihood value for the less complex model and $\hat{L}_2$ is the maximized likelihood value for the more complex model. Thus, this statistic is a function of the ratio $\hat{L}_1/\hat{L}_2$ of maximized likelihoods. Since $\hat{L}_2$ corresponds to the more complex model, we have $\hat{L}_2 > \hat{L}_1 > 0$; so $0 < \hat{L}_1/\hat{L}_2 < 1$, $-\infty < \ln(\hat{L}_1/\hat{L}_2) < 0$, and hence $0 < -2\ln(\hat{L}_1/\hat{L}_2) < +\infty$. Clearly, the larger $\hat{L}_2$ is relative to $\hat{L}_1$; and hence the better the more complex model agrees with the data in comparison to the less complex model, the larger will be the value of the test statistic $-2\ln(\hat{L}_1/\hat{L}_2)$, and so the likelier this value will be large enough to fall in the rejection region (i.e., the specified area in the *upper* tail of the appropriate chi-square distribution).

The likelihood ratio statistic $-2\ln(\hat{L}_1/\hat{L}_2)$ can be written equivalently, using algebra about logarithms,[10] as follows:

$$-2\ln(\hat{L}_1/\hat{L}_2) = -2\ln\hat{L}_1 - (-2\ln\hat{L}_2) \tag{22.18}$$

The expressions $-2\ln\hat{L}_1$ and $-2\ln\hat{L}_2$ are called *log-likelihood statistics*. When a computer program is used to carry out the ML estimation procedure, these two log-likelihood statistics are provided in separate outputs for the two models being compared. To carry out the likelihood ratio test, the investigator simply finds $-2\ln\hat{L}_1$ and $-2\ln\hat{L}_2$ from the output and then subtracts one log-likelihood from the other to obtain the value of the test statistic.

Let us study some numerical examples of likelihood ratio tests. Again, we will consider models 0, 1, and 2 and the data set used earlier. First we will compare models 0 and 1 via a likelihood ratio test. Since model 0 is a special case of model 1 when $\beta_1 = 0$, this likelihood ratio test addresses $H_0: \beta_1 = 0$ versus $H_A: \beta_1 \neq 0$. The appropriate chi-square distribution for the likelihood ratio statistic under this null hypothesis has 1 degree of freedom (since one parame-

---

[9]The reader should keep in mind that a given likelihood ratio test will always entail a comparison between two models that are members of a hierarchical class, so that the terms *more complex* and *less complex* are well defined.

[10]$\ln(a/b) = \ln a - \ln b$ for any two positive values a and b.

ter is restricted to being equal to 0 under H_0). Using the previous notation (see expression (22.15) and footnote 7), we are claiming, for large n, that

$$-2\ln\left[\frac{L\left(\mathbf{Y}; \hat{\beta}_0, \hat{\sigma}_0^2\right)}{L\left(\mathbf{Y}; \hat{\beta}_0, \hat{\beta}_1, \hat{\sigma}_1^2\right)}\right] = n\ln\left(\frac{\hat{\sigma}_0^2}{\hat{\sigma}_1^2}\right) \qquad \chi_1^2 \qquad (22.19)$$

has approximately a χ_1^2 distribution under H_0: $\beta_1 = 0$.

Based on our previous numerical work, we can compute the numerical value of (22.19) for our particular data set as

$$-2\ln\left[\frac{(121.8426)^{-4}}{(10.7397)^{-4}}\right] = 8\ln\left(\frac{121.8426}{10.7397}\right) = 8\ln 11.3451 = 19.43$$

which corresponds to a P-value of less than .0005 (based on upper-tail χ_1^2-values) and thus provides strong evidence in favor of H_A: $\beta_1 \neq 0$.[11] The discrepancy between the likelihood ratio statistic of 19.43 and the Wald statistic of 82.14 obtained earlier is alarming. However, this discrepancy is entirely plausible because of the small sample size ($n = 8$). Because the two test statistics are only asymptotically equivalent, their numerical values are reasonably close only when n is large. Thus, although we have chosen this data set for pedagogical purposes, it is actually an inappropriate one for the application of large-sample statistical procedures!

As another illustration, a test of H_0: $\beta_2 = 0$ versus H_A: $\beta_2 \neq 0$ involves a comparison of the maximized likelihoods for models 1 and 2. In particular, we will assume that

$$-2\ln\left[\frac{L\left(\mathbf{Y}; \hat{\beta}_0, \hat{\beta}_1, \hat{\sigma}_1^2\right)}{L\left(\mathbf{Y}; \hat{\beta}_0, \hat{\beta}_1, \hat{\beta}_2, \hat{\sigma}_2^2\right)}\right] = n\ln\left(\frac{\hat{\sigma}_1^2}{\hat{\sigma}_2^2}\right)$$

has approximately a χ_1^2 distribution for large samples under H_0: $\beta_2 = 0$. For our data, the numerical value of this test statistic is

$$-2\ln\left[\frac{(10.7397)^{-4}}{(0.4270)^{-4}}\right] = 25.80$$

which strongly favors H_A: $\beta_2 \neq 0$.[12]

As a third illustration, a likelihood ratio test involving the maximized likelihood values for models 0 and 2 provides a test of H_0: $\beta_1 = \beta_2 = 0$ versus H_A: "At least one of the parameters β_1 and β_2 differs from 0." The appropriate likelihood ratio test statistic is

$$-2\ln\left[\frac{L\left(\mathbf{Y}; \hat{\beta}_0, \hat{\sigma}_0^2\right)}{L\left(\mathbf{Y}; \hat{\beta}_0, \hat{\beta}_1, \hat{\beta}_2, \hat{\sigma}_2^2\right)}\right] = n\ln\left(\frac{\hat{\sigma}_0^2}{\hat{\sigma}_2^2}\right)$$

---

[11]A computer program would output the log-likelihood statistics $-2\ln\hat{L}_0 = -2\ln(121.8426)^{-4} = 38.42$ and $-2\ln\hat{L}_1 = -2\ln(10.7397)^{-4} = 18.99$, so that the likelihood ratio statistic $-2\ln(\hat{L}_0/\hat{L}_1)$ can be computed by subtraction as $-2\ln\hat{L}_0 - (-2\ln\hat{L}_1) = 38.42 - 18.99 = 19.43$.

[12]A computer program would output the log-likelihood statistics $-2\ln\hat{L}_1 = -2\ln(10.7397)^{-4} = 18.99$ and $-2\ln\hat{L}_2 = -2\ln(0.4270)^{-4} = -6.81$; the likelihood ratio statistic $-2\ln(\hat{L}_1/\hat{L}_2)$ can be computed by subtraction as $-2\ln\hat{L}_1 - (-2\ln\hat{L}_2) = 18.99 - (-6.81) = 25.80$.

which, for large samples and under H_0: $\beta_1 = \beta_2 = 0$, has approximately a chi-square distribution with *2 degrees of freedom*. (The null hypothesis restricts *two* independent model parameters to being 0, thus leading to the 2 degrees of freedom for the chi-square statistic.) The computed likelihood ratio test statistic is

$$-2 \ln \left[\frac{(121.8426)^{-4}}{(0.4270)^{-4}} \right] = -2 \ln (121.8426)^{-4} - (-2 \ln (0.4270)^{-4}) = 45.23$$

which is highly significant. This test is completely analogous to an overall F test in standard regression analysis.

Even though a sample size of $n = 8$ is too small to justify using statistical inference-making procedures whose desirable statistical properties are asymptotic (i.e., they hold only for very large samples), the decisions made about the importance of the linear (β_1) and quadratic (β_2) effects in the data agree with the conclusions drawn based on the standard regression analysis given in section 13-6.

One way to assess the goodness of fit of this second-degree model in X is to employ a likelihood ratio statistic to examine whether adding a cubic term in X (i.e., the term $\beta_3 X^3$) to the second-degree model significantly improves prediction of Y (i.e., a test of H_0: $\beta_3 = 0$ versus H_A: $\beta_3 \neq 0$). This particular statistic has the form $n \ln (\hat{\sigma}_2^2 / \hat{\sigma}_3^2)$, where $\hat{\sigma}_3^2 = (1/n) \sum_{i=1}^{n} [Y_i - (\hat{\beta}_0 + \hat{\beta}_1 X_i + \hat{\beta}_2 X_i^2 + \hat{\beta}_3 X_i^3)]^2$. For our data, it can be seen from Table 13-5 in Chapter 13 that $\hat{\sigma}_3^2 = 0.056/8 = 0.007$; hence, $n \ln (\hat{\sigma}_2^2 / \hat{\sigma}_3^2) = 8 \ln (0.025/0.007) = 10.18$, which corresponds to a P-value of between .0005 and .005, based on the chi-square distribution with 1 degree of freedom. Although this result argues for adding the term $\beta_3 X^3$ to the quadratic model, other information suggests the contrary. Specifically, the change in R^2 in going from a quadratic to a cubic model is negligible ($\triangle R^2 = .999 - .997 = .002$); a plot of the data clearly suggests no more than a second-degree function in X; and the likelihood ratio test under discussion is based on a data set with only $n = 8$ observations and so is not completely reliable. In light of these considerations, the correct conclusion is that a second-degree model in X suffices to describe the $X - Y$ relationship with high precision.

22-4 Summary

In this chapter, we have discussed the maximum likelihood (ML) method of estimating parameters and associated inference-making procedures. In ML estimation, the likelihood function to be maximized must be specified. For linear regression models involving independent normally distributed response variables, the ML estimates of regression coefficients are identical to the least-squares estimates. Nevertheless, ML estimation is particularly useful for estimating regression coefficients in nonlinear models. The estimation procedure for such models typically requires a computer program that employs an iterative algorithm.

Procedures for testing hypotheses and constructing confidence intervals use maximized likelihood values and estimated covariance matrices for the various models under study. Two alternative large-sample test procedures for testing hypotheses involve the Wald statistic and the likelihood ratio statistic. Both statistics frequently yield similar (though not identical) numerical results in small samples; when doubt exists, the likelihood ratio statistic is preferred.[13] Upper and lower limits, respectively, of large sample confidence intervals for (linear functions of) regression coefficients are obtained by adding to and subtracting from an ML point estimate a percentile of the standard normal distribution multiplied by the standard error of the point estimate.

Problems

1. A certain drug is suspected of lowering blood pressure as a side effect. A clinical trial is conducted to investigate this suspicion. Thirty-two patients are randomized into drug and placebo groups (16 per group). Their initial and posttreatment systolic blood pressures (SBP, measured in mm Hg), and their body size as measured by the Quetelet index (QUET), are recorded.

   The researchers conduct a linear regression analysis to investigate whether the mean changes in SBP differ between the placebo and drug groups, controlling for initial SBP and QUET. The dependent variable is posttreatment SBP. One model under consideration involves three predictors: DRUG status, initial SBP, and QUET as main effects only (together with a constant term). Suppose that a computer program that calculates least-squares estimates of the regression coefficients (e.g., the REGRESSION procedure in SAS) is used to fit this model.

   a. Are the least-squares estimates of the regression coefficients identical to ML estimates of these same coefficients for the above model? Explain briefly.

   b. Assuming that the estimation procedure uses ML estimation, how would you carry out a Wald test for the effect of the DRUG status variable, controlling for initial SBP and QUET? (Include in your answer the null hypothesis being tested, the form of the test statistic, and its large-sample distribution under the null hypothesis.)

   c. Is the Wald test procedure described in part (b) equivalent to a partial F test obtained from least-squares estimation? Explain briefly.

   d. How would you carry out a likelihood ratio test for the effect of the DRUG status variable, controlling for initial SBP and QUET? (Include in your answer the null hypothesis being tested, the form of the test statistic, and its large-sample distribution under the null hypothesis.)

   e. Should you expect to obtain the same P-value for the likelihood ratio test described in part (d) as for the Wald test described in part (c)? Explain briefly.

   f. Assuming ML estimation, state the formula for a large-sample 95% confidence interval for the effect of the DRUG status variable controlling for initial SBP and QUET.

2. The data for this question consist of a sample of 50 persons from the 1967–1980 Evans County Study (Schoenbach et al. 1986). Two basic independent variables are of interest: AGE and chronic disease status (CHR), where CHR is coded as 0 = none and 1 = presence of chronic disease. A product term of the form AGE × CHR is also considered. The dependent variable is time until death, a continuous variable. The primary question of interest is

---

[13]One statistical advantage of the likelihood ratio statistic over the Wald statistic is that the former is less likely than the latter to be influenced by collinearity problems for the model being fit.

whether CHR, considered as the exposure variable, is related to survival time, controlling for AGE. The computer results, based on ML estimation,[14] are as follows:

Model 1:

Variable	Coeff	S.E.	Chisq	P-value
CHR	0.8595	0.3116	7.61	.0058

$-2 \ln \hat{L} = 285.74$

Model 2:

Variable	Coeff	S.E.	Chisq	P-value
CHR	0.8051	0.3252	6.13	.0133
AGE	0.0856	0.0193	19.63	.0000

$-2 \ln \hat{L} = 264.90$

Model 3:

Variable	Coeff	S.E.	Chisq	P-value
CHR	1.0009	2.2556	0.20	.6572
AGE	0.0874	0.0276	10.01	.0016
CHR×AGE	-0.0030	0.0345	0.01	.9301

$-2 \ln \hat{L} = 264.89$

a. Assuming a regression model that contains the main effects of CHR and AGE, as well as the interaction effect of CHR with AGE, carry out a Wald test for significant interaction. (Include in your answer the null hypothesis, the form of the test statistic, and its distribution under the null hypothesis.) What are your conclusions about interaction, based on this test?

b. For the same model considered in part (a), how would you carry out a likelihood ratio test for significant interaction? (Include in your answer the null hypothesis, the form of the test statistic, and its distribution under the null hypothesis.) What are your conclusions about interaction, based on this test? How do these results compare with those in part (a)?

c. Assuming no interaction, carry out a Wald test for the significance of the CHR variable, controlling for AGE. (As in the preceding parts of this problem, state the null hypothesis, the form of the test statistic, and its distribution under the null hypothesis.) What are your conclusions, based on this test?

d. For the same no-interaction model considered in part (c), how would you carry out a likelihood ratio test for significance of the CHR variable, controlling for AGE? (Include in

[14]The analysis described in Problem 2 is an example of a *survival analysis,* and the model considered is called the *Cox proportional hazards model.* See Kleinbaum (1996) for a detailed discussion of survival analysis.

your answer the null hypothesis, the form of the test statistic, and its distribution under the null hypothesis.) Why can't you actually carry out this test, given the output information provided earlier?

e. For the no-interaction model considered in part (c), compute a 95% confidence interval for the coefficient of the CHR variable, controlling for AGE. What does the computed confidence interval say about the reliability of the point estimate of the effect of CHR in this model?

f. What is your overall conclusion about the effect of CHR on survival time, based on the computer results from this study?

References

Kleinbaum, D. G. 1996. *Survival Analysis—A Self-Learning Text.* New York: Springer-Verlag.

Schoenbach, V. J.; Kaplan, B. H.; Fredman, L.; and Kleinbaum, D. G. 1986. "Social Ties and Mortality in Evans County, Georgia." *American Journal of Epidemiology* 123(4): 577–91.

Logistic Regression Analysis

[handwritten annotations:]

$y = 0, 1$

$P(Y=1) = P$

$P\left(\frac{Y}{1}\right) = P^y(1-P)^Y$

$P(Y=y) = \left(\frac{P}{1-P}\right)^y (1-p) = 1-p \, e^{y \log P/1-p}$

$\log \frac{P}{1-p} = B_0 + B_1 X$

$P = \frac{1}{1+e^{-B_0 + B_1 x}}$

$E(y) = p$

23-1 Preview

Logistic regression analysis is the most popular regression technique available for modeling dichotomous dependent variables. This chapter describes the logistic model form and several of its key features, particularly how an odds ratio can be estimated from it. We also demonstrate how logistic regression may be applied, using a real-life data set.

Maximum likelihood procedures are used to estimate the model parameters of a logistic model. Therefore, the general principles and inference-making procedures described in Chapter 22 on ML estimation directly carry over to the likelihood functions appropriate for logistic regression analysis. We will examine two alternative ML procedures for logistic regression— called unconditional and conditional—which involve different likelihood functions to be maximized. The latter (conditional) method is recommended when the amount of data available for analysis is not large relative to the number of parameters in the model, as is often the case when matching on potential confounders is carried out in the selection of subjects.

23-2 The Logistic Model

Logistic regression is a mathematical modeling approach that can be used to describe the relationship of several predictor variables $X_1, X_2,\ldots, X_k$ to a *dichotomous* dependent variable Y, where Y is typically coded as 1 or 0 for its two possible categories. The logistic model describes the expected value of Y (i.e., $E(Y)$) in terms of the following "logistic" formula:

$$E(Y) = \frac{1}{1 + \exp\left[-\left(\beta_0 + \sum_{j=1}^{k} \beta_j X_j\right)\right]}$$

For (0, 1) random variables such as Y, it follows from basic statistical principles about expected values[1] that $E(Y)$ is equivalent to the probability $\mathrm{pr}(Y = 1)$; so the formula for the logistic model

---

[1] For a (0, 1) random variable Y, $E(Y) = [0 \times \mathrm{pr}(Y = 0)] + [1 \times \mathrm{pr}(Y = 1)] = \mathrm{pr}(Y = 1)$.

can be written in a form that describes the probability of occurrence of one of the two possible outcomes of Y, as follows:

$$\text{pr}(Y = 1) = \cfrac{1}{1 + \exp\left[-\left(\beta_0 + \sum_{j=1}^{k} \beta_j X_j\right)\right]}. \tag{23.1}$$

The logistic model (23.1) is useful in many important practical situations where the response variable takes only one of two possible values. For example, a study of the development of a particular disease in some human population could employ a logistic model to describe in probabilistic terms whether a given individual in the study group will ($Y = 1$) or will not ($Y = 0$) develop the disease in question during a follow-up period of interest.

The first step of a logistic regression analysis is to postulate (based on knowledge about and experience with the process under study) a mathematical model describing the mean of Y as a function of the X_j and the β_j values. The model is then fitted to the data by maximum likelihood, and eventually appropriate statistical inferences are made (after the model's adequacy of fit is verified, including consideration of the relevant regression diagnostic indices).

The mathematical expression given on the right side of the logistic model formula given by (23.1) is of the general mathematical form

$$f(z) = \frac{1}{1 + e^{-z}}$$

where $z = \beta_0 + \sum_{j=1}^{k} \beta_j X_j$. The function $f(z)$ is called the *logistic function*. This function is well-suited to modeling a probability, since the values of $f(z)$ range from 0 to 1 as z varies from $-\infty$ to $+\infty$. In epidemiologic studies, such a probability can be used to state an individual's risk of developing a disease. The logistic model, therefore, is set up to ensure that, whatever estimate of risk we get, it always falls between 0 and 1. This is not true for other possible models, which is why the logistic model is often used when a probability must be estimated.

Another reason why the logistic model is popular relates to the general sigmoid shape of the logistic function (see Figure 23-1). A sigmoid shape is particularly appealing to epidemiologists if the variable z is viewed as representing an index that combines the contributions of several risk factors, so that $f(z)$ represents the risk for a given value of z. In this context, the risk is

FIGURE 23-1   The logistic function $f(z) = \dfrac{1}{1 + e^{-z}}$.

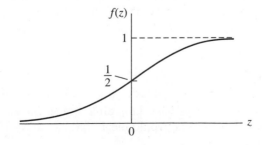

minimal for low z values, rises over a range of intermediate values of z, and remains close to 1 once z gets large enough. Epidemiologists believe that this sigmoid shape applies to a variety of disease conditions.

23-3 Estimating the Odds Ratio Using Logistic Regression

As in any regression model, the regression coefficients β_j in the logistic model given by (23.1) play an important role in providing information about the relationships of the predictors in the model to the dependent variable. For the logistic model, quantification of these relationships involves a parameter called the *odds ratio*.

The odds ratio is a widely used measure of effect in epidemiologic studies. By "measure of effect," we mean a measure that compares two or more groups in predicting the outcome (dependent) variable. To describe an odds ratio, we first define *odds* as the ratio of the probability that some event (e.g., developing lung cancer) will occur divided by the probability that the same event will not occur (e.g., not developing lung cancer), Thus, the odds for some event D is given by the formula

$$\text{odds}(D) = \frac{\text{pr}(D)}{\text{pr}(\text{not } D)} = \frac{\text{pr}(D)}{1 - \text{pr}(D)}$$

For example, if $\text{pr}(D) = .25$, then

$$\text{odds}(D) = \frac{.25}{1 - .25} = 1/3$$

An odds of one-third can be interpreted to mean that the probability of the event's occurring is $1/3$ the probability of the event's not occurring; equivalently, the odds are "3 to 1" that the event will not happen.

Any odds ratio (OR), by definition, is a ratio of two odds; that is,

$$\text{OR}_{A \text{ vs. } B} = \frac{\text{odds}(D_A)}{\text{odds}(D_B)} = \frac{\text{pr}(D_A)}{1 - \text{pr}(D_A)} \bigg/ \frac{\text{pr}(D_B)}{1 - \text{pr}(D_B)}$$

in which the subscripts A and B denote two groups of individuals being compared. For example, suppose that $A = S$ denotes a group of smokers and $B = NS$ denotes a group of nonsmokers; then D_S describes smokers who develop lung cancer, and D_{NS} describes nonsmokers who develop lung cancer. If $\text{pr}(D_S) = .25$ and $\text{pr}(D_{NS}) = .10$, the odds ratio that compares the odds of developing lung cancer for smokers with the odds of developing lung cancer for nonsmokers is given by

$$\text{OR}_{S \text{ vs. } NS} = \frac{.25}{1 - .25} \bigg/ \frac{.10}{1 - .10} = \frac{1}{3} \bigg/ \frac{1}{9} = 3$$

In other words, the odds of developing lung cancer for smokers is three times the corresponding odds for nonsmokers, suggesting roughly that smokers have a threefold higher risk of developing lung cancer than nonsmokers. An odds ratio of one would mean that the odds for the two groups are the same; that is, it would indicate that there is no effect of smoking on the development of lung cancer.

Where does logistic regression fit in here? To answer this, we must consider an equivalent way to write the logistic regression model, called the *logit form* of the model. The "logit" is a transformation of the probability $\text{pr}(Y = 1)$, defined as the natural log odds of the event $D = \{Y = 1\}$. In other words,

$$\text{logit}[\text{pr}(Y = 1)] = \log_e[\text{odds}(Y = 1)] = \log_e\left[\frac{\text{pr}(Y = 1)}{1 - \text{pr}(Y = 1)}\right] \tag{23.2}$$

If we then substitute the logistic model formula (23.1) for $\text{pr}(Y = 1)$ into equation (23.2), it follows that

$$\text{logit}[\text{pr}(Y = 1)] = \beta_0 + \sum_{j=1}^{k} \beta_j X_j \tag{23.3}$$

Equation (23.3) is called the *logit form* of the model. The logit form is given by the linear function $\beta_0 + \sum_{j=1}^{k} \beta_j X_j$. For convenience, many authors describe the logistic model in its logit form given by (23.3), rather than in its original form defined by (23.1).

For example, if Y denotes lung cancer status ($1 = \text{yes}, 0 = \text{no}$) and there is only one (i.e., $k = 1$) predictor X_1—say, smoking status ($1 = \text{smoker}, 0 = \text{nonsmoker}$)—then the logistic model (23.1) can be written equivalently in logit form (23.3) as

$$\text{logit}[\text{pr}(Y = 1)] = \beta_0 + \beta_1 X_1 = \beta_0 + \beta_1(\text{smoking status})$$

To obtain an expression for the odds ratio from a logistic model, we must compare the odds for two groups of individuals. For the preceding example involving one predictor, the two groups are smokers ($X_1 = 1$) and nonsmokers ($X_1 = 0$). Thus, for this example, the log odds for smokers and nonsmokers can be written as

$$\log_e \text{odds}(\text{smokers}) = \beta_0 + (\beta_1 \cdot 1) = \beta_0 + \beta_1$$

and

$$\log_e \text{odds}(\text{nonsmokers}) = \beta_0 + (\beta_1 \cdot 0) = \beta_0$$

respectively. It follows that the odds ratio comparing smokers to nonsmokers is given by

$$\text{OR}_{S\text{ vs. }NS} = \frac{\text{odds}(\text{smokers})}{\text{odds}(\text{nonsmokers})} = \frac{e^{(\beta_0 + \beta_1)}}{e^{\beta_0}} = e^{\beta_1}$$

In other words, for the simple example involving one (0–1) predictor, the odds ratio comparing the two categories of the predictor is obtained by exponentiating the coefficient of the predictor in the logistic model.

Generally, when computing an odds ratio, we can define the two groups (or individuals) that are to be compared in terms of two different specifications of the set of predictors $X_1, X_2, \ldots, X_k$. We do this by letting $\mathbf{X_A} = (X_{A1}, X_{A2}, \ldots, X_{Ak})$ and $\mathbf{X_B} = (X_{B1}, X_{B2}, \ldots, X_{Bk})$ denote the collection of X's for groups (or individuals) A and B, respectively. For example, if $k = 3$, and X_1 is smoking status ($1 = \text{yes}, 0 = \text{no}$), X_2 is age (continuous) and X_3 is race ($1 = \text{black}, 0 = \text{white}$), then $\mathbf{X_A}$ and $\mathbf{X_B}$ are two specifications of these three variables—say,

$$\mathbf{X_A} = (1, 45, 1) \text{ and } \mathbf{X_B} = (0, 45, 1).$$

Here, $\mathbf{X_A}$ denotes the group of 45-year-old black smokers, whereas $\mathbf{X_B}$ denotes the group of 45-year-old black nonsmokers.

To obtain a general formula for the odds ratio, we must divide the odds for group (or individual) A by the odds for group (or individual) B and then substitute the logit form of the logistic model given by (23.1), to obtain an expression involving the logistic model parameters. From algebra, the following result is obtained:

$$OR_{\mathbf{X_A} \text{ vs. } \mathbf{X_B}} = \frac{\text{odds for } \mathbf{X_A}}{\text{odds for } \mathbf{X_B}} = \frac{e^{(\beta_0 + \sum_{j=1}^{k} \beta_j X_{Aj})}}{e^{(\beta_0 + \sum_{j=1}^{k} \beta_j X_{Bj})}}$$

We can simplify the above expression further[2] to obtain the following general formula for the odds ratio:

$$OR_{\mathbf{X_A} \text{ vs. } \mathbf{X_B}} = e^{\sum_{j=1}^{k} (X_{Aj} - X_{Bj})\beta_j} \tag{23.4}$$

The constant term β_0 in the logistic model (23.1) drops out of the odds ratio expression (23.4), so the odds ratio depends only on exponentiating the sum of the β_j coefficients multiplied by the difference in corresponding values of the X_j's for groups (or individuals) A and B. Expression (23.4) describes a "population" odds ratio parameter because the β_j terms in this expression are themselves unknown population parameters. An estimate of this population odds ratio can be obtained by fitting the logistic model using maximum likelihood estimation and substituting the ML estimates $\hat{\beta}_j$, together with values of X_{Aj} and X_{Bj}, into the formula (23.4) to obtain a numerical value for the odds ratio.

For example, if $\mathbf{X_A}$ and $\mathbf{X_B}$ are two specifications of the three variables smoking status, age, and race, so that $\mathbf{X_A} = (1, 45, 1)$ and $\mathbf{X_B} = (0, 45, 1)$, as described earlier, then

$$OR_{\mathbf{X_A} \text{ vs. } \mathbf{X_B}} = e^{(1-0)\beta_1 + (45-45)\beta_2 + (1-1)\beta_3} = e^{\beta_1}$$

If the estimate of the β_1 coefficient from ML estimation turns out to be, say, $\hat{\beta}_1 = 1.32$, then the estimated odds ratio will be $e^{1.32} = 3.74$. As in the previous example (which involved only the smoking status variable), the odds ratio expression is computed by exponentiating the coefficient of the smoking status variable. However, the logistic model in this example involves three variables, as follows (in logit form):

$$\text{logit}[\text{pr}(Y = 1)] = \beta_0 + \beta_1(\text{smoking}) + \beta_2(\text{age}) + \beta_3(\text{race})$$

so the value of the estimate $\hat{\beta}_1$ for this three-variable model may be numerically different from the value obtained for $\hat{\beta}_1$ in the model containing only smoking status.

In this latest example, the smoking status variable changes from 1 in group A to 0 in group B, whereas the other variables remain the same for each group—namely, age is 45 and race is 1. In general, whenever only one variable (e.g., smoking) changes, while the other variables are fixed, we say that the odds ratio comparing two categories of the changing variable (e.g., smokers versus nonsmokers) is an *adjusted odds ratio* that *controls* for the other variables (i.e., those that are fixed at specific values) in the model. The variable of interest—in this case, smoking

---

[2]For any two values a and b, it follows that $\dfrac{e^a}{e^b} = e^{(a-b)}$.

status—is often referred to as the *exposure* (or *study*) variable; the other variables in the model are called *control* (or *confounder*) variables. Thus, we have as an important special case of the odds ratio expression (23.4) the following rule:

☐ An adjusted odds ratio can be obtained by exponentiating the coefficient of a (0–1) exposure variable in the logistic model (provided that there are no cross-product terms in the model involving the exposure variable).

The examples that we have considered so far have involved only *main effect variables* like smoking, age, and race; we have not considered product terms like "smoking $\times$ age" or "smoking $\times$ race," nor have we considered exposure variables other than (0–1) variables. When the model contains product terms, like "smoking $\times$ age" or exposure variables that are not (0–1) variables, the preceding simple rule for obtaining an adjusted odds ratio does not work. In such instances, we must use the general formula given by (23.4).

For example, suppose, as before, that smoking is a (0–1) exposure variable; and that age (continuous) and race (0–1) are control variables; but that the logistic model we want to fit is given (in logit form) as

$$\text{logit}[\text{pr}(Y = 1)] = \beta_0 + \beta_1(\text{smoking}) + \beta_2(\text{age}) + \beta_3(\text{race})$$
$$+\beta_4(\text{smoking} \times \text{age}) + \beta_5(\text{smoking} \times \text{race}) \qquad (23.5)$$

Then, to obtain an adjusted odds ratio for the effect of smoking adjusted for age and race, we need to specify two sets of values $\mathbf{X_A}$ and $\mathbf{X_B}$ for the collection of predictor variables defined by

$$\mathbf{X} = (\text{smoking, age, race, smoking} \times \text{age, smoking} \times \text{race})$$

If the two groups being compared are, as before, 45-year-old black smokers (i.e., group A), and 45-year-old black nonsmokers (i.e., group B), then $\mathbf{X_A}$ and $\mathbf{X_B}$ are given as

$$\mathbf{X_A} = (1, 45, 1, 1 \times 45, 1 \times 1) = (1, 45, 1, 45, 1)$$

and

$$\mathbf{X_B} = (0, 45, 1, 0 \times 45, 0 \times 1) = (0, 45, 1, 0, 0)$$

Applying the odds ratio formula (23.4) to model (23.5) with the given values for $\mathbf{X_A}$ and $\mathbf{X_B}$, we obtain the following expression for the adjusted odds ratio:

$$\text{OR}_{\mathbf{X_A} \text{ vs. } \mathbf{X_B}} = e^{(1-0)\beta_1 + (45-45)\beta_2 + (1-1)\beta_3 + (45-0)\beta_4 + (1-0)\beta_5} = e^{\beta_1 + 45\beta_4 + \beta_5}$$

Since this odds ratio describes the effect of smoking adjusted for age = 45 and race = 1, we can alternatively denote the odds ratio as

$$\text{OR}_{(S \text{ vs. } NS \mid \text{age}=45, \text{race}=1)}$$

where the | sign is common mathematical notation for "given." Thus, we have

$$\text{OR}_{(S \text{ vs. } NS \mid \text{age}=45, \text{race}=1)} = e^{\beta_1 + 45\beta_4 + \beta_5}$$

Rather than involving only β_1, the preceding expression involves the three coefficients β_1, β_4, and β_5, each of which is a coefficient of a variable in the model that contains the exposure variable (i.e., smoking) in some form. The coefficients β_4 and β_5 are included because they are coefficients of "interaction terms" in the model being fit. The model essentially says that the value of the odds ratio for the effect of smoking should vary depending on the values of the variables age and race (which are components of the two interaction variables in the model) and of their coefficients β_4 and β_5. Since the "fixed" value of age is 45 and the "fixed" value of race is 1, these two values multiply their corresponding coefficients in the odds ratio expression. The variables age and race in model (23.5) are examples of *effect modifiers* (see Kleinbaum, Kupper, and Morgenstern 1982, Chapter 13) since the *effect* of the exposure variable smoking status, as quantified by the odds ratio $OR_{(S \text{ vs. } NS \mid \text{age, race})}$, changes (i.e., *is modified*) depending on the values of age and race.

To illustrate how age and race are effect modifiers of smoking status in model (23.5), suppose that we let the values of age and race in $\mathbf{X_A}$ and $\mathbf{X_B}$ be fixed but unspecified, so we can write $\mathbf{X_A}$ and $\mathbf{X_B}$ as

$$\mathbf{X_A} = (1, \text{age, race}, 1 \times \text{age}, 1 \times \text{race}) = (1, \text{age, race, age, race})$$

and

$$\mathbf{X_B} = (0, \text{age, race}, 0 \times \text{age}, 0 \times \text{race}) = (0, \text{age, race}, 0, 0)$$

Then, substituting $\mathbf{X_A}$ and $\mathbf{X_B}$ into formula (23.4) for the odds ratio, we obtain

$$OR_{(S \text{ vs. } NS \mid \text{age, race})} = e^{(1-0)\beta_1 + (\text{age}-\text{age})\beta_2 + (\text{race}-\text{race})\beta_3 + (\text{age}-0)\beta_4 + (\text{race}-0)\beta_5}$$

$$= e^{\beta_1 + \beta_4(\text{age}) + \beta_5(\text{race})}$$

This formula says that the value of the odds ratio for the effect of smoking varies depending on the values of the variables age and race (which are components of the two interaction variables in the model) and of their coefficients β_4 and β_5. So, for example, if we have age = 45 and race = 1, as in our earlier example, the adjusted odds ratio is given by

$$OR_{(S \text{ vs. } NS \mid \text{age}=45, \text{race}=1)} = e^{\beta_1 + 45\beta_4 + \beta_5}$$

Similarly, if we have age = 35 and race = 0, the adjusted odds ratio is given by

$$OR_{(S \text{ vs. } NS \mid \text{age}=35, \text{race}=0)} = e^{\beta_1 + 35\beta_4}$$

And if age = 20 and race = 1, then

$$OR_{(S \text{ vs. } NS \mid \text{age}=20, \text{race}=1)} = e^{\beta_1 + 20\beta_4 + \beta_5}$$

The preceding examples illustrate another general rule that describes an important special case of the general odds ratio formula (23.4):

> ☐ For a logistic model that contains a (0–1) exposure variable together with interaction terms that are products of the exposure variable with other control variables, the adjusted odds ratio is obtained by exponentiating a linear function of the regression coefficients involving the exposure alone and the product terms in the model involving exposure. Moreover, since the model being fit contains interaction terms, the numerical value of the adjusted odds ratio will vary depending on the values of the control variables (i.e., effect modifiers) that are components of the product terms involving exposure.

We now consider one other important special case of the general odds ratio formula (23.4). Suppose that the exposure variable is a continuous variable, like systolic blood pressure (SBP), so that, for example, our model contains SBP and the control variables age and race; that is, the logit form of the no-interaction model is given by

$$\text{logit}[\text{pr}(Y = 1)] = \beta_0 + \beta_1(\text{SBP}) + \beta_2(\text{age}) + \beta_3(\text{race}) \tag{23.6}$$

To obtain an adjusted odds ratio for the effect of SBP, controlling for age and race, we must specify two values of the exposure variable (SBP) to be compared. Two specified values are needed, even when the exposure variable (like SBP) is continuous, because an odds ratio by definition compares the odds for two groups (or individuals). For example, if the two values of SBP are 160 and 120, and age and race are considered fixed, the general odds ratio expression (23.4) simplifies to

$$OR_{(SBP=160\,vs.\,SBP=120\,|\,age,\,race)} = e^{(160-120)\beta_1 + (age-age)\beta_2 + (race-race)\beta_3} = e^{40\beta_1}$$

More generally, if we specify the two values of SBP to be SBP_1 and SBP_0, then the odds ratio is

$$OR_{(SBP_1\,vs.\,SBP_0\,|\,age,\,race)} = e^{(SBP_1-SBP_0)\beta_1}$$

This odds ratio expression simplifies to e^{β_1} when the difference $(SBP_1 - SBP_0)$ equals one. Thus, exponentiating the coefficient of the exposure variable gives an odds ratio for comparing any two groups that differ by one unit of SBP. A one-unit difference in SBP is rarely of interest, however. A more typical choice of SBP values to be compared would be clinically meaningful values of blood pressure, such as SBP values of 160 and 120. One possible strategy for choosing values of SBP to compare is to categorize the distribution of SBP in a data set into clinically different categories—say, quintiles. Then, using the mean or median SBP in each quintile, we could compute odds ratios for all possible pairs of mean or median SBP values. We would then obtain a table of odds ratios to consider in assessing the relationship of SBP to the outcome variable. (For a more thorough treatment of important special cases for computing the odds ratio, see chapter 3 of Kleinbaum (1994).)

ML estimates of the regression coefficients are typically obtained by using standard computer packages for logistic regression. These estimates can then be used in appropriate odds ratio formulas based on the general expression (23.4) to obtain numerical values for adjusted odds ratios. Since these are point estimates, researchers typically carry out statistical inferences about the odds ratios that are being estimated. For example, if the adjusted odds ratio is given by the simple expression e^{β_1}, involving a single coefficient, then the null hypothesis that this odds ratio equals 1 can be stated equivalently as $H_0: \beta_1 = 0$, since, under H_0, $e^{\beta_1} = e^0 = 1$. The test for this null hypothesis can be carried out by using either the Wald test or the likelihood ratio test described in Chapter 22 on ML estimation methods. A confidence interval for the adjusted odds ratio can be obtained by first calculating a confidence interval for β_1, as described in section 22-3, and then exponentiating the lower and upper limits, which for this simple situation is given by the formula

$$\exp\left(\hat{\beta}_1 \pm Z_{1-\alpha/2} s_{\hat{\beta}_1}\right)$$

where $\hat{\beta}_1$ is the ML estimate of β_1, $s_{\hat{\beta}_1}$ is the standard error of $\hat{\beta}_1$, and $(1 - \alpha)$ is the confidence level.

More detailed discussion of the properties and applications of logistic regression may be found in several other textbooks, including Kleinbaum (1994), Hosmer and Lemeshow (1989), Collett (1991), and Kleinbaum, Kupper, and Morgenstern (1982). In particular, a general analysis strategy for selecting the variables to retain in a logistic model is described in Kleinbaum (1994), chapters 6 and 7, and in Kleinbaum, Kupper, and Morgenstern (1982), chapters 21–24. The goal of this strategy is to obtain a *valid* estimate (in the context of the analysis of epidemiologic research data) of the relationship between a specified exposure variable and a particular disease variable, while controlling for or adjusting other covariates that, if not correctly taken into account, can lead to an incorrect assessment of the strength of the exposure–disease relationship of interest. These covariates can act as confounders and/or effect modifiers, terms we discussed in Chapter 11.

23-4 A Numerical Example of Logistic Regression

Dengue fever is an acute infectious disease caused by a virus transmitted by the *Aedes* mosquito. A retrospective survey of an epidemic of dengue fever[3] was carried out in three Mexican cities in 1984. In this section, we review the analyses of a subset of the data from this study obtained on a two-stage stratified random sample of 196 persons from the city of Puerto Vallarta, 57 of whom were determined to be suffering cases of dengue fever. The goal of the analyses was to identify risk factors associated with having the disease, especially the effect of the absence of mosquito netting on a subject's bed as a determinant of the disease. The following variables were tracked in a computer file called DENGUE.DAT[4]:

Column 1: Subject ID

Column 2: Dengue fever status (DENGUE): 1 = yes, 2 = no

Column 3: AGE in years

Column 4: Use of mosquito netting (MOSNET): 0 = yes, 1 = no

Column 5: Geographical sector in which the subject lived (SECTOR): 1–5.

The variable SECTOR was treated as a categorical variable in the regression analysis, so four dummy variables had to be created to distinguish the five geographical sectors. These variables were defined so that sector 5 was the referent group, as follows:

$$SECTOR1 = \begin{cases} 1 & \text{if sector 1} \\ 0 & \text{if other} \end{cases} \qquad SECTOR2 = \begin{cases} 1 & \text{if sector 2} \\ 0 & \text{if other} \end{cases}$$

$$SECTOR3 = \begin{cases} 1 & \text{if sector 3} \\ 0 & \text{if other} \end{cases} \qquad SECTOR4 = \begin{cases} 1 & \text{if sector 4} \\ 0 & \text{if other} \end{cases}$$

The following block of edited computer output comes from fitting SAS's LOGISTIC procedure for a logistic regression model that regresses the dichotomous outcome variable DENGUE on the predictors MOSNET, AGE, and SECTORj, for $j = 1, 2, 3, 4$. The logit form of the model being fit is given as

$$\text{logit}[\text{pr}(Y = 1)] = \beta_0 + \beta_1(\text{AGE}) + \beta_2(\text{MOSNET}) + \beta_3(\text{SECTOR1})$$
$$+ \beta_4(\text{SECTOR2}) + \beta_5(\text{SECTOR3}) + \beta_6(\text{SECTOR4}) \qquad (23.7)$$

where Y denotes the dependent variable DENGUE.

---

[3]Dantes, Koopman, et al. (1988).

[4]The data file DENGUE.DAT is stored on a computer disk (accompanying this text) that contains other data files considered in examples and problems throughout the text.

```
                    The LOGISTIC Procedure
                      Response Profile
                 Ordered
                  Value      DENGUE           Count
                    1          1               57
                    2          2              139
                Criteria for Assessing Model Fit

                               Intercept
                   Intercept      and
Criterion            Only      Covariates   Chi-Square for Covariates
AIC                238.329      217.706      .
SC                 241.607      240.653      .
-2 LOG L           236.329      203.706     32.623 with 6 DF (p=0.0001)
Score                 .            .         28.775 with 6 DF (p=0.0001)

               Analysis of Maximum Likelihood Estimates
                                    Wald      Pr >
                 Parameter  Standard Chi-     Chi-    Standardized  Odds
Variable    DF   Estimate    Error   Square   Square  Estimate      Ratio
INTERCPT    1    -1.9001    1.3254   2.0551   0.1517                0.150
AGE         1     0.0243    0.00906  7.1778   0.0074   0.252890     1.025
MOSNET      1     0.3335    1.2718   0.0688   0.7931   0.034212     1.396
SECTOR1     1    -2.2200    1.0723   4.2861   0.0384  -0.441811     0.109
SECTOR2     1    -0.6589    0.5536   1.4164   0.2340  -0.142513     0.517
SECTOR3     1     0.8121    0.4750   2.9235   0.0873   0.173824     2.253
SECTOR4     1     0.5310    0.4502   1.3911   0.2382   0.121456     1.701
```

Using the given computer output, we now focus on the information provided under the heading "Analysis of Maximum Likelihood Estimates." From this information, we can see that the ML coefficients obtained for the fitted model are

$$\hat{\beta}_0 = -1.9001, \hat{\beta}_1 = 0.0243, \hat{\beta}_2 = 0.3335, \hat{\beta}_3 = -2.2200,$$
$$\hat{\beta}_4 = -0.6589, \hat{\beta}_5 = 0.8121, \hat{\beta}_6 = 0.5310$$

so the fitted model is given (in logit form) by

$$\text{logit}[\widehat{\text{pr}}(Y = 1)] = -1.9001 + 0.0243(\text{AGE}) + 0.3335(\text{MOSNET})$$
$$- 2.2200(\text{SECTOR1}) - 0.6589(\text{SECTOR2})$$
$$+ 0.8121(\text{SECTOR3}) + 0.5310(\text{SECTOR4})$$

Based on this fitted model and the information provided in the computer output, we can compute the estimated odds ratio for contracting dengue fever for persons who did not use mosquito netting relative to persons who did use mosquito netting, controlling for age and sector. We do this using the previously stated rule for adjusted odds ratios for (0–1) variables (when there is no interaction), by exponentiating the estimated coefficient ($\hat{\beta}_2$) of the MOSNET variable in the fitted model. We then obtain the following value for the adjusted odds ratio:

$$\widehat{\text{OR}}_{(\text{MOSNET}=1 \text{ vs. MOSNET}=0 \mid \text{age, sector})} = e^{0.3335} = 1.396$$

A 95% confidence interval for e^{β_2} can be obtained by computing

$$\exp[0.3335 \pm 1.96(1.2718)]$$

where 1.2718 is the estimated standard error of the estimator $\hat{\beta}_2$ of β_2 in model (23.7). The resulting lower and upper confidence limits are 0.115 and 16.89, respectively. The Wald (chi-square) statistic for testing the null hypothesis that the adjusted odds ratio e^{β_2} is equal to 1 (or equivalently, that the coefficient β_2 of the MOSNET variable equals 0), is shown in the output to be 0.0688, with a P-value equal to 0.7931.

From these results, it can be concluded that the odds of contracting dengue is about 1.4 times higher for a person who did not use mosquito netting than for a person who did use mosquito netting. The Wald statistic is not statistically significant, however, and the 95% confidence interval is very wide (and includes the null value of 1). Thus, there is no statistical evidence in these data that the nonuse of mosquito netting significantly increases the probability (or risk) of contracting dengue fever.

The preceding computer output contains, under the heading "Criteria for Assessing Model Fit," the log likelihood statistic $(-2 \log \hat{L})$ of 203.706 for the fitted model (in the column labeled "Intercept and Covariates"). To compute the likelihood ratio test for the null hypothesis H_0: $\beta_2 = 0$, we must compare this value of $-2 \log \hat{L}$ to the corresponding value of $-2 \log \hat{L}$ for the "reduced" model, which is obtained under the null hypothesis by dropping the exposure variable (MOSNET) from the "full" model given by (23.7). The reduced model is written in logit form as

$$\text{logit}[\text{pr}(Y = 1)] = \beta_0 + \beta_1(\text{AGE}) + \beta_3(\text{SECTOR1}) + \beta_4(\text{SECTOR2})$$
$$+ \beta_5(\text{SECTOR3}) + \beta_6(\text{SECTOR4})$$

The following block of edited computer output for the reduced model indicates that the log likelihood statistic for the fitted model has changed slightly to 203.778. The likelihood ratio statistic for comparing these two models is given by the difference:

$$\text{LR} = -2 \log \hat{L}_R - (-2 \log \hat{L}_F) = 203.778 - 203.706 = 0.072$$

Because the null hypothesis H_0: $\beta_2 = 0$ involves only one parameter, the preceding LR statistic is distributed as approximately a chi-square variable with 1 degree of freedom under the null hypothesis. The test is nonsignificant, which agrees with the conclusion from the Wald test described earlier, thereby supporting the previous conclusion that using mosquito netting did not significantly affect a person's probability of contracting dengue fever.

```
                    The LOGISTIC Procedure
                      Response Profile
                 Ordered
                  Value        DENGUE              Count
                    1            1                   57
                    2            2                  139
              Criteria for Assessing Model Fit

                              Intercept
                 Intercept       and
Criterion          Only       Covariates    Chi-Square for Covariates
AIC              238.329       215.778        .
SC               241.607       235.447        .
-2 LOG L         236.329       203.778       32.551 with 5 DF (p=0.0001)
Score               .             .          28.766 with 5 DF (p=0.0001)
```

```
              Analysis of Maximum Likelihood Estimates
                                      Wald      Pr >
                     Parameter  Standard  Chi-      Chi-    Standardized    Odds
  Variable    DF     Estimate     Error  Square    Square     Estimate     Ratio
  INTERCPT    1      -1.5717     0.4229  13.8144    0.0002                  0.208
  AGE         1       0.0240    0.00901   7.1129    0.0077     0.250447     1.024
  SECTOR1     1      -2.2333     1.0715   4.3444    0.0371    -0.444446     0.107
  SECTOR2     1      -0.6650     0.5531   1.4455    0.2293    -0.143843     0.514
  SECTOR3     1       0.8224     0.4735   3.0169    0.0824     0.176013     2.276
  SECTOR4     1       0.5439     0.4478   1.4755    0.2245     0.124412     1.723
```

Returning to model (23.7), involving the predictors AGE, MOSNET, and SECTOR1 through SECTOR4, we can also evaluate the effect of predictors other than MOSNET on the probability of getting dengue fever. For example, let us examine the effect of AGE, controlling for MOSNET and the SECTOR j variables. From the first block of computer output, we find that the Wald chi-square statistic corresponding to the AGE variable is 7.1778. The null hypothesis being tested here is H_0: $\beta_1 = 0$, where β_1, is the coefficient of the AGE variable in model (23.7). The P-value for this test, as shown in the output, is .0074, so we can reject the null hypothesis at less than the .01 significance level and conclude that the AGE variable is a significant predictor of dengue fever status in model (23.7). In particular, the positive estimated coefficient for AGE suggests that the risk of dengue fever increases with increasing age.

The likelihood ratio test of the same null hypothesis (i.e., H_0: $\beta_1 = 0$) would require subtracting the log likelihood statistic for model (23.7) from the log likelihood statistic for the "reduced" model, which is obtained by dropping the AGE variable from model (23.7). This reduced model has not been presented in either of the blocks of computer output, so an additional computer run would be required obtain the log likelihood statistic for it.[5] The resulting likelihood ratio test does in fact give a significant result, too, corresponding (though not numerically identical) to the significant result obtained from the Wald test.

Since the (Wald or likelihood ratio) test just described measures the effect of a continuous variable (AGE), its specific purpose is to assess whether the effect of the AGE variable in model (23.7) describes a linear relationship between AGE and the log odds of dengue fever, controlling for the other variables in the model; that is, when a variable being tested is continuous and no other variables in the model are squared terms (e.g., AGE^2) or product terms (e.g., AGE $\times$ MOSNET) involving this variable, we are testing whether a linear effect in AGE is more plausible than no effect (i.e., H_0: $\beta_1 = 0$) of AGE. Such a test is typically referred to as a (*linear*) *trend test*. Thus, using the results from the first block of computer output considered previously, we conclude that a linear trend test for AGE in model (23.7) is significant.

Since the AGE variable is continuous in model (23.7), the quantity $e^{\hat{\beta}_1}$, where $\hat{\beta}_1 = 0.0243$ is the estimated coefficient of the AGE variable, describes the odds ratio comparing two persons whose age differs by only one year. As mentioned earlier, if the exposure variable of interest is continuous, a one-unit difference in the variable is rarely of interest. Instead, we need to compare meaningfully (e.g., clinically) different categories of the continuous variable. For example, we might consider two persons whose age differs by five years—say, a 35-year-old person and a 30-year-old person—as having a meaningful difference in age. Using the general odds ratio

---

[5]Under SAS's logistic procedure, the log likelihood statistic for the reduced model is 211.143.

expression (23.4), we can compute the estimated odds ratio for two persons with a five-year age difference, controlling for mosquito netting and sector, as follows:

$$\widehat{OR}_{(AGE_1-AGE_0=5\,|\,\text{mosnet, sector})}$$
$$= e^{[5\hat{\beta}_1+(\text{mosnet}-\text{mosnet})\hat{\beta}_2+(\text{sector1}-\text{sector1})\hat{\beta}_3+\cdots+(\text{sector4}-\text{sector4})\hat{\beta}_6]}$$
$$= e^{5\hat{\beta}_1} = e^{5(0.0243)} = 1.13$$

Since we are controlling for mosquito netting and sector, we assume that the MOSNET variable has the same value (= mosnet) and that the four sector variables (SECTOR1 through SECTOR4) have the same value for the two persons being compared. The coefficients of these five variables then drop out of the odds ratio expression, and we have only the quantity $5\hat{\beta}_1$ to exponentiate. The adjusted odds ratio is therefore given by $\exp[5\hat{\beta}_1]$ rather than by $\exp[\hat{\beta}_1]$, since we are considering the effect of a five-year rather than a one-year difference in age. Thus, for two persons whose age differs by five years, the adjusted odds ratio that controls for use of mosquito netting and sector equals 1.13, which is quite close to the null value of 1. To obtain a 95% confidence interval for the adjusted odds ratio $e^{5\hat{\beta}_1}$, we make the following calculation:

$$95\% \text{ CI for } e^{5\hat{\beta}_1}: \exp\left[5\hat{\beta}_1 \pm 1.96(5s_{\hat{\beta}_1})\right] = \exp\left[5(0.0243) \pm 1.96(5)(0.0091)\right]$$

which yields 95% confidence limits of 1.03 and 1.23. Since the confidence limits do not include the null value of 1, the effect of a five-year difference in age is statistically significant, though small (i.e., the point estimate of the odds ratio is 1.13).

If, instead of a five-year difference in age, we consider the effect of a ten-year difference in age, the formulas for the estimated adjusted odds ratio and 95% confidence interval become

$$\exp\left[10\hat{\beta}_1\right]$$

and

$$\exp\left[10\hat{\beta}_1 \pm 1.96(10s_{\hat{\beta}_1})\right]$$

respectively. The corresponding computed values are 1.28 for the adjusted odds ratio and 1.07 and 1.52 for the confidence limits.

Table 23-1 presents point and interval estimates of the adjusted odds ratio for age differences of from five to forty years in five-year increments. The adjusted odds ratio for the effect

TABLE 23-1   Point and interval estimates of the adjusted odds ratio for the effect of age based on the fit of model (23.7) from the dengue fever study.

$AGE_1 - AGE_0$	$\widehat{OR}_{(AGE_1-AGE_0\,|\,\text{mosnet, sector})} =$ $\exp[(AGE_1 - AGE_0)\hat{\beta}_1]$	95% Confidence Limits for $OR_{(AGE_1-AGE_0\,|\,\text{mosnet, sector})}$
5	1.13	(1.03, 1.23)
10	1.28	(1.07, 1.52)
15	1.44	(1.10, 1.88)
20	1.63	(1.14, 2.32)
25	1.84	(1.18, 2.87)
30	2.07	(1.21, 3.54)
35	2.34	(1.25, 4.37)
40	2.64	(1.30, 5.39)

of age increases from 1.13 for an age difference of five years to 2.64 for an age difference of forty years. The width of the 95% confidence interval also increases as the age difference increases.

We now consider one more illustration of numerical calculations using the dengue fever data. This time we consider a model involving an interaction term. The following block of edited computer output applies to a model that contains the product term MSA = MOSNET $\times$ AGE in addition to the predictors AGE, MOSNET, and SECTOR1 through SECTOR4. This "interaction" model is written in logit form as

$$\text{logit}[\text{pr}(Y = 1)] = \beta_0 + \beta_1(\text{AGE}) + \beta_2(\text{MOSNET}) + \beta_3(\text{SECTOR1})$$
$$+ \beta_4(\text{SECTOR2}) + \beta_5(\text{SECTOR3}) + \beta_6(\text{SECTOR4}) \qquad (23.8)$$
$$+ \beta_7(\text{MSA})$$

```
              The LOGISTIC Procedure
                 Response Profile
            Ordered
             Value       DENGUE          Count
               1            1              57
               2            2             139

          Criteria for Assessing Model Fit

                           Intercept
                              and
               Intercept   Covariates   Chi-Square for Covariates
Criterion        Only
AIC            238.329      218.995        .
SC             241.607      245.220        .
-2 LOG L       236.329      202.995      33.334 with 7 DF (p=0.0001)
Score            .            .           29.492 with 7 DF (p=0.0001)

          Analysis of Maximum Likelihood Estimates
                                    Wald      Pr >
                   Parameter  Standard  Chi-   Chi-  Standardized  Odds
Variable   DF      Estimate   Error    Square  Square  Estimate    Ratio
INTERCPT   1       -0.8080    1.6311   0.2454  0.6203              0.446
AGE        1       -0.00434   0.0362   0.0143  0.9048  -0.045185   0.996
MOSNET     1       -0.8043    1.6433   0.2396  0.6245  -0.082505   0.447
SECTOR1    1       -2.2929    1.0804   4.5042  0.0338  -0.456309   0.101
SECTOR2    1       -0.6813    0.5541   1.5118  0.2189  -0.147362   0.506
SECTOR3    1        0.8153    0.4756   2.9388  0.0865   0.174497   2.260
SECTOR4    1        0.5115    0.4515   1.2830  0.2573   0.116992   1.668
MSA        1        0.0306    0.0374   0.6689  0.4134   0.316912   1.031
```

Suppose that we wish to use the information in the computer output to compute the estimated odds ratio for contracting dengue fever for persons who did not use mosquito netting relative to persons who did use mosquito netting, controlling for age and sector. We previously considered this question using the no-interaction model (23.7), but now we are using model (23.8) instead. We again use the general odds ratio formula (23.4) to compute the adjusted odds ratio, but this time we must take into account the coefficient of the product term MSA. Nevertheless, since the exposure variable of interest (MOSNET) is a (0–1) variable, the general odds ratio formula (23.4) simplifies to the previously stated (see section 23-3) rule for calculating an adjusted odds ratio when the logistic model contains interaction terms: the adjusted odds ratio is obtained by exponentiating a linear function of those regression coefficients involving the exposure alone and those product terms in the model involving exposure. Thus, for interaction model (23.8), we compute the estimated adjusted odds ratio for the effect of mosquito net use, controlling for age and sector as follows:

$$\widehat{OR}_{(MOSNET=1 \text{ vs. } MOSNET=0 \,|\, age, sector)} = \exp\left[\hat{\beta}_2 + \hat{\beta}_7(age)\right]$$
$$= \exp\left[-0.8043 + 0.0306(age)\right]$$

where $\hat{\beta}_2 = -0.8043$ is the estimated coefficient of the MOSNET variable and $\hat{\beta}_7 = 0.0306$ is the estimated coefficient of the MSA (= MOSNET $\times$ AGE) variable obtained from the last batch of computer output. The adjusted odds ratio formula says that the value of the ratio depends on the value we specify for age, which is exactly what we mean when we assume (using model 23.8) that age is an effect modifier of the relationship between mosquito net use and risk of contracting dengue fever. Table 23-2 shows computed values for this adjusted odds ratio for various specifications of the effect modifier age.

Table 23-2 indicates that, based on the interaction model (23.8), the adjusted odds ratio for the effect of mosquito net use varies from values below the null value of 1 when age is 25 or less to values increasingly above 1 (though below 3) when age is over 30. These results suggest, for example, that at age 20 a person who does not use a mosquito net (i.e., MOSNET = 1) is 0.83 times as likely to get dengue fever as a person who uses a mosquito net, whereas at age 40 a person who does not use a mosquito net is 1.52 times as likely to get dengue fever as a person who uses a mosquito net. The precision of these point estimates has not been taken into account by Table 23-2, however, though any calculation of confidence intervals would take precision into account.

We can obtain confidence intervals for any of the preceding adjusted odds ratios by first computing a confidence interval for the linear function $\beta_2 + \beta_7(age)$, which is the log of the (population) adjusted odds ratio $\exp[\beta_2 + \beta_7(age)]$ and then exponentiating the lower and upper limits of this confidence interval. For a 95% confidence interval, we need to use the following formula:

$$\exp\left(\hat{L} \pm 1.96 S_{\hat{L}}\right)$$

TABLE 23-2   Effect of age on the value of the estimated adjusted odds ratio.

Age	$\exp[-0.8043 + 0.0306(age)]$
10	0.61
15	0.71
20	0.83
25	0.96
26.28	1.00
30	1.12
35	1.31
40	1.52
45	1.77
50	2.07
55	2.41
60	2.81

where $\hat{L} = \hat{\beta}_2 + \hat{\beta}_7(\text{age})$ and $S_{\hat{L}} = \sqrt{\widehat{\text{Var}}(\hat{L})}$, and where $\widehat{\text{Var}}(\hat{L})$ is computed using the formula for the variance of a linear combination of random variables, which for this example turns out to be

$$\widehat{\text{Var}}(\hat{L}) = \widehat{\text{Var}}(\hat{\beta}_2) + (\text{age})^2\widehat{\text{Var}}(\hat{\beta}_7) + 2(\text{age})\widehat{\text{Cov}}(\hat{\beta}_2, \hat{\beta}_7)$$

The numerical values of the variances and the covariances involving the $\hat{\beta}_1$ are typically printed out as an option by the computer program used. For model (23.8), the numerical values obtained for the preceding variance formula are:

$$\widehat{\text{Var}}(\hat{\beta}_2) = 2.7004, \qquad \widehat{\text{Var}}(\hat{\beta}_7) = 0.001399, \qquad \text{and} \qquad \widehat{\text{Cov}}(\hat{\beta}_2, \hat{\beta}_7) = -0.0435$$

If, for example, we let age $= 40$, then

$$\hat{L} = -0.8043 + 0.0306(40) = 0.4197$$

$$\widehat{\text{Var}}(\hat{L}) = 2.7004 + (40)^2(0.001399) + 2(40)(-0.0435) = 1.4588$$

and

$$S_{\hat{L}} = \sqrt{\widehat{\text{Var}}(\hat{L})} = 1.2078$$

The 95% confidence interval for the adjusted odds ratio is then given by

$$\exp[0.4197 \pm (1.96)(1.2078)]$$

which yields lower and upper confidence limits of 0.14 and 16.23, respectively. This confidence interval is extremely wide, which implies that the point estimate $e^{0.4197} = 1.52$ when age is 40 is very unreliable. Furthermore, since the confidence interval includes the null value of 1, a test of significance for the adjusted odds ratio is not significant at the 5% level. Computations of 95% intervals at values of age other than 40 also yield very wide intervals and nonsignificant findings.

Finally, we may wish to compare model (23.8), which contains the interaction term MSA, with the no-interaction model (23.7). We can do this by using a likelihood ratio test that takes the following form:

$$\text{LR} = -2 \log \hat{L}_R - (-2 \log \hat{L}_F) = 203.706 - 202.995 = 0.711$$

The full model (F) in this case is model (23.8), and the reduced model (R) is model (23.7). The null hypothesis is $H_0: \beta_7 = 0$, which involves only one parameter, so that the LR statistic is approximately chi-square with 1 d.f. under H_0. The test statistic value of 0.711 is nonsignificant, indicating that the no-interaction model (23.7) is preferable to the interaction model (23.8). In other words, the odds ratio for the effect of MOSNET adjusted for age and sector is best expressed by the single value $\exp(0.3335) = 1.396$, based on the fit of model (23.7), rather than by the expression $\exp(\hat{L})$, where $\hat{L} = \hat{\beta}_2 + \hat{\beta}_7(\text{age})$, which is based on on model (23.8).

23-5 Theoretical Considerations

In this section, we consider the form of the likelihood function to be maximized for a logistic regression model. In particular, we will distinguish between two alternative ML procedures for estimation, called *unconditional* and *conditional*, that involve different likelihood functions. To describe these two procedures, we must first discuss the distributional properties of the dependent (i.e., outcome) variable underlying the logistic model.

For logistic regression, the basic dependent random variable of interest is a dichotomous variable Y taking the value 1 with probability θ and the value 0 with probability $1 - \theta$. Such a random variable is called *Bernoulli* (or point-binomial) and has the simple discrete probability distribution

$$\text{pr}(Y; \theta) = \theta^Y (1 - \theta)^{1-Y}, \quad Y = 0, 1 \tag{23.9}$$

The name *point-binomial* arises because (23.9) is a special case of the binomial distribution $_nC_Y\theta^Y(1 - \theta)^{n-Y}$ in which $n = 1$, where $_nC_Y$ denotes the number of combinations of n distinct objects selected Y-at-a-time.

In general, for a study sample of n subjects, suppose for $i = 1, 2,..., n$ that Y_i is the value of Y for the ith subject and has the Bernoulli distribution

$$\text{pr}(Y_i; \theta_i) = \theta_i^{Y_i} (1 - \theta_i)^{1-Y_i}, \quad Y_i = 0, 1 \tag{23.10}$$

For example, θ_i could represent the probability that individual i in a random sample of n individuals from some population will develop some particular disease during the follow-up period in question.

23-5-1 Unconditional ML Estimation

Given that $Y_1, Y_2,..., Y_n$ are mutually independent, the likelihood function based on (23.10) is obtained as the product of the marginal distributions for the Y_i's—namely,

$$L(\mathbf{Y}; \boldsymbol{\theta}) = \prod_{i=1}^{n} \text{pr}(Y_i; \theta_i) = \prod_{i=1}^{n} \left[\theta_i^{Y_i} (1 - \theta_i)^{1-Y_i} \right] \tag{23.11}$$

where $\boldsymbol{\theta} = (\theta_1, \theta_2,..., \theta_n)$. Now suppose, without loss of generality, that the first n_1 out of the n individuals in our random sample actually develop the disease in question (so that $Y_1 = Y_2 = \cdots = Y_{n_1} = 1$), and that the remaining $n - n_1$ individuals do not (so that $Y_{n_1+1} = Y_{n_1+2} = \cdots = Y_n = 0$). Given this set of observed outcomes, the likelihood expression (23.11) takes the specific form

$$L(\mathbf{Y}; \boldsymbol{\theta}) = \left(\prod_{i=1}^{n_1} \theta_i \right) \left[\prod_{i=n_1+1}^{n} (1 - \theta_i) \right] \tag{23.12}$$

We can work with expression (23.12) to write the likelihood function in terms of the regression coefficients β_j in the logistic model. We let $\mathbf{X}_i = (X_{i1}, X_{i2},..., X_{ik})$ denote the set of values of the k predictors $X_1, X_2,..., X_k$ specific to individual i. Then the logistic model assumes that the relationship between θ_i and the X_{ij}'s is of the specific form

$$\theta_i = \frac{1}{1 + \exp\left[-\left(\beta_0 + \sum_{j=1}^{k} \beta_j X_{ij} \right) \right]}, \quad i = 1, 2,..., n \tag{23.13}$$

where $\beta_j, j = 0, 1,..., k$, are unknown regression coefficients that must be estimated. (The right side of (23.13) has the same form as the right side of the logistic model form (23.1), where the X_{ij} in (23.13) have been substituted for the X_j in (23.1).)

Now, if we replace θ_i in the likelihood (23.12) with the logistic function expression (23.13), we obtain the so-called *unconditional likelihood function* characterizing standard logistic regression analysis—namely,

$$
L(\mathbf{Y}; \boldsymbol{\beta}) = \prod_{i=1}^{n_1} \left\{ 1 + \exp\left[-\left(\beta_0 + \sum_{j=1}^{k} \beta_j X_{ij} \right) \right] \right\}^{-1}
$$

$$
\times \prod_{i=n_1+1}^{n} \left(\exp\left[-\left(\beta_0 + \sum_{j=1}^{k} \beta_j X_{ij} \right) \right] \left\{ 1 + \exp\left[-\left(\beta_0 + \sum_{j=1}^{k} \beta_j X_{ij} \right) \right] \right\}^{-1} \right)
$$

where $\boldsymbol{\beta} = (\beta_0, \beta_1,...,\beta_k)$. The term *unconditional likelihood* refers to the unconditional probability of obtaining the particular set of data under consideration. More, specifically, the unconditional likelihood function is the joint probability distribution for discrete data or the joint density function for continuous data. A *conditional likelihood,* which we discuss in the next subsection, gives the conditional probability of obtaining the data configuration *actually observed,* given all possible configurations (i.e., permutations) of the data values.

With a little algebraic manipulation, we can verify that an equivalent expression for the preceding likelihood function is[6]

$$
L(\mathbf{Y}; \boldsymbol{\beta}) = \frac{\prod_{i=1}^{n_1} \exp\left(\beta_0 + \sum_{j=1}^{k} \beta_j X_{ij} \right)}{\prod_{i=1}^{n} \left[1 + \exp\left(\beta_0 + \sum_{j=1}^{k} \beta_j X_{ij} \right) \right]}
\tag{23.14}
$$

Because (23.14) is a complex nonlinear function of the elements of $\boldsymbol{\beta}$, maximizing of (23.14) to find the ML estimator $\hat{\boldsymbol{\beta}}$ of $\boldsymbol{\beta}$ must involve using appropriate computer algorithms. Such programs (e.g., SAS's LOGISTIC procedure) also produce the maximized likelihood value $L(\mathbf{Y}; \hat{\boldsymbol{\beta}})$ and the estimated (large-sample) covariance matrix $\hat{\mathbf{V}}(\hat{\boldsymbol{\beta}})$ for a given model, which can then be used to make appropriate statistical inferences.

23-5-2 Conditional ML Estimation

An alternative to using the unconditional likelihood function (23.14) for estimating the elements of $\boldsymbol{\beta} = (\beta_0, \beta_1, \beta_2,..., \beta_k)$ is to employ a conditional likelihood function. The primary reason for using conditional likelihood methods is that unconditional methods can lead to seri-

---

[6]The likelihood (23.14) is based on the responses of *individual* subjects, with (23.10) pertaining to the *i*th subject. In contrast, a categorical data analysis involving the binomial distribution is based on the responses of *groups* of subjects, with, say, Y_i subjects out of n_i in the *i*th group contracting the disease in question. In this situation, the underlying distribution of Y_i is binomial; that is,

$$
\text{pr}(Y_i; \theta_i) = {}_{n_i}C_{Y_i} \theta_i^{Y_i} (1 - \theta_i)^{n_i - Y_i}
$$

and the validity of certain categorical data analyses requires that n_i be fairly large in each group. The effects of small samples on the validity of ML analyses based on (23.14) are briefly discussed in the next subsection.

ously biased estimates of the elements of β when the amount of data available for analysis is not large. Although "not large" is admittedly a vague term, it acknowledges the potential problem of using large-sample-based statistical procedures such as maximum likelihood when the number of parameters to be estimated constitutes a fair proportion of the available data. This is often the situation when data involving *matching* must be analyzed.

In some matched case–control studies, for example, each case (i.e., a person with the disease in question) is matched with one or more controls (i.e., persons not having the disease in question) that have the same values (or are in the same categories) as the cases for the covariates involved in the matching. Analyzing such (matched) data requires that the data be cast into strata corresponding to the matched sets (e.g., pairs) with stratum-specific sample sizes that are small (i.e., reflecting sparse data). In particular, for pair-matched case–control data, each stratum contains only two subjects. A logistic model for analyzing matched data requires that the matching be taken into account by including indicator (i.e., dummy) variables in the model to reflect the matching strata. Thus, the model will have as many parameters as there are matched sets (plus parameters for the exposure variable, any other unmatched variables, and possibly even product terms of unmatched variables with exposure); so the total number of parameters in the model is large relative to the number of subjects in the study. See Kleinbaum (1994), chapter 8, and Kleinbaum, Kupper, and Morgenstern (1982), chapter 24, for further discussion of the principles of matching and modeling.

To see why using a conditional likelihood function is appropriate for matched data, let us consider a *pair-matched case–control study* to assess the effect of a (0–1) exposure variable E on a (0–1) health outcome variable D. Suppose that the matching involves the variables age, race, and sex; then, for each case, a control subject is found who has the same age (or age category), race, and sex as the corresponding case. Suppose, too, that the study involves 100 matched pairs, so the study size n is 200. If no predictor variables were considered in the study other than E and the matching variables age, race, and sex, then a (no-interaction) logistic regression model appropriate for analyzing these data is

$$\text{logit}[\text{pr}(D = 1)] = \beta_0 + \beta_1 E + \sum_{i=1}^{99} \gamma_i V_i \qquad (23.15)$$

where V_i, $i = 1,\ldots,99$, denote a set of dummy variables distinguishing the collection of 100 matched pairs; for example, the V_i may be defined as

$$V_i = \begin{cases} 1 & \text{for the } i\text{th matched set} \\ 0 & \text{otherwise} \end{cases}$$

Model (23.15) allows us to predict case–control status as a function of exposure status (E) and the matching variables age, race, and sex, where the matching variables are incorporated into the model as dummy variables.[7] The number of parameters in model (23.15) is 101. This is an

---

[7]The analysis of a matched-pair study can (equivalently) be carried out without using logistic regression if no variables are controlled other than those involved in the matching. Such an analysis is a "stratified" analysis, and involves using a Mantel-Haenszel test, odds ratio, and confidence interval. Equivalently, the stratified analysis can be carried out using McNemar's test and estimation procedure (see Kleinbaum, Kupper, and Morgenstern (1982), chapter 18).

example of a model whose number of parameters (101) is "large" relative to the number of subjects (200), so use of a conditional likelihood function is appropriate.

If an unconditional likelihood is used to fit model (23.15), the resulting (biased) estimated odds ratio for the exposure effect is the squared value of the estimated odds ratio obtained from using a conditional likelihood; that is, if $\hat{\beta}_U$ and $\hat{\beta}_C$ denote the estimates of β_1 using unconditional and conditional likelihoods, respectively, then

$$\widehat{OR}_U = e^{\hat{\beta}_U} \equiv (\widehat{OR}_C)^2 = e^{2\hat{\beta}_C}$$

Thus, for example, if a pair-matched analysis using a conditional likelihood results in an estimated odds ratio of, say, 3, a corresponding analysis (involving the same model on the same data) using an unconditional likelihood would yield a biased estimate of 3^2, or 9.

Consider the data in the following table, which comes from the "Agent Orange Study" (Donovan, MacLennan, and Adena, 1984), a pair-matched case–control study involving 8502 matched pairs (i.e., $n = 17,004$):

		$D = 0$	
		$E = 1$	$E = 0$
$D = 1$	$E = 1$	2	125
	$E = 0$	121	8254

In this table, the data layout separates the 8502 case–control pairs into four cells depending on whether both the case and the control were exposed (2 *concordant* pairs), the case was exposed and the control was unexposed (125 *discordant* pairs), the case was unexposed and the control was exposed (121 *discordant* pairs), or both the case and the control are unexposed (8254 *concordant* pairs). The D and E variables are defined D = case-control status (1 = baby born with genetic anomaly, 0 = baby without anomaly), and E = father's status (1 = Vietnam vet, 0 = non-Vietnam vet). The matching variables are M_1 = time period of birth, M_2 = mother's age, M_3 = health insurance status, and M_4 = hospital. Since only the matching variables are being controlled, the analysis can be carried out using Mantel-Haenszel (or McNemar) statistics from a stratified analysis (see Kleinbaum, Kupper, and Morgenstern 1982, chapter 17) without the need to perform a logistic regression.

For these data, the Mantel-Haenszel (i.e., McNemar) test of H_0: "No (E, D) association" gives a 1-d.f. chi-square statistic of 0.0650 ($P = 0.80$) and an estimated Mantel-Haenszel odds ratio of 1.033.[8] An equivalent analysis (yielding the same results) based on logistic regression with a conditional likelihood uses the following logistic model:

$$\text{logit}[\text{pr}(D = 1)] = \beta_0 + \beta_1 E + \sum_{i=1}^{8501} \gamma_i V_i \qquad (23.16)$$

---

[8]McNemar's test is computed from the data in the preceding table using only the information from the (125 + 121) discordant pairs. The test statistic is computed as $(125 - 121)^2/(125 + 121) = 0.0650$. The (Mantel-Haenszel) odds ratio is computed as the ratio of discordant pairs: $125/121 = 1.033$.

where the V_i denote dummy variables that distinguish the 8502 matched sets. Model (23.16) contains 8503 parameters, which is a large number (over 50%) relative to the number of subjects in the study ($n = 17,004$).

Summarized output from fitting model (23.16) using a conditional likelihood is presented in Table 23-3. The output indicates that the odds ratio estimate is exp (0.0325) = 1.033 and that the Wald test statistic is 0.0650 with a P-value of 0.80. These are the same results we obtained from the stratified analysis. The output does not show estimated coefficients for β_0 and for the $\{\gamma_i\}$; this is because these parameters, which distinguish the different matching strata, drop out of the conditional likelihood function (given by equation 23.17) for this model, and therefore cannot be estimated.

The output in Table 23-3 was obtained by using SAS's PHREG procedure, a program for fitting Cox's Proportional Hazards (PH) model—a model for assessing survival analysis data, where the response variable, "time to development of disease" (i.e., "time to failure" or "survival time"), is continuous rather than discrete.[9] PHREG was employed because the structure of the conditional likelihood function for logistic regression is a special case of Cox's (1975) partial likelihood, which is used to fit a PH model. In particular, the conditional likelihood function for the general logistic model

$$pr(Y = 1) = \cfrac{1}{1 + \exp\left[-\left(\beta_0 + \sum_{j=1}^{k} \beta_j X_j\right)\right]}$$

takes the form

$$L_C(\mathbf{Y}; \boldsymbol{\beta}) = \cfrac{\prod_{i=1}^{n_1} \exp\left(\beta_0 + \sum_{j=1}^{k} \beta_j X_{ij}\right)}{\sum_u \left(\prod_{l=1}^{n_1} \exp\left(\beta_0 + \sum_{j=1}^{k} \beta_j X_{ulj}\right)\right)} \tag{23.17}$$

where the sum in the denominator is over all partitions of the set $\{1, 2,..., n\}$ into two subsets, the first of which contains n_1 elements. This formula assumes, without loss of generality, that

TABLE 23-3 Edited computer output from SAS's PROC PHREG for logistic regression of the "Agent Orange Data," using conditional maximum likelihood estimation of model (23.16).

Variable	d.f.	Parameter Estimate	Standard Error	Wald Chi-Square	Pr > Chi-Square	Odds Ratio
E	1	0.0325	0.1275	0.0650	0.7987	1.033

[9]Although we will not discuss the statistical methodology known as survival analysis here (see Kleinbaum 1996), estimating survival models such as the Cox PH model requires a maximum likelihood–based procedure to which the general principles discussed in Chapter 22 apply and for which appropriate computer programs (e.g., SAS's PHREG) are available.

the first n_1 of the n subjects actually develop the disease; so X_{ij} denotes the value of variable X_j for the ith of these first n_1 subjects. The X_{ulj} term in the denominator, on the other hand, denotes the value of X_j for the lth person in the uth partition of the data into n_1 cases and $n - n_1$ non-cases. There are $_nC_{n_1} = n!/(n_1)!(n - n_1)!$ such partitions and hence that many terms in the summation.[10]

In words, the conditional likelihood function (23.17) can be considered analogous to a conditional probability. More specifically, if we let $\mathbf{X}_l$ denote the set of predictor values observed on the lth subject, $L_C(\mathbf{Y}; \boldsymbol{\beta})$ is the conditional probability (based on the underlying logistic model assumption) that the first n_1 members of the observed set $\mathbf{X}_1, \mathbf{X}_2, ..., \mathbf{X}_n$ actually go with the n_1 subjects who developed the disease in question, *given* (or conditional on) the observed set $\mathbf{X}_1, \mathbf{X}_2, ..., \mathbf{X}_n$ and *given* the fact that exactly n_1 of the n subjects under study actually developed the disease in question. Expressed in yet another way, (23.17) compares the likelihood of what was actually observed (the numerator) relative to the likelihood of all possible arrangements of the given data set (the denominator). The reason that (23.17) is called a *conditional likelihood* is that it is completely analogous to a conditional probability and is conditional on (arrangements of) the data actually observed.

With regard to data analysis, the conditional likelihood (23.17) is employed just like any other (e.g., unconditional) likelihood used with maximum likelihood procedures: Wald and/or likelihood ratio tests are used to test hypotheses about regression coefficients in the model, and large-sample confidence intervals are constructed using a percentage point of the $N(0,1)$ distribution and estimates of the variances and covariances of the estimated regression coefficients. The computational aspects for a conditional likelihood are, nevertheless, somewhat more involved than for an unconditional likelihood, because of the permutations of the data required to evaluate the denominator in (23.17). For example, if $n = 20$ and $n_1 = 3$, then $_{20}C_3 = 1140$, so the denominator in (23.17) involves a sum of more than a thousand terms. The computational complexity of the conditional likelihood is not a computational handicap for data analysis, however, since many computer packages are available (e.g., SAS, EGRET) for carrying out the required computations.

23-6 An Example of Conditional ML Estimation Involving Pair-matched Data with Unmatched Covariates

In this section, we describe an application of a conditional likelihood in a logistic regression analysis of case–control data involving both matched and unmatched covariates. The data derive from a pair-matched case–control study of the effect of estrogen use on the development of endometrial cancer. This study was conducted on 63 matched pairs of women living in a Los

---

[10]In the likelihood (23.17), the constant term β_0 drops out, since **exp** (β_0) can be factored out of both the numerator and the denominator. For the pair-matched model given by (23.16), the γ_i parameters can be factored out similarly; thus, in addition to β_0, the γ_i coefficients drop out of the conditional likelihood and so do not have to be estimated.

Angeles retirement village from 1971 to 1975 (Breslow and Day 1980; McNeil 1996). The following variables were involved in the study (and are analyzed here):

Outcome variable: D = CASE (endometrial cancer status: 1 = case, 0 = control),

Exposure variable: E = EST (estrogen use: 1 = yes, 0 = no),

Matching variables: M_1 = age, M_2 = marital status, M_3 = date of entry into retirement village,

Covariates not matched on: GALL (gall bladder disease status: 1 = present, 0 = absent).

Table 23-4 summarizes the data compiled for use by SAS's PHREG procedure. Three variables included in this table but not previously mentioned are WGT, STRATUM, and SURVT. The WGT variable is used to combine in a single line of data the information from several pairs that have the same combination of covariate values (including exposure status). For example, the information on lines 7 and 8 of Table 23-4 identifies 21 pairs for which the value of the variable

TABLE 23-4   Data summary for SAS's PHREG from a pair-matched case–control study of the relationship of estrogen use to endometrial cancer status on 63 matched pairs of women living in a Los Angeles retirement village from 1971 to 1975 (Breslow and Day 1980; McNeil 1996).

WGT	STRATUM	CASE	EST	GALL	SURVT
1	1	1	0	0	1
1	1	0	0	0	2
1	2	1	0	0	1
1	2	0	1	0	2
1	3	1	0	0	1
1	3	0	1	1	2
21	4	1	1	0	1
21	4	0	0	0	2
18	5	1	1	0	1
18	5	0	1	0	2
1	6	1	1	0	1
1	6	0	0	1	2
3	7	1	1	0	1
3	7	0	1	1	2
2	8	1	0	1	1
2	8	0	0	0	2
1	9	1	0	1	1
1	9	0	1	0	2
1	10	1	0	1	1
1	10	0	0	1	2
6	11	1	1	1	1
6	11	0	0	0	2
4	12	1	1	1	1
4	12	0	1	0	2
1	13	1	1	1	1
1	13	0	0	1	2
2	14	1	1	1	1
2	14	0	1	1	2

EST is 1 for the case and 0 for the control and for which the value of the variable GALL is 0 for both case and control; each of these 21 case–control pairs has the following layout:

EST GALL

$D = 1$	1	0
$D = 0$	1	0

The WGT variable lets the user avoid listing a separate pair of lines for each case–control pair in the entire data set—provided, of course, that some pairs have the same values for all covariates.[11]

The STRATUM variable distinguishes the different strata. The cases in a given stratum are listed on a separate line from the controls. Thus, the listing shows two lines with the same stratum number; for example, for stratum number 4, the information for the 21 cases in the 21 pairs that have the same values of the covariates is given on one line, followed by a line of information for the 21 controls. In the (PHREG) program, the 21 pairs identified by the two lines for stratum 4 are treated as (21) separate strata, rather than as being *pooled* into one stratum.[12]

The SURVT variable shown in Table 23-4 is needed to tell the survival analysis program (in this case, PHREG) which variable to use for survival time, as required for carrying out a survival analysis. Since we are not actually carrying out a survival analysis here, the purpose of the SURVT variable is to "trick" the survival analysis program into fitting a logistic regression model using a conditional likelihood function. In the table, the value of SURVT is 1 for the cases and 2 for the controls. As a general rule (for carrying out the "trick"), the survival time for the controls (who have "artificially" survived) must be longer than the survival time for the cases (who have "artificially" failed).

In analyzing the data of Table 23-4, we first describe the results obtained when the control of the variable GALL is ignored. The analysis then simplifies to a stratified analysis for matched data with no unmatched covariates. The corresponding frequencies describing the numbers of concordant and discordant case–control pairs are given in Table 23-5.

TABLE 23-5   Frequencies of case–control pairs by exposure status, ignoring the variable GALL, in the case–control study of the relationship of estrogen use to endometrial cancer status.

		CASE = 0	
		EST = 1	EST = 0
CASE = 1	EST = 1	27	29
	EST = 0	3	4

63

---

[11]The use of such a WGT variable is particularly advantageous for pair matching without unmatched covariates; e.g., in the Agent Orange study data (described in section 23-5), only four lines of data are required, corresponding to the two types of concordant pairs and the two types of discordant pairs.

[12]Pooling is a reasonable alternative (and may lead to more precise estimates) for pairs that have the same values on all matching variables.

For the data in Table 23-5, the Mantel-Haenszel test statistic is computed as

$$\chi^2_{MH} = \frac{(29-3)^2}{(29+3)} = 21.125 \quad (P < .01)$$

and the Mantel-Haenszel odds ratio estimated is computed as

$$\widehat{mOR} = \frac{29}{3} = 9.67$$

These results indicate a strong and significant association between estrogen use and endometrial cancer, based on these study data. However, this analysis does not control for the variable GALL; we now have to control for GALL, using a logistic model with a conditional likelihood.

Table 23-6 summarizes the PHREG output for fitting the following logistic model:

$$\text{logit}[\text{pr}(\text{CASE} = 1)] = \beta_0 + \beta_1(\text{EST}) + \beta_2(\text{GALL}) + \sum_{i=1}^{62} \gamma_i V_i \qquad (23.18)$$

where the V_i, $i = 1,\dots, 62$, denote 62 dummy variables that distinguish the 63 matched pairs. Model (23.18) is a no-interaction model that estimates the effect of the exposure variable EST on the outcome CASE, controlling for the confounding effects of the unmatched covariate GALL and the matching variables reflected by the V_i. If we wanted to consider the possibility of interaction between EST and GALL, we could add the product term EST $\times$ GALL to the preceding model. We leave the assessment of interaction to the interested reader, however, and report only on results for the no-interaction model (23.18).

TABLE 23-6   Edited output from SAS's PHREG[13] for model (23.18) from the pair-matched case–control study of the relationship of estrogen use to endometrial cancer status on 63 matched pairs of women living in a Los Angeles retirement village (1971–1975).

Variable	d.f.	Parameter Estimate	Standard Error	Wald Chi-Square	Pr > Chi-Square	Odds Ratio
EST	1	2.209	0.610	13.127	0.000	9.107
GALL	1	0.695	0.616	1.128	0.259	2.003

---

[13]For pair-matched data, an alternative method for fitting model (23.18) with a conditional likelihood uses an unconditional logistic regression program (e.g., SAS's LOGISTIC) with a no-intercept (i.e., noint) option. This approach requires the data file to contain a line of information for each case–control pair (rather than for each subject), with independent variables defined as difference scores between corresponding covariate values (including exposure) for each case–control pair. The dependent variable is defined as a column of 1's. This alternative yields the same output shown in Table 23-6 because the likelihood function being maximized by the unconditional (LOGISTIC) program is a constant multiple of the likelihood being maximized by the conditional (PHREG) program. The model being fit by the unconditional program is identical to model (23.18), although it can be stated in terms of difference scores as follows:

$$\text{logit}[\text{pr}(\text{OUT} = 1)] = \beta_1 * (\text{ESTD}) + \beta_2 * (\text{GALLD})$$

where OUT denotes the dependent variable (all 1's), and ESTD and GALLD denote variables for the difference scores corresponding to variables EST and GALL, respectively.

From Table 23-6, the odds ratio for the effect of EST, controlling for GALL and the matching variables (not shown in the table), is estimated as exp (2.209) = 9.107. The Wald statistic for testing whether this odds ratio differs significantly from one is given by

$$\chi^2_{1\,d.f.} = \left(\frac{2.209}{0.610}\right)^2 = 13.127 \quad (P = 0.000)$$

which is highly significant. The estimated odds ratio and corresponding test statistic previously obtained from a stratified analysis that ignored the variable GALL were 9.67 and 21.125, respectively, versus 9.107 and 13.127 when controlling for GALL. Nevertheless, both analyses lead to the same conclusion: estrogen use has a strong effect on the development of endometrial cancer. We can also compute a confidence interval for the (adjusted) odds ratio, using the output of Table 23-6. To obtain a 95% confidence interval for the odds ratio, we calculate

$$\exp [2.209 \pm 1.96 \times (0.610)]$$

which yields lower and upper limits of 2.755 and 30.102, respectively.

Table 23-6 does not provide estimates of the coefficients of the V_i variables that control for the matching. This is because these coefficients drop out of the conditional likelihood (23.17) and therefore cannot be estimated. Although the V_i variables are not explicitly specified in the data listing given by Table 23-4, the matched pairs are identified in the listing by the STRATUM variable, which is used by the (PHREG) program to control for the matching.

23-7 Summary

This chapter described the key features of logistic regression analysis, the most popular regression technique available for modeling dichotomous dependent variables. The *logistic model* is defined as a probability for the occurrence of one of two possible outcomes, using the following formula:

$$\mathrm{pr}(Y = 1) = \frac{1}{1 + \exp\left[-\left(\beta_0 + \sum_{j=1}^{k} \beta_j X_j\right)\right]}$$

The most important reason for the popularity of the logistic model is that the right-hand side of the preceding expression ensures that the predicted value of Y will always lie between 0 and 1. Using the logit form of the logistic model defined by the expression

$$\mathrm{logit}[\mathrm{pr}(Y = 1)] = \beta_0 + \sum_{j=1}^{k} \beta_j X_j$$

we can estimate an odds ratio: the general formula for an odds ratio comparing two specifications of the set of predictors $\mathbf{X_A}$ and $\mathbf{X_B}$ is

$$\mathrm{OR}_{\mathbf{X_A}\,vs.\,\mathbf{X_B}} = e^{\sum_{j=1}^{k} (X_{Aj} - X_{Bj})\beta_j}$$

Maximum likelihood estimates and associated standard errors of the regression coefficients in a logistic model are typically obtained by using computer packages for logistic regression. These statistics can then be used to obtain numerical values for estimated adjusted odds ratios, to test hypotheses, and to obtain confidence intervals for population odds ratios based on standard maximum likelihood techniques.

When performing a logistic regression analysis, we must decide between two alternative ML procedures, called *unconditional* and *conditional*. The key distinction between the two involves whether the number of parameters in the model constitutes a "large" proportion of the total study size. If so, the situation is typical of matched data, requiring the use of the conditional approach to ensure validity of the odds ratio estimate.

The chapter included examples of using both the unconditional approach (based on SAS's LOGISTIC procedure) and the conditional approach (based on SAS's PHREG procedure). The PHREG program is a survival analysis program for fitting a Cox Proportional Hazards model. Since the conditional likelihood for logistic regression has the same structure as the partial likelihood used for the Cox model, we can "trick" the survival analysis program (PHREG) into carrying out the conditional estimation procedure by adding appropriate STRATUM and SURVT variables to the data set.

Problems

1. A researcher was interested in determining risk factors for high blood pressure (hypertension) among women. Data from a sample group of 680 women were collected. The following table gives the observed relationship between hypertension and smoking:

	Hypertension	
	Yes	No
Smokers	$a = 28$	$b = 271$
Nonsmokers	$c = 13$	$d = 368$
	41	639

Let π be the probability of having hypertension, and suppose that the researcher used logistic regression to model the relationship between smoking and hypertension.

One model the researcher considered is

$$\log_e \frac{\pi}{1 - \pi} = \beta_0 + \beta_1 X_1 \qquad \text{(model 0)}$$

where X_1 = smoking status (1 = smoker, 0 = nonsmoker). For this model, the estimated odds ratio for the effect of smoking on hypertension status can be computed by using the simple formula for an odds ratio in a 2×2 table—namely, ad/bc, where a, b, c, and d are the cell frequencies in the preceding table.

a. Based on this information, compute the point estimate $\hat{\beta}_1$ of β_1.

The researcher ultimately decided to use the following logistic regression model:

$$\log_e \frac{\pi}{1 - \pi} = \beta_0 + \beta_1 X_1 + \beta_2 X_2 + \beta_3 X_1 X_2 \qquad \text{(model 1)}$$

where X_1 = smoking status (1 = smoker, 0 = nonsmoker) and X_2 = age. The following information was obtained:

Parameter	Estimate	SE
β_0	-2.8	1.2
β_1	0.706	0.311
β_2	0.0004	0.0001
β_3	0.0006	0.0003

$-2 \ln \hat{L} = 303.84$

The estimated variance covariance matrix for $\hat{\beta}_0, \hat{\beta}_1, \hat{\beta}_2$, and $\hat{\beta}_3$ is estimated as follows:

$$
\begin{array}{cccc}
\hat{\beta}_0 & \hat{\beta}_1 & \hat{\beta}_2 & \hat{\beta}_3 \\
\begin{bmatrix}
1.44 & 0.0001 & 0.0001 & 0.0001 \\
 & 0.0967 & 0.1 \times 10^{-8} & 2.0 \times 10^{-8} \\
 & & 1.0 \times 10^{-8} & 3.0 \times 10^{-8} \\
 & & & 9.0 \times 10^{-8}
\end{bmatrix}
\end{array}
$$

b. What is the estimated logistic regression model for the relationship between age and hypertension for nonsmokers?

c. What is a 20-year-old smoker's predicted probability of having hypertension?

d. Estimate the odds ratio comparing a 20-year-old smoker to a 21-year-old smoker. Interpret this estimate.

e. Find a 95% confidence interval for the population odds ratio being estimated in part (d).

f. The log likelihood $(-2 \ln \hat{L})$ for the model consisting of the intercept, age, and smoking status was 308.00. Use this information plus other information provided earlier to perform a likelihood ratio test of the null hypothesis that $\beta_3 = 0$ in model 1.

2. A five-year follow-up study on 600 disease-free subjects was carried out to assess the effect of a (0–1) exposure variable E on the development or not of a certain disease. The variables AGE (continuous) and obesity status (OBS), the latter a (0–1) variable, were determined at the start of follow-up and were to be considered as control variables in analyzing the data. For this study, answer the following questions.

a. State the logit form of a logistic model that assesses the effect of the (0, 1) exposure variable E, controlling for the confounding effects of AGE and OBS and for the interaction effects of AGE with E and OBS with E.

b. Given the model described in part (a), give a formula for the odds ratio for the exposure-disease relationship that controls for the confounding and interactive effects of AGE and OBS.

c. Use the formula described in part (b) to derive an expression for the estimated odds ratio for the exposure–disease relationship that considers both confounding and interaction when AGE = 40 and OBS = 1.

d. Give a formula for a 95% confidence interval for the population adjusted odds ratio being estimated in part (c). In stating this confidence interval formula, write out the formula for the estimated variance of the log of the estimated adjusted odds ratio in terms of estimated variances and covariances of appropriate regression coefficients.

e. State the null hypothesis (in terms of model parameters) for simultaneously testing for no interaction of AGE with E and of OBS with E.

f. State the formula for the likelihood ratio statistic that tests the null hypothesis described in part (e). What is the large-sample distribution of this statistic, including its degrees of freedom, under the null hypothesis?

g. Give the formula for the Wald statistic for testing the null hypothesis described in part (e). What is the large-sample distribution of this statistic under the null hypothesis?

h. Assuming that both the likelihood ratio statistic described in part (f) and the Wald statistic described in part (g) are nonsignificant, state the logit form of the no-interaction model that is appropriate to consider at this point in the analysis.

i. For the no-interaction model given in part (h), give the formula for the (adjusted) odds ratio for the effect of exposure, controlling for AGE and OBS.

j. For the no-interaction model given in part (h), give the formula for a 95% confidence interval for the adjusted odds ratio for the effect of exposure, controlling for AGE and OBS.

k. Describe the Wald test for the effect of exposure, controlling for AGE and OBS, in the no-interaction model given in part (h). Include a description of the null hypothesis being tested, the form of the test statistic, and its distribution under the null hypothesis.

3. A study was conducted on a sample of 53 patients presenting with prostate cancer who had also undergone a laparotomy to ascertain the extent of nodal involvement (Collett 1991). The result of the laparotomy is a binary response variable, where 0 signifies the absence of, and 1 the presence of, nodal involvement. The purpose of the study was to determine variables that could be used to forecast whether the cancer has spread to the lymph nodes. Five predictor variables were considered, each measurable without surgery. The following printout provides information for fitting two logistic models based on these data. The five predictor variables were: age of patient at diagnosis, level of serum acid phosphatase, result of an X-ray examination (0 = negative, 1 = positive), size of the tumor as determined by a rectal exam (0 = small, 1 = large), and summary of the pathological grade of the tumor as determined from a biopsy (0 = less serious, 1 = more serious).

a. Which method of estimation do you think was used to obtain estimates of parameters for both models—conditional or unconditional ML estimation? Explain briefly.

b. For model I, test the null hypothesis of no effect of X-ray status on response. State the null hypothesis in terms of an odds ratio parameter; give the formula for the test statistic; state the distribution of the test statistic under the null hypothesis; and finally, carry out the test, using the printout data for model I. Is the test significant?

Model I

Variable Name	Coeff	StErr	p-value	$\widehat{OR}$	95% CI	
0 constant	0.057	3.460	0.987			
1 age	−0.069	0.058	0.232	0.933	0.833	1.045
2 acid	2.434	1.316	0.064	11.409	0.865	150.411
3 xray	2.045	0.807	0.011	7.731	1.589	37.611
4 tsize	1.564	0.774	0.043	4.778	1.048	21.783
5 tgrad	0.761	0.771	0.323	2.141	0.473	9.700

d.f.: 47   Dev: 48.126*

* The deviance statistic is a likelihood ratio statistic that compares a current model of interest to the baseline model containing as many parameters as there are data points. The difference in deviance statistics obtained for two (hierarchically ordered) models being compared is equivalent to the difference in log-likelihood statistics for each model. Thus a likelihood ratio test can equivalently be carried out by using differences in deviance statistics. See section 24-5 for further details about deviances.

Model II

Variable Name	Coeff	StErr	p-value	$\widehat{OR}$		95% CI
0 constant	2.928	4.044	0.469			
1 age	−0.101	0.067	0.132	0.904	0.792	1.031
2 acid	1.462	1.559	0.349	4.314	0.203	91.678
3 xray	−19.171	20.027	0.338	0.000	0.000	5.259×10^8
4 tsize	1.285	0.894	0.151	3.613	0.626	20.845
5 tgrad	0.421	0.923	0.648	1.523	0.250	9.291
7 age*xray	0.257	0.278	0.356	1.292	0.750	2.228
8 acid*xray	7.987	9.197	0.385	2943.008	0.000	1.982×10^{11}
9 tsiz*xray	2.236	2.930	0.445	9.356	0.030	2920.383
10 tgrd*xray	−0.624	2.376	0.793	0.536	0.005	56.403

d.f.: 43   Dev: 44.474

c. Using the printout data for model I, compute the point estimate and 95% confidence interval for the odds ratio for the effect of X-ray status on response for a person of age 50, with phosphatase acid level .50, tsize equal to 0, and tgrad equal to 0.

d. State the logit form of model II given in the accompanying printout.

e. Using the results for model II, give an expression for the odds ratio that describes the effect of X-ray status on the response, controlling for age, phosphatase acid level, tsize, and tgrad. Using this expression, compute the odds ratio of the effect of X-ray status on the response for a person of age 50, with phosphatase acid level .50, tsize equal to 0, and tgrad equal to 0.

f. For model II, give an expression for the estimated variance of the estimated adjusted odds ratio relating X-ray status to response for a person of age 50, with phosphatase acid level .50, tsize equal to 0, and tgrad equal to 0. Write this expression for the estimated variance in terms of variances and covariances obtained from the variance-covariance matrix.

g. Using your answer to part (f), give an expression for a 95% confidence interval for the odds ratio relating X-ray status to response, for a person of age 50, with phosphatase acid level .50, tsize equal to 0, and tgrad equal to 0.

h. For model II, carry out a "chunk" test for the combined interaction of X-ray status with each of the variables age, phosphatase acid level, tsize, and tgrad. State the null hypothesis in terms of one or more model coefficients; give the formula for the test statistic and its distribution and degrees of freedom under the null hypothesis; and report the P-value. Is the test significant?

i. If you had to choose between model I and model II, which would you pick as the "better" model? Explain.

Assume that the following model has been defined as the initial model to be considered, using a strategy to obtain a "best" model:

$$logit[pr(Y = 1)] = \alpha + \beta_1(xray) + \beta_2(age) + \beta_3(acid) + \beta_4(tsize) + \beta_5(tgrad)$$
$$+ \beta_6(age*acid) + \beta_7(xray*age) + \beta_8(xray*acid)$$
$$+ \beta_9(xray*tsize) + \beta_{10}(xray*tgrad) + \beta_{11}(xray*age*acid)$$

    j. For this model—and considering the variable xray to be the only "exposure" variable of interest, with the variables age, acid, tsize, and tgrad considered for control—which β's in the above model are coefficients of (potential) effect modifiers? Explain briefly.

    k. Assume that the only interaction term found significant is the product term xray*age. What variables are left in the model at the end of the interaction stage?

    l. Based on the (reduced) model described in part (k) (where the only significant interaction term is xray*age), what expression for the odds ratio describes the effect of xray on nodal involvement status?

    m. Suppose that, as a result of confounding and precision assessment, the variables age* acid, acid, tgrad, and tsize are dropped from the model described in part (k). What is your final model, and what expression for the odds ratio describes the effect of xray on nodal involvement status?

References

Breslow, N., and Day, N. 1980. *Statistical Methods in Cancer Research. Vol. I: The Analysis of Case-Control Studies.* Lyon, France: IIARC Scientific Publications, No. 32.

Collett, D. 1991. *Modeling Binary Data.* London: Chapman & Hall.

Cox, D. R. 1975. "Partial Likelihood." *Biometrika* 62: 269–76.

Dantes, H. G.; Koopman, J. S.; et al. 1988. "Dengue Epidemics on the Pacific Coast of Mexico." *International Journal of Epidemiology* 17(1): 178–86.

Donovan, J. W.; MacLennan, R.; and Adena, M. 1984. "Vietnam Service and the Risk of Congenital Anomalies; a Case-Control Study," *Medical Journal of Australia* 140: 394–97.

Hosmer, D. W., and Lemeshow, S. 1989. *Applied Logistic Regression.* New York: John Wiley & Sons.

Kleinbaum, D. G. 1994. *Logistic Regression—A Self-Learning Text.* New York: Springer-Verlag.

———. 1996. *Survival Analysis—A Self-Learning Text.* New York: Springer-Verlag.

Kleinbaum, D. G.; Kupper, L. L.; and Morgenstern, H. 1982. *Epidemiologic Research—Principles and Quantitative Methods.* New York: Van Nostrand Reinhold.

McNeil, D. 1996. *Epidemiologic Research Methods.* New York: John Wiley & Sons.

24

Poisson Regression Analysis

24-1 Preview

Poisson regression analysis is a regression technique available for modeling dependent variables that describe *count* (i.e., *discrete*) *data.* The purpose of this chapter is to describe the Poisson regression model form and several key features of the model, particularly how a rate ratio can be estimated using Poisson regression. We also use real-life data to demonstrate how Poisson regression may be applied. Maximum likelihood procedures are used to estimate the model parameters of a Poisson regression model. Therefore, the general principles and inference-making procedures described in Chapter 22 on ML estimation carry over directly to the likelihood functions appropriate for Poisson regression analysis.

24-2 The Poisson Distribution

The methodology of Poisson regression analysis assumes that the underlying distribution of the response variable Y under consideration is Poisson. The Poisson probability distribution with parameter μ is given by the formula

$$\text{pr}(Y; \mu) = \frac{\mu^Y e^{-\mu}}{Y!}, \quad Y = 0, 1, \ldots, \infty \tag{24.1}$$

Theoretically, a Poisson random variable can take any nonnegative integer value. From (24.1), for example, the probability that Y takes the value 10 is

$$\text{pr}(Y = 10; \mu) = \frac{\mu^{10} e^{-\mu}}{10!} = \frac{\mu^{10} e^{-\mu}}{3,628,800}$$

This probability changes as a function of the value of μ.

The Poisson distribution is often used to model the occurrence of rare events, such as the number of new cases of lung cancer developing in some population over a certain period of time or the number of automobile accidents occurring at a certain location per year. The Poisson distribution possesses an interesting statistical attribute: $E(Y) = \text{Var}(Y) = \mu$ when Y has the Poisson distribution (24.1).

24-3 An Example of Poisson Regression

To illustrate the utility of Poisson regression analysis, let us consider a data analysis situation where Poisson regression has been used quite successfully. Table 24-1 gives nonmelanoma skin cancer data for women stratified by age in two metropolitan areas: Dallas–Ft. Worth and Minneapolis–St. Paul (Scotto, Kopf, and Urbach 1974). In this example, the dependent variable Y is a count, the number of cases of skin cancer. Since eight age strata and two metropolitan areas are involved, we let Y_{ij} denote the count for the ith age stratum in the jth area, where i ranges from 1 to 8 for the eight age groups and $j = 0$ (Minneapolis–St. Paul) or $j = 1$ (Dallas–Ft. Worth). We also let ℓ_{ij} denote the population size for the ith age stratum in the jth area. For these data, one analysis goal is to determine whether the risk for skin cancer adjusted for age is higher in one metropolitan area than in the other. The term *risk* in this context essentially means the probability associated with an event of interest—for example, the probability of developing skin cancer. We will let λ_{ij} denote the true (i.e., population) risk in the (i, j)th group. The ratio

$$RR_i = \frac{\lambda_{i1}}{\lambda_{i0}}$$

is commonly referred to as the *relative risk* or *risk ratio,* which in this case is the population risk for Dallas–Ft. Worth in the ith age group *divided by* the risk for Minneapolis–St. Paul in the ith age group. If $RR_i = 1$, then the population risks are the same in the ith age group; if $RR_i > 1$, however, then the risk for Dallas–Ft. Worth is higher than the risk for Minneapolis–St. Paul in this age group.

As indicated in the last column of Table 24-1, the estimated risk ratios in all age groups are greater than 1, which clearly suggests that the Dallas–Ft. Worth area has a higher overall incidence of skin cancer than Minneapolis–St. Paul. Our objective here is to use Poisson regression analysis to determine whether such a data pattern is statistically significant and to obtain an estimate of the overall risk ratio that adjusts for the effect of age.

TABLE 24-1   Comparison of incidence of nonmelanoma skin cancer among women in Minneapolis–St. Paul and Dallas–Ft. Worth.

Age Group (yr)	Minneapolis–St. Paul		Dallas–Ft. Worth		Estimated Risk Ratio*
	No. of Cases	Population Size	No. of Cases	Population Size	
15–24	1	172,675	4	181,343	3.81
25–34	16	123,065	38	146,207	2.00
35–44	30	96,216	119	121,374	3.14
45–54	71	92,051	221	111,353	2.57
55–64	102	72,159	259	83,004	2.21
65–74	130	54,722	310	55,932	2.33
75–84	133	32,185	226	29,007	1.89
85 +	40	8,328	65	7,538	1.80

Source: Adapted from Scotto, Kopf, and Urbach (1974).

* With Minneapolis–St. Paul as the reference group.

Where does the Poisson distribution enter into this problem? Notice, first, that the count Y_{ij} is, in theory, a binomial random variable with mean $\mu_{ij} = \ell_{ij}\lambda_{ij}$. We know from statistical theory that the binomial distribution can be approximated by a Poisson distribution with the same mean, provided that the population size is large and the binomial probability parameter is small, so that the expected binomial count (i.e., the mean μ) is small relative to the population size, In other words, the Poisson distribution provides a good approximation to the binomial distribution for rare events. The data in Table 24-1 satisfy this requirement reasonably well, since all stratum-specific counts are quite small relative to the corresponding population sizes.

To develop a Poisson regression model for the above situation, we need to define a model for the expected number of skin cancer cases, $E(Y_{ij})$, in terms of the variables of interest. Here, two underlying variables are of interest: "age" and "area." Since "age" has been categorized into eight groups, we will use seven dummy variables to index them.[1] The variable "area," which contains two categories, requires only one dummy variable. Thus, one possible model for the expected number of skin cancer cases in the (i, j)th group can be written as

$$E(Y_{ij}) = \mu_{ij} = \ell_{ij}\lambda_{ij}, \qquad i = 1, 2,\ldots,8; \quad j = 0, 1$$

where

$$\ln \lambda_{ij} = \alpha + \sum_{k=1}^{7} \alpha_k U_k + \beta E$$

with

$$U_k = \begin{cases} 1 & \text{if } k = i, \\ 0 & \text{otherwise} \end{cases} \qquad k = 1, 2,\ldots, 7$$

$$E = \begin{cases} 1 & \text{if } j = 1 \quad \text{(Dallas–Ft. Worth)} \\ 0 & \text{if } j = 0 \quad \text{(Minneapolis–St. Paul)} \end{cases}$$

Using this model, we can write the risks λ_{ij} in terms of the parameters α_i and β to obtain

$$\ln \lambda_{i0} = \alpha + \alpha_i \qquad \text{and} \qquad \ln \lambda_{i1} = \alpha + \alpha_i + \beta, \qquad i = 1, 2,\ldots, 7$$

and for $i = 8$,

$$\ln \lambda_{80} = \alpha \qquad \text{and} \qquad \ln \lambda_{81} = \alpha + \beta$$

since $U_k = 0$, $k = 1, 2,\ldots, 7$ for $i = 8$. Hence,

$$\ln \lambda_{i1} - \ln \lambda_{i0} = (\alpha + \alpha_i + \beta - \alpha - \alpha_i) = \beta, \qquad i = 1, 2,\ldots, 7$$

Also,

$$\ln \lambda_{81} - \ln \lambda_{80} = (\alpha + \beta - \alpha) = \beta$$

In other words,

$$RR_i = \frac{\lambda_{i1}}{\lambda_{i0}} = \mathbf{exp}\left[\ln\left(\frac{\lambda_{i1}}{\lambda_{i0}}\right)\right] = \mathbf{exp}[\ln \lambda_{i1} - \ln \lambda_{i0}] = \mathbf{exp}[\beta] = e^{\beta}, \quad i = 1, 2,\ldots, 8.$$

Thus, using this model, we can estimate the risk ratio for any age group by fitting the model, estimating the coefficient of the E-variable, and then exponentiating this estimate. Since the

---

[1] Alternatively, the model can be defined by using eight dummy variables for "age" and one dummy variable for "area"; when eight dummy variables are used for "age," using a constant term is redundant. The eight-dummy-variable alternative was used in the published analysis of this data set.

estimated risk ratio $e^{\hat{\beta}}$ is independent of i, we can interpret $e^{\hat{\beta}}$ as being an estimate of an overall risk ratio adjusted for age.

The example just described illustrates the type of model used in performing a Poisson regression analysis. In general, instead of having two variables (like age and area) to consider, we may have several (say k) predictor variables $x_1, x_2,\ldots, x_k$ to examine. Nevertheless, the general method of fitting a Poisson regression model is still to use the Poisson model formulation to derive a likelihood function that can then be maximized so that parameter estimates, estimated standard errors, maximized likelihood statistics, and other information can be produced. Since packaged programs can now carry out such analyses, a user need only specify the model to be fit; the program then determines the likelihood function, maximizes it, and computes relevant statistics. We shall return later to the same example to illustrate methods of Poisson regression analysis numerically.

The preceding example (strictly speaking) involves a model for estimating the *risk* of developing a disease. A more general and popular application of Poisson regression involves modeling *failure rates* for different subgroups of interest. The estimated failure rate, or more simply the estimated rate, is generally defined as

$$\hat{\lambda} = \frac{Y}{\ell}$$

where Y is the observed count of health failures (e.g., the number of cases of skin cancer or the number of new cases of heart disease) for a subgroup of interest, and ℓ denotes the accumulated length of (disease-free) follow-up time for all persons in the subgroup. Thus, $\hat{\lambda}$ measures the number of failures relative to the total amount of follow-up time for all persons in a given subgroup. If, for example, the data in Table 24-1 were based on a one-year follow-up study of the Minneapolis–St. Paul and Dallas–Ft. Worth populations, then the numbers in the table giving population size might be considered as *person-years* of follow-up time for the age–area subgroups. The ratio of two rates (e.g., $\lambda_{i1}/\lambda_{i0}$) is commonly referred to as a *rate ratio*. Other terms used are *incidence density ratio* (abbreviated *IDR*) and *hazard ratio*.

24-4 Poisson Regression: General Considerations

We are now ready to describe the general Poisson regression analysis framework. The dependent variable Y is, as already mentioned, typically a count of health failures obtained for each of a number of subgroups that are described by a set of predictor variables $X_1, X_2,\ldots, X_k$. For subgroup i, $i = 1, 2,\ldots, n$, let Y_i denote the observed number of failures, and let ℓ_i denote the total length of follow-up time for all persons in that subgroup. Let $\mathbf{X}_i = (X_{i1}, X_{i2},\ldots, X_{ik})$ denote the set of values of $X_1, X_2,\ldots, X_k$ specific to subgroup i, let $\boldsymbol{\beta} = (\beta_0, \beta_1,\ldots, \beta_k)$ be a set of unknown parameters, and let $\lambda(\mathbf{X}_i, \boldsymbol{\beta})$ denote some specific function of $\mathbf{X}_i$ and $\boldsymbol{\beta}$ (e.g., $\exp(\beta_0 + \sum_{j=1}^{k}\beta_j X_{ij})$) that represents the failure rate for subgroup i (i.e., $\lambda(\mathbf{X}_i, \boldsymbol{\beta})$ measures the rate at which failures occur per unit of follow-up time). Then the expected number of failures in the ith subgroup is

$$E(Y_i) = \mu_i = \ell_i\, \lambda(\mathbf{X}_i, \boldsymbol{\beta}), \qquad i = 1, 2,\ldots, n \tag{24.2}$$

where Y_i is a Poisson random variable. It is required that $\lambda(\mathbf{X}_i, \boldsymbol{\beta}) > 0$.[2]

---

[2]Formula (24.2) is analogous (but not equivalent) to the formula for the mean $\mu = np$ of a binomial random variable; here, ℓ_i is similar to n, and $\lambda(\mathbf{X}_i, \boldsymbol{\beta})$ is similar to p.

Under the assumption that Y_i is Poisson with mean μ_i,[3] so that

$$\text{pr}(Y_i; \mu_i) = \frac{\mu_i^{Y_i} e^{-\mu_i}}{Y_i!}, \qquad i = 1, 2, \ldots, n \tag{24.3}$$

it follows from (24.2) and (24.3) that

$$\text{pr}(Y_i; \boldsymbol{\beta}) = \frac{[\ell_i \lambda(\mathbf{X}_i, \boldsymbol{\beta})]^{Y_i} e^{-\ell_i \lambda(\mathbf{X}_i, \boldsymbol{\beta})}}{Y_i!} \tag{24.4}$$

where $Y_i = 0, 1, \ldots, \infty$ and $i = 1, 2, \ldots, n$.

Note that the only real conceptual difference between Poisson regression and standard multiple regression is that the former involves the Poisson distribution whereas the latter involves the normal distribution. In each instance, the analysis goal is the same—namely, to fit to the data a regression equation that will accurately model $E(Y)$ as a function of a set of predictor variables $X_1, X_2, \ldots X_k$.[4] This is exactly what expression (24.2) is saying!

In the most general sense, then, regression analysis pertains to modeling the mean of the dependent variable under consideration as a function of certain predictor variables. The form of likelihood function that is used to estimate the regression coefficient set $\boldsymbol{\beta}$ is determined by the assumptions made about the distribution of that dependent variable.

As we did earlier to obtain the likelihood function (22.12), let us assume that $Y_1, Y_2, \ldots, Y_n$ constitute a mutually independent set of Poisson random variables, with Y_i having the probability distribution (24.4). Then, the *likelihood function for Poisson regression analysis* is of the general form

$$
\begin{aligned}
L(\mathbf{Y}; \boldsymbol{\beta}) &= \prod_{i=1}^{n} \text{pr}(Y_i; \boldsymbol{\beta}) = \prod_{i=1}^{n} \left\{ \frac{[\ell_i \lambda(\mathbf{X}_i, \boldsymbol{\beta})]^{Y_i} e^{-\ell_i \lambda(\mathbf{X}_i, \boldsymbol{\beta})}}{Y_i!} \right\} \\[2ex]
&= \frac{\left\{ \prod_{i=1}^{n} [\ell_i \lambda(\mathbf{X}_i, \boldsymbol{\beta})]^{Y_i} \right\} \exp\left[-\sum_{i=1}^{n} \ell_i \lambda(\mathbf{X}_i, \boldsymbol{\beta}) \right]}{\prod_{i=1}^{n} Y_i!}
\end{aligned}
\tag{24.5}
$$

where $E(Y_i) = \mu_i = \ell_i \lambda(\mathbf{X}_i, \boldsymbol{\beta})$, $i = 1, 2, \ldots, n$.

To utilize the likelihood function (24.5) in practice, the investigator must specify a particular form for the rate function $\lambda(\mathbf{X}_i, \boldsymbol{\beta})$. Such a specification should be based on the process under study

---

[3]That the Poisson distribution is useful for modeling certain types of health count data can be loosely argued on the basis of the well-known Poisson approximation to the binomial distribution (see, e.g., Remington and Schork (1985), chapter 5). If $Y \frown \text{Bin}(n, \pi)$, and if n is large and π is very small, then $Y \frown \text{Poi}(\mu = n\pi)$. For many health outcomes (e.g., the development of a rare disease), the length ℓ_i of follow-up time (analogous to n) is large, and the rate $\lambda(\mathbf{X}_i, \boldsymbol{\beta})$ of occurrence of the health outcome in question (analogous to π) is small, thus suggesting the Poisson model.

[4]If, in the likelihood (22.12) of Chapter 22 we replace $E(Y_i) = \beta_0 + \beta_1 X_i$ with, say, $E(Y_i) = \beta_0 + e^{\beta_1 X_i}$, then we change from a *linear* (in the coefficients) model to a *nonlinear* model, and hence from a *linear* regression analysis to a *nonlinear* one. The major effect of this change is that we have to solve a set of nonlinear (as opposed to linear) likelihood equations in the β's. This solution generally requires some sort of computer-assisted iteration procedure.

and on previous knowledge of and experience with the relationships among the variables under consideration. Examples of possible choices for $\lambda(\mathbf{X}_i, \boldsymbol{\beta})$ are $e^{\lambda_i^*}$ when $\lambda_i^* = \beta_0 + \sum_{j=1}^{k} \beta_j X_{ij}$, λ_i^* when $\lambda_i^* > 0$, and $\ln \lambda_i^*$ when $\lambda_i^* > 1$.

Recall that the maximum likelihood estimators $\hat{\beta}_0, \hat{\beta}_1, \ldots, \hat{\beta}_k$ of $\beta_0, \beta_1, \ldots, \beta_k$ are obtained from (24.5) as the solutions of the $k + 1$ equations

$$\frac{\partial}{\partial \beta_j} [\ln L(\mathbf{Y}; \boldsymbol{\beta})] = 0 \qquad j = 0, 1, \ldots, k \tag{24.6}$$

The solution to the set of ML equations given by (24.6) must generally be obtained by a computer-based iteration procedure. Frome (1983) discusses the use of algorithms for solving the system of equations (24.6). In particular, he argues for the use of a computational algorithm referred to as *iteratively reweighted least squares* (IRLS).[5] Several statistical packages, such as SAS (using PROC GENMOD), can be utilized to find the ML estimator $\hat{\boldsymbol{\beta}}$ of $\boldsymbol{\beta}$ based on the likelihood (24.5). In addition, the estimated covariance matrix $\hat{\mathbf{V}}(\hat{\boldsymbol{\beta}})$ of $\boldsymbol{\beta}$, measures of goodness of fit of the model under consideration, and certain regression diagnostic statistics (i.e., indices useful for detecting influential observations and multicollinearity) can be obtained as part of the computer output.

For an application of the above procedures, we return to the data in Table 24-1 describing nonmelanoma skin cancer data for women stratified by age in Minneapolis–St. Paul and Dallas–Ft. Worth (adapted from Scotto, Kopf, and Urbach (1974), and reanalyzed by Frome and Checkoway (1985)). We previously considered the following Poisson regression model for the expected number of skin cancer cases in subgroup (i, j), $i = 1, 2, \ldots, 8$ and $j = 0, 1$:

$$E(Y_{ij}) = \mu_{ij} = \ell_{ij} \lambda_{ij}$$

where

$$\ln \lambda_{ij} = \alpha + \sum_{k=1}^{7} \alpha_k U_k + \beta E$$

Here, the U_k's were 0–1 dummy variables indexing the age strata, and E was a 0–1 variable delineating metropolitan area (1 = Dallas–Ft. Worth, 0 = Minneapolis–St. Paul). For this model, the risk (or more generally, rate) ratio

$$RR_i = \frac{\lambda_{i1}}{\lambda_{i0}}$$

reduced to the expression

$$RR_i = e^{\beta}$$

where e^{β} is independent of i and represents an overall rate ratio estimate adjusted for age.

---

[5]The fact that $E(Y_i) = \text{Var}(Y_i) = \mu_i = \ell_i \lambda(\mathbf{X}_i, \boldsymbol{\beta})$, means that the variance of the response variable is *not* constant (i.e., it varies as a function of ℓ_i and $\mathbf{X}_i$, thus requiring a weighted-least-squares regression analysis). And because this variance is a mathematical function of $\boldsymbol{\beta}$, the weights in such a weighted regression analysis necessarily change as a function of the change in the estimate $\hat{\boldsymbol{\beta}}$ at each step of the iteration process (i.e., a reweighting is required at each step). This is the reason for the terminology "iteratively reweighted least squares," or IRLS for short.

The likelihood function L for the preceding model, based on the assumption that the count Y_{ij} follows the Poisson distribution with mean $\mu_{ij} = \ell_{ij}\lambda_{ij}$, is given by the expression

$$L = \prod_{i=1}^{8} \left\{ \left[\frac{(\ell_{i0}\lambda_{i0})^{Y_{i0}} e^{-\ell_{i0}\lambda_{i0}}}{Y_{i0}!} \right] \left[\frac{(\ell_{i1}\lambda_{i1})^{Y_{i1}} e^{-\ell_{i1}\lambda_{i1}}}{Y_{i1}!} \right] \right\}$$

where $\lambda_{i0} = \exp(\alpha + \alpha_i)$, $\lambda_{i1} = \exp(\alpha + \alpha_i + \beta)$ for $i = 1,\ldots, 7$, $\lambda_{80} = \exp(\alpha)$, and $\lambda_{81} = \exp(\alpha + \beta)$.

The use of a Poisson regression computer package would then maximize this likelihood function to produce the nine parameter estimates

$$\{\hat{\alpha}, \hat{\alpha}_1, \hat{\alpha}_2, \ldots, \hat{\alpha}_7, \hat{\beta}\}$$

along with a 9×9 estimated covariance matrix. The computer output for these data, using SAS's GENMOD procedure is given next. From this output, we see that the estimates of β and its standard error are

$$\hat{\beta} = 0.806, \text{ s.e.}(s_{\hat{\beta}}) = 0.0522$$

Thus, the point estimate of the adjusted rate ratio is given by

$$e^{\hat{\beta}} = e^{0.806} = 2.2389$$

```
                  The GENMOD Procedure
           Criteria For Assessing Goodness Of Fit

Criterion                DF        Value       Value/DF
Deviance                  7       8.3426        1.1918
Scaled Deviance           7       8.3426        1.1918
Pearson Chi-Square        7       8.2189        1.1741
Scaled Pearson X2         7       8.2189        1.1741
Log Likelihood            .    7201.7897          .

              Analysis Of Parameter Estimates

Parameter    DF     Estimate     Std Err    ChiSquare     Pr > Chi
INTERCEPT     1      -5.4851      0.1037    2798.4771       0.0000
CITY          1       0.8064      0.0522     238.5764       0.0000
U1            1      -6.1743      0.4577     181.9416       0.0000
U2            1      -3.5441      0.1675     447.7882       0.0000
U3            1      -2.3270      0.1275     333.2712       0.0000
U4            1      -1.5791      0.1138     192.4421       0.0000
U5            1      -1.0870      0.1109      96.0623       0.0000
U6            1      -0.5221      0.1086      23.1044       0.0000
U7            1      -0.1156      0.1109       1.0866       0.2972
```

An approximate large-sample 95% confidence interval for e^{β} is calculated as

$$\exp[\hat{\beta} \pm 1.96\widehat{\text{s.e.}}(s_{\hat{\beta}})] = \exp[0.806 \pm 1.96(0.0522)]$$
$$= \exp(0.806 \pm 0.1023)$$

which gives the 95% confidence limits

$$(e^{0.7037}, e^{0.9083}) = (2.0212, 2.4801)$$

A large-sample test of $H_0: \beta = 0$ versus $H_A: \beta \neq 0$ can be based on the Wald statistic

$$Z = \frac{\hat{\beta} - 0}{\widehat{\text{s.e.}}(s_{\hat{\beta}})}$$

which is approximately $N(0, 1)$ under $H_0: \beta = 0$.

For our example,

$$Z = \frac{0.806 - 0}{0.0522} = 15.44 \quad (P \approx 0)$$

Thus, the foregoing Poisson regression analysis indicates that a statistically significant effect is due to area and that the overall (adjusted for age) rate for nonmelanoma skin cancer in women in Dallas–Ft. Worth is approximately 2.2 times the corresponding adjusted rate in women in Minneapolis–St. Paul; a 95% confidence interval for the (adjusted) rate ratio is (2.0212, 2.4801). We will return to this example to illustrate how to evaluate confounding and goodness of fit.

24-5 Measures of Goodness of Fit

Measures of the goodness of fit of Poisson regression models are obtained from comparisons of maximized likelihood values. Suppose that Y_i has the Poisson distribution (24.3) and that $Y_1, Y_2, \ldots, Y_n$ are mutually independent; then, expressed as a general function of $\mu_1, \mu_2, \ldots, \mu_n$ (i.e., ignoring the predictors $X_1, X_2, \ldots, X_k$ completely), the likelihood function takes the form

$$L(\mathbf{Y}; \boldsymbol{\mu}) = \prod_{i=1}^{n} \frac{\mu_i^{Y_i} e^{-\mu_i}}{Y_i!} = \frac{\left(\prod_{i=1}^{n} \mu_i^{Y_i}\right) \exp\left(-\sum_{i=1}^{n} \mu_i\right)}{\prod_{i=1}^{n} Y_i!} \tag{24.7}$$

where $\boldsymbol{\mu} = (\mu_1, \mu_2, \ldots, \mu_n)$. The system of ML equations

$$\frac{\partial}{\partial \mu_i}[\ln L(\mathbf{Y}; \boldsymbol{\mu})] = 0, \qquad i = 1, 2, \ldots, n$$

leads to the solution $\hat{\mu}_i = Y_i$, $i = 1, 2, \ldots, n$. Thus, the maximized likelihood value for the likelihood function (24.7) is

$$L(\mathbf{Y}; \hat{\boldsymbol{\mu}}) = \frac{\left(\prod_{i=1}^{n} Y_i^{Y_i}\right) \exp\left(-\sum_{i=1}^{n} Y_i\right)}{\prod_{i=1}^{n} Y_i!} \tag{24.8}$$

where $\hat{\boldsymbol{\mu}} = (\hat{\mu}_1, \hat{\mu}_2, \ldots, \hat{\mu}_n) = (Y_1, Y_2, \ldots, Y_n)$.

The value of the maximized likelihood $L(\mathbf{Y}; \hat{\boldsymbol{\mu}})$ based on (24.7) will be larger (for any set of data) than that achieved by maximizing a likelihood such as (24.5) when $(k + 1) < n$. This is because (24.7) imposes no restrictions on the structure of μ_i, whereas (24.5) imposes the restriction $\mu_i = \ell_i \lambda(\mathbf{X}_i, \boldsymbol{\beta})$. In other words, (24.5) can be thought of as the likelihood function under

H_0: $\mu_i = \ell_i \lambda(\mathbf{X}_i, \boldsymbol{\beta})$, $i = 1, 2,..., n$, whereas (24.7) is the likelihood under H_A: "μ_i is unrestricted in structure, $i = 1, 2,..., n$."

Thus, if $L(\mathbf{Y}; \hat{\boldsymbol{\beta}})$ is the maximized likelihood value under (24.5), where $\hat{\boldsymbol{\beta}}$ is the ML estimator of β, then

$$-2 \ln \left[\frac{L(\mathbf{Y}; \hat{\boldsymbol{\beta}})}{L(\mathbf{Y}; \hat{\boldsymbol{\mu}})} \right] \tag{24.9}$$

is a likelihood-ratio-type statistic reflecting the goodness of fit of the model $\mu_i = \ell_i \lambda(\mathbf{X}_i, \boldsymbol{\beta})$ relative to the model where no structure has been imposed on μ_i. Since the objective of any regression analysis is to obtain a parsimonious description of the data, the model $\mu_i = \ell_i \lambda(\mathbf{X}_i, \boldsymbol{\beta})$ involving $k + 1$ parameters will (we hope) provide a maximized likelihood value almost as large as can be obtained by the baseline (and uninformative) model that involves as many parameters (namely, n) as data points. By "almost as large," we mean that $L(\mathbf{Y}; \hat{\boldsymbol{\beta}})$ will not be significantly smaller than $L(\mathbf{Y}; \hat{\boldsymbol{\mu}})$ based on a likelihood ratio test using (24.9).

The quantity

$$D(\hat{\boldsymbol{\beta}}) = -2 \ln \left[\frac{L(\mathbf{Y}; \hat{\boldsymbol{\beta}})}{L(\mathbf{Y}; \hat{\boldsymbol{\mu}})} \right] \tag{24.10}$$

is the goodness-of-fit statistic employed to assess whether $L(\mathbf{Y}; \hat{\boldsymbol{\beta}})$ is significantly less than $L(\mathbf{Y}; \hat{\boldsymbol{\mu}})$ and thus to suggest meaningful lack of fit to the data of the assumed regression model $\mu_i = \ell_i \lambda(\mathbf{X}_i, \boldsymbol{\beta})$. The quantity $D(\hat{\boldsymbol{\beta}})$ is also called the *deviance* for the Poisson regression model $\mu_i = \ell_i \lambda(\mathbf{X}_i, \boldsymbol{\beta})$, and it can be thought of as a measure of residual variation about (or deviation from) the fitted model. Under H_0: $\mu_i = \ell_i \lambda(\mathbf{X}_i, \boldsymbol{\beta})$, the deviance $D(\hat{\boldsymbol{\beta}})$ is typically (although not strictly legitimately) assumed to have (for large samples) an approximate chi-square distribution with $n - k - 1$ degrees of freedom, where n is the number of parameters (i.e., the number of subgroups, cells, or categories) specified in the likelihood (24.7), and $k + 1$ is the number of parameters (i.e., β_j's) in the likelihood (24.5). Thus, a very approximate test for goodness-of-fit of the model $\mu_i = \ell_i \lambda(\mathbf{X}_i, \boldsymbol{\beta})$ to a given data set can be performed by comparing the calculated value of $D(\hat{\boldsymbol{\beta}})$ to an appropriate upper-tail value of the chi-square distribution with $n - k - 1$ degrees of freedom.

With $\hat{Y}_i = \ell_i \lambda(\mathbf{X}_i, \boldsymbol{\beta})$ denoting the predicted response in cell i under model (24.2), the quantity (24.10) can be written in the form

$$D(\hat{\boldsymbol{\beta}}) = 2 \sum_{i=1}^{n} \left[Y_i \ln \left(\frac{Y_i}{\hat{Y}_i} \right) - (Y_i - \hat{Y}_i) \right] \tag{24.11}$$

Hence, $D(\hat{\boldsymbol{\beta}})$ behaves like SSE $= \sum_{i=1}^{n}(Y_i - \hat{Y}_i)^2$ in standard multiple linear regression analysis. When the fitted model exactly predicts the observed data (i.e., $Y_i = \hat{Y}_i$, $i = 1, 2,..., n$), then $D(\hat{\boldsymbol{\beta}}) = 0$; and the larger the discrepancy between observed and predicted responses, the larger is the value of $D(\hat{\boldsymbol{\beta}})$.

When the predicted values are all of reasonable size (i.e., $\hat{Y}_i > 3$, $i = 1, 2,..., n$), then (24.11) can be reasonably approximated by the more familiar Pearson-type observed-versus-predicted chi-square statistic of the form

$$\chi^2 = \sum_{i=1}^{n} \frac{(Y_i - \hat{Y}_i)^2}{\hat{Y}_i} \tag{24.12}$$

As a word of caution, the statistic (24.12) can be misleadingly large when certain $\hat{Y}_i$-values are very small.

The deviances for various models in a hierarchical class can be used to produce likelihood ratio tests. In particular, consider again the likelihood (24.5) involving the parameter set $\boldsymbol{\beta} = (\beta_0, \beta_1,\ldots, \beta_k)$, with deviance $D(\hat{\boldsymbol{\beta}})$ given by (24.10). Now, for $0 < r < k$, suppose that we wish to test whether the last $k - r$ parameters in $\boldsymbol{\beta}$ are equal to 0; i.e., our null hypothesis is H_0: $\beta_{r+1} = \beta_{r+2} = \cdots = \beta_k = 0$. Under H_0, the (null hypothesis) likelihood can be obtained by replacing $\boldsymbol{\beta}$ in (24.5) with $\boldsymbol{\beta}_r$, where

$$\boldsymbol{\beta}_r = (\beta_0, \beta_1,\ldots, \beta_r; 0, 0,\ldots, 0)$$

If we denote this likelihood function by $L(\mathbf{Y}; \boldsymbol{\beta}_r)$, and if $\hat{\boldsymbol{\beta}}_r$ is the maximum likelihood estimator of $\boldsymbol{\beta}_r$ using $L(\mathbf{Y}; \boldsymbol{\beta}_r)$, then the likelihood ratio test of H_0 is performed using the test statistic

$$-2 \ln \left[\frac{L(\mathbf{Y}; \hat{\boldsymbol{\beta}}_r)}{L(\mathbf{Y}; \hat{\boldsymbol{\beta}})} \right] \qquad (24.13)$$

which has approximately a chi-square distribution with $k - r$ degrees of freedom for large samples when H_0 is true.

Furthermore, expression (24.13) is exactly equal to the deviance difference

$$D(\hat{\boldsymbol{\beta}}_r) - D(\hat{\boldsymbol{\beta}}) \qquad (24.14)$$

To see this, recall the general definition of $D(\hat{\boldsymbol{\beta}})$ given by expression (24.10). Using (24.10) and (24.14), we have

$$
\begin{aligned}
D(\hat{\boldsymbol{\beta}}_r) - D(\hat{\boldsymbol{\beta}}) &= -2 \ln \left[\frac{L(\mathbf{Y}; \hat{\boldsymbol{\beta}}_r)}{L(\mathbf{Y}; \hat{\boldsymbol{\mu}})} \right] + 2 \ln \left[\frac{L(\mathbf{Y}; \hat{\boldsymbol{\beta}})}{L(\mathbf{Y}; \hat{\boldsymbol{\mu}})} \right] \\
&= -2 \ln \left[\frac{L(\mathbf{Y}; \hat{\boldsymbol{\beta}}_r)}{L(\mathbf{Y}; \hat{\boldsymbol{\beta}})} \right]
\end{aligned}
$$

which is exactly the likelihood ratio test statistic (24.13). Under H_0: $\beta_{r+1} = \beta_{r+2} = \cdots = \beta_k = 0$, the difference $D(\hat{\boldsymbol{\beta}}_r) - D(\hat{\boldsymbol{\beta}})$ has approximately a chi-square distribution with $k - r$ degrees of freedom in large samples.

Thus, when we use Poisson regression to analyze a set of data, members of a set of candidate models within a hierarchical class can be compared by considering differences between pairs of deviances for these models.

24-6 Continuation of Skin Cancer Data Example

We again consider the data in Table 24-1 giving skin cancer counts for women stratified by age in Minneapolis–St. Paul and Dallas–Ft. Worth. For these data, we used ML estimation to fit the following Poisson regression model for the expected number of skin cancer cases:

$$E(Y_{ij}) = \mu_{ij} = \ell_{ij} \lambda_{ij}$$

where

$$\ln \lambda_{ij} = \alpha + \sum_{k=1}^{7} \alpha_k U_k + \beta E$$

The set $\{U_i\}$ are 0–1 dummy variables indexing the age strata, and E contrasts metropolitan areas (1 = Dallas–Ft. Worth, 0 = Minneapolis–St. Paul). We will refer to this model as model 1. For this model, the estimated rate ratio adjusted for age was $e^{\hat{\beta}} = 2.2389$, and a 95% confidence interval for the true adjusted rate ratio was (2.0212, 2.4801). Also, a large-sample test for $H_0: \beta = 0$ versus $H_A: \beta \neq 0$ yielded a Z-statistic of 15.44 ($P \approx 0$), which is highly significant.

Two additional questions of interest for these data is:

1. Is "age" an effect modifier? That is, does the "area" effect (as measured by a rate ratio parameter) differ for different age strata?

2. If the answer to question 1 is no, is "age" a confounder? That is, does "age" need to be in the model in some form in order to produce a valid estimate of the "area" effect?

At this point, readers may wish to review the definitions and properties of effect modifiers and confounders discussed in Chapter 11.

To answer question 1 directly, we could modify model 1 to include interaction terms, as follows:

$$\text{model 2: } \ln \lambda_{ij} = \alpha + \sum_{k=1}^{7} \alpha_k U_k + \beta E + \sum_{k=1}^{7} \delta_k (E U_k)$$

To avoid a singularity (i.e., perfect collinearity), we have added product terms involving only seven of the eight U_i's.

With the preceding revised interaction model, we can test for effect modification by testing $H_0: \delta_1 = \delta_2 = \cdots = \delta_7 = 0$, using a likelihood ratio χ^2 statistic with 7 degrees of freedom. This test involves comparing model 1 (without interaction terms) to model 2 (with seven EU_i terms). Earlier, we looked at the computer output from fitting model 1. The computer output from fitting model 2 (again using SAS's GENMOD procedure) is presented next. From this latter block of output, we see that the deviance for model 2 is exactly zero, because model 2 fits the data perfectly (i.e., we have fit a model with 16 parameters to a data set of size $n = 16$). Thus, the likelihood ratio test statistic for testing $H_0: \delta_1 = \delta_2 = \cdots = \delta_7 = 0$ is obtained by subtracting the deviance for model 2 (i.e., zero) from the deviance for model 1, which is just the deviance for model 1. Therefore, an equivalent way to carry out this particular interaction test is to use the deviance previously obtained for model 1, which has the value 8.34. When compared with upper-tail chi-square values for which

d.f. = [Number of Y_{ij}'s] − [Number of parameters in model 1]
      = 16 − 9 = 7

there is clearly no lack of fit in model 1 (i.e., there are no large deviations of observed Y_{ij}-values from predicted $\hat{Y}_{ij}$-values). This indicates that adding more terms (e.g., of the form EU_i) to model 1 will not significantly improve the fit of that model.

```
                        The GENMOD Procedure
                Criteria For Assessing Goodness Of Fit
Criterion                   DF           Value           Value/DF
Deviance                     0          0.0000              .
Scaled Deviance              0          0.0000              .
Pearson Chi-Square           0          0.0000              .
Scaled Pearson X2            0          0.0000              .
Log Likelihood               .       7205.9610              .
                  Analysis of Parameter Estimates
```

Parameter	DF	Estimate	Std Err	ChiSquare	Pr > Chi
INTERCEPT	1	-5.3385	0.1581	1139.9829	0.0000
CITY	1	0.5792	0.2010	8.3076	0.0039
U1	1	-6.7207	1.0124	44.0657	0.0000
U2	1	-3.6094	0.2958	148.8871	0.0000
U3	1	-2.7347	0.2415	128.2000	0.0000
U4	1	-1.8289	0.1977	85.5824	0.0000
U5	1	-1.2232	0.1866	42.9868	0.0000
U6	1	-0.7040	0.1808	15.1595	0.0001
U7	1	-0.1504	0.1803	0.6957	0.4042
CU1	1	0.7581	1.1360	0.4454	0.5045
CU2	1	0.1135	0.3594	0.0996	0.7523
CU3	1	0.5664	0.2866	3.9068	0.0481
CU4	1	0.3659	0.2429	2.2694	0.1320
CU5	1	0.2126	0.2325	0.8364	0.3604
CU6	1	0.2776	0.2265	1.5026	0.2203
CU7	1	0.0549	0.2288	0.0577	0.8102

The braces grouping CU1–CU7 are labeled EU_k, $k = 1, 2, \ldots, 7$.

To answer question 2 about confounding, we need to see whether $\hat{\beta}$, or $e^{\hat{\beta}}$, changes meaningfully if we ignore (i.e., don't control for) "age." In particular, we need to drop the age terms (i.e., the "$\sum_{k=1}^{7} \alpha_k U_k$" component) from model 1 to see whether the estimated coefficient of E changes meaningfully from its value of 0.806 (or whether the estimate of the rate ratio changes from its value of $e^{\hat{\beta}} = 2.2389$). If we fit

model 3: $\ln \lambda_{ij} = \alpha + \beta_0 E$

to these data, we obtain $\hat{\beta}_0 = 0.743$ and a crude rate ratio estimate of

$$\widehat{RR}_c = e^{\hat{\beta}_0} = e^{0.744} = 2.1043$$

There is enough change here to suggest that "age" is a confounder and so should be controlled at the analysis stage.

To this point, we have treated age as a *categorical* variable, using seven terms of the form $\alpha_k U_k$ ($k = 1, 2, \ldots, 7$) plus a constant α in model 1 to reflect the eight age strata. An alternative analysis that treats age as an interval variable is suggested by plotting $\ln \hat{\lambda}_{ij}$ on $\ln T_i$ where

$$T_i = \frac{[\text{Midpoint of } i\text{th age interval}] - 15}{35}, \qquad i = 1, 2, \ldots, 8$$

These plots are presented in Figure 24-1 and the values used for the plots are given in Table 24-2. Figure 24-1 shows that the plots graph as a straight line for each j, with the two lines essentially parallel. These results suggest using a parsimonious model involving only a single linear effect of "age," with no interaction terms involving "age" and E.

FIGURE 24-1   Plots of $\ln \hat{\lambda}_{ij}$ on $\ln T_i$, where $T_i = \lfloor(\text{midpoint of age group } i) - 15\rfloor/35$, using values in Table 24-2 from the study of nonmelanoma skin cancer in two cities (Scotto, Kopf, and Urbach 1974).

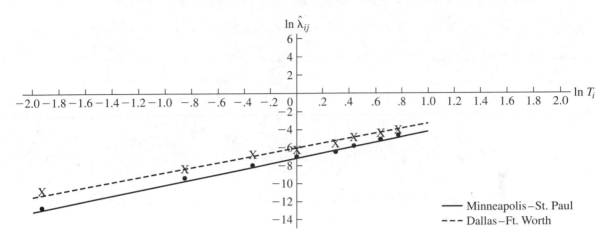

TABLE 24-2   Values of $\ln \hat{\lambda}_{ij}$ and $\ln T_i$, where $T_i = [(\text{midpoint of age group } i) - 15]/35$, using the data in Table 24-1 from the study of nonmelanoma skin cancer in two cities (Scotto, Kopf, and Urbach 1974).

Age Group	Midpoint	T_i	$\ln T_i$	$\ln \hat{\lambda}_{i0}$	$\ln \hat{\lambda}_{i1}$
15–24	20	0.1429	−1.94591	−12.0592	−10.7219
25–34	30	0.4286	−0.84730	−8.9479	−8.2552
35–44	40	0.7143	−0.33647	−8.0732	−6.9275
45–54	50	1.0000	0.00000	−7.1674	−6.2223
55–64	60	1.2857	0.25131	−6.5617	−5.7698
65–74	70	1.5714	0.45199	−6.0425	−5.1953
75–84	80	1.8571	0.61904	−5.4889	−4.8548
85+	90	2.1429	0.76214	−5.3385	−4.7533

$\hat{\lambda}_{i0} = Y_{i0}/\ell_{i0} = $ Crude rate for Minneapolis–St. Paul
$\hat{\lambda}_{i1} = Y_{i1}/\ell_{i1} = $ Crude rate for Dallas–Ft. Worth

In particular, consider the following model:

model 4: $\ln \lambda_{ij} = \alpha + \theta \ln T_i + \beta E$

Model 4 says that

$$\lambda_{i0} = e^\alpha T_i^\theta \qquad \text{and} \qquad \lambda_{i1} = e^\alpha T_i^\theta e^\beta$$

so

$$RR_i = \frac{\lambda_{i1}}{\lambda_{i0}} = e^\beta$$

When $i = 4$ (so that $T_4 = 1$) and $j = 0$, we have $\lambda_{40} = e^\alpha T_4^\theta = e^\alpha (1)^\theta = e^\alpha$, so $\ln \lambda_{40} = \alpha$. Hence, α (the intercept term in model 4) is the natural logarithm of the rate for the 45- to 54-year-old age group ($i = 4$) in Minneapolis–St. Paul ($j = 0$). The following block of computer output (using SAS's GENMOD procedure) comes from fitting model 4.

```
                    The GENMOD Procedure
            Criteria For Assessing Goodness Of Fit
Criterion               DF          Value        Value/DF
Deviance                13         14.8476        1.1421
Scaled Deviance         13         14.8476        1.1421
Pearson Chi-Square      13         14.6475        1.1267
Scaled Pearson X2       13         14.6475        1.1267
Log Likelihood           .       7198.5372
             Analysis of Parameter Estimates

Parameter    DF    Estimate    Std Err     ChiSquare     Pr > Chi
INTERCEPT     1     -7.0764     0.0476    22071.7488      0.0000
CITY          1      0.8051     0.0522      238.0716      0.0000
LOGT          1      2.2894     0.0627     1334.0067      0.0000
```

From this latest block of output, the following point estimates and standard errors can be obtained:

$$\hat{\alpha} = -7.076 \qquad \text{s.e.}(s_{\hat{\alpha}}) = 0.048$$

$$\hat{\beta} = 0.805 \qquad \text{s.e.}(s_{\hat{\beta}}) = 0.052$$

$$\hat{\theta} = 2.289 \qquad \text{s.e.}(s_{\hat{\theta}}) = 0.063$$

The deviance value for model 4 is 14.85, with $16 - 3 = 13$ degrees of freedom, indicating a good fit to the data. The estimated adjusted rate ratio for model 4 is

$$e^{\hat{\beta}} = e^{0.805} = 2.2367$$

The 95% confidence interval for e^{β} is given by

$$\exp[0.805 \pm 1.96(0.052)] = \exp(0.805 \pm 0.1019)$$
$$= (e^{0.7031}, e^{0.9069})$$
$$= (2.020, 2.477)$$

Thus, using model 4 leads to the same answers obtained from model 1, with a very small gain in precision (since the confidence interval for model 4 is slightly narrower than that for model 1). Since model 4 contains fewer parameters than model 1, without compromising on validity or precision, and since model 4 describes the linear effect of "age" in a clear-cut manner, it is the model of choice!

A summary of the results of fitting various models to the data set under consideration can be presented in a Poisson ANOVA table, as illustrated in Table 24-3. As this table indicates, models 1 and 4 clearly have the best deviance values; but model 4 becomes our final choice. Model 1 has a deviance of 8.3, whereas model 4 has a deviance of 14.8—but model 1 *should* have a smaller deviance (i.e., should fit the data better) than model 4, since it contains nine parameters, whereas model 4 contains only three. The issue is really whether model 1 fits the data *significantly* better than model 4. To address this issue, we can look at the difference in the two deviance values (namely, $14.8 - 8.3 = 6.5$), just as we would look at the difference in SSE values in standard multiple regression analysis.

Specifically, under H_0: $\alpha_i = \theta \ln T_i$, where $i = 1, 2, \ldots, 8$,

Deviance for model 4 $-$ Deviance for model 1

TABLE 24-3   Poisson ANOVA table for the skin cancer data of Table 24-1.

	Model for $\ln \lambda_{ij}$	Number of Parameters	$D(\beta)$	d.f.
	α	1	2794.3	15
Model 3:	$\alpha + \beta E$	2	2573.3	14
	$\alpha + \theta \ln T_i$	2	272.7	14
Model 4:	$\alpha + \theta \ln T_i + \beta E$	3	14.8	13
	$\alpha + \sum_{k=1}^{7} \alpha_k U_k$	8	268.4	8
Model 1:	$\alpha + \sum_{k=1}^{7} \alpha_k U_k + \beta E$	9	8.3	7
	α_{ij}	16	0.0	0

is approximately a chi-square random variable for large n with $13 - 7 = 6$ degrees of freedom. The alternative hypothesis is H_A: "$\{\alpha_i\}$ have unspecified structure." Since $\chi^2_{6,0.60} = 6.211$, the P-value for this test is about .40. Hence, there is absolutely no evidence that model 1 provides a better fit to the data than model 4.

Finally, a pseudo-R^2 for model 4 can be computed as

$$\text{Pseudo-}R^2 = \frac{2794.3 - 14.8}{2794.3} = .9947$$

which indicates a superb fit to the data.

24-7 A Second Illustration of Poisson Regression Analysis

We now consider an example given by Frome (1983). The data to be used appear in Table 24-4. The basic response variable Y is "number of lung cancer deaths," which is assumed to have a Poisson distribution. More specifically, Y_{ij} denotes the observed number of lung cancer deaths in row i and column j, $i = 1, 2,..., 9$ and $j = 1, 2,..., 7$; thus there are $n = 9 \times 7 = 63$ subgroups. The rows represent "years of smoking" (defined as age minus 20 years) in five-year categories from 15 through 19 to 55 through 59; the columns represent "number of cigarettes smoked per day," starting at 0 for nonsmokers and going up to 35 or more for the heaviest smokers. The variable ℓ_{ij} denotes the number of man-years at risk for cell (i, j). The variable T_i, defined as the midpoint of the ith "years of smoking" category divided by 42.5, will be employed when we fit some dose–response models to the data in Table 24-4; the variable D_j will denote the dosage level variable for the jth "number of cigarettes smoked per day" category.

One particular form of failure rate model $\lambda(\mathbf{X}_{ij}, \boldsymbol{\beta})$ to be fit to these data is the standard two-way cross-classification model *in exponentiated form*. (Failure rates are always nonnegative, and using an exponential function ensures that this is reflected in the model.) In particular, consider modeling the rate $\lambda(\mathbf{X}_{ij}, \boldsymbol{\beta}) \equiv \lambda_{ij}$ for cell (i, j) as

$$\lambda_{ij} = e^{\mu + \alpha_i + \delta_j} \tag{24.15}$$

TABLE 24-4   Man-years at risk (ℓ_{ij}) and observed number (Y_{ij}) of lung cancer deaths (in parentheses).

Years of Smoking*	42.5T_i	Number of Cigarettes Smoked per Day						
		0 ($D_1 = 0$)	1–9 ($D_2 = 5.2$)	10–14 ($D_3 = 11.2$)	15–19 ($D_4 = 15.9$)	20–24 ($D_5 = 20.4$)	25–34 ($D_6 = 27.4$)	35+ ($D_7 = 40.8$)
15–19	17.5	10,366 (1)	3,121 (0)	3,577 (0)	4,317 (0)	5,683 (0)	3,042 (0)	670 (0)
20–24	22.5	8,162 (0)	2,937 (0)	3,286 (1)	4,214 (0)	6,385 (1)	4,050 (1)	1,166 (0)
25–29	27.5	5,969 (0)	2,288 (0)	2,546 (1)	3,185 (0)	5,483 (1)	4,290 (4)	1,482 (0)
30–34	32.5	4,496 (0)	2,015 (0)	2,219 (2)	2,560 (4)	4,687 (6)	4,268 (9)	1,580 (4)
35–39	37.5	3,512 (0)	1,648 (1)	1,826 (0)	1,893 (0)	3,646 (5)	3,529 (9)	1,336 (6)
40–44	42.5	2,201 (0)	1,310 (2)	1,386 (1)	1,334 (2)	2,411 (12)	2,424 (11)	924 (10)
45–49	47.5	1,421 (0)	927 (0)	988 (2)	849 (2)	1,567 (9)	1,409 (10)	556 (7)
50–54	52.5	1,121 (0)	710 (3)	684 (4)	470 (2)	857 (7)	663 (5)	255 (4)
55–59	57.5	826 (2)	606 (0)	449 (3)	280 (5)	416 (7)	284 (3)	104 (1)

Note: If ℓ_{ij} really were the number of people in cell (i, j) from which the observed number Y_{ij} of lung cancer cases developed, then Y_{ij} could be treated as a binomial random variable with sample size ℓ_{ij} and unknown probability (or risk) of lung cancer death π_{ij}, in which case categorical data analysis methods might be utilized (although having several cells with no deaths can be problematic).

The quantity T_i is the midpoint of the ith "years of smoking" category divided by 42.5; D_j is the dosage level variable for the jth "cigarettes per day" category.

* Age minus 20 years.

where μ is the overall mean, α_i is the fixed effect of the ith row, and δ_j is the fixed effect of the jth column. Here, as in standard two-way ANOVA, we impose the constraints $\sum_{i=1}^{9}\alpha_i = \sum_{j=1}^{7}\delta_j = 0$, so a total of $1 + (9-1) + (7-1) = 15$ parameters must be estimated using model (24.15).

As with standard regression model representations of ANOVA-type data (see elsewhere in this book), the "x" variables underlying model (24.15) are the usual dummy variables used to index the various rows and columns; their appearance has been suppressed for notational convenience. An equivalent way to write (24.15) in dummy variable regression notation is $\lambda_{ij} = \exp(\mu + \sum_{k=1}^{8}\alpha_k R_k + \sum_{\ell=1}^{6}\delta_\ell C_\ell)$, where the R_k's and C_ℓ's denote dummy variables indexing the rows and columns, respectively.

The Poisson-model-based likelihood function for the data in Table 22-1 and under model (24.15) is

$$\prod_{i=1}^{9}\prod_{j=1}^{7}\left[\frac{(\lambda_{ij})^{Y_{ij}}e^{-\lambda_{ij}}}{(Y_{ij})!}\right]$$

where $\lambda_{ij} = \exp(\mu + \alpha_i + \delta_j)$, $\alpha_9 = -\sum_{i=1}^{8}\alpha_i$, and $\delta_7 = -\sum_{j=1}^{6}\delta_j$. For the data in Table 24-4, using IRLS methods leads to the following estimates (in exponentiated form):

$$e^{\hat{\mu}} = 7.69 \times 10^{-5}$$

$$e^{\hat{\alpha}_1} = 0.039, \quad e^{\hat{\alpha}_2} = 0.117, \quad e^{\hat{\alpha}_3} = 0.247, \quad e^{\hat{\alpha}_4} = 1.105, \quad e^{\hat{\alpha}_5} = 1.144,$$

$$e^{\hat{\alpha}_6} = 3.017, \quad e^{\hat{\alpha}_7} = 3.823, \quad e^{\hat{\alpha}_8} = 6.047, \quad e^{\hat{\alpha}_9} = 10.052$$

$$e^{\hat{\delta}_1} = 1.00, \quad e^{\hat{\delta}_2} = 3.39, \quad e^{\hat{\delta}_3} = 8.16, \quad e^{\hat{\delta}_4} = 10.10, \quad e^{\hat{\delta}_5} = 18.20,$$

$$e^{\hat{\delta}_6} = 22.60, \quad e^{\hat{\delta}_7} = 36.80$$

Given these estimates, the predicted number $\hat{Y}_{ij}$ of cancer deaths in cell (i, j) is $\hat{Y}_{ij} = \ell_{ij}e^{\hat{\mu}+\hat{\alpha}_i+\hat{\delta}_j}$. For example, when $i = 4$ and $j = 5$, then

$$\hat{Y}_{45} = \ell_{45}e^{\hat{\mu}}e^{\hat{\alpha}_4}e^{\hat{\delta}_5}$$
$$= (4{,}687)(7.69 \times 10^{-5})(1.105)(18.20)$$
$$= 7.25$$

which should be compared with the actual observed value $Y_{45} = 6$. It can be shown that the deviance for this fitted model, as calculated using (24.11), has the numerical value of 51.47. (Formula (24.12) is not appropriate for these data, since several $\hat{Y}_{ij}$'s are close to 0 in value.) When compared with critical values of the chi-square distribution with $63 - 15 = 48$ degrees of freedom, the value 51.47 does not suggest any significant lack of fit in the cross-classification model (24.15).

However, using model (24.15) involves estimating *15* parameters (whereas n is only 63); hence, it does not provide a parsimonious description of the data. In addition, it is of considerable interest with these data to fit a model whose parameters can realistically be interpreted in terms of the mathematical theory of carcinogenesis.

In what follows, we will consider the *four-parameter nonlinear* model described by Frome (1983)—namely,

$$\lambda_{ij} = \lambda(T_i, D_j; \gamma, \alpha, \theta, \delta) = (\gamma + \alpha D_j^{\theta})T_i^{\delta} \qquad (24.16)$$

where T_i and D_j are as defined earlier.[6]

By using the mathematical identity $e^{\ln a} = a$ for $a > 0$, we can write (24.16) in an equivalent form considered by Frome (1983):

$$\lambda_{ij} = [e^{\ln\gamma} + e^{(\ln\alpha+\theta\ln D_j)}]e^{\delta\ln T_i} = [e^{\beta_3} + e^{(\beta_1+\beta_2 X_{2j})}]e^{\beta_0 X_{1i}} \equiv \lambda(\mathbf{X}_{ij}, \boldsymbol{\beta}) \qquad (24.17)$$

where $\mathbf{X}_{ij} = (X_{1i}, X_{2j}) = (\ln T_i, \ln D_j)$ and $\boldsymbol{\beta} = (\beta_0, \beta_1, \beta_2, \beta_3) = (\delta, \ln\alpha, \theta, \ln\gamma)$. Expressing (24.16) in the exponential form (24.17) ensures that predicted rates are always positive.

The fitting of model (24.17) by IRLS gives the estimates $\hat{\beta}_0 = 4.46, \hat{\beta}_1 = 1.82, \hat{\beta}_2 = 1.29$, and $\hat{\beta}_3 = 2.94$. The estimated standard errors for these four estimators (obtained as the square roots of the appropriate diagonal elements of the estimated covariance matrix) are, respectively, 0.33, 0.66, 0.20, and 0.58. For example, an approximate large-sample 95% confidence interval for $\alpha(= e^{\beta_i})$ would be

$$\exp\left(\hat{\beta}_1 \pm 1.96\sqrt{\widehat{\text{Var }}\hat{\beta}_1}\right)$$

giving $\exp[1.82 \pm 1.96(0.66)] = \exp(1.82 \pm 1.29)$, and thus the interval $(e^{0.53}; e^{3.11}) = (1.70, 22.40)$.

Finally, a summary of the results of fitting various subset models of (24.16) is provided in Table 24-5.

As discussed earlier, calculating the various differences between deviances in the ANOVA-type table presented in Table 24-5 will provide likelihood ratio tests for the importance of the

---

[6]In model (24.16), γ represents the background (i.e., $D_j = 0$ for a nonsmoker) rate at age 62.5 (i.e., $T_i = [\text{age} - 20]/42.5 = 1$ at age 62.5), αD_j^{θ} describes the effect of dosage (i.e., amount smoked) on lung cancer death rates, and T_i^{δ} is the multiplicative effect (on $\gamma + \alpha D_j^{\theta}$) of the time elapsed since the smoking habit was started. Frome (1983) provides further discussion of scientific evidence in favor of using a model like (24.16) to describe the incidence of lung cancer.

TABLE 24-5   Summary of analyses of data in Table 24-4.

Model for λ_{ij}	Number of Parameters	$D(\beta)$	d.f.*
γ	1	445.10	62
γT_i^δ	2	180.82	61
$(\gamma + \alpha D_j) T_i^\delta$	3	61.84	60
$(\gamma + \alpha D_j^\theta) T_i^\delta$	4	59.58	59

* d.f. = 63 − (Number of parameters in fitted model).

parameters γ, δ, α, and θ. First of all, the difference $445.10 - 180.82 = 264.28$, when compared with upper-tail values of the χ_1^2 distribution, argues strongly for rejecting $H_0: \delta = 0$ in favor of $H_A: \delta \neq 0$. This means that the (multiplicative) effect of time since first exposure is important. Second, the difference $180.82 - 61.84 = 118.98$ demands rejection of $H_0: \alpha = 0$ in favor of $H_A: \alpha \neq 0$, suggesting that the amount smoked is an important variable. Finally, the difference $61.84 - 59.58 = 2.26$ does not lead to rejection of $H_0: \theta = 1$, so the first power of dosage seems most appropriate. Since the deviance value of 61.84 with 60 degrees of freedom for the model $\lambda_{ij} = (\gamma + \alpha D_j) T_i^\delta$ does not suggest any lack of fit, this model seems preferable.[7]

24-8 Summary

This chapter described the general form and several key features of the Poisson regression model. The "typical" Poisson regression model expresses the log *failure* (e.g., *disease*) *rate* as a linear function of a set of predictors. The Poisson regression method, nevertheless, allows for more complicated nonlinear models as well. The dependent variable is a count of the number of occurrences of an event of interest, such as the number of cases of a disease that occur over a given follow-up time period. For the "typical" Poisson regression model, the natural measure of effect that is estimated is a *rate ratio* associated with a particular predictor (i.e., exposure) of interest.

Maximum likelihood estimation is used to estimate the regression coefficients of a Poisson regression model. The likelihood function assumes that the underlying response (a count) has the Poisson distribution. A measure of goodness of fit of a Poisson regression model is obtained by using the deviance statistic, which is a likelihood-ratio-type statistic reflecting the fit of a current model of interest relative to the baseline (uninformative) model, which involves as many parameters as there are data points. The difference in deviance statistics obtained for two (hierarchically ordered) models being compared is equivalent to the difference in log-likelihood statistics for each model, so a likelihood ratio test for the Poisson model can be carried out equivalently by using differences in deviance statistics. Finally, a tabular summary of the results of fitting various models to the same data set can be presented in a *Poisson ANOVA table*, which contains deviance and corresponding d.f. (degrees of freedom) information for each model being fit.

---

[7]The reduction in the deviance due to adding a parameter (or a group of parameters) is *order-dependent*; consequently, a different order of parameter additions can lead to a different final model.

Problems

1. Suppose that, for each of three age groups (25–34, 35–44, and 45–54), we have recorded yearly sex-specific lung cancer mortality rates for the five-year period 1990 through 1994. These data are to be analyzed by Poisson regression methods to see whether the change (if any) in log rate over time varies by age–sex group and, if so, to quantify that variation. In what follows, we will index the six age–sex groups, as follows:

   group 1 ($i = 1$): 25–34-year-old females

   group 2 ($i = 2$): 35–44-year-old females

   group 3 ($i = 3$): 45–54-year-old females

   group 4 ($i = 4$): 25–34-year-old males

   group 5 ($i = 5$): 35–44-year-old males

   group 6 ($i = 6$): 45–54-year-old males

   Consider the following model for the expected cell count $E(Y_{ik})$ for age–sex group i in year $(1990 + k)$, where $i = 1, 2, 3, 4, 5, 6$ and $k = 0, 1, 2, 3, 4$:

   $$E(Y_{ik}) = \ell_{ik} \lambda_{ik}$$

   where

   $$\ln \lambda_{ik} = \sum_{i=1}^{6} \alpha_i A_i + \beta k \tag{1}$$

   Here, ℓ_{ik} and λ_{ik} are, respectively, the person-years at risk and the (unknown) population lung cancer mortality rate in cell (i, k). The independent variables in model (1) are defined as follows:

   $$A_i = \begin{cases} 1 & \text{if age–sex group } i \\ 0 & \text{otherwise} \end{cases}$$

   $$k = \text{year} - 1990$$

   a. What is the total number n of data points, or pairs (ℓ_{ik}, Y_{ik}), for this data set?

   b. Based on model (1), what is the expected cell count for a 40-year-old male in 1992, written as a function of α_5 and β?

   c. For the ith age–sex group, how does model (1) describe the *change* in log rate over time?

   d. What does model (1) assume about the effect of age–sex group on the *change* in log rate over time?

   e. Find a general expression for

   $$\text{IDR}_{ik} = \frac{\lambda_{ik}}{\lambda_{10}}$$

   the incidence density ratio (IDR) comparing the mortality rate for age group i and year $1990 + k$ to the mortality rate for the reference category "25–34-year-old females in 1990."

   f. Suppose that it is of interest to assess whether the *change* in log rate over time actually varies by age–sex group (i.e., whether there is a group-by-time interaction). By adding appropriate cross-product terms to model (1), construct a model that will permit such an assessment and then discuss how you would interpret this model.

Consider the following Poisson regression ANOVA table, based on fitting various models to the data under study:

Model for $\ln \lambda_{ik}$	Number of Parameters	$D(\hat{\beta})$	d.f.
(1) α	1	300	29
(2) $\alpha + \beta k$	2	200	28
(3) $\sum_{i=1}^{6} \alpha_i A_i$	6	175	24
(4) $\sum_{i=1}^{6} \alpha_i A_i + \beta k$	7	60	23
(5) $\sum_{i=1}^{6} \alpha_i A_i + \beta_1 k + \beta_2 k^2$	8	59	22
(6) $\sum_{i=1}^{6} \alpha_i A_i + \beta k + \gamma(A_1 k)$	8	25	22
(7) $\sum_{i=1}^{6} \alpha_i A_i + \beta k + \sum_{i=1}^{5} \gamma_1(A_i k)$	12	20	18
(8) α_{ik}	30	0	0

Using this table, answer the following questions:

g. Ignoring (for now) the variable "time," is there evidence that average mortality rates differ among the six age–sex groups?

h. Is adding the linear time term (βk) to model (3) worthwhile?

i. Assuming that the change in log rate over time is the same for all age–sex groups, do the data argue for adding a quadratic time term to a model that already contains a linear component of time?

j. Is there evidence in the data that the change in log rate over time differs for different age–sex groups?

k. Carry out a test of H_0: "All six age–sex groups have the same slope" versus H_A: "All age–sex groups except group 1 have the same slope."

l. Of the models that fit the data well (use $\alpha = .1$ for any test of lack of fit that you perform), which one would you choose as your final model? Why?

m. For your final model chosen in part (l), calculate a pseudo-R^2-value.

n. Now, assume that model (6) has been fit to the data, resulting in the following point estimates and standard errors:

Parameter	Point Estimate	Standard Error
$\hat{\alpha}_1$	0.50	0.25
$\hat{\alpha}_2$	1.00	0.40
$\hat{\alpha}_3$	1.50	0.30
$\hat{\alpha}_4$	1.25	0.30
$\hat{\alpha}_5$	1.50	0.50
$\hat{\alpha}_6$	1.75	0.40
$\hat{\beta}$	0.50	0.20
$\hat{\gamma}_1$	-3.00	0.50

o. Find and interpret approximate 95% confidence intervals for the following parameters: (1) the common slope for age–sex groups 2 through 6, and (2) the slope for age–sex group 1. [*Note:* The estimated covariance between $\hat{\beta}$ and $\hat{\gamma}_1$ is $\widehat{\text{Cov}}(\hat{\beta}, \hat{\gamma}_1) = -.10$.]

2. A five-year follow-up study was carried out to assess the relationship of diet and weight to the incidence of stomach cancer in 40- to 50-year-old males in a certain metropolitan area. Let Y_{ij} denote the number of cases of stomach cancer found in the ith weight category of diet type j, based on the following table:

	Low-cholesterol Diet ($j = 1$)	High-cholesterol Diet ($j = 2$)
Low Weight ($i = 1$)	Y_{11}	Y_{12}
Medlow Weight ($i = 2$)	Y_{21}	Y_{22}
Medhigh Weight ($i = 3$)	Y_{31}	Y_{32}
High Weight ($i = 4$)	Y_{41}	Y_{42}

Consider the following Poisson regression model for these data:

Model 1: $E(Y_{ij}) = \ell_{ij} \lambda_{ij}$, $i = 1, 2, 3, 4$ and $j = 1, 2$

where ℓ_{ij} =man-years at risk in the ijth category, λ_{ij} =incidence density (i.e., rate) in the ijth category, and λ_{ij} is modeled as follows:

$$\ln \lambda = \sum_{i=1}^{4} \alpha_i V_i + \beta C \quad \text{where} \quad V_i = \begin{cases} 1 & \text{if Weight group } i \\ 0 & \text{otherwise} \end{cases}$$

$$C = \begin{cases} 1 & \text{if Low-cholesterol Diet} \\ 0 & \text{if High-cholesterol Diet} \end{cases}$$

a. Based on this model, give an expression for the rate ratio (RR) of low-cholesterol diet subjects to high-cholesterol diet subjects, adjusting for weight group.

b. Give a large-sample formula for the 95% interval estimate of the RR described in part (a).

c. How would you test whether there is a significant interaction between diet type and weight group?

d. Assuming that there is no interaction between weight group and diet type, suppose that you decide to consider the following model as an alternative to model 1:

Model 2: $\ln \lambda = \alpha + \gamma W_i + \beta C$

where W_i denotes the midpoint (in pounds) of weight group i.

Describe how to use a "difference of deviances" approach to evaluate whether model 2 fits the data significantly better than model 1. (In your answer, state the null hypothesis, describe the test statistic in terms of deviances, and state the distribution of the test statistic, including degrees of freedom, under the null hypothesis.) How might you criticize the use of this approach?

3. The following set of questions relates to using Poisson regression methods to analyze data from an in vitro study of human chromosome damage. In this study, using Poisson regression is appropriate because it models "count" response data, where the counts represent the number of broken chromosomes in a sample of 100 cells taken from each of $n = 40$ individuals. (A Poisson distribution is appropriate for describing counts per unit of *space*—e.g., 100 cells—as well as for describing counts per unit of *time*.)

The design of the study described here involved randomly assigning each of the cell samples for the 40 individuals to one of four treatment groups consisting of ten individuals each. The cells in each treatment group were exposed to different combinations of two drugs, A and B, as follows:

Treatment Group	Number of Individuals n_j	Drugs
1	10	neither A nor B
2	10	A only
3	10	B only
4	10	both A and B
	40	

The exposed cells were then examined for chromosome breakage, and the number of broken chromosomes in the 100 cells was counted for each individual. *The investigator wishes to assess the effects of drug use and of drug interaction on the rate of chromosome damage.*

Let Y_{ij} denote the random variable for the number of chromosome breaks counted for each individual i ($i = 1, 2,..., 10$) within treatment group j ($j = 1, 2, 3, 4$). Also, let $\ell_{ij} = 100$, where ℓ_{ij} denotes the amount of person-space of observation. Assume that the drugs act in a multiplicative fashion on the rate of chromosome breakage (ℓ_{ij}). That is, the model being fitted can be written as

$$\ln \lambda_{ij} = \beta_0 + \sum_{k=1}^{p} \beta_k X_{ijk}$$

where X_{ijk} denotes the value of the kth predictor on the ith individual in the jth treatment group.

a. Based on the preceding scenario, write out the form of the Poisson regression model for $\ln \lambda_{ij}$ as a function of the predictors of interest in this study. In doing so, *make sure to explicitly define a set of categorical predictors to include in this model; these predictors should reflect the primary objective of the study.*

b. Based on your model in part (a), describe *two* alternative ways you can test the null hypothesis of no interaction among the drugs. Include in your answer the null hypothesis (in terms of regression coefficients), the formula for your test statistic(s), the degrees of freedom for each test, and the distribution of the test statistic under the null hypothesis.

c. Fill in the blanks in the following Poisson regression ANOVA table, which is based on fitting various models for the rate of chromosome breakage:

	Model for $\ln \lambda$	Number of Parameters	Deviance	Dev. d.f.
(I)	Baseline (i.e., β_0)	1	90	39
(II)	Main effect of drug A only	—	85	—
(III)	Main effect of drug B only	—	84	—
(IV)	Main effects of both drug A and drug B	—	40	—
(V)	Main effects and interaction	—	30	—

d. According to the ANOVA table, which models provide good fit to the data? Explain briefly.

e. According to the ANOVA table, does drug A have a significant main effect? Explain briefly.

f. According to the ANOVA table, does drug B have a significant effect over and above the effect of drug A? Explain briefly.

g. According to the ANOVA table, do drugs A and B have a significant interaction effect? Explain briefly.

h. According to the ANOVA table (and in light of your answers to parts (d) through (g)), formulate an expression for the rate ratio for the effect of drug A given drug B status and an expression for the rate ratio for the effect of drug B given drug A status.

References

Baker, R. J., and Nelder, J. A. 1978. *Generalized Linear Interactive Modeling (GLIM), Release 3,* Oxford: Numerical Algorithms Group.

Frome, E. L. 1983. "The Analysis of Rates Using Poisson Regression Models." *Biometrics* 39: 665–74.

Frome, E. L., and Checkoway, H. 1985. "Use of Poisson Regression Models in Estimating Incidence Rates and Ratios." *American Journal of Epidemiology* 121(2): 309–23.

Remington, R. D., and Schork, M. A. 1985. *Statistics with Applications to the Biological and Health Sciences.* Englewood Cliffs, N.J.: Prentice-Hall.

Scotto, J.; Kopf, A. W.; and Urbach, F. 1974. "Non-Melanoma Skin Cancer among Caucasians in Four Areas of the United States." *Cancer* 34: 1333–38.

Stokes, M. E., and Koch, G. G. 1983. "A Macro for Maximum Likelihood Fitting of Log-Linear Models to Poisson and Multinomial Counts with Contrast Matrix Capability for Hypothesis Testing." *Proceedings of the Eighth Annual SAS Users Group International Conference,* pp. 795–800.

A

Appendix—Tables

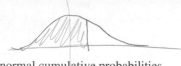

Table A-1   Standard normal cumulative probabilities

z	0.00	0.01	0.02	0.03	0.04	0.05	0.06	0.07	0.08	0.09
-3.8	0.0001	0.0001	0.0001	0.0001	0.0001	0.0001	0.0001	0.0001	0.0001	0.0001
-3.7	0.0001	0.0001	0.0001	0.0001	0.0001	0.0001	0.0001	0.0001	0.0001	0.0001
-3.6	0.0002	0.0002	0.0001	0.0001	0.0001	0.0001	0.0001	0.0001	0.0001	0.0001
-3.5	0.0002	0.0002	0.0002	0.0002	0.0002	0.0002	0.0002	0.0002	0.0002	0.0002
-3.4	0.0003	0.0003	0.0003	0.0003	0.0003	0.0003	0.0003	0.0003	0.0003	0.0002
-3.3	0.0005	0.0005	0.0005	0.0004	0.0004	0.0004	0.0004	0.0004	0.0004	0.0003
-3.2	0.0007	0.0007	0.0006	0.0006	0.0006	0.0006	0.0006	0.0005	0.0005	0.0005
-3.1	0.0010	0.0009	0.0009	0.0009	0.0008	0.0008	0.0008	0.0008	0.0007	0.0007
-3.0	0.0014	0.0013	0.0013	0.0012	0.0012	0.0011	0.0011	0.0011	0.0010	0.0010
-2.9	0.0019	0.0018	0.0018	0.0017	0.0016	0.0016	0.0015	0.0015	0.0014	0.0014
-2.8	0.0026	0.0025	0.0024	0.0023	0.0023	0.0022	0.0021	0.0021	0.0020	0.0019
-2.7	0.0035	0.0034	0.0033	0.0032	0.0031	0.0030	0.0029	0.0028	0.0027	0.0026
-2.6	0.0047	0.0045	0.0044	0.0043	0.0041	0.0040	0.0039	0.0038	0.0037	0.0036
-2.5	0.0062	0.0060	0.0059	0.0057	0.0055	0.0054	0.0052	0.0051	0.0049	0.0048
-2.4	0.0082	0.0080	0.0078	0.0076	0.0073	0.0071	0.0069	0.0068	0.0066	0.0064
-2.3	0.0107	0.0104	0.0102	0.0099	0.0096	0.0094	0.0091	0.0089	0.0087	0.0084
-2.2	0.0139	0.0136	0.0132	0.0129	0.0125	0.0122	0.0119	0.0116	0.0113	0.0110
-2.1	0.0179	0.0174	0.0170	0.0166	0.0162	0.0158	0.0154	0.0150	0.0146	0.0143
-2.0	0.0228	0.0222	0.0217	0.0212	0.0207	0.0202	0.0197	0.0192	0.0188	0.0183
-1.9	0.0287	0.0281	0.0274	0.0268	0.0262	0.0256	0.0250	0.0244	0.0239	0.0233
-1.8	0.0359	0.0351	0.0344	0.0336	0.0329	0.0322	0.0314	0.0307	0.0301	0.0294
-1.7	0.0446	0.0436	0.0427	0.0418	0.0409	0.0401	0.0392	0.0384	0.0375	0.0367
-1.6	0.0548	0.0537	0.0526	0.0516	0.0505	0.0495	0.0485	0.0475	0.0465	0.0455
-1.5	0.0668	0.0655	0.0643	0.0630	0.0618	0.0606	0.0594	0.0582	0.0571	0.0559
-1.4	0.0808	0.0793	0.0778	0.0764	0.0749	0.0735	0.0721	0.0708	0.0694	0.0681
-1.3	0.0968	0.0951	0.0934	0.0918	0.0901	0.0885	0.0869	0.0853	0.0838	0.0823
-1.2	0.1151	0.1131	0.1112	0.1093	0.1075	0.1057	0.1038	0.1020	0.1003	0.0985
-1.1	0.1357	0.1335	0.1314	0.1292	0.1271	0.1251	0.1230	0.1210	0.1190	0.1170
-1.0	0.1587	0.1562	0.1539	0.1515	0.1492	0.1469	0.1446	0.1423	0.1401	0.1379
-0.9	0.1841	0.1814	0.1788	0.1762	0.1736	0.1711	0.1685	0.1660	0.1635	0.1611
-0.8	0.2119	0.2090	0.2061	0.2033	0.2005	0.1977	0.1949	0.1922	0.1894	0.1867
-0.7	0.2420	0.2389	0.2358	0.2327	0.2297	0.2266	0.2236	0.2206	0.2177	0.2148
-0.6	0.2743	0.2709	0.2676	0.2643	0.2611	0.2578	0.2546	0.2514	0.2483	0.2451
-0.5	0.3085	0.3050	0.3015	0.2981	0.2946	0.2912	0.2877	0.2843	0.2810	0.2776
-0.4	0.3446	0.3409	0.3372	0.3336	0.3300	0.3264	0.3228	0.3192	0.3156	0.3121
-0.3	0.3821	0.3783	0.3745	0.3707	0.3669	0.3632	0.3594	0.3557	0.3520	0.3483
-0.2	0.4207	0.4168	0.4129	0.4090	0.4052	0.4013	0.3974	0.3936	0.3897	0.3859
-0.1	0.4602	0.4562	0.4522	0.4483	0.4443	0.4404	0.4364	0.4325	0.4286	0.4247
-0.0	0.5000	0.4960	0.4920	0.4880	0.4840	0.4801	0.4761	0.4721	0.4681	0.4641

Note: Table entry is the area under the standard normal curve to the left of the indicated z-value, thus giving $P(Z < z)$.

z	0.00	0.01	0.02	0.03	0.04	0.05	0.06	0.07	0.08	0.09
0.0	0.5000	0.5040	0.5080	0.5120	0.5160	0.5199	0.5239	0.5279	0.5319	0.5359
0.1	0.5398	0.5438	0.5478	0.5517	0.5557	0.5596	0.5636	0.5675	0.5714	0.5753
0.2	0.5793	0.5832	0.5871	0.5910	0.5948	0.5987	0.6026	0.6064	0.6103	0.6141
0.3	0.6179	0.6217	0.6255	0.6293	0.6331	0.6368	0.6406	0.6443	0.6480	0.6517
0.4	0.6554	0.6591	0.6628	0.6664	0.6700	0.6736	0.6772	0.6808	0.6844	0.6879
0.5	0.6915	0.6950	0.6985	0.7019	0.7054	0.7088	0.7123	0.7157	0.7190	0.7224
0.6	0.7257	0.7291	0.7324	0.7357	0.7389	0.7422	0.7454	0.7486	0.7517	0.7549
0.7	0.7580	0.7611	0.7642	0.7673	0.7703	0.7734	0.7764	0.7794	0.7823	0.7852
0.8	0.7881	0.7910	0.7939	0.7967	0.7995	0.8023	0.8051	0.8078	0.8106	0.8133
0.9	0.8159	0.8186	0.8212	0.8238	0.8264	0.8289	0.8315	0.8340	0.8365	0.8389
1.0	0.8413	0.8438	0.8461	0.8485	0.8508	0.8531	0.8554	0.8577	0.8599	0.8621
1.1	0.8643	0.8665	0.8686	0.8708	0.8729	0.8749	0.8770	0.8790	0.8810	0.8830
1.2	0.8849	0.8869	0.8888	0.8907	0.8925	0.8943	0.8962	0.8980	0.8997	0.9015
1.3	0.9032	0.9049	0.9066	0.9082	0.9099	0.9115	0.9131	0.9147	0.9162	0.9177
1.4	0.9192	0.9207	0.9222	0.9236	0.9251	0.9265	0.9279	0.9292	0.9306	0.9319
1.5	0.9332	0.9345	0.9357	0.9370	0.9382	0.9394	0.9406	0.9418	0.9429	0.9441
1.6	0.9452	0.9463	0.9474	0.9484	0.9495	0.9505	0.9515	0.9525	0.9535	0.9545
1.7	0.9554	0.9564	0.9573	0.9582	0.9591	0.9599	0.9608	0.9616	0.9625	0.9633
1.8	0.9641	0.9649	0.9656	0.9664	0.9671	0.9678	0.9686	0.9693	0.9699	0.9706
1.9	0.9713	0.9719	0.9726	0.9732	0.9738	0.9744	0.9750	0.9756	0.9761	0.9767
2.0	0.9772	0.9778	0.9783	0.9788	0.9793	0.9798	0.9803	0.9808	0.9812	0.9817
2.1	0.9821	0.9826	0.9830	0.9834	0.9838	0.9842	0.9846	0.9850	0.9854	0.9857
2.2	0.9861	0.9864	0.9868	0.9871	0.9875	0.9878	0.9881	0.9884	0.9887	0.9890
2.3	0.9893	0.9896	0.9898	0.9901	0.9904	0.9906	0.9909	0.9911	0.9913	0.9916
2.4	0.9918	0.9920	0.9922	0.9924	0.9927	0.9929	0.9931	0.9932	0.9934	0.9936
2.5	0.9938	0.9940	0.9941	0.9943	0.9945	0.9946	0.9948	0.9949	0.9951	0.9952
2.6	0.9953	0.9955	0.9956	0.9957	0.9959	0.9960	0.9961	0.9962	0.9963	0.9964
2.7	0.9965	0.9966	0.9967	0.9968	0.9969	0.9970	0.9971	0.9972	0.9973	0.9974
2.8	0.9974	0.9975	0.9976	0.9977	0.9977	0.9978	0.9979	0.9979	0.9980	0.9981
2.9	0.9981	0.9982	0.9982	0.9983	0.9984	0.9984	0.9985	0.9985	0.9986	0.9986
3.0	0.9986	0.9987	0.9987	0.9988	0.9988	0.9989	0.9989	0.9989	0.9990	0.9990
3.1	0.9990	0.9991	0.9991	0.9991	0.9992	0.9992	0.9992	0.9992	0.9993	0.9993
3.2	0.9993	0.9993	0.9994	0.9994	0.9994	0.9994	0.9994	0.9995	0.9995	0.9995
3.3	0.9995	0.9995	0.9995	0.9996	0.9996	0.9996	0.9996	0.9996	0.9996	0.9997
3.4	0.9997	0.9997	0.9997	0.9997	0.9997	0.9997	0.9997	0.9997	0.9997	0.9998
3.5	0.9998	0.9998	0.9998	0.9998	0.9998	0.9998	0.9998	0.9998	0.9998	0.9998
3.6	0.9998	0.9998	0.9999	0.9999	0.9999	0.9999	0.9999	0.9999	0.9999	0.9999
3.7	0.9999	0.9999	0.9999	0.9999	0.9999	0.9999	0.9999	0.9999	0.9999	0.9999
3.8	0.9999	0.9999	0.9999	0.9999	0.9999	0.9999	0.9999	0.9999	0.9999	0.9999
3.9	1.0000									

Table A-1 Standard normal cumulative probabilities (*continued*)

z	$P(Z < z)$
-4.265	0.00001
-3.891	0.00005
-3.719	0.0001
-3.291	0.0005
-3.090	0.001
-2.576	0.005
-2.326	0.01
-2.054	0.02
-1.960	0.025
-1.881	0.03
-1.751	0.04
-1.645	0.05
-1.555	0.06
-1.476	0.07
-1.405	0.08
-1.341	0.09
-1.282	0.10
-1.036	0.15
-0.842	0.20
-0.674	0.25
-0.524	0.30
-0.385	0.35
-0.253	0.40
-0.126	0.45
0	0.50

z	$P(Z < z)$
0	0.50
0.126	0.55
0.253	0.60
0.385	0.65
0.524	0.70
0.674	0.75
0.842	0.80
1.036	0.85
1.282	0.90
1.341	0.91
1.405	0.92
1.476	0.93
1.555	0.94
1.645	0.95
1.751	0.96
1.881	0.97
1.960	0.975
2.054	0.98
2.326	0.99
2.576	0.995
3.090	0.999
3.291	0.9995
3.719	0.9999
3.891	0.99995
4.265	0.99999

Table A-2   Percentiles of the _t_ distribution

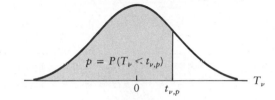

$$p = P(T_\nu < t_{\nu,p})$$

Student's _t_ distribution

% df	55	65	75	85	90	95	97.5	99	99.5	99.95
1	0.158	0.510	1.000	1.963	3.078	6.314	12.706	31.821	63.657	636.619
2	0.142	0.445	0.816	1.386	1.886	2.920	4.303	6.965	9.925	31.599
3	0.137	0.424	0.765	1.250	1.638	2.353	3.182	4.541	5.841	12.924
4	0.134	0.414	0.741	1.190	1.533	2.132	2.776	3.747	4.604	8.610
5	0.132	0.408	0.727	1.156	1.476	2.015	2.571	3.365	4.032	6.869
6	0.131	0.404	0.718	1.134	1.440	1.943	2.447	3.143	3.707	5.959
7	0.130	0.402	0.711	1.119	1.415	1.895	2.365	2.998	3.499	5.408
8	0.130	0.399	0.706	1.108	1.397	1.860	2.306	2.896	3.355	5.041
9	0.129	0.398	0.703	1.100	1.383	1.833	2.262	2.821	3.250	4.781
10	0.129	0.397	0.700	1.093	1.372	1.812	2.228	2.764	3.169	4.587
11	0.129	0.396	0.697	1.088	1.363	1.796	2.201	2.718	3.106	4.437
12	0.128	0.395	0.695	1.083	1.356	1.782	2.179	2.681	3.055	4.318
13	0.128	0.394	0.694	1.079	1.350	1.771	2.160	2.650	3.012	4.221
14	0.128	0.393	0.692	1.076	1.345	1.761	2.145	2.624	2.977	4.140
15	0.128	0.393	0.691	1.074	1.341	1.753	2.131	2.602	2.947	4.073
16	0.128	0.392	0.690	1.071	1.337	1.746	2.120	2.583	2.921	4.015
17	0.128	0.392	0.689	1.069	1.333	1.740	2.110	2.567	2.898	3.965
18	0.127	0.392	0.688	1.067	1.330	1.734	2.101	2.552	2.878	3.922
19	0.127	0.391	0.688	1.066	1.328	1.729	2.093	2.539	2.861	3.883
20	0.127	0.391	0.687	1.064	1.325	1.725	2.086	2.528	2.845	3.850
21	0.127	0.391	0.686	1.063	1.323	1.721	2.080	2.518	2.831	3.819
22	0.127	0.390	0.686	1.061	1.321	1.717	2.074	2.508	2.819	3.792
23	0.127	0.390	0.685	1.060	1.319	1.714	2.069	2.500	2.807	3.768
24	0.127	0.390	0.685	1.059	1.318	1.711	2.064	2.492	2.797	3.745
25	0.127	0.390	0.684	1.058	1.316	1.708	2.060	2.485	2.787	3.725
26	0.127	0.390	0.684	1.058	1.315	1.706	2.056	2.479	2.779	3.707
27	0.127	0.389	0.684	1.057	1.314	1.703	2.052	2.473	2.771	3.690
28	0.127	0.389	0.683	1.056	1.313	1.701	2.048	2.467	2.763	3.674
29	0.127	0.389	0.683	1.055	1.311	1.699	2.045	2.462	2.756	3.659
30	0.127	0.389	0.683	1.055	1.310	1.697	2.042	2.457	2.750	3.646
35	0.127	0.388	0.682	1.052	1.306	1.690	2.030	2.438	2.724	3.591
40	0.126	0.388	0.681	1.050	1.303	1.684	2.021	2.423	2.704	3.551
45	0.126	0.388	0.680	1.049	1.301	1.679	2.014	2.412	2.690	3.520
50	0.126	0.388	0.679	1.047	1.299	1.676	2.009	2.403	2.678	3.496
60	0.126	0.387	0.679	1.045	1.296	1.671	2.000	2.390	2.660	3.460
70	0.126	0.387	0.678	1.044	1.294	1.667	1.994	2.381	2.648	3.435
80	0.126	0.387	0.678	1.043	1.292	1.664	1.990	2.374	2.639	3.416
90	0.126	0.387	0.677	1.042	1.291	1.662	1.987	2.368	2.632	3.402
100	0.126	0.386	0.677	1.042	1.290	1.660	1.984	2.364	2.626	3.390
120	0.126	0.386	0.677	1.041	1.289	1.658	1.980	2.358	2.617	3.373
140	0.126	0.386	0.676	1.040	1.288	1.656	1.977	2.353	2.611	3.361
160	0.126	0.386	0.676	1.040	1.287	1.654	1.975	2.350	2.607	3.352
180	0.126	0.386	0.676	1.039	1.286	1.653	1.973	2.547	2.603	3.345
200	0.126	0.386	0.676	1.039	1.286	1.653	1.972	2.345	2.601	3.340
∞	0.126	0.385	0.674	1.036	1.282	1.645	1.960	2.326	2.576	3.291

Table A-3   Percentiles of the chi-square distribution

χ^2_v distribution

$$p = P(\chi^2_v < \chi^2_{v,p})$$

% df	0.5	1	2.5	5	10	20	30	40	50	60	70	80	90	95	97.5	99	99.5	99.95
1	0.0001	0.0002	0.001	0.004	0.016	0.064	0.148	0.275	0.455	0.708	1.074	1.642	2.706	3.841	5.024	6.635	7.879	12.116
2	0.010	0.020	0.051	0.103	0.211	0.446	0.713	1.022	1.386	1.833	2.408	3.219	4.605	5.991	7.378	9.210	10.597	15.202
3	0.072	0.115	0.216	0.352	0.584	1.005	1.424	1.869	2.366	2.946	3.665	4.642	6.251	7.815	9.348	11.345	12.838	17.730
4	0.207	0.297	0.484	0.711	1.064	1.649	2.195	2.753	3.357	4.045	4.878	5.989	7.779	9.488	11.143	13.277	14.860	19.997
5	0.412	0.554	0.831	1.145	1.610	2.343	3.000	3.655	4.351	5.132	6.064	7.289	9.236	11.070	12.833	15.086	16.750	22.105
6	0.676	0.872	1.237	1.635	2.204	3.070	3.828	4.570	5.348	6.211	7.231	8.558	10.645	12.592	14.449	16.812	18.548	24.103
7	0.989	1.239	1.690	2.167	2.833	3.822	4.671	5.493	6.346	7.283	8.383	9.803	12.017	14.067	16.013	18.475	20.278	26.018
8	1.344	1.646	2.180	2.733	3.490	4.594	5.527	6.423	7.344	8.351	9.524	11.030	13.362	15.507	17.535	20.090	21.955	27.868
9	1.735	2.088	2.700	3.325	4.168	5.380	6.393	7.357	8.343	9.414	10.656	12.242	14.684	16.919	19.023	21.666	23.589	29.666
10	2.156	2.558	3.247	3.940	4.865	6.179	7.267	8.295	9.342	10.473	11.781	13.442	15.987	18.307	20.483	23.209	25.188	31.420
11	2.603	3.053	3.816	4.575	5.578	6.989	8.148	9.237	10.341	11.530	12.899	14.631	17.275	19.675	21.920	24.725	26.757	33.137
12	3.074	3.571	4.404	5.226	6.304	7.807	9.034	10.182	11.340	12.584	14.011	15.812	18.549	21.026	23.337	26.217	28.300	34.821
13	3.565	4.107	5.009	5.892	7.042	8.634	9.926	11.129	12.340	13.636	15.119	16.985	19.812	22.362	24.736	27.688	29.819	36.478
14	4.075	4.660	5.629	6.571	7.790	9.467	10.821	12.078	13.339	14.685	16.222	18.151	21.064	23.685	26.119	29.141	31.319	38.109
15	4.601	5.229	6.262	7.261	8.547	10.307	11.721	13.030	14.339	15.733	17.322	19.311	22.307	24.996	27.488	30.578	32.801	39.719
16	5.142	5.812	6.908	7.962	9.312	11.152	12.624	13.983	15.338	16.780	18.418	20.465	23.542	26.296	28.845	32.000	34.267	41.308
17	5.697	6.408	7.564	8.672	10.085	12.002	13.531	14.937	16.338	17.824	19.511	21.615	24.769	27.587	30.191	33.409	35.718	42.879
18	6.265	7.015	8.231	9.390	10.865	12.857	14.440	15.893	17.338	18.868	20.601	22.760	25.989	28.869	31.526	34.805	37.156	44.434
19	6.844	7.633	8.907	10.117	11.651	13.716	15.352	16.850	18.338	19.910	21.689	23.900	27.204	30.144	32.852	36.191	38.582	45.973
20	7.434	8.260	9.591	10.851	12.443	14.578	16.266	17.809	19.337	20.951	22.775	25.038	28.412	31.410	34.170	37.566	39.997	47.498
21	8.034	8.897	10.283	11.591	13.240	15.445	17.182	18.768	20.337	21.991	23.858	26.171	29.615	32.671	35.479	38.932	41.401	49.011
22	8.643	9.542	10.982	12.338	14.041	16.314	18.101	19.729	21.337	23.031	24.939	27.301	30.813	33.924	36.781	40.289	42.796	50.511
23	9.260	10.196	11.689	13.091	14.848	17.187	19.021	20.690	22.337	24.069	26.018	28.429	32.007	35.172	38.076	41.638	44.181	52.000
24	9.886	10.856	12.401	13.848	15.659	18.062	19.943	21.752	23.337	25.106	27.096	29.553	33.196	36.415	39.364	42.980	45.559	53.479
25	10.520	11.524	13.120	14.611	16.473	18.940	20.867	22.616	24.337	26.143	28.172	30.675	34.382	37.652	40.646	44.314	46.928	54.947
26	11.160	12.198	13.844	15.379	17.292	19.820	21.792	23.579	25.336	27.179	29.246	31.795	35.563	38.885	41.923	45.642	48.290	56.407
27	11.808	12.879	14.573	16.151	18.114	20.703	22.719	24.544	26.336	28.214	30.319	32.912	36.741	40.113	43.195	46.963	49.645	57.858
28	12.461	13.565	15.308	16.928	18.939	21.588	23.647	25.509	27.336	29.249	31.391	34.027	37.916	41.337	44.461	48.278	50.993	59.300
29	13.121	14.256	16.047	17.708	19.768	22.475	24.577	26.475	28.336	30.283	32.461	35.139	39.087	42.557	45.722	49.588	52.336	60.735
30	13.787	14.953	16.791	18.493	20.599	23.364	25.508	27.442	29.336	31.316	33.530	36.250	40.256	43.773	46.979	50.892	53.672	62.162
35	17.192	18.509	20.569	22.465	24.797	27.836	30.178	32.282	34.336	36.475	38.859	41.778	46.059	49.802	53.203	57.342	60.275	69.199
40	20.707	22.164	24.433	26.509	29.051	32.345	34.872	37.134	39.335	41.622	44.165	47.269	51.805	55.758	59.342	63.691	66.766	76.095
45	24.311	25.901	28.366	30.612	33.350	36.884	39.585	41.995	44.335	46.761	49.452	52.729	57.505	61.656	65.410	69.957	73.166	82.876
50	27.991	29.707	32.357	34.764	37.689	41.449	44.313	46.864	49.335	51.892	54.723	58.164	63.167	67.505	71.420	76.154	79.490	89.561
60	35.534	37.485	40.482	43.188	46.459	50.641	53.809	56.620	59.335	62.135	65.227	68.972	74.397	79.082	83.298	88.379	91.952	102.695
70	43.275	45.442	48.758	51.739	55.329	59.898	63.346	66.396	69.334	72.358	75.689	79.715	85.527	90.531	95.023	100.425	104.215	115.578
80	51.172	53.540	57.153	60.391	64.278	69.207	72.915	76.188	79.334	82.566	86.120	90.405	96.578	101.879	106.629	112.329	116.321	128.261
90	59.196	61.754	65.647	69.126	73.291	78.558	82.511	85.993	89.334	92.761	96.524	101.054	107.565	113.145	118.136	124.116	128.299	140.782
100	67.328	70.065	74.222	77.929	82.358	87.945	92.129	95.808	99.334	102.946	106.906	111.667	118.498	124.342	129.561	135.807	140.169	153.167
120	83.852	86.923	91.573	95.705	100.624	106.806	111.419	115.465	119.334	123.289	127.616	132.806	140.233	146.567	152.211	158.950	163.648	177.603
140	100.655	104.034	109.137	113.659	119.029	125.758	130.766	135.149	139.334	143.604	148.269	153.854	161.827	168.613	174.648	181.840	186.847	201.683
160	117.679	121.346	126.870	131.756	137.546	144.783	150.158	154.856	159.334	163.898	168.876	174.828	183.311	190.516	196.915	204.530	209.824	225.481
180	134.884	138.820	144.741	149.969	156.153	163.868	169.588	174.580	179.334	184.173	189.446	195.743	204.704	212.304	219.044	227.056	232.620	249.048
200	152.241	156.432	162.728	168.279	174.835	183.003	189.049	194.319	199.334	204.434	209.985	216.609	226.021	233.994	241.058	249.445	255.264	272.423

Table A-4 Percentiles of the F distribution

Upper 25% point of the F distribution

$$p = P(F_{\nu_1, \nu_2} < F_{\nu_1, \nu_2, p})$$

F distribution

DEGREES OF FREEDOM FOR NUMERATOR

ν_2 \\ ν_1	1	2	3	4	5	6	7	8	9	10	11	12	13	14	15	16	17	18	19	20	25	30	40	50	100	150	200
1	5.83	7.50	8.20	8.58	8.82	8.98	9.10	9.19	9.26	9.32	9.37	9.41	9.44	9.47	9.49	9.52	9.53	9.55	9.57	9.58	9.63	9.67	9.71	9.74	9.80	9.81	9.82
2	2.57	3.00	3.15	3.23	3.28	3.31	3.34	3.35	3.37	3.38	3.39	3.39	3.40	3.41	3.41	3.41	3.42	3.42	3.43	3.43	3.44	3.44	3.45	3.46	3.47	3.47	3.47
3	2.02	2.28	2.36	2.39	2.41	2.42	2.43	2.44	2.44	2.44	2.45	2.45	2.45	2.45	2.46	2.46	2.46	2.46	2.46	2.46	2.46	2.47	2.47	2.47	2.47	2.47	2.47
4	1.81	2.00	2.05	2.06	2.07	2.08	2.08	2.08	2.08	2.08	2.08	2.08	2.08	2.08	2.08	2.08	2.08	2.08	2.08	2.08	2.08	2.08	2.08	2.08	2.08	2.08	2.08
5	1.69	1.85	1.88	1.89	1.89	1.89	1.89	1.89	1.89	1.89	1.89	1.89	1.89	1.89	1.89	1.88	1.88	1.88	1.88	1.88	1.88	1.88	1.88	1.88	1.87	1.87	1.87
6	1.62	1.76	1.78	1.79	1.79	1.78	1.78	1.78	1.77	1.77	1.77	1.77	1.77	1.76	1.76	1.76	1.76	1.76	1.76	1.76	1.75	1.75	1.75	1.75	1.74	1.74	1.74
7	1.57	1.70	1.72	1.72	1.71	1.71	1.70	1.70	1.69	1.69	1.69	1.68	1.68	1.68	1.68	1.68	1.67	1.67	1.67	1.67	1.66	1.66	1.66	1.66	1.65	1.65	1.65
8	1.54	1.66	1.67	1.66	1.66	1.65	1.64	1.64	1.63	1.63	1.63	1.62	1.62	1.62	1.62	1.62	1.61	1.61	1.61	1.61	1.60	1.60	1.59	1.59	1.58	1.58	1.58
9	1.51	1.62	1.63	1.63	1.62	1.61	1.60	1.60	1.59	1.59	1.58	1.58	1.58	1.57	1.57	1.57	1.57	1.56	1.56	1.56	1.55	1.55	1.54	1.54	1.53	1.53	1.53
10	1.49	1.60	1.60	1.59	1.59	1.58	1.57	1.56	1.56	1.55	1.55	1.54	1.54	1.54	1.53	1.53	1.53	1.53	1.53	1.52	1.52	1.51	1.51	1.50	1.49	1.49	1.49
11	1.47	1.58	1.58	1.57	1.56	1.55	1.54	1.53	1.53	1.52	1.52	1.51	1.51	1.51	1.50	1.50	1.50	1.50	1.49	1.49	1.49	1.48	1.47	1.47	1.46	1.46	1.46
12	1.46	1.56	1.56	1.55	1.54	1.53	1.52	1.51	1.51	1.50	1.49	1.49	1.49	1.48	1.48	1.48	1.47	1.47	1.47	1.47	1.46	1.45	1.45	1.44	1.43	1.43	1.43
13	1.45	1.55	1.55	1.53	1.52	1.51	1.50	1.49	1.49	1.48	1.47	1.47	1.47	1.46	1.46	1.46	1.45	1.45	1.45	1.45	1.44	1.43	1.42	1.42	1.41	1.41	1.40
14	1.44	1.53	1.53	1.52	1.51	1.50	1.48	1.48	1.47	1.46	1.46	1.45	1.45	1.44	1.44	1.44	1.44	1.43	1.43	1.43	1.42	1.41	1.41	1.40	1.39	1.39	1.38
15	1.43	1.52	1.52	1.51	1.49	1.48	1.47	1.46	1.46	1.45	1.44	1.44	1.43	1.43	1.43	1.42	1.42	1.42	1.41	1.41	1.40	1.40	1.39	1.38	1.37	1.37	1.37
16	1.42	1.51	1.51	1.50	1.48	1.47	1.46	1.45	1.44	1.44	1.43	1.43	1.42	1.42	1.41	1.41	1.41	1.40	1.40	1.40	1.39	1.38	1.37	1.37	1.36	1.35	1.35
17	1.42	1.51	1.50	1.49	1.47	1.46	1.45	1.44	1.43	1.43	1.42	1.41	1.41	1.41	1.40	1.40	1.39	1.39	1.39	1.39	1.38	1.37	1.36	1.36	1.34	1.34	1.34
18	1.41	1.50	1.49	1.48	1.46	1.45	1.44	1.43	1.42	1.42	1.41	1.40	1.40	1.40	1.39	1.39	1.38	1.38	1.38	1.38	1.37	1.36	1.35	1.35	1.33	1.33	1.32
19	1.41	1.49	1.49	1.47	1.46	1.44	1.43	1.42	1.41	1.41	1.40	1.40	1.39	1.39	1.38	1.38	1.37	1.37	1.37	1.37	1.36	1.35	1.34	1.33	1.32	1.31	1.31
20	1.40	1.49	1.48	1.47	1.45	1.44	1.43	1.42	1.41	1.40	1.39	1.39	1.38	1.37	1.37	1.37	1.36	1.36	1.36	1.36	1.35	1.34	1.33	1.33	1.31	1.30	1.30
21	1.40	1.48	1.48	1.46	1.44	1.43	1.42	1.41	1.40	1.39	1.39	1.38	1.37	1.37	1.37	1.36	1.36	1.35	1.35	1.35	1.34	1.33	1.32	1.32	1.30	1.30	1.29
22	1.40	1.48	1.47	1.45	1.44	1.42	1.41	1.40	1.39	1.39	1.38	1.37	1.37	1.36	1.36	1.36	1.35	1.35	1.35	1.34	1.33	1.32	1.31	1.31	1.30	1.29	1.28
23	1.39	1.47	1.47	1.45	1.43	1.42	1.41	1.40	1.39	1.38	1.37	1.37	1.36	1.36	1.35	1.35	1.34	1.34	1.34	1.34	1.33	1.32	1.31	1.30	1.29	1.28	1.28
24	1.39	1.47	1.46	1.44	1.43	1.41	1.40	1.39	1.38	1.38	1.37	1.36	1.36	1.35	1.35	1.34	1.34	1.34	1.33	1.33	1.32	1.31	1.31	1.30	1.28	1.27	1.27
25	1.39	1.47	1.46	1.44	1.42	1.41	1.40	1.39	1.38	1.37	1.36	1.36	1.35	1.35	1.34	1.34	1.33	1.33	1.33	1.33	1.31	1.31	1.30	1.29	1.27	1.27	1.26
26	1.38	1.46	1.45	1.44	1.42	1.41	1.39	1.38	1.37	1.37	1.36	1.35	1.35	1.34	1.34	1.33	1.33	1.33	1.32	1.32	1.31	1.30	1.29	1.28	1.27	1.26	1.26
27	1.38	1.46	1.45	1.43	1.42	1.40	1.39	1.38	1.37	1.36	1.35	1.35	1.34	1.34	1.33	1.33	1.32	1.32	1.32	1.32	1.30	1.30	1.28	1.28	1.26	1.25	1.25
28	1.38	1.46	1.45	1.43	1.41	1.40	1.39	1.38	1.37	1.36	1.35	1.34	1.34	1.33	1.33	1.32	1.32	1.32	1.31	1.31	1.30	1.29	1.28	1.27	1.25	1.25	1.24
29	1.38	1.45	1.45	1.43	1.41	1.40	1.38	1.37	1.36	1.35	1.35	1.34	1.33	1.33	1.32	1.32	1.32	1.31	1.31	1.31	1.30	1.29	1.27	1.27	1.25	1.24	1.24
30	1.38	1.45	1.44	1.42	1.41	1.39	1.38	1.37	1.36	1.35	1.34	1.34	1.33	1.32	1.32	1.32	1.31	1.31	1.30	1.30	1.29	1.28	1.27	1.26	1.25	1.24	1.24
32	1.37	1.45	1.44	1.42	1.40	1.39	1.37	1.36	1.35	1.34	1.34	1.33	1.32	1.32	1.31	1.31	1.30	1.30	1.30	1.30	1.28	1.28	1.26	1.25	1.24	1.23	1.23
34	1.37	1.44	1.44	1.42	1.40	1.38	1.37	1.36	1.35	1.34	1.33	1.33	1.32	1.31	1.31	1.30	1.30	1.30	1.29	1.29	1.28	1.27	1.26	1.25	1.23	1.22	1.22
36	1.37	1.44	1.43	1.41	1.39	1.38	1.36	1.35	1.34	1.34	1.33	1.32	1.32	1.31	1.30	1.30	1.30	1.29	1.29	1.28	1.27	1.26	1.25	1.24	1.22	1.22	1.21
38	1.37	1.44	1.43	1.41	1.39	1.38	1.36	1.35	1.34	1.33	1.32	1.32	1.31	1.31	1.30	1.29	1.29	1.29	1.28	1.28	1.27	1.26	1.24	1.24	1.22	1.21	1.21
40	1.36	1.44	1.42	1.40	1.39	1.37	1.36	1.35	1.34	1.33	1.32	1.31	1.31	1.30	1.30	1.29	1.29	1.28	1.28	1.28	1.26	1.25	1.24	1.23	1.21	1.21	1.20
42	1.36	1.43	1.42	1.40	1.38	1.37	1.35	1.34	1.33	1.33	1.32	1.31	1.30	1.30	1.29	1.29	1.28	1.28	1.27	1.27	1.26	1.25	1.23	1.23	1.21	1.20	1.20
44	1.36	1.43	1.42	1.40	1.38	1.37	1.35	1.34	1.33	1.32	1.31	1.31	1.30	1.29	1.29	1.28	1.28	1.28	1.27	1.27	1.25	1.24	1.23	1.22	1.20	1.19	1.19
46	1.36	1.43	1.42	1.40	1.38	1.36	1.35	1.34	1.33	1.32	1.31	1.30	1.30	1.29	1.28	1.28	1.28	1.27	1.27	1.26	1.25	1.24	1.23	1.22	1.20	1.19	1.19
48	1.36	1.43	1.41	1.39	1.38	1.36	1.35	1.34	1.33	1.32	1.31	1.30	1.29	1.29	1.28	1.28	1.27	1.27	1.27	1.26	1.25	1.24	1.22	1.22	1.20	1.19	1.18
50	1.35	1.42	1.41	1.39	1.37	1.36	1.34	1.33	1.32	1.31	1.30	1.30	1.29	1.28	1.28	1.27	1.27	1.27	1.26	1.26	1.25	1.23	1.22	1.21	1.19	1.18	1.18
60	1.35	1.42	1.41	1.38	1.37	1.35	1.33	1.32	1.31	1.30	1.29	1.29	1.28	1.27	1.27	1.26	1.26	1.26	1.25	1.25	1.24	1.22	1.21	1.20	1.18	1.17	1.16
70	1.35	1.41	1.40	1.38	1.36	1.35	1.33	1.32	1.31	1.30	1.29	1.28	1.28	1.27	1.26	1.26	1.25	1.25	1.24	1.24	1.23	1.22	1.20	1.19	1.16	1.16	1.15
80	1.34	1.41	1.40	1.38	1.36	1.34	1.33	1.31	1.30	1.29	1.28	1.27	1.27	1.26	1.25	1.25	1.25	1.24	1.24	1.23	1.22	1.21	1.19	1.18	1.15	1.15	1.14
90	1.34	1.41	1.39	1.37	1.35	1.34	1.32	1.31	1.29	1.28	1.28	1.27	1.26	1.26	1.25	1.24	1.24	1.23	1.23	1.23	1.21	1.20	1.18	1.17	1.14	1.13	1.13
100	1.34	1.41	1.39	1.37	1.35	1.34	1.32	1.30	1.29	1.28	1.27	1.27	1.26	1.25	1.25	1.24	1.24	1.23	1.23	1.22	1.21	1.20	1.18	1.17	1.14	1.13	1.13
125	1.34	1.40	1.39	1.36	1.34	1.33	1.31	1.30	1.29	1.28	1.27	1.26	1.25	1.24	1.24	1.24	1.23	1.23	1.22	1.22	1.20	1.19	1.18	1.16	1.14	1.12	1.12
150	1.33	1.40	1.38	1.36	1.34	1.32	1.31	1.29	1.28	1.27	1.26	1.26	1.25	1.24	1.23	1.23	1.22	1.22	1.21	1.21	1.20	1.19	1.16	1.15	1.13	1.11	1.11
200	1.33	1.40	1.38	1.36	1.34	1.32	1.30	1.29	1.28	1.27	1.26	1.25	1.24	1.24	1.23	1.23	1.22	1.21	1.21	1.20	1.19	1.18	1.16	1.14	1.11	1.10	1.10
300	1.33	1.39	1.38	1.35	1.33	1.32	1.30	1.28	1.27	1.26	1.26	1.25	1.24	1.23	1.23	1.22	1.22	1.21	1.21	1.20	1.19	1.17	1.15	1.14	1.10	1.09	1.09
500	1.33	1.39	1.38	1.35	1.33	1.31	1.30	1.28	1.27	1.26	1.25	1.24	1.24	1.23	1.22	1.22	1.21	1.21	1.20	1.20	1.18	1.17	1.15	1.13	1.10	1.09	1.08
1000	1.32	1.39	1.37	1.35	1.33	1.31	1.29	1.28	1.27	1.26	1.25	1.24	1.23	1.23	1.22	1.21	1.21	1.20	1.20	1.20	1.18	1.16	1.14	1.13	1.10	1.08	1.07

DEGREES OF FREEDOM FOR DENOMINATOR

Table A-4   Percentiles of the F distribution (continued)

Upper 10% point of the F distribution

DEGREES OF FREEDOM FOR NUMERATOR

denom	1	2	3	4	5	6	7	8	9	10	11	12	13	14	15	16	17	18	19	20	25	30	40	50	100	150	200
1	39.9	49.5	53.6	55.8	57.2	58.2	58.9	59.4	59.9	60.2	60.5	60.7	60.9	61.1	61.2	61.3	61.5	61.6	61.7	61.7	62.1	62.3	62.5	62.7	63.0	63.1	63.2
2	8.53	9.00	9.16	9.24	9.29	9.33	9.35	9.37	9.38	9.39	9.40	9.41	9.41	9.42	9.42	9.43	9.43	9.44	9.44	9.44	9.45	9.46	9.47	9.47	9.48	9.48	9.49
3	5.54	5.46	5.39	5.34	5.31	5.28	5.27	5.25	5.24	5.23	5.22	5.22	5.21	5.20	5.20	5.20	5.19	5.19	5.19	5.18	5.17	5.17	5.16	5.15	5.14	5.14	5.14
4	4.54	4.32	4.19	4.11	4.05	4.01	3.98	3.95	3.94	3.92	3.91	3.90	3.89	3.88	3.87	3.86	3.86	3.85	3.85	3.84	3.83	3.82	3.80	3.80	3.78	3.77	3.77
5	4.06	3.78	3.62	3.52	3.45	3.40	3.37	3.34	3.32	3.30	3.28	3.27	3.26	3.25	3.24	3.23	3.22	3.22	3.21	3.21	3.19	3.17	3.16	3.15	3.13	3.12	3.12
6	3.78	3.46	3.29	3.18	3.11	3.05	3.01	2.98	2.96	2.94	2.92	2.90	2.89	2.88	2.87	2.86	2.85	2.85	2.84	2.84	2.81	2.80	2.78	2.77	2.75	2.74	2.73
7	3.59	3.26	3.07	2.96	2.88	2.83	2.78	2.75	2.72	2.70	2.68	2.67	2.65	2.64	2.63	2.62	2.61	2.61	2.60	2.59	2.57	2.56	2.54	2.52	2.50	2.49	2.48
8	3.46	3.11	2.92	2.81	2.73	2.67	2.62	2.59	2.56	2.54	2.52	2.50	2.49	2.48	2.46	2.45	2.45	2.44	2.43	2.42	2.40	2.38	2.36	2.35	2.32	2.31	2.31
9	3.36	3.01	2.81	2.69	2.61	2.55	2.51	2.47	2.44	2.42	2.40	2.38	2.36	2.35	2.34	2.33	2.32	2.31	2.30	2.30	2.27	2.25	2.23	2.22	2.19	2.18	2.17
10	3.29	2.92	2.73	2.61	2.52	2.46	2.41	2.38	2.35	2.32	2.30	2.28	2.27	2.26	2.24	2.23	2.22	2.22	2.21	2.20	2.17	2.16	2.13	2.12	2.09	2.08	2.07
11	3.23	2.86	2.66	2.54	2.45	2.39	2.34	2.30	2.27	2.25	2.23	2.21	2.19	2.18	2.17	2.16	2.15	2.14	2.13	2.12	2.10	2.08	2.05	2.04	2.01	1.99	1.99
12	3.18	2.81	2.61	2.48	2.39	2.33	2.28	2.24	2.21	2.19	2.17	2.15	2.13	2.12	2.10	2.09	2.08	2.08	2.07	2.06	2.04	2.01	1.99	1.97	1.94	1.93	1.92
13	3.14	2.76	2.56	2.43	2.35	2.28	2.23	2.20	2.16	2.14	2.12	2.10	2.08	2.07	2.05	2.04	2.03	2.02	2.01	2.01	1.98	1.96	1.93	1.92	1.88	1.87	1.86
14	3.10	2.73	2.52	2.39	2.31	2.24	2.19	2.15	2.12	2.10	2.08	2.05	2.04	2.02	2.01	2.00	1.99	1.98	1.97	1.96	1.93	1.91	1.89	1.87	1.83	1.82	1.82
15	3.07	2.70	2.49	2.36	2.27	2.21	2.16	2.12	2.09	2.06	2.04	2.02	2.00	1.99	1.97	1.96	1.95	1.94	1.93	1.92	1.89	1.87	1.85	1.83	1.79	1.78	1.77
16	3.05	2.67	2.46	2.33	2.24	2.18	2.13	2.09	2.06	2.03	2.01	1.99	1.97	1.95	1.94	1.93	1.92	1.91	1.90	1.89	1.86	1.84	1.81	1.79	1.76	1.74	1.74
17	3.03	2.64	2.44	2.31	2.22	2.15	2.10	2.06	2.03	2.00	1.98	1.96	1.94	1.93	1.91	1.90	1.89	1.88	1.87	1.86	1.83	1.81	1.78	1.76	1.73	1.71	1.71
18	3.01	2.62	2.42	2.29	2.20	2.13	2.08	2.04	2.00	1.98	1.95	1.93	1.92	1.90	1.89	1.87	1.86	1.85	1.84	1.84	1.80	1.78	1.75	1.74	1.70	1.68	1.68
19	2.99	2.61	2.40	2.27	2.18	2.11	2.06	2.02	1.98	1.96	1.93	1.91	1.89	1.88	1.86	1.85	1.84	1.83	1.82	1.81	1.78	1.76	1.73	1.71	1.67	1.66	1.65
20	2.97	2.59	2.38	2.25	2.16	2.09	2.04	2.00	1.96	1.94	1.91	1.89	1.87	1.86	1.84	1.83	1.82	1.81	1.80	1.79	1.76	1.74	1.71	1.69	1.65	1.64	1.63
21	2.96	2.57	2.36	2.23	2.14	2.08	2.02	1.98	1.95	1.92	1.90	1.87	1.86	1.84	1.83	1.81	1.80	1.79	1.78	1.78	1.74	1.72	1.69	1.67	1.63	1.62	1.61
22	2.95	2.56	2.35	2.22	2.13	2.06	2.01	1.97	1.93	1.90	1.88	1.86	1.84	1.83	1.81	1.80	1.79	1.78	1.77	1.76	1.73	1.70	1.67	1.65	1.61	1.60	1.59
23	2.94	2.55	2.34	2.21	2.11	2.05	1.99	1.95	1.92	1.89	1.87	1.84	1.83	1.81	1.80	1.78	1.77	1.76	1.75	1.74	1.71	1.69	1.66	1.64	1.59	1.58	1.57
24	2.93	2.54	2.33	2.19	2.10	2.04	1.98	1.94	1.91	1.88	1.85	1.83	1.81	1.80	1.78	1.77	1.76	1.75	1.74	1.73	1.70	1.67	1.64	1.62	1.58	1.56	1.56
25	2.92	2.53	2.32	2.18	2.09	2.02	1.97	1.93	1.89	1.87	1.84	1.82	1.80	1.79	1.77	1.76	1.75	1.74	1.73	1.72	1.68	1.66	1.63	1.61	1.56	1.55	1.54
26	2.91	2.52	2.31	2.17	2.08	2.01	1.96	1.92	1.88	1.86	1.83	1.81	1.79	1.77	1.76	1.75	1.73	1.72	1.71	1.71	1.67	1.65	1.61	1.59	1.55	1.54	1.53
27	2.90	2.51	2.30	2.17	2.07	2.00	1.95	1.91	1.87	1.85	1.82	1.80	1.78	1.76	1.75	1.74	1.72	1.71	1.70	1.70	1.66	1.64	1.60	1.58	1.54	1.52	1.52
28	2.89	2.50	2.29	2.16	2.06	2.00	1.94	1.90	1.87	1.84	1.81	1.79	1.77	1.75	1.74	1.73	1.71	1.70	1.69	1.69	1.65	1.63	1.59	1.57	1.53	1.51	1.50
29	2.89	2.50	2.28	2.15	2.06	1.99	1.93	1.89	1.86	1.83	1.80	1.78	1.76	1.75	1.73	1.72	1.71	1.69	1.68	1.68	1.64	1.62	1.58	1.56	1.52	1.50	1.49
30	2.88	2.49	2.28	2.14	2.05	1.98	1.93	1.88	1.85	1.82	1.79	1.77	1.75	1.74	1.72	1.71	1.70	1.69	1.68	1.67	1.63	1.61	1.57	1.55	1.51	1.49	1.48
32	2.87	2.48	2.26	2.13	2.04	1.97	1.91	1.87	1.83	1.81	1.78	1.76	1.74	1.72	1.71	1.69	1.68	1.67	1.66	1.65	1.62	1.59	1.56	1.53	1.49	1.47	1.46
34	2.86	2.47	2.25	2.12	2.02	1.96	1.90	1.86	1.82	1.79	1.77	1.75	1.73	1.71	1.69	1.68	1.67	1.66	1.65	1.64	1.60	1.58	1.54	1.52	1.47	1.46	1.45
36	2.85	2.46	2.24	2.11	2.01	1.94	1.89	1.85	1.81	1.78	1.75	1.73	1.71	1.70	1.68	1.67	1.66	1.65	1.64	1.63	1.59	1.56	1.53	1.51	1.46	1.44	1.43
38	2.84	2.45	2.23	2.10	2.01	1.94	1.88	1.84	1.80	1.77	1.75	1.72	1.70	1.69	1.67	1.66	1.65	1.63	1.62	1.61	1.58	1.55	1.52	1.49	1.45	1.43	1.42
40	2.84	2.44	2.23	2.09	2.00	1.93	1.87	1.83	1.79	1.76	1.74	1.71	1.70	1.68	1.66	1.65	1.64	1.62	1.61	1.61	1.57	1.54	1.51	1.48	1.44	1.42	1.41
42	2.83	2.43	2.22	2.08	1.99	1.92	1.86	1.82	1.78	1.75	1.73	1.71	1.69	1.67	1.65	1.64	1.63	1.62	1.61	1.60	1.56	1.53	1.50	1.47	1.42	1.40	1.40
44	2.82	2.43	2.21	2.08	1.98	1.91	1.86	1.81	1.77	1.75	1.72	1.70	1.68	1.66	1.65	1.63	1.62	1.61	1.60	1.59	1.55	1.52	1.49	1.46	1.41	1.39	1.39
46	2.82	2.42	2.21	2.07	1.98	1.91	1.85	1.81	1.77	1.74	1.71	1.69	1.67	1.65	1.64	1.62	1.61	1.60	1.59	1.58	1.54	1.52	1.48	1.46	1.40	1.39	1.38
48	2.81	2.42	2.20	2.07	1.97	1.90	1.85	1.80	1.77	1.74	1.71	1.69	1.67	1.65	1.63	1.62	1.61	1.59	1.58	1.57	1.54	1.51	1.48	1.45	1.40	1.38	1.37
50	2.81	2.41	2.20	2.06	1.97	1.90	1.84	1.80	1.76	1.73	1.70	1.68	1.66	1.64	1.63	1.61	1.60	1.59	1.58	1.57	1.53	1.50	1.46	1.44	1.39	1.37	1.36
60	2.79	2.39	2.18	2.04	1.95	1.87	1.82	1.77	1.74	1.71	1.68	1.66	1.64	1.62	1.60	1.59	1.58	1.56	1.55	1.54	1.50	1.48	1.44	1.41	1.36	1.34	1.33
70	2.78	2.38	2.16	2.03	1.93	1.86	1.80	1.76	1.72	1.69	1.66	1.64	1.62	1.60	1.59	1.57	1.56	1.55	1.54	1.53	1.49	1.46	1.42	1.39	1.34	1.31	1.30
80	2.77	2.37	2.15	2.02	1.92	1.85	1.79	1.75	1.71	1.68	1.65	1.63	1.61	1.59	1.57	1.56	1.55	1.53	1.52	1.51	1.47	1.44	1.40	1.38	1.32	1.30	1.28
90	2.76	2.36	2.15	2.01	1.91	1.84	1.78	1.74	1.70	1.67	1.64	1.62	1.60	1.58	1.56	1.55	1.54	1.52	1.51	1.50	1.46	1.43	1.39	1.36	1.30	1.28	1.27
100	2.76	2.36	2.14	2.00	1.91	1.83	1.78	1.73	1.69	1.66	1.64	1.61	1.59	1.57	1.56	1.54	1.53	1.52	1.50	1.49	1.45	1.42	1.38	1.35	1.29	1.27	1.26
125	2.75	2.35	2.13	1.99	1.89	1.82	1.77	1.72	1.68	1.65	1.62	1.60	1.58	1.56	1.54	1.53	1.51	1.50	1.49	1.48	1.44	1.41	1.36	1.34	1.27	1.25	1.23
150	2.74	2.34	2.12	1.98	1.89	1.81	1.76	1.71	1.67	1.64	1.61	1.59	1.57	1.55	1.53	1.52	1.50	1.49	1.48	1.47	1.43	1.40	1.35	1.33	1.26	1.23	1.22
200	2.73	2.33	2.11	1.97	1.88	1.80	1.75	1.70	1.66	1.63	1.60	1.58	1.56	1.54	1.52	1.51	1.49	1.48	1.47	1.46	1.41	1.38	1.34	1.31	1.24	1.21	1.20
300	2.72	2.32	2.10	1.96	1.87	1.79	1.74	1.69	1.65	1.62	1.59	1.57	1.55	1.53	1.51	1.49	1.48	1.47	1.46	1.45	1.40	1.37	1.32	1.29	1.21	1.19	1.18
500	2.72	2.31	2.09	1.96	1.86	1.79	1.73	1.68	1.64	1.61	1.58	1.56	1.54	1.52	1.50	1.49	1.47	1.46	1.45	1.44	1.39	1.36	1.31	1.28	1.21	1.18	1.16
1000	2.71	2.31	2.09	1.95	1.85	1.78	1.72	1.68	1.64	1.61	1.58	1.55	1.53	1.51	1.49	1.48	1.46	1.45	1.44	1.43	1.38	1.35	1.30	1.27	1.20	1.16	1.15

DEGREES OF FREEDOM FOR DENOMINATOR

Table A-4 Percentiles of the *F* distribution (*continued*)

Upper 5% point of the F distribution

DEGREES OF FREEDOM FOR NUMERATOR

den \\ num	1	2	3	4	5	6	7	8	9	10	11	12	13	14	15	16	17	18	19	20	25	30	40	50	100	150	200
1	161	200	216	225	230	234	237	239	241	242	243	244	245	245	246	246	247	247	248	248	249	250	251	252	253	253	254
2	18.5	19.0	19.2	19.2	19.3	19.3	19.4	19.4	19.4	19.4	19.4	19.4	19.4	19.4	19.4	19.4	19.4	19.4	19.4	19.4	19.4	19.5	19.5	19.5	19.5	19.5	19.5
3	10.1	9.55	9.28	9.12	9.01	8.94	8.89	8.85	8.81	8.79	8.76	8.74	8.73	8.71	8.70	8.69	8.68	8.67	8.67	8.66	8.63	8.62	8.59	8.58	8.55	8.54	8.54
4	7.71	6.94	6.59	6.39	6.26	6.16	6.09	6.04	6.00	5.96	5.94	5.91	5.89	5.87	5.86	5.84	5.83	5.82	5.81	5.80	5.77	5.75	5.72	5.70	5.66	5.65	5.65
5	6.61	5.79	5.41	5.19	5.05	4.95	4.88	4.82	4.77	4.74	4.70	4.68	4.66	4.64	4.62	4.60	4.59	4.58	4.57	4.56	4.52	4.50	4.46	4.44	4.41	4.39	4.39
6	5.99	5.14	4.76	4.53	4.39	4.28	4.21	4.15	4.10	4.06	4.03	4.00	3.98	3.96	3.94	3.92	3.91	3.90	3.88	3.87	3.83	3.81	3.77	3.75	3.71	3.70	3.69
7	5.59	4.74	4.35	4.12	3.97	3.87	3.79	3.73	3.68	3.64	3.60	3.57	3.55	3.53	3.51	3.49	3.48	3.47	3.46	3.44	3.40	3.38	3.34	3.32	3.27	3.26	3.25
8	5.32	4.46	4.07	3.84	3.69	3.58	3.50	3.44	3.39	3.35	3.31	3.28	3.26	3.24	3.22	3.20	3.19	3.17	3.16	3.15	3.11	3.08	3.04	3.02	2.97	2.96	2.95
9	5.12	4.26	3.86	3.63	3.48	3.37	3.29	3.23	3.18	3.14	3.10	3.07	3.05	3.03	3.01	2.99	2.97	2.96	2.95	2.94	2.89	2.86	2.83	2.80	2.76	2.74	2.73
10	4.96	4.10	3.71	3.48	3.33	3.22	3.14	3.07	3.02	2.98	2.94	2.91	2.89	2.86	2.85	2.83	2.81	2.80	2.79	2.77	2.73	2.70	2.66	2.64	2.59	2.57	2.56
11	4.84	3.98	3.59	3.36	3.20	3.09	3.01	2.95	2.90	2.85	2.82	2.79	2.76	2.74	2.72	2.70	2.69	2.67	2.66	2.65	2.60	2.57	2.53	2.51	2.46	2.44	2.43
12	4.75	3.89	3.49	3.26	3.11	3.00	2.91	2.85	2.80	2.75	2.72	2.69	2.66	2.64	2.62	2.60	2.58	2.57	2.56	2.54	2.50	2.47	2.43	2.40	2.35	2.33	2.32
13	4.67	3.81	3.41	3.18	3.03	2.92	2.83	2.77	2.71	2.67	2.63	2.60	2.58	2.55	2.53	2.51	2.50	2.48	2.47	2.46	2.41	2.38	2.34	2.31	2.26	2.24	2.23
14	4.60	3.74	3.34	3.11	2.96	2.85	2.76	2.70	2.65	2.60	2.57	2.53	2.51	2.48	2.46	2.44	2.43	2.41	2.40	2.39	2.34	2.31	2.27	2.24	2.19	2.17	2.16
15	4.54	3.68	3.29	3.05	2.90	2.79	2.71	2.64	2.59	2.54	2.51	2.48	2.45	2.42	2.40	2.38	2.37	2.35	2.34	2.33	2.28	2.25	2.20	2.18	2.12	2.10	2.10
16	4.49	3.63	3.24	3.01	2.85	2.74	2.66	2.59	2.54	2.49	2.46	2.42	2.40	2.37	2.35	2.33	2.32	2.30	2.29	2.28	2.23	2.19	2.15	2.12	2.07	2.05	2.04
17	4.45	3.59	3.20	2.96	2.81	2.70	2.61	2.55	2.49	2.45	2.41	2.38	2.35	2.33	2.31	2.29	2.27	2.26	2.24	2.23	2.18	2.15	2.10	2.08	2.02	2.00	1.99
18	4.41	3.55	3.16	2.93	2.77	2.66	2.58	2.51	2.46	2.41	2.37	2.34	2.31	2.29	2.27	2.25	2.23	2.22	2.20	2.19	2.14	2.11	2.06	2.04	1.98	1.96	1.95
19	4.38	3.52	3.13	2.90	2.74	2.63	2.54	2.48	2.42	2.38	2.34	2.31	2.28	2.26	2.23	2.21	2.20	2.18	2.17	2.16	2.11	2.07	2.03	2.00	1.94	1.92	1.91
20	4.35	3.49	3.10	2.87	2.71	2.60	2.51	2.45	2.39	2.35	2.31	2.28	2.25	2.22	2.20	2.18	2.17	2.15	2.14	2.12	2.07	2.04	1.99	1.97	1.91	1.89	1.88
21	4.32	3.47	3.07	2.84	2.68	2.57	2.49	2.42	2.37	2.32	2.28	2.25	2.22	2.20	2.18	2.16	2.14	2.12	2.11	2.10	2.05	2.01	1.96	1.94	1.88	1.86	1.84
22	4.30	3.44	3.05	2.82	2.66	2.55	2.46	2.40	2.34	2.30	2.26	2.23	2.20	2.17	2.15	2.13	2.11	2.10	2.08	2.07	2.02	1.98	1.94	1.91	1.85	1.83	1.82
23	4.28	3.42	3.03	2.80	2.64	2.53	2.44	2.37	2.32	2.27	2.24	2.20	2.18	2.15	2.13	2.11	2.09	2.08	2.06	2.05	2.00	1.96	1.91	1.88	1.82	1.80	1.79
24	4.26	3.40	3.01	2.78	2.62	2.51	2.42	2.36	2.30	2.25	2.22	2.18	2.15	2.13	2.11	2.09	2.07	2.05	2.04	2.03	1.97	1.94	1.89	1.86	1.80	1.78	1.77
25	4.24	3.39	2.99	2.75	2.60	2.49	2.40	2.34	2.28	2.24	2.20	2.16	2.14	2.11	2.09	2.07	2.05	2.04	2.02	2.01	1.96	1.92	1.87	1.84	1.78	1.76	1.75
26	4.23	3.37	2.98	2.74	2.59	2.47	2.39	2.32	2.27	2.22	2.18	2.15	2.12	2.09	2.07	2.05	2.03	2.02	2.00	1.99	1.94	1.90	1.85	1.82	1.76	1.74	1.73
27	4.21	3.35	2.96	2.73	2.57	2.46	2.37	2.31	2.25	2.20	2.17	2.13	2.10	2.08	2.06	2.04	2.02	2.00	1.99	1.97	1.92	1.88	1.84	1.81	1.74	1.72	1.71
28	4.20	3.34	2.95	2.71	2.56	2.45	2.36	2.29	2.24	2.19	2.15	2.12	2.09	2.06	2.04	2.02	2.00	1.99	1.97	1.96	1.91	1.87	1.82	1.79	1.73	1.70	1.69
29	4.18	3.33	2.93	2.70	2.55	2.43	2.35	2.28	2.22	2.18	2.14	2.10	2.08	2.05	2.03	2.01	1.99	1.97	1.96	1.94	1.89	1.85	1.81	1.77	1.71	1.69	1.67
30	4.17	3.32	2.92	2.69	2.53	2.42	2.33	2.27	2.21	2.16	2.13	2.09	2.06	2.04	2.01	1.99	1.98	1.96	1.95	1.93	1.88	1.84	1.79	1.76	1.70	1.67	1.66
32	4.15	3.29	2.90	2.67	2.51	2.40	2.31	2.24	2.19	2.14	2.10	2.07	2.04	2.01	1.99	1.97	1.95	1.94	1.92	1.91	1.85	1.82	1.77	1.74	1.67	1.64	1.63
34	4.13	3.28	2.88	2.65	2.49	2.38	2.29	2.23	2.17	2.12	2.08	2.05	2.02	1.99	1.97	1.95	1.93	1.92	1.90	1.89	1.83	1.80	1.75	1.71	1.65	1.62	1.61
36	4.11	3.26	2.87	2.63	2.48	2.36	2.28	2.21	2.15	2.11	2.07	2.03	2.00	1.98	1.95	1.93	1.92	1.90	1.88	1.87	1.81	1.78	1.73	1.69	1.62	1.60	1.59
38	4.10	3.24	2.85	2.62	2.46	2.35	2.26	2.19	2.14	2.09	2.05	2.02	1.99	1.96	1.94	1.92	1.90	1.88	1.87	1.85	1.80	1.76	1.71	1.68	1.61	1.58	1.57
40	4.08	3.23	2.84	2.61	2.45	2.34	2.25	2.18	2.12	2.08	2.04	2.00	1.97	1.95	1.92	1.90	1.89	1.87	1.85	1.84	1.78	1.74	1.69	1.66	1.59	1.56	1.55
42	4.07	3.22	2.83	2.59	2.44	2.32	2.24	2.17	2.11	2.06	2.03	1.99	1.96	1.94	1.91	1.89	1.87	1.86	1.84	1.83	1.77	1.73	1.68	1.65	1.57	1.55	1.53
44	4.06	3.21	2.82	2.58	2.43	2.31	2.23	2.16	2.10	2.05	2.01	1.98	1.95	1.92	1.90	1.88	1.86	1.84	1.83	1.81	1.76	1.72	1.67	1.63	1.56	1.53	1.52
46	4.05	3.20	2.81	2.57	2.42	2.30	2.22	2.15	2.09	2.04	2.00	1.97	1.94	1.91	1.89	1.87	1.85	1.83	1.82	1.80	1.75	1.71	1.65	1.62	1.55	1.52	1.51
48	4.04	3.19	2.80	2.57	2.41	2.29	2.21	2.14	2.08	2.03	1.99	1.96	1.93	1.90	1.88	1.86	1.84	1.82	1.81	1.79	1.74	1.70	1.64	1.61	1.54	1.51	1.49
50	4.03	3.18	2.79	2.56	2.40	2.29	2.20	2.13	2.07	2.03	1.99	1.95	1.92	1.89	1.87	1.85	1.83	1.81	1.80	1.78	1.73	1.69	1.63	1.60	1.52	1.50	1.48
60	4.00	3.15	2.76	2.53	2.37	2.25	2.17	2.10	2.04	1.99	1.95	1.92	1.89	1.86	1.84	1.82	1.80	1.78	1.76	1.75	1.69	1.65	1.59	1.56	1.48	1.45	1.44
70	3.98	3.13	2.74	2.50	2.35	2.23	2.14	2.07	2.02	1.97	1.93	1.89	1.86	1.84	1.81	1.79	1.77	1.75	1.74	1.72	1.66	1.62	1.57	1.53	1.45	1.42	1.40
80	3.96	3.11	2.72	2.49	2.33	2.21	2.13	2.06	2.00	1.95	1.91	1.88	1.84	1.82	1.79	1.77	1.75	1.73	1.72	1.70	1.64	1.60	1.54	1.51	1.43	1.39	1.38
90	3.95	3.10	2.71	2.47	2.32	2.20	2.11	2.04	1.99	1.94	1.90	1.86	1.83	1.80	1.78	1.76	1.74	1.72	1.70	1.69	1.63	1.59	1.53	1.49	1.41	1.38	1.36
100	3.94	3.09	2.70	2.46	2.31	2.19	2.10	2.03	1.97	1.93	1.89	1.85	1.82	1.79	1.77	1.75	1.73	1.71	1.69	1.68	1.62	1.57	1.52	1.48	1.39	1.36	1.34
125	3.92	3.07	2.68	2.44	2.29	2.17	2.08	2.01	1.96	1.91	1.87	1.83	1.80	1.77	1.75	1.73	1.71	1.69	1.67	1.66	1.60	1.55	1.49	1.45	1.36	1.33	1.31
150	3.90	3.06	2.66	2.43	2.27	2.16	2.07	2.00	1.94	1.89	1.85	1.82	1.79	1.76	1.73	1.71	1.69	1.67	1.66	1.64	1.58	1.54	1.48	1.44	1.34	1.31	1.29
200	3.89	3.04	2.65	2.42	2.26	2.14	2.06	1.98	1.93	1.88	1.84	1.80	1.77	1.74	1.72	1.69	1.67	1.66	1.64	1.62	1.56	1.52	1.46	1.41	1.32	1.28	1.26
300	3.87	3.03	2.63	2.40	2.24	2.13	2.04	1.97	1.91	1.86	1.82	1.78	1.75	1.72	1.70	1.68	1.66	1.64	1.62	1.61	1.54	1.50	1.43	1.39	1.30	1.26	1.23
500	3.86	3.01	2.62	2.39	2.23	2.12	2.03	1.96	1.90	1.85	1.81	1.77	1.74	1.71	1.69	1.66	1.64	1.62	1.61	1.59	1.53	1.48	1.42	1.38	1.28	1.23	1.21
1000	3.85	3.00	2.61	2.38	2.22	2.11	2.02	1.95	1.89	1.84	1.80	1.76	1.73	1.70	1.68	1.65	1.63	1.61	1.60	1.58	1.52	1.47	1.41	1.36	1.26	1.22	1.19

DEGREES OF FREEDOM FOR DENOMINATOR

Upper 2.5% point of the F distribution

DEGREES OF FREEDOM FOR NUMERATOR

	1	2	3	4	5	6	7	8	9	10	11	12	13	14	15	16	17	18	19	20	25	30	40	50	100	150	200
1	648	800	864	900	922	937	948	957	963	969	973	977	980	983	985	987	989	990	992	993	998	1001	1006	1008	1013	1015	1016
2	38.5	39.0	39.2	39.2	39.3	39.3	39.4	39.4	39.4	39.4	39.4	39.4	39.4	39.4	39.4	39.4	39.4	39.4	39.4	39.4	39.5	39.5	39.5	39.5	39.5	39.5	39.5
3	17.4	16.0	15.4	15.1	14.9	14.7	14.6	14.5	14.5	14.4	14.4	14.3	14.3	14.3	14.3	14.2	14.2	14.2	14.2	14.2	14.1	14.1	14.0	14.0	14.0	13.9	13.9
4	12.2	10.6	9.98	9.60	9.36	9.20	9.07	8.98	8.90	8.84	8.79	8.75	8.71	8.68	8.66	8.63	8.61	8.59	8.58	8.56	8.50	8.46	8.41	8.38	8.32	8.30	8.29
5	10.0	8.43	7.76	7.39	7.15	6.98	6.85	6.76	6.68	6.62	6.57	6.52	6.49	6.46	6.43	6.40	6.38	6.36	6.34	6.33	6.27	6.23	6.18	6.14	6.08	6.06	6.05
6	8.81	7.26	6.60	6.23	5.99	5.82	5.70	5.60	5.52	5.46	5.41	5.37	5.33	5.30	5.27	5.24	5.22	5.20	5.18	5.17	5.11	5.07	5.01	4.98	4.92	4.89	4.88
7	8.07	6.54	5.89	5.52	5.29	5.12	4.99	4.90	4.82	4.76	4.71	4.67	4.63	4.60	4.57	4.54	4.52	4.50	4.48	4.47	4.40	4.36	4.31	4.28	4.21	4.19	4.18
8	7.57	6.06	5.42	5.05	4.82	4.65	4.53	4.43	4.36	4.30	4.24	4.20	4.16	4.13	4.10	4.08	4.05	4.03	4.02	4.00	3.94	3.89	3.84	3.81	3.74	3.72	3.70
9	7.21	5.71	5.08	4.72	4.48	4.32	4.20	4.10	4.03	3.96	3.91	3.87	3.83	3.80	3.77	3.74	3.72	3.70	3.68	3.67	3.60	3.56	3.51	3.47	3.40	3.38	3.37
10	6.94	5.46	4.83	4.47	4.24	4.07	3.95	3.85	3.78	3.72	3.66	3.62	3.58	3.55	3.52	3.50	3.47	3.45	3.44	3.42	3.35	3.31	3.26	3.22	3.15	3.13	3.12
11	6.72	5.26	4.63	4.28	4.04	3.88	3.76	3.66	3.59	3.53	3.47	3.43	3.39	3.36	3.33	3.30	3.28	3.26	3.24	3.23	3.16	3.12	3.06	3.03	2.96	2.93	2.92
12	6.55	5.10	4.47	4.12	3.89	3.73	3.61	3.51	3.44	3.37	3.32	3.28	3.24	3.21	3.18	3.15	3.13	3.11	3.09	3.07	3.01	2.96	2.91	2.87	2.80	2.78	2.76
13	6.41	4.97	4.35	4.00	3.77	3.60	3.48	3.39	3.31	3.25	3.20	3.15	3.12	3.08	3.05	3.03	3.00	2.98	2.96	2.95	2.88	2.84	2.78	2.74	2.67	2.65	2.63
14	6.30	4.86	4.24	3.89	3.66	3.50	3.38	3.29	3.21	3.15	3.09	3.05	3.01	2.98	2.95	2.92	2.90	2.88	2.86	2.84	2.78	2.73	2.67	2.64	2.56	2.54	2.53
15	6.20	4.77	4.15	3.80	3.58	3.41	3.29	3.20	3.12	3.06	3.01	2.96	2.92	2.89	2.86	2.84	2.81	2.79	2.77	2.76	2.69	2.64	2.59	2.55	2.47	2.45	2.44
16	6.12	4.69	4.08	3.73	3.50	3.34	3.22	3.12	3.05	2.99	2.93	2.89	2.85	2.82	2.79	2.76	2.74	2.72	2.70	2.68	2.61	2.57	2.51	2.47	2.40	2.37	2.36
17	6.04	4.62	4.01	3.66	3.44	3.28	3.16	3.06	2.98	2.92	2.87	2.82	2.79	2.75	2.72	2.70	2.67	2.65	2.63	2.62	2.55	2.50	2.44	2.41	2.33	2.30	2.29
18	5.98	4.56	3.95	3.61	3.38	3.22	3.10	3.01	2.93	2.87	2.81	2.77	2.73	2.70	2.67	2.64	2.62	2.60	2.58	2.56	2.49	2.44	2.38	2.35	2.27	2.24	2.23
19	5.92	4.51	3.90	3.56	3.33	3.17	3.05	2.96	2.88	2.82	2.76	2.72	2.68	2.65	2.62	2.59	2.57	2.55	2.53	2.51	2.44	2.39	2.33	2.30	2.22	2.19	2.18
20	5.87	4.46	3.86	3.51	3.29	3.13	3.01	2.91	2.84	2.77	2.72	2.68	2.64	2.60	2.57	2.55	2.52	2.50	2.48	2.46	2.40	2.35	2.29	2.25	2.17	2.14	2.13
21	5.83	4.42	3.82	3.48	3.25	3.09	2.97	2.87	2.80	2.73	2.68	2.64	2.60	2.56	2.53	2.51	2.48	2.46	2.44	2.42	2.36	2.31	2.25	2.21	2.13	2.10	2.09
22	5.79	4.38	3.78	3.44	3.22	3.05	2.93	2.84	2.76	2.70	2.65	2.60	2.56	2.53	2.50	2.47	2.45	2.43	2.41	2.39	2.32	2.27	2.21	2.17	2.09	2.06	2.05
23	5.75	4.35	3.75	3.41	3.18	3.02	2.90	2.81	2.73	2.67	2.62	2.57	2.53	2.50	2.47	2.44	2.42	2.39	2.37	2.36	2.29	2.24	2.18	2.14	2.06	2.03	2.01
24	5.72	4.32	3.72	3.38	3.15	2.99	2.87	2.78	2.70	2.64	2.59	2.54	2.50	2.47	2.44	2.41	2.39	2.36	2.35	2.33	2.26	2.21	2.15	2.11	2.02	2.00	1.98
25	5.69	4.29	3.69	3.35	3.13	2.97	2.85	2.75	2.68	2.61	2.56	2.51	2.48	2.44	2.41	2.38	2.36	2.34	2.32	2.30	2.23	2.18	2.12	2.08	2.00	1.97	1.95
26	5.66	4.27	3.67	3.33	3.10	2.94	2.82	2.73	2.65	2.59	2.54	2.49	2.45	2.42	2.39	2.36	2.34	2.31	2.29	2.28	2.21	2.16	2.09	2.05	1.97	1.94	1.92
27	5.63	4.24	3.65	3.31	3.08	2.92	2.80	2.71	2.63	2.57	2.51	2.47	2.43	2.39	2.36	2.34	2.31	2.29	2.27	2.25	2.18	2.13	2.07	2.03	1.94	1.91	1.90
28	5.61	4.22	3.63	3.29	3.06	2.90	2.78	2.69	2.61	2.55	2.49	2.45	2.41	2.37	2.34	2.32	2.29	2.27	2.25	2.23	2.16	2.11	2.05	2.01	1.92	1.89	1.88
29	5.59	4.20	3.61	3.27	3.04	2.88	2.76	2.67	2.59	2.53	2.48	2.43	2.39	2.36	2.32	2.30	2.27	2.25	2.23	2.21	2.14	2.09	2.03	1.99	1.90	1.87	1.86
30	5.57	4.18	3.59	3.25	3.03	2.87	2.75	2.65	2.57	2.51	2.46	2.41	2.37	2.34	2.31	2.28	2.26	2.23	2.21	2.20	2.12	2.07	2.01	1.97	1.88	1.85	1.84
32	5.53	4.15	3.56	3.22	3.00	2.84	2.71	2.62	2.54	2.48	2.43	2.38	2.34	2.31	2.28	2.25	2.22	2.20	2.18	2.16	2.09	2.04	1.98	1.93	1.85	1.82	1.80
34	5.50	4.12	3.53	3.19	2.97	2.81	2.69	2.59	2.52	2.45	2.40	2.35	2.31	2.28	2.25	2.22	2.20	2.17	2.15	2.13	2.06	2.01	1.95	1.90	1.82	1.78	1.77
36	5.47	4.09	3.50	3.17	2.94	2.78	2.66	2.57	2.49	2.43	2.37	2.33	2.29	2.25	2.22	2.20	2.17	2.15	2.13	2.11	2.04	1.99	1.92	1.88	1.79	1.76	1.74
38	5.45	4.07	3.48	3.15	2.92	2.76	2.64	2.55	2.47	2.41	2.35	2.31	2.27	2.23	2.20	2.17	2.15	2.13	2.11	2.09	2.01	1.96	1.90	1.85	1.76	1.73	1.71
40	5.42	4.05	3.46	3.13	2.90	2.74	2.62	2.53	2.45	2.39	2.33	2.29	2.25	2.21	2.18	2.15	2.13	2.11	2.09	2.07	1.99	1.94	1.88	1.83	1.74	1.71	1.69
42	5.40	4.03	3.45	3.11	2.89	2.73	2.61	2.51	2.43	2.37	2.32	2.27	2.23	2.20	2.16	2.14	2.11	2.09	2.07	2.05	1.98	1.92	1.86	1.81	1.72	1.69	1.67
44	5.39	4.02	3.43	3.09	2.87	2.71	2.59	2.50	2.42	2.36	2.30	2.26	2.22	2.18	2.15	2.12	2.10	2.07	2.05	2.03	1.96	1.91	1.84	1.80	1.70	1.67	1.65
46	5.37	4.00	3.42	3.08	2.86	2.70	2.58	2.48	2.41	2.34	2.29	2.24	2.20	2.17	2.13	2.11	2.08	2.06	2.04	2.02	1.94	1.89	1.82	1.78	1.69	1.65	1.63
48	5.35	3.99	3.40	3.07	2.84	2.69	2.56	2.47	2.39	2.33	2.27	2.23	2.19	2.15	2.12	2.09	2.07	2.05	2.02	2.01	1.93	1.88	1.81	1.77	1.67	1.64	1.62
50	5.34	3.97	3.39	3.05	2.83	2.67	2.55	2.46	2.38	2.32	2.26	2.22	2.18	2.14	2.11	2.08	2.06	2.03	2.01	1.99	1.92	1.87	1.80	1.75	1.66	1.62	1.60
60	5.29	3.93	3.34	3.01	2.79	2.63	2.51	2.41	2.33	2.27	2.22	2.17	2.13	2.09	2.06	2.03	2.01	1.98	1.96	1.94	1.87	1.82	1.74	1.70	1.60	1.56	1.54
70	5.25	3.89	3.31	2.97	2.75	2.59	2.47	2.38	2.30	2.24	2.18	2.14	2.10	2.06	2.03	2.00	1.97	1.95	1.93	1.91	1.83	1.78	1.71	1.66	1.56	1.52	1.50
80	5.22	3.86	3.28	2.95	2.73	2.57	2.45	2.35	2.28	2.21	2.16	2.11	2.07	2.03	2.00	1.97	1.95	1.92	1.90	1.88	1.81	1.75	1.68	1.63	1.53	1.49	1.47
90	5.20	3.84	3.26	2.93	2.71	2.55	2.43	2.34	2.26	2.19	2.14	2.09	2.05	2.02	1.98	1.95	1.93	1.91	1.88	1.86	1.79	1.73	1.66	1.61	1.50	1.46	1.44
100	5.18	3.83	3.25	2.92	2.70	2.54	2.42	2.32	2.24	2.18	2.12	2.08	2.04	2.00	1.97	1.94	1.91	1.89	1.87	1.85	1.77	1.71	1.64	1.59	1.48	1.44	1.42
125	5.15	3.80	3.22	2.89	2.67	2.51	2.39	2.30	2.22	2.15	2.10	2.05	2.01	1.97	1.94	1.91	1.89	1.86	1.84	1.82	1.74	1.68	1.61	1.56	1.45	1.40	1.38
150	5.13	3.78	3.20	2.87	2.65	2.49	2.37	2.28	2.20	2.13	2.08	2.03	1.99	1.95	1.92	1.89	1.87	1.84	1.82	1.80	1.72	1.67	1.59	1.54	1.42	1.38	1.35
200	5.10	3.76	3.18	2.85	2.63	2.47	2.35	2.26	2.18	2.11	2.06	2.01	1.97	1.93	1.90	1.87	1.84	1.82	1.80	1.78	1.70	1.64	1.56	1.51	1.39	1.35	1.32
300	5.07	3.73	3.16	2.83	2.61	2.45	2.33	2.23	2.16	2.09	2.04	1.99	1.95	1.91	1.88	1.85	1.82	1.80	1.77	1.75	1.67	1.62	1.54	1.48	1.36	1.31	1.28
500	5.05	3.72	3.14	2.81	2.59	2.43	2.31	2.22	2.14	2.07	2.02	1.97	1.93	1.89	1.86	1.83	1.80	1.78	1.76	1.74	1.65	1.60	1.52	1.46	1.34	1.28	1.25
1000	5.04	3.70	3.13	2.80	2.58	2.42	2.30	2.20	2.13	2.06	2.01	1.96	1.92	1.88	1.85	1.82	1.79	1.77	1.74	1.72	1.64	1.58	1.50	1.45	1.32	1.26	1.23

DEGREES OF FREEDOM FOR DENOMINATOR

Table A-4   Percentiles of the F distribution (continued)

Upper 1% point of the F distribution

DEGREES OF FREEDOM FOR NUMERATOR

df denom	1	2	3	4	5	6	7	8	9	10	11	12	13	14	15	16	17	18	19	20	25	30	40	50	100	150	200
1	4052	5000	5403	5625	5764	5859	5928	5981	6022	6056	6083	6106	6126	6143	6157	6170	6181	6192	6201	6209	6240	6261	6287	6303	6334	6345	6350
2	98.5	99.0	99.2	99.2	99.3	99.3	99.4	99.4	99.4	99.4	99.4	99.4	99.4	99.4	99.4	99.4	99.4	99.4	99.4	99.4	99.5	99.5	99.5	99.5	99.5	99.5	99.5
3	34.1	30.8	29.5	28.7	28.2	27.9	27.7	27.5	27.3	27.2	27.1	27.1	27.0	26.9	26.9	26.8	26.8	26.8	26.7	26.7	26.6	26.5	26.4	26.4	26.2	26.2	26.2
4	21.2	18.0	16.7	16.0	15.5	15.2	15.0	14.8	14.7	14.5	14.5	14.4	14.3	14.2	14.2	14.2	14.1	14.1	14.0	14.0	13.9	13.8	13.7	13.7	13.6	13.5	13.5
5	16.3	13.3	12.1	11.4	11.0	10.7	10.5	10.3	10.2	10.1	9.96	9.89	9.84	9.77	9.72	9.68	9.64	9.61	9.58	9.55	9.45	9.38	9.29	9.24	9.13	9.09	9.08
6	13.7	10.9	9.78	9.15	8.75	8.47	8.26	8.10	7.98	7.87	7.79	7.72	7.66	7.60	7.56	7.52	7.48	7.45	7.42	7.40	7.30	7.23	7.14	7.09	6.99	6.95	6.93
7	12.2	9.55	8.45	7.85	7.46	7.19	6.99	6.84	6.72	6.62	6.54	6.47	6.41	6.36	6.31	6.28	6.24	6.21	6.18	6.16	6.06	5.99	5.91	5.86	5.75	5.72	5.70
8	11.3	8.65	7.59	7.01	6.63	6.37	6.18	6.03	5.91	5.81	5.73	5.67	5.61	5.56	5.52	5.48	5.44	5.41	5.38	5.36	5.26	5.20	5.12	5.07	4.96	4.93	4.91
9	10.6	8.02	6.99	6.42	6.06	5.80	5.61	5.47	5.35	5.26	5.18	5.11	5.05	5.01	4.96	4.92	4.89	4.86	4.83	4.81	4.71	4.65	4.57	4.52	4.41	4.38	4.36
10	10.0	7.56	6.55	5.99	5.64	5.39	5.20	5.06	4.94	4.85	4.77	4.71	4.65	4.60	4.56	4.52	4.49	4.46	4.43	4.41	4.31	4.25	4.17	4.12	4.01	3.98	3.96
11	9.65	7.21	6.22	5.67	5.32	5.07	4.89	4.74	4.63	4.54	4.46	4.40	4.34	4.29	4.25	4.21	4.18	4.15	4.12	4.10	4.01	3.94	3.86	3.81	3.71	3.67	3.66
12	9.33	6.93	5.95	5.41	5.06	4.82	4.64	4.50	4.39	4.30	4.22	4.16	4.10	4.05	4.01	3.97	3.94	3.91	3.88	3.86	3.76	3.70	3.62	3.57	3.47	3.43	3.41
13	9.07	6.70	5.74	5.21	4.86	4.62	4.44	4.30	4.19	4.10	4.02	3.96	3.91	3.86	3.82	3.78	3.75	3.72	3.69	3.66	3.57	3.51	3.43	3.38	3.27	3.24	3.22
14	8.86	6.51	5.56	5.04	4.69	4.46	4.28	4.14	4.03	3.94	3.86	3.80	3.75	3.70	3.66	3.62	3.59	3.56	3.53	3.51	3.41	3.35	3.27	3.22	3.11	3.08	3.06
15	8.68	6.36	5.42	4.89	4.56	4.32	4.14	4.00	3.89	3.80	3.73	3.67	3.61	3.56	3.52	3.49	3.45	3.42	3.40	3.37	3.28	3.21	3.13	3.08	2.98	2.94	2.92
16	8.53	6.23	5.29	4.77	4.44	4.20	4.03	3.89	3.78	3.69	3.62	3.55	3.50	3.45	3.41	3.37	3.34	3.31	3.28	3.26	3.16	3.10	3.02	2.97	2.86	2.83	2.81
17	8.40	6.11	5.19	4.67	4.34	4.10	3.93	3.79	3.68	3.59	3.52	3.46	3.40	3.35	3.31	3.27	3.24	3.21	3.19	3.16	3.07	3.00	2.92	2.87	2.76	2.73	2.71
18	8.29	6.01	5.09	4.58	4.25	4.01	3.84	3.71	3.60	3.51	3.43	3.37	3.32	3.27	3.23	3.19	3.16	3.13	3.10	3.08	2.98	2.92	2.84	2.78	2.68	2.64	2.62
19	8.18	5.93	5.01	4.50	4.17	3.94	3.77	3.63	3.52	3.43	3.36	3.30	3.24	3.19	3.15	3.12	3.08	3.05	3.03	3.00	2.91	2.84	2.76	2.71	2.60	2.57	2.55
20	8.10	5.85	4.94	4.43	4.10	3.87	3.70	3.56	3.46	3.37	3.29	3.23	3.18	3.13	3.09	3.05	3.02	2.99	2.96	2.94	2.84	2.78	2.69	2.64	2.54	2.50	2.48
21	8.02	5.78	4.87	4.37	4.04	3.81	3.64	3.51	3.40	3.31	3.24	3.17	3.12	3.07	3.03	2.99	2.96	2.93	2.90	2.88	2.79	2.72	2.64	2.58	2.48	2.44	2.42
22	7.95	5.72	4.82	4.31	3.99	3.76	3.59	3.45	3.35	3.26	3.18	3.12	3.07	3.02	2.98	2.94	2.91	2.88	2.85	2.83	2.73	2.67	2.58	2.53	2.42	2.38	2.36
23	7.88	5.66	4.76	4.26	3.94	3.71	3.54	3.41	3.30	3.21	3.14	3.07	3.02	2.97	2.93	2.89	2.86	2.83	2.80	2.78	2.69	2.62	2.54	2.48	2.37	2.34	2.32
24	7.82	5.61	4.72	4.22	3.90	3.67	3.50	3.36	3.26	3.17	3.09	3.03	2.98	2.93	2.89	2.85	2.82	2.79	2.76	2.74	2.64	2.58	2.49	2.44	2.33	2.29	2.27
25	7.77	5.57	4.68	4.18	3.85	3.63	3.46	3.32	3.22	3.13	3.06	2.99	2.94	2.89	2.85	2.81	2.78	2.75	2.72	2.70	2.60	2.54	2.45	2.40	2.29	2.25	2.23
26	7.72	5.53	4.64	4.14	3.82	3.59	3.42	3.29	3.18	3.09	3.02	2.96	2.90	2.86	2.81	2.78	2.75	2.72	2.69	2.66	2.57	2.50	2.42	2.36	2.25	2.21	2.19
27	7.68	5.49	4.60	4.11	3.78	3.56	3.39	3.26	3.15	3.06	2.99	2.93	2.87	2.82	2.78	2.75	2.71	2.68	2.66	2.63	2.54	2.47	2.38	2.33	2.22	2.18	2.16
28	7.64	5.45	4.57	4.07	3.75	3.53	3.36	3.23	3.12	3.03	2.96	2.90	2.84	2.79	2.75	2.72	2.68	2.65	2.63	2.60	2.51	2.44	2.35	2.30	2.19	2.15	2.13
29	7.60	5.42	4.54	4.04	3.73	3.50	3.33	3.20	3.09	3.00	2.93	2.87	2.81	2.77	2.73	2.69	2.66	2.63	2.60	2.57	2.48	2.41	2.33	2.27	2.16	2.12	2.10
30	7.56	5.39	4.51	4.02	3.70	3.47	3.30	3.17	3.07	2.98	2.91	2.84	2.79	2.74	2.70	2.66	2.63	2.60	2.57	2.55	2.45	2.39	2.30	2.25	2.13	2.09	2.07
32	7.50	5.34	4.46	3.97	3.65	3.43	3.26	3.13	3.02	2.93	2.86	2.80	2.74	2.70	2.65	2.62	2.58	2.55	2.53	2.50	2.41	2.34	2.25	2.20	2.08	2.04	2.02
34	7.44	5.29	4.42	3.93	3.61	3.39	3.22	3.09	2.98	2.89	2.82	2.76	2.70	2.66	2.61	2.58	2.54	2.51	2.49	2.46	2.37	2.30	2.21	2.16	2.04	2.00	1.98
36	7.40	5.25	4.38	3.89	3.57	3.35	3.18	3.05	2.95	2.86	2.79	2.72	2.67	2.62	2.58	2.54	2.51	2.48	2.45	2.43	2.33	2.26	2.18	2.12	2.00	1.96	1.94
38	7.35	5.21	4.34	3.86	3.54	3.32	3.15	3.02	2.92	2.83	2.75	2.69	2.64	2.59	2.55	2.51	2.48	2.45	2.42	2.40	2.30	2.23	2.14	2.09	1.97	1.93	1.90
40	7.31	5.18	4.31	3.83	3.51	3.29	3.12	2.99	2.89	2.80	2.73	2.66	2.61	2.56	2.52	2.48	2.45	2.42	2.39	2.37	2.27	2.20	2.11	2.06	1.94	1.90	1.87
42	7.28	5.15	4.29	3.80	3.49	3.27	3.10	2.97	2.86	2.78	2.70	2.64	2.59	2.54	2.50	2.46	2.43	2.40	2.37	2.34	2.25	2.18	2.09	2.03	1.91	1.87	1.85
44	7.25	5.12	4.26	3.78	3.47	3.24	3.08	2.95	2.84	2.75	2.68	2.62	2.56	2.52	2.47	2.44	2.40	2.37	2.35	2.32	2.22	2.15	2.07	2.01	1.89	1.84	1.82
46	7.22	5.10	4.24	3.76	3.44	3.22	3.06	2.93	2.82	2.73	2.66	2.60	2.54	2.50	2.45	2.42	2.38	2.35	2.33	2.30	2.20	2.13	2.04	1.99	1.86	1.82	1.80
48	7.19	5.08	4.22	3.74	3.43	3.20	3.04	2.91	2.80	2.71	2.64	2.58	2.53	2.48	2.44	2.40	2.37	2.33	2.31	2.28	2.18	2.12	2.02	1.97	1.84	1.80	1.78
50	7.17	5.06	4.20	3.72	3.41	3.19	3.02	2.89	2.78	2.70	2.63	2.56	2.51	2.46	2.42	2.38	2.35	2.32	2.29	2.27	2.17	2.10	2.01	1.95	1.82	1.78	1.76
60	7.08	4.98	4.13	3.65	3.34	3.12	2.95	2.82	2.72	2.63	2.56	2.50	2.44	2.39	2.35	2.31	2.28	2.25	2.22	2.20	2.10	2.03	1.94	1.88	1.75	1.70	1.68
70	7.01	4.92	4.07	3.60	3.29	3.07	2.91	2.78	2.67	2.59	2.51	2.45	2.40	2.35	2.31	2.27	2.23	2.20	2.18	2.15	2.05	1.98	1.89	1.83	1.70	1.65	1.62
80	6.96	4.88	4.04	3.56	3.26	3.04	2.87	2.74	2.64	2.55	2.48	2.42	2.36	2.31	2.27	2.23	2.20	2.17	2.14	2.12	2.01	1.94	1.85	1.79	1.65	1.61	1.58
90	6.93	4.85	4.01	3.53	3.23	3.01	2.84	2.72	2.61	2.52	2.45	2.39	2.33	2.29	2.24	2.21	2.17	2.14	2.11	2.09	1.99	1.92	1.82	1.76	1.62	1.57	1.55
100	6.90	4.82	3.98	3.51	3.21	2.99	2.82	2.69	2.59	2.50	2.43	2.37	2.31	2.27	2.22	2.19	2.15	2.12	2.09	2.07	1.97	1.89	1.80	1.74	1.60	1.55	1.52
125	6.84	4.78	3.94	3.47	3.17	2.95	2.79	2.66	2.55	2.47	2.39	2.33	2.28	2.23	2.19	2.15	2.11	2.08	2.05	2.03	1.93	1.85	1.76	1.69	1.55	1.50	1.47
150	6.81	4.75	3.91	3.45	3.14	2.92	2.76	2.63	2.53	2.44	2.37	2.31	2.25	2.20	2.16	2.12	2.09	2.06	2.03	2.00	1.90	1.83	1.73	1.66	1.52	1.46	1.43
200	6.76	4.71	3.88	3.41	3.11	2.89	2.73	2.60	2.50	2.41	2.34	2.27	2.22	2.17	2.13	2.09	2.06	2.03	2.00	1.97	1.87	1.79	1.69	1.63	1.48	1.42	1.39
300	6.72	4.68	3.85	3.38	3.08	2.86	2.70	2.57	2.47	2.38	2.31	2.24	2.19	2.14	2.10	2.06	2.03	1.99	1.97	1.94	1.84	1.76	1.66	1.59	1.44	1.38	1.35
500	6.69	4.65	3.82	3.36	3.05	2.84	2.68	2.55	2.44	2.36	2.28	2.22	2.17	2.12	2.07	2.04	2.00	1.97	1.94	1.92	1.81	1.74	1.63	1.57	1.41	1.34	1.31
1000	6.66	4.63	3.80	3.34	3.04	2.82	2.66	2.53	2.43	2.34	2.27	2.20	2.15	2.10	2.06	2.02	1.98	1.95	1.92	1.90	1.79	1.72	1.61	1.54	1.38	1.32	1.28

DEGREES OF FREEDOM FOR DENOMINATOR

Table A-4 Percentiles of the *F* distribution (*continued*)

Upper 0.5% point of the F distribution

DEGREES OF FREEDOM FOR NUMERATOR

Denom \ Num	1	2	3	4	5	6	7	8	9	10	11	12	13	14	15	16	17	18	19	20	25	30	40	50	100	150	200
1	****	****	****	****	****	****	****	****	****	****	****	****	****	****	****	****	****	****	****	****	****	****	****	****	****	****	****
2	199	199	199	199	199	199	199	199	199	199	199	199	199	199	199	199	199	199	199	199	199	199	199	199	199	199	199
3	55.6	49.8	47.5	46.2	45.4	44.8	44.4	44.1	43.9	43.7	43.5	43.4	43.3	43.2	43.1	43.0	42.9	42.9	42.8	42.8	42.6	42.5	42.3	42.2	42.0	42.0	41.9
4	31.3	26.3	24.3	23.2	22.5	22.0	21.6	21.4	21.1	21.0	20.8	20.7	20.6	20.5	20.4	20.4	20.3	20.3	20.2	20.2	20.0	19.9	19.8	19.7	19.5	19.4	19.4
5	22.8	18.3	16.5	15.6	14.9	14.5	14.2	14.0	13.8	13.6	13.5	13.4	13.3	13.2	13.1	13.1	13.0	13.0	12.9	12.9	12.8	12.7	12.5	12.5	12.3	12.2	12.2
6	18.6	14.5	12.9	12.0	11.5	11.1	10.8	10.6	10.4	10.3	10.1	10.0	9.95	9.88	9.81	9.76	9.71	9.66	9.62	9.59	9.45	9.36	9.24	9.17	9.03	8.98	8.95
7	16.2	12.4	10.9	10.1	9.52	9.16	8.89	8.68	8.51	8.38	8.27	8.18	8.10	8.03	7.97	7.91	7.87	7.83	7.79	7.75	7.62	7.53	7.42	7.35	7.22	7.17	7.15
8	14.7	11.0	9.60	8.81	8.30	7.95	7.69	7.50	7.34	7.21	7.10	7.01	6.94	6.87	6.81	6.76	6.72	6.68	6.64	6.61	6.48	6.40	6.29	6.22	6.09	6.04	6.02
9	13.6	10.1	8.72	7.96	7.47	7.13	6.88	6.69	6.54	6.42	6.31	6.23	6.15	6.09	6.03	5.98	5.94	5.90	5.86	5.83	5.71	5.62	5.52	5.45	5.32	5.28	5.26
10	12.8	9.43	8.08	7.34	6.87	6.54	6.30	6.12	5.97	5.85	5.75	5.66	5.59	5.53	5.47	5.42	5.38	5.34	5.31	5.27	5.15	5.07	4.97	4.90	4.77	4.73	4.71
11	12.2	8.91	7.60	6.88	6.42	6.10	5.86	5.68	5.54	5.42	5.32	5.24	5.16	5.10	5.05	5.00	4.96	4.92	4.89	4.86	4.74	4.65	4.55	4.49	4.36	4.31	4.29
12	11.8	8.51	7.23	6.52	6.07	5.76	5.52	5.35	5.20	5.09	4.99	4.91	4.84	4.77	4.72	4.67	4.63	4.59	4.56	4.53	4.41	4.33	4.23	4.17	4.04	3.99	3.97
13	11.4	8.19	6.93	6.23	5.79	5.48	5.25	5.08	4.94	4.82	4.72	4.64	4.57	4.51	4.46	4.41	4.37	4.33	4.30	4.27	4.15	4.07	3.97	3.91	3.78	3.74	3.71
14	11.1	7.92	6.68	6.00	5.56	5.26	5.03	4.86	4.72	4.60	4.51	4.43	4.36	4.30	4.25	4.20	4.16	4.12	4.09	4.06	3.94	3.86	3.76	3.70	3.57	3.53	3.50
15	10.8	7.70	6.48	5.80	5.37	5.07	4.85	4.67	4.54	4.42	4.33	4.25	4.18	4.12	4.07	4.02	3.98	3.95	3.91	3.88	3.77	3.69	3.58	3.52	3.39	3.35	3.33
16	10.6	7.51	6.30	5.64	5.21	4.91	4.69	4.52	4.38	4.27	4.18	4.10	4.03	3.97	3.92	3.87	3.83	3.80	3.76	3.73	3.62	3.54	3.44	3.37	3.25	3.20	3.18
17	10.4	7.36	6.16	5.50	5.07	4.78	4.56	4.39	4.25	4.14	4.05	3.97	3.90	3.84	3.79	3.75	3.71	3.67	3.64	3.61	3.49	3.41	3.31	3.25	3.12	3.07	3.05
18	10.2	7.21	6.03	5.37	4.96	4.66	4.44	4.28	4.14	4.03	3.94	3.86	3.79	3.73	3.68	3.64	3.60	3.56	3.53	3.50	3.38	3.30	3.20	3.14	3.01	2.96	2.94
19	10.1	7.09	5.92	5.27	4.85	4.56	4.34	4.18	4.04	3.93	3.84	3.76	3.70	3.64	3.59	3.54	3.50	3.46	3.43	3.40	3.29	3.21	3.11	3.04	2.91	2.87	2.85
20	9.94	6.99	5.82	5.17	4.76	4.47	4.26	4.09	3.96	3.85	3.76	3.68	3.61	3.55	3.50	3.46	3.42	3.38	3.35	3.32	3.20	3.12	3.02	2.96	2.83	2.78	2.76
21	9.83	6.89	5.73	5.09	4.68	4.39	4.18	4.01	3.88	3.77	3.68	3.60	3.54	3.48	3.43	3.38	3.34	3.31	3.27	3.24	3.13	3.05	2.95	2.88	2.75	2.71	2.68
22	9.73	6.81	5.65	5.02	4.61	4.32	4.11	3.94	3.81	3.70	3.61	3.54	3.47	3.41	3.36	3.31	3.27	3.24	3.21	3.18	3.06	2.98	2.88	2.82	2.69	2.64	2.62
23	9.63	6.73	5.58	4.95	4.54	4.26	4.05	3.88	3.75	3.64	3.55	3.47	3.41	3.35	3.30	3.25	3.21	3.18	3.15	3.12	3.00	2.92	2.82	2.76	2.62	2.58	2.56
24	9.55	6.66	5.52	4.89	4.49	4.20	3.99	3.83	3.69	3.59	3.50	3.42	3.35	3.30	3.25	3.20	3.16	3.12	3.09	3.06	2.95	2.87	2.77	2.70	2.57	2.52	2.50
25	9.48	6.60	5.46	4.84	4.43	4.15	3.94	3.78	3.64	3.54	3.45	3.37	3.30	3.25	3.20	3.15	3.11	3.08	3.04	3.01	2.90	2.82	2.72	2.65	2.52	2.47	2.45
26	9.41	6.54	5.41	4.79	4.38	4.10	3.89	3.73	3.60	3.49	3.40	3.33	3.26	3.20	3.15	3.11	3.07	3.03	3.00	2.97	2.85	2.77	2.67	2.61	2.47	2.43	2.40
27	9.34	6.49	5.36	4.74	4.34	4.06	3.85	3.69	3.56	3.45	3.36	3.28	3.22	3.16	3.11	3.07	3.03	2.99	2.96	2.93	2.81	2.73	2.63	2.57	2.43	2.38	2.36
28	9.28	6.44	5.32	4.70	4.30	4.02	3.81	3.65	3.52	3.41	3.32	3.25	3.18	3.12	3.07	3.03	2.99	2.95	2.92	2.89	2.77	2.69	2.59	2.53	2.39	2.35	2.32
29	9.23	6.40	5.28	4.66	4.26	3.98	3.77	3.61	3.48	3.38	3.29	3.21	3.15	3.09	3.04	2.99	2.95	2.92	2.88	2.86	2.74	2.66	2.56	2.49	2.36	2.31	2.29
30	9.18	6.35	5.24	4.62	4.23	3.95	3.74	3.58	3.45	3.34	3.25	3.18	3.11	3.06	3.01	2.96	2.92	2.89	2.85	2.82	2.71	2.63	2.52	2.46	2.32	2.28	2.25
32	9.09	6.28	5.17	4.56	4.17	3.89	3.68	3.52	3.39	3.29	3.20	3.12	3.06	3.00	2.95	2.90	2.86	2.83	2.80	2.77	2.65	2.57	2.47	2.40	2.26	2.22	2.19
34	9.01	6.22	5.11	4.50	4.11	3.84	3.63	3.47	3.34	3.24	3.15	3.07	3.01	2.95	2.90	2.85	2.81	2.78	2.75	2.72	2.60	2.52	2.42	2.35	2.21	2.16	2.14
36	8.94	6.16	5.06	4.46	4.06	3.79	3.58	3.42	3.30	3.19	3.10	3.03	2.96	2.90	2.85	2.81	2.77	2.73	2.70	2.67	2.56	2.48	2.37	2.30	2.17	2.12	2.09
38	8.88	6.11	5.02	4.41	4.02	3.75	3.54	3.39	3.26	3.15	3.06	2.99	2.92	2.87	2.82	2.77	2.73	2.70	2.66	2.63	2.52	2.44	2.33	2.27	2.12	2.08	2.05
40	8.83	6.07	4.98	4.37	3.99	3.71	3.51	3.35	3.22	3.12	3.03	2.95	2.89	2.83	2.78	2.74	2.70	2.66	2.63	2.60	2.48	2.40	2.30	2.23	2.09	2.04	2.01
42	8.78	6.03	4.94	4.34	3.95	3.68	3.48	3.32	3.19	3.09	3.00	2.92	2.86	2.80	2.75	2.71	2.67	2.63	2.60	2.57	2.45	2.37	2.26	2.20	2.06	2.00	1.98
44	8.74	5.99	4.91	4.31	3.92	3.65	3.45	3.29	3.16	3.06	2.97	2.89	2.83	2.77	2.72	2.68	2.64	2.60	2.57	2.54	2.42	2.34	2.24	2.17	2.03	1.97	1.95
46	8.70	5.96	4.88	4.28	3.90	3.62	3.42	3.26	3.14	3.03	2.94	2.87	2.80	2.75	2.70	2.65	2.61	2.58	2.54	2.51	2.40	2.32	2.21	2.14	2.00	1.95	1.92
48	8.66	5.93	4.85	4.25	3.87	3.60	3.40	3.24	3.11	3.01	2.92	2.85	2.78	2.72	2.67	2.63	2.59	2.55	2.52	2.49	2.37	2.29	2.19	2.12	1.97	1.92	1.90
50	8.63	5.90	4.83	4.23	3.85	3.58	3.38	3.22	3.09	2.99	2.90	2.82	2.76	2.70	2.65	2.61	2.57	2.53	2.50	2.47	2.35	2.27	2.16	2.10	1.95	1.90	1.87
60	8.49	5.79	4.73	4.14	3.76	3.49	3.29	3.13	3.01	2.90	2.82	2.74	2.68	2.62	2.57	2.53	2.49	2.45	2.42	2.39	2.27	2.19	2.08	2.01	1.86	1.81	1.78
70	8.40	5.72	4.66	4.08	3.70	3.43	3.23	3.08	2.95	2.85	2.76	2.68	2.62	2.56	2.51	2.47	2.43	2.39	2.36	2.33	2.21	2.13	2.02	1.95	1.80	1.74	1.71
80	8.33	5.67	4.61	4.03	3.65	3.39	3.19	3.03	2.91	2.80	2.72	2.64	2.58	2.52	2.47	2.43	2.39	2.35	2.32	2.29	2.17	2.08	1.97	1.90	1.75	1.69	1.66
90	8.28	5.62	4.57	3.99	3.62	3.35	3.15	3.00	2.87	2.77	2.68	2.61	2.54	2.49	2.44	2.39	2.35	2.32	2.28	2.25	2.13	2.05	1.94	1.87	1.71	1.65	1.62
100	8.24	5.59	4.54	3.96	3.59	3.33	3.13	2.97	2.85	2.74	2.66	2.58	2.52	2.46	2.41	2.37	2.33	2.29	2.26	2.23	2.11	2.02	1.91	1.84	1.68	1.62	1.59
125	8.17	5.53	4.49	3.91	3.54	3.28	3.08	2.93	2.80	2.70	2.61	2.54	2.47	2.42	2.37	2.32	2.28	2.24	2.21	2.18	2.06	1.98	1.86	1.79	1.63	1.56	1.53
150	8.12	5.49	4.45	3.88	3.51	3.25	3.05	2.89	2.76	2.67	2.58	2.51	2.44	2.38	2.33	2.29	2.25	2.21	2.18	2.15	2.03	1.94	1.83	1.76	1.59	1.53	1.49
200	8.06	5.44	4.41	3.84	3.47	3.21	3.01	2.86	2.73	2.63	2.54	2.47	2.40	2.35	2.30	2.25	2.21	2.18	2.14	2.11	1.99	1.91	1.79	1.71	1.54	1.48	1.44
300	8.00	5.39	4.36	3.80	3.43	3.17	2.97	2.82	2.69	2.59	2.51	2.43	2.37	2.31	2.26	2.21	2.17	2.14	2.10	2.07	1.95	1.87	1.75	1.67	1.50	1.43	1.39
500	7.95	5.35	4.33	3.76	3.40	3.14	2.94	2.79	2.66	2.56	2.48	2.40	2.34	2.28	2.23	2.19	2.14	2.11	2.07	2.04	1.92	1.84	1.72	1.64	1.46	1.39	1.35
1000	7.91	5.33	4.30	3.74	3.37	3.11	2.92	2.77	2.64	2.54	2.45	2.38	2.32	2.26	2.21	2.16	2.12	2.09	2.05	2.02	1.90	1.81	1.69	1.61	1.43	1.36	1.31

DEGREES OF FREEDOM FOR DENOMINATOR

Table A-4 Percentiles of the F distribution (continued)

Upper 0.1% point of the F distribution

DEGREES OF FREEDOM FOR NUMERATOR

Denom \ Num	1	2	3	4	5	6	7	8	9	10	11	12	13	14	15	16	17	18	19	20	25	30	40	50	100	150	200
1	****	****	****	****	****	****	****	****	****	****	****	****	****	****	****	****	****	****	****	****	****	****	****	****	****	****	****
2	999	999	999	999	999	999	999	999	999	999	999	999	999	999	999	999	999	999	999	999	999	999	999	999	999	999	999
3	167	148	141	137	135	133	132	131	130	129	129	128	128	128	127	127	127	127	127	126	126	125	125	125	124	124	124
4	74.1	61.2	56.2	53.4	51.7	50.5	49.7	49.0	48.5	48.1	47.7	47.4	47.2	46.9	46.8	46.6	46.5	46.3	46.2	46.1	45.7	45.4	45.1	44.9	44.5	44.3	44.3
5	47.2	37.1	33.2	31.1	29.8	28.8	28.2	27.6	27.2	26.9	26.6	26.4	26.2	26.1	25.9	25.8	25.7	25.6	25.5	25.4	25.1	24.9	24.6	24.4	24.1	24.0	24.0
6	35.5	27.0	23.7	21.9	20.8	20.0	19.5	19.0	18.7	18.4	18.2	18.0	17.8	17.7	17.6	17.4	17.4	17.3	17.2	17.1	16.9	16.7	16.4	16.3	16.0	15.9	15.9
7	29.2	21.7	18.8	17.2	16.2	15.5	15.0	14.6	14.3	14.1	13.9	13.7	13.6	13.4	13.3	13.2	13.1	13.1	13.0	12.9	12.7	12.5	12.3	12.2	12.0	11.9	11.8
8	25.4	18.5	15.8	14.4	13.5	12.9	12.4	12.0	11.8	11.5	11.4	11.2	11.1	10.9	10.8	10.8	10.7	10.6	10.5	10.5	10.3	10.1	9.92	9.80	9.57	9.49	9.45
9	22.9	16.4	13.9	12.6	11.7	11.1	10.7	10.4	10.1	9.89	9.72	9.57	9.44	9.33	9.24	9.15	9.08	9.01	8.95	8.90	8.69	8.55	8.37	8.26	8.04	7.96	7.93
10	21.0	14.9	12.6	11.3	10.5	9.93	9.52	9.20	8.96	8.75	8.59	8.45	8.32	8.22	8.13	8.05	7.98	7.91	7.86	7.80	7.60	7.47	7.30	7.19	6.98	6.91	6.87
11	19.7	13.8	11.6	10.3	9.58	9.05	8.66	8.35	8.12	7.92	7.76	7.63	7.51	7.41	7.32	7.24	7.17	7.11	7.06	7.01	6.81	6.68	6.52	6.42	6.21	6.14	6.10
12	18.6	13.0	10.8	9.63	8.89	8.38	8.00	7.71	7.48	7.29	7.14	7.00	6.89	6.79	6.71	6.63	6.57	6.51	6.45	6.40	6.22	6.09	5.93	5.83	5.63	5.56	5.52
13	17.8	12.3	10.2	9.07	8.35	7.86	7.49	7.21	6.98	6.80	6.65	6.52	6.41	6.31	6.23	6.16	6.09	6.03	5.98	5.93	5.75	5.63	5.47	5.37	5.17	5.10	5.07
14	17.1	11.8	9.73	8.62	7.92	7.44	7.08	6.80	6.58	6.40	6.26	6.13	6.02	5.93	5.85	5.78	5.71	5.66	5.60	5.56	5.38	5.25	5.10	5.00	4.81	4.74	4.71
15	16.6	11.3	9.34	8.25	7.57	7.09	6.74	6.47	6.26	6.08	5.94	5.81	5.71	5.62	5.54	5.46	5.40	5.35	5.29	5.25	5.07	4.95	4.80	4.70	4.51	4.44	4.41
16	16.1	11.0	9.01	7.94	7.27	6.80	6.46	6.19	5.98	5.81	5.67	5.55	5.44	5.35	5.27	5.20	5.14	5.09	5.04	4.99	4.82	4.70	4.54	4.45	4.26	4.19	4.16
17	15.7	10.7	8.73	7.68	7.02	6.56	6.22	5.96	5.75	5.58	5.44	5.32	5.22	5.13	5.05	4.99	4.92	4.87	4.82	4.78	4.60	4.48	4.33	4.24	4.05	3.98	3.95
18	15.4	10.4	8.49	7.46	6.81	6.35	6.02	5.76	5.56	5.39	5.25	5.13	5.03	4.94	4.87	4.80	4.74	4.68	4.63	4.59	4.42	4.30	4.15	4.06	3.87	3.80	3.77
19	15.1	10.2	8.28	7.27	6.62	6.18	5.85	5.59	5.39	5.22	5.08	4.97	4.87	4.78	4.70	4.64	4.58	4.52	4.47	4.43	4.26	4.14	3.99	3.90	3.71	3.65	3.61
20	14.8	9.95	8.10	7.10	6.46	6.02	5.69	5.44	5.24	5.08	4.94	4.82	4.72	4.64	4.56	4.49	4.44	4.38	4.33	4.29	4.12	4.00	3.86	3.77	3.58	3.51	3.48
21	14.6	9.77	7.94	6.95	6.32	5.88	5.56	5.31	5.11	4.95	4.81	4.70	4.60	4.51	4.44	4.37	4.31	4.26	4.21	4.17	4.00	3.88	3.74	3.64	3.46	3.39	3.36
22	14.4	9.61	7.80	6.81	6.19	5.76	5.44	5.19	4.99	4.83	4.70	4.58	4.49	4.40	4.33	4.27	4.20	4.15	4.10	4.06	3.89	3.78	3.63	3.54	3.35	3.28	3.25
23	14.2	9.47	7.67	6.70	6.08	5.65	5.33	5.09	4.89	4.73	4.60	4.48	4.39	4.30	4.23	4.16	4.10	4.05	4.00	3.96	3.79	3.68	3.53	3.44	3.25	3.19	3.16
24	14.0	9.34	7.55	6.59	5.98	5.55	5.23	4.99	4.80	4.64	4.51	4.39	4.30	4.21	4.14	4.07	4.02	3.96	3.92	3.87	3.71	3.59	3.45	3.36	3.17	3.10	3.07
25	13.9	9.22	7.45	6.49	5.89	5.46	5.15	4.91	4.71	4.56	4.42	4.31	4.22	4.13	4.06	3.99	3.94	3.88	3.84	3.79	3.63	3.52	3.37	3.28	3.09	3.03	2.99
26	13.7	9.12	7.36	6.41	5.80	5.38	5.07	4.83	4.64	4.48	4.35	4.24	4.14	4.06	3.99	3.92	3.86	3.81	3.77	3.72	3.56	3.44	3.30	3.21	3.02	2.95	2.92
27	13.6	9.02	7.27	6.33	5.73	5.31	5.00	4.76	4.57	4.41	4.28	4.17	4.08	3.99	3.92	3.86	3.80	3.75	3.70	3.66	3.49	3.38	3.23	3.14	2.96	2.89	2.86
28	13.5	8.93	7.19	6.25	5.66	5.24	4.93	4.69	4.50	4.35	4.22	4.11	4.01	3.93	3.86	3.80	3.74	3.69	3.64	3.60	3.43	3.32	3.18	3.09	2.90	2.83	2.80
29	13.4	8.85	7.12	6.19	5.59	5.18	4.87	4.64	4.45	4.29	4.16	4.05	3.96	3.88	3.80	3.74	3.68	3.63	3.59	3.54	3.38	3.27	3.12	3.03	2.84	2.78	2.74
30	13.3	8.77	7.05	6.12	5.53	5.12	4.82	4.58	4.39	4.24	4.11	4.00	3.91	3.82	3.75	3.69	3.63	3.58	3.53	3.49	3.33	3.22	3.07	2.98	2.79	2.73	2.69
32	13.1	8.64	6.94	6.01	5.43	5.02	4.72	4.48	4.30	4.14	4.02	3.91	3.81	3.73	3.66	3.60	3.54	3.49	3.44	3.40	3.24	3.13	2.98	2.89	2.70	2.64	2.60
34	13.0	8.52	6.83	5.92	5.34	4.93	4.63	4.40	4.22	4.06	3.94	3.83	3.74	3.65	3.58	3.52	3.46	3.41	3.37	3.33	3.16	3.05	2.91	2.82	2.63	2.56	2.52
36	12.8	8.42	6.74	5.84	5.26	4.86	4.56	4.33	4.14	3.99	3.87	3.76	3.67	3.59	3.51	3.45	3.40	3.34	3.30	3.26	3.10	2.98	2.84	2.75	2.56	2.49	2.46
38	12.7	8.33	6.66	5.76	5.19	4.79	4.49	4.26	4.08	3.93	3.80	3.70	3.60	3.52	3.45	3.39	3.34	3.28	3.24	3.20	3.04	2.92	2.78	2.69	2.50	2.43	2.40
40	12.6	8.25	6.59	5.70	5.13	4.73	4.44	4.21	4.02	3.87	3.75	3.64	3.55	3.47	3.40	3.34	3.28	3.23	3.19	3.14	2.98	2.87	2.73	2.64	2.44	2.38	2.34
42	12.5	8.18	6.53	5.64	5.07	4.68	4.38	4.16	3.97	3.83	3.70	3.59	3.50	3.42	3.35	3.29	3.23	3.18	3.14	3.10	2.94	2.83	2.68	2.59	2.40	2.33	2.29
44	12.4	8.12	6.48	5.59	5.02	4.63	4.34	4.11	3.93	3.78	3.66	3.55	3.46	3.38	3.31	3.25	3.19	3.14	3.10	3.06	2.89	2.78	2.64	2.55	2.35	2.28	2.25
46	12.4	8.06	6.42	5.54	4.98	4.59	4.30	4.07	3.89	3.74	3.62	3.51	3.42	3.34	3.27	3.21	3.15	3.10	3.06	3.02	2.86	2.74	2.60	2.51	2.31	2.24	2.21
48	12.3	8.00	6.38	5.50	4.94	4.55	4.26	4.03	3.85	3.70	3.58	3.48	3.38	3.31	3.24	3.17	3.12	3.07	3.02	2.98	2.82	2.71	2.56	2.47	2.28	2.21	2.17
50	12.2	7.96	6.34	5.46	4.90	4.51	4.22	4.00	3.82	3.67	3.55	3.44	3.35	3.27	3.20	3.14	3.09	3.04	2.99	2.95	2.79	2.68	2.53	2.44	2.25	2.18	2.14
60	12.0	7.77	6.17	5.31	4.76	4.37	4.09	3.86	3.69	3.54	3.42	3.32	3.23	3.15	3.08	3.02	2.96	2.91	2.87	2.83	2.67	2.55	2.41	2.32	2.12	2.05	2.01
70	11.8	7.64	6.06	5.20	4.66	4.28	3.99	3.77	3.60	3.45	3.33	3.23	3.14	3.06	2.99	2.93	2.87	2.83	2.78	2.74	2.58	2.47	2.32	2.23	2.03	1.95	1.92
80	11.7	7.54	5.97	5.12	4.58	4.20	3.92	3.70	3.53	3.39	3.27	3.16	3.07	3.00	2.93	2.87	2.81	2.76	2.72	2.68	2.52	2.41	2.26	2.16	1.96	1.89	1.85
90	11.6	7.47	5.91	5.06	4.53	4.15	3.87	3.65	3.48	3.34	3.22	3.11	3.02	2.95	2.88	2.82	2.76	2.71	2.67	2.63	2.47	2.36	2.21	2.11	1.91	1.83	1.79
100	11.5	7.41	5.86	5.02	4.48	4.11	3.83	3.61	3.44	3.30	3.18	3.07	2.99	2.91	2.84	2.78	2.73	2.68	2.63	2.59	2.43	2.32	2.17	2.08	1.87	1.79	1.75
125	11.4	7.30	5.77	4.93	4.40	4.03	3.75	3.54	3.37	3.23	3.11	3.00	2.92	2.84	2.77	2.71	2.66	2.61	2.56	2.52	2.36	2.25	2.10	2.01	1.79	1.71	1.67
150	11.3	7.24	5.71	4.88	4.35	3.98	3.71	3.49	3.32	3.18	3.06	2.96	2.87	2.80	2.73	2.67	2.61	2.56	2.52	2.48	2.32	2.21	2.06	1.96	1.74	1.66	1.62
200	11.2	7.15	5.63	4.81	4.29	3.92	3.65	3.43	3.26	3.12	3.00	2.90	2.81	2.74	2.67	2.61	2.55	2.51	2.46	2.42	2.26	2.15	2.00	1.90	1.68	1.60	1.55
300	11.0	7.07	5.56	4.75	4.22	3.86	3.59	3.38	3.21	3.07	2.95	2.85	2.76	2.69	2.62	2.56	2.50	2.46	2.41	2.37	2.21	2.10	1.94	1.85	1.62	1.53	1.48
500	11.0	7.00	5.51	4.69	4.18	3.81	3.54	3.33	3.16	3.02	2.91	2.81	2.72	2.64	2.58	2.52	2.46	2.41	2.37	2.33	2.17	2.05	1.90	1.80	1.57	1.48	1.43
1000	10.9	6.96	5.46	4.65	4.14	3.78	3.51	3.30	3.13	2.99	2.87	2.77	2.69	2.61	2.54	2.48	2.43	2.38	2.34	2.30	2.14	2.02	1.87	1.77	1.53	1.44	1.38

DEGREES OF FREEDOM FOR DENOMINATOR

Table A-5   Values of $\frac{1}{2}\ln\frac{1+r}{1-r}$

r	0.000	0.001	0.002	0.003	0.004	0.005	0.006	0.007	0.008	0.009
0.000	0.0000	0.0010	0.0020	0.0030	0.0040	0.0050	0.0060	0.0070	0.0080	0.0090
0.010	0.0100	0.0110	0.0120	0.0130	0.0140	0.0150	0.0160	0.0170	0.0180	0.0190
0.020	0.0200	0.0210	0.0220	0.0230	0.0240	0.0250	0.0260	0.0270	0.0280	0.0290
0.030	0.0300	0.0310	0.0320	0.0330	0.0340	0.0350	0.0360	0.0370	0.0380	0.0390
0.040	0.0400	0.0410	0.0420	0.0430	0.0440	0.0450	0.0460	0.0470	0.0480	0.0490
0.050	0.0501	0.0511	0.0521	0.0531	0.0541	0.0551	0.0561	0.0571	0.0581	0.0591
0.060	0.0601	0.0611	0.0621	0.0631	0.0641	0.0651	0.0661	0.0671	0.0681	0.0691
0.070	0.0701	0.0711	0.0721	0.0731	0.0741	0.0751	0.0761	0.0771	0.0782	0.0792
0.080	0.0802	0.0812	0.0822	0.0832	0.0842	0.0852	0.0862	0.0872	0.0882	0.0892
0.090	0.0902	0.0912	0.0922	0.0933	0.0943	0.0953	0.0963	0.0973	0.0983	0.0993
0.100	0.1003	0.1013	0.1024	0.1034	0.1044	0.1054	0.1064	0.1074	0.1084	0.1094
0.110	0.1105	0.1115	0.1125	0.1135	0.1145	0.1155	0.1165	0.1175	0.1185	0.1195
0.120	0.1206	0.1216	0.1226	0.1236	0.1246	0.1257	0.1267	0.1277	0.1287	0.1297
0.130	0.1308	0.1318	0.1328	0.1338	0.1348	0.1358	0.1368	0.1379	0.1389	0.1399
0.140	0.1409	0.1419	0.1430	0.1440	0.1450	0.1460	0.1470	0.1481	0.1491	0.1501
0.150	0.1511	0.1522	0.1532	0.1542	0.1552	0.1563	0.1573	0.1583	0.1593	0.1604
0.160	0.1614	0.1624	0.1634	0.1644	0.1655	0.1665	0.1676	0.1686	0.1696	0.1706
0.170	0.1717	0.1727	0.1737	0.1748	0.1758	0.1768	0.1779	0.1789	0.1799	0.1810
0.180	0.1820	0.1830	0.1841	0.1851	0.1861	0.1872	0.1882	0.1892	0.1903	0.1913
0.190	0.1923	0.1934	0.1944	0.1954	0.1965	0.1975	0.1986	0.1996	0.2007	0.2017
0.200	0.2027	0.2038	0.2048	0.2059	0.2069	0.2079	0.2090	0.2100	0.2111	0.2121
0.210	0.2132	0.2142	0.2153	0.2163	0.2174	0.2184	0.2194	0.2205	0.2215	0.2226
0.220	0.2237	0.2247	0.2258	0.2268	0.2279	0.2289	0.2300	0.2310	0.2321	0.2331
0.230	0.2342	0.2353	0.2363	0.2374	0.2384	0.2395	0.2405	0.2416	0.2427	0.2437
0.240	0.2448	0.2458	0.2469	0.2480	0.2490	0.2501	0.2511	0.2522	0.2533	0.2543
0.250	0.2554	0.2565	0.2575	0.2586	0.2597	0.2608	0.2618	0.2629	0.2640	0.2650
0.260	0.2661	0.2672	0.2682	0.2693	0.2704	0.2715	0.2726	0.2736	0.2747	0.2758
0.270	0.2769	0.2779	0.2790	0.2801	0.2812	0.2823	0.2833	0.2844	0.2855	0.2866
0.280	0.2877	0.2888	0.2898	0.2909	0.2920	0.2931	0.2942	0.2953	0.2964	0.2975
0.290	0.2986	0.2997	0.3008	0.3019	0.3029	0.3040	0.3051	0.3062	0.3073	0.3084
0.300	0.3095	0.3106	0.3117	0.3128	0.3139	0.3150	0.3161	0.3172	0.3183	0.3195
0.310	0.3206	0.3217	0.3228	0.3239	0.3250	0.3261	0.3272	0.3283	0.3294	0.3305
0.320	0.3317	0.3328	0.3339	0.3350	0.3361	0.3372	0.3384	0.3395	0.3406	0.3417
0.330	0.3428	0.3439	0.3451	0.3462	0.3473	0.3484	0.3496	0.3507	0.3518	0.3530
0.340	0.3541	0.3552	0.3564	0.3575	0.3586	0.3597	0.3609	0.3620	0.3632	0.3643
0.350	0.3654	0.3666	0.3677	0.3689	0.3700	0.3712	0.3723	0.3734	0.3746	0.3757
0.360	0.3769	0.3780	0.3792	0.3803	0.3815	0.3826	0.3838	0.3850	0.3861	0.3873
0.370	0.3884	0.3896	0.3907	0.3919	0.3931	0.3942	0.3954	0.3966	0.3977	0.3989
0.380	0.4001	0.4012	0.4024	0.4036	0.4047	0.4059	0.4071	0.4083	0.4094	0.4106
0.390	0.4118	0.4130	0.4142	0.4153	0.4165	0.4177	0.4189	0.4201	0.4213	0.4225
0.400	0.4236	0.4248	0.4260	0.4272	0.4284	0.4296	0.4308	0.4320	0.4332	0.4344
0.410	0.4356	0.4368	0.4380	0.4392	0.4404	0.4416	0.4429	0.4441	0.4453	0.4465
0.420	0.4477	0.4489	0.4501	0.4513	0.4526	0.4538	0.4550	0.4562	0.4574	0.4587
0.430	0.4599	0.4611	0.4623	0.4636	0.4648	0.4660	0.4673	0.4685	0.4697	0.4710
0.440	0.4722	0.4735	0.4747	0.4760	0.4772	0.4784	0.4797	0.4809	0.4822	0.4835
0.450	0.4847	0.4860	0.4872	0.4885	0.4897	0.4910	0.4923	0.4935	0.4948	0.4061
0.460	0.4973	0.4986	0.4999	0.5011	0.5024	0.5037	0.5049	0.5062	0.5075	0.5088
0.470	0.5101	0.5114	0.5126	0.5139	0.5152	0.5165	0.5178	0.5191	0.5204	0.5217
0.480	0.5230	0.5243	0.5256	0.5279	0.5282	0.5295	0.5308	0.5321	0.5334	0.5347
0.490	0.5361	0.5374	0.5387	0.5400	0.5413	0.5427	0.5440	0.5453	0.5466	0.5480

Table A-5 Values of $\frac{1}{2}\ln\frac{1+r}{1-r}$ (continued)

r	0.000	0.001	0.002	0.003	0.004	0.005	0.006	0.007	0.008	0.009
0.500	0.5493	0.5506	0.5520	0.5533	0.5547	0.5560	0.5573	0.5587	0.5600	0.5614
0.510	0.5627	0.5641	0.5654	0.5668	0.5681	0.5695	0.5709	0.5722	0.5736	0.5750
0.520	0.5763	0.5777	0.5791	0.5805	0.5818	0.5832	0.5846	0.5860	0.5874	0.5888
0.530	0.5901	0.5915	0.5929	0.5943	0.5957	0.5971	0.5985	0.5999	0.6013	0.6027
0.540	0.6042	0.6056	0.6070	0.6084	0.6098	0.6112	0.6127	0.6141	0.6155	0.6170
0.550	0.6184	0.6198	0.6213	0.6227	0.6241	0.6256	0.6270	0.6285	0.6299	0.6314
0.560	0.6328	0.6343	0.6358	0.6372	0.6387	0.6401	0.6416	0.6431	0.6446	0.6460
0.570	0.6475	0.6490	0.6505	0.6520	0.6535	0.6550	0.6565	0.6579	0.6594	0.6610
0.580	0.6625	0.6640	0.6655	0.6670	0.6685	0.6700	0.6715	0.6731	0.6746	0.6761
0.590	0.6777	0.6792	0.6807	0.6823	0.6838	0.6854	0.6869	0.6885	0.6900	0.6916
0.600	0.6931	0.6947	0.6963	0.6978	0.6994	0.7010	0.7026	0.7042	0.7057	0.7073
0.610	0.7089	0.7105	0.7121	0.7137	0.7153	0.7169	0.7185	0.7201	0.7218	0.7234
0.620	0.7250	0.7266	0.7283	0.7299	0.7315	0.7332	0.7348	0.7364	0.7381	0.7398
0.630	0.7414	0.7431	0.7447	0.7464	0.7481	0.7497	0.7514	0.7531	0.7548	0.7565
0.640	0.7582	0.7599	0.7616	0.7633	0.7650	0.7667	0.7684	0.7701	0.7718	0.7736
0.650	0.7753	0.7770	0.7788	0.7805	0.7823	0.7840	0.7858	0.7875	0.7893	0.7910
0.660	0.7928	0.7946	0.7964	0.7981	0.7999	0.8017	0.8035	0.8053	0.8071	0.8089
0.670	0.8107	0.8126	0.8144	0.8162	0.8180	0.8199	0.8217	0.8236	0.8254	0.8273
0.680	0.8291	0.8310	0.8328	0.8347	0.8366	0.8385	0.8404	0.8423	0.8442	0.8461
0.690	0.8480	0.8499	0.8518	0.8537	0.8556	0.8576	0.8595	0.8614	0.8634	0.8653
0.700	0.8673	0.8693	0.8712	0.8732	0.8752	0.8772	0.8792	0.8812	0.8832	0.8852
0.710	0.8872	0.8892	0.8912	0.8933	0.8953	0.8973	0.8994	0.9014	0.9035	0.9056
0.720	0.9076	0.9097	0.9118	0.9139	9.9160	0.9181	0.9202	0.9223	0.9245	0.9266
0.730	0.9287	0.9309	0.9330	0.9352	0.9373	0.9395	0.9417	0.9439	0.9461	0.9483
0.740	0.9505	0.9527	0.9549	0.9571	0.9594	0.9616	0.9639	0.9661	0.9684	0.9707
0.750	0.9730	0.9752	0.9775	0.9799	0.9822	0.9845	0.9868	0.9892	0.9915	0.9939
0.760	0.9962	0.9986	1.0010	1.0034	1.0058	1.0082	1.0106	1.0130	1.0154	1.0179
0.770	1.0203	1.0228	1.0253	1.0277	1.0302	1.0327	1.0352	1.0378	1.0403	1.0428
0.780	1.0454	1.0479	1.0505	1.0531	1.0557	1.0583	1.0609	1.0635	1.0661	1.0688
0.790	1.0714	1.0741	1.0768	1.0795	1.0822	1.0849	1.0876	1.0903	1.0931	1.0958
0.800	1.0986	1.1014	1.1041	1.1070	1.1098	1.1127	1.1155	1.1184	1.1212	1.1241
0.810	1.1270	1.1299	1.1329	1.1358	1.1388	1.1417	1.1447	1.1477	1.1507	1.1538
0.820	1.1568	1.1599	1.1630	1.1660	1.1692	1.1723	1.1754	1.1786	1.1817	1.1849
0.830	1.1870	1.1913	1.1946	1.1979	1.2011	1.2044	1.2077	1.2111	1.2144	1.2178
0.840	1.2212	1.2246	1.2280	1.2315	1.2349	1.2384	1.2419	1.2454	1.2490	1.2526
0.850	1.2561	1.2598	1.2634	1.2670	1.2708	1.2744	1.2782	1.2819	1.2857	1.2895
0.860	1.2934	1.2972	1.3011	1.3050	1.3089	1.3129	1.3168	1.3209	1.3249	1.3290
0.870	1.3331	1.3372	1.3414	1.3456	1.3498	1.3540	1.3583	1.3626	1.3670	1.3714
0.880	1.3758	1.3802	1.3847	1.3892	1.3938	1.3984	1.4030	1.4077	1.4124	1.4171
0.890	1.4219	1.4268	1.4316	1.4366	1.4415	1.4465	1.4516	1.4566	1.4618	1.4670
0.900	1.4722	1.4775	1.4828	1.4883	1.4937	1.4992	1.5047	1.5103	1.5160	1.5217
0.910	1.5275	1.5334	1.5393	1.5453	1.5513	1.5574	1.5636	1.5698	1.5762	1.5825
0.920	1.5890	1.5956	1.6022	1.6089	1.6157	1.6226	1.6296	1.6366	1.6438	1.6510
0.930	1.6584	1.6659	1.6734	1.6811	1.6888	1.6967	1.7047	1.7129	1.7211	1.7295
0.940	1.7380	1.7467	1.7555	1.7645	1.7736	1.7828	1.7923	1.8019	1.8117	1.8216
0.950	1.8318	1.8421	1.8527	1.8635	1.8745	1.8857	1.8972	1.9090	1.9210	1.9333
0.960	1.9459	1.9588	1.9721	1.9857	1.9996	2.0140	2.0287	2.0439	2.0595	2.0756
0.970	2.0923	2.1095	2.1273	2.1457	2.1649	2.1847	2.2054	2.2269	2.2494	2.2729
0.980	2.2976	2.3223	2.3507	2.3796	2.4101	2.4426	2.4774	2.5147	2.5550	2.5988
0.990	2.6467	2.6996	2.7587	2.8257	2.9031	2.9945	3.1063	3.2504	3.4534	3.8002

r	z
0.9999	4.95172
0.99999	6.10303

Source: Albert E. Waugh, *Statistical Tables and Problems,* McGraw-Hill Book Company, New York, 1952, Table A11, pp. 40–41, with the kind permission of the author and publisher.

Note: To obtain $\frac{1}{2}\ln(1+r)/(1-r)$ when r is negative, use the negative of the value corresponding to the absolute value of r; e.g., if $r = 0.242$, $\frac{1}{2}\ln(1+0.242)/(1-0.242) = -0.2469$.

725

Table A-6 Upper α point of Studentized range, $q_{k,v} = R/S$, k = sample size for range R, v = number of degrees of freedom for S

(entry $= q_{k,v,1-\alpha}$, where $P(q_{k,v} > q_{k,v,1-\alpha}) = \alpha$)

$\alpha = .05$

v \ k	2	3	4	5	6	7	8	9	10	11	12	13	14	15	16	17	18	19	20
1	18.0	27.0	32.8	37.1	40.4	43.1	45.4	47.4	49.1	50.6	52.0	53.2	54.3	55.4	56.3	57.2	58.0	58.8	59.6
2	6.08	8.33	9.80	10.9	11.7	12.4	13.0	13.5	14.0	14.4	14.7	15.1	15.4	15.7	15.9	16.1	16.4	16.6	16.8
3	4.50	5.91	6.82	7.50	8.04	8.48	8.85	9.18	9.46	9.72	9.95	10.2	10.3	10.5	10.7	10.8	11.0	11.1	11.2
4	3.93	5.04	5.76	6.29	6.71	7.05	7.35	7.60	7.83	8.03	8.21	8.37	8.52	8.66	8.79	8.91	9.03	9.13	9.23
5	3.64	4.60	5.22	5.67	6.03	6.33	6.58	6.80	6.99	7.17	7.32	7.47	7.60	7.72	7.83	7.93	8.03	8.12	8.21
6	3.46	4.34	4.90	5.30	5.63	5.90	6.12	6.32	6.49	6.65	6.79	6.92	7.03	7.14	7.24	7.34	7.43	7.51	7.59
7	3.34	4.16	4.68	5.06	5.36	5.61	5.82	6.00	6.16	6.30	6.43	6.55	6.66	6.76	6.85	6.94	7.02	7.10	7.17
8	3.26	4.04	4.53	4.89	5.17	5.40	5.60	5.77	5.92	6.05	6.18	6.29	6.39	6.48	6.57	6.65	6.73	6.80	6.87
9	3.20	3.95	4.41	4.76	5.02	5.24	5.43	5.59	5.74	5.87	5.98	6.09	6.19	6.28	6.36	6.44	6.51	6.58	6.64
10	3.15	3.88	4.33	4.65	4.91	5.12	5.30	5.46	5.60	5.72	5.83	5.93	6.03	6.11	6.19	6.27	6.34	6.40	6.47
11	3.11	3.82	4.26	4.57	4.82	5.03	5.20	5.35	5.49	5.61	5.71	5.81	5.90	5.98	6.06	6.13	6.20	6.27	6.33
12	3.08	3.77	4.20	4.51	4.75	4.95	5.12	5.27	5.39	5.51	5.61	5.71	5.80	5.88	5.95	6.02	6.09	6.15	6.21
13	3.06	3.73	4.15	4.45	4.69	4.88	5.05	5.19	5.32	5.43	5.53	5.63	5.71	5.79	5.86	5.93	5.99	6.05	6.11
14	3.03	3.70	4.11	4.41	4.64	4.83	4.99	5.13	5.25	5.36	5.46	5.55	5.64	5.71	5.79	5.85	5.91	5.97	6.03
15	3.01	3.67	4.08	4.37	4.59	4.78	4.94	5.08	5.20	5.31	5.40	5.49	5.57	5.65	5.72	5.78	5.85	5.90	5.96
16	3.00	3.65	4.05	4.33	4.56	4.74	4.90	5.03	5.15	5.26	5.35	5.44	5.52	5.59	5.66	5.73	5.79	5.84	5.90
17	2.98	3.63	4.02	4.30	4.52	4.70	4.86	4.99	5.11	5.21	5.31	5.39	5.47	5.54	5.61	5.67	5.73	5.79	5.84
18	2.97	3.61	4.00	4.28	4.49	4.67	4.82	4.96	5.07	5.17	5.27	5.35	5.43	5.50	5.57	5.63	5.69	5.74	5.79
19	2.96	3.59	3.98	4.25	4.47	4.65	4.79	4.92	5.04	5.14	5.23	5.31	5.39	5.46	5.53	5.59	5.65	5.70	5.75
20	2.95	3.58	3.96	4.23	4.45	4.62	4.77	4.90	5.01	5.11	5.20	5.28	5.36	5.43	5.49	5.55	5.61	5.66	5.71
24	2.92	3.53	3.90	4.17	4.37	4.54	4.68	4.81	4.92	5.01	5.10	5.18	5.26	5.32	5.38	5.44	5.49	5.55	5.59
30	2.89	3.49	3.85	4.10	4.30	4.46	4.60	4.72	4.82	4.92	5.00	5.08	5.15	5.21	5.27	5.33	5.38	5.43	5.47
40	2.86	3.44	3.79	4.04	4.23	4.39	4.52	4.63	4.73	4.82	4.90	4.98	5.04	5.11	5.16	5.22	5.27	5.31	5.36
60	2.83	3.40	3.74	3.98	4.16	4.31	4.44	4.55	4.65	4.73	4.81	4.88	4.94	5.00	5.06	5.11	5.15	5.20	5.24
120	2.80	3.36	3.68	3.92	4.10	4.24	4.36	4.47	4.56	4.64	4.71	4.78	4.84	4.90	4.95	5.00	5.04	5.09	5.13
∞	2.77	3.31	3.63	3.86	4.03	4.17	4.29	4.39	4.47	4.55	4.62	4.68	4.74	4.80	4.85	4.89	4.93	4.97	5.01

Source: From pp. 176–177 of *Biometrika Tables for Statisticians*, Vol. I, by E. S. Pearson and H. O. Hartley, published by the Biometrika Trustees, Cambridge University Press, Cambridge, 1954. Reproduced with the kind permission of the authors and the publisher. Corrections of ±1 in the last figure, supplied by James Pacheres, have been incorporated in 41 entries.

Table A-6 Upper α point of Studentized range (continued)

$\alpha = .01$

v \ k	2	3	4	5	6	7	8	9	10	11	12	13	14	15	16	17	18	19	20
1	90.0	135	164	186	202	216	227	237	246	253	260	266	272	277	282	286	290	294	298
2	14.0	19.0	22.3	24.7	26.6	28.2	29.5	30.7	31.7	32.6	33.4	34.1	34.8	35.4	36.0	36.5	37.0	37.5	37.9
3	8.26	10.6	12.2	13.3	14.2	15.0	15.6	16.2	16.7	17.1	17.5	17.9	18.2	18.5	18.8	19.1	19.3	19.5	19.8
4	6.51	8.12	9.17	9.96	10.6	11.1	11.5	11.9	12.3	12.6	12.8	13.1	13.3	13.5	13.7	13.9	14.1	14.2	14.4
5	5.70	6.97	7.80	8.42	8.91	9.32	9.67	9.97	10.2	10.5	10.7	10.9	11.1	11.2	11.4	11.6	11.7	11.8	11.9
6	5.24	6.33	7.03	7.56	7.97	8.32	8.61	8.87	9.10	9.30	9.49	9.65	9.81	9.95	10.1	10.2	10.3	10.4	10.5
7	4.95	5.92	6.54	7.01	7.37	7.68	7.94	8.17	8.37	8.55	8.71	8.86	9.00	9.12	9.24	9.35	9.46	9.55	9.65
8	4.74	5.63	6.20	6.63	6.96	7.24	7.47	7.68	7.87	8.03	8.18	8.31	8.44	8.55	8.66	8.76	8.85	8.94	9.03
9	4.60	5.43	5.96	6.35	6.66	6.91	7.13	7.32	7.49	7.65	7.78	7.91	8.03	8.13	8.23	8.32	8.41	8.49	8.57
10	4.48	5.27	5.77	6.14	6.43	6.67	6.87	7.05	7.21	7.36	7.48	7.60	7.71	7.81	7.91	7.99	8.07	8.15	8.22
11	4.39	5.14	5.62	5.97	6.25	6.48	6.67	6.84	6.99	7.13	7.25	7.36	7.46	7.56	7.65	7.73	7.81	7.88	7.95
12	4.32	5.04	5.50	5.84	6.10	6.32	6.51	6.67	6.81	6.94	7.06	7.17	7.26	7.36	7.44	7.52	7.59	7.66	7.73
13	4.26	4.96	5.40	5.73	5.98	6.19	6.37	6.53	6.67	6.79	6.90	7.01	7.10	7.19	7.27	7.34	7.42	7.48	7.55
14	4.21	4.89	5.32	5.63	5.88	6.08	6.26	6.41	6.54	6.66	6.77	6.87	6.96	7.05	7.12	7.20	7.27	7.33	7.39
15	4.17	4.83	5.25	5.56	5.80	5.99	6.16	6.31	6.44	6.55	6.66	6.76	6.84	6.93	7.00	7.07	7.14	7.20	7.26
16	4.13	4.78	5.19	5.49	5.72	5.92	6.08	6.22	6.35	6.46	6.56	6.66	6.74	6.82	6.90	6.97	7.03	7.09	7.15
17	4.10	4.74	5.14	5.43	5.66	5.85	6.01	6.15	6.27	6.38	6.48	6.57	6.66	6.73	6.80	6.87	6.94	7.00	7.05
18	4.07	4.70	5.09	5.38	5.60	5.79	5.94	6.08	6.20	6.31	6.41	6.50	6.58	6.65	6.72	6.79	6.85	6.91	6.96
19	4.05	4.67	5.05	5.33	5.55	5.73	5.89	6.02	6.14	6.25	6.34	6.43	6.51	6.58	6.65	6.72	6.78	6.84	6.89
20	4.02	4.64	5.02	5.29	5.51	5.69	5.84	5.97	6.09	6.19	6.29	6.37	6.45	6.52	6.59	6.65	6.71	6.76	6.82
24	3.96	4.54	4.91	5.17	5.37	5.54	5.69	5.81	5.92	6.02	6.11	6.19	6.26	6.33	6.39	6.45	6.51	6.56	6.61
30	3.89	4.45	4.80	5.05	5.24	5.40	5.54	5.65	5.76	5.85	5.93	6.01	6.08	6.14	6.20	6.26	6.31	6.36	6.41
40	3.82	4.37	4.70	4.93	5.11	5.27	5.39	5.50	5.60	5.69	5.77	5.84	5.90	5.96	6.02	6.07	6.12	6.17	6.21
60	3.76	4.28	4.60	4.82	4.99	5.13	5.25	5.36	5.45	5.53	5.60	5.67	5.73	5.79	5.84	5.89	5.93	5.98	6.02
120	3.70	4.20	4.50	4.71	4.87	5.01	5.12	5.21	5.30	5.38	5.44	5.51	5.56	5.61	5.66	5.71	5.75	5.79	5.83
∞	3.64	4.12	4.40	4.60	4.76	4.88	4.99	5.08	5.16	5.23	5.29	5.35	5.40	5.45	5.49	5.54	5.57	5.61	5.65

Table A-7 Orthogonal polynomial coefficients

k	POLYNOMIAL	1	2	3	4	5	6	7	8	9	10	(Σp_i^2)
3	Linear	−1	0	1								2
	Quadratic	1	−2	1								6
4	Linear	−3	−1	1	3							20
	Quadratic	1	−1	−1	1							4
	Cubic	−1	3	−3	1							20
5	Linear	−2	−1	0	1	2						10
	Quadratic	2	−1	−2	−1	2						14
	Cubic	−1	2	0	−2	1						10
	Quartic	1	−4	6	−4	1						70
6	Linear	−5	−3	−1	1	3	5					70
	Quadratic	5	−1	−4	−4	−1	5					84
	Cubic	−5	7	4	−4	−7	5					180
	Quartic	1	−3	2	2	−3	1					28
	Quintic	−1	5	−10	10	−5	1					252
7	Linear	−3	−2	−1	0	1	2	3				28
	Quadratic	5	0	−3	−4	−3	0	5				84
	Cubic	−1	1	1	0	−1	−1	1				6
	Quartic	3	−7	1	6	1	−7	3				154
	Quintic	−1	4	−5	0	5	−4	1				84
	Sextic	1	−6	15	−20	15	−6	1				924
8	Linear	−7	−5	−3	−1	1	3	5	7			168
	Quadratic	7	1	−3	−5	−5	−3	1	7			168
	Cubic	−7	5	7	3	−3	−7	−5	7			264
	Quartic	7	−13	−3	9	9	−3	−13	7			616
	Quintic	−7	23	−17	−15	15	17	−23	7			2184
	Sextic	1	−5	9	−5	−5	9	−5	1			264
	Septic	−1	7	−21	35	−35	21	−7	1			3,432
9	Linear	−4	−3	−2	−1	0	1	2	3	4		60
	Quadratic	28	7	−8	−17	−20	−17	−8	7	28		2,772
	Cubic	−14	7	13	9	0	−9	−13	−7	14		990
	Quartic	14	−21	−11	9	18	9	−11	−21	14		2,002
	Quintic	−4	11	−4	−9	0	9	4	−11	4		468
	Sextic	4	−17	22	1	−20	1	22	−17	4		1,980
	Septic	−1	6	−14	14	0	−14	14	−6	1		858
	Octic	1	−8	28	−56	70	−56	28	−8	1		12,870
10	Linear	−9	−7	−5	−3	−1	1	3	5	7	9	330
	Quadratic	6	2	−1	−3	−4	−4	−3	−1	2	6	132
	Cubic	−42	14	35	31	12	−12	−31	−35	−14	42	8,580
	Quartic	18	−22	−17	3	18	18	3	−17	−22	18	2,860
	Quintic	−6	14	−1	−11	−6	6	11	1	−14	6	780
	Sextic	3	−11	10	6	−8	−8	6	10	11	3	660
	Septic	−9	47	−86	92	56	−56	−42	86	−47	9	29,172
	Octic	1	−7	20	−28	14	14	−28	20	−7	1	2,860
	Novic	−1	9	−36	84	−126	126	−84	36	−9	1	48,620

Table A-8A Bonferroni corrected jackknife residual critical values

α = 0.1

k	n=5	10	15	20	25	50	100	200	400	800
1	6.96	3.50	3.27	3.22	3.21	3.27	3.39	3.54	3.70	3.86
2	31.82	3.71	3.33	3.25	3.23	3.28	3.39	3.54	3.70	3.86
3		4.03	3.41	3.29	3.25	3.28	3.40	3.54	3.70	3.86
4		4.60	3.51	3.33	3.27	3.29	3.40	3.54	3.70	3.86
5		5.84	3.63	3.37	3.30	3.29	3.40	3.54	3.70	3.86
6		9.92	3.81	3.43	3.33	3.30	3.40	3.54	3.70	3.86
7		63.66	4.06	3.50	3.36	3.30	3.40	3.54	3.70	3.86
8			4.46	3.58	3.39	3.31	3.40	3.54	3.70	3.86
9			5.17	3.69	3.44	3.31	3.40	3.54	3.70	3.86
10			6.74	3.83	3.49	3.32	3.40	3.54	3.70	3.86
15				7.45	3.99	3.36	3.41	3.54	3.70	3.86
20					8.05	3.41	3.42	3.55	3.70	3.86
40						4.50	3.47	3.55	3.70	3.86
80							3.92	3.58	3.70	3.86

α = 0.05

k	n=5	10	15	20	25	50	100	200	400	800
1	9.92	4.03	3.65	3.54	3.50	3.51	3.60	3.73	3.87	4.02
2	63.66	4.32	3.73	3.58	3.53	3.51	3.60	3.73	3.87	4.02
3		4.77	3.83	3.62	3.55	3.52	3.60	3.73	3.87	4.02
4		5.60	3.95	3.67	3.58	3.53	3.61	3.73	3.87	4.02
5		7.45	4.12	3.73	3.61	3.53	3.61	3.73	3.87	4.02
6		14.09	4.36	3.81	3.65	3.54	3.61	3.73	3.88	4.02
7		127.32	4.70	3.89	3.69	3.55	3.61	3.73	3.88	4.02
8			5.25	4.00	3.73	3.56	3.61	3.73	3.88	4.02
9			6.25	4.15	3.79	3.57	3.61	3.73	3.88	4.02
10			8.58	4.33	3.85	3.61	3.61	3.73	3.88	4.03
15				9.46	4.50	3.67	3.62	3.74	3.88	4.03
20					10.21	3.67	3.63	3.74	3.88	4.03
40						5.04	3.69	3.75	3.88	4.03
80							4.23	3.78	3.88	4.03

α = 0.01

k	n=5	10	15	20	25	50	100	200	400	800
1	22.33	5.41	4.55	4.29	4.17	4.03	4.06	4.15	4.27	4.40
2	318.31	5.96	4.68	4.35	4.20	4.04	4.06	4.15	4.27	4.40
3		6.87	4.85	4.42	4.24	4.05	4.06	4.15	4.27	4.40
4		8.61	5.08	4.50	4.28	4.06	4.06	4.15	4.27	4.40
5		12.92	5.37	4.60	4.33	4.07	4.07	4.15	4.27	4.40
6		31.60	5.80	4.72	4.39	4.07	4.07	4.15	4.27	4.40
7		636.62	6.43	4.86	4.45	4.08	4.07	4.15	4.27	4.40
8			7.50	5.05	4.53	4.09	4.07	4.15	4.27	4.40
9			9.57	5.29	4.62	4.10	4.07	4.15	4.27	4.40
10			14.82	5.62	4.72	4.12	4.08	4.15	4.27	4.40
15				16.33	5.81	4.18	4.09	4.15	4.27	4.40
20					17.60	4.28	4.10	4.16	4.27	4.40
40						6.44	4.18	4.17	4.27	4.40
80							4.97	4.21	4.28	4.40

Table A-8B Bonferroni corrected studentized residual critical values

α = 0.1

k	n=5	10	15	20	25	50	100	200	400	800
1	1.70	2.26	2.48	2.61	2.71	2.98	3.23	3.44	3.64	3.82
2	1.41	2.21	2.46	2.60	2.70	2.98	3.22	3.44	3.64	3.82
3		2.14	2.43	2.59	2.69	2.98	3.22	3.44	3.64	3.82
4		2.05	2.40	2.57	2.69	2.98	3.22	3.44	3.64	3.82
5		1.92	2.37	2.56	2.68	2.98	3.22	3.44	3.64	3.82
6		1.71	2.32	2.54	2.66	2.97	3.22	3.44	3.64	3.82
7		1.41	2.27	2.51	2.65	2.97	3.22	3.44	3.64	3.82
8			2.19	2.49	2.64	2.96	3.22	3.44	3.64	3.82
9			2.09	2.45	2.62	2.96	3.22	3.44	3.64	3.82
10			1.94	2.41	2.60	2.94	3.22	3.44	3.64	3.82
15				1.95	2.45	2.92	3.21	3.44	3.64	3.82
20					1.96	2.54	3.18	3.44	3.64	3.82
40							2.96	3.43	3.64	3.82
80								3.41	3.63	3.82

α = 0.05

k	n=5	10	15	20	25	50	100	200	400	800
1	1.71	2.36	2.61	2.77	2.87	3.16	3.40	3.61	3.81	3.99
2	1.41	2.30	2.59	2.75	2.86	3.15	3.40	3.61	3.81	3.99
3		2.22	2.56	2.73	2.85	3.15	3.40	3.61	3.31	3.99
4		2.11	2.52	2.71	2.84	3.15	3.40	3.61	3.31	3.99
5		1.95	2.47	2.69	2.82	3.15	3.40	3.61	3.81	3.99
6		1.72	2.42	2.67	2.81	3.14	3.40	3.61	3.81	3.99
7		1.41	2.35	2.64	2.79	3.13	3.39	3.61	3.81	3.99
8			2.25	2.60	2.78	3.13	3.39	3.61	3.81	3.99
9			2.13	2.56	2.76	3.13	3.39	3.61	3.81	3.99
10			1.96	2.51	2.73	3.13	3.39	3.61	3.81	3.99
15				1.97	2.54	3.10	3.38	3.61	3.81	3.99
20					1.97	3.07	3.35	3.60	3.81	3.99
40						2.62	3.08	3.60	3.80	3.99
80								3.58	3.80	3.99

α = 0.01

k	n=5	10	15	20	25	50	100	200	400	800
1	1.73	2.54	2.87	3.06	3.19	3.51	3.77	3.99	4.18	4.35
2	1.41	2.45	2.83	3.03	3.17	3.51	3.77	3.99	4.18	4.35
3		2.33	2.78	3.01	3.15	3.51	3.77	3.99	4.18	4.35
4		2.18	2.72	2.98	3.14	3.50	3.77	3.99	4.18	4.35
5		1.98	2.65	2.94	3.11	3.50	3.77	3.99	4.18	4.35
6		1.73	2.57	2.91	3.09	3.49	3.77	3.99	4.18	4.35
7		1.41	2.47	2.86	3.07	3.49	3.76	3.99	4.18	4.35
8			2.35	2.81	3.04	3.48	3.76	3.98	4.18	4.35
9			2.19	2.75	3.01	3.47	3.76	3.98	4.18	4.35
10			1.99	2.68	2.97	3.47	3.76	3.98	4.18	4.35
15				1.99	2.70	3.43	3.75	3.98	4.17	4.35
20					1.99	3.38	3.74	3.98	4.17	4.35
40						2.75	3.69	3.97	4.17	4.35
80							3.31	3.94	4.17	4.34

Table A-9 Critical values for leverages, n = sample size, k = number of predictors

$\alpha = .10$

n \ k	1	2	3	4	5	6	7	8	9	10	15	20	40	80
10	0.626	0.759	0.847	0.911	0.956	0.984	0.997	1.000						
15	0.481	0.595	0.679	0.748	0.806	0.855	0.897	0.932	0.959	0.980				
20	0.394	0.491	0.565	0.627	0.682	0.731	0.775	0.815	0.851	0.883	0.988			
25	0.335	0.419	0.484	0.540	0.589	0.635	0.676	0.715	0.751	0.784	0.918	0.992		
30	0.293	0.366	0.424	0.474	0.519	0.560	0.599	0.635	0.669	0.701	0.837	0.937		
40	0.236	0.295	0.342	0.383	0.420	0.455	0.487	0.518	0.547	0.576	0.701	0.806		
60	0.172	0.214	0.248	0.279	0.306	0.332	0.356	0.380	0.402	0.424	0.524	0.612	0.888	
80	0.137	0.170	0.197	0.221	0.242	0.263	0.283	0.301	0.319	0.337	0.418	0.491	0.737	
100	0.114	0.141	0.164	0.183	0.201	0.219	0.235	0.250	0.266	0.280	0.348	0.410	0.625	0.941
200	0.064	0.079	0.091	0.102	0.111	0.121	0.130	0.138	0.146	0.155	0.192	0.227	0.353	0.568
400	0.036	0.043	0.050	0.055	0.060	0.065	0.070	0.075	0.079	0.083	0.104	0.122	0.190	0.311
800	0.020	0.024	0.027	0.030	0.032	0.035	0.037	0.040	0.042	0.044	0.055	0.065	0.100	0.164

$\alpha = .05$

n \ k	1	2	3	4	5	6	7	8	9	10	15	20	40	80
10	0.683	0.802	0.879	0.933	0.969	0.990	0.999	1.000						
15	0.531	0.639	0.719	0.782	0.835	0.880	0.916	0.946	0.969	0.986				
20	0.436	0.531	0.602	0.662	0.714	0.761	0.802	0.839	0.872	0.901	0.991			
25	0.372	0.454	0.518	0.573	0.621	0.665	0.705	0.742	0.776	0.807	0.931	0.994		
30	0.325	0.398	0.455	0.505	0.549	0.589	0.627	0.662	0.695	0.726	0.855	0.947		
40	0.261	0.321	0.368	0.409	0.446	0.480	0.512	0.543	0.572	0.600	0.722	0.823		
60	0.190	0.233	0.268	0.298	0.326	0.352	0.376	0.400	0.422	0.444	0.543	0.630	0.898	
80	0.151	0.185	0.212	0.236	0.258	0.279	0.299	0.318	0.336	0.353	0.435	0.508	0.751	
100	0.126	0.154	0.176	0.196	0.215	0.232	0.248	0.264	0.279	0.294	0.363	0.425	0.638	0.946
200	0.070	0.085	0.098	0.108	0.119	0.128	0.137	0.146	0.154	0.162	0.201	0.236	0.362	0.570
400	0.039	0.047	0.053	0.059	0.064	0.069	0.074	0.079	0.083	0.088	0.108	0.127	0.196	0.317
800	0.021	0.025	0.029	0.032	0.034	0.037	0.039	0.042	0.044	0.046	0.057	0.067	0.103	0.168

$\alpha = .01$

n \ k	1	2	3	4	5	6	7	8	9	10	15	20	40	80
10	0.785	0.875	0.930	0.965	0.986	0.997	1.000	1.000						
15	0.629	0.724	0.792	0.844	0.887	0.921	0.948	0.969	0.984	0.994				
20	0.524	0.612	0.677	0.731	0.777	0.817	0.852	0.883	0.910	0.933	0.996			
25	0.450	0.529	0.589	0.640	0.685	0.724	0.761	0.794	0.824	0.851	0.953	0.997		
30	0.394	0.466	0.521	0.568	0.610	0.648	0.683	0.716	0.746	0.774	0.889	0.964		
40	0.318	0.377	0.424	0.464	0.501	0.534	0.565	0.595	0.622	0.649	0.763	0.855		
60	0.231	0.275	0.310	0.341	0.369	0.395	0.420	0.443	0.465	0.487	0.584	0.668	0.917	
80	0.183	0.218	0.246	0.271	0.293	0.314	0.334	0.353	0.372	0.389	0.471	0.543	0.778	
100	0.152	0.181	0.205	0.225	0.244	0.262	0.279	0.295	0.310	0.325	0.394	0.456	0.666	0.956
200	0.085	0.100	0.113	0.124	0.135	0.145	0.154	0.163	0.172	0.180	0.219	0.255	0.383	0.598
400	0.046	0.054	0.061	0.067	0.073	0.078	0.083	0.088	0.092	0.097	0.118	0.138	0.208	0.330
800	0.025	0.029	0.033	0.036	0.039	0.041	0.044	0.046	0.049	0.051	0.062	0.073	0.110	0.175

Table A-10 Critical values for the maximum of N values of Cook's $d(i) \times (n - k - 1)$
(Bonferroni correction used) n observations and k predictors

$\alpha = 0.1$

k	n=5	10	15	20	25	50	100	200	400	800
1	14.96	11.13	11.84	12.68	13.46	16.39	19.97	23.94	28.70	33.80
2	40.53	12.21	12.09	12.63	13.22	15.65	18.64	22.09	25.96	30.12
3		13.30	12.09	12.35	12.79	14.84	17.48	20.52	23.86	27.50
4		15.21	12.18	12.14	12.45	14.23	16.62	19.36	22.30	25.97
5		19.33	12.44	12.03	12.21	13.76	15.95	18.49	21.39	24.51
6		31.06	12.94	12.01	12.04	13.39	15.43	17.81	20.36	23.51
7		96.01	13.79	12.08	11.94	13.10	15.02	17.27	19.75	22.42
8			15.26	12.26	11.90	12.85	14.70	16.83	19.20	21.73
9			18.00	12.55	11.91	12.66	14.40	16.52	18.62	21.45
10			23.93	13.02	11.97	12.50	14.16	16.16	18.43	20.55
15				27.66	13.60	12.01	13.39	15.16	17.00	19.34
20					30.94	11.83	12.92	14.53	16.31	18.35
40						15.95	12.26	13.56	15.10	16.83
80							13.49	13.05	14.39	15.85

$\alpha = 0.05$

k	n=5	10	15	20	25	50	100	200	400	800
1	24.97	15.24	15.55	16.37	17.18	20.41	24.31	28.83	33.88	40.15
2	82.06	16.56	15.63	16.01	16.56	19.08	22.33	26.05	30.20	33.96
3		18.16	15.50	15.49	15.85	17.93	20.72	24.14	27.57	32.06
4		21.28	15.59	15.14	15.33	17.06	19.63	22.49	25.83	29.31
5		28.40	15.94	14.95	14.96	16.41	18.70	21.39	24.42	28.24
6		50.22	16.70	14.91	14.70	15.91	17.97	20.54	23.48	26.68
7		192.90	17.99	15.00	14.55	15.50	17.49	20.00	22.35	25.67
8			20.32	15.25	14.48	15.19	17.05	19.31	22.06	24.44
9			24.78	15.69	14.49	14.92	16.69	18.85	21.34	24.29
10			34.72	16.38	14.58	14.70	16.38	18.42	20.49	23.33
15				39.98	16.94	14.03	15.36	17.16	19.39	21.75
20					44.63	13.79	14.81	16.52	18.46	20.32
40						19.50	13.92	15.22	16.83	18.76
80							15.55	14.58	15.99	17.52

$\alpha = 0.01$

p	n=5	10	15	20	25	50	100	200	400	800
1	77.29	28.72	26.88	27.24	27.92	31.46	36.10	41.22	49.42	68.39
2	415.27	30.97	26.13	25.65	25.81	28.12	32.61	37.34	44.99	57.70
3		35.12	25.66	24.22	24.33	26.17	29.15	34.23	37.55	52.58
4		44.09	25.82	23.58	23.20	24.56	27.31	31.26	35.28	40.60
5		66.83	26.66	23.20	22.49	23.39	25.84	29.44	34.14	36.91
6		150.47	28.48	23.12	22.00	22.55	24.35	28.42	31.04	36.91
7		964.09	31.80	23.34	21.71	21.79	24.19	26.87	31.04	33.55
8			37.84	23.93	21.59	21.26	23.28	25.83	29.31	33.55
9			50.10	24.93	21.64	20.76	22.23	25.62	28.21	30.50
10			80.67	26.54	21.83	20.37	22.11	24.53	28.21	30.50
15				92.09	27.02	19.16	20.22	22.40	25.64	27.73
20					102.32	18.82	19.18	21.32	23.31	25.21
40						29.95	18.04	19.32	21.17	22.91
80							20.67	18.57	20.12	22.90

Appendix—Matrices and Their Relationship to Regression Analysis

B-1 Preview

Statisticians have found matrix mathematics to be a very useful vehicle for compactly presenting the concepts, methods, and formulae of regression analysis and other multivariable methods. Moreover, matrix formulation of such topics has had the important practical implication of permitting extremely efficient and accurate use of the computer for carrying out multivariable analyses on large data sets.

This appendix will summarize some of the more elementary but important notions and manipulations of matrix algebra and will use this tool to describe the general least-squares procedures of multiple regression. Admittedly, the material in this appendix is more mathematical than in the main body of the text and is not absolutely necessary knowledge for the applied user of the multivariable methods we describe. Nevertheless, the reader who is comfortable with the mathematical level used here should find the matrix formulation of regression analysis a powerful and unifying supplement that may facilitate the learning of more advanced multivariable methods.

B-2 Definitions

A *matrix* may be simply defined as a rectangular array of numbers. For example,

$$A = \begin{bmatrix} 2 & 3 & 1 \\ 1 & 1 & 2 \end{bmatrix}, \qquad B = \begin{bmatrix} 2 & 1 \\ 3 & 2 \end{bmatrix}, \qquad C = \begin{bmatrix} 1 \\ 1 \\ 3 \end{bmatrix}$$

are all matrices.

The *dimensions* of a matrix are the number of rows and the number of columns that it has. For example, the matrix A above has two rows and three columns:

$$
\begin{array}{ccc}
\text{column} & \text{column} & \text{column} \\
1 & 2 & 3 \\
\downarrow & \downarrow & \downarrow
\end{array}
$$

$$
\begin{array}{c}
\text{row 1} \rightarrow \\
\\
\text{row 2} \rightarrow
\end{array}
\begin{bmatrix}
2 & 3 & 1 \\
1 & 1 & 2
\end{bmatrix}
$$

It is customary to say that A is a 2×3 matrix or to write $A_{2 \times 3}$. The dimensions of the matrices B and C above are 2×2 and 3×1, respectively. An example of a 1×4 matrix is any matrix with one row and four columns, such as $D_{1 \times 4} = [-2 \quad 3 \quad 3 \quad 0]$. Incidentally, matrices that contain only one row or only one column are often referred to as *vectors*; thus the matrices

$$
C = \begin{bmatrix} 1 \\ 1 \\ 3 \end{bmatrix} \quad \text{and} \quad D = [-2 \quad 3 \quad 3 \quad 0]
$$

are examples of a column vector and a row vector, respectively. Also, the matrix B is called a *square matrix* because it has the same number of rows and columns.

The numbers forming the rectangular array of a matrix are called the *elements* of the matrix. If we let a_{ij} denote the element in the ith row and jth column of the matrix A above,

$$
a_{11} = 2, \quad a_{12} = 3, \quad a_{13} = 1
$$
$$
a_{21} = 1, \quad a_{22} = 1, \quad a_{23} = 2
$$

It is often informative to write

$$
A_{2 \times 3} = ((a_{ij}))
$$

indicating that the matrix A (represented by a capital letter) with two rows and three columns has typical element a_{ij} (represented by the corresponding lowercase letter). Thus, if you were given that

$$
B_{3 \times 2} = ((b_{ij}))
$$

where $b_{21} = 3$, $b_{31} = 2$, $b_{11} = -2$, $b_{12} = 6$, $b_{22} = 0$, $b_{32} = 1$, you should construct B to be

$$
B = \begin{bmatrix} -2 & 6 \\ 3 & 0 \\ 2 & 1 \end{bmatrix}
$$

B-3 Matrices in Regression Analysis

Given any set of multivariable data suitable for a regression analysis, a number of key matrices can be defined that directly correspond to the basic components of the regression model being postulated. For example, consider the following $n = 12$ pairs of observations on $Y = $ WGT and $X = $ HGT:

						Child						
Variable	1	2	3	4	5	6	7	8	9	10	11	12
Y (WGT)	64	71	53	67	55	58	77	57	56	51	76	68
X (HGT)	57	59	49	62	51	50	55	48	42	42	61	57

We can, in correspondence with the straight-line regression model $Y = \beta_0 + \beta_1 X + E$, define four matrices: $\mathbf{Y}$, the vector of observations on Y; $\mathbf{X}$, the matrix of independent variables; $\boldsymbol{\beta}$, the vector of parameters to be estimated; and $\mathbf{E}$, the vector of errors.

For the data given, these matrices are defined as follows:

$$
\mathbf{Y}_{12\times1} = \begin{bmatrix} 64 \\ 71 \\ 53 \\ 67 \\ 55 \\ 58 \\ 77 \\ 57 \\ 56 \\ 51 \\ 76 \\ 68 \end{bmatrix}, \quad
\mathbf{X}_{12\times2} = \begin{bmatrix} 1 & 57 \\ 1 & 59 \\ 1 & 49 \\ 1 & 62 \\ 1 & 51 \\ 1 & 50 \\ 1 & 55 \\ 1 & 48 \\ 1 & 42 \\ 1 & 42 \\ 1 & 61 \\ 1 & 57 \end{bmatrix}, \quad
\boldsymbol{\beta}_{2\times1} = \begin{bmatrix} \beta_0 \\ \beta_1 \end{bmatrix}, \quad
\mathbf{E}_{12\times1} = \begin{bmatrix} E_1 \\ E_2 \\ E_3 \\ E_4 \\ E_5 \\ E_6 \\ E_7 \\ E_8 \\ E_9 \\ E_{10} \\ E_{11} \\ E_{12} \end{bmatrix}
$$

Notice that the first column of the $\mathbf{X}$ matrix of independent variables contains only 1's. This is the general convention used for any regression model containing a constant term β_0; motivation for adopting this convention follows by imagining the β_0 term to be of the form $\beta_0 X_0$, where X_0 is a dummy variable always taking the value 1. The vectors of errors $\mathbf{E}$ contains random (and unobservable) error values, one for each pair of observations, which represent the differences between the observed Y-values and their (unknown) expected values under the given model.

B-4 Transpose of a Matrix

The transpose $\mathbf{A}'$ of a matrix $\mathbf{A}$ is defined to be that matrix whose (i, j)th element a'_{ij} is equal to the (j, i)th element of $\mathbf{A}$. For example, if

$$
\mathbf{A} = \begin{bmatrix} 2 & 3 & 1 \\ 1 & 1 & 2 \end{bmatrix}
$$

then

$$
\mathbf{A}' = \begin{bmatrix} 2 & 1 \\ 3 & 1 \\ 1 & 2 \end{bmatrix}
$$

since $a'_{11} = a_{11} = 2$, $a'_{12} = a_{21} = 1$, $a'_{21} = a_{12} = 3$, $a'_{22} = a_{22} = 1$, $a'_{31} = a_{13} = 1$, $a'_{32} = a_{23} = 2$.

Another way of looking at this is that the first column of $\mathbf{A}$ becomes the first row of $\mathbf{A}'$, the second column of $\mathbf{A}$ becomes the second row of $\mathbf{A}'$, and so on. Thus, if $\mathbf{A}$ is $r \times c$, then $\mathbf{A}'$ is $c \times r$.

As examples, the transposes of

$$\mathbf{A}_{3\times 2} = \begin{bmatrix} 1 & 2 \\ 3 & 1 \\ 1 & 1 \end{bmatrix}, \qquad \mathbf{B}_{3\times 3} = \begin{bmatrix} 1 & 0 & 2 \\ 0 & 4 & -5 \\ 2 & -5 & 3 \end{bmatrix}, \qquad \mathbf{C}_{4\times 1} = \begin{bmatrix} 0 \\ 1 \\ 2 \\ -2 \end{bmatrix}$$

are

$$\mathbf{A}'_{2\times 3} = \begin{bmatrix} 1 & 3 & 1 \\ 2 & 1 & 1 \end{bmatrix}, \qquad \mathbf{B}'_{3\times 3} = \begin{bmatrix} 1 & 0 & 2 \\ 0 & 4 & -5 \\ 2 & -5 & 3 \end{bmatrix}, \qquad \mathbf{C}'_{1\times 4} = \begin{bmatrix} 0 & 1 & 2 & -2 \end{bmatrix}$$

Also, the transpose of the matrix $\mathbf{X}_{12\times 2}$ of the previous section is given by

$$\mathbf{X}'_{2\times 12} = \begin{bmatrix} 1 & 1 & 1 & 1 & 1 & 1 & 1 & 1 & 1 & 1 & 1 & 1 \\ 57 & 59 & 49 & 62 & 51 & 50 & 55 & 48 & 42 & 42 & 61 & 57 \end{bmatrix}$$

Note that in the examples above, the matrix $\mathbf{B}$ is such that $\mathbf{B} = \mathbf{B}'$ (equality here means that corresponding elements are equal). A matrix satisfying this condition is said to be a *symmetric matrix*. Note that a symmetric matrix $\mathbf{A}$ must always be square, since otherwise $\mathbf{A}$ and $\mathbf{A}'$ would have different dimensions and so could not possibly be equal. A necessary and sufficient condition for the square matrix $\mathbf{A} = ((a_{ij}))$ to be symmetric is that $a_{ij} = a_{ji}$ for every $i \neq j$.

Correlation matrices such as

$$\mathbf{R}_1 = \begin{bmatrix} 1 & r_{xy} \\ r_{xy} & 1 \end{bmatrix} \qquad \text{or} \qquad \mathbf{R}_2 = \begin{bmatrix} 1 & r_{12} & r_{13} \\ r_{12} & 1 & r_{23} \\ r_{13} & r_{23} & 1 \end{bmatrix}$$

are always symmetric.

An important special case where the above condition for symmetry is satisfied is when $a_{ij} = a_{ji} = 0$ for every $i \neq j$. A square matrix having this property is said to be a *diagonal matrix*, the general form of which is given (for the 3×3 case) by

$$\begin{bmatrix} a_{11} & 0 & 0 \\ 0 & a_{22} & 0 \\ 0 & 0 & a_{33} \end{bmatrix}$$

Diagonal

The most often used diagonal matrix is the *identity matrix* $\mathbf{I}$, which has 1's on the diagonal; for example, the 3×3 identity matrix is

$$\mathbf{I} = \begin{bmatrix} 1 & 0 & 0 \\ 0 & 1 & 0 \\ 0 & 0 & 1 \end{bmatrix}$$

We will see shortly that an identity matrix serves the same algebraic function for matrix multiplication that the number 1 serves for ordinary scalar multiplication.

B-5 Matrix Addition

The sum of two matrices, say **A** and **B**, is obtained by adding together the corresponding elements of each matrix. Clearly, such addition can be performed only when the two matrices have the same dimensions. Thus, for example, we can add the matrices

$$\mathbf{A}_{2\times3} = \begin{bmatrix} 2 & 3 & 1 \\ 1 & 1 & 2 \end{bmatrix} \quad \text{and} \quad \mathbf{B}_{2\times3} = \begin{bmatrix} 4 & 1 & 5 \\ 1 & 3 & 1 \end{bmatrix}$$

(since they have the same dimensions) to get

$$\mathbf{A} + \mathbf{B}_{2\times3} = \begin{bmatrix} 2+4 & 3+1 & 1+5 \\ 1+1 & 1+3 & 2+1 \end{bmatrix} = \begin{bmatrix} 6 & 4 & 6 \\ 2 & 4 & 3 \end{bmatrix}$$

We could not, however, add the matrices **A** and **C**, where

$$\mathbf{C}_{3\times2} = \begin{bmatrix} 5 & 4 \\ 1 & 4 \\ 2 & 6 \end{bmatrix}$$

since the dimensions of these matrices are not the same.

An example of a more abstract use of matrix addition would be to sum the two 12×1 vectors

$$\mathbf{D}_{12\times1} = \begin{bmatrix} \beta_0 + 57\beta_1 \\ \beta_0 + 59\beta_1 \\ \vdots \\ \beta_0 + 57\beta_1 \end{bmatrix} \quad \text{and} \quad \mathbf{E}_{12\times1} = \begin{bmatrix} E_1 \\ E_2 \\ \vdots \\ E_{12} \end{bmatrix}$$

to obtain

$$\mathbf{D} + \mathbf{E}_{12\times1} = \begin{bmatrix} \beta_0 + 57\beta_1 + E_1 \\ \beta_0 + 59\beta_1 + E_2 \\ \vdots \\ \beta_0 + 57\beta_1 + E_{12} \end{bmatrix}$$

Actually, you may recognize that each element of the matrix $\mathbf{D} + \mathbf{E}$ is obtained by substituting for X in the right side of the straight-line regression equation

$$Y = \beta_0 + \beta_1 X + E$$

each of the 12 X (HGT) values given in the data set of Section B-3.

B-6 Matrix Multiplication

Multiplication of two matrices is somewhat more complicated than addition. The first rule to remember is that the product **AB** of two matrices **A** and **B** can exist if and only if the *number of columns* of **A** *is equal to the number of rows of* **B**. Thus, if **A** is 2×3 and **B** is

3×4, the product **AB** exists since **A** has 3 columns and **B** has 3 rows. However, the product **BA** does not exist, since the number of columns of **B** (i.e., 4) is not equal to the number of rows of **A** (i.e., 2). Notationally, therefore, a matrix product can exist only if the dimensions of the matrices can be represented as follows:

Equal numbers

$$\mathbf{A}_{m \times n} \quad \times \quad \mathbf{B}_{n \times p} \quad = \quad \mathbf{AB}_{m \times p}$$

Product dimensions

Note also from the expression above that the dimensions $m \times p$ of the product matrix **AB** are given by the number of rows of the *pre*multiplier (i.e., **A**) and the number of columns of the *post*multiplier (i.e., **B**). For example, if **A** is 2×3 and **B** is 3×4, the dimensions of the product **AB** are 2×4.

Now, to carry out matrix multiplication, consider the two matrices

$$\mathbf{A}_{2 \times 3} = \begin{bmatrix} 2 & 1 & 0 \\ 0 & 3 & 1 \end{bmatrix} \quad \text{and} \quad \mathbf{B}_{3 \times 2} = \begin{bmatrix} 1 & -2 \\ 1 & 0 \\ 3 & 2 \end{bmatrix}$$

Since **A** is 2×3 and **B** is 3×2, the product **AB** will be a 2×2 matrix. If we let the elements of **AB** be denoted by

$$\mathbf{AB}_{2 \times 2} = ((c_{ij})) = \begin{bmatrix} c_{11} & c_{12} \\ c_{21} & c_{22} \end{bmatrix}$$

we can obtain the upper-left-hand corner element c_{11} by working with the first row of **A** and the first column of **B**, as follows:

$$\begin{array}{ccc} \mathbf{A} & \mathbf{B} & \mathbf{AB} \\ \begin{bmatrix} 2 & 1 & 0 \\ 0 & 3 & 1 \end{bmatrix} & \begin{bmatrix} 1 & -2 \\ 1 & 0 \\ 3 & 2 \end{bmatrix} & = \begin{bmatrix} (2 \times 1) + (1 \times 1) \\ + (0 \times 3) = 3 & c_{12} \\ c_{21} & c_{22} \end{bmatrix} \end{array}$$

What we have done here is to calculate the product of each element in row 1 of **A** with the corresponding element in column 1 of **B**, and then add up these three products to obtain $c_{11} = 3$:

Column 1 of **B**

$$c_{11} = (2 \times 1) + (1 \times 1) + (0 \times 3) = 3$$

Row 1 of **A**

To find the element in the second row and first column of **AB** (i.e., c_{21}), we work with the second row of **A** and the first column of **B**, as follows:

$$
\overset{\mathbf{A}}{\begin{bmatrix} 2 & 1 & 0 \\ 0 & 3 & 1 \end{bmatrix}} \overset{\mathbf{B}}{\begin{bmatrix} 1 & -2 \\ 1 & 0 \\ 3 & 2 \end{bmatrix}} = \overset{\mathbf{AB}}{\begin{bmatrix} 3 & c_{12} \\ (0 \times 1) + (3 \times 1) & \\ + (1 \times 3) = 6 & c_{22} \end{bmatrix}}
$$

Thus, for the element c_{21}, we find

Column 1 of **B**

$$c_{21} = (0 \times 1) + (3 \times 1) + (1 \times 3) = 6$$

Row 2 of **A**

Continuing this process, we find

$$c_{12} = (2 \times -2) + (1 \times 0) + (0 \times 2) = -4$$
$$c_{22} = (0 \times -2) + (3 \times 0) + (1 \times 2) = 2$$

Thus,

$$
\mathbf{AB}_{2 \times 2} = \begin{bmatrix} 2 & 1 & 0 \\ 0 & 3 & 1 \end{bmatrix} \begin{bmatrix} 1 & -2 \\ 1 & 0 \\ 3 & 2 \end{bmatrix} = \begin{bmatrix} 3 & -4 \\ 6 & 2 \end{bmatrix}
$$

In general, if $\mathbf{A}_{m \times n} = ((a_{ij}))$ and $\mathbf{B}_{n \times p} = ((b_{ij}))$, the (i, j)th element c_{ij} of the product

$$\mathbf{AB}_{m \times p} = ((c_{ij}))$$

is defined to be

$$c_{ij} = \sum_{l=1}^{n} a_{il}b_{lj} \qquad i = 1, 2, \ldots, m; \quad j = 1, 2, \ldots, p$$

Thus, as another example, if

$$\mathbf{A}_{2 \times 2} = \begin{bmatrix} -1 & 3 \\ 2 & 2 \end{bmatrix} \quad \text{and} \quad \mathbf{B}_{2 \times 1} = \begin{bmatrix} 0 \\ 1 \end{bmatrix}$$

then $m = 2$, $p = 1$, $n = 2$, and

$$c_{11} = \sum_{l=1}^{2} a_{1l}b_{l1} = (-1 \times 0) + (3 \times 1) = 3$$

$$c_{21} = \sum_{l=1}^{2} a_{2l}b_{l1} = (2 \times 0) + (2 \times 1) = 2$$

so that

$$\mathbf{AB}_{2 \times 2} = \begin{bmatrix} -1 & 3 \\ 2 & 2 \end{bmatrix} \begin{bmatrix} 0 \\ 1 \end{bmatrix} = \begin{bmatrix} 3 \\ 2 \end{bmatrix}$$

Here are a few other examples for additional practice:

1. Find **AI** and **IA**, where

$$A_{2\times2} = \begin{bmatrix} -1 & 3 \\ 2 & 2 \end{bmatrix} \quad \text{and} \quad I_{2\times2} = \begin{bmatrix} 1 & 0 \\ 0 & 1 \end{bmatrix}$$

2. Find **X'X**, where

$$X_{3\times2} = \begin{bmatrix} 1 & 10 \\ 1 & 15 \\ 1 & 20 \end{bmatrix}$$

The answer to problem 1 is

$$\mathbf{AI} = \mathbf{IA} = \mathbf{A}$$

which indicates why **I** is generally referred to as the identity matrix, since, like the scalar identity 1, the product of any square matrix (e.g., **A**) with an appropriate identity matrix (of the right dimensions) will always yield the original matrix **A**, whether **I** premultiplies or postmultiplies **A**.

The answer to problem 2 is

$$\mathbf{X'X}_{2\times2} = \begin{bmatrix} 1 & 1 & 1 \\ 10 & 15 & 20 \end{bmatrix} \begin{bmatrix} 1 & 10 \\ 1 & 15 \\ 1 & 20 \end{bmatrix} = \begin{bmatrix} 3 & 45 \\ 45 & 725 \end{bmatrix}$$

which is a symmetric matrix, as will be the case whenever any matrix is multiplied (pre or post) by its own transpose.

B-7 Inverse of a Matrix

The definition of an *inverse matrix* parallels the basic property of the *reciprocal* of an ordinary (scalar) number. That is, for any nonzero number a, its reciprocal $1/a$ satisfies the equation

$$a \times \frac{1}{a} = \frac{1}{a} \times a = 1$$

In words, when a is either pre- or postmultiplied by its reciprocal, the result is the scalar 1. Analogously, we say that a square matrix **A** has an inverse $\mathbf{A}^{-1}$ if and only if

$$\mathbf{AA}^{-1} = \mathbf{A}^{-1}\mathbf{A} = \mathbf{I}$$

That is, *the product of* **A** *by its inverse must be equal to an identity matrix.* We hasten to add at this point that *only the inverses of square matrices are being considered.* Thus, if **A** is $n \times n$, $\mathbf{A}^{-1}$ must also be $n \times n$.

We shall not describe here any of the many algorithms available for actually computing the inverse of a matrix.[1] For most practical applications, one can use standard computer packages for such computations. We wish to emphasize instead that one important attribute of inverses in statistical analyses is that their use permits the solution of matrix equations for a matrix of unknowns (e.g., the vector β of unknown regression parameters), in a manner similar to how division is used in ordinary algebra. Most specifically, in regression analysis, the use of an inverse is crucial because it provides a means for efficiently solving the least-squares equations, as well as providing compact formulae for additional components of the analysis (e.g., the variances of and covariances among the estimated regression coefficients).

Some examples of matrix inverses are given as follows:

$$\mathbf{A} = \begin{bmatrix} 2 & 0 & 1 \\ 1 & -1 & 2 \\ 1 & 0 & 0 \end{bmatrix} \quad \text{and} \quad \mathbf{A}^{-1} = \begin{bmatrix} 0 & 0 & 1 \\ 2 & -1 & -3 \\ 1 & 0 & -2 \end{bmatrix}$$

(The reader can check this out by multiplying $\mathbf{A}$ and $\mathbf{A}^{-1}$ together to obtain $\mathbf{I}_{3\times 3}$.) The inverse of

$$\mathbf{X'X} = \begin{bmatrix} n & \sum_{i=1}^{n} X_i \\ \sum_{i=1}^{n} X_i & \sum_{i=1}^{n} X_i^2 \end{bmatrix}$$

is

$$(\mathbf{X'X})^{-1} = \begin{bmatrix} \dfrac{\sum_{i=1}^{n} X_i^2}{n \sum_{i=1}^{n} (X_i - \bar{X})^2} & \dfrac{-\bar{X}}{\sum_{i=1}^{n} (X_i - \bar{X})^2} \\ \dfrac{-\bar{X}}{\sum_{i=1}^{n} (X_i - \bar{X})^2} & \dfrac{1}{\sum_{i=1}^{n} (X_i - \bar{X})^2} \end{bmatrix}$$

B-8 Matrix Formulation of Regression Analysis

We have previously seen in Section B-3 that when fitting a straight-line model

$$Y = \beta_0 + \beta_1 X + E$$

to a set of data consisting of n pairs of observations on the variables X and Y, we can define several matrices to characterize the regression problem under consideration: $\mathbf{Y}$, the $n \times 1$

---

[1] For further details, see N. R. Draper and H. Smith, *Applied Regression Analysis* (New York: Wiley, 1966); W. Mendenhall, *Introduction to Linear Models and the Design and Analysis of Experiments* (Belmont, Calif.: Wadsworth, 1968).

vector of observations on Y; $\mathbf{X}$, the $n \times 2$ matrix of independent variables; $\boldsymbol{\beta}$, the 2×1 vector of parameters; and $\mathbf{E}$, the $n \times 1$ vector of random errors.

In general, whenever we are considering a regression problem involving p independent variables using a model such as

$$Y = \beta_0 + \beta_1 X_1 + \beta_2 X_2 + \cdots + \beta_p X_p + E$$

we can analogously construct appropriate matrices based on the multivariable data set being considered. Thus, if the data on the ith individual consists of the $p + 1$ values

$$Y_i, X_{i1}, X_{i2}, \ldots, X_{ip} \qquad i = 1, 2, \ldots, n$$

the following matrices can be constructed:

$$\mathbf{Y}_{n \times 1} = \begin{bmatrix} Y_1 \\ Y_2 \\ \vdots \\ Y_n \end{bmatrix} = \begin{array}{l} \text{vector of} \\ \text{observations on } Y \end{array}$$

$$\mathbf{X}_{n \times (p+1)} = \begin{bmatrix} 1 & X_{11} & X_{12} & \cdots & X_{1p} \\ 1 & X_{21} & X_{22} & \cdots & X_{2p} \\ \vdots & \vdots & \vdots & & \vdots \\ 1 & X_{n1} & X_{n2} & \cdots & X_{np} \end{bmatrix} = \begin{array}{l} \text{matrix of} \\ \text{independent} \\ \text{variables} \end{array}$$

$$\boldsymbol{\beta}_{(p+1) \times 1} = \begin{bmatrix} \beta_0 \\ \beta_1 \\ \vdots \\ \beta_p \end{bmatrix} = \text{vector of parameters}$$

$$\mathbf{E}_{n \times 1} = \begin{bmatrix} E_1 \\ E_2 \\ \vdots \\ E_n \end{bmatrix} = \text{vector of random errors}$$

Using the matrices above in conjunction with the notions of matrix addition and multiplication, we can formulate the general regression model in matrix terms as follows:

$$\mathbf{Y}_{n \times 1} = \mathbf{X}_{n \times (p+1)} \boldsymbol{\beta}_{(p+1) \times 1} + \mathbf{E}_{n \times 1} \tag{B.1}$$

This compact equation summarizes in a single statement the n equations

$$Y_i = \beta_0 + \beta_1 X_{i1} + \beta_2 X_{i2} + \cdots + \beta_p X_{ip} + E_i \qquad i = 1, 2, \ldots, n$$

Note that this equivalence follows from the following matrix calculations:

$$
\mathbf{X}\boldsymbol{\beta} + \mathbf{E} =
\begin{bmatrix}
1 & X_{11} & X_{12} & \cdots & X_{1p} \\
1 & X_{21} & X_{22} & \cdots & X_{2p} \\
\vdots & \vdots & & & \vdots \\
1 & X_{n1} & X_{n2} & \cdots & X_{np}
\end{bmatrix}
\begin{bmatrix}
\beta_0 \\ \beta_1 \\ \vdots \\ \beta_p
\end{bmatrix}
+
\begin{bmatrix}
E_1 \\ E_2 \\ \vdots \\ E_n
\end{bmatrix}
$$

$$
=
\begin{bmatrix}
\beta_0 + \beta_1 X_{11} + \beta_2 X_{12} + \cdots + \beta_p X_{1p} \\
\beta_0 + \beta_1 X_{21} + \beta_2 X_{22} + \cdots + \beta_p X_{2p} \\
\vdots \\
\beta_0 + \beta_1 X_{n1} + \beta_2 X_{n2} + \cdots + \beta_p X_{np}
\end{bmatrix}
+
\begin{bmatrix}
E_1 \\ E_2 \\ \vdots \\ E_n
\end{bmatrix}
$$

$$
=
\begin{bmatrix}
\beta_0 + \beta_1 X_{11} + \beta_2 X_{12} + \cdots + \beta_p X_{1p} + E_1 \\
\beta_0 + \beta_1 X_{21} + \beta_2 X_{22} + \cdots + \beta_p X_{2p} + E_2 \\
\vdots \\
\beta_0 + \beta_1 X_{n1} + \beta_2 X_{n2} + \cdots + \beta_p X_{np} + E_n
\end{bmatrix}
$$

Based on the general matrix equation given by (B.1), a description of all the essential features of regression analysis can be expressed in matrix notation. In particular, the least-squares solution for the estimates of the regression coefficients in the parameter vector $\boldsymbol{\beta}$ can now be compactly written. This least-squares solution, in matrix terms, is that vector $\hat{\boldsymbol{\beta}}$ which minimizes the error sum of squares (given in matrix terms)

$$(\mathbf{Y} - \mathbf{X}\hat{\boldsymbol{\beta}})'(\mathbf{Y} - \mathbf{X}\hat{\boldsymbol{\beta}})$$

The solution to this minimization problem, which is obtained via the use of matrix calculus, yields the following easy-to-remember matrix formula:

$$\hat{\boldsymbol{\beta}} = (\mathbf{X}'\mathbf{X})^{-1}\mathbf{X}'\mathbf{Y}$$

where $\hat{\boldsymbol{\beta}}'_{1 \times (p+1)} = [\hat{\beta}_0 \quad \hat{\beta}_1 \quad \ldots \quad \hat{\beta}_p]$ denotes the vector of estimated regression coefficients.

Thus, although we have pointed out in the text the futility of *explicitly* giving the solutions to the least-squares equations for models of more complexity than a straight line, we can at least *implicitly* express these solutions in matrix notation and, because of modern computer technology, conveniently carry through with the computation of the least-squares solutions using this matrix representation.

At this point it is not our intention to carry through completely with the matrix formulation of every other aspect of a regression analysis. The reader is referred to Draper and Smith (1966; see footnote 1) for a fuller treatment of this matrix approach. However, some additional matrix results will be summarized below to give more of an indication of the utility of the matrix approach:

1. The *vector of predicted responses* is given by $\hat{\mathbf{Y}} = \mathbf{X}\hat{\boldsymbol{\beta}}$.

2. The *error sum of squares* SSE is given by SSE $= \mathbf{Y}'\mathbf{Y} - \hat{\boldsymbol{\beta}}'\mathbf{X}'\mathbf{Y}$.

3. The *regression sum of squares* is given by $\hat{\boldsymbol{\beta}}'\mathbf{X}'\mathbf{Y} - n\bar{Y}^2$.

4. The *variances of the regression coefficients*, that is, $\sigma_{\hat{\beta}_j}^2$, $j = 0, 1, 2, \ldots, p$, are given by the diagonal elements of the matrix $(\mathbf{X}'\mathbf{X})^{-1}\sigma^2$.

Finally, although we do not present the details here, any test of a (linear) statistical hypothesis concerning some subset of the regression coefficients can be formulated and carried out entirely by means of matrix operations.

Appendix—ANOVA Information for Four Common Balanced Repeated Measures Designs

Section 21-4 of Chapter 21 describes the data layout, ANOVA model, and corresponding partitioning of variability for four relatively common repeated measures designs:

1. A balanced repeated measures design with one crossover factor

2. A balanced repeated measures design with two crossover factors

3. A balanced repeated measures design with one nest factor

4. A balanced repeated measures design with one crossover factor and one nest factor

This appendix provides the ANOVA tables, expected mean squares formulas, and appropriate F statistics for these four designs.

C-1 Balanced Repeated Measures Design with One Crossover Factor (Treatments)

ANOVA model with Treatments as a fixed factor[1]:

$$Y_{ij} = \mu + S_i + \tau_j + E_{ij}$$

where

$$i = 1,\ldots,s$$
$$j = 1,\ldots,t$$
$$s = \text{Number of subjects}$$
$$t = \text{Number of treatments}$$

---

[1] If Treatments is taken as a random factor, we replace the fixed effect τ_j with the random effect T_j in the ANOVA and regression models, and we assume that T_j is distributed as $N(0, \sigma_T^2)$ and is independent of the other random effects.

$\tau_j =$ Fixed effect of treatment j (and $\displaystyle\sum_{j=1}^{t} \tau_j = 0$)

$S_i =$ Random effect of subject i

$E_{ij} =$ Random error for treatment j within subject i

and we assume that

$\{S_i\}$ and $\{E_{ij}\}$ are mutually independent

S_i is distributed as $N(0, \sigma_S^2)$

E_{ij} is distributed as $N(0, \sigma^2)$

Regression model:

$$Y = \mu + \sum_{i=1}^{s-1} S_i X_i + \sum_{j=1}^{t-1} \tau_j Z_j + E$$

where

$$X_i = \begin{cases} 1 \text{ for subject } i \\ 0 \text{ if not subjects } i \text{ or } s \\ -1 \text{ for subject } s \end{cases} \qquad Z_j = \begin{cases} 1 \text{ for treatment } j \\ 0 \text{ if not treatments } j \text{ or } t \\ -1 \text{ for treatment } t \end{cases}$$

for $i = 1, 2,\ldots, s - 1$ $\qquad\qquad$ for $j = 1, 2,\ldots, t - 1$

TABLE C-1   ANOVA table for balanced repeated measures design with one crossover factor.

Source	d.f.	MS	$F\left(\begin{smallmatrix}\text{Fixed or Random}\\\text{Treatments Factor}\end{smallmatrix}\right)$
Between Subjects	$s - 1$	MS_S	MS_S/MS_{TS}
Within Subjects	$s(t - 1)$	MS_W	
Treatments	$t - 1$	MS_T	MS_T/MS_{TS}
Treatments $\times$ Subjects (i.e., Error)	$(s - 1)(t - 1)$	MS_{TS}	
Total (corrected)	$st - 1$		

TABLE C-2   Expected mean squares for balanced repeated measures design with one crossover factor.

Source	d.f.	MS	Fixed Factor $E(MS)$	Random Factor $E(MS)$
Between Subjects	$s - 1$	MS_S	$\sigma^2 + t\sigma_S^2$	$\sigma^2 + t\sigma_S^2$
Within Subjects	$s(t - 1)$	MS_W	$\sigma^2 + \dfrac{1}{t-1}\displaystyle\sum_{j=1}^{t} \tau_j^2$	$\sigma^2 + \sigma_T^2$
Treatments	$t - 1$	MS_T	$\sigma^2 + \dfrac{s}{t-1}\displaystyle\sum_{j=1}^{t} \tau_j^2$	$\sigma^2 + s\sigma_T^2$
Treatments $\times$ Subjects	$(s - 1)(t - 1)$	MS_{TS}	σ^2	σ^2

TABLE C-3 F statistics, null hypotheses, and ratios of expected mean squares for balanced repeated measures design with one crossover factor.

Source	H_0	F	d.f.$_1$	d.f.$_2$	Ratio of $E(MS)$
Treatments (Fixed)	$\tau_j = 0$ $j = 1,\ldots,t$	MS_T / MS_{TS}	$t-1$	$(s-1)(t-1)$	$\dfrac{\sigma^2 + \dfrac{s}{t-1}\sum_{j=1}^{t} \tau_j^2}{\sigma^2}$
Treatments (Random)	$\sigma_T^2 = 0$	MS_T / MS_{TS}	$t-1$	$(s-1)(t-1)$	$\dfrac{\sigma^2 + s\sigma_T^2}{\sigma^2}$

C-2 Balanced Repeated Measures Design with Two Crossover Factors

ANOVA model treating both Factor A and Factor B as fixed factors[2]:

$$Y_{ijk} = \mu + S_i + \alpha_j + \beta_k + \delta_{jk} + S_{ij} + S_{ik} + E_{ijk}$$

where

$i = 1,\ldots,s$

$j = 1,\ldots,a$

$k = 1,\ldots,b$

$s = $ Number of subjects

$a = $ Number of levels of Factor A

$b = $ Number of levels of Factor B

$\mu = $ Overall mean

$\alpha_j = $ Fixed effect of level j of Factor A (and $\sum_{j=1}^{a} \alpha_j = 0$)

$\beta_k = $ Fixed effect of level k of Factor B (and $\sum_{k=1}^{b} \beta_k = 0$)

---

[2]If Factor A is treated as a random factor, we replace the fixed effect α_j with the random effect A_j in the ANOVA and regression models, and we assume that A_j is distributed as $N(0, \sigma_A^2)$. If Factor B is treated as a random factor, we replace the fixed effect β_k with the random effect B_k, and we assume that B_k is distributed as $N(0, \sigma_B^2)$. If either Factor A or Factor B is treated as a random factor, we replace the fixed interaction effect δ_{jk} with the random effect D_{jk}, and we assume that D_{jk} is distributed as $N(0, \sigma_{AB}^2)$. We also assume, when using A_j, B_k, and/or D_{jk} random effects in the model, that $\{A_j\}$, $\{B_k\}$, and $\{D_{jk}\}$ are mutually independent and independent of $\{S_i\}$, $\{S_{ij}\}$, $\{S_{ik}\}$, and $\{E_{ijk}\}$.

δ_{jk} = Fixed interaction effect of level j of Factor A with level k of Factor B

$$\left(\text{and } \sum_{j=1}^{a} \delta_{jk} = 0 \text{ for each } k, \sum_{k=1}^{b} \delta_{jk} = 0 \text{ for each } j\right)$$

S_i = Random effect of subject i

S_{ij} = Random effect of level j of Factor A within subject i

S_{ik} = Random effect of level k of Factor B within subject i

E_{ijk} = Random error for level j of Factor A and level k of Factor B within subject i

and we assume that

$\{S_i\}$, $\{S_{ij}\}$, $\{S_{ik}\}$, and $\{E_{ijk}\}$ are mutually independent,

S_i is distributed as $N(0, \sigma_S^2)$

S_{ij} is distributed as $N(0, \sigma_{SA}^2)$

S_{ik} is distributed as $N(0, \sigma_{SB}^2)$

E_{ijk} is distributed as $N(0, \sigma^2)$

Regression model:

$$Y = \mu + \sum_{i=1}^{s-1} S_i X_i + \sum_{j=1}^{a-1} \alpha_j U_j + \sum_{k=1}^{b-1} \beta_k V_k + \sum_{j=1}^{a-1}\sum_{k=1}^{b-1} \delta_{jk} U_j V_k + \sum_{i=1}^{s-1}\sum_{j=1}^{a-1} S_{ij} X_i U_j$$

$$+ \sum_{i=1}^{s-1}\sum_{k=1}^{b-1} S_{ik} X_i V_k + E$$

where

$$X_i = \begin{cases} 1 \text{ for subject } i \\ 0 \text{ if not subjects } i \text{ or } s \\ -1 \text{ for subject } s \end{cases} \qquad U_j = \begin{cases} 1 \text{ if level } j \text{ of } A \\ 0 \text{ if not levels } j \text{ or } a \text{ of } A \\ -1 \text{ if level } a \text{ of } A \end{cases}$$

for $i = 1, 2, \ldots, s - 1$ \qquad\qquad for $j = 1, 2, \ldots, a - 1$

$$V_k = \begin{cases} 1 \text{ if level } k \text{ of } B \\ 0 \text{ if not levels } k \text{ or } b \text{ of } B \\ -1 \text{ if level } b \text{ of } B \end{cases}$$

for $k = 1, 2, \ldots, b - 1$

TABLE C-4   ANOVA table for balanced repeated measures design with two crossover factors.[3]

Source	d.f.	MS	$F\left(\begin{array}{c}\text{Both Factors}\\\text{Fixed}\end{array}\right)$	$F\left(\begin{array}{c}\text{One or Both}\\\text{Factors Random}\end{array}\right)$[†]
Between Subjects	$s-1$	MS_S		
Within Subjects	$(s-1)ab$	MS_W		
Factor A	$a-1$	MS_A	MS_A/MS_{SA}	$MS_A/MS_{SA(\text{adj})}$
Subjects $\times$ Factor A	$(s-1)(a-1)$	MS_{SA}		
Factor B	$b-1$	MS_B	MS_B/MS_{SB}	$MS_B/MS_{SB(\text{adj})}$
Subjects $\times$ Factor B	$(s-1)(b-1)$	MS_{SB}		
$A \times B$	$(a-1)(b-1)$	MS_{AB}	$MS_{AB}/MS_{\text{Error}}$	$MS_{AB}/MS_{\text{Error}}$
Subjects $\times A \times B$ (i.e., Error)	$(s-1)(a-1)(b-1)$	MS_{Error}		
Total (corrected)	sab			

[†] Using Satterthwaite's approximation procedure,

$$MS_{SA(\text{adj})} = MS_{AB} + MS_{SA} - MS_{\text{Error}} \quad \text{and} \quad MS_{SB(\text{adj})} = MS_{AB} + MS_{SB} - MS_{\text{Error}}$$

TABLE C-5   Expected mean squares for balanced repeated measures design with two crossover factors.

Source	MS	(Both Factors Fixed) $E(MS)$	(Both Factors Random)[†] $E(MS)$
Factor A	MS_A	$\sigma^2 + b\sigma_{SA}^2 + \dfrac{sb}{a-1}\sum_{j=1}^{a}\alpha_j^2$	$\sigma^2 + b\sigma_{SA}^2 + sb\sigma_A^2 + s\sigma_{AB}^2$
Subjects $\times$ Factor A	MS_{SA}	$\sigma^2 + b\sigma_{SA}^2$	$\sigma^2 + b\sigma_{SA}^2$
Factor B	MS_B	$\sigma^2 + a\sigma_{SB}^2 + \dfrac{sa}{b-1}\sum_{k=1}^{b}\beta_k^2$	$\sigma^2 + a\sigma_{SB}^2 + sa\sigma_B^2 + s\sigma_{AB}^2$
Subjects $\times$ Factor B	MS_{SB}	$\sigma^2 + a\sigma_{SB}^2$	$\sigma^2 + a\sigma_{SB}^2$
$A \times B$	MS_{AB}	$\sigma^2 + \dfrac{s}{(a-1)(b-1)}\sum_{j=1}^{a}\sum_{k=1}^{b}\delta_{jk}^2$	$\sigma^2 + s\sigma_{AB}^2$
Subjects $\times A \times B$	MS_{Error}	σ^2	σ^2

[†] If Factor A is fixed and Factor B is random, the $E(MS)$ for Factor A in the last column of the table (i.e., both factors random) is modified by substituting the term $\dfrac{sb}{a-1}\sum_{j=1}^{a}\alpha_j^2$ for $sb\sigma_A^2$, with the other terms remaining the same. If Factor B is fixed and Factor A is random, the $E(MS)$ for Factor B in the last column of the table (i.e., both factors random) is modified by substituting the term $\dfrac{sa}{b-1}\sum_{k=1}^{b}\beta_k^2$ for $sa\sigma_B^2$, with the other terms remaining the same. Also, $E(MS_{SA(\text{adj})}) = \sigma^2 + b\sigma_{SA}^2 + s\sigma_{AB}^2$ and $E(MS_{SB(\text{adj})}) = \sigma^2 + a\sigma_{SB}^2 + s\sigma_{AB}^2$.

[3]When one or both factors are random, SAS's GLM procedure calculates the F tests for Factors A and B using Satterthwaite's approximation procedure (see Burdick and Graybill 1992; Gaylor and Hopper 1969). Satterthwaite's approximation involves replacing MS_{SA} and MS_{SB} with $MS_{SA(\text{adj})}$ and $MS_{SB(\text{adj})}$, respectively, in the denominators of the F tests for Factors A and B. The denominator degrees of freedom (d.f.2) in the F tests for Factors A and B are also modified by using formulas provided in Table C-6.

TABLE C-6 F statistics, null hypotheses, and ratios of expected mean squares for balanced repeated measures design with two crossover factors.

Source	H_0	F	d.f.1	d.f.2[†]	Ratio of E(MS)	
A	$\alpha_j = 0$ $j = 1, \ldots, a$	MS_A / MS_{SA}	$a - 1$	$(s-1)(a-1)$	$\dfrac{\sigma^2 + b\sigma_{SA}^2 + \dfrac{sb}{a-1}\sum\limits_{j=1}^{a}\alpha_j^2}{\sigma^2 + b\sigma_{SA}^2}$	$\left(\begin{array}{c}\text{Both}\\ \text{Factors}\\ \text{Fixed}\end{array}\right)$
	$\sigma_A^2 = 0$	$MS_A / MS_{SA(\text{adj})}$	$a - 1$	d.f.$2_{A(\text{adj})}$	$\dfrac{\sigma^2 + b\sigma_{SA}^2 + sb\sigma_A^2 + s\sigma_{AB}^2}{\sigma^2 + b\sigma_{SA}^2 + s\sigma_{AB}^2}$	$\left(\begin{array}{c}\text{Factor}\\ A\\ \text{Random}\end{array}\right)$
B	$\beta_k = 0$ $k = 1, \ldots, b$	MS_B / MS_{SB}	$b - 1$	$(s-1)(b-1)$	$\dfrac{\sigma^2 + a\sigma_{SB}^2 + \dfrac{sa}{b-1}\sum\limits_{k=1}^{b}\beta_k^2}{\sigma^2 + a\sigma_{SB}^2}$	$\left(\begin{array}{c}\text{Both}\\ \text{Factors}\\ \text{Fixed}\end{array}\right)$
	$\sigma_B^2 = 0$	$MS_B / MS_{SB(\text{adj})}$	$b - 1$	d.f.$2_{B(\text{adj})}$	$\dfrac{\sigma^2 + b\sigma_{SB}^2 + sb\sigma_B^2 + s\sigma_{AB}^2}{\sigma^2 + b\sigma_{SB}^2 + s\sigma_{AB}^2}$	$\left(\begin{array}{c}\text{Factor}\\ B\\ \text{Random}\end{array}\right)$
$A \times B$	$\delta_{jk} = 0$ all j, k	$MS_{AB} / MS_{\text{Error}}$	$(a-1)(b-1)$	$(s-1)(a-1)$ $\times (b-1)$	$\dfrac{\sigma^2 + \dfrac{s}{(a-1)(b-1)}\sum\limits_{j=1}^{a}\sum\limits_{k=1}^{b}\delta_{jk}^2}{\sigma^2}$	$\left(\begin{array}{c}\text{Both}\\ \text{Factors}\\ \text{Fixed}\end{array}\right)$
	$\sigma_{AB}^2 = 0$	$MS_{AB} / MS_{\text{Error}}$	$(a-1)(b-1)$	$(s-1)(a-1)$ $\times (b-1)$	$\dfrac{\sigma^2 + s\sigma_{AB}^2}{\sigma^2}$	$\left(\begin{array}{c}\text{Factors}\\ A \text{ or } B\\ \text{Random}\end{array}\right)$

[†] Using Satterthwaite's approximation procedure,[4] d.f.$2_{A(\text{adj})}$ and d.f.$2_{B(\text{adj})}$ are computed as

$$\text{d.f.}2_{A(\text{adj})} = \frac{(MS_{AB} + MS_{SA} - MS_{\text{Error}})^2}{\dfrac{(MS_{AB})^2}{(a-1)(b-1)} + \dfrac{(MS_{SA})^2}{(s-1)(a-1)} + \dfrac{(MS_{\text{Error}})^2}{(s-1)(a-1)(b-1)}}$$

$$\text{d.f.}2_{B(\text{adj})} = \frac{(MS_{AB} + MS_{SB} - MS_{\text{Error}})^2}{\dfrac{(MS_{AB})^2}{(a-1)(b-1)} + \dfrac{(MS_{SB})^2}{(s-1)(b-1)} + \dfrac{(MS_{\text{Error}})^2}{(s-1)(a-1)(b-1)}}$$

---

[4]When using Satterthwaite's approximation procedure for any linear combination of MS terms of the general form

$$MS_{\text{adj}} = c_1 MS_1 + c_2 MS_2 + \cdots + c_h MS_h$$

we compute the corresponding (approximate) denominator degrees of freedom by using the general formula

$$\text{d.f.}2_{\text{adj}} = \frac{(c_1 MS_1 + c_2 MS_2 + \cdots + c_h MS_h)^2}{\dfrac{(c_1 MS_1)^2}{\text{d.f.}_1} + \dfrac{(c_2 MS_2)^2}{\text{d.f.}_2} + \cdots + \dfrac{(c_h MS_h)^2}{\text{d.f.}_h}}$$

where d.f.$_1, \ldots,$ d.f.$_h$ are the degrees of freedom corresponding to $MS_1, \ldots, MS_h$, respectively.

C-3 Balanced Repeated Measures Design with One Nest Factor (Treatments)

ANOVA model with Treatments as a fixed factor[5]:

$$Y_{ijk} = \mu + S_{i(j)} + \tau_j + E_{k(ij)}$$

where

$$i = 1,\ldots,s$$

$$j = 1,\ldots,t$$

$$k = 1,\ldots,r$$

$s =$ Number of subjects given each treatment

$t =$ Number of treatments

$r =$ Number of repeats measured on the ith subject for a given treatment

$\mu =$ Overall mean

$\tau_j =$ Fixed effect of treatment j (and $\sum \tau_j = 0$)

$S_{i(j)} =$ Random effect of subject i within subject j

$E_{k(ij)} =$ Random error for repeat k on subject i at treatment j

and it is assumed that

$\{S_{i(j)}\}$ and $\{E_{ijk}\}$ are mutually independent

$S_{i(j)}$ is distributed as $N(0, \sigma_S^2)$

$E_{k(ij)}$ is distributed as $N(0, \sigma^2)$

Regression model:

$$Y = \mu + \sum_{j=1}^{t}\sum_{i=1}^{s-1} S_{i(j)} X_{i(j)} + \sum_{j=1}^{t-1} \tau_j Z_j + E$$

---

[5]If Treatments is treated as a random factor, we replace the fixed effect τ_j with the random effect T_j in the ANOVA and regression models, and we assume that T_j is distributed as $N(0, \sigma_T^2)$.

where

$$X_{i(j)} = \begin{cases} 1 \text{ if subject } i \text{ within treatment } j \\ 0 \text{ if not subjects } i \text{ or } s \text{ within treatment } j \\ -1 \text{ if subject } s \text{ within treatment } j \\ 0 \text{ if otherwise} \end{cases} \qquad Z_j = \begin{cases} 1 \text{ if treatment } j \\ 0 \text{ if not treatments } j \text{ or } t \\ -1 \text{ if treatment } t \end{cases}$$

for $i = 1, 2,..., s - 1$ and $j = 1, 2,..., t$ for $j = 1, 2,..., t - 1$

TABLE C-7   ANOVA table for balanced repeated measures design with one nest factor.

Source	d.f.	MS	$F\left(\begin{array}{c}\text{Fixed or Random}\\ \text{Treatments Factor}\end{array}\right)$
Between Subjects	$ts - 1$	MS_S	
Treatments	$t - 1$	MS_T	$MS_T/MS_{S(T)}$
Subjects within Treatments	$t(s - 1)$	$MS_{S(T)}$	
Within Subjects (i.e., Error)	$ts(r - 1)$	MS_W	
Total (corrected)	$str - 1$		

TABLE C-8   Expected mean squares for balanced repeated measures design with one nest factor.

Source	d.f.	MS	Fixed Factor $E(MS)$	Random Factor $E(MS)$
Between Subjects	$ts - 1$	MS_S		
Treatments	$t - 1$	MS_T	$\sigma^2 + r\sigma_S^2 + \dfrac{sr}{t-1}\sum_{j=1}^{t}\tau_j^2$	$\sigma^2 + r\sigma_S^2 + sr\sigma_T^2$
Subjects within Treatments	$t(s - 1)$	$MS_{S(T)}$	$\sigma^2 + r\sigma_S^2$	$\sigma^2 + r\sigma_S^2$
Within Subjects (i.e., Error)	$ts(r - 1)$	MS_W	σ^2	σ^2

TABLE C-9   F statistics, null hypotheses, and ratios of expected mean squares for balanced repeated measures design with one nest factor.

Source	H_0	F	d.f.1	d.f.2	Ratio of $E(MS)$
Treatments (Fixed)	$\tau_j = 0$ $j = 1,...,t$	$MS_T/MS_{S(T)}$	$t - 1$	$(s - 1)(t - 1)$	$\dfrac{\sigma^2 + r\sigma_S^2 + \dfrac{sr}{t-1}\sum_{j=1}^{t}\tau_j^2}{\sigma^2 + r\sigma_S^2}$
Treatments (Random)	$\sigma_T^2 = 0$	$MS_T/MS_{S(T)S}$	$t - 1$	$(s - 1)(t - 1)$	$\dfrac{\sigma^2 + r\sigma_S^2 + sr\sigma_T^2}{\sigma^2 + r\sigma_S^2}$

C-4 Balanced Repeated Measures Design with One Crossover Factor and One Nest Factor

ANOVA model treating both Factor A and Factor B fixed factors[6]:

$$Y_{ijk} = \mu + S_{i(j)} + \alpha_j + \beta_k + \delta_{jk} + E_{k(ij)}$$

where

$i = 1,\ldots,s$

$j = 1,\ldots,a$

$k = 1,\ldots,b$

$s =$ Number of subjects observed at each level of Factor A

$a =$ Number of levels of Factor A (with s different subjects per level)

$b =$ Number of levels of Factor B (with each subject observed at all levels of Factor B)

$\mu =$ Overall mean

$\alpha_j =$ Fixed effect of level j of Factor A (and $\displaystyle\sum_{j=1}^{a} \alpha_j = 0$)

$\beta_k =$ Fixed effect of level k of Factor B (and $\displaystyle\sum_{k=1}^{b} \beta_k = 0$)

$\delta_{jk} =$ Fixed interaction effect of level j of Factor A with level k of Factor B

(and $\displaystyle\sum_{j=1}^{a} \delta_{jk} = 0$ for each k, $\displaystyle\sum_{k=1}^{b} \delta_{jk} = 0$ for each j)

$S_{i(j)} =$ Random effect of subject i within level j of Factor A

$E_{k(ij)} =$ Random error for kth level of Factor B on subject i within level j of Factor A

---

[6]If Factor A is treated as a random factor, we replace the fixed effect α_j with the random effect A_j in the ANOVA and regression models, and we assume that A_j is distributed as $N(0, \sigma_A^2)$. If Factor B is treated as a random factor, we replace the fixed effect β_k with the random effect B_k, and we assume that B_k is distributed as $N(0, \sigma_B^2)$. If either Factor A or Factor B is treated as a random factor, we replace the fixed interaction effect δ_{jk} with the random effect D_{jk}, and we assume that D_{jk} is distributed as $N(0, \sigma_{AB}^2)$. We also assume, when using A_j, B_k, and/or D_{jk} random effects in the model, that $\{A_j\}$, $\{B_k\}$, and $\{D_{jk}\}$ are mutually independent and independent of $\{S_{i(j)}\}$ and $\{E_{k(ij)}\}$.

and it is assumed that

$\{S_{i(j)}\}$ and $\{E_{k(ij)}\}$ are mutually independent

$S_{i(j)}$ is distributed as $N(0, \sigma_S^2)$

$E_{k(ij)}$ is distributed as $N(0, \sigma^2)$

Regression model:

$$Y = \mu + \sum_{i=1}^{s-1}\sum_{j=1}^{a} S_{i(j)}X_{i(j)} + \sum_{j=1}^{a-1} \alpha_j U_j + \sum_{k=1}^{b-1} \beta_k V_k + \sum_{j=1}^{a-1}\sum_{k=1}^{b-1} \delta_{jk}U_jV_k + E$$

where

$$X_{i(j)} = \begin{cases} 1 \text{ if subject } i \text{ within level } j \text{ of Factor } A \\ 0 \text{ if not subjects } i \text{ or } s \text{ within level } j \text{ of Factor } A \\ -1 \text{ if subject } s \text{ within level } j \text{ of Factor } A \\ 0 \text{ if otherwise} \end{cases}$$

for $i = 1, 2,..., s - 1$ and $j = 1, 2,...,a$

$$U_j = \begin{cases} 1 \text{ if level } j \text{ of Factor } A \\ 0 \text{ if not levels } j \text{ or } a \text{ of Factor } A \\ -1 \text{ if level } a \text{ Factor } A \end{cases} \qquad V_k = \begin{cases} 1 \text{ if level } k \text{ of Factor } B \\ 0 \text{ if not levels } k \text{ or } b \text{ of Factor } B \\ -1 \text{ if level } b \text{ of Factor } B \end{cases}$$

for $j = 1, 2,..., a - 1$ for $k = 1, 2,..., b - 1$

TABLE C-10   ANOVA table for balanced repeated measures design with one crossover factor and one nest factor.

Source	d.f.	MS	$F\left(\begin{array}{c}\text{Both Factors}\\\text{Fixed}\end{array}\right)$	$F\left(\begin{array}{c}\text{One or Both}\\\text{Factors Random}\end{array}\right)$[†]
Between Subjects	$sa - 1$	MS_S		
⎰ Factor A	$a - 1$	MS_A	$MS_A/MS_{S(A)}$	$MS_A/MS_{S(A)\text{adj}}$
⎱ Subjects (within A)	$a(s - 1)$	$MS_{S(A)}$		
Within Subjects	$sa(b - 1)$	MS_W		
⎧ Factor B	$b - 1$	MS_B	MS_B/MS_{Error}	MS_B/MS_{AB}
⎨ $A \times B$	$(a - 1)(b - 1)$	MS_{AB}	$MS_{AB}/MS_{\text{Error}}$	$MS_{AB}/MS_{\text{Error}}$
⎩ Subjects $\times B$ (within A) (i.e., Error)	$a(b - 1)(s - 1)$	MS_{Error}		
Total (corrected)	sab			

[†] Using Satterthwaite's approximation procedure,
$$MS_{S(A)\text{adj}} = MS_{AB} + MS_{S(A)} - MS_{\text{Error}}$$

TABLE C-11 Expected mean squares for balanced repeated measures design with one crossover factor and one nest factor.

Source	MS	(Both Factors Fixed) E(MS)	(Both Factors Random)[†] E(MS)
Factor A	MS_A	$\sigma^2 + b\sigma_S^2 + \dfrac{bs}{a-1}\sum\limits_{j=1}^{a}\alpha_j^2$	$\sigma^2 + b\sigma_S^2 + bs\sigma_A^2 + s\sigma_{AB}^2$
Factor B	MS_B	$\sigma^2 + \dfrac{as}{b-1}\sum\limits_{k=1}^{b}\beta_k^2$	$\sigma^2 + as\sigma_B^2 + s\sigma_{AB}^2$
$A \times B$	MS_{AB}	$\sigma^2 + \dfrac{s}{(a-1)(b-1)}\sum\limits_{j=1}^{a}\sum\limits_{k=1}^{b}\delta_{jk}^2$	$\sigma^2 + s\sigma_{AB}^2$
Subjects (within A)	$MS_{S(A)}$	$\sigma^2 + b\sigma_S^2$	$\sigma^2 + b\sigma_S^2$
Subjects $\times B$ (within A) i.e., Error	MS_{Error}	σ^2	σ^2

[†] If Factor A is fixed and Factor B is random, the E(MS) for Factor A in the last column of the table (i.e., both factors random) is modified by substituting the term $\dfrac{bs}{a-1}\sum\limits_{j=1}^{a}\alpha_j^2$ for $bs\sigma_A^2$, with the other terms remaining the same. If Factor B is fixed and Factor A is random, the E(MS) for Factor B in the last column of the table (i.e., both factors random) is modified by substituting the term $\dfrac{as}{b-1}\sum\limits_{k=1}^{b}\beta_k^2$ for $as\sigma_B^2$, with the other terms remaining the same. Also $E(MS_{S(A)adj}) = \sigma^2 + b\sigma_S^2 + s\sigma_{AB}^2$.

TABLE C-12 F statistics, null hypotheses, and ratios of expected mean squares for balanced repeated measures design with one crossover factor and one nest factor.

Source	H_0	F	d.f.1	d.f.2[†]	Ratio of E(MS)	
A	$\alpha_j = 0$ $j = 1,\ldots,a$	$MS_A/MS_{S(A)}$	$a-1$	$a(s-1)$	$\dfrac{\sigma^2 + b\sigma_S^2 + \dfrac{bs}{a-1}\sum\limits_{j=1}^{a}\alpha_j^2}{\sigma^2 + b\sigma_S^2}$	$\begin{pmatrix}\text{Both}\\\text{Factors}\\\text{Fixed}\end{pmatrix}$
	$\sigma_A^2 = 0$	$MS_A/MS_{S(A)adj}$	$a-1$	$\text{d.f.2}_{A(adj)}$	$\dfrac{\sigma^2 + b\sigma_S^2 + bs\sigma_A^2 + s\sigma_{AB}^2}{\sigma^2 + b\sigma_S^2 + s\sigma_{AB}^2}$	$\begin{pmatrix}\text{Factor }A\\\text{Random}\end{pmatrix}$
B	$\beta_k = 0$ $k = 1,\ldots,b$	MS_B/MS_{Error}	$b-1$	$a(b-1)(s-1)$	$\dfrac{\sigma^2 + \dfrac{as}{b-1}\sum\limits_{k=1}^{b}\beta_k^2}{\sigma^2}$	$\begin{pmatrix}\text{Both}\\\text{Factors}\\\text{Fixed}\end{pmatrix}$
	$\sigma_B^2 = 0$	MS_B/MS_{AB}	$b-1$	$(a-1)(b-1)$	$\dfrac{\sigma^2 + as\sigma_B^2 + s\sigma_{AB}^2}{\sigma^2 + s\sigma_{AB}^2}$	$\begin{pmatrix}\text{Factor }B\\\text{Random}\end{pmatrix}$
$A \times B$	$\delta_{jk} = 0$ all j, k	MS_{AB}/MS_{Error}	$(a-1)(b-1)$	$a(b-1)(s-1)$	$\dfrac{\sigma^2 + \dfrac{s}{(a-1)(b-1)}\sum\limits_{j=1}^{a}\sum\limits_{k=1}^{b}\delta_{jk}^2}{\sigma^2}$	$\begin{pmatrix}\text{Both}\\\text{Factors}\\\text{Fixed}\end{pmatrix}$
	$\sigma_{AB}^2 = 0$	MS_{AB}/MS_{Error}	$(a-1)(b-1)$	$a(b-1)(s-1)$	$\dfrac{\sigma^2 + s\sigma_{AB}^2}{\sigma^2}$	$\begin{pmatrix}\text{Factors}\\A\text{ or }B\\\text{Random}\end{pmatrix}$

[†] Using Satterthwaite's approximation procedure, d.f.$2_{A(adj)}$ is computed as $\text{d.f.}2_{A(adj)} = \dfrac{(MS_{AB} + MS_{S(A)} - MS_{Error})^2}{\dfrac{(MS_{AB})^2}{(a-1)(b-1)} + \dfrac{(MS_{S(A)})^2}{a(s-1)} + \dfrac{(MS_{Error})^2}{a(b-1)(s-1)}}$

C-5 Balanced Two-Group Pre/Posttest Design

ANOVA model treating both Groups (Factor A) and Tests (Factor B) as fixed factors[7]:

$$Y_{ijk} = \mu + S_{i(j)} + \alpha_j + \beta_k + \delta_{jk} + E_{k(ij)}$$

where

$i = 1,\ldots,s$

$j = 1, 2$

$k = 1, 2$

$s = $ Number of subjects observed at each level of Factor A

$\mu = $ Overall mean

$\alpha_j = $ Fixed effect of level j of Factor A

$\beta_k = $ Fixed effect of level k of Factor B

$\delta_{jk} = $ Fixed interaction effect of level j of Factor A with level k of Factor B

$S_{i(j)} = $ Random effect of subject i within level j of Factor A

$E_{k(ij)} = $ Random error for kth level of Factor B on subject i within level j of Factor A

and it is assumed that

$\{S_{i(j)}\}$ and $\{E_{k(ij)}\}$ are mutually independent

$S_{i(j)}$ is distributed as $N(0, \sigma_S^2)$

$E_{k(ij)}$ is distributed as $N(0, \sigma^2)$

and that the usual summation restrictions hold separately for α_j, β_k, and δ_{jk}.

Regression model:

$$Y = \mu + \sum_{i=1}^{s-1} S_{i(1)}X_{i(1)} + \sum_{i=1}^{s-1} S_{i(2)}X_{i(2)} + \alpha G + \beta T + \delta GT + E$$

where (using effect coding):

$$G = \begin{cases} 1 \text{ if group 1} \\ -1 \text{ if group 2} \end{cases} \qquad T = \begin{cases} 1 \text{ if pretest} \\ -1 \text{ if posttest} \end{cases}$$

---

[7]If Groups is treated as a random factor, we replace the fixed effect α_j with the random effect G_j in the ANOVA and regression models, and we assume that G_j is distributed as $N(0, \sigma_G^2)$. If Tests is treated as a random factor, we replace the fixed effect β_k with the random effect T_k, and we assume that T_k is distributed as $N(0, \sigma_T^2)$. If either Groups or Tests is treated as a random factor, we replace the fixed interaction effect δ_{jk} with the random effect I_{jk}, and we assume that I_{jk} is distributed as $N(0, \sigma_{GT}^2)$. We also assume, when using G_j, T_k, and/or I_{jk} random effects in the model, that $\{G_j\}$, $\{T_k\}$, and $\{I_{jk}\}$ are mutually independent and independent of $\{S_{i(j)}\}$ and $\{E_{k(ij)}\}$.

$$X_{i(1)} = \begin{cases} 1 \text{ if subject } k \text{ within group 1} \\ 0 \text{ if not subjects } k \text{ or } s \text{ within group 1} \\ -1 \text{ if subject } s \text{ within group 1} \end{cases} \quad \text{for } i = 1,\ldots, s$$

$$X_{i(2)} = \begin{cases} 1 \text{ if subject } k \text{ within group 2} \\ 0 \text{ if not subjects } k \text{ or } s \text{ within group 2} \\ -1 \text{ if subject } s \text{ within group 2} \end{cases} \quad \text{for } i = 1,\ldots, s$$

TABLE C-13   ANOVA table for balanced two-group pre/posttest design.

Source	d.f.	MS	$F\left(\begin{array}{c}\text{Both Factors}\\\text{Fixed}\end{array}\right)$	$F\left(\begin{array}{c}\text{One or Both}\\\text{Factors Random}\end{array}\right)^{\dagger}$
Between Subjects	$2s-1$	MS_S		
Groups	1	MS_G	$MS_G/MS_{S(G)}$	$MS_G/MS_{S(G)(\text{adj})}$
Subjects (within Groups)	$2(s-1)$	$MS_{S(G)}$		
Within Subjects	$2s$	MS_W		
Tests	1	MS_T	MS_T/MS_{Error}	MS_T/MS_{GT}
Groups × Tests	1	MS_{GT}	$MS_{GT}/MS_{\text{Error}}$	$MS_{GT}/MS_{\text{Error}}$
Subjects × Tests (within Groups) i.e., Error	$2(s-1)$	MS_{Error}		
Total (corrected)	$4s$			

$\dagger$ Using Satterthwaite's approximation procedure,
$$MS_{S(G)\text{adj}} = MS_{GT} + MS_{S(T)} - MS_{\text{Error}}$$

TABLE C-14   Expected mean squares for balanced two-group pre/posttest design.

Source	MS	(Both Factors Fixed) $E(MS)$	(Both Factors Random)† $E(MS)$
Groups	MS_G	$\sigma^2 + 2\sigma_S^2 + 2s(\alpha_1^2 + \alpha_2^2)$	$\sigma^2 + 2\sigma_S^2 + 2s\sigma_G^2 + s\sigma_{GT}^2$
Tests	MS_T	$\sigma^2 + 2(\beta_1^2 + \beta_2^2)$	$\sigma^2 + 2s\sigma_T^2 + s\sigma_{GT}^2$
Groups × Tests	MS_{GT}	$\sigma^2 + s\sum_{j=1}^{2}\sum_{k=1}^{2}\delta_{jk}^2$	$\sigma^2 + s\sigma_{GT}^2$
Subjects (within Groups)	$MS_{S(T)}$	$\sigma^2 + 2\sigma_S^2$	$\sigma^2 + 2\sigma_S^2$
Subjects × B (within Groups) (i.e., Error)	MS_{Error}	σ^2	σ^2

$\dagger$ If Groups is fixed and Tests is random, the $E(MS)$ for Groups in the last column of the table (i.e., both factors random) is modified by substituting the term $2s(\alpha_1^2 + \alpha_2^2)$ for $2s\sigma_G^2$, with the other terms remaining the same. If Tests is fixed and Groups is random, the $E(MS)$ for Tests in the last column of the table (i.e., both factors random) is modified by substituting the term $2s(\beta_1^2 + \beta_2^2)$ for $2s\sigma_T^2$, with the other terms remaining the same. Also, $E(MS_{S(A)\text{adj}}) = \sigma^2 + 2\sigma_S^2 + s\sigma_{GT}^2$.

TABLE C-15 *F* statistics, null hypotheses, and ratios of expected mean squares for balanced two-group pre/posttest design.

Source	H_0	F	d.f.1	d.f.2[†]	Ratio of $E(MS)$	
Groups	$\alpha_j = 0$ $j = 1, 2$	$MS_G/MS_{S(G)}$	1	$2(s-1)$	$\dfrac{\sigma^2 + 2\sigma_S^2 + 2s(\alpha_1^2 + \alpha_2^2)}{\sigma^2 + 2\sigma_S^2}$	$\begin{pmatrix}\text{Both}\\\text{Factors}\\\text{Fixed}\end{pmatrix}$
	$\sigma_G^2 = 0$	$MS_G/MS_{S(G)\text{adj}}$	1	$\text{d.f.2}_{G(\text{adj})}$	$\dfrac{\sigma^2 + 2\sigma_S^2 + 2s\sigma_G^2 + s\sigma_{GT}^2}{\sigma^2 + 2\sigma_S^2 + s\sigma_{GT}^2}$	$\begin{pmatrix}\text{Factor}\\A\\\text{Random}\end{pmatrix}$
Tests	$\beta_k = 0$ $k = 1, 2$	MS_T/MS_{Error}	1	$2(s-1)$	$\dfrac{\sigma^2 + 2s(\beta_1^2 + \beta_2^2)}{\sigma^2}$	$\begin{pmatrix}\text{Both}\\\text{Factors}\\\text{Fixed}\end{pmatrix}$
	$\sigma_T^2 = 0$	MS_T/MS_{GT}	1	1	$\dfrac{\sigma^2 + 2s\sigma_T^2 + s\sigma_{GT}^2}{\sigma^2 + s\sigma_{GT}^2}$	$\begin{pmatrix}\text{Factor}\\B\\\text{Random}\end{pmatrix}$
Groups × Tests	$\delta_{jk} = 0$ all j, k	$MS_{GT}/MS_{\text{Error}}$	1	$2(s-1)$	$\dfrac{\sigma^2 + s\displaystyle\sum_{j=1}^{2}\sum_{k=1}^{2}\delta_{jk}^2}{\sigma^2}$	$\begin{pmatrix}\text{Both}\\\text{Factors}\\\text{Fixed}\end{pmatrix}$
	$\sigma_{GT}^2 = 0$	$MS_{GT}/MS_{\text{Error}}$	1	$2(s-1)$	$\dfrac{\sigma^2 + s\sigma_{GT}^2}{\sigma^2}$	$\begin{pmatrix}\text{Factors}\\A \text{ or } B\\\text{Random}\end{pmatrix}$

[†] Using Satterthwaite's approximation procedure, $\text{d.f.2}_{G(\text{adj})}$ is computed as

$$\text{d.f.2}_{G(\text{adj})} = \frac{(MS_{GT} + MS_{S(G)} - MS_{\text{Error}})^2}{\dfrac{(MS_{GT})^2}{1} + \dfrac{(MS_{S(G)})^2}{2(s-1)} + \dfrac{(MS_{\text{Error}})^2}{2(s-1)}}$$

References

Burdick, R. K., and Graybill, F. A. 1992. *Confidence Intervals on Variance Components*. Marcel Decker, New York.

Gaylor, D. W., and Hopper, F. N. 1969. "Estimating the Degrees of Freedom for Linear Combinations of Mean Squares by Satterthwaite's Formula." *Technometrics* 11: 691–706.

Appendix—Solutions to Selected Exercises

Chapter 3

2. nominal, ordinal, interval, ratio

3. a. 0.8413   **b.** −0.842

4. a. 18.475   **b.** 0.699

5. a. −1.350   **b.** 0.05

6. a. 2.51   **b.** 0.025

7. a. 0   **b.** 0   **c.** 0

8. standard normal

9. a. 3.0   **b.** 3   **c.** 2.8 or 3.11

10. e

11. a. 5.0   **b.** (187.44, 192.56)

12. $t_{0.975, 27} = 2.052$

13. (24.66, 35.33)

14. 3.6858

15. a. significant difference   **b.** significant difference

16. nonsignificant difference

17. $t_{0.995, 10} = 3.619$

18. b

19. a. Type I error   **b.** correct decision   **c.** correct decision   **d.** Type II error

20. a, b

21. c

22. b

23. a. $1 - \alpha$ **b.** α **c.** β **d.** $1 - \beta$

24. Accept H_0.

25. b

Chapter 5

1. a. Dry Weight (Y) does increase with increasing Age (X), but the relationship may not be linear. An exponential relationship between X and Y may better fit the data. Log Dry Weight (Z) increases linearly with increasing Age (X).

   b. $Y = \beta_0 + \beta_1 X + E$ $Z = \beta_0' + \beta_1' X + E$

   c. $\hat{Y} = -1.885 + 0.235X$ $\hat{Z} = -2.689 + 0.196X$

   d. The regression line for Log_{10} Dry Weight regressed on Age has a better fit. It is more appropriate to run a linear regression of Z on X.

   e. 95% confidence intervals: for β_1': $(0.189, 0.203)$, for β_0': $(-2.759, -2.619)$

   f. $(-1.149, -1.096)$

3. a. The relationship between Time (Y) and Inc (X) does not appear to be linear.

   b. $\hat{\beta}_0 = 19.626$ $\hat{\beta}_1 = 0.0007$

   c. $\hat{Y} = 19.626 + 0.0007X$. The regression line fits the data poorly.

   d. The linearity assumption is not met.

   e. TS: $T = 2.023$, $p = 0.0582$ (from SAS output). We do not reject H_0, since the p-value > 0.05.

   f. The scatter plot suggests that a parabola would better fit the data.

5. a. $\hat{Y} = 2.174 + 1.177X$. The line fits the data well.

   b. No.

   c. TS: $T = 13.5$ $p = 0.0001$ (from SAS output).
Since the p-value < 0.05, we reject H_0.

   d. TS: $T = 0.954$ $0.15 < p < 0.25$.
Since the p-value > 0.05, we do not reject H_0.

   e. $(44.146, 47.276)$

7. a. $\hat{Y}_1 = -122.345 + 6.227X$ $\hat{Y}_2 = -1.697 + 0.299X$

   b. Y_2 regressed on X.

   c. TS: $T = -57.934$. Critical value: $t_{17} \sim 2.898$ under H_a at $\alpha = 0.01$. We see that $|T| = 57.934 > 2.898$, so we reject H_0 at $\alpha = 0.01$.

   d. $(0.264, 0.334)$.

   e. $(11.333, 12.163)$

9. a. $\hat{\beta}_0 = 2.936$ $\hat{\beta}_1 = -1.785$

   b. $\hat{Y} = 2.936 - 1.785X$. The line fits the data well.

   d. $\hat{Y} = 862.979\hat{X}^{-1.785}$

   e. $(6.266, 9.311)$, $(321.366, 528.445)$

   f. One could plot the transformed data (X', Y') and then draw the estimated regression line on the plot. Then one could compare the fit of the estimated line of (X, Y) versus that of (X', Y'). If one did this comparison, the straight-line regression of (X, Y) gives a better fit.

11. a. $\hat{Y} = 3.707 - 0.012X$.

   b. TS: $T = -8.684$ $p = 0.001$ (from SAS output).
Since the p-value < 0.05, we reject H_0.

   c. Including the data from the three experiments, rather than just using the average values, would provide more information and might improve the sensitivity of the analysis.

   d. $(1.243, 3.737)$

e. It is inappropriate, since the estimated model relies on data that do not include any information for average growth rate when exposed to a gas with a molecular weight of 200.

f. The chosen X values are not uniformly distributed in the experiment. There are large gaps between the X values of 39.9, 83.8, and 131.3. This may result in a fitted line subject to inaccuracies for predicting Y based on X.

13. b. From the SAS output: $\hat{B}_0 = 0.116$ $\hat{B}_1 = 0.005$

c. TS: $T = 0.811$ $p = 0.433$ (from SAS output).
Since the p-value > 0.10, we do not reject H_0.

d. TS: $T = 0.046$ $p = 0.964$ (from SAS output).
Since the p-value > 0.10, we do not reject H_0.

e. $\hat{Y} = 0.116 + 0.005X$

f. The line does not differ from the line plotted in part (e). The evidence suggests that there is no significant linear relationship. Determining a well-fitting line is difficult, given the dispersion of the data.

15. a. As X increases, the dispersion of Y increases.

b. $\hat{B}_0 = -2.546$ $\hat{B}_1 = 0.032$ $\hat{Y} = -2.546 + 0.032X$

c. This chart implies a better linear relationship between X and Y.

d. $\hat{B}_0 = -6.532$ $\hat{B}_1 = 1.430$ $\hat{Y} = -6.532 + 1.430X$

e. The natural log transformation provides the best representation. The natural log plot illustrates the linear relationship better, and the dispersion of the data is more similar at each level of toluene exposure. The first plot indicates that there may be a violation of homoscedasticity for the untransformed data.

17. a. Yes.

b. $\hat{Y} = 1.643 + 1.057X$. The line appears to fit the data well.

c. 95% CI for β_1: (0.580, 1.534). The 95% CI does not include the null value of zero, indicating that there is a significant linear relationship at $\alpha = 0.05$.

d. No, it is not appropriate, since the data used to estimate the regression line do not inlude a \$10 million advertising expenditure in its range.

19. a. There may be a slight negative linear relationship.

b. $Y = \hat{B}_0 + \hat{B}_1 X + E$ $\hat{B}_0 = 76.008$ $\hat{B}_1 = -0.015$.
The baseline OWNEROCC $= 76\%$, and as OWNCOST increases by \$1,000, the percentage of OWNEROCC decreases by $\sim 2\%$.

c. $\hat{Y} = 76.008 - 0.015X$. The line fits the data well.

d. TS: $T = -2.607$ $p = 0.0155$ (from SAS output).
Since the p-value < 0.05, we reject H_0.

e. $(-0.027, -0.003)$. We are 95% confident that the true slope is between -0.027 and -0.003. Since the interval does not contain zero, we conclude that the slope is not equal to zero (at $\alpha = 0.05$).

Chapter 6

1. a. (1) $r = 0.863$ **(2)** $r = 0.999$

b. (1) (0.546, 0.964)   **(2)** (0.996, 1.000)

c. (1) $r^2 = 0.744$. 74% of the variation in Y is explained with the help of X.
(2) $r^2 = 0.998$. 99% of the variation in Z is explained with the help of X.

d. The regression using Log_{10} Dry Weight appears to fit better. This agrees with Chapter 5, Problem 1(d).

3. a. $r = 0.980$

   b. Substitute $r(S_y/S_x)$ and $\dfrac{(n-1)}{(n-2)}(S_Y^2 - \hat{\beta}_1^2 S_X^2)$ for $\hat{\beta}_1$ and $S_{Y|X}^2$ in formula 5.9.

   c. $T' = 13.929$; $T \sim t_8$ under H_0: $\rho = 0$. The p-value for this test is less than 0.001; therefore, we reject H_0.

   d. The graph of Y vs. X does not illustrate a linear relationship.

5. a. $r^2 = 0.601$, $r = 0.775$. 60% of the variation in SBP (Y) is explained by AGE (X).

   b. (0.554, 0.907). Since $\rho = 0$ is not included in the interval, we reject H_0: $\rho = 0$ at $\alpha = 0.01$.

7. a. $r^2 = 0.1853$, $r = 0.430$. 19% of the variation in SBP (Y) is explained by AGE (X).

   b. $T = 2.02$; Critical value: $t_{18,0.975} = 2.101$. At $\alpha = 0.05$, since $|T| <$ critical value, we would not reject H_0.

   c. $(-0.015, 0.733)$. Since $\rho = 0$ is included in the interval, we do not reject H_0: $\rho = 0$ at $\alpha = 0.05$.

9. a. $r^2 = 0.9101$, $r = 0.954$. 91% of the variation in Y is explained by X.

   b. $T = 13.49$; Critical value: $t_{18,0.975} = 2.101$. At $\alpha = 0.05$, since $|T| >$ critical value, we would reject H_0.

   c. (0.885, 0.982). Since $\rho = 0$ is not included in the interval, we reject H_0: $\rho = 0$ at $\alpha = 0.05$.

11. a. $r^2 = 0.9728$, $r = 0.986$. 98% of the variation in Y_2 is explained by X.

   b. $T = 24.640$; Critical value: $t_{17,0.975} = 2.110$. At $\alpha = 0.05$, since $|T| >$ critical value, we would reject H_0.

   c. (0.963, 0.995). Since $\rho = 0$ is not included in the interval, we reject H_0: $\rho = 0$ at $\alpha = 0.05$.

13. a. $r = 0.977$

   b. (0.849, 0.997). Since $\rho = 0$ is not included in the interval, we reject H_0: $\rho = 0$ at $\alpha = 0.05$.

15. $Z = 3.62$; Critical value $= 1.96$. Since $|Z|$ is $>$ critical value, we reject the null hypothesis.

17. a. $r^2 = 0.9558$, $r = 0.978$. 96% of the variation in Y is explained by X. *includes 16th obs.

   b. $T = 17.424$; Critical value: $t_{14,0.975} = 2.145$. At $\alpha = 0.05$, we would reject H_0.

   c. (0.936, 0.993). Since $\rho = 0$ is not included in the interval, we reject H_0: $\rho = 0$ at $\alpha = 0.05$.

19. a. $r^2 = 0.0310$, $r = -0.176$. 3% of the variation in Y is explained by X.

   b. $T = 0.759$; Critical value: $t_{18,0.975} = 2.101$. At $\alpha = 0.05$, we would not reject H_0.

   c. $(-0.574, 0.289)$. Since $\rho = 0$ is included in the interval, we do not reject H_0: $\rho = 0$ at $\alpha = 0.05$.

21. a. $r^2 = 0.9813$, $r = 0.991$. 98% of the variation in Y is explained by X.

   b. $T = 55.088$; Critical value: $t_{38,0.975} = 2.0$. At $\alpha = 0.05$, we would reject H_0.

   c. (0.985, 0.995). Since $\rho = 0$ is not included in the interval, we reject H_0: $\rho = 0$ at $\alpha = 0.05$.

23. a. $r^2 = 0.9044$, $r = 0.951$. 90% of the variation in Y is explained by X.

   b. $T = 6.155$; Critical value: $t_{4,0.975} = 2.776$. At $\alpha = 0.05$, we would reject H_0.

   c. (0.611, 0.995). Since $\rho = 0$ is not included in the interval, we reject H_0: $\rho = 0$ at $\alpha = 0.05$.

25. a. $r^2 = 0.2207$, $r = 0.470$. 22% of the variation in Y is explained by X.
   b. $T = 2.608$; Critical value: $t_{24, 0.975} = 2.064$. At $\alpha = 0.05$, we would reject H_0.
   c. (0.101, 0.725). Since $\rho = 0$ is not included in the interval, we reject H_0: $\rho = 0$ at $\alpha = 0.05$.

Chapter 7

1. a. (1)

	d.f.	Sum of Squares	Mean Sum of Squares	Variance Ratio
Regression	1	6.080	6.080	26.607
Residual	9	2.088	0.232	
	10	8.168		

a. (2)

	d.f.	Sum of Squares	Mean Sum of Squares	Variance Ratio
Regression	1	4.222	4.222	6031.429
Residual	9	0.006	0.0007	
	10	4.228		

b. (1) H_0: $\beta_1 = 0$ H_a: $\beta_1 \neq 0$; Critical value: $F_{1, 9, 0.95} = 5.12$. Since $26.607 > 5.12$, we reject H_0 at $\alpha = 0.05$.
b. (2) H_0: $\beta_1 = 0$ H_a: $\beta_1 \neq 0$; Critical value: $F_{1, 9, 0.95} = 5.12$. Since $6031.429 > 5.12$, we reject H_0 at $\alpha = 0.05$.

5. a.

	d.f.	Sum of Squares	Mean Sum of Squares	Variance Ratio
Regression	1	450.865	450.865	4.093
Residual	18	1982.916	110.162	
	19	2433.781		

b. H_0: $\beta_1 = 0$ H_a: $\beta_1 \neq 0$; Critical value: $F_{1, 18, 0.95} = 4.41$. $4.093 < 4.41$, so we do not reject H_0 at $\alpha = 0.05$.
c. $T^2 = (2.023)^2 = 4.093$. The values are virtually the same.
d. The hypotheses for each test are equivalent. As mentioned in the text, the F statistic and T statistic are equivalent after squaring T. Using the information that the tests of hypotheses are equivalent (as are the test statistics), one may infer that the resulting p-values for each test should also be equivalent.

9. a.

	d.f.	Sum of Squares	Mean Sum of Squares	Variance Ratio
Regression	1	76858486.06	76858486.06	60.758
Residual	28	35419546.54	1264983.805	
	29	112278032.670		

b. H_0: $\beta_1 = 0$ H_a: $\beta_1 \neq 0$; Critical value: $F_{1, 28, 0.95} = 4.20$.
Since $60.758 > 4.20$, we reject H_0 at $\alpha = 0.05$.

13. a.

	d.f.	Sum of Squares	Mean Sum of Squares	Variance Ratio
Regression	1	133806354.1	133806354.1	3155.890
Residual	14	593584.922	42398.923	
	15	134399939.0		

b. H_0: $\beta_1 = 0$ H_a: $\beta_1 \neq 0$; Critical value: $F_{1, 14, 0.95} = 4.60$

Since $3155.89 > 4.60$, we would reject H_0 and conclude that there is a significant linear relationship of Y on X at $\alpha = 0.05$.

17. a.

	d.f.	Sum of Squares	Mean Sum of Squares	Variance Ratio
Regression	1	177.547	177.547	3061.155
Residual	58	3.364	0.058	
	59	180.911		

b. $H_0: \beta_1 = 0$ $H_a: \beta_1 \neq 0$; Critical value: $F_{1, 58, 0.95} = 4.00$. Since $3061.155 > 4.00$, we reject H_0 at $\alpha = 0.05$.

21. a.

	d.f.	Sum of Squares	Mean Sum of Squares	Variance Ratio
Regression	1	132.626	132.626	6.796
Residual	24	468.336	19.514	
	25	600.962		

b. $H_0: \beta_1 = 0$ $H_a: \beta_1 \neq 0$; Critical value: $F_{1, 24, 0.95} = 4.26$. Since $6.796 > 4.26$, we reject H_0 at $\alpha = 0.05$.

Chapter 8

1. a. i $\hat{Y} = 145.771$ **ii** $\hat{Y} = 135.825$ **iii** $\hat{Y} = 141.475$

As QUET increases from 3.0 to 3.5, average SBP increases by an estimated 4.296 points, from 141.475 to 145.771.

b. SBP on AGE: $R^2 = 0.601$; SBP on AGE and SMK: $R^2 = 0.730$; SBP on AGE, SMK, and QUET: $R^2 = 0.761$.

The model using AGE and SMK to predict SBP appears to be the best choice.

5. a.

	d.f.	Sum of Squares	Mean Sum of Squares	F
Regression	3	25974	8658.0	80.871
Residual	21	2248.23	107.059	
Total	24	28222.23		

b. $R^2 = 0.920$. There is a strong positive linear relationship between education resources and student performance.

9. a. As temperature increases from 20 to 25, the average oxygen consumption increases by an estimated 0.197 units.

b. As weight increases from 0.25 to 0.5, the average oxygen consumption increases by an estimated 0.148 units.

c. i $R^2 = 0.019$ **ii** $R^2 = 0.814$ **iii** $R^2 = 0.943$

13. a. $\hat{Y} = 6.874 - 0.004X_2 - 0.234X_3$ **b.** $\hat{Y} = 3.734$

c. $R^2 = 0.553$. The model explains about half of the variation in Yield (Y). The model has a limited ability to predict the yield for a company using 1989 ranking and P-E ratio as predictors.

Chapter 9

1. a. i $H_0: \beta_1 = 0$ $H_a: \beta_1 \neq 0$ (Full model: $Y = \beta_0 + \beta_1 X_1 + E$).
$F(X_1)_{1, 30} = 45.18$, $p = 0.0001$. At $\alpha = 0.05$, we reject H_0.

ii $H_0: \beta_1 = \beta_2 = 0$ $H_a:$ at least one $\beta_i \neq 0$ ($i = 1, 2$)
(Full model: $Y = \beta_0 + \beta_1 X_1 + \beta_2 X_2 + E$)
$F(X_1, X_2)_{2, 29} = 39.16$, $p = 0.0001$. At $\alpha = 0.05$, we reject H_0.

iii $H_0: \beta_1 = \beta_2 = \beta_3 = 0$ H_a: at least one $\beta_i \neq 0$ ($i = 1, 2, 3$)
(Full model $Y = \beta_0 + \beta_1 X_1 + \beta_2 X_2 + \beta_3 X_3 + E$).
$F(X_1, X_2, X_3)_{3,28} = 29.71, p = 0.0001$. At $\alpha = 0.05$, we reject H_0.

b. Using the criterion of favoring a more parsimonious model, one would choose the model tested in a(i).

5. a. i $H_0: \beta_1 = 0$ $H_a: \beta_1 \neq 0$ (Full model: $Y = \beta_0 + \beta_1 X_1 + E$).
$F(X_1)_{1,40} = 0.405, p > 0.25$. At $\alpha = 0.05$, we do not reject H_0.

ii $H_0: \beta_2 = 0$ $H_a: \beta_2 \neq 0$ (Full model: $Y = \beta_0 + \beta_2 X_2 + E$)
$F(X_2)_{1,40} = 0.19, p > 0.25$. At $\alpha = 0.05$, we do not reject H_0.

iii $H_0: \beta_3 = 0$ $H_a: \beta_3 \neq 0$ (Full model: $Y = \beta_0 + \beta_3 X_3 + E$).
$F(X_3)_{1,40} = 7.58, p = 0.009$. At $\alpha = 0.05$, we reject H_0.

b. $H_0: \beta_1 = \beta_2 = \beta_3 = 0$ H_a: at least one $\beta_i \neq 0$ ($i = 1, 2, 3$)
(Full model: $Y = \beta_0 + \beta_1 X_1 + \beta_2 X_2 + \beta_3 X_3 + E$).
$F(X_1, X_2, X_3)_{3,38} = 2.737, 0.05 < p < 0.10$. At $\alpha = 0.05$, we do not reject H_0.

c. $H_0: \beta_4 = \beta_5 = 0$ H_a: at least one $\beta_i \neq 0$ ($i = 4, 5$)
(Full model: $Y = \beta_0 + \beta_1 X_1 + \beta_2 X_2 + \beta_3 X_3 + \beta_4 X_1 X_3 + \beta_5 X_2 X_3 + E$)
$F(X_1 X_3, X_2 X_3 \mid X_1, X_2, X_3)_{2,36} = 0.362, p > 0.25$. At $\alpha = 0.05$, we do not reject H_0.

d. $H_0: \beta_3 = 0$ $H_a: \beta_3 \neq 0$ (Full model $Y = \beta_0 + \beta_1 X_1 + \beta_2 X_2 + \beta_3 X_3 + E$).
$F(X_3 \mid X_1, X_2)_{1,38} = 6.855, 0.01 < p < 0.025$. At $\alpha = 0.05$, we reject H_0.

e. X_3 is associated with Y, but the other two independent variables are not.

9. a. $H_0: \beta_1 = \beta_2 = 0$ H_a: At least one $\beta_i \neq 0$ ($i = 1, 2$)
(Full model: $Y = \beta_0 + \beta_1 X_1 + \beta_2 X_2 + E$).
$F(X_1, X_2)_{2,4} = 20.03, p = 0.0082$. At $\alpha = 0.05$, we reject H_0.

b. i $H_0: \beta_1 = 0$ $H_a: \beta_1 \neq 0$ in the model $Y = \beta_0 + \beta_1 X_1 + E$.
$F(X_1)_{1,5} = 40.06, p = 0.0032$. At $\alpha = 0.05$, we reject H_0.

ii $H_0: \beta_2 = 0$ $H_a: \beta_2 \neq 0$ in the model $Y = \beta_0 + \beta_1 X_1 + \beta_2 X_2 + E$.
$F(X_2 \mid X_1)_{1,4} = 0.01, p = 0.9344$. At $\alpha = 0.05$, we do not reject H_0.

c. i $H_0: \beta_2 = 0$ $H_a: \beta_2 \neq 0$ in the model $Y = \beta_0 + \beta_2 X_2 + E$.
$F(X_2)_{1,5} = 37.832, 0.001 < p < 0.005$. At $\alpha = 0.05$, we reject H_0.

ii $H_0: \beta_1 = 0$ $H_a: \beta_1 \neq 0$ (Full model: $Y = \beta_0 + \beta_1 X_1 + \beta_2 X_2 + E$).
$F(X_1 \mid X_2)_{1,4} = 9.80, p = 0.0352$. At $\alpha = 0.05$, we reject H_0.

d.

Source	d.f.	SS	MS	F	R^2
$X_1 \mid X_2$	1	1402.315	1402.315	9.8	0.909
$X_2 \mid X_1$	1	1.098	1.098	0.01	
Residual	4	572.393	143.098		
Total	6	6305.714			

e. X_1 is the only necessary predictor.

13. a. $H_0: \beta_1 = \beta_2 = 0$ H_a: at least one $\beta_i \neq 0$ ($i = 1, 2$)
($X_1 = $ OWNCOST, $X_2 = $ URBAN)
(Full model: $Y = \beta_0 + \beta_1 X_1 + \beta_2 X_2 + E$).
$F(X_1, X_2)_{2,23} = 8.521, p = 0.017$. At $\alpha = 0.05$, we reject H_0.

b. i $H_0: \beta_1 = 0$ $H_a: \beta_1 \neq 0$ (Full model: $Y = \beta_0 + \beta_1 X_1 + E$).
$F(X_1)_{1,24} = 8.84, p = 0.0068$. At $\alpha = 0.05$, we reject H_0.

ii $H_0: \beta_2 = 0$ $H_a: \beta_2 \neq 0$ (Full model: $Y = \beta_0 + \beta_1 X_1 + \beta_2 X_2 + E$).
$F(X_2 \mid X_1)_{1,23} = 8.21, p = 0.0088$. At $\alpha = 0.05$, we reject H_0.

c. i $H_0: \beta_2 = 0$ $H_a: \beta_2 \neq 0$ (Full model: $Y = \beta_0 + \beta_2 X_2 + E$).
$F(X_1)_{1,24} = 13.17, 0.001 < p < 0.005$. At $\alpha = 0.05$, we reject H_0.

ii $H_0: \beta_1 = 0$ $H_a: \beta_1 \neq 0$ (Full model: $Y = \beta_0 + \beta_1 X_1 + \beta_2 X_2 + E$).
$F(X_1 \mid X_2)_{1,23} = 4.42, p = 0.0467$. At $\alpha = 0.05$, we reject H_0.

d.

Source	d.f.	SS	MS	F	R^2
$X_1 \mid X_2$	1	66.334	66.334	4.42	0.4256
$X_3 \mid X_2$	1	123.158	123.158	8.21	
Residual	23	345.183	15.008		
Total	25	600.962			

e. Both predictors are necessary.

Chapter 10

1. a. Age with an r value of 0.7752.

b. i $r_{SBP,SMK \mid AGE} = 0.568$.   **ii** $r_{SBP,QUET \mid AGE} = 0.318$

c. $H_0: \rho_{SBP,SMK \mid AGE} = 0$ $H_a: \rho_{SBP,SMK \mid AGE} \neq 0$
$F(SMK \mid AGE)_{1,29} = 13.83, p < 0.001$. At $\alpha = 0.05$, we reject H_0.

d. $H_0: \rho_{SBP,QUET \mid AGE,SMK} = 0$ $H_a: \rho_{SBP,QUET \mid AGE,SMK} \neq 0$
$T_{28} = 1.91, p = 0.066$. At $\alpha = 0.05$, we do not reject H_0.

e. Based on the results for a–d, we find that the following variables (ranked in order of their significance) helped explain the variation in SBP : (1) AGE, (2) SMK, (3) QUET.

f. $r^2_{SBP(QUET,SMK) \mid AGE} = 0.401$
$H_0: \rho_{SBP(QUET,SMK) \mid AGE} = 0$ $H_a: \rho_{SBP(QUET,SMK) \mid AGE} \neq 0$
$F(QUET, SMK \mid AGE)_{2,28} = 9.371, p < 0.001$.
The highly significant p-value suggests that both SMK and QUET are important variables, but there is room for debate since the increase in r^2 going from model 1, with only AGE (0.601), to model 3, with all 3 variables (0.761), is small (0.160).

5. a. i $r^2_{YX_1 \mid X_3} = 0.076$ **ii** $r^2_{YX_2 \mid X_3} = 0.214$
iii $r^2_{YX_2 \mid X_3} = 0.215$ The computations are nearly equivalent with the difference being due to round-off error.

b. X_2 should be considered next for entry into the model because X_2 has a higher partial correlation than does X_1.

c. $H_0: \rho_{YX_2 \mid X_3} = 0$ $H_a: \rho_{YX_2 \mid X_3} \neq 0$
$T_{17} = 2.154, 0.02 < p < 0.05$. At $\alpha = 0.05$, we reject H_0.

d. $r^2_{YX_1 \mid X_2,X_3} = 0.082$
$H_0: \rho_{YX_1 \mid X_2,X_3} = 0$ $H_a: \rho_{YX_1 \mid X_2,X_3} \neq 0$
$T_{16} = 1.199, p = 0.248$. At $\alpha = 0.05$, we do not reject H_0.

e. $r^2_{Y(X_1,X_2) \mid X_3} = 0.279$
$H_0: \rho_{Y(X_1,X_2) \mid X_3} = 0$ $H_a: r_{Y(X_1,X_2) \mid X_3} \neq 0$
$F(X_1, X_2 X_3)_{2,16} = 3.098, 0.05 < p < 0.10$. At $\alpha = 0.05$, we do not reject H_0.

f. Based on the above results, only X_3 should be included in the model at $\alpha = 0.05$.

9. a. i $H_0: \rho_{YX_1} = 0$ $H_a: \rho_{YX_1} \neq 0$
$F(X_1)_{1,45} = 0.89, p = 0.35$. At $\alpha = 0.05$, we do not reject H_0.

    ii H_0: $\rho_{YX_2} = 0$ H_a: $\rho_{YX_2} \neq 0$

        $F(X_2)_{1,45} = 197.58$, $p = 0.0001$. At $\alpha = 0.05$, we reject H_0.

  b. i H_0: $\rho_{YX_1|X_2} = 0$ H_a: $\rho_{YX_1|X_2} \neq 0$

        $F(X_1, X_2)_{1,44} = 98.605$, $p < 0.001$. At $\alpha = 0.05$, we reject H_0.

   ii H_0: $\rho_{YX_2|X_1} = 0$ H_a: $\rho_{YX_2|X_1} \neq 0$

        $F(X_2 \mid X_1)_{1,44} = 709.641$, $p < 0.001$. At $\alpha = 0.05$, we reject H_0.

  c. Both X_1 and X_2 should be included in the model with X_2 being the more important than X_1.

13. a. $r^2_{Y|X_1,X_2} = 0.909$ **b.** $r_{YX_2|X_1} = -\sqrt{0.002} = -0.045$

  c. $r_{YX_1|X_2} = \sqrt{0.710} = 0.843$

  d. H_0: $\rho_{YX_2|X_1} = 0$ H_a: $\rho_{YX_2|X_1} \neq 0$

    $T_4 = -0.09$, $p > 0.90$. At $\alpha = 0.05$, we do not reject H_0.

  e. H_0: $\rho_{YX_1|X_2} = 0$ H_a: $\rho_{YX_1|X_2} \neq 0$

    $T_4 = 3.131$, $0.02 < p < 0.05$. At $\alpha = 0.05$, we reject H_0.

  f. X_1 should be included in the model, while X_2 should not be included.

17. a. $r^2_{Y|X_1,X_2} = 0.426$ **b.** $r_{YX_2|X_1} = -\sqrt{0.263} = -0.513$

  c. $r_{YX_1|X_2} = -\sqrt{0.161} = -0.401$

  d. H_0: $\rho_{YX_2|X_1} = 0$ H_a: $\rho_{YX_2|X_1} \neq 0$

    $T_{23} = -2.865$, $0.001 < p < 0.01$. At $\alpha = 0.05$, we reject H_0.

  e. H_0: $\rho_{YX_1|X_2} = 0$ H_a: $\rho_{YX_1|X_2} \neq 0$

    $T_{23} = -2.1$, $0.02 < p < 0.05$. At $\alpha = 0.05$, we reject H_0.

  f. Both variables should be included in the model with X_2 being the more important predictor of Y.

Chapter 11

1. a. $WGT = \beta_0 + \beta_1 HGT + \beta_2 AGE + \beta_3 AGE^2 + E$

  b. $\hat{\beta}_1$ does not change when either AGE or AGE^2 are removed from the model. However, $\hat{\beta}_1$ changes "significantly" when both AGE and AGE^2 are removed from the model. Thus, there is confounding due to AGE and AGE^2.

  c. AGE^2 can be dropped from the model because $\hat{\beta}_1$ does not change significantly.

  d. AGE^2 should not be retained in the model, because the 95% C.I. for β_1 is narrower when AGE^2 is absent from the model.

  e. Considering the change in $\hat{\beta}_1$ and the width of the 95% C.I., the final model should be $WGT = \beta_0 + \beta_1 HGT + \beta_2 AGE + E$.

  f. Revise the initial model as

    $WGT = \beta_0 + \beta_1 HGT + \beta_2 AGE + \beta_3 AGE^2 + \beta_4 HGT * AGE + \beta_5 HGT * AGE^2 + E$.

  g. We would test for interaction by performing a multiple-partial F test for H_0: $\beta_4 = \beta_5 = 0$. If this test proved significant, we would perform separate partial F tests to assess H_0: $\beta_4 = 0$ and H_0: $\beta_5 = 0$.

3. a. There is no confounding due to X_2, because $\hat{\beta}_1$ does not change when X_2 is removed from the model.

  b. $r_{YX_1} = 0.265$ $r_{YX_1|X_2} = \sqrt{0.5} = 0.707$

    Since the two correlation coefficients are significantly different, we conclude that confounding exists.

  c. The conclusions for confounding depend on the definition of confounding.

d. Since $H_0: \beta_2 = 0$ is rejected ($p = 0.0005$), we conclude that confounding exists, which is contradictory to part (a).

5. a. i $H_0: \beta_1 = 0$ $H_a: \beta_1 \neq 0$ $F(X_1)_{1,40} = 0.4$, $p = 0.528$. At $\alpha = 0.05$, we do not reject H_0.

    ii $H_0: \beta_2 = 0$ $H_a: \beta_2 \neq 0$ $F(X_2)_{1,40} = 0.19$, $p = 0.669$. At $\alpha = 0.05$, we do not reject H_0.

    iii $H_0: \beta_3 = 0$ $H_a: \beta_3 \neq 0$ $F(X_3)_{1,40} = 7.58, p = 0.009$. At $\alpha = 0.05$, we reject H_0.

b. $H_0: \beta_1 = \beta_2 = \beta_3 = 0$ H_a: At least one $\beta_i \neq 0$ ($i = 1, 2, 3$)
$F(X_1, X_2, X_3)_{3,38} = 2.737, 0.05 < p < 0.1$. At $\alpha = 0.05$, we do not reject H_0.

c. $H_0: \rho_{Y(X_1X_3, X_2X_3)|X_1,X_2,X_3} = 0$ $H_a: \rho_{Y(X_1X_3, X_2X_3)|X_1,X_2,X_3} \neq 0$
$F(X_1X_3, X_2X_3 \mid X_1, X_2, X_3)_{2,36} = 0.362, p > 0.25$. At $\alpha = 0.05$, we do not reject H_0.
The two regression lines are parallel.

d. $H_0: \rho_{YX_3|X_1,X_2} = 0$ $H_a: \rho_{YX_3|X_1,X_2} \neq 0$
$F(X_3 \mid X_1, X_2)_{1,38} = 6.855, p = 0.014$. At $\alpha = 0.05$, we reject H_0.

e. First fit the full model
$$Y = \beta_0 + \beta_1 X_1 + \beta_2 X_2 + \beta_3 X_3 + E \quad (1)$$
Next, fit the reduced models
$$Y = \beta_0 + \beta_1 X_1 + \beta_3 X_3 + E \quad (2)$$
$$Y = \beta_0 + \beta_2 X_2 + \beta_3 X_3 + E \quad (3)$$
$$Y = \beta_0 + \beta_3 X_3 + E \quad (4)$$
We assess confounding by noting how $\hat{\beta}_3$ changes for the different models. In particular, if $\hat{\beta}_3$ from model (2), (3), or (4) differs from $\hat{\beta}_3$ from model (1), then X_1, X_2, or X_1 and X_2, respectively, are confounders. To assess precision, we note how the $100(1 - \alpha)\%$ C.I.'s for $\hat{\beta}_3$ change. We only eliminate potential confounders from the model if the width of the C.I. for $\hat{\beta}_3$ does not widen significantly.

f. From the information provided, we can assess the confounding effects of X_1 or X_2 alone with respect to X_3 but not for X_1 and X_2 taken together.

7. a. No, there is no meaningful change in the estimate for $\hat{\beta}_1$ when X_2 is added to the model.

b. No, the confidence interval for $\hat{\beta}_1$ is narrower when only X_1 is in the model.

c. No, there is not evidence suggesting that including X_2 to the model improves the precision and/or the validity of the estimated relationship between X_1 and Y.

9. a. Let OWNCOST $= X_1$ and INCOME $= X_2$
$H_0: \beta_1 = \beta_2 = 0$ H_a: at least one β_i does not equal 0.
(Full model: $Y = \beta_0 + \beta_1 X_1 + \beta_2 X_2 + E$)
$F(X_1, X_2)_{2,23} = 6.38, p = 0.006$. At $\alpha = 0.05$, we reject H_0.

b. $H_0: \beta_1 = 0$ $H_a: \beta_1 \neq 0$ (Full model: $Y = \beta_0 + \beta_1 X_1 + \beta_2 X_2 + E$).
$F(X_1 \mid X_2)_{1,23} = 11.47, p = 0.003$. At $\alpha = 0.05$, we reject H_0.

c. $H_0: \beta_2 = 0$ $H_a: \beta_2 \neq 0$ (Full model $Y = \beta_0 + \beta_1 X_1 + \beta_2 X_2 + E$).
$F(X_2 \mid X_1)_{1,23} = 4.87, p = 0.038$. At $\alpha = 0.05$, we reject H_0.

d. Including X_2 does meaningfully change $\hat{\beta}_1$, and it should therefore be included in the model as a confounder, assuming there is no interaction between X_1 and X_2.

Chapter 12

1. a. $Y = \beta_0 + \beta_1(\text{AGE}) + E$, where Y denotes dry weight.

   d. The largest (absolute value) jackknife residual is $r_{(-11)} = 3.568$ (not significant).

e. The primary problem is illustrated by the plot of the jackknife residuals vs. AGE. The systematic pattern of the residuals suggests that the straight-line regression model is not appropriate.

f. The jackknife residuals follow an exact t-distribution. For samples less than 30, one should expect heavier tails than for a normal curve. At sample sizes above 30, the distribution should reflect sampling from a standard normal distribution.

6. a. $Y = \beta_0 + \beta_1(\text{AGE}) + E$, where Y denotes average total sleep time.

    d. The largest (absolute value) jackknife residual is $r_{(-5)} = 2.190$ (not significant).

    e. In general the residuals appear to be in good shape.

13. a. $Y = \beta_0 + \beta_1(\text{BODY WEIGHT}) + E$, where Y denotes latency to seizure.

    d. The largest (absolute value) jackknife residual is $r_{(-14)} = 2.558$ (not significant).

    e. The plots of the jackknife residuals do not suggest any troublesome observations.

20. a. $Y = \beta_0 + \beta_1(\text{HEIGHT}) + \beta_2(\text{WEIGHT}) + \beta_3(\text{AGE}) + \beta_4(\text{FEMALE}) + E$, where Y denotes forced expiratory volume in one second.

    b.   Variable-added-last tests

Variable	SS	F
INTERCEPT	0.013	0.016
HEIGHT	0.312	0.395
WEIGHT	0.478	0.603
AGE	0.047	0.059
FEMALE	1.605	2.028

The critical value is $F_{1,14,0.95} = 4.60$, and hence none of the predictors significantly explains the variation in FEV_1.

    h. Collinearity of the predictors does not appear to be a problem.

25. a. $Y = \beta_0 + \beta_1(\text{ADVERTISING}) + E$, where Y denotes sales.

    d. The largest studentized residual (absolute value) = 1.527, which is no cause for alarm.

    e. The plot of the jackknife residuals versus the predictor does not suggest any troublesome problems.

Chapter 13

1. b. From the computer output, we find:

    (1) Degree 1: $\hat{Y} = -1.932 + 0.246X$

    (2) Degree 2: $\hat{Y} = 3.172 - 0.781X + 0.047X^2$

    (3) $\ln Y$ on X: $\ln \hat{Y} = -6.21 + 0.451X$

    (4) The above fitted equations are plotted on the graphs presented for 1(a).

    c.

Source	d.f.	SS	MS	F
Regression	1	12.705	12.705	43.69
Lack of fit	4	4.419	1.105	57.03
Residual	16	4.651	0.2908	
Pure error	12	0.232	0.0194	
Total	17	17.357		

d.

Source	d.f.	SS	MS	F
Degree 1 (X)	1	12.705	12.705	43.69
Regression	2	16.61	8.305	
Degree 2 ($X^2 \mid X$)	1	3.905	3.905	78.46
Lack of fit	3	0.514	0.171	8.85
Residual	15	0.746	0.0497	
Pure error	12	0.232	0.0194	
Total	17	17.357		

e. $r_{XY}^2 = 0.732$; $r^2(quadratic) = 0.957$

f. Test for significance of straight-line regression of Y on X

H_0: The straight-line regression is not significant.

$F_{1,16} = 43.69$, $p < 0.001$. At $\alpha = 0.05$, we reject H_0 and conclude that the straight-line regression is significant.

Test for adequacy of straight-line model

H_0: The straight-line model is adequate.

$F_{4,12} = 56.959$, $p < 0.001$. At $\alpha = 0.05$, we reject H_0 and conclude that the straight-line model is not adequate.

g. Test for significance of quadratic regression

H_0: The quadratic regression is not significant.

$F_{2,15} = 166.77$, $p = 0.0001$. At $\alpha = 0.05$, we reject H_0 and conclude that the quadratic regression is significant.

Test for addition of X^2 term

H_0: The addition of X^2 to a model already containing X is not significant.

Partial $F(X^2 \mid X)_{1,15} = 78.41$, $p = 0.0001$. At $\alpha = 0.05$, we reject H_0 and conclude that the addition of X^2 is significant.

Test for adequacy of quadratic model

H_0: The quadratic model is adequate.

$F_{3,12} = 8.84$, $p = 0.002$. At $\alpha = 0.05$, we reject H_0 and conclude that the quadratic model is not adequate.

h. Test for significance of straight-line regression of $\ln Y$ on X

H_0: The straight-line regression is not significant.

$F_{1,16} = 4277.167$, $p = 0.0001$. At $\alpha = 0.05$, we reject H_0 and conclude that the straight-line regression is significant.

Test for adequacy of straight-line model of $\ln Y$ on X

H_0: The straight-line model is adequate.

$F_{4,12} = 0.896$, $p = 0.471$. At $\alpha = 0.05$, we do not reject H_0 and conclude that the straight-line model is adequate.

i. R^2 (straight-line regression of $\ln Y$ on X) $= 0.9965$

R^2 (quadratic regression of Y on X) $= 0.957$

A comparison of the above two R^2 shows that the straight-line fit of $\ln Y$ on X provides a better fit.

j. **(1)** Homoscedasticity assumption appears to be much more reasonable when using $\ln Y$ on X than when using Y on X.

(2) The straight-line regression of $\ln Y$ on X is preferred.

k. The independence assumption is violated.

5. b. <u>Test for significance of straight-line regression</u>
H_0: The straight-line regression is not significant.
$F_{1,24} = 978.04$, $p < 0.001$. At $\alpha = 0.05$, we reject H_0 and conclude that the straight-line regression is significant.
<u>Test for adequacy of straight-line model</u>
H_0: The straight-line model is adequate.
$F_{16,8} = 1.57$, $p > 0.25$. At $\alpha = 0.05$, we do not reject H_0 and conclude that the straight-line model is adequate.

c. <u>Test for addition of X^2 to the model</u>
H_0: The addition of X^2 is not significant.
Partial $F(X^2 \mid X)_{1,23} = 0.55$, $p > 0.25$. At $\alpha = 0.05$, we do not reject H_0.

d. The straight-line model is most appropriate.

9. a. Fit the model VOC_SIZE $= \beta_0 + \beta_1(\text{AGE}^*) + \beta_2(\text{AGE}^*)^2 + \beta_3(\text{AGE}^*)^3 + E$, where AGE* = AGE − 2.867.
$\hat{\beta}_0 = 741.84$ $\hat{\beta}_1 = 645.60$ $\hat{\beta}_2 = 70.43$ $\hat{\beta}_3 = -31.18$

b. Using variables-added-in-order tests, the best model includes AGE*, $(\text{AGE}^*)^2$, and $(\text{AGE}^*)^3$.

c. Using variables-added-last tests, the best model includes AGE*, $(\text{AGE}^*)^2$, and $(\text{AGE}^*)^3$.

d. The only large predictor correlation is between AGE* and $(\text{AGE}^*)^3$. The largest condition index ($\text{CI}_3 = 6.05$) suggests that the centered data do not have any serious collinearity problems.

f. The estimated regression coefficients differ from those in problem 8, but the best model includes the linear, quadratic, and cubic terms as in problem 8. Also, the sums of squares for the variables-added-in-order test are the same as in problem 8. The centering of AGE greatly reduced the previous collinearity problems. Centering does not affect the residual diagnostics.

13. a. The estimated equation is
$$\hat{Y} = 10.64 + 94.42(\text{LIN\_TOL}) + 14.59(\text{QUAD\_TOL}) - 0.73(\text{CUB\_TOL}),$$
where LN_TOL, QUAD_TOL, and COL_TOL denote the centered PPM_TOLU orthogonal polynomials.

b. Using the variable-added-in-order tests, the best model includes LIN_TOL and QUAD_TOL .

c. Using the variable-added-last tests, the best model includes LIN_TOL and QUAD_TOL, which is the same model as in part (b).

d. The orthogonal polynomials are uncorrelated with each other, which implies that any collinearities are eliminated as shown by the condition indices.

e. The residual plots suggest that the variances increase as the predicted values increase.

Chapter 14

1. a. For smokers: $\hat{Y} = 79.225 + 20.118X$. For nonsmokers: $\hat{Y} = 49.312 + 26.303X$.

b. H_0: $\beta_{1SMK} = \beta_{1\overline{SMK}}$ H_a: $\beta_{1SMK} < \beta_{1\overline{SMK}}$
$T_{28} = 0.892$, $0.15 < p < 0.25$. At $\alpha = 0.05$, we do not reject H_0 that the slopes for smokers and nonsmokers are the same.

c. H_0: $\beta_{0SMK} = \beta_{0\overline{SMK}}$ H_a: $\beta_{0SMK} \neq \beta_{0\overline{SMK}}$

$T_{28} = -1.24$, $0.20 < p < 0.30$. At $\alpha = 0.05$, we do not reject H_0, and we conclude that the two intercepts are equal.

d. The straight lines for smokers and nonsmokers are coincident since both tests failed to reject H_0.

5. a. For NY: $\hat{Y} = 2.174 + 1.177X$.   For CA: $\hat{Y} = 8.030 + 1.036X$.

   b. $H_0: \beta_{1NY} = \beta_{1CA}$ $H_a: \beta_{1NY} > \beta_{1CA}$
   $T_{33} = 1.115$, $0.10 < p < 0.15$. At $\alpha = 0.05$, we do not reject H_0, and we conclude that the slopes are the same for NY and CA.

   c. $H_0: \beta_{0NY} = \beta_{0CA}$ $H_a: \beta_{0NY} > \beta_{0CA}$
   $T_{33} = 1.219$, $0.85 < p < 0.90$. At $\alpha = 0.05$, we do not reject H_0 that the two intercepts are equal for NY and CA.

   d. Since the tests for equal slopes and equal intercepts did not lead to rejection, we can conclude that the lines are coincident.

   e. $H_0: \rho_{NY} = \rho_{CA}$ $H_a: \rho_{NY} \neq \rho_{CA}$
   $Z = 0.252$, $p > 0.80$. At $\alpha = 0.05$, we do not reject H_0, and we conclude that the correlation coefficients for each straight line regression are not significantly different.

9. a. $SBP = \beta_0 + \beta_1 AGE + \beta_2 QUET + \beta_3 SMK + \beta_4 AGE*SMK + \beta_5 QUET*SMK + E$
   Smokers: $SBP = (\beta_0 + \beta_3) + (\beta_1 + \beta_4)AGE + (\beta_2 + \beta_5)QUET + E$
   Nonsmokers: $SBP = \beta_0 + \beta_1 AGE + \beta_2 QUET + E$

   b. Smokers: $\hat{SBP} = 48.076 + 1.466(AGE) + 6.744(QUET)$
   Nonsmokers: $\hat{SBP} = 48.613 + 1.029(AGE) + 10.451(QUET)$

   c. $H_0: \beta_4 = \beta_5 = 0$ H_a: at least one $\beta_i \neq 0$ (Full Model given in part (a)).
   $F(QUET * SMK, AGE * SMK \mid AGE, QUET, SMK)_{2,26} = 0.222$, $p > 0.25$
   At $\alpha = 0.05$, we do not reject H_0 and conclude the two lines are coincident.

   d. $H_0: \beta_3 = \beta_4 = \beta_5 = 0$ H_a: At least one $\beta_i \neq 0$ (Full Model given in part (a)).
   $F(SMK, QUET * SMK, AGE * SMK \mid AGE, QUET)_{3,26} = 4.562$, $0.01 < p < 0.025$.
   At $\alpha = 0.05$, we reject H_0 and conclude the two lines are not coincident.

13. a. $R = 1$ and $TD = 1$: $\hat{SBPL} = (\hat{\beta}_0 + \hat{\beta}_2 + \hat{\beta}_4 + \hat{\beta}_9) + \hat{\beta}_1(SBP1) + (\hat{\beta}_6 + \hat{\beta}_{11})RW$
   $R = 0$ and $TD = 1$: $\hat{SBPL} = (\hat{\beta}_0 + \hat{\beta}_3 + \hat{\beta}_4 + \hat{\beta}_7) + \hat{\beta}_1(SBP1) + (\hat{\beta}_6 + \hat{\beta}_{13})RW$

   b. $H_0: \beta_{11} = \beta_{12} = \beta_{13} = \beta_{14} = 0$ H_a: At least one $\beta_i \neq 0$.
   $F(X_{11}, X_{12}, X_{13}, X_{14} \mid X_1..., X_{10})_{4,89} = 1.410$, $0.1 < p < 0.25$. At $\alpha = 0.05$, we do not reject H_0, and we conclude that the lines are parallel.

   c. H_0: The three regression lines corresponding to rural, town, and urban background are parallel. (i.e., $H_0: \beta_7 = \beta_8 = \beta_9 = \beta_{10} = \beta_{11} = \beta_{12} = \beta_{13} = \beta_{14} = 0$).
   H_a: At least one $\beta_i \neq 0$.
   $F(X_7,..., X_{14} \mid X_1..., X_6)_{8,89} = 1.103$ with $p > 0.25$. At $\alpha = 0.05$, we do not reject H_0, and we conclude that the three regression lines are parallel.

   d. $H_0: \beta_2 = \beta_3 = \beta_4 = \beta_5 = \beta_7 = \beta_8 = \beta_9 = \beta_{10} = \beta_{11} = \beta_{12} = \beta_{13} = \beta_{14} = 0$
   H_a: At least one $\beta_i \neq 0$.
   $$F(X_2, X_3, X_4, X_5, X_7..., X_{14} \mid X_1, X_6)_{12,89} = \frac{\dfrac{\text{SS Regression (full model)} - \text{SS Regression }(X_1, X_6)}{12}}{\text{MS Residual (full model)}}$$

17. a. $Y = \beta_0 + \beta_1 X + \beta_2 Z + \beta_3 XZ + E$ where $Z = 1$ if cool, 0 if warm.

b. For cool: $\hat{Y} = 104.003 + 2.465X$.   For warm: $\hat{Y} = 96.830 + 3.485X$.

c. $H_0: \beta_2 = \beta_3 = 0$ H_a: At least one $\beta_i \neq 0$ (Full model given in part (a))

$F(Z, XZ \mid X)_{2,14} = 41.875$, $p = 0.0076$. At $\alpha = 0.05$, we reject H_0 and conclude that the lines do not coincide.

d. $H_0: \beta_3 = 0$ H_a: $\beta_3 \neq 0$ (Full model given in part (a))

$F(XZ \mid X, Z)_{1,14} = 9.699$, $p < 0.01$. At $\alpha = 0.05$, we reject H_0 and conclude that the lines are not parallel.

e. Baseline sales are higher during the warm season than in the cool season. Advertising expenditures are higher in the cool season than in the warm season. By spending more money in advertising during the cool season, retailers are able to surpass the sales revenue of the warm season.

21. a. $Y = \beta_0 + \beta_1 X + \beta_2 Z + \beta_3 XZ + E$

b. $H_0: \beta_2 = \beta_3 = 0$ H_a: At least one $\beta_i \neq 0$ (Full model given in part (a))

$F(Z, X_1 Z \mid X_1)_{2,50} = 2.704$, $0.05 < p < 0.10$. At $\alpha = 0.05$, we do not reject H_0, and we conclude that the lines coincide.

d. $H_0: \beta_3 = 0$ H_a: $\beta_3 \neq 0$ (Full model given in part (a))

$F(X_1 Z \mid X_1, Z)_{1,50} = 3.587$, $p = 0.064$, At $\alpha = 0.05$, we do not reject H_0, and we conclude that the lines are parallel.

e. The change in refraction-baseline refractive relationship is the same for males and females.

Chapter 15

1. a. $SBP = \beta_0 + \beta_1(QUET) + \beta_2 SMK + E$

b. For smokers: $SBP = (\beta_0 + \beta_2) + \beta_1(QUET) + E; \overline{SBP}_{(adj)} = 148.548$

For nonsmokers: $SBP = \beta_0 + \beta_1(QUET) + E; \overline{SBP}_{(adj)} = 139.977$

c. $H_0: \beta_2 = 0$ H_a: $\beta_2 \neq 0$ in the model $SBP = \beta_0 + \beta_1(QUET) + \beta_2 SMK + E$

$T_{29} = 2.707$, $p = 0.011$. At $\alpha = 0.05$, we reject H_0 and conclude that mean SBP differs for smokers and for nonsmokers, after adjusting for QUET.

d. Finding the 95% confidence interval for the true difference in adjusted mean SBP is equivalent to finding the 95% confidence interval for $\hat{\beta}_2$. The 95% confidence interval for $\hat{\beta}_2$ is (2.094, 15.048).

5. a. $VIAD = \beta_0 + \beta_1 IQM + \beta_2 IQF + \beta_3 Z + E$ where $Z_1 = 1$ if female, 0 if male.

b. For males: VIAD (adj) $= -3.307$ vs. -3.00 unadjusted.

For females: VIAD (adj) $= 1.889$ vs. 1.60 unadjusted.

c. $H_0: \beta_3 = 0$ H_a: $\beta_3 \neq 0$ in the model $VIAD = \beta_0 + \beta_1 IQM + \beta_2 IQF + \beta_3 Z + E$

$T_{16} = 1.659$, $p = 0.117$. At $\alpha = 0.05$, we do not reject H_0, and we conclude that the mean scores do not significantly differ by gender, after adjusting for IQM and IQF.

d. 95% CI: $(-1.446, 11.838)$

9. a. $LN\_BRNTL = \beta_0 + \beta_1 WGT + \beta_2 Z_1 + \beta_3 Z_2 + \beta_4 Z_3 + E$, where $Z_1 = 1$ if 100 ppm, 0 otherwise; $Z_2 = 1$ if 500 ppm, 0 otherwise; $Z_3 = 1$ if 1000 ppm, 0 otherwise.

b. $\hat{\beta}_0 = -0.764$ $\hat{\beta}_1 = 0.0006$ $\hat{\beta}_2 = 0.828$ $\hat{\beta}_3 = 3.571$ $\hat{\beta}_4 = 4.214$

c.

PPM_TOLU	Adjusted Means	Unadjusted Means
50	−0.537	−0.548
100	0.291	0.282
500	3.034	3.019
1000	3.677	3.668

  d. $H_0: \beta_2 = \beta_3 = \beta_4 = 0$ H_a: At least one $\beta_i \neq 0$ ($i = 2, 3, 4$)
    (Full model: LN_BRNTL $= \beta_0 + \beta_1 WGT + \beta_2 Z_1 + \beta_3 Z_2 + \beta_4 Z_3 + E$)
    $F(Z_1, Z_2, Z_3 \mid WGT)_{3,55} = 1662.526$, $p < 0.001$. At $\alpha = 0.05$, we reject H_0 and conclude that the adjusted means significantly differ.

13. a. $Y = \beta_0 + \beta_1 AGE + \beta_2 Z + E$
  b. $Y = \beta_0 + \beta_1 AGE + \beta_2 Z + \beta_3 AGE * Z + E$
    $H_0: \beta_3 = 0$ $H_a: \beta_3 \neq 0$ (Full model: $Y = \beta_0 + \beta_1 AGE + \beta_2 Z + \beta_3 AGE * Z + E$)
    $T_{26} = -1.677$, $p = 0.106$. At $\alpha = 0.05$, we do not reject H_0, and we conclude that the ANACOVA model in part (a) is appropriate.

  c.

Location	Adjusted Means	Unadjusted Means
Intown/inner	81.526	82.187
Outer	85.46	84.807

    $H_0: \beta_2 = 0$ $H_a: \beta_2 \neq 0$
    $T_{27} = 1.212$, $p = 0.236$. At $\alpha = 0.05$, we do not reject H_0, and we conclude that the adjusted means do not significantly differ.

Chapter 16

1. a. The final model by forward selection is $\hat{WGT} = 37.600 + 0.053(AGE*HGT)$
  b. The final model by backward elimination is $\hat{WGT} = 37.600 + 0.053(AGE*HGT)$
  c. The best model resulting from this approach is:
    $\hat{WGT} = 6.553 + 0.722(HGT) + 2.050(AGE)$
  d. The first two approaches result in the model including only the AGE*HGT interaction. It is difficult to interpret results for such a model, since the variables that make up the interaction term are not included. The third approach results in a model that is easier to interpret. It is possible to conduct modified forward and backward stepwise strategies in which interaction terms are only considered for inclusion if the terms that make up the interaction term are already in the model.

3. For smokers: the final estimated model is $\hat{SBP} = 102.220 + 0.252(AGE*QUET)$
  For nonsmokers: the final estimated model is $\hat{SBP} = 93.073 + 0.250(AGE*QUET)$
  These two models are different from those obtained from 2(d) by putting SMK = 1 (for smokers) and SMK = 0 (for nonsmokers).

5. The best regression models using the sequential procedure of adding AGE first for Females and Males are
  Females: $\hat{DEP} = 190.012 - 1.099AGE - 1.217MC$
  Males: $\hat{DEP} = 270.056 - 2.514AGE - 1.065MC$
  For females, the above model was selected as best because of its high r^2 (0.401), satisfactory $C(p)$ (2.238), and low MSE (3274.364). The next best model was the full model with $r^2 = 0.413$, $C(p) = 4.0$, MSE = 3478.318. Emphasizing parsimony, we find that the above model is more favorable than the full model.
  The reasoning is similar for selecting the model shown above for males

7. a. Using the $C(p)$ criterion exclusively, we find that the three models with the most favorable $C(p)$ values contain AGE alone (1.23), AGE and WGT (2.63), or all three variables (4). None of these models is particularly impressive, and since it would be hard to argue that either of the multiple-variable models is better than the model with AGE alone, we could take AGE alone as the best model. Upon further investigation, the difficulty is seen to be partly due to none of these models having a significant overall F test.

b. Since there seems to be no rationale for grouping the variables (age, weight, and height) in any way, chunks are taken to be the three variable specific pairs of linear/quadratic terms (i.e., AGE_C and AGE_CSQ; HGT_C and HGT_CSQ; WGT_C and WGT_CSQ, where the _C terms are the centered variables, and the _CSQ terms are the squared centered terms). A plausible forward chunkwise strategy is to treat each chunk/pair as a distinct entity that cannot be split and then proceed in the usual forward manner (use $\alpha = 0.10$):

Step 1: WGT_C, WGT_CSQ added to the model. F test $= 2.78$
Step 2: AGE_C, AGE_CSQ added to the model. F test $= 3.49$
Step 3: Stop HGT_C, HGT_CSQ not significant. F test $= 0.16$

c. The all possible regressions method yields the following "best" models for each of the model sizes:

Number in Model	Variables	$C(p)$	R^2	MSE
1	WGT_CSQ	8.026	0.078	0.771
2	WGT_C, WGT_CSQ	5.213	0.300	0.631
3	WGT_C, WGT_CSQ, AGE_CSQ	4.337	0.432	0.554
4	WGT_C, WGT_CSQ, AGE_C, AGE_CSQ	3.312	0.571	0.456
5	WGT_C, WGT_CSQ, AGE_C, AGE_CSQ, HGT_CSQ	5.225	0.576	0.497
6	Full Model	7.00	0.586	0.539

The best model is most likely the four-variable model including WGT_C, WGT_CSQ, AGE_C, and AGE_CSQ. The R^2, $C(p)$, and MSE for this model are better than for any of the smaller models; and they are similar to, if not better than, the statistics for the larger models, which of course are less parsimonious.

d. Any model containing only first-order terms (part (a)) is seriously deficient. The model in parts (b) and (c) is the best one.

9. a. A model containing X_1, X_3, and X_4 is the best model by this method. The model has a relatively high R^2, a satisfactory $C(p)$, and one of the lowest MSEs. The model also has the benefit of parsimony compared to the larger models.

b. The selected model contains X_1, X_3, and X_4.

c. The selected model contains X_1, X_3, and X_4.

d. All three methods selected the same model, and it appears to be the best model for the reasons cited in (a).

11. a. The model containing X_1 and X_3 would be the recommended model. Its R^2, $C(p)$, and MSE are clearly are superior to the best one-variable model statistics, and they are similar to the statistics for the less parsimonious full model.

b. The results of the stepwise regression show that the model containing X_1 and X_3 is the best model.

c. The model: $Y = \beta_0 + \beta_1 X_1 + \beta_3 X_3 + E$ appears to be best. This model is chosen, given the results in parts (a) and (b).

Chapter 17

1. a.

Treatment	Mean	S
1	7.5	1.643
2	5	1.265
3	4.333	1.033
4	5.167	1.472
5	6.167	2.041

$\overline{Y} = 5.633$

b. ANOVA table:

Source	d.f.	SS	MS	F
Treatment	4	36.467	9.117	3.90
Error	25	58.50	2.34	
Total	29	94.967		

c. H_0: $\mu_1 = \mu_2 = \mu_3 = \mu_4 = \mu_5 = 0$

H_a: At least two treatments have different population means.

$$F_{4,25} = \frac{9.117}{2.34} = 3.896, \text{ with } p = 0.0136.$$

At $\alpha = 0.05$, we reject H_0 and conclude that at least two treatments have different population means.

d. Estimates of true effects ($\mu_i - \mu$) where μ is the overall mean:

Treatment

i	$(\overline{Y}_i - \overline{Y})$
1	1.8667
2	−0.6333
3	−1.3000
4	−0.4667
5	0.5333
Total $\sum_{i=1}^{5} (\overline{Y}_i - \overline{Y})$	0.000

e. We define X_i such that

$X_i = 1$, for $R_x i$; 0, otherwise

where $i = 1, 2, 3, 4$. Then the appropriate regression model is

$Y = \beta_0 + \alpha_1 X_1 + \alpha_2 X_2 + \alpha_3 X_3 + \alpha_4 X_4 + E$

where the regression coefficients are as follows:

$\beta_0 = \mu_5$

$\alpha_1 = \mu_2 - \mu_5, \alpha_2 = \mu_2 - \mu_5, \alpha_3 = \mu_3 - \mu_5, \alpha_4 = \mu_4 - \mu_5$

For $X_i = -1$, for $R_x 5$; 1, for $R_x i$; 0, otherwise

the regression coefficients are:

$\beta_0 = \mu, \alpha_1 = \mu_1 - \mu, \alpha_2 = \mu_2 - \mu, \alpha_3 = \mu_3 - \mu, \alpha_4 = \mu_4 - \mu$, and also

$-(\alpha_1 + \alpha_2 + \alpha_3 + \alpha_4) = \mu_5 - \mu$

f. Hand calculated 95% confidence intervals for $(\mu_1 - \mu_3)$ and $(\mu_1 - \mu_2)$ are as follows:

Comparison	Scheffé	Tukey	LSD
$(\mu_1 - \mu_3)$	(0.238, 6.096)	(0.57, 5.764)	(0.34, 5.994)
$(\mu_1 - \mu_2)$	(−0.429, 5.429)	(−0.097, 5.097)	(−0.327, 5.327)

Conclusions: Treatment 1 and 3 significantly differ; treatments 1 and 2 do not significantly differ. We conclude that all other remaining comparisons, which involve smaller (in absolute value) pairwise sample mean differences, are not significant using $\alpha = 0.05$. Of the three methods, Scheffé's method gives the widest interval, the LSD method gives the next widest, and Tukey's method gives the narrowest interval.

5. a.

Source	d.f.	SS	MS	F
Parties	3	5625	1875.00	37.31
Error	16	804	50.25	
Total	19	6429		

b. H_0: $\mu_1 = \mu_2 = \mu_3 = \mu_4$

H_a: at least two parties have different mean authoritarianism scores.

$F_{(3, 16)} = 37.31$, with $p < 0.001$. At $\alpha = 0.05$, we reject H_0 and conclude that the authoritarianism scores of the members of different political parties significantly differ.

c. Regression model:

$Y = \beta_0 + \alpha_1 X_1 + \alpha_2 X_2 + \alpha_3 X_3 + E$ where $X_i = 1$ if party #i, 0 otherwise; $i = 1, 2, 3$

d. Tukey's method confidence intervals for different comparisons are then given as follows:

Comparison	Confidence Interval	Remark
P_3 vs P_4:	$(\overline{Y}_3 - \overline{Y}_4) \pm 12.93 = (32.07, 57.93)$	significant
P_3 vs P_2:	$(\overline{Y}_3 - \overline{Y}_2) \pm 12.93 = (2.07, 27.93)$	significant
P_3 vs P_1:	$(\overline{Y}_3 - \overline{Y}_1) \pm 12.93 = (-2.93, 22.93)$	not significant, since 0 in interval
P_1 vs P_4:	$(\overline{Y}_1 - \overline{Y}_4) \pm 12.93 = (22.07, 47.93)$	significant
P_1 vs P_2:	$(\overline{Y}_1 - \overline{Y}_2) \pm 12.93 = (-7.93, 17.93)$	not significant
P_2 vs P_4:	$(\overline{Y}_2 - \overline{Y}_4) \pm 12.93 = (17.07, 42.93)$	significant

At $\alpha = 0.05$, all differences except for $(\overline{Y}_1 - \overline{Y}_2)$ and $(\overline{Y}_3 - \overline{Y}_1)$ are significant.

9. a. H_0: $\mu_1 = \mu_2 = \mu_3$ H_a: At least two mean generation times differ.

$F_{(3, 20)} = 46.99$, with $p < 0.001$. At $\alpha = 0.05$, we reject H_0 and conclude that between strains significantly differ.

b.

Source	d.f.	SS	MS	F
Strains	3	134713.00	44904.33	46.99
Error	20	19113.244	955.662	
Total	23	153826.244		

c. Using Tukey's method confidence intervals for pairwise comparisons, we conclude that

$\mu_C > (\mu_D = \mu_A) > \mu_B$

where μ_A, μ_B, μ_C, and μ_D are population means from Strains A, B, C, and D respectively.

13. a.

Source	d.f.	SS	MS	F
Dosage	3	566.628	188.876	14.71
w/in Dosage	44	564.96	12.84	
Total	47	1131.588		

17. a. $Y_{ij} = \mu + \alpha_i + E_{ij}$, $i = 1, 2, 3; j = 1, ..., 6$.

b. MSM $= 4.667$ $F = 5.53$ MSE $= 0.844$

c. H_0: $\mu_1 = \mu_2 = \mu_3$

H_a: There is a significant difference in attitude toward advertising by practice type.

$F_{2, 15} = 5.53$, with $p = 0.0159$. At $\alpha = 0.05$, we reject H_0 and conclude that there are significant differences in attitude toward advertising by practice type.

d. GP vs IM: $3.6667 - 2.00 \pm 1.4398$ $(0.2269, 3.1065)$ significant
  GP vs FP: $3.6667 - 2.333 \pm 1.4398$ $(-0.1061, 2.7735)$ not significant
  FP vs IM: $2.333 - 2.00 \pm 1.4398$ $(-1.1068, 1.7728)$ not significant

e. Same as 17(a).

f.

Source	d.f.	SS	MS	F
Model	2	26.333	13.167	20.43
Error	15	9.667	0.6444	
Total	17	35.999		

g. $H_0: \mu_1 = \mu_2 = \mu_3$ H_a: At least two means differ by practice type. $F_{2,15} = 20.43$, $p = 0.0001$. At $\alpha = 0.05$, we reject H_0 and conclude that there is a significant difference in influence on prescription writing habits by practice type.

h. GP vs IM: $4.3333 - 1.5000 \pm 1.2578$ $(1.5755, 4.0911)$ significant
  GP vs FP: $4.3333 - 2.1667 \pm 1.2578$ $(0.9088, 3.4244)$ significant
  FP vs IM: $2.1667 - 1.500 \pm 1.2578$ $(-0.5911, 1.9245)$ N.S.

21. a. $Y_{ij} = \mu + \alpha_i + E_{ij}$, $i = 1,\dots, 3; j = 1,\dots, n_i$. Clear zone sizes are fixed-effect factors.

b.

Source	d.f.	SS	MS	F	p-value
Model	2	14.704	7.352	5.462	0.0073
Error	48	64.592	1.346		
Total	50	79.296			

c. $H_0: \mu_1 = \mu_2 = \mu_3$ H_a: At least two mean five-year changes differ by clear zone. $F_{2,48} = 5.462$, with $p = 0.0073$. At $\alpha = 0.05$, we reject H_0 and conclude that mean five-year changes significantly differ by clear zone size.

d. μ_1 vs μ_3: $(0.3368, 2.2188)$ significant
  μ_1 vs μ_2: $(-0.2562, 1.6603)$ not significant
  μ_2 vs μ_3: $(-1.6603, 0.2562)$ not significant
  The mean five-year refractive changes significantly differ between 3.0 mm and 4.0 mm.

Chapter 18

1. a. The rats are the blocks, and the three chemicals are the treatments.

b. SSE $= 25.0$ MSE $= 1.786$ Type I SS (chem) $= 25.0$ MS (chem) $= 12.5$

c. $H_0: \mu_1 = \mu_2 = \mu_3$ H_a: The mean irritation scores differ by chemical type. $F_{2,14} = 7.00$ with $p < 0.0078$. At $\alpha = 0.05$, we reject H_0 and conclude the toxic effects of the three chemicals significantly differ.

d. 98% CI on $\mu_I - \mu_{II}$: $(-3.0, 0.5)$ **e.** $R^2 = 0.635$

f. Fixed-effects ANOVA model: $Y_{ij} = \mu + \tau_i + \beta_j + E_{ij}$, $i = 1, 2, 3; j = 1, 2,\dots, 8$
where
$Y_{ij} =$ observation on the jth rat for the ith chemical effect
$\mu =$ overall mean
$\tau_i = i$th chemical effect
$\beta_j = j$th rat effect
$E_{ij} =$ error due to observation on the jth rat for the ith chemical effect (E_{ij}'s are independent and assumed to be normally distributed).

Regression model: $Y = \beta_0 + \alpha_1 X_1 + \alpha_2 X_2 + \sum_{j=1}^{7} \beta_j Z_j + E$

$$\text{where } X_1 = \begin{cases} 1 \text{ if chemical I} \\ 0 \text{ if chemical II,} \\ -1 \text{ if chemical III} \end{cases} X_2 = \begin{cases} 0 \text{ if chemical I} \\ 1 \text{ if chemical II,} \\ -1 \text{ if chemical III} \end{cases} Z_j = \begin{cases} -1 \text{ if rat 8} \\ 1 \text{ if rat } j \ (j = 1, 2, \ldots, 7) \\ 0 \text{ otherwise} \end{cases}$$

g. Assumptions underlying the model: (i) additivity of the model (no interaction), (ii) homogeneity of variance, (iii) normality of the errors, (iv) independence of the errors.

5. $H_0: \mu_1 = \mu_2 = \mu_3$ H_a: At least two mean ESP scores differ by person.
$F_{2,8} = 15.12$ with $p = 0.0019$. At $\alpha = 0.05$, we reject H_0 and conclude that there are significant differences in ESP ability by person. Note that, since the F test for blocking on days is not significant, one may conclude that blocking on days is not necessary.

9. a.–f.

Source	d.f.	SS	MS	F
Treatments	4	160	40	5.00
Blocks	5	240	48	6.00
Error	20	160	8	

g. Test of treatments: $F_{4,20} = 5.00$ with $0.005 < p < 0.01$. At $\alpha = 0.05$, we reject H_0 and conclude that there is a significant main effect of treatments.

Test of blocks: $F_{5,20} = 6.00$ with $0.001 < p < 0.005$. At $\alpha = 0.05$, we reject H_0 and conclude that there is a significant main effect of blocks.

Chapter 19

1. a. The two factors are levorphanol and epinephrine.
b. Both factors should be considered fixed.
c. Rearrangement of the data into a two-way ANOVA layout:

Levels of Epinephrine

	Absence −	Presence +
Absence −	(Control) 1.90, 1.80, 1.54, 4.10, 1.89	(Epinephrine only) 5.33, 4.85, 5.26, 4.92, 6.07
Presence +	(Levporphanol only) 0.82, 3.36, 1.64, 1.74, 1.21	(Levorphanol and Epinephrine) 3.08, 1.42, 4.54, 1.25, 2.57

Levels of Levorphanol

d. Table of sample means:

	Epinephrine −	+	Total (Row Mean)
Levorphanol −	2.25	5.28	3.77
+	1.75	2.57	2.16
Total (Col. Mean)	2.00	3.93	2.96

The presence of levorphanol appears to reduce stress, whereas the presence of epinephrine appears to increase stress. In the presence of epinephrine, levorphanol reduces stress by an average of 2.71 units, whereas in the absence of epinephrine, levorphanol reduces stress by only 0.50 units, suggesting the possibility of an interaction.

e.

Source	d.f.	SS	MS	F
Levorphenol	1	12.832	12.932	12.60
Epinephrine	1	18.586	18.586	18.25
Interaction	1	6.161	6.161	6.05
Error	16	16.298	1.019	
Total	19	53.877		

f. <u>Main effect for levorphanol</u>

H_0: Levorphenol has no significant main effect on stress.

H_a: Levorphenol has a significant main effect on stress.

$F_{1,16} = 12.60, p = 0.0027$

At $\alpha = 0.05$, we reject H_0 and conclude that levorphanol has a significant main effect on stress.

<u>Main effect for epinephrine</u>

H_0: Epinephrine has no significant main effect on stress.

H_a: Epinephrine has a significant main effect on stress.

$F_{1,16} = 18.25, p = 0.0066$

At $\alpha = 0.05$, we reject H_0 and conclude that epinephrine has a significant main effect on stress.

<u>Interaction</u>

H_0: There is no significant interaction between levorphanol and epinephrine on stress.

H_a: There is a significant interaction between levorphanol and epinephrine on stress.

$F_{1,16} = 6.05, p = 0.0267$

At $\alpha = 0.05$, we reject H_0 and conclude that there is significant interaction between levorphanol and epinephrine on stress.

5. a. Yes. There is an apparent same-direction interaction, which is reflected in the fact that the difference in mean waiting times between suburban and rural court locations is larger for State 1 than for State 2.

b. <u>Main effect of State:</u> $F_{1,594} = 217.45, p < 0.001$

<u>Main effect of Court Location:</u> $F_{2,594} = 184.86, p < 0.001$

<u>Interaction effect:</u> $F_{2,594} = 10.96, p < 0.001$

All effects are highly statistically significant.

c. Regression model:

$$Y = \beta_0 + \beta_1 S + \beta_2 C_1 + \beta_3 C_2 + \beta_4 SC_1 + \beta_5 SC_2 + E$$

where

$S = 1$ if State 1, -1 if State 2 $C_1 = -1$ if Urban, 1 if Rural, 0 if Suburban

$C_2 = -1$ if Urban, 1 if Suburban, 0 if Rural

d. We might consider a model of the form

$$Y = \beta'_0 + \beta'_1 S + \beta'_2 C + \beta'_3 SC + E$$

where S is as defined in (c) and where C is a variable taking on three values (i.e., 0, 1, and 2) that increase directly with the degree of urbanization. One difficulty here is how to determine the appropriate values for C.

9. a. "Species": fixed factor; "Locations": random factor, unless only the four locations chosen are of interest.

b.

Source	F (Mixed)	p-value	F (Both fixed)	p-value
Species	9.109 (d.f. = 2,6)	$0.01 < p < 0.025$	9.455 (d.f. = 2,48)	$p < 0.001$
Locations	0.811 (d.f. = 3,6)	$p > 0.25$	0.841 (d.f. = 3,48)	$p > 0.25$
Interaction	1.038 (d.f. = 6,48)	$p > 0.25$	1.038 (d.f. = 6,48)	$p > 0.25$

13. a.

Source	d.f.	SS	F	p
AIRTEMP	3	0.03	0.32	0.813
DOSE	2	0.06	1.04	0.370
A*D	6	0.05	0.29	0.936
Error	24	0.70		
TOTAL	35			

b. From the ANOVA table, we see that the main effects are not significant, and neither is the interaction term.

c. An appropriate multiple regression model is:

$$Y = \mu + \sum_{i=2}^{4} \alpha_i X_i + \sum_{j=2}^{3} \beta_j Z_j + \sum_{i=2}^{4}\sum_{j=2}^{3} X_i Z_j \gamma_{ij} + E$$

in which

$$X_i = \begin{cases} -1 & \text{for AIRTEMP} = 21 \\ 1 & \text{for level } i \text{ of AIRTEMP, } i = 2, 3, 4 \\ 0 & \text{otherwise} \end{cases}$$

$$Z_j = \begin{cases} -1 & \text{for DOSE} = 0 \\ 1 & \text{for level } j \text{ of DOSE, } j = 2, 3 \\ 0 & \text{otherwise} \end{cases}$$

$\alpha_i = \mu_i. - \mu.., i = 2, 3, 4$
$\beta_j = \mu_j. - \mu.., j = 2, 3$
$\gamma_{ij} = \mu_{ij} - \mu_i + \mu.., i = 2, 3, 4; j = 2, 3$
AIRTEMP = 21 and DOSE = 0 correspond to control levels and hence are of least interest. The larger AIRTEMPS and DOSES are of primary interest, so they are directly parameterized in the model.

d. A natural polynomial model is
$$Y = \beta_0 + \beta_1(\text{AIRTEMP}) + \beta_2(\text{DOSE}) + \beta_3(\text{AIRTEMP} * \text{DOSE}) + E.$$
Note that higher-order terms can be added to the model if deemed reasonable.

17. a. The ANOVA model:
$Y_{ijk} = \mu + \alpha_i + \beta_j + \gamma_{ij} + E_{ijk}, i = 1, 2; j = 1, 2; k = 1,\ldots, 6$
α_i = effect for the ith school type
β_j = effect for the jth reputation category
γ_{ij} = interaction effect for the ith school type and jth reputation category
E_{ijk} = error term for the observation ijk
The factors are fixed.

b. d.f. = 3 $F_{3, 20} = 2.572$, with $0.05 < p < 0.10$.

c. Main effect of school type:
H_0: There is no significant main effect of school type on starting salary.

H_a: There is a significant main effect of school type on starting salary.

$F_{1,20} = 0.52$, $p = 0.4786$. At $\alpha = 0.05$, we do not reject H_0, and we conclude that school type does not have a significant main effect on starting salary.

Main effect of reputation rank:

H_0: There is no significant main effect of reputation rank on starting salary.

H_a: There is a significant main effect of reputation rank on starting salary.

$F_{1,20} = 6.21$, $p = 0.0216$. At $\alpha = 0.05$, we reject H_0 and conclude that reputation rank has a significant main effect on starting salary.

Test of interaction:

H_0: There is no significant interaction between school type and reputation rank with respect to starting salary.

H_a: There is significant interaction between school type and reputation rank with respect to starting salary.

$F_{1,20} = 0.99$, $p = 0.3323$. At $\alpha = 0.05$, we do not reject H_0, and we conclude that there is no significant interaction between school type and reputation rank.

Chapter 20

1. a. Table of sample means:

Traditional Rank

		HI	MED	LO	Total
	HI	135	147.5	165	152.5
Modern	MED	145	155	163	155.91
Rank	LO	161.67	143.33	121.67	142.22
	Total	147.22	148.75	154.23	

The preceding table indicates than the sample mean blood pressure for males with low modern rank is lower than for other modern rank categories. It also illustrates that the sample mean blood pressure for males with low traditional rank is higher than the other traditional rank categories. Finally, the table hints at an interaction effect: for persons with high modern rank, mean blood pressure increases with decreasing traditional rank, whereas mean blood pressure for persons with low modern rank decreases with decreasing traditional rank. In other words, this interaction suggests that persons with incongruous cultural roles (i.e., HI-LO, LO-HI) tend to have higher blood pressures than persons with more congruent cultural roles.

b. Regression model:

$$Y = \beta_0 + \alpha_1 X_1 + \alpha_2 X_2 + \beta_1 Z_1 + \beta_2 Z_2 + \gamma_{11} X_1 Z_1 + \gamma_{12} X_1 Z_2 + \gamma_{21} X_2 Z_1 + \gamma_{22} X_2 Z_2 + E$$

where

$$X_i = \begin{cases} -1 & \text{if LO modern rank} \\ 1 & \text{if modern rank } i \ (i = 1 \text{ for HI, } i = 2 \text{ for MED}) \\ 0 & \text{otherwise} \end{cases}$$

and

$$Z_i = \begin{cases} -1 & \text{if LO traditional rank} \\ 1 & \text{if traditional rank } i \ (i = 1 \text{ for HI, } i = 2 \text{ for MED}) \\ 0 & \text{otherwise} \end{cases}$$

c. Modern main effect:

(i) $F(X_1, X_2)_{2,27} = 2.076$, $0.10 < p < 0.25$, which is not significant.

(ii) $F(X_1, X_2 \mid Z_1, Z_2)_{2,25} = 1.768$, $0.10 < p < 0.25$, which is not significant.

Traditional main effect:

(i) $F(Z_1, Z_2)_{2,27} = 0.578$, $p > 0.25$, which is not significant.

(ii) $F(Z_1, Z_2 \mid X_1, X_2)_{2,25} = 0.397$, $p > 0.25$, which is not significant.

Interaction:

$F(X_1 Z_1, X_1 Z_2, X_2 Z_1, X_2 Z_2 \mid X_1, X_2, Z_1, Z_2)_{4,21} = 15.069$, $p < 0.001$, which is highly significant.

d. $Y = \beta_0 + \beta_1 X_1 + \beta_2 X_2 + \beta_3 X_1 X_2 + E$, where

$$X_1 = \begin{cases} 0 & \text{if LO modern rank} \\ 1 & \text{if MED modern rank} \\ 2 & \text{if HI modern rank} \end{cases} \quad X_2 = \begin{cases} 0 & \text{if LO traditional rank} \\ 1 & \text{if MED traditional rank} \\ 2 & \text{if HI traditional rank} \end{cases}$$

The difficulty arises with regard to assigning numerical values to the categories of each factor. The coding scheme for X_1 and X_2 given here assumes that the categories are "equally spaced," which may not really be the case.

5. a. Table of means:

		Social Class			
		Lo	Med	Hi	Row means
Number	0	14.571	12.00	13.50	13.5
of times	1	9.00	12.25	7.40	9.4
victimized	2+	8.00	2.33	5.75	5.8
Col means		11.3	9.7	8.8	

The preceding table suggests a downward trend in confidence with increasing number of victimizations and a slight downward trend with increasing social class status score.

b. From the ANOVA table given in the question, we obtain

$F(\text{Victim})_{2,31} = 8.81$, $p < 0.001$, which is significant.

$F(\text{SCLS})_{2,31} = 0.501$, $p > 0.25$, which is not significant.

$F(\text{Interaction})_{4,31} = 1.21$, $p > 0.25$, which is not significant.

c. We can use the following regression model:

$$Y = \beta_0 + \alpha_1 X_1 + \alpha_2 X_2 + \beta_1 Z_1 + \beta_2 Z_2 + \gamma_{11} X_1 Z_1 + \gamma_{12} X_1 Z_2 + \gamma_{21} X_2 Z_1 + \gamma_{22} X_2 Z_2 + E$$

where

$$X_1 = \begin{cases} -1 & \text{if no. of times victimized} = 0 \\ 1 & \text{if no. of times victimized} = 1 \\ 0 & \text{otherwise} \end{cases}$$

$$X_2 = \begin{cases} -1 & \text{if no. of times victimized} = 0 \\ 1 & \text{if no. of times victimized} = 2+ \\ 0 & \text{if no. of times victimized} = 1 \end{cases}$$

$$Z_j = \begin{cases} -1 & \text{if social class status} = \text{LO} \\ 1 & \text{if social class status} = j \ (j = 1 \text{ for MED}, j = 2 \text{ for HI}) \\ 0 & \text{otherwise} \end{cases}$$

We then obtain the following F values from the regression results given in the question:

Main effect of VICTIM:

$F(X_1, X_2)_{2,37} = 9.03, p < 0.001$, which is significant.

$F(X_1, X_2 \mid Z_1, Z_2)_{2,35} = 8.61, p < 0.001$, which is significant.

Main effect of SCLS:

$F(Z_1, Z_2)_{2,37} = 0.695, p > 0.25$, which is not significant.

$F(Z_1, Z_2 \mid X_1, X_2)_{2,35} = 0.707, p > 0.25$, which is not significant.

Interaction:

$F(X_1Z_1, X_1Z_2, X_2Z_1, X_2Z_2 \mid X_1, X_2, Z_1, Z_2)_{4,31} = 1.10$,

$p > 0.25$, which is not significant.

  d. The error term may have a nonconstant variance.

9. a. A tabulation of DOSAGE * SEX sample sizes reveals that all combinations of these factors have 10 subjects except for 9 in the DOSAGE = 0, SEX = Male combination. We could call this a nearly orthogonal design since we would only need one more observation in the cell with 9 to make a perfectly orthogonal design (equal cell sizes).

b.

	TYPE I		TYPE III	
Source	F	p-value	F	p-value
DOSAGE	2.65	$0.05 < p < 0.10$	2.64	$0.05 < p < 0.10$
SEX	0.22	$p > 0.25$	0.20	$p > 0.25$
DOSAGE*SEX	0.14	$p > 0.25$	0.14	$p > 0.25$

  c. At $\alpha = 0.05$, no effects are significant.

  d. Since no effects are significant, no multiple-comparison tests are justified.

13. a $Y_{ijk} = \mu + \alpha_i + \beta_j + \gamma_{ij} + E_{ijk}, i = 1, 2, 3; j = 1, 2, 3; k = 1, \ldots, n_{ij};$

$n_{11} = 3, n_{12} = 6, n_{13} = 11, n_{21} = 4, n_{22} = 5, n_{23} = 6, n_{31} = 4, n_{32} = 4, n_{33} = 8$

α_i = effect for the ith clear zone ($i = 1$: 3.0mm, $i = 2$: 3.5mm, $i = 3$: 4.0mm)

β_j = effect for the jth baseline curvature category

γ_{ij} = interaction effect for the ith clear zone and jth curvature

E_{ijk} = error term for the observation ijk

  b. d.f. = 8 $F_{8,42} = 3.595$ $0.001 < p < 0.005$

  c. Using the SAS output: At $\alpha = 0.10$, we would conclude that there is a significant interaction between CLRZONE and BASECURV. We would also conclude that CLRZONE and BASECURV have significant main effects on refractive error.

Chapter 21

1. a. Days is a crossover factor, since every subject is measured at every level of this factor.

  b. ANOVA model: $Y_{ij} = \mu + S_i + \tau_j + E_{ij}$, where $i = 1, \ldots, 19$ and $j = 1, 2, 3$.

μ is the overall mean.

S_i is the random effect of subject i.

τ_j is the fixed effect of day j.

E_{ij} is the random error for Day j within subject i.

Assumptions: $\{S_i\}$ and $\{E_{ij}\}$ are mutually independent; S_i is distributed as $N(0, \sigma_S^2)$; and E_{ij} is distributed as $N(0, \sigma^2)$.

  c. $H_0: \tau_1 = \tau_2 = \tau_3 = 0$

  d. There seems to be a larger difference between Friday (15.78), and the other days (18.36 and 18.25).

e. $F_{2,36} = 1.56$, p-value $= 0.22$, which indicates nonsignificance.

f. $E[MS_T] = \sigma^2 + \frac{19}{2}\sum_{j=1}^{3} \tau_j^2$. Under H_0, this reduces to σ^2.

g. $\hat{\rho} = 0.86$

5. a. Factor A is a nest factor, since each rat is observed at only one level of Factor A.

b. The sample means range from .76 to 2.34, which indicates a fairly large difference between different levels of Factor A, leading to the suggestion that there is an effect of Factor A.

c. The ANOVA model is $Y_{ijk} = \mu + S_{i(j)} + \tau_j + E_{k(ij)}$, where $i = 1,\ldots, 24$; $j = 1,\ldots, 6$; and $k = 1, 2, 3, 4$.

μ is the overall mean.

τ_j is the fixed effect of Factor A level j.

$S_{i(j)}$ is the random effect of rat i within level j of Factor A.

$E_{k(ij)}$ is the random error for repeat k on rat i within level j of Factor A.

This model assumes $\{S_{(i)j}\}$ and $\{E_{k(ij)}\}$ are mutually independent, that $S_{i(j)}$ is distributed as $N(0, \sigma_S^2)$ and that $E_{k(ij)}$ is distributed as $N(0, \sigma^2)$.

d. Test for Factor A: $F_{5,18} = 143.0641$, which has a p-value of 0.0001 (highly significant).

e. $E[MS_A] = \sigma^2 + 4\sigma_S^2 + \frac{(24)(4)}{5}\sum_{j=1}^{6} \tau_j^2$. Under H_0: all $\tau_j = 0$, $E[MS_A]$ reduces to $\sigma^2 + 4\sigma_S^2$.

9. a. ANOVA model: $Y_{ijk} = \mu + S_i + \alpha_j + \beta_k + \delta_{jk} + S_{ij} + S_{ik} + E_{ijk}$. This model assumes that $\{S_i\}, \{S_{ij}\}, \{S_{ik}\}$, and $\{E_{ijk}\}$ are mutually independent, that S_i is distributed as $N(0, \sigma_S^2)$, that S_{ij} is distributed as $N(0, \sigma_{SD}^2)$, that S_{ik} is distributed as $N(0, \sigma_{ST}^2)$, and that E_{ijk} is distributed as $N(0, \sigma^2)$, where D denotes the Days factor and T denotes the Time factor.

b. Neither Days ($F = 1.8226$, $p = 0.1762$) nor Time ($F = 2.2986$, $p = 0.1469$) nor their interaction ($F = 0.3324$, $p = 0.7194$) is significant.

d. The correlation structure for this model is not exchangeable. The correlation between responses on different days but at the same time of day differs from the correlation between responses on different days at different times, and each of these differs from the correlation between responses on the same day but at different times.

Chapter 22

1. a. Yes, they are identical if we assume that the linear regression model is fit to normally distributed data.

b. $H_0: \beta_1 = 0$ $H_a: \beta_1 \neq 0$

(Full Model: $Y = \beta_0 + \beta_1 DRUG + \beta_2 SBP + \beta_3 QUET + E$ where DRUG $= 1$ if drug, 0 if placebo)

Test Statistic: $Z = \frac{\hat{\beta}_1}{\sqrt{Var_{\hat{\beta}_1}}}$, which is approximately standard normal under H_0.

Critical Value: If $Z > 1.96$, then reject H_0 at $\alpha = 0.05$.

c. No; however, as N increases, they approach one another.

d. $H_0: \beta_1 = 0$ $H_a: \beta_1 \neq 0$

Test statistic: $-2 \log L$ (reduced) $- (-2 \log L$ (full)) $= \chi^2$ with 1 d.f. where

Full model equals: $Y = \beta_0 + \beta_1 DRUG + \beta_2 SBP + \beta_3 QUET + E$

Reduced model equals: $Y = \beta_0 + \beta_2 SBP + \beta_3 QUET + E$

Under H_0 the test statistic is approximately chi-square with 1 d.f.
Critical value: if $\chi^2 > 3.841$, then reject H_0 at $\alpha = 0.05$.

e. Yes, since $\chi^2 = Z^2$ when Z is normally distributed.

f. 95% confidence interval for β_1: $\hat{\beta}_1 \pm 1.96 * \sqrt{\text{Var}_{\hat{\beta}_1}}$

Chapter 23

1. a. $\hat{OR} = 2.925 \Rightarrow \hat{\beta}_1 = \ln(2.925) = 1.073$

b. logit $[\text{pr}\,(Y = 1)] = -2.8 + 0.706X_1 + 0.0004X_2 + 0.0006X_3$

c. $[\text{pr}\,(Y = 1)] = 0.112$

d. $\hat{OR}_{20\,\text{vs}\,21} = 0.999$

The odds for hypertension for a 20-year-old smoker are essentially equal to those for a 21-year-old smoker.

e. 95% CI for $\hat{OR}_{20\,\text{yr old smoker vs 21 yr old smoker}}$: $(0.9982, 0.9998)$

f. $H_0: \beta_3 = 0 \qquad H_a: \beta_3 \neq 0$

Test Statistic: $-2 \log L\,(\text{reduced}) - (-2 \log L\,(\text{full})) = \chi^2$ with 1 d.f.

$\chi^2 = 308.00 - 303.84 = 4.16$, with $0.025 < p < 0.05$.

At $\alpha = 0.05$, we reject H_0 and conclude that there is significant interaction between age and smoking.

Chapter 24

1. a. $n = 30$ (6 age-sex groups * 5 years)

b. Here, $i = 5$ and $k = 1992 - 1990 = 2$, so $E(Y_{52}) = l_{52}\lambda_{52} = l_{52}e^{(\alpha_5 + 2\beta)}$

c. Log rate changes linearly with time. In particular, for the ith group,

$$\ln \lambda_{ik} = \alpha_i + \beta k$$

so α_i is the intercept and β is the slope of the straight line relating the response $\ln \lambda_{ik}$ to the time variable $k = [\text{year}] - 1960$.

d. Model (1) assumes no interaction between age-sex group and time in the sense that the change in log rate over time (as measured by β) does not depend on i. Since $\ln \lambda_{ik} = \alpha_i + \beta k$, a graph of $\ln \lambda_{ik}$ versus k for each i would plot as a series of parallel straight lines—that is, lines all with the same slope (β) but possibly different intercepts (the α_i's). A lack of parallelism would reflect interaction between age-sex groups and time because the change in log rate over time would differ for different age-sex groups.

e. $\ln \text{IDR}_{ik} = \ln \lambda_{ik} - \ln \lambda_{10} = (\alpha_i + \beta k) - (\alpha_1 + \beta * 0) = (\alpha_i - \alpha_1) + \beta k$, so that

$$\text{IDR}_{ik} = e^{\alpha_i - \alpha_1}e^{\beta K}$$

Note that this is a function of both age-sex group (i) and time (k).

f. An appropriate model is

$$E(Y_{ik}) = l_{ik}\lambda_{ik}$$

where

$$\ln \lambda_{ik} = \sum_{i=1}^{6} \alpha_i A_i + \beta k + \sum_{i=1}^{5} \gamma_i(A_i k)$$

For age-sex group i, then,

$$\ln \lambda_{ik} = \alpha_i + \beta k + \gamma_i k$$
$$= \alpha_i + (\beta + \gamma_i)k$$
$$= \alpha_i + \delta_i k$$

where $\delta_i = \beta + \gamma_i$. Hence the slope for group i (namely, δ_i) is now a function of i. Only when all six δ_i's (or equivalently, all six γ_i's) are equal will the straight lines be parallel.

g. Yes, since $D(\hat{\beta})_{(1)} - D(\hat{\beta})_{(3)} = 300 - 175 = 125$, which is highly significant when compared to appropriate upper-tail χ^2-values with $29 - 24 = 5$ d.f.

h. Yes, since $D(\hat{\beta})_{(3)} - D(\hat{\beta})_{(4)} = 175 - 60 = 115$, which is highly significant when compared to appropriate upper-tail χ^2-values with $24 - 23 = 1$ d.f.

i. No, since $D(\hat{\beta})_{(4)} - D(\hat{\beta})_{(5)} = 60 - 59 = 1$, which is clearly not significant when compared to appropriate upper-tail χ^2-values with $23 - 22 = 1$ d.f.

j. Yes, since $D(\hat{\beta})_{(4)} - D(\hat{\beta})_{(7)} = 60 - 20 = 40$, which is highly significant when compared to appropriate upper-tail χ^2-values with $23 - 18 = 5$ d.f.

k. H_0 is rejected, since $D(\hat{\beta})_{(4)} - D(\hat{\beta})_{(6)} = 60 - 25 = 35$, which is highly significant when compared to appropriate upper-tail χ^2-values with $23 - 22 = 1$ d.f.

l. Only models (6) and (7) are candidates to be the final model. Model (6) has a deviance of 25 based on 22 d.f., indicating a good fit to the data. Model (7) has a deviance of 20 based on 18 d.f., also indicating a good fit. All other candidate models have significant lack of fit. Note that $D(\hat{\beta})_{(6)} - D(\hat{\beta})_{(7)} = 25 - 20 = 5$, which is clearly not significant when compared to appropriate upper-tail χ^2-values with $22 - 18 = 4$ d.f. Hence, model (6) certainly fits the data as well as model (7), and it also characterizes very specifically the type of interaction present in the data (namely, that the group 1 slope differs from the slope common to the other five groups). Model (6) is our choice as the final model.

m. For model (6),
$$\text{pseudo } R^2 = \frac{300 - 25}{300} = 0.917$$
which is indicative of a good model.

n. $\hat{\beta} \pm 1.96 * SE_{\hat{\beta}} = 0.50 \pm 1.96(0.20) = (0.108, 0.892)$

o. The point estimate of δ_1 is $\hat{\delta}_1 = \hat{\beta} + \hat{\gamma}_1 = 0.50 - 3.00 = -2.50$. The variance of the estimator $\hat{\delta}_1$ is $\text{Var}(\hat{\delta}_1) = \text{Var}(\hat{\beta}) + \text{Var}(\hat{\gamma}_1) + 2\,\text{Cov}(\hat{\beta}_1, \hat{\gamma}_1)$, which equals
$$\text{Var}(\hat{\delta}_1) = (0.20)^2 + (0.50)^2 + 2(-0.10) = 0.09$$
Finally, an approximate 95% CI for δ_1 is
$$\hat{\delta}_1 \pm 1.96\sqrt{\widehat{\text{Var}}(\hat{\delta}_1)} = -2.50 \pm 1.96\sqrt{0.90} = (-3.088, -1.912).$$

These two confidence intervals suggest that log rate increases linearly ($\beta > 0$) for groups 2 through 6, but decreases ($\delta_1 < 0$) for group 1.

Index